Tumor Suppressor Genes

METHODS IN MOLECULAR BIOLOGY™

John M. Walker, SERIES EDITOR

METHODS IN MOLECULAR BIOLOGY™

Tumor Suppressor Genes

Volume 2
Regulation, Function, and Medicinal Applications

Edited by

Wafik S. El-Deiry, MD, PhD

Departments of Medicine, Genetics, and Pharmacology,
University of Pennsylvania School of Medicine, Philadelphia, PA

Humana Press ✳ Totowa, New Jersey

Production Editor: Tracy Catanese

Cover illustration: Fig. 3B from Volume 2, Chapter 5, "*In Situ* Hybridization in Cancer and Normal Tissue," by Shrihari Kadkol, Jeremy Juang, and Tzyy-choou Wu.

Cover design by Patricia F. Cleary

For additional copies, pricing for bulk purchases, and/or information about other Humana titles, contact Humana at the above address or at any of the following numbers: Tel.: 973-256-1699; Fax: 973-256-8341; E-mail: humana@humanapr.com; or visit our Website: humanapress.com

Printed in the United States of America. 10 9 8 7 6 5 4 3 2 1

Library of Congress Cataloging in Publication Data

Tumor suppressor genes / edited by Wafik S. El-Deiry.
      p. ; cm. -- (Methods in molecular biology ; v. 223)
   Includes bibliographical references and index.
   Contents: v. 1. Pathways and isolation strategies -- v. 2. Regulation, function, and medicinal applications.
   ISBN 0-89603-986-2 (v. 1 : alk. paper); 1-59259-328-3 (v. 1 : e-book) -- ISBN 0-89603-987-0 (v. 2 : alk. paper);1-59259-329-1 (v. 2 : e-book)
    1. Antioncogenes--Laboratory manuals. I. El-Deiry, Wafik S. II. Methods in molecular biology (Totowa, N.J.) ; v. 222-223.
   [DNLM: 1. Genes, Tumor Suppressor--physiology--Laboratory Manuals. 2. Neoplasms--genetics--Laboratory Manuals. 3. Neoplasms--therapy--Laboratory Manuals. 4. Signal Transduction--physiology--Laboratory Manuals. 5. Tumor Suppressor Proteins--physiology--Laboratory Manuals. QZ 25 T925 2003]
   RC268.43 .T82 2003
   616.99'4042--dc21
                                                                              2002027501

Preface

It has become clear that tumors result from excessive cell proliferation and a corresponding reduction in cell death caused by the successive accumulation of mutations in key regulatory target genes over time. During the 1980s, a number of oncogenes were characterized, whereas from the 1990s to the present, the emphasis has shifted to tumor suppressor genes (TSGs). It has become clear that oncogenes and TSGs function in the same pathways, providing positive and negative growth regulatory activities. The signaling pathways controlled by these genes involve virtually every process in cell biology, including nuclear events, cell cycle, cell death, cytoskeletal, cell membrane, angiogenesis, and cell adhesion effects. Mutations in tumor suppressor genes have been identified in familial cancer syndromes, and the same genes in many cases have been found to be mutationally inactivated in sporadically occurring cancers. In their normal state, TSGs control cancer development and progression, as well as contribute to the sensitivity of cancers to a variety of therapeutics. Understanding the classes of TSGs, the biochemical pathways they function in, and how they are regulated provides an essential lesson in cancer biology. We cannot hope to advance our current knowledge and to develop new and more effective therapies without understanding the relevant pathways and how they influence the present approaches to therapy. Moreover, it is important to be able to access not only the powerful tools now available to discover these genes, but also their links to cell biology and growth control.

The scope of this two volume work, *Tumor Suppressor Genes, Volume 1: Pathways and Isolation Strategies* and *Volume 2: Regulation, Function, and Medicinal Applications*, is broad in the sense that it covers all the known tumor suppressor pathways and provides key information on the road to their discovery, analysis, and uses in cancer therapeutics. The aim of the first volume, *Pathways and Isolation Strategies*, is to educate the reader about known TSGs and the relevance of the biochemical pathways they regulate to human cancer. The reader has an opportunity in Volume 1 to access state-of-the-art protocols that have been successful in the identification of TSGs in the past, and that can be applied to isolate novel TSGs. With a novel TSG in hand, the reader has an opportunity in Volume 2, *Regulation, Function, and Medicinal Applications*, to explore the cell biology and biochemical function of the encoded protein, as well as its physiological role in vivo. Finally, in Volume 2, the reader is exposed to strategies for the use of information on TSGs to develop diagnostic and therapeutic strategies for cancer.

The two volumes of *Tumor Suppressor Genes* bring together many of the world's experts in the identification and characterization of TSGs. The work is thus intended to become the core reference and compilation of the emerging pathways and the growing number of molecules that suppress cancer. Importantly, it should also serve as a wide-ranging source of protocols useful in understanding and characterizing the function of TSGs. One of the challenges facing cancer researchers and clinicians is to bring forward and develop active therapeutics. This book, by example, puts forward highly useful paradigms for rational drug design, based on our dramatic new understanding of molecular pathogenesis.

Tumor Suppressor Genes thus lays down a firm and timely foundation for understanding cancer. In this age of expression profiling and proteomics, there has already been revealed a remarkable complexity and interrelatedness of seemingly diverse processes and signal transduction pathways. For the student, this book provides a reference to the basics concerning the identity of the major TSGs and the signaling pathways they use to inhibit tumors. For the investigator, it provides not only a critical update, but also an extremely useful compendium of newly assembled research protocols, including both classical methods and state-of-the-art techniques. For the translational scientist, the book provides fertile ground for the development of therapeutic strategies based on understanding the mechanisms of action and appreciating the existing preclinical data. One of the criticisms of an effort leading to such a book is that the field is moving very quickly and material is likely to be outdated. However, with many of the world's leading experts providing a comprehensive overview of all tumor suppressing pathways, along with their detailed protocols, we believe we have provided an invaluable resource for continued learning and discovery. Finally, *Tumor Suppressor Genes* provides a bridge to those interested in translational research by giving examples of the rationale for many of the most promising manipulations that may lead to novel therapeutic agents.

Tumor Suppressor Genes is targeted at a broad audience including medical and graduate students, postdoctoral fellows, physician scientists, academics, and principal investigators. The text provides the critical information needed to rapidly gain appreciation of important TSGs in human cancer, as well as modern methods for their discovery, analysis, and clinical application. The reader is enabled to learn the background and then access the literature in which studies designed to define the biology and biochemistry of known TSGs have been performed. Details of protocols with examples of their previous uses allow the researcher to apply current technologies to novel or known genes whose role has not yet been defined.

In summary, *Tumor Suppressor Genes* is a comprehensive compilation of the known tumor suppressing pathways, and the key molecular approaches to their discovery, analysis, and clinical applications. It should be of enduring value to students at all levels of experimental biology and medicine, as well as those clinicians who want to better understand the molecular biology of cancer.

Wafik S. El-Deiry, MD, PhD

Contents

CONTENTS OF THE COMPANION VOLUME
Volume 1: Pathways and Isolation Strategies

Contributors

MAMOUN AHRAM • *Molecular Biosciences, Pacific Northwest National Laboratory, Battelle, Richland, WA*

TERESA ACOSTA ALMEIDA • *Institute of Cancer Genetics, Department of Pathology, Columbia University, New York, NY*

DARIO C. ALTIERI • *Department of Cancer Biology and the Cancer Center, University of Massachusetts Medical School, Worcester, MA*

SALLY A. AMUNDSON • *National Cancer Institute, National Institutes of Health, Bethesda, MD*

ANDREW M. ARSHAM • *Abramson Family Cancer Research Institute and Howard Hughes Medical Institute, University of Pennsylvania School of Medicine, Philadelphia, PA*

JENNIFER J. ASCAÑO • *Division of Gastroenterology, University of Virginia Health System, Charlottesville, VA*

EVANS C. BAILEY • *Department of Biochemistry and Molecular Genetics, The University of Alabama at Birmingham, Birmingham, AL*

ALBERT S. BALDWIN, JR. • *Lineberger Comprehensive Cancer Center, University of North Carolina, Chapel Hill, NC*

VIMLA BAND • *Department of Radiation Oncology, New England Medical Center, Boston, MA*

BRYAN L. BETZ • *Department of Pathology and Laboratory Medicine, University of North Carolina, Chapel Hill, NC*

KAPIL BHALLA • *Department of Interdisciplinary Oncology, Moffitt Cancer Center and Research Institute, Tampa, FL*

SIHAM BIADE • *Department of Pharmacology, Fox Chase Cancer Center, Philadelphia, PA*

MICHAEL BIRRER • *Biomarkers Branch, Division of Clinical Sciences, National Cancer Institute, Rockville, MD*

MIKHAIL V. BLAGOSKLONNY • *Medicine Branch, National Institutes of Health, Bethesda, MD*

KATHY BOON • *Duke University Medical Center, Durham, NC*

RAINER K. BRACHMANN • *Departments of Medicine and Biological Chemistry, University of California at Irvine, Irvine, CA*

JAMES M. BUGNI • *McArdle Laboratory for Cancer Research, University of Wisconsin Medical School, Madison, WI*

JUDITH CAMPISI • *Life Sciences Division, Lawrence Berkeley National Laboratory, Berkeley, CA, and Buck Institute for Age Research, Novato, CA*

XINBIN CHEN • *Department of Cell Biology, University of Alabama at Birmingham, Birmingham, AL*

SANJEEV DAS • *Department of Microbiology and Cell Biology, Indian Institute of Science, Bangalore, India*

PAUL DENT • *Department of Radiation Oncology, Virginia Commonwealth University, Medical College of Virginia, Richmond, VA*

THEODORE L. DEWEESE • *Johns Hopkins Oncology Center, Baltimore, MD*

DAVID T. DICKER • *Howard Hughes Medical Institute, Departments of Medicine, Genetics, and Pharmacology, University of Pennsylvania School of Medicine, Philadelphia, PA*

MICHAEL DOHN • *Department of Cell Biology, University of Alabama at Birmingham, Birmingham, AL*

LAWRENCE A. DONEHOWER • *Division of Molecular Virology, Baylor College of Medicine, Houston, TX*

NORMAN R. DRINKWATER • *McArdle Laboratory for Cancer Research, University of Wisconsin Medical School, Madison, WI*

ANNIE DUTRIAUX • *Department of Molecular Biology, Cell Biology, and Biochemistry, Brown University, Providence, RI*

WAFIK S. EL-DEIRY • *Departments of Medicine, Genetics, and Pharmacology, University of Pennsylvania School of Medicine, Philadelphia, PA*

MICHAEL R. EMMERT-BUCK • *Pathogenetics Unit, Laboratory of Pathology, National Cancer Institute, National Institutes of Health, Bethesda, MD*

GREG H. ENDERS • *Departments of Medicine and Genetics, University of Pennsylvania School of Medicine, Philadelphia, PA*

ERVIN EPSTEIN, JR. • *Department of Dermatology, UCSF Medical Center, San Francisco, CA*

KERI FAIR • *Section of Hematology/Oncology, University of Chicago, Chicago, IL*

ANDREW P. FEINBERG • *Johns Hopkins University School of Medicine, Baltimore, MD*

ARMANDO FELSANI • *CNR, Istituto di Neurobiologia e Medicina Molecolare, Rome, Italy*

PAUL B. FISHER • *Department of Urology and Pathology, College of Physicians and Surgeons, Columbia University, New York, NY*

ALBERT J. FORNACE, JR. • *National Cancer Institute, National Institutes of Health, Bethesda, MD*

ADI F. GAZDAR • *Hamon Cancer Center, UT Southwestern Medical Center, Dallas, TX*

AMATO J. GIACCIA • *Department of Radiation Oncology, Stanford University School of Medicine, Stanford, CA*

DONNA GILFOR • *Department of Radiation Oncology, Medical College of Virginia, Virginia Commonwealth University, Richmond, VA*

ANTONIO GIORDANO • *Sbarro Institute for Cancer Research and Molecular Medicine, Temple University, Philadelphia, PA*

EYAL GOTTLIEB • *Abramson Family Cancer Research Institute, Department of Cancer Biology, University of Pennsylvania, Philadelphia, PA*

WILLIAM M. GRADY • *Division of Gastroenterology, Vanderbilt University Medical Center, Nashville, TN*

SUSAN F. GRAMMER • *Biotechwrite: Biomedical and Science Communications, Kalamazoo, MI*

STEPHEN GRANT • *Department of Medicine/Hematology, Medical College of Virginia, Virginia Commonwealth University, Richmond, VA*

ANDREI V. GUDKOV • *Lerner Research Institute, Department of Molecular Biology, Cleveland Clinic Foundation, Cleveland, OH*

RUTH HEMMER • *Department of Molecular Biology, Cell Biology, and Biochemistry, Brown University, Providence, RI*

MEENHARD HERLYN • *The Wistar Institute, Philadelphia, PA*

TYLER JACKS • *Center for Cancer Research and Howard Hughes Medical Institute, Massachusetts Institute of Technology, Cambridge, MA*

LESLEY J. JARDINE • *Department of Cancer Cell Biology, Harvard School of Public Health, Boston, MA*

STEPHEN N. JONES • *Department of Cell Biology, University of Massachusetts Medical Center, Worcester, MA*

JEREMY JUANG • *Department of Pathology, The Johns Hopkins Medical Institutions, Baltimore, MD*

SHRIHARI KADKOL • *Department of Pathology, The Johns Hopkins Medical Institutions, Baltimore, MD*

WILLIAM G. KAELIN, JR. • *Dana-Farber Cancer Institute, Boston, MA*

KUM KUM KHANNA • *Queensland Institute of Medical Research, Brisbane, Queensland, Australia*

ADI KIMCHI • *Department of Molecular Genetics and Virology, The Weizmann Institute of Science, Rehovot, Israel*

TAKAHIKO KOBAYASHI • *Hokkaido University Medical Hospital, Sapporo, Japan*

ELENA KOMAROVA • *Lerner Research Institute, Department of Molecular Biology, Cleveland Clinic Foundation, Cleveland, OH*

SMITA LAKHOTIA • *Department of Microbiology and Cell Biology, Indian Institute of Science, Bangalore, India*

FREDRICK S. LEACH • *Departments of Urology and Molecular and Human Genetics, Baylor College of Medicine, Houston, TX*

MICHELLE M. LE BEAU • *Section of Hematology/Oncology, University of Chicago, Chicago, IL*

CRISTINA T. LEWIS • *Pfizer Global Research and Development, San Diego, CA*

IURI D. LOURO • *Division of Hematology/Oncology, Department of Medicine, The University of Alabama at Birmingham, Birmingham, AL*

TIMOTHY K. MACLACHLAN • *CuraGen Corporation, Branford, CT*

LEE V. MADRID • *Lineberger Comprehensive Cancer Center, University of North Carolina, Chapel Hill, NC*

CARL G. MAKI • *Department of Cancer Cell Biology, Harvard School of Public Health, Boston, MA*

JAMES J. MANFREDI • *Derald H. Ruttenberg Cancer Center, Mount Sinai School of Medicine, New York, NY*

FRANCESCA M. MARASSI • *The Burnham Institute, La Jolla, CA*

SANFORD D. MARKOWITZ • *Howard Hughes Medical Institute, and Department of Medicine and Ireland Cancer Center, Case Western Reserve University and University Hospitals of Cleveland, Cleveland, OH*

MICHAEL R. MATTERN • *Department of Oncology Research, GlaxoSmithKline, King of Prussia, PA*

MATTHEW MAURER • *Derald H. Ruttenberg Cancer Center, Mount Sinai School of Medicine, New York, NY*

E. ROBERT McDONALD III • *Howard Hughes Medical Institute, Departments of Medicine, Genetics, and Pharmacology, University of Pennsylvania School of Medicine, Philadelphia, PA*

MARGARET E. McLAUGHLIN • *Center for Cancer Research and Howard Hughes Medical Institute, Massachusetts Institute of Technology, Cambridge, MA; Department of Pathology, Brigham and Women's Hospital, Boston, MA*

PATRICE J. MORIN • *Laboratory of Biological Chemistry, National Institute on Aging, National Institutes of Health, Baltimore, MD*

MAUREEN E. MURPHY • *Department of Pharmacology, The Fox Chase Cancer Center, Philadelphia, PA*

KIICHIRO NAKAJIMA • *Peptide Institute Co. Ltd., Osaka, Japan*

MARK A. NELSON • *Department of Pathology, College of Medicine, University of Arizona, Tucson, AZ*

WILLIAM G. NELSON • *Johns Hopkins Oncology Center, Baltimore, MD*

RAMADEVI NIMMANAPALLI • *Department of Interdisciplinary Oncology, Moffitt Cancer Center and Research Institute, Tampa, FL*

SUSAN NOZELL • *Department of Cell Biology, University of Alabama at Birmingham, Birmingham, AL*

PATRICK M. O'CONNOR • *Pfizer Global Research and Development, San Diego, CA*

MICHAEL OHH • *Dana-Farber Cancer Institute, Boston, MA*

JUN-ICHI OKANO • *Division of Gastroenterology, Department of Medicine, University of Pennsylvania School of Medicine, Philadelphia, PA*

MICHAEL S. O'REILLY • *Department of Radiation Oncology and Cancer Biology, M. D. Anderson Cancer Center, University of Texas, Houston, TX*

MARCO G. PAGGI • *Center for Experimental Research, Regina Elena Cancer Institute, Rome, Italy*

RAYMOND A. PAGLIARINI • *Boyer Center for Molecular Medicine, Yale University School of Medicine, New Haven, CT*

NICKOLAS PAPADOPOULOS • *Department of Pathology, Institute of Cancer Genetics, Columbia University, New York, NY; Present address: GMP Genetics, Waltham, MA*

RAMON PARSONS • *Department of Pathology, Columbia University, New York, NY*

STEVEN M. POWELL • *Division of Gastroenterology and Hepatology, University of Virginia Health System, Charlottesville, VA*

NADINE PUGET • *Beth Israel Deaconess Medical Center and Harvard Medical School, Boston, MA*

HUA QIAN • *Department of Surgery, Shanghai Second Medical University, Shangai, P. R. China*

LIANG QIAO • *Department of Radiation Oncology, Medical College of Virginia, Virginia Commonwealth University, Richmond, VA*

ANTHONY G. QUINN • *Experimental Medicine, AstraZeneca R&D Charnwood, Leicestershire, UK*

ANA T. QUIÑONES • *Yale University School of Medicine, New Haven, CT*

TIMOTHY R. REBBECK • *Department of Biostatistics and Epidemiology, University of Pennsylvania School of Medicine, Philadelphia, PA*

MARK REDSTON • *Department of Pathology, Brigham and Women's Hospital, Boston, MA*

M. STACEY RICCI • *Howard Hughes Medical Institute, Departments of Medicine, Genetics, and Pharmacology, University of Pennsylvania School of Medicine, Philadelphia, PA*

GREGORY J. RIGGINS • *Duke University Medical Center, Durham, NC*

GAVIN P. ROBERTSON • *Department of Pharmacology, Penn State College of Medicine, Hershey, PA*

IGOR B. RONINSON • *Department of Molecular Genetics, University of Illinois, Chicago, IL*

JACK A. ROTH • *Department of Thoracic and Cardiovascular Surgery, University of Texas M. D. Anderson Cancer Center, Houston, TX*

J. MICHAEL RUPPERT • *Department of Medicine, University of Alabama at Birmingham, Birmingham, AL*

ANIL K. RUSTGI • *Division of Gastroenterology, Department of Medicine, University of Pennsylvania School of Medicine, Philadelphia, PA*

LINDA SARGENT • *Genetic Susceptibility Team, Toxicology and Molecular Biology Branch, Health Effects Laboratory Division, NIOSH, Morgantown, WV*

KAPAETTU SATYAMOORTHY • *The Wistar Institute, Philadelphia, PA*

RALPH SCULLY • *Beth Israel Deaconess Medical Center, and Harvard Medical School, Boston, MA*

JOHN M. SEDIVY • *Department of Molecular Biology, Cell Biology, and Biochemistry, Brown University, Providence, RI*

JULIAN A. SIMON • *Program in Molecular Pharmacology, Fred Hutchinson Cancer Research Center, Seattle, WA*

M. CELESTE SIMON • *Abramson Family Cancer Research Institute and Howard Hughes Medical Institute, University of Pennsylvania School of Medicine, Philadelphia, PA*

LAURA SIMPSON • *Institute of Cancer Genetics, College of Physicians and Surgeons, Columbia University, New York, NY*

LORRAINE SNYDER • *Division of Gastroenterology, Department of Medicine, University of Pennsylvania School of Medicine, Philadelphia, PA*

KUMARAVEL SOMASUNDARAM • *Department of Microbiology and Cell Biology, Indian Institute of Science, Bangalore, India*

LI-KUO SU • *Department of Molecular and Cellular Oncology, M.D. Anderson Cancer Center, The University of Texas, Houston, TX*

KATSUYUKI TAMAI • *Cyclex Co. Ltd., Nagano, Japan*

YOICHI TAYA • *National Cancer Center Research Institute, Tokyo, Japan*

CRAIG B. THOMPSON • *Abramson Family Cancer Research Institute, Department of Cancer Biology, University of Pennsylvania, Philadelphia, PA*

EDWARD C. THORNBORROW • *Derald H. Ruttenberg Cancer Center, Mount Sinai School of Medicine, New York, NY*

MEREDITH UNGER • *Abramson Family Cancer Research Institute, University of Pennsylvania Cancer Center, Philadelphia PA*

JENNY VARLEY • *CR-UK Department of Cancer Genetics, Paterson Institute of Cancer Research, Manchester, UK*

ARVIND K. VIRMANI • *Hamon Cancer Center, UT Southwestern Medical Center, Dallas, TX*

NARENDRA WAJAPEYEE • *Department of Microbiology and Cell Biology, Indian Institute of Science, Bangalore, India*

TING WANG • *Department of Genetics, Washington University School of Medicine, St. Louis, MO*

VALERIE M. WEAVER • *Department of Pathology, The Institute for Medicine and Engineering, University of Pennsylvania, Philadelphia, PA*

ASHANI T. WEERARATNA • *Cancer Genetics Branch, National Human Genome Research Institute, National Institutes of Health, Bethesda, MD*

WENYI WEI • *Department of Molecular Biology, Cell Biology, and Biochemistry, Brown University, Providence, RI*

BERNARD E. WEISSMAN • *Department of Pathology and Laboratory Medicine, University of North Carolina, Chapel Hill, NC*

BRYAN R. G. WILLIAMS • *Lerner Research Institute, Department of Cancer Biology, The Cleveland Clinic Foundation, Cleveland, OH*

AMY WILLIS • *Department of Cell Biology, University of Alabama at Birmingham, Birmingham, AL*

GEN SHENG WU • *Karmanos Cancer Institute, Department of Pathology, Wayne State University School of Medicine, Detroit, MI*

TZYY-CHOOU WU • *Department of Pathology, The Johns Hopkins Medical Institution, Baltimore, MD*

TIAN XU • *Boyer Center for Molecular Medicine, Yale University School of Medicine, New Haven, CT*

TIMOTHY J. YEN • *Fox Chase Cancer Center, Philadelphia, PA*

KUMIKO YOSHIZAWA-KUMAGAYE • *Peptide Institute Co. Ltd., Osaka, Japan*

ZHONG YUN • *Division of Radiation and Cancer Biology, Department of Radiation Oncology, Stanford University School of Medicine, Stanford, CA*

BIN-BING S. ZHOU • *Incyte Genomics, Newark, DE*

I

UNDERSTANDING THE FUNCTION AND REGULATION OF TUMOR SUPPRESSOR GENES

1

Utilizing NMR to Study the Structure of Growth-Inhibitory Proteins

Francesca M. Marassi

1. Introduction

The underlying premise of structural biology is that the fundamental understanding of biological functions lies in the three-dimensional structures of proteins and other biopolymers. The two well-established experimental methods for determining the high-resolution structures of proteins have both contributed to the wealth of structural information available for the tumor suppressor genes. The tumor suppressor proteins whose structures have been determined by nuclear magnetic resonance (NMR) spectroscopy are listed in **Table 1**. Although X-ray crystallography plays a central role in high-throughput structure determination in the current structural genomics efforts, several features of NMR spectroscopy make it extremely well suited for three-dimensional structure determination as well as for the structure–function analysis of proteins *(1,2)*.

An important advantage of NMR spectroscopy is that it does not require crystals for structure determination, so that NMR structural studies can be carried out in samples that are similar to physiologic conditions in which the protein is normally functional. Indeed, NMR spectroscopy can be applied to a wide variety of samples, ranging from isotropic solutions to crystalline powders, including those with slowly reorienting or immobile macromolecules, such as membrane proteins in lipid environments *(3,4)*. This is especially significant because many proteins are insoluble and do not provide crystals of suitable quality for crystallographic analysis.

NMR is capable of resolving signals from all atomic sites in proteins, and each site has several well-characterized nuclear spin interactions that can be used as sources of information about molecular structure and dynamics, as well as chemical interactions. The spin interactions can be probed through radiofrequency irradiations and sample manipulations, and provide an immediate characterization of the foldedness of proteins in solution, even prior to complete three-dimensional structure determination. They also provide the basis for a structure-guided approach to the design and optimization of high-affinity ligands and to screen libraries of potential drugs *(5,6)*.

From: *Methods in Molecular Biology, Vol. 223: Tumor Suppressor Genes: Regulation, Function, and Medicinal Applications.* Edited by: Wafik S. El-Deiry © Humana Press Inc., Totowa, NJ

Table 1
Tumor Suppressor Proteins Whose Structures Have Been Determined by NMR Spectroscopy in Solution, with Protein Data Bank Identification (PDB ID) Codes (http://www.rcsb.org/pdb/)

Tumor suppressor structures determined by NMR spectroscopy	PDB ID
Refined solution structure of the oligomerization domain of the tumour suppressor p53 *(39,40)*	1SAE, 1SAF, 1SAG, 1SAH, 1SAI, 1SAJ, 1SAK, 1SAL
Solution structure determination of a p53 mutant dimerization domain *(44)*	1AU1
NMR solution structure of designed p53 dimer *(63)*	1HS5
Solution structure of a conserved C-terminal domain of p73 with structural homology to the Sam domain *(64)*	1COK
Solution structure of P18-Ink4C, 21 structures *(56)*	1BU9
Tumor suppressor P16Ink4A: determination of solution structure and analyses of its interaction with cyclin-dependent kinase 4 *(58)*	1A5E, 2A5E
Solution NMR structure of tumor suppressor P16Ink4A *(59)*	1DC2
Tumor suppressor P15(Ink4B) structure by comparative modeling and NMR data *(59)*	1D9S
High-resolution solution structure of human pNR-2/pS2: a single trefoil motif protein *(65)*	1PS2
NMR solution structure of the disulfide-linked homo-dimer of human Tff1 *(66)*	1HI7

1.1. NMR of Soluble Proteins

Solution NMR methods rely on rapid molecular reorientation for line narrowing, and standard multidimensional solution NMR methods can be successfully applied to proteins in the size range of 25–35 kDa *(7)*. A recent analysis estimates that at least 25% of open reading frames in a genome will be suitable for NMR structure determination, and that 15–20% of new protein structures are determined by NMR methods *(1)*. The recent advances in high-field-magnet technology, cryogenic probes, partial sample deuteration *(8)*, and transverse relaxation optimized spectroscopy (TROSY) *(9)* all increase the sensitivity of the NMR experiments and extend the size limit of proteins that can have their structures determined by solution NMR to 50 kDa. Although large multidomain proteins are not generally suitable for structure determination by solution NMR spectroscopy, these also tend to exhibit flexibility in the domain linker regions, which can impede crystallization. As a result, the majority of structural information for these systems, including many tumor suppressor proteins and their complexes, comes from studies of their individual domains using both NMR and X-ray.

1.2. NMR of Membrane Proteins

NMR spectroscopy can also be extended to the study of membrane proteins, which do not easily yield high quality crystals, and thus pose a considerable problem for crystallographic analysis *(3,4)*. Solution NMR methods can be successfully applied to relatively small membrane proteins in micelles, although in this case the size limitation is substantially more severe because the many lipid molecules associated with each protein slow its overall reorientation rate *(4)*. Using currently available instruments and methods, it is difficult to resolve, assign, and measure the long-range Nuclear Overhauser Effects (NOEs) between hydrogens on hydrophobic side chains that are needed to determine tertiary structures based on distance constraints. However, the ability to weakly align membrane proteins in micelles enables the measurement of residual dipolar couplings, and improves the feasibility of determining the structures of membrane proteins using solution NMR methods *(10,11)*.

Nonetheless, it is highly desirable to determine the structures of membrane proteins in the definitive environment of lipid bilayer membranes, where solution NMR methods fail completely. Solid-state NMR spectroscopy is well suited for this task, with both oriented-sample and magic-angle spinning methods providing approaches to measure orientational and distance parameters for structure determination *(3,4)*. Several tumor suppressor genes encode membrane-bound proteins, for example, the deleted in colon cancer (DCC) and the neurofibromatosis type 2 (NF-2) genes, and solid-state NMR provides an important approach toward their structure determination.

2. NMR of Proteins in Solution

The determination of protein structures by multidimensional solution NMR spectroscopy is straightforward in principle, and for globular proteins that are soluble and do not aggregate in aqueous solution, the application of this approach is generally straightforward in practice as well, especially if uniformly ^{15}N- and/or uniformly ^{15}N- and ^{13}C-labeled samples can be prepared by expression in bacteria *(12,13)*. The strategy for protein structure determination by NMR spectroscopy is outlined in **Fig. 1** and described below.

2.1. Expression of Isotopically Labeled Proteins

The development of expression systems for the production of isotopically labeled proteins is as important as that of pulse sequences or instrumentation for the success of NMR structural studies. The expression of isotopically labeled proteins can be obtained in several organisms including bacteria, insect, yeast or human cells, and in cell-free expression systems *(14)*, however the most commonly used expression strategy is bacterial expression via an inducible T7 RNA polymerase promoter *(4,15,16)*.

Several *Escherichia coli* expression systems are available, all of which involve the use of fusion proteins. The incorporation of engineered affinity tags, such as poly-His tags for metal affinity chromatography, is often used to simplify protein isolation and purification. This process can be further facilitated by selecting fusion partners that form inclusion bodies After inclusion body isolation, and fusion protein affinity purification and cleavage, the resulting target protein is purified and then dissolved in the appropriate buffer for NMR studies.

The ability to express proteins in bacteria provides the opportunity to incorporate a variety of isotopic labeling schemes into the overall experimental strategy, since it allows both selective and uniform labeling. For selective labeling by amino acid type, the bacteria harboring the protein gene are grown on defined media, where only the amino acid of interest is labeled and the others are not. Uniform labeling, where all the nuclei of one or several types (^{15}N, ^{13}C, ^{2}H) are incorporated in the protein, is accomplished by growing the bacteria on defined media containing ^{15}N-labeled ammonium

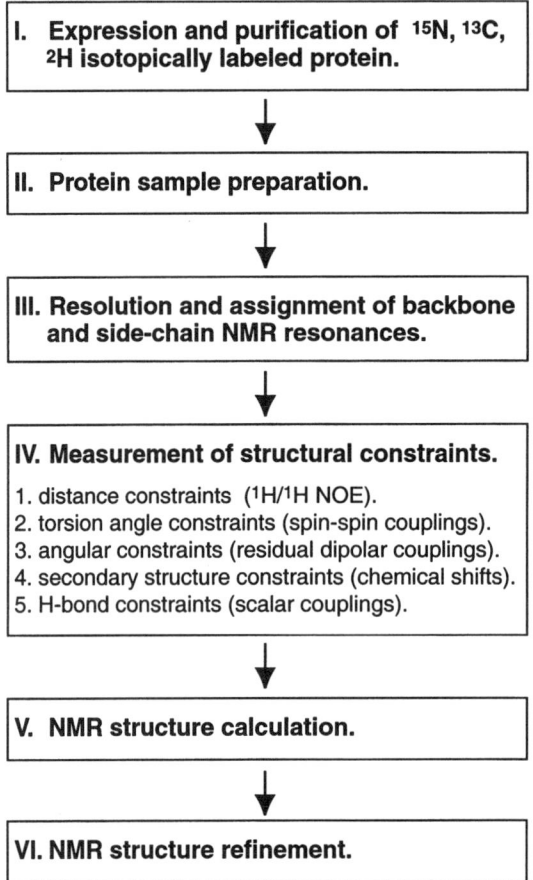

Fig. 1. Strategy for protein structure determination by solution NMR spectroscopy.

sulfate, or ^{13}C-labeled glucose, or D_2O, or a combination of these. The availability of uniformly labeled samples is a prerequisite for triple-resonance $^{13}C/^{15}N/^1H$ spectroscopy, which is essential for the structure determination of larger proteins and protein complexes in solution.

2.2. Protein Sample Preparation

The primary goal in NMR sample preparation is to reduce the effective rotational correlation time of the protein as much as possible, so that resonances will have the narrowest achievable line widths. Careful handling of the protein throughout the purification is essential, since subtle changes in the protocol can have a significant impact on the quality of the resulting spectra. It is essential to optimize protein concentrations, counterions, pH, and temperature, in order to obtain well-resolved two-dimensional heteronuclear correlation NMR spectra with narrow ^1H and ^{15}N resonance line width. Narrow line widths in both frequency dimensions, and the presence of one well-defined resonance for each amide site in the protein, reflect a high-quality sample *(4,16)*. As the protein size increases, solubilization generally becomes more difficult and aggregation more likely.

2.3. Protein Structure Determination

2.3.1. Resolution and Assignment of Backbone and Side-Chain Resonances

The resolution and assignment of backbone and side-chain resonances are based on both through-bond and through-space spin interactions, and are observed in two- and three-dimensional NMR spectra. There are basically two strategies for assigning resolved resonances to specific residues in a protein. One involves short-range homonuclear ^{1}H/^{1}H NOEs *(12,13)*, and the other relies on spin–spin couplings in uniformly ^{15}N- and ^{13}C-labeled proteins *(17–19)*. The procedure starts with heteronuclear edited TOCSY experiments, supplemented with triple-resonance ^{13}C/^{15}N/^{1}H experiments. Selective isotopic labeling may be necessary in order to resolve and assign some of the resonances, especially in cases of limited chemical shift dispersion. Further, the incorporation of ^{2}H is often needed in studies of larger proteins or protein complexes, in order to limit spin diffusion and line broadening.

2.3.2. Measurement of Structural Constraints

The measurements of as many homonuclear ^{1}H/^{1}H NOEs as possible among the assigned resonances provide the short-range and long-range distance constraints required for structure determination. The cross-peaks between pairs of ^{1}H nuclei in the protein structure are grouped into three classes of strong, medium and weak intensity, corresponding to interhydrogen distances of 1.9–2.5 Å, 1.9–3.5 Å, and 3.0–5.0 Å, respectively. These are supplemented by other structural constraints, such as spin–spin coupling constants and chemical shifts, in order to assign resonances, obtain torsion angle and H-bond constraints, and to characterize the secondary structure of the protein. The ^{13}Cα and ^{13}Cβ chemical shifts are particularly useful for characterizing secondary structure in the early stages of structure determination *(20,21)*. The amide resonances detected in a two-dimensional ^{1}H/^{15}N correlation spectrum at different times after the addition of D$_2$O to the sample can be used to assign hydrogen bond constraints.

The measurements of residual dipolar couplings from weakly aligned protein samples provide direct long-range angular constraints with respect to a molecule-fixed reference frame, which can be used for structure determination *(22,23)*. Aqueous solutions containing bicelles *(24)*, purple membrane fragments *(25)*, or rod-shaped viruses *(26,27)* have all been successfully employed to obtain residual couplings in soluble proteins and other macromolecules, although these media can also complicate studies of large proteins and complexes, since the increased solvent viscosity leads to reorientation rates that are too slow to give adequately resolved spectra. In addition, lanthanide ions can be used to weakly align membrane proteins in lipid micelles *(10,11)*.

2.3.3. Structure Calculation and Refinement

Structure determination involves the interpretation of the distance and angular constraints in terms of secondary and tertiary protein structure. This is achieved through a combination of distance geometry, simulated annealing, molecular dynamics, and other calculations, and yields a family of energy-minimized, three-dimensional protein structures *(13)*. This final stage of the structure determination procedure requires essentially complete assignment of the protein resonances. The lack of a significant number of unambiguously assigned long-range NOEs has limited the ability of solution NMR spectroscopy to determine the tertiary structures of larger proteins, protein complexes, and membrane proteins. Fortunately, the measurement of residual dipolar couplings from weakly aligned protein samples offers an additional set of constraints for structure determination. These couplings can be used to overcome limitations resulting from having few long-range NOE distance restraints. Structures are calculated by inclusion of all available distance and orientational constraints *(28,29)*.

3. NMR Structural Studies of Tumor Suppressor Proteins

3.1. Structure of the p53 Tumor Suppressor

The p53 tumor suppressor protein is a 393-residue transcription factor that activates genes involved in the control of the cell cycle and apoptosis, in response to DNA damage *(30)*. Because over one-half of all human cancers involve mutations or deletions of p53, this molecule has been the subject of several structural studies aimed at understanding the differences between the wild-type and mutant molecules *(31)*. The full-length protein comprises an acidic trans-activation domain (residues 1–70), a DNA-binding domain (residues 90–300), a homo-tetramerization domain (residues 324–355), and basic regulatory domain (residues 355–393). The structures of several domains of p53 have been determined by NMR and/or X-ray crystallography. Recently, the NMR spectrum of a 67-kDa dimer of p53, comprising the DNA-binding and oligomerization domains, has been assigned for structure determination *(32)*. This was possible through the use of triple resonance and TROSY spectroscopy of $^{15}N^-$, $^{13}C^-$ and ^2H-labeled protein.

Structures of the DNA-binding domain in complex with target DNA and with p53-binding protein 2 *(33,34)* have been determined by X-ray crystallography. The structure of the trans-activation domain complexed with the MDM2 oncoprotein *(35)* was determined by X-ray crystallography, and multidimensional NMR spectroscopy was utilized to identify chalcone derivative MDM2 inhibitors that bind to a subsite of the p53 tumor suppressor-binding cleft of human MDM2 *(36)*. Solution NMR spectroscopy was utilized to compare the structure of the p53 DNA-binding domain in wild-type and mutant p53, and monitor the structural changes introduced by hot-spot mutations. By following changes in chemical shifts, the mutation R248Q, which was believed to affect only interactions with DNA, was shown to introduce structural changes that perturb the structure of the p53 DNA-binding domain *(37)*.

The structure of the tetramerization domain has been determined by both NMR spectroscopy *(38–40)* and crystallography *(41,42)*. The tetramerization domain is required for tumor suppressor activity *(43)*, and since it is only 30 residues long and its function can be easily assayed, it well suited for structural studies. Its solution structure, shown in **Fig. 2**, consists of a dimer of two primary dimers, with a well-defined globular hydrophobic core, whose subunits form a β-strand, followed by a tight turn and an α-helix. NMR studies demonstrate that conservative hydrophobic amino acid mutations influence the helix packing and disrupt tetramerization of the p53 complex *(44)*.

Recently, two new p53 homologs, p63 and p73, have been identified (reviewed in **ref. *31***). The high level of sequence identity in critical functional regions of the p53, p63, and p73 molecules suggests that the three-dimensional structures of their respective domains will be very similar. In addition, the new family members have a conserved C-terminal domain with a predicted regulatory function. The solution structure of this

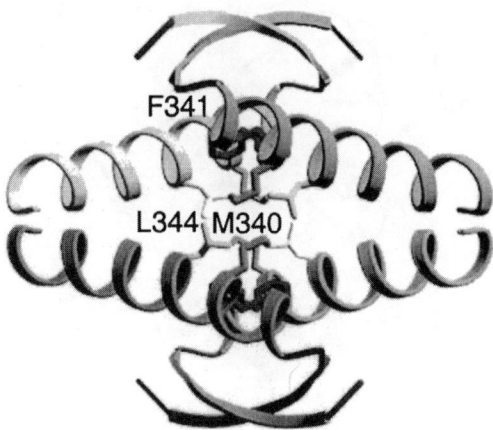

Fig. 2. Solution NMR structure of the p53 tetramerization domain (PDB ID 3SAK) *(40)*. The residues that switch the domain packing and stoichiometry upon substitution are shown *(44)*. The letters N and C respectively identify the amino and carboxy termini of the protein.

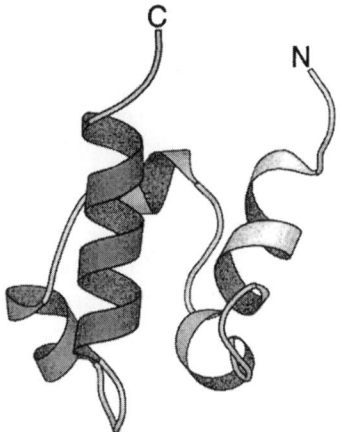

Fig. 3. Solution NMR structure of the p73 SAM domain (PDB ID 1COK) *(64)*. The letters N and C respectively identify the amino and carboxy termini of the protein.

domain has been determined by NMR spectroscopy and is shown in **Fig. 3** *(31)*. It forms a 5-helix bundle similar to those of sterile α-motif (SAM) domains from Ephrin tyrosine kinases, suggesting that it is a protein–protein interaction module, possibly involved in developmental processes.

Finally, the structure of the Ca^{2+} signaling protein S100B in complex with p53 has been determined using NMR spectroscopy *(45,46)*. Upon Ca^{2+} binding to its EF hand, S100B undergoes a large conformational change that is a prerequisite for its interaction with p53 *(47,48)*. This, in turn, inhibits protein kinase C-dependent phosphorylation of p53 at residues Ser376 and Thr377 in its C-terminal regulatory domain, and provides a mechanism for regulating the cellular functions of the tumor suppressor. S100B inhibits p53 tetramerization, and promotes dissociation of the p53 tetramer *(49)*. In addition, it has been shown to protect p53 from thermal denaturation and aggregation in vitro. The solution structure shows that the S100B homo-dimer recognizes two molecules of p53 and inhibits its posttranslational modification.

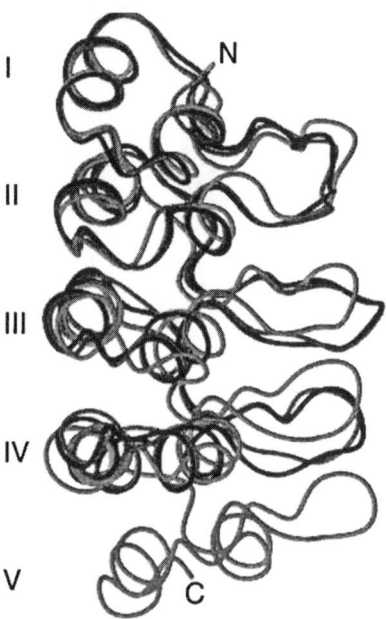

Fig. 4. Superposition of the solution NMR structures of the tumor suppressor INK4 p15, p16 and p18 proteins (PDB IDs 1D9S, 1DC2, 1BU9) *(56,59)*. The helix–turn–helix ankyrin repeats are numbered I through V. The letters N and C respectively identify the amino and carboxy termini of the protein.

3.2. Structures of the Tumor Suppressors INK4

The cyclin-dependent kinase (CDK) inhibitors bind to CDKs and inhibit their kinase activity, thus regulating some of the most fundamental decisions in the cell cycle. The INK4 (inhibitor of cyclin-dependent kinase 4) family consists of four tumor suppressor proteins, p15, p16, p18, and p19, ranging in size from 13.7 to 18 kDa *(50–53)*. Among these, mutations in p16 have been tied to the development of cancer, and the tumor suppressor function is well established for p16 and to a lesser extent for p15. Three-dimensional structures of the INK4 proteins have been determined using both X-ray crystallography and NMR spectroscopy, with the following structures reported in recent years: the solution *(54)* and crystal *(55)* structures of p19; the solution *(56)* and crystal *(57)* structures of p18; the solution structure of p16 *(58,59)*; and the solution structure of p15 *(59)*.

All the INK4 family members are highly homologous in sequences and structures, and fold as ankyrin repeats, arrays of four (p15, p16) or five (p18, p19) 33-residue helix–turn–helix motifs connected by long loops, as shown in **Fig. 4**. Despite their considerable homology, they also show appreciable differences in conformational flexibility, stability, and aggregation tendency. Because the smaller INK4 proteins, p15 and p16, display the highest degree of conformational flexibility and instability, no crystal structures have been reported for their free forms. However, their NMR structures could be determined in solution, and were refined at high resolution through the use of high-field spectroscopy at 800 MHz *(59)*.

3.3. Structural Studies of the Wilms Tumor Suppressor Protein

NMR spectroscopy has been used to study the structural changes resulting from post-transcriptional modification of the Wilms tumor suppressor protein (WT1) in the 4-zinc finger DNA-binding domain *(60)*. WT1 is a transcription factor that contains a C-terminal DNA-binding domain with four Cys2His2 zinc fingers, a Pro/Glu-rich N-terminus, an activation and a repressor domain, nuclear localization signals, and self-association domains. Its function is modulated by a posttranscriptional modification that adds three amino acids into one of the linker regions between the DNA-binding zinc fingers. NMR resonance assignments and chemical shift changes were used to characterize the structural differences between two isoforms of the WT1 DNA-binding domain, with a (Lys-Thr-Ser) sequence insertion and without it. These studies were carried out both with WT1 free in solution and in complex with a 14-base DNA duplex corresponding to the WT1 recognition element. In the absence of the DNA, the two isoforms are nearly identical in structure; however, the linker regions become more structured upon DNA binding, and insertion of the Lys-Thr-Ser sequence disrupts important interactions of the linker region with the adjacent zinc fingers, thus lowering the stability of the complex with DNA *(60)*. Using NMR, it was also shown that DNA binding induces a conformational change and helix capping in the conserved zinc finger-linker region of WT1 *(61)*.

3.4. Binding of Elongin C to a von Hippel–Lindau Tumor Suppressor Peptide

NMR spectroscopy was used to study the structural basis for the interaction of Elongin A, an F-box-containing protein, with Elongin C, a SKP1 homolog, and the modulation of this interaction by the tumor suppressor von Hippel-Lindau protein (VHL) *(62)*. Elongin is a hetero-trimeric transcription elongation factor composed of subunits A, B, and C in mammals. Complexes of elongin C with elongin A and with a peptide from the VHL tumor suppressor were analyzed by NMR. Elongin C was shown to oligomerize in solution and to undergo significant structural rearrangements upon binding of its two partner proteins.

4. Conclusions

NMR spectroscopy is extremely well suited to determine the structures and dynamics of tumor suppressor proteins and to study their interactions in complexes with proteins, DNA, or drug molecules. The methods for expression and purification of proteins from bacteria and the preparation of samples are as important as the instrumentation and methods for the NMR experiments. Recent technological advances in NMR spectroscopy enhance the sensitivity of the experiments, and extend the size range of molecules that can have their structures determined by NMR. Thus, the prospects for expanding the current tumor suppressor gene structure database are excellent, as structural studies are extended beyond the single domain, to multiple domains or full-length proteins and their complexes *(1,32)*, and as solid-state NMR spectroscopy is used to determine the structures of membrane-bound tumor suppressor proteins *(3,4)*.

Acknowledgments

The author thanks the Department of Defense Breast Cancer Research Program (DAMD-17-00-1–0506) and the W.W. Smith Charitable Trust (H9804) for grant support.

References

1. Montelione, G. T., Zheng, D., Huang, Y. J., Gunsalus, K. C., and Szyperski, T. (2000) Protein NMR spectroscopy in structural genomics. *Nature Struct. Biol., Struct. Genomics Suppl.* **7**, 982–985.
2. Wuthrich, K. (1998) The second decade into the third millenium. *Nat. Struct. Biol., NMR Suppl.* **5**, 492–495.
3. Marassi, F. M. and Opella, S. J. (1998) NMR structural studies of membrane proteins. *Curr. Opin. Struct. Biol.* **8**, 640–648.
4. Opella, S. J., Ma, C., and Marassi, F. M. (2001). NMR of membrane associated peptides and proteins, *Meth. Enzymol.* **339**, in press.
5. Schuker, S. B., Hajduk, P. J., Meadows, R. P., and Fesik, S. W. (1996) Discovering high affinity ligands for proteins: SAR by NMR. *Science* **274**, 1531–1534.
6. Moore, J. M. (1999) NMR screening in drug discovery. *Curr. Opin. Biotechnol.* **10**, 54–58.
7. Clore, G. M. and Gronenborn, A. M. (1997) NMR structures of proteins and protein complexes beyond 20,000 Mr. *Nat. Struct. Biol. NMR Suppl.* **4**, 849–853.
8. Gardner, K. H. and Kay, L. E. (1998) The use of ^{2}H, ^{13}C, ^{15}N multidimensional NMR to study the structure and dynamics of proteins. *Annu. Rev. Biophys. Biomol. Struct.* **27**, 357–406.
9. Pervushin, K., Riek, R., Wider, G., and Wuthrich K. (1997) Attenuated T2 relaxation by mutual cancellation of dipole-dipole coupling and chemical shift anisotropy indicates an avenue to NMR structures of very large biological macromolecules in solution. *Proc. Natl. Acad. Sci. USA* **94**, 12366–12371.
10. Veglia, G. and Opella, S. J. (2000) Lanthanide ion binding to adventitious sites aligns membrane proteins in micelles for solution NMR spectroscopy. *J. Am. Chem. Soc.* **122**, 11733–11734.
11. Ma, C. and Opella, S. J. (2000) Lanthanide ions bind specifically to an added EF-hand and orient a membrane protein in micelles for solution NMR spectroscopy. *J. Magn. Reson.* **146**, 381–384.
12. Wuthrich, K. (1986) *NMR of Proteins and Nucleic Acids.* Wiley, New York.
13. Clore, G. M. and Gronenborn, A. M. (1989) Determination of three-dimensional structures of proteins and nucleic acids in solution by nuclear magnetic resonance spectroscopy. *Crit. Rev. Biochem. Mol. Biol.* **24**, 479–564.
14. Kigawa, T., Yabuki, T., Yoshida, Y., et al. (1999) Cell-free production and stable-isotope labeling of milligram quantities of proteins. *FEBS Lett.* **442**, 15–19.
15. Studier, F. W. and Moffat, B. A. (1986) Use of bacteriophage T7 RNA polymerase to direct selective high-level expression of cloned genes. *J. Mol. Biol.* **189**, 113–130.
16. Edwards, A. M., Arrowsmith, C. H., Christendat, D., et al. (2000) Protein production:feeding the crystallographers and NMR spectroscopists. *Nat. Struct. Biol. Struct. Genomics Suppl.* **7**, 970–972.
17. Ikura, M., Krinks, M., Torchia, D. A., and Bax, A. (1990) An efficient NMR approach for obtaining sequence-specific resonance assignments of larger proteins based on multiple isotopic labeling. *FEBS Lett.* **266**, 155–158.
18. Ikura, M., Kay, L. E., and Bax, A. (1990) A novel approach for sequential assignment of ^{1}H, ^{13}C, and ^{15}N spectra of proteins: heteronuclear triple-resonance three-dimensional NMR spectroscopy. Application to calmodulin. *Biochemistry* **29**, 4659–4667.
19. Moseley, H. N. and Montelione, G. T. (1999) Automated analysis of NMR assignments and structures for proteins. *Curr. Opin. Struct. Biol.* **9**, 635–642.
20. Wishart, D. S., Sykes, B. D., and Richards, F. M. (1991) Relationship between nuclear magnetic resonance chemical shift and protein secondary structure. *J. Mol. Biol.* **222**, 311–333.
21. Wishart, D. S., Sykes, B. D., and Richards, F. M. (1992) The chemical shift index: a fast and simple method for the assignment of protein secondary structure through NMR spectroscopy. *Biochemistry* **31**, 1647–1651.

22. Tolman, J. R., Flanagan, J. M., Kennedy, M. A., and Prestegard, J. H. (1995) Nuclear magnetic dipole interactions in field-oriented proteins: information for structure determination in solution. *Proc. Natl. Acad. Sci. USA* **92**, 9279–9283.
23. Tjandra, N., Grzesiek, S. and Bax, A. (1996) Magnetic field dependence of nitrogen-proton J splittings in ^{15}N-enriched human ubiquitin resulting from relaxation interference and residual dipolar coupling. *J. Am. Chem. Soc.* **118**, 6264–6272.
24. Tjandra, N. and Bax, A. (1997). Direct measurement of distances and angles in biomolecules by NMR in a dilute liquid crystalline medium. *Science* **278**, 1111–1114.
25. Sass, J., Cordier, F., Hoffmann, A., et al. (1999) Purple membrane induced alignment of biological macromolecules in the magnetic field. *J. Am. Chem. Soc.* **121**, 2047–2055.
26. Hansen, M. R., Mueller, L., and Pardi, A. (1998) Tunable alignment of macromolecules by filamentous phage yields dipolar coupling interactions. *Nat. Struct. Biol.* **5**, 1065–1074.
27. Clore, G. M., Starich, M. R., and Gronenborn, A. M. (1998) Measurement of residual dipolar couplings of macromolecules aligned in the nematic phase of a colloidal suspension of rod-shaped viruses. *J. Am. Chem. Soc.* **120**, 10571–10572.
28. Brunger, A. T., Adams, P. D., Clore, G. M., et al. (1998) Crystallography & NMR system: a new software suite for macromolecular structure determination. *Acta Crystallogr.* **D54**, 905–921.
29. Prestegard, J. H. (1998) New techniques in structural NMR—anisotropic interactions. *Nat. Struct. Biol. NMR Suppl.* **5**, 517–522.
30. Levine, A. J. (1997) p53, the cellular gatekeeper for growth and division. *Cell* **88**, 323–331.
31. Arrowsmith, C.H. (1999) Structure and function of the p53 family. *Cell Death Diff.* **6**, 1169–1173.
32. Mulder, F. A., Ayed, A., Yang, D., Arrowsmith, C. H., and Kay, L. E. (2000) Assignment of ^{1}H(N), ^{15}N, ^{13}C(alpha), ^{13}CO and ^{13}C(beta) resonances in a 67 kDa p53 dimer using 4D-TROSY NMR spectroscopy. *J. Biomol. NMR* **18**, 173–176.
33. Cho, Y., Gorina, S., Jeffrey, P. D., and Pavletich, N. P. (1994) Crystal structure of a p53 tumor suppressor-DNA complex: understanding tumorigenic mutations. *Science* **265**, 346–355.
34. orina, S. and Pavletich, N. P. (1996) Structure of the p53 tumor suppressor bound to the ankyrin and SH3 domains of 53BP2. *Science* **274**, 1001.
35. Kussie, P. H., Gorina, S., Marechal, V., et al. (1996). Structure of the MDM2 oncoprotein bound to the p53 tumor suppressor transactivation domain. *Science* **274**, 948–953.
36. Stoll, R., Renner, C., Hansen, S., et al. (2001) Chalcone derivatives antagonize interactions between the human oncoprotein MDM2 and p53. *Biochemistry* **40**, 336–344.
37. Wong K. B., DeDecker, B. S., Freund, S. M., Proctor M. R., Bycroft M., and Fersht A. R. (1999) Hot-spot mutants of p53 core domain evince characteristic local structural changes. *Proc. Natl. Acad. Sci. USA* **96**, 8438–8442.
38. Lee, W., Harvey, T. S., Yin, Y., Yau, P., Litchfield, D., and Arrowsmith C. H. (1994) Solution structure of the tetrameric minimum transforming domain of p53. *Nat. Struct. Biol.* **1**, 877–890.
39. Clore, G. M., Ernst, J., Clubb, R., et al. (1995) Refined solution structure of the oligomerization domain of the tumour suppressor p53. *Nat. Struct. Biol.* **2**, 321–333.
40. Kuszewski, J., Gronenborn, A. M., and Clore, G. M. (1999) Improving the packing and accuracy of NMR structure with a pseudopotential for the radius of gyration. *J. Am. Chem. Soc.* **121**, 2337–2338.
41. Jeffrey, P. D., Gorina, S., and Pavletich, N. P. (1995) Crystal structure of the tetramerization domain of the P53 tumor suppressor at 1.7 angstroms. *Science* **267**, 1498.
42. Mittl, P. R., Chene, P., and Grutter, M. G. (1998) Crystallization and structure solution of p53 (residues 326-356) by molecular replacement using an NMR model as template. *Acta Crystallogr.* **D54**, 86–89.
43. Pietenpol, J. A., Tokino, T., Thiagalingam, S., El-Deiry, W. S., Kinzler, K. W., and Vogelstein, B. (1994) Sequence-specific transcriptional activation is essential for growth suppression by p53. *Proc. Natl. Acad. Sci USA* **91**, 1998–2002.

44. McCoy, M., Stavridi, E. S., Waterman, J. L., Wieczorek, A. M., Opella, S. J., and Hala-zonetis, T. D. (1997) Hydrophobic side-chain size is a determinant of the three-dimensional structure of the p53 oligomerization domain. *EMBO J.* **16**, 6230.

45. Rustandi, R.R., Drohat, A.C., Baldisseri, D.M. Wilder, P.T., and Weber, D.J. (1998) The Ca(2+)-dependent interaction of S100B(βββ) with a peptide derived from p53. *Biochemistry* **37**, 1951–1960.

46. Rustandi, R. R., Baldisseri, D. M., and Weber D. J. (2000) Structure of the negative regulatory domain of p53 bound to S100B(ββ). *Nat. Struct. Biol.* **7**, 570–574.

47. Baudier, J., Delphin, C., Grunwald, D., Khochbin, S., and Lawrence, J. J. (1992) Characterization of the tumor suppressor protein p53 as a protein kinase C substrate and a S100b-binding protein. *Proc. Natl. Acad. Sci. USA* **89**, 11627–11631.

48. Delphin, C., Ronjat, M., Deloulme, J. C., et al. (1999) Calcium-dependent interaction of S100B with the C-terminal domain of the tumor suppressor p53. *J. Biol. Chem.* **274**, 10539–10544.

49. Scotto, C., Deloulme, J. C., Rousseau, D., Chambaz, E., and Baudier, J. (1998) Calcium and S100B regulation of p53-dependent cell growth arrest and apoptosis. *Mol. Cell Biol.* **18**, 4272–4281.

50. Hannon, G. J. and Beach, D. (1994) p15 INK4B is a potential effector of TGF-beta-induced cell cycle arrest. *Nature* **371**, 257–261.

51. Serrano, M., Hannon, G. J., and Beach, D. (1993) A new regulatory motif in cell-cycle control causing specific inhibition of cyclinD0CDK4. *Nature* **366**, 704–707.

52. Guan, K. L., Jenkins, C. W., Li, Y., et al. (1994) Growth suppression by p18, a p16 INK40MTS1—and p14 INK4B0MTS2—related CDK6 inhibitor, correlates with wild-type pRb function. Genes Dev. 8, 2939–2352.

53. Guan, K. L., Jenkins, C. W., Li, Y., et al. (1996) Isolation and characterization of p19 INK4d, a p16-related inhibitor specific to CDK6 and CDK4. *Mol. Biol. Cell.* **7**, 57–70.

54. Luh, F. Y., Archer, S. J., Domaille, P. J., et al. (1997) Structure of the cyclin-dependent kinase inhibitor p19 INK4d. *Nature* **389**, 999–1003.

55. Baumgartner, R., Fernandez-Catalan, C., Winoto, A., Huber, R., Engh, R. A., and Holak, T. A. (1998) Structure of human cyclin-dependent kinase inhibitor p19INK4d: comparison to known ankyrin-repeat-containing structures and implications for the dysfunction of tumor suppressor p16INK4a. *Structure* **6**, 1279–1290.

56. Li, J., Byeon, I. J. Ericson, K., Poi, M. J., O'Maille, P., Selby, T., and Tsai M. D. (1999) Tumor suppressor INK4: determination of the solution structure of p18INK4C and demonstration of the functional significance of loops in p18INK4C and p16INK4A. *Biochemistry* **38**, 2930–2940.

57. Venkataramani, R., Swaminathan, K., and Marmorstein, R. (1998) Crystal structure of the CDK406 inhibitory protein p18 INK4c provides insights into ankyrin-like repeat structure/function and tumor-derived p16 INK4 mutations. *Nat. Struct. Biol.* 5, 74–81.

58. Byeon, I. J., Li, J., Ericson, K., et al. (1998) Tumor suppressor p16INK4A: determination of solution structure and analyses of its interaction with cyclin-dependent kinase 4. *Mol. Cell* **1**, 421–431.

59. Yuan, C., Selby, T. L., Li, J., Byeon, I. J., and L. Tsai, M. D. (2000) Tumor suppressor INK4: refinement of p16INK4A structure and determination of p15INK4B structure by comparative modeling and NMR data. *Protein Sci.* **9**, 1120–1128.

60. Laity, J. H., Chung, J., Dyson, H. J., and Wright, P. E. (2000) Alternative splicing of Wilms' tumor suppressor protein modulates DNA binding activity through isoform-specific DNA-induced conformational changes. *Biochemistry* **39**, 5341–5348.

61. Laity, J. H., Dyson, H. J., and Wright, P. E. (2000) DNA-induced alpha-helix capping in conserved linker sequences is a determinant of binding affinity in Cys(2)-His(2) zinc fingers. *J. Mol. Biol.* **295**, 719–727.

62. Botuyan, M.V., Koth, C.M., Mer, G., et al. (1999) Binding of elongin A or a von Hippel-Lindau peptide stabilizes the structure of yeast elongin C. *Proc. Natl. Acad. Sci. USA* **96**, 9033–9038.
63. Davison, T. S., Nie, X., Ma, W., et al. (2001) Structure and functionality of a designed p53 dimer. *J. Mol. Biol.* **307,** 605–617.
64. Chi, S.-W., Ayed, A., and Arrowsmith, C. H. (1999) Solution structure of a conserved C-terminal domain of p73 with structural homology to the Sam domain. *EMBO J.* **18**, 4438–4445.
65. Polshakov, V. I., Williams, M. A., Gargaro, A. R., et al. (1997) High-resolution solution structure of human pNR-2/pS2: a single trefoil motif protein. *J. Mol. Biol.* **267**, 418–432.
66. Williams, M. A., Westley, B. R., May, F. E. B., and Feeney, J. (2001) The solution structure of the disulphide-linked homodimer of the human trefoil protein TFF1. *FEBS Lett.* **493**, 70–74.

2

Generation and Application of Phospho-specific Antibodies for p53 and pRB

Yoichi Taya, Kiichiro Nakajima, Kumiko Yoshizawa-Kumagaye, and Katsuyuki Tamai

1. Introduction

The functions of many proteins are likely to be regulated by phosphorylation. Thus, antibodies that can recognize specifically phosphorylated sites on proteins have a wide variety of uses for studying the function and regulation of phosphoproteins. We have improved methods for generation of phosphorylation site-specific antibodies and have successfully obtained antibodies for the analysis of most of the phosphorylation sites on p53 and RB proteins.

The RB protein (pRB) was first shown by Taya and colleagues to be phosphorylated by a cyclin-dependent kinase (Cdk) at multiple sites in vitro (reviewed in **ref. *1***). Subsequently a variety of novel Cdks and their inhibitory proteins have been identified, and it is commonly understood now that phosphorylation of pRB by Cdks plays a key role in the regulation of cellular proliferation and in cancer *(1–3)*. In an attempt to elucidate the physiologic relevance of phosphorylation of pRB, it was decided to generate antibodies to recognize specific phosphorylation sites of pRB using chemically synthesized phosphopeptides as antigens. However, it was not easy to synthesize peptides containing stably phosphorylated serine or threonine. Therefore, we have improved methods for obtaining such peptides.

Taking advantage of this improved methodology, the production of antibodies to specific phosphorylation sites of pRB was initiated. After demonstrating that this approach is successful for pRB, we also embarked upon generation of a series of antibodies specific for specific phosphorylation sites of p53.

We generally synthesize phosphopeptides as shown in **Fig. 1** for immunization of rabbits or mice. Cys is coupled to the N-terminus of most peptides to allow conjugation with KLH for more effective immunization. Because epitopes of antibodies can comprise as few as three or four amino acid residues, it is recommended that, when possible, only three residues be placed C-terminal to phosphoserine/threonine. In our experience we have observed that if there are more residues on this external side of the phosphorylated residue, antibodies directed against unphosphorylated epitopes are pref-

From: *Methods in Molecular Biology, Vol. 223: Tumor Suppressor Genes: Regulation, Function, and Medicinal Applications.* Edited by: Wafik S. El-Deiry © Humana Press Inc., Totowa, NJ

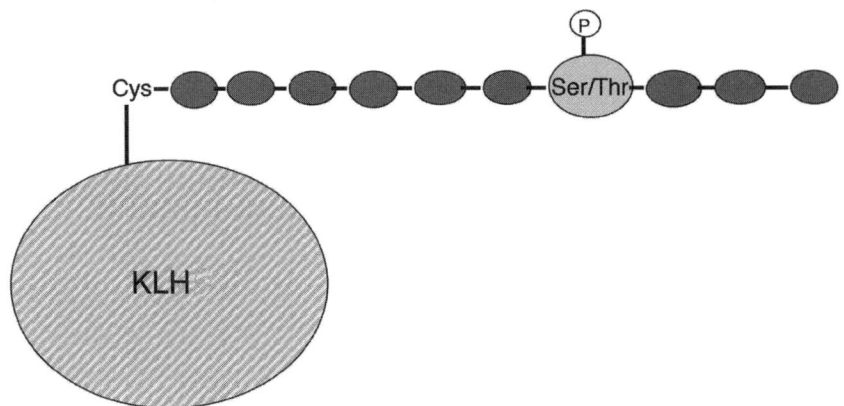

Fig. 1. Schematic presentation of a phosphopeptide antigen.

erentially produced. It is acceptable to include 6 amino acids plus N-terminal Cys on the other (internal) side of the phosphorylated residue.

If the peptide is too hydrophobic this will render it insoluble, so it is better to replace 2 or 3 amino acids on the internal side (Cys side) with 2 or 3 Arg residues. For the purpose of chemical synthesis, internal Met or Cys residues should be avoided. When an inappropriate amino acid residue is located on the N-terminal side, it is recommended to exchange the length of the N-terminal side with that of the C-terminal side, putting the Cys residue for conjugation with KLH at the C-terminus.

1.1. Chemical Synthesis of Phosphopeptides

Establishing methods for generating the most effective phosphopeptides is very important. To synthesize phosphopeptides, there are two basic strategies: (a) The prephosphorylation method, in which the protected phosphorylated amino acid derivatives are used as the building blocks of peptides; or (2) the postphosphorylation method, in which the unprotected hydroxy groups are phosphorylated on the completely assembled peptide chain. In this chapter we focus on the method we have used most extensively, which is the prephosphorylration procedure for the synthesis of desired phosphorylated peptides involving the *t*-butoxycarbonyl (Boc), and the 9-fluorenyl-methoxycarbonyl (Fmoc) strategies.

2. Materials

2.1. Recommended Standard Synthesis of Phosphopeptides by the Boc Strategy (4,5)

1. Automated peptide synthesizer: ABI-430A (Applied Biosystems, Foster City, CA, USA).
2. Resin: Boc-amino acid preloaded PAM resin (Applied Biosystems, Foster City, CA, USA)
3. All of the Boc-amino acid derivatives Arg (Mts), Glu (OBzl), Asp (OBzl), Cys (MeOBzl), His (Bom), Lys (Cl-Z), Ser (Bzl), Thr (Bzl), Tyr (Br-Z), Trp (Hoc) were purchased from Peptide Institute (Osaka, Japan), and the phosphorylated, Boc-Ser (ALL2) and Boc-Thr (PO3All2) were synthesized for our own uses *(6,7)*.

4. Preparative reversed-phase (RP)-HPLC: YMC-Pack ODS-AM (30 × 250 mm, YMC, Kyoto, Japan) using Shimdzu LC-8 A HPLC Apparatus.

2.2. Recommended Standard Synthesis of Phosphopeptides by the Fmoc Strategy (8)

1. Automated peptide synthesizer: ABI-433A (Applied Biosystems, Foster City, CA, USA).
2. Resin: Fmoc-amino acid preloaded *p*-hydroxymethyl-phenoxymethylated resin (Wang resin) (0.25-mmol sacle) (Applied Biosystems, Foster City, CA, USA).
3. All of the Fmoc-amino acid derivatives were purchased from Peptide Institute (Osaka, Japan).
4. Arg (Pbf), Glu (OtBu), Asp (OtBu), Cys (Trt), His (Trt), Lys (Boc), Ser (tBu), Thr (tBu), Tyr (tBu), Trp (Boc), and the phosphorylated Fmoc-amino acids, Fmoc-Ser(PO(OH)OBzl) and Thr(PO(OH)OBzl), were purchased from Watanabe Chemical Industries (Hiroshima, Japan) or Novabiochem (Laufelfingen, Switzerland).
5. Preparative RP-HPLC: YMC-Pack ODS-AM (30 × 250 mm, YMC, Kyoto, Japan) using Shimdzu LC-8 A HPLC apparatus.

3. Methods

3.1. Recommended Standard Synthesis of Phosphopeptides by the Boc Strategy (4,5)

3.1.1. Synthesis of Protected Peptide Resin by the Boc Strategy

The elongation of the desired protected peptide resin was carried out by the 0.5-mmol scale standard protocol of the benzotriazol active ester method in the system software version 1.40. The synthesis starts with the first Boc-amino acid attached to the PAM resin. After deprotection (50% trifluorocetic acid [TFA]/dichloromethane [DCM] (v/v) of the Boc group of first amino function, the next Boc-amino acid was coupled (N,N′-dicyclo-hexylcarbodiimide [DCC]/N-hydroxybenzotriazole [HOBt]). Consecutive deprotection/coupling steps accomplished chain elongation. The finally obtained protected peptide resin was dried and used for the following deprotection procedure (*see* **Note 1**).

3.1.2. Deprotection and Cleavage Procedure by TFMSA

To the peptide resin (0.3-mmol scale) were added thioanisole (36 mmol, 120 eq), *p*-cresol (36 mmol, 120 eq), trifluoromethanesulfonic acid (TFMSA) (36 mmol, 120 eq), and TFA (30 mL) in a round-bottom flask (500 mL) with stirring at −2−0° (*see* **Note 2**).

After the reaction mixture was stirring for 2.5 h at −2−0°, the reaction mixture was diluted by dry ether (300 mL) at −2−0°. Thus obtained precipitate was filtered off by the aid of a microfilter, and the crude phosphopeptide was extracted by TFA to separate the resin. Dry ether was added again to the TFA solution obtained, and the crude phosphopeptide precipitate was filtered off and dried over NaOH *in vacuo*.

3.1.3. Purification Procedure

To a solution of thus obtained crude compound (0.3 mmol) in water (20 mL) was added dithiothreitol (DTT) (1.5 mmol), and the pH of the solution was adjusted to pH 8–9 with dilute ammonia aqueous to prepare the thiol-free peptide (*see* **Note 3**). After the mixture was allowed to stand at room temperature for 10–20 min, the reduction reaction was stopped by addition of TFA. The mixture was immediately purified by preparative RP-HPLC (CH₃CN/0.1% TFA aq. linear gradient system). The desired fractions were combined and lyophylized to obtain the purified titred phosphopeptide.

Isolated peptide was identified by amino acid analysis and mass spectrometry. The purity of the final product was inspected by analytical RP-HPLC.

3.2. Recommended Standard Synthesis of Phosphopeptides by the Fmoc Strategy (8)

3.2.1. Synthesis of the Protected Peptide Resin by the Fmoc Strategy

The elongation of the desired protected peptide resin was carried out by the 0.25-mmol-scale standard protocol of the 2-(1H-benzotriazole-1-yl)-1,1,3,3-tetramethyluronium hexafluorophosphate (HBTU)/diisopropylethylamine (DIPEA) method in the system software. The synthesis starts with the first Fmoc-amino acid atached to the Wang resin (0.25 mmol).

After deprotection (piperidine/dimethyl sulfoxide (DMF), (1/4 v/v) of the Fmoc group of the first amino function, the next Fmoc amino acid was coupled (HBTU/ DIPEA). Consecutive deprotection/coupling steps accomplished chain elongation. The finally obtained protected peptide resin was dried and used for the following deprotection procedure.

3.2.2. Deprotection and Cleavage Procedure by TFA

To the peptide resin (0.25 mmol) were added TFA (20 mL), triisopropylsilane (TIS) (0.54 mL), H_2O (0.54 mL), and thiophenol (0.54 mL) (92.5/2.5/2.5/2.5, v/v) in a round-bottom flask (500 mL) with stirring at room temperature.

After the reaction mixture was stirring for 1.5–2 h at room temperature, the reaction mixture was diluted by dry ether (300 mL). The precipitate obtained was filtered off with a microfilter, and the crude phosphopeptide was extracted with 0.1% TFA to separate the resin. The extracted solution was lyophylized to obtain the desired crude phosphopetide. The crude product was purified in the same way as described for the Boc Strategy.

3.3. Generation of Phospho-specific Antibodies

3.3.1. Coupling Peptide to Carrier Protein with MBS

1. Dissolve keyhole limpet hemocyanin (KLH) to a concentration of 16 mg/mL in 1 mL of 10 mM sodium phosphate buffer, pH 7.2 (*see* **Note 4**).
2. Prepare 280 mg/mL of *m*-maleimidobenzoyl-N-hydroxysuccinimide ester (MBS) in dimethylformamide freshly.
3. Add 10 μL of MBS with stirring to avoid a high local concentration and continue to stir for 30 min at room temperature.
4. Centriguge at 15,000 rpm at 4°C for 5 min. Take a supernatant.
5. Separate the MBS-activated KLH from the free MBS by gel filtration on Sephadex G25 equilibrated with 50 mM sodium phosphate buffer, pH 6.0.
6. Pool MBS-activated KLH and divide equally into 10 microcentrifuge tubes. Store at −80°C until use.
7. Thaw MBS-activated KLH in one tube.
8. Add 0.5 vol of 0.2 M Na₂HPO₄ (pH should be 7.3–7.5).

9. Dissolve 1 mg of the peptide containing cysteine residue at the N- or C-terminus in 0.5 mL of distilled water, and add to MBS-activated KLH.
10. Stir the reaction for 3 h at room temperature.
11. Divide equally into 10 microcentrifuge tubes. Store at −80°C until use.

3.3.1.1. DEOXIDATION OF PEPTIDE

When free sulfhydryl residue in the peptide seems to be oxidized, the peptide should be deoxidized by the following protocol. (Ellman's reagent can be used to determine if there are free sulfhydryls available on the terminus of the peptide.)

1. Dissolve 1 mg of the peptide containing cysteine residue at the N- or C-terminus in 0.5 mL of 20 mM sodium phosphate buffer, pH 8.5.
2. Add $NaBH_4$ to a final concentration of 0.1 M.
3. Stand the reaction for 45 min at room temperature.
4. Adjust the final pH to 4.0 by adding 0.1 N HCl.
5. Stand the reaction for 5 min at room temperature.
6. Adjust the final pH to 7.5 by adding 0.2 M Na_2HPO_4.
7. Add to MBS-activated KLH.
8. Stir the reaction for 3 h at room temperature.
9. Dialyze against phosphate-buffered saline (PBS) at 4°C overnight.
10. Divide equally into 10 microcentrifuge tubes. Store at −80°C until use.

3.3.2. Production of Anti-phosphopeptide Polyclonal Antibody

3.3.2.1. IMMUNIZATION OF RABBITS

Two or more female rabbits (weight ~2.5 kg) were immunized subcutaneously by standard protocol with ~100 μg of phosphopeptide-KLH conjugate emulsified with Freund's complete adjuvant (FCA) 4 times at biweekly intervals, and boosted with phospho-peptide-KLH conjugate in Freund's incomplete adjuvant (*see* **Note 5**). One week after boost, bleed approx. 70 mL from marginal ear vein of rabbits.

3.3.2.2. AMMONIUM SULFATE PRECIPITATION

1. Incubate rabbit antiserum (~30 mL) for 30 min at 56°C (heat inactivation).
2. Centrifuge at 15,000 rpm at 4°C for 5 min.
3. Transfer the supernatant to an appropriate vessel.
4. Determine the volume of serum and add an equal volume of PBS.
5. Add slowly 0.8 vol of saturated ammonium sulfate solution.
6. Stir gently for 15 min at room temperature.
7. Centrifuge at 15,000 rpm at 4°C for 15 min.
8. Discard the supernatant.
9. Resuspend the precipitate with PBS (2 vol of initial antiserum).
10. Repeat **steps 4–7.**
11. Resuspend the precipitate with PBS (0.5 vol of initial antiserum)
12. Dialyze against 50 vol of PBS using dialyzing tubing at 4°C overnight (3 changes).
13. Centrifuge at 15,000 rpm at 4°C for 15 min for remove any remaining debris.

3.3.2.3. COUPLING THE PEPTIDE-CONTAINING CYSTEINE RESIDUE TO THE GEL

1. Pack 1 mL of iodeacetyl immobilized gel (SulfoLink coupling gel: Pierce, Rockford, IL, USA) into appropriate-sized column (*see* **Note 6**).
2. Equilibrate the column with 6 column vols of 50 mM Tris-HCl, pH 8.5, and 5 mM EDTA-Na.

3. Dissolve 2 mg of the peptide containing cysteine residue at the N- or C-terminus in 1 mL of 50 mM Tris-HCl, pH 8.5, and 5 mM EDTA-Na.
4. Apply the peptide solution on the column and rotate the column for 1 h at room temperature or overnight at 40°C.
5. Wash the column with 3 column vols of 50 mM Tris-HCl, pH 8.5, and 5 mM EDTA-Na.
6. Add 1 mL of 50 mM cysteine solution to the column to block any remaining active iodeacetyl groups, and rotate the column for 1 h at room temperature.
7. Wash the column with at least 15 column vols of 1 M NaCl and 15 column vols of 0.05% NaN$_3$.
8. Equilibrate the column with 6 column vols of PBS containing 0.3 M NaCl.

3.3.2.4. AFFINITY CHROMATOGRAPHY

1. Equilibrate the phosphopeptide coupled gel in a column with 5 gel vols of PBS containing 0.3 M NaCl.
2. Filter the γ-globulin fraction of the antiserum using a Millipore filter (pore size: 0.45 μm).
3. Apply the antibody solution onto the column and pass through the column twice.
4. Wash the column with 10 column vols of PBS containing 0.3 M NaCl and 0.1% Triton X-100.
5. Wash the column with 10 column vols of PBS containing 0.3 M NaCl.
6. Elute with 0.17 M glycine-HCl (pH 2.5) containing 10% glycerol. Collect 2 mL of each eluate in appropriate tubes containing 400 μL of 1 M Tris-HCl (pH 8.0) on ice.
7. Identify the immunoglobulin-containing fractions by measuring absorbance at 280 nm (1 OD = 0.7 mg/mL).
8. Dialyze these fractions against PBS at 4°C overnight (3 changes).
9. Determine the antibody concentration by measuring absorbance at 280 nm.
10. Apply the antibody solution onto non-phosphopeptide-coupled gel column equilibrated with PBS containing 0.3 M NaCl for absorption of anti-non-phosphopeptide antibody and pass-through the column twice. Collect the pass through fraction.
11. Determine the antibody concentration by measuring absorbance at 280 nm.
12. Confirm the reactivity against phosphopeptide and the absence of reactivity against non-phosphopeptide by ELISA using non-phosphopeptide- and phosphopeptide-coated microplates.

3.3.2.5. ELISA FOR MEASUREMENT OF SPECIFICITY

1. Coat a 96-well microtiter plate with 50 μL/well of each synthetic peptide at the concentration of 5 μg/mL in 0.1 M NaHCO$_3$ (pH 9.0).
2. Incubate the plate at 4°C overnight. Wells are emptied, and uncoated sites on the plates are blocked by the addition of 200 mL of 2% bovine serum albumin (BSA) and 1% sucrose in PBS, followed by incubation for 2 h at room temperature.
3. Wells are emptied and incubated with 50 μL/well of serially diluted anti-phosphopeptide antibody for 1 h at room temperature.
4. Wash microtiter plate 3 times with PBS containing 0.05% Tween-20.
5. Add 50 μL/well of 1:30,000-diluted horseradish peroxidase (HRP)-conjugated anti-rabbit IgG (MBL) or anti-mouse IgG (MBL) in PBS containing 1% BSA.
6. Incubates for 1 h at room temperature.
7. Wash microtiter plate 3 times with PBS containing 0.05% Tween-20.
8. Add 50 μL/well of 1 mM tetramethyl benzidine (TMB) in 0.1 M sodium acetate, pH 6.0, containing 0.01% H$_2$O$_2$ for 15 min at room temperature.
9. The reaction is stopped by adding 50 μL/mL of 2 N H$_2$SO$_4$, and absorbance at 450 nm is determined using a microplate reader.

3.3.3. Production of Anti-phosphopeptide Mouse Monoclonal Antibody

3.3.3 1. IMMUNIZING MICE

Four or more female Balb/c mice (4–5 wk) were immunized at the footpad with 10–20 µg of phosphopeptide-KLH conjugate emulsified with Freund's complete adjuvant (FCA) twice at 4-d intervals, 3 d after last immunization, boosted with ~20 µg of phospho-peptide-KLH conjugate in Freund's incomplete adjuvant (*see* **Notes 7** and **8**).

3.3.3.2. PRODUCING HYBRIDOMAS

Production of hybridoma is according to standard protocol using polyethylene glycol except that the lymphocytes from two leg lymph nodes of immunized mice are fused to myeloma cells. This protocol is very efficient for producing hybridomas that secret anti-phospho specific monoclonal antibody when an appropriate mouse strain was selected.

3.4. Applications

3.4.1. p53

When cells are exposed to a variety of stresses including DNA damage, heat shock, hypoxia, and osmotic shock, p53 protein accumulates rapidly and is also activated as a transcriptional factor, which then leads to growth arrest or apoptosis. We have shown that phosphorylation plays an important role in regulating stabilization and activation of p53 using phospho-specific antibodies.

At least 8 in-vivo phosphorylation sites are known in human p53 *(9,10)*. We have generated rabbit polyclonal antibodies to recognize each phosphorylation site. In addition, we have also made antibodies for several other potential phosphorylation sites. Although monoclonal antibodies for several sites were obtained, polyclonal antibodies are much stronger and normally they have been used for most of our experiments.

3.4.1.1. SER-15

We showed that phosphorylation of Ser-15 occurs after DNA damage and is important for its release from its negative regulator MDM2 *(11,12)*. Phosphorylation of Ser-20 is also involved in this release. Thereafter, Ser-15 was shown to be phosphorylated by ATM *(13–15)* or ATR *(16)*.

3.4.1.2. SER-20

We showed that Chk1 and/or Chk2 directly phosphorylate this site *(17)*.

3.4.1.3. SER-33

Two different kinases have been shown to phosphorylate Ser-33: the cyclin H CDK7/MAT1 complex *(18)* and the p38MAP kinase *(19,20)*.

3.4.1.4. SER-46

We showed that phosphorylation of this site regulates apoptosis induction by p53 *(21)*. It was also shown that p53DINP1 is involved in this phosphorylation *(22)*.

3.4.1.5. SER-392

This site was found to be specifically phosphorylated after DNA damage by UV but not γ-radiation or etoposide-induced DNA damage *(23)*.

3.4.2. pRB

There are about 13 in-vivo phosphorylation sites on pRB *(1)*. In an attempt to elucidate the physiologic function of pRB, we generated all antibodies to recognize each phosphorylation site. Using these antibodies, we showed that Ser-780 is phosphorylated by Cdk4-cyclin D1 but not by Cdk2-cyclin E or by Cdk2-cyclin A *(24)*. This was the first clear demonstration of different substrate specificities of two cyclin-dependent kinases, Cdk4 and Cdk2. This study also established the presence of a Cdk4-specific phosphorylation site on pRB. Subsequently, we showed that 13 phosphorylated sites can be classified to Cdk4- or Cdk2-specific sites using these antibodies (**ref. *25*** and unpublished data).

In contrast to p53, strong and good monoclonal antibodies were obtained for phosphorylation sites of pRB. Probably, this is at least partially due to the abundance of pRB in cells compared to p53.

4. Notes

1. Caution: DCC is an aggressive allergen. To avoid the chemical hazard, you have to avoid direct contact with the DCC solution, and it is necessary to perform all handling in a well-ventilated hood.
2. Caution: TFMSA is a very strong acid. It must be slowly added with careful handling in a well-ventilated hood.
3. Cys residue is necessary for the conjugation with corresponding carrier protein at the N- or C-terminus position of your peptide sequence. If Cys residue is not in your peptide sequence, this DTT reduction procedure is not necessary. Then, the TFMSA-treated products can be used directly for the purification procedure.
4. Alternatively, Bovine Serum Albumin (BSA) or Ovalbumin (OVA) can be used as a carrier protein.
5. TiterMax Gold (CytRx Corp., Atlanta, GA, USA) or RIBI adjuvant (RIBI ImmunoChem Research, Hamilton, MT, USA) can be substituted for FCA.
6. NHS-activated Sepharose 4 Fast Flow, Thiopropyl Sepharose 6B (Amersham Pharmacia Biotech, UK) can be used for this purpose.
7. If any positive clone could not be obtained using Balb/c mice, another strain, A/J, DBA2, or C57Black, has to be used.
8. TiterMax Gold (CytRx Corp., Atlanta, GA, USA) or RIBI adjuvant (RIBI ImmunoChem Research, Hamilton, MT, USA) can be substituted for FCA.

References

1. Taya, Y. (1997) RB-kinases and RB-binding proteins: new points of view. *Trends Biochem. Sci.*, **22**, 14–17.
2. Sherr, C. J. (1994) The ins and outs of RB: coupling gene expression to the cell cycle clock. *Trends Cell Biol.* **4**, 15–18.

3. Weinberg, R. A. (1995) The retinoblastoma protein and cell cycle control. *Cell* **81**, 323–330.

4. Wakamiya, T., Saruta, K., Kusumoto, S., et al. (1993) An effcient procedure for synthesis of phosphopeptides through the benzyl phosphate-protection by the Boc-mode solid-phase method. *Chem. Lett.* **1993**, 401–1404T

5. Wakamiya, T., Saruta, K., Kusumoto, S., Aimoto, S., Kumagaye, K.- Y., and Nakajima, K. (1994) Synthetic study of phosphopeptides by solid-phase method in *Peptide Chemistry 1993*, Okada, Y., ed.). Protein Research Foundation, Osaka, Japan, pp. 17–20 .

6. Ueno, Y., Suda, F., Taya, Y., Noyori, R., Hayakawa, H., and Hata, T. (1995) Allyl derivative synthesis: a synthetic study of phosphopeptides by solid-phase. *Bioorg. Med. Chem. Lett.* **5**, 823–826.

7. Kitas, E., Knorr, R., Treciak, A., and Bannwarth, W. (1991) Alternative strategies for Fmoc solid-phase synthesis of O-phospho-L-tyrosine-containing peptides. *Helv. Chim. Acta* **74**, 1314–1328.

8. Wakamiya, T., Togashi, R., Nishida, T., et al. (1997) Synthetic study of phosphopeptides related to heat shock protein HSP27. *Bioorg. Med. Chem.* **5**, 135–145.

9. Prives, C. (1998) Signaling to p53: breaking the MDM2-p53 circuit. *Cell* **95**, 5–8.

10. Lyungman, M. (2000) Dial 9-1-1 for p53: mechanism of p53 activation by cellular stress. *Neoplasia* **2**, 208–225.

11. Shieh, S.-Y., Ikeda, M., Taya, Y., and Prives, C. (1997) DNA damage-induced phosphorylation of p53 alleviates inhibition by MDM2. *Cell* **91**, 325–334.

12. Siliciano, J.D., Canman, C.E., Taya, Y., Sakaguchi, K., Appella, E., and Kastan, M. B. (1997) DNA damage induces phosphorylation of the amino terminus of p53. *Genes Dev.* **11**, 3471–3481.

13. Banin, S., Moyal, S., Shieh, S. Y., et al. (1998) Enhanced phosphorylation of p53 by ATM in response to DNA damage. *Science* **281**, 1674–1677.

14. Canman, C. E., Lim, D. S., Cimprich, K. A., et al. (1998) Activation of the ATM kinase by ionizing radiation and phosphorylation of p53. *Science* **281**, 1677–1679.

15. Nakagawa, K., Taya, Y., Tamai, K., and Yamaizumi, M. (1999) Requirement of ATM in the phosphorylation of the human p53 protein at serine 15 following DNA double-strand breaks. *Mol. Cell. Biol.* **19**, 2828–2834.

16. Tibbetts, R. S., Williams, J. M., Taya, Y., Shieh, S.-Y., Prives, C., and Abraham, R. T. (1999) A role for ATR in the DNA damage-induced phosphorylation of p53. *Genes Dev.* **13**, 152–157.

17. Shieh, S.-Y, Ahn, J, Tamai, K., Taya, Y., and Prives, C. (2000) The human homologues of checkpoint kinases Chk1 and Cds1 (Chk2) phosphorylate p53 at multiple DNA damage inducible sites. *Genes Dev.* **14**, 289–300.

18. Ko, L. J., Chen, X., Shieh, S.-Y., et al. (1997) p53 is phosphorylated by CDK7/cyclin H in a p36/MAT1 dependent manner. *Mol. Cell. Biol.* **17**, 7220–7229.

19. Takekawa, M., Adachi, M., Nakahata, A., et al. (2000) p53-inducible Wip1 phosphatasemediates a negative feedback regulation of p38 MAPK-p53 signaling in response to UV radiation. *EMBO J.* **19**, 6517–6521.

20. Kishi, H., Nakagawa, K., Matsumoto, M., et al. (2001) Osmotic shock induces G1-arrest through p53 phosphorylation at Ser33 by activated p38MAPK without phosphorylation at Ser15 and Ser20. *J. Biol. Chem.* **276**, 39115–391122.

21. Oda, K., Arakawa, H., Tanaka, T., et al. (2000) *p53AIP1*, a potential mediator of p53-dependent apoptosis, and its regulation by Ser46-phosphorylated p53. *Cell* **102**, 849–862.

22. Okamura, S., Arakawa, H., Tanaka, T., et al. (2001) p53DINP1, a novel p53-inducible gene, regulates p53-dependent apoptosis. *Mol. Cell* **8**, 85–94.

23. Lu, H., Taya, Y., Ikeda, M., and Levine, A. J. (1998) Ultraviolet radiation, but not γ radiation or etoposide-induced DNA damage, results in the phosphorylation of the murine p53 protein at serine-389. *Proc. Natl. Acad. Sci. USA* **95**, 6399–6402.

24. Kitagawa, M., Higashi, H., Jung, H.-K., et al. (1996) The consensus motif for phosphorylation by cyclin D1-Cdk4 is different from that for phosphorylation by cyclin A/E-Cdk2. *EMBO J.* 15, 7060–7069.

25. Adams, P. D., Li, X., Sellers, W. R., et al. (1999) The retinoblastoma protein contains a C-terminal motif that targets it for phosphorylation by cyclin/cdk2 complexes. *Mol. Cell. Biol.* 19, 1068–1080.

3

Stability and Ubiquitination of the Tumor Suppressor Protein p53

Lesley J. Jardine and Carl G. Maki

1. Introduction

The ubiquitin-dependent proteolysis pathway constitutes a major pathway in the cell for selective protein degradation. The covalent attachment of multiple ubiquitin molecules to lysine residues of a target protein serves to signal its recognition and rapid degradation by the 26S proteasome. The process of ubiquitination requires the concerted action of three classes of cellular enzymes, designated E1, E2, and E3 (reviewed in **refs. 1** and **2**). Ubiquitin is first activated in an ATP-dependent manner through its covalent thio-ester linkage to the E1 ubiquitin-activating enzyme. The activated ubiquitin is then transferred to the E2 enzyme, also known as a ubiquitin conjugating enzyme (UBC), again in the form of a high-energy thio-ester bond. In some cases the E2 enzyme can transfer the ubiquitin directly to a substrate, while the E3 enzyme (also known as a ubiquitin protein ligase) assists in the recognition of the substrate *(3,4)*. In other cases, the E2 enzyme transfers the ubiquitin to the E3, and the E3 then transfers the ubiquitin to the substrate protein *(5)*. In either case, the ubiquitin is transferred to the substrate through formation of an isopeptide bond between the carboxyl terminus of ubiquitin and the e-amino group of lysine residues on the target protein. Additional ubiquitin moieties may be linked sequentially to each other, leading to the formation of multiubiquitin chains. The precise nature of a multiubiquitination reaction is unclear, though recent studies suggest the presence of yet a fourth enzymatic activity, designated E4, which may play a role in this process *(6)*. The multiubiquitinated substrate is then recognized and degraded by the 26S proteasome. In order for this process to be efficient, it is likely that the E1, E2, and E3 enzymes involved form multiprotein complexes to allow rapid thiol ester transfer of ubiquitin molecules.

Wild-type p53 is a tumor suppressor protein and potent inhibitor of cell growth. The activity most associated with the growth suppressive function of p53 is its ability to bind DNA in a sequence-specific manner and activate gene transcription *(7–9)*. Wild-type p53 is expressed at low levels in most cell types due to a short protein half-life *(10,11)*. In contrast, p53 is stabilized and its levels increase in response to various stresses, including DNA damage, hypoxia, and aberrant oncogene signaling

From: *Methods in Molecular Biology, Vol. 223: Tumor Suppressor Genes: Regulation, Function, and Medicinal Applications.* Edited by: Wafik S. El-Deiry © Humana Press Inc., Totowa, NJ

(reviewed in **ref.** *12*). The stress-induced stabilization of p53 is believed to result from an inhibition of p53 ubiquitination *(11)*. The effect of increasing p53 is to inhibit cell growth through either a G1 cell cycle arrest or apoptosis *(13)*. These activities are mediated by proteins such as p21 and bax, whose genes are transcriptionally activated by p53 *(14,15)*.

The first evidence that p53 is a target of the ubiquitin system came from studies of the oncogenic human papillomaviruses (HPVs), which showed that the HPV E6 onco-protein could promote the ubiquitin-dependent degradation of p53 *(16)*. E6 forms a complex with the cellular factor E6AP, and the E6:E6AP complex functions as an E3 enzyme in HPV-infected cells that can transfer ubiquitin directly to p53 *(5,17,18)*. Proof that p53 is targeted for ubiquitination in the absence of E6 came with the demonstration of p53:ubiquitin conjugates in cells that lack E6 expression *(19)*. Factors that can participate in the E6-independent ubiquitination of p53 have now been identified, and one of these factors is MDM2. MDM2 binds to the N-terminal transactivation of p53 and inhibits the activity of p53 as a transcription factor. Kubbutat et al. (1997) and Haupt et al. (1997) independently demonstrated that MDM2 binding to p53 promoted the rapid degradation of p53 in a proteasome-dependent manner *(20,21)*. Later studies confirmed that MDM2 can promote the ubiquitination of p53 in vitro and in transiently transfected cells *(22–24)*. There is some evidence that, at least in vitro, MDM2 can function as an E3 enzyme and transfer ubiquitin directly to p53 *(22)*. However, it has not been fully clarified if, in cells, MDM2 alone transfers ubiquitin directly to p53 or is part of a larger ubiquitination complex that transfers ubiquitin to p53. Interestingly, the MDM2 gene is transcriptionally activated by p53. Thus, MDM2 functions in an autoregulatory feedback loop in which p53 activates MDM2 expression, and increased levels of MDM2 then bind p53 and inhibit p53 activity.

Our laboratory can readily detect ubiquitin:p53 conjugates in normal cells, and in cells transiently overexpressing wild-type p53 and MDM2. This chapter outlines some of the methods used to assess the stability and ubiquitination of p53.

2. Materials
2.1. Cell Lines and Plasmid DNAs

RKO cells are a human colon cancer cell line that expresses wild-type p53. U2OS cells are a human osteosarcoma cell line that expresses wild-type p53. Both cell types are maintained in Modified Eagle's Medium supplemented with 50 µg/mL penicillin and streptomycin. DNA encoding HA-tagged wild-type p53 was previously described *(25)*, and was obtained from Christine Jost and William Kaelin (Dana Farber Cancer Institute, USA). DNA encoding MDM2 was obtained from Steve Grossmann (Dana Farber Cancer Institute, USA). DNA encoding myc-tagged ubiquitin has been described *(26)*, and was obtained from Ronald Kopito (Stanford University, USA).

2.2. Lysis Solutions
1. Lysis buffer: 50 mM Tris-HCl, pH 7.5, 150 mM NaCl, 5 mM EDTA, 0.5% NP40, 1 mM phenylmethylsulfonyl fluoride (PMSF), 2 µg/mL aprotinin, 5 µg/mL leupeptin.
2. Low-salt lysis buffer: 50 mM Tris-HCl, pH 7.5, 10 mM NaCl, 5 mM EDTA, 0.5% NP40, 1 mM PMSF, 2 µg/mL aprotinin, 5 µg/mL leupeptin.

3. $1\times$ RIPA buffer: 20 mM Tris-HCl, pH 7.5, 2 mM EDTA, 150 mM NaCl, 0.25% sodium dodecyl sulfate (SDS), 1% NP40, 1% deoxycholic acid, 1 mM PMSF, 2 μg/mL aprotinin, 5 μg/mL leupeptin.

4. $5\times$ RIPA buffer: 100 mM Tris-HCl, pH 7.5, 10 mM EDTA, 750 mM NaCl, 1.25% SDS, 5% NP40, 5% deoxycholic acid, 1 mM PMSF, 2 μg/mL aprotinin, 5 μg/mL leupeptin.

5. $4\times$ loading buffer: 2.5 mL of 1 M Tris-HCl, pH 6.8, are combined with 4 mL of 20% SDS. 0.6 g of dithiothreotol are added and mixed until dissolved. 4 mL of glycerol are then added to bring the volume to 10.5 mL. A speck of bromophenol blue is added to give a blue color. 1-mL aliquots are stored at –20°C.

2.3. Immunoblot Solutions

1. TNET: 10 mM Tris-HCl, pH 7.5, 2.5 mM EDTA, 50 mM NaCl, 0.1% Tween-20.
2. Gel running buffer: 190 mM glycine, 24 mM Tris-HCl base (MW 121.1), 0.1% SDS.
3. Transfer buffer: 48 mM Tris-HCl base (MW 121.1), 38 mM glycine, 0.37% SDS, 20% methanol.
4. Blocking solution: 5% powdered milk in TNET.
5. Blot stripping solution: 68 mM Tris-HCl, pH 6.8, 20% SDS, 0.7% 2-mercaptoethanol.

2.4. Transfection Solutions

1. $2\times$ HBS: 280 mM NaCl, 1.5 mM Na_2HPO_4, 50 mM HEPES adjusted to pH 6.8 with NaOH, filter-sterilized, and stored at 4°C.
2. 2.5 M $CaCl_2$ was prepared by dissolving 3.675 g of $CaCl_2$ in 10 mL of H_2O. The 2.5 M $CaCl_2$ was filter-sterilized and stored at 4°C.
3. Antibodies: The p53 monoclonal antibody Ab-6 was purchased from Oncogene Science (Cambridge, MA, USA). The MDM2 monoclonal antibody SMP14 was purchased from Santa Cruz. Anti-ubiquitin monoclonal antibody (MAB1510) was purchased from Zymed. Anti-HA monoclonal antibody (HA.11) was purchased from Babco. Anti-HA polyclonal antibody (Y-11) was purchased from Santa Cruz. Anti-myc monoclonal antibody (9E10) can be purchased from Oncogene Science.
4. Protein A–agarose beads: Protein A–agarose beads were purchased from Gibco BRL. Immediately prior to use, the bead slurry was washed 3 times with ice-cold $1\times$ RIPA buffer and resuspended in a volume of $1\times$ RIPA buffer equivalent to 30 μL/sample.
5. Proteasome inhibitors: MG132 was purchased from Peptides International (Lexington, KY, USA). MG132 was dissolved in DMSO to a final concentration of 10 mM. Lactacystin was purchased from Biomol Research Labs, Inc. (Plymouth Meeting, PA, USA). Lactacystin was dissolved in H_2O to a final concentration of 10 mM.
6. Cyclohexamide: Cyclohexamide (CHX) powder was purchased from Sigma. A 2% stock solution of CHX was prepared by dissolving 0.02 g CHX in 1 mL of ethanol.

3. Methods

3.1. Stabilization of p53 by Proteasome Inhibitor Treatment

Wild-type p53 is degraded through the ubiquitin-proteasome degradation pathway. Accordingly, inhibitors of the proteasome are expected to inhibit p53 degradation and thus stabilize the p53 protein. This section outlines a method used to demonstrate the effect of proteasome inhibitors on p53 levels and stability.

1. Cells are plated such that at the time of the experiment the cells are actively growing and are 80–90% confluent. U2OS cells (a human osteosarcoma cell line that expresses wild-type

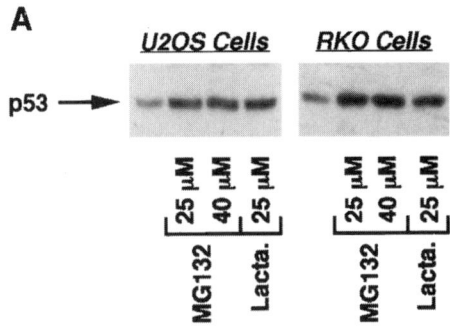

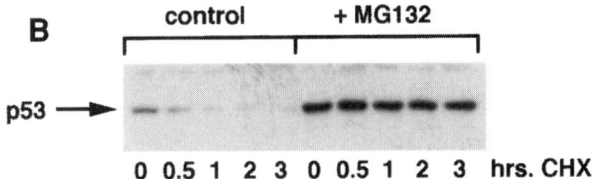

Fig. 1. Proteasome inhibition stabilizes p53. **(A)** U2OS and RKO were either untreated (no tr.) or exposed to the indicated concentration of the proteasome inhibitors MG132 or lactacystin (Lacta.) for 7 h. p53 levels were determined by immunoblot analysis with the p53 monoclonal antibody Ab-6, as described in the text. **(B)** U2OS cells were exposed to 25 mM MG132 for 6 h. The cells were then treated with the 25 ng/mL of cyclohexamide (CHX) to inhibit *de novo* protein synthesis. Cell extracts were prepared at the indicated time after CHX treatment and p53 steady-state levels were determined by immunoblot analysis with the p53 monoclonal antibody Ab-6. The half-life ($t_{1/2}$) of p53 in untreated U2OS cells was approximately 30 min. The half-life of p53 in MG132-treated cells was greater than 3 h.

p53) and RKO cells (a human colon cancer cell line that expresses wild-type p53) were used in the experiment illustrated in **Fig. 1**.

2. Proteasome inhibitors are then added directly to the cell growth media. A proteasome inhibitor that is commonly used is MG132, which can be purchased from Peptides International (Lexington, KY, USA). MG132 is dissolved in DMSO at a concentration of 10 mM. In addition to inhibition of the proteasome, MG132 also has minor inhibitory activity against nonproteasomal degradation pathways *(27)*. Lactacystin is a highly specific inhibitor of the proteasome *(28)*, and can be purchased from Biomol (Plymouth Meeting, PA, USA). Lactacystin is dissolved in H_2O to a final concentration of 10 mM. MG132 and lactacystin are added directly to the cell growth media such that their final concentrations in the media are between 10 and 40 μM. Control cells are treated with either DMSO or H_2O only.

3. P53 levels are determined by immunoblot analysis at various time points following treatment. In our experiments, a clear increase in p53 levels is observed in U2OS and RKO cells exposed to the inhibitors MG132 and lactacystin for 5–10 h (**Fig. 1A**).

3.2. Immunoblot Analysis for p53 Protein Levels

1. Cell growth media is aspirated from the untreated or proteasome inhibitor-treated cells and replaced with PBS. The PBS is then aspirated from the plate.

2. Ice-cold lysis buffer is added directly to the cells on the plate. Typically, 500 μL of lysis buffer are used if the cells are grown in a 35-mm or 60-mm cell culture dish, and 0.8–1.0 mL of lysis buffer are used if cells are grown in a 10-cm culture dish.

3. The cells are removed from the dish in the lysis buffer using a cell scraper. The cells in lysis buffer are then transferred to a 1.5-mL microfuge tube and placed on ice for 30 min with light vortexing every 5–10 min. Cell lysis occurs during this 30 min.

4. The lysed cells are then spun at 16,000*g* for 15 minutes in a refrigerated microfuge to pellet the cellular debris. The supernatant (protein extract) is removed from the debris pellet and transferred to a second microfuge tube on ice. The concentration of cellular protein can now be determined. Our laboratory uses the Bio-Rad Protein Assay solution to determine the protein concentration.

5. For immunoblot analysis of p53, cellular protein is resolved by sodium dodecyl sulfate–polyacrylamide gel electrophoresis (SDS-PAGE) through a 9% polyacrylamide gel. For analysis of endogenous p53 levels, 50–70 μg of cellular protein are typically loaded per lane if using a minigel (8 × 5 × 0.1 cm). 100 μg are typically loaded if using a 14 × 10 × 0.15-cm gel.

6. Proteins are transferred from the gel to a PVDF membrane or nitrocellulose membrane. The membrane is then immersed in blocking solution and placed on a shaker for 1 h.

7. The membrane is rinsed 2× quickly with TNET, then placed in TNET on a shaker for one 15-min wash, and two 5-min washes.

8. The TNET is then replaced with the p53 antibody solution. The p53 antibody solutions consist of the p53 antibodies Ab-6 or Ab-2 (both from Oncogene Science) diluted 1:1000 in TNET. The blot in antibody solution is placed on a shaker for 90 min. This antibody solution can be used in multiple experiments.

9. The blot is then washed with TNET as described in **step 7.**

10. The TNET is then replaced with the horseradish peroxidase (HRP)-conjugated secondary antibody solution. This solution consists of HRP-conjugated anti-mouse IgG from Jackson Laboratories, with the antibody present at a 1:10,000 dilution in TNET. The blot in secondary antibody is placed on a shaker for 1 h.

11. The blot is rinsed 3× quickly with TNET, then placed in TNET on a shaker for one 15-min wash, followed by three 5-min washes.

12. The blot is then exposed to enhanced chemiluminescence (ECL) reagents and exposed to film to visualize p53. Exposure times may vary from anywhere from 2 s to 15 min.

3.3. Determination of p53 Stability

p53 stability measurements can be performed to confirm that the increased p53 protein levels observed in proteasome inhibitor-treated cells results from stabilization of the p53 protein. This section outlines a method that can be used to measure the stability of the p53 protein in untreated and proteasome inhibitor-treated cells.

1. Five separate dishes of cells are exposed to proteasome inhibitors for 5–10 h, as described earlier. Control cells are again exposed to either DMSO or H_2O only.

2. Cyclohexamide (CHX) is added to each dish to inhibit *de novo* protein synthesis. Our laboratory adds CHX such that its final concentration in the growth media is 25 μg/mL. Higher concentrations of CHX may have undesired secondary effects on protein degradation, and are therefore not recommended.

3. Protein extracts from the control and proteasome inhibitor-treated cells are harvested at various time points after CHX treatment as described earlier. The steady-state levels of p53 are determined by immunoblot analysis as described earlier. Because CHX inhibits new protein synthesis, the rate at which p53 steady-state levels decrease following CHX treatment is a

measure of p53 protein stability. In untreated control cells, the half-life ($t_{1/2}$) of p53 is between 30 and 60 min. In contrast, proteasome inhibitor treatment increases the $t_{1/2}$ of p53 to greater than 3 h (**Fig. 1B**).

3.4. Ubiquitination of the Endogenous p53 Protein

The fact that p53 levels increase in response to proteasome inhibitors is consistent with the hypothesis that p53 is degraded through the ubiquitin-proteasome pathway. To confirm that p53 is targeted for ubiquitination, it is necessary to demonstrate the presence of ubiquitin:p53 conjugates. This section outlines a method by which this can be accomplished.

1. The abundance of ubiquitinated p53 species in a cell is very low. It is therefore necessary to start with a large amount of protein extract to visualize the low amount of ubiquitinated p53. Our laboratory typically starts with 1–2 confluent and actively growing 10-cm dishes of RKO or U2OS cells.
2. The cells can be either untreated or exposed in various ways to increase p53 levels. In the experiment shown in **Fig. 2**, RKO cells were untreated or exposed to a dose of ionizing radiation (IR) of 6.8 Gy. The p53 protein is stabilized in IR-treated cells, and p53 steady-state levels increase.
3. Cell lysates are harvested 5–9 h after IR treatment as described below.

3.5. Harvesting Cells to Visualize Ubiquitinated p53

We have found that an efficient way to visualize ubiquitinated p53 is first to extract cellular proteins in low-salt lysis buffer. Ubiquitin-p53 conjugates and a portion of nonubiquitinated p53 can be extracted from the cells under these conditions. A fraction of nonubiquitinated p53 is not extracted under these conditions.

1. Cell growth medium is aspirated from the cells and the cells are rinsed with PBS. 1.0 mL of ice-cold, low-salt lysis buffer is then added directly to the cells in a 10-cm dish. The cells in low-salt lysis buffer are gently scraped from the dish using a cell scraper, and placed in a 1.5-mL microfuge tube on ice for 5 min with occasional light mixing (not vortexing).
2. The cells are then spun at 380g for 2 min in a refrigerated microfuge. The supernatant (S) is removed from the pellet (P) and transferred to a second microfuge tube on ice. 250 µL of 5× RIPA buffer is added to the supernatant (S) fraction. The pellet (P) is resuspended in 1250 mL of 1× RIPA buffer.
3. The S and P fractions are transferred to separate 15-mL conical tubes and sonicated on ice using a Branson 450 sonifier, setting 5, 50% output. This sonnication disrupts the DNA component of the P fraction.
4. Following sonication, the samples are transferred to microfuge tubes, and spun down at 16,000g in a refrigerated microfuge for 15 min. The supernatant of each fraction is removed and transferred to a microfuge tube on ice. 100 µL of the S and P fractions are removed to new microfuge tubes and set aside.
5. 10 µL of the p53 antibody Ab-6 (Oncogene Science) is added to the remaining 1150 µL S and P fractions and rocked overnight at 4°C to immunoprecipitate p53.
6. 30 µL of protein A–agarose beads are added to the immunoprecipitations. Rocking is continued at 4°C for 1 h.
7. The beads are spun down for 15 s in a refrigerated microfuge. The beads are then washed 3× with 1.0 mL of ice-cold 1× RIPA buffer. Following the final rinse, the beads are resuspended in approximately 40 µL of 1× RIPA buffer. 4× loading buffer is added to 1×. The

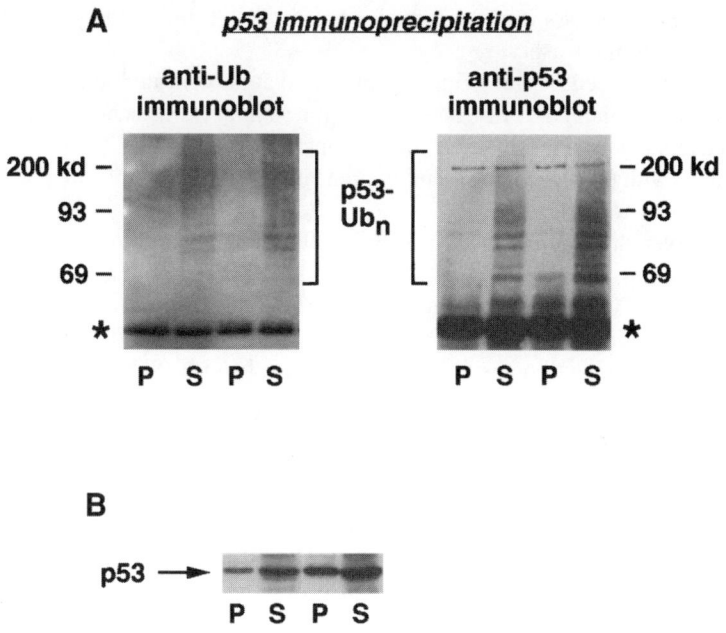

Fig. 2. Detection of p53:ubiquitin conjugates in cells that express wild-type p53. **(A)** Two confluent and actively growing 10-cm^2 dishes of RKO cells were either untreated (no tr.) or exposed to ionizing radiation (IR). The cells were harvested for detection of ubiquitinated p53 7 h after IR treatment, as described in the text. Briefly, the cells were lysed in low-salt lysis buffer and centrifuged at low speed. Following centrifugation, the supernatant (S) and pellet (P) were sonicated in $1 \times$ RIPA buffer and immunoprecipitated with the p53 monoclonal antibody Ab-6. The immunoprecipitates were examined by immunoblot analysis with an anti-ubiquitin monoclonal antibody, or with the p53 antibody Ab-6. A ladder of bands that are ubiquitin:p53 conjugates are detected in the (S) fractions from untreated and IR-treated cells when the immunoprecipitates are probed with Ab-6 (*right*). A series of bands and a high-molecular-weight smear are also detected in the (S) fractions from untreated and IR-treated cells when the immunoprecipitates are probed with an antibody against ubiquitin (*left*). The asterisk (*) marks the position of non-ubiquitinated p53 and the IgG heavy-chain used in the immunoprecipitation. **(B)** p53 levels in the (S) and (P) fractions were determined by immunoblot analysis with the p53 antibody Ab-6 without prior immunoprecipitation.

immunoprecipitations are then resolved through a 9% polyacrylamide gel and transferred to a PVDF membrane.

8. Following transfer, the membrane is immersed in distilled H_2O in a Pyrex baking dish. The dish is covered with foil and autoclaved on slow exhaust for 15 min. The blot may turn a slight yellow color during this time.

9. The blot is then placed in blocking solution for 1 h, followed by rinsing in TNET as described earlier.

10. To probe for ubiquitinated p53, the blot is probed with an anti-ubiquitin antibody. Of the currently available anti-ubiquitin antibodies, Zymed monoclonal anti-ubiquitin (MAB1510) seems to work the best (*see* **Note 1**). The blot is placed in the anti-ubiquitin antibody for 90 min, and the blot is washed and processed with the secondary antibody as described earlier. Under these conditions, the ubiquitin antibody recognizes a series of high-molecular-weight bands and a high-molecular smear in S fraction of untreated and IR-treated RKO cells, but not in the P fraction of untreated or IR-treated cells (**Fig. 2**, left).

11. To confirm that the bands recognized by the anti-ubiquitin antibody are p53–ubiquitin conjugates, the blot can be stripped of the ubiquitin antibody, and reprobed with antibodies spe-

cific to p53. First, the blot is given 3 quick rinses in TNET, followed by one 15-min wash in TNET and two 5-min washes.

12. The blot is then placed in blot stripping solution that is preheated to 50°C. If possible, the blot should be maintained on a shaker at 50°C for 30 min. If this is not possible, the blot can be covered and placed on a shaker at room temperature for 15 min. The stripping solution is then replaced with fresh stripping solution preheated to 50°C, and the blot placed on a shaker for an additional 15 min.

13. The blot is given three quick rinses in TNET, followed by three 5-min washes in TNET, and placed in blocking solution for 1 h. The blot is then probed with an anti-p53 antibody, as described earlier. Under these conditions, the p53 antibody recognizes a series of bands in the S fraction, but not in the P fraction (**Fig. 2**, right). Some protein bands are recognized by both the ubiquitin and p53 antibodies, indicating that these bands are ubiquitin–p53 conjugates.

3.6. MDM2-Mediated Ubiquitination of P53

MDM2 can promote the ubiquitination and degradation of p53 when p53 and MDM2 are coexpressed in transiently transfected cells *(23,24)*. Our laboratory has found that the degradation of p53 is most readily observed with coexpression of epitope-tagged (myc-tagged) ubiquitin *(26)*. Moreover, our ability to see p53 ubiquitination depends on the amount of p53 expression DNA included in the transfection. **Figure 3** illustrates the ubiquitination and degradation of p53 that is mediated by MDM2 in cells transiently transfected with p53 and MDM2 expression DNAs.

1. U2OS cells were transfected by the calcium phosphate transfection method. First, 0.5 µg of an expression DNA encoding myc-tagged ubiquitin is combined in a microfuge tube with an expression DNA encoding HA-tagged wild-type p53 alone, or in combination with 3 µg of an expression plasmid encoding MDM2. In the experiments shown in **Fig. 3,** the amount of HA-tagged p53 DNA was varied between 0.05 µg and 0.5 µg. In each case, the total amount of DNA is adjusted to 4 mg by the addition of sonnicated salmon sperm DNA. H_2O is added to bring the total volume to 90 µL.

2. 100 µL of 2× HBS was added to bring the volume to 190 µL.

3. 10 µL of 2.5 *M* $CaCl_2$ is then added to each tube one tube at a time. Immediately following the addition of $CaCl_2$ to a tube, the tube is vortexed for 5 s. This process is repeated for each transfection. When $CaCl_2$ has been added to each sample, the tubes are spun down for 10 s and placed at room temperature for 30 min, during which time a DNA precipitate is formed.

4. During this 30 min, U2OS cells or other cells are trypsinized and plated for transfection. Typically, one confluent and actively growing 10-cm dish of U2OS cells are plated into nine 3.5-cm dishes. This trypsinization and plating usually takes 15–20 min to complete.

5. Each 200-µL DNA precipitate is added directly to the culture medium of the trypsinized and plated U2OS cells. The precipitate and cells are mixed in the dish by gentle swirling. The cells are cultured in the presence of the DNA precipitate for 16–20 h.

6. The medium is next aspirated from the cells and the cells are rinsed 2× with 2 mL of pre-warmed growth medium, and subsequently incubated in the presence of growth media for 6–10 h. Cells are harvested in lysis buffer for immunoblot analysis of p53, as described earlier.

3.6.1. Results

As shown in **Fig. 3A,** HA p53 levels were decreased with MDM2 expression under all conditions tested, indicating that MDM2 can promote p53 degradation. The degradation of p53 was most evident when relatively low amounts of HA p53 DNA were included in the transfection. A high-molecular-weight ladder of HA-p53 bands that are

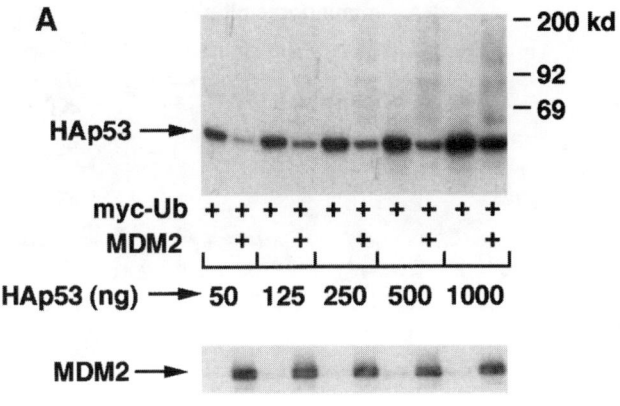

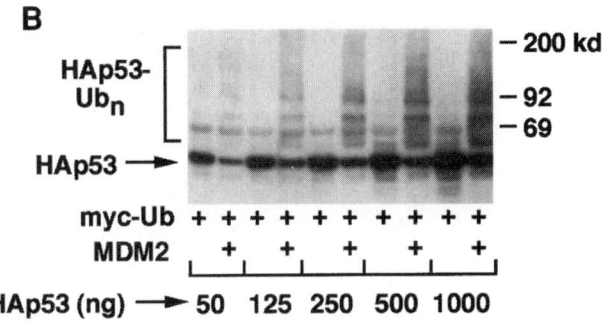

Fig. 3. MDM2-mediated ubiquitination and degradation of p53. U2OS cells were transfected with expression DNAs encoding MDM2 (3 mg), myc-tagged ubiquitin (0.5 mg), and the indicated amount of HAp53. In each case, the total amount of DNA was adjusted to 4 mg by the addition of salmon sperm DNA. Cell lysates were prepared 20 h after transfection and examined by immunoblot analysis with the anti-HA monoclonal antibody HA.11, and the anti-MDM2 monoclonal antibody SMP14. A short exposure (5 s, **A**) and long exposure (3 min, **B**) of the HA immunoblot were obtained. HAp53 levels decrease with MDM2 expression under all conditions, indicative of MDM2-mediated p53 degradation. A ladder of high-molecular-weight HAp53:ubiquitin conjugates is only observed when relatively high levels of DNA encoding HAp53 are included in the transfection.

ubiquitin:p53 conjugates were evident only under conditions when relatively high amounts of HA p53 DNA were included in the transfection (long exposure, **Fig. 3B**). These results indicate that the ability to see the ubiquitination of p53 that is mediated by MDM2 depends on the amount of input p53 DNA included in the transfection.

3.7. Detection of HAp53:myc-Ubiquitin Conjugates

To confirm that the ladder of p53 bands observed with p53 and MDM2 cotransfection are ubiquitin:p53 conjugates, an experiment like the one performed in **Fig. 4** can be performed. Saos-2 cells (p53-null) were used in the experiment shown in **Fig. 4**. However, a similar experiment could be performed in any cell type.

1. Cells are transfected with the indicated combinations of HA p53, MDM2, and myc-tagged ubiquitin.

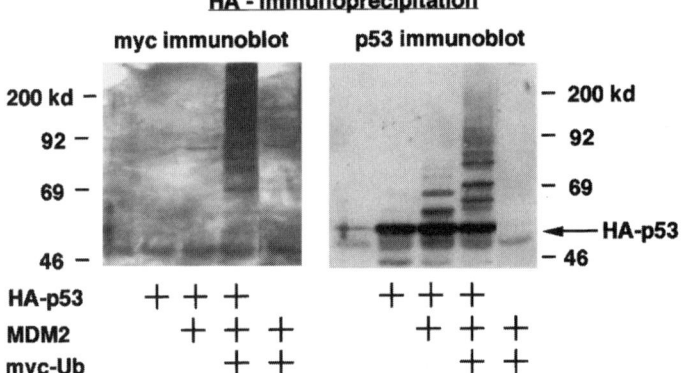

Fig. 4. Recognition of HAp53:ubiquitin conjugates in transfected cells. Saos-2 cells were transfected with DNAs encoding HAp53, MDM2, and myc-tagged ubiquitin as indicated. Cell extracts were immunoprecipitated with the anti-HA polyclonal antibody HA.11, and the immunoprecipitates were examined by immunoblot analysis with the anti-myc monoclonal antibody 9E10 (*left*). The blot was then stripped and reprobed with the p53 antibody Ab-6 (*right*). A ladder of bands that are ubiquitin:p53 conjugates is detected by both the myc and p53 antibodies. Note that the size of ubiquitinated p53 shifts to a slightly higher molecular weight when myc-ubiquitin is included in the transfection (*right*), consistent with the addition of the myc-tag.

2. 200 μg of lysate from the transfected cells were immunoprecipitated with a polyclonal anti-HA antibody (HA.11, Santa Cruz, cat no. sc-805). We typically use 2 μL of the 200-μg/mL antibody stock to immunoprecipitate p53 from 200 μg of protein extract.
3. The HA immunoprecipitates are resolved through SDS-PAGE, and analyzed by immunoblot analysis with antibodies against either the myc or p53. As shown in **Fig. 4,** a ladder of high-molecular-weight bands is detected by the anti-myc antibody in the HA immunoprecipitates, and these same bands are also recognized by the antibody against p53. These results confirm that these bands are ubiquitinated species of HA-tagged wild-type p53, and therefore confirm that MDM2 can promote the ubiquitination of p53.

4. Notes

1. Blots probed with commercially available anti-ubiquitin antibodies often have high levels of background staining. Several things can be done to overcome this problem. First, 5% powdered milk can be added to the anti-ubiquitin antibody dilution in TNET. Second, 5% powdered milk can be added to the TNET washes after probing the blot with both the primary (anti-ubiquitin) and secondary (HRP conjugated anti-mouse IgG) antibodies, and more extensive washing can be done. Third, the ECL chemiluminescence reagents can be diluted 1:2 in H_2O immediately prior to use.

References

1. Ciechanover, A., Orian, A., and Schwartz, A. L. (2000) The ubiquitin-mediated proteolytic pathway: mode of action and clinical implications. *J. Cell Biochem.* **77,** 40–51.
2. Hochstrasser, M. (1996) Ubiquitin-dependent protein degradation. *Annu. Rev. Genet.* **30,** 405–439.
3. Tyers, M. and Jorgensen, P. (2000) Proteolysis and the cell cycle: with this RING I do thee destroy. *Curr. Opin. Genet. Dev.* **10,** 54–64.

4. Jackson, P. K., Eldridge, A. G., Freed E., et al. (2000) The lore of the RINGs: substrate recognition and catalysis by ubiquitin ligases. *Trends Cell Biol.* **10,** 429–439.
5. Schneffner, M., Nuber, U., Huibregtse, J. M. (1995) Protein ubiquitination involving an E1-E2-E3 enzyme ubiquitin thioester cascade. *Nature* **373,** 81–83.
6. Koegl, M., Hoppe, T., Schlenker, S., Ulrich, H. D., Mayer, T. U., and Jentsch, S. (1999) A novel ubiquitination factor, E4, involved in multiubiquitin chain assembly. *Cell* **96,** 635–644.
7. Funk, W., Pak, D. T., Karas, R. H., Wright, W.E., and Shay, J. W. (1992) Transcriptionally active DNA binding site for human p53 protein complexes. *Mol. Cell. Biol.* **12,** 2866–2871.
8. Kern, S. E., Pietenpol, J. A., Thiagalingam, S., Seymour, A., Kinzler, K. W., and Vogelstein, B. (1992) Oncogenic forms of p53 inhibit p53-regulated gene expression. *Science* **256,** 827–830.
9. Pietenpol, J. A., Tokino, T., Thiagalingman, S., El-Deiry, W. S., Kinzler, K. W., and Vogelstein, B. (1994) Sequence-specific transcriptional activation is essential for growth suppression by p53. *Proc. Natl. Acad. Sci. USA* **91,** 1998–2002.
10. Maltzman, W. and Czyzyk, L. (1984) UV irradiation stimulates levels of p53 cellular tumor antigen in nontransformed mouse cells. *Mol. Cell. Biol.* **4,** 1689–1694.
11. Maki, C. G. and Howley, P. M. (1997) Ubiquitination of p53 and p21 is differentially affected by ionizing and UV radiation. *Mol. Cell. Biol.* **17,** 355–363.
12. Giaccia, A. J. and Kastan, M. B. (1998) The complexity of p53 modulation: emerging patterns from divergent signals. *Genes Dev.* **12,** 2973–2983.
13. Levine, A. J. (1997) P53, the cellular gatekeeper for growth and division. *Cell* **88,** 323–331.
14. El-Deiry, W. S., Tokino, T., Velculesu, V. E., et al. (1993) WAF1, a potential mediator of p53 tumor suppression. *Cell* **75,** 817–825.
15. Han, J., Sabbatini, P., Perez, D., Rao, L., Modha, D., and White, E. (1996) The E1B 19K protein blocks apoptosis by interacting with and inhibiting the p53-inducible and death promoting BAX protein. *Genes Dev.* **10,** 461–477.
16. Scheffner, M., Werness, B. A., Huibregtse, J. M., Levine, A. J., and Howley, P. M. (1990) The E6 oncoprotein encoded by human papillomavirus types 16 and 18 promotes the degradation of p53. *Cell* **63,** 1129–1136.
17. Huibregtse, J. M., Scheffner, M., and Howley, P. M. (1991) A cellular protein mediates association of p53 with the E6 oncoprotein of human papillomavirus types 16 or 18. *EMBO J.* **10,** 4129–4135.
18. Scheffner, M., Huibregtse, J. M., Vierstra, R. D., and Howley, P. M. The HPV16 E6 and E6AP complex functions as a ubiquitin protein ligase in the ubiquitination of p53. *Cell* **75,** 495–505.
19. Maki, C. G., Huibregtse, J. M., and Howley, P. M. In vivo ubiquitination and proteasome-mediated degradation of p53. *Cancer Res.* **56,** 2649–2654.
20. Kubbutat, M. H., Jones, S. N., and Vousden, K. H. Regulation of p53 stability by MDM2. *Nature* **387,** 299–303.
21. Haupt, Y., Maya, R., Kazaz, A., and Oren, M. Mdm2 promotes the rapid degradation of p53. *Nature* **387,** 296–299.
22. Honda, R., Tanaka, H,, Yasuda, H. Oncoprotein MDM2 is a ubiquitin ligase E3 for tumor suppressor p53. *FEBS Lett.* **420,** 25–27.
23. Fuchs, S. Y., Adler, V., Buschmann, T., Wu, X., and Ronai, Z. (1998) Mdm2 association with p53 targets its ubiquitination. *Oncogene* **17,** 2543–2547.
24. Maki, C. G. (1999) Oligomerization is required for p53 to be efficiently ubiquitinated by MDM2. *J. Biol. Chem.* **274,** 16531–16535.
25. Jost, C. A., Marin, M. C., and Kaelin, W. G. (1997) P73 is a simian p53-related protein that can induce apoptosis. *Nature* **389,** 191–194.
26. Ward, C. L., Omura, S., and Kopito, R. R. (1995) Degradation of CFTR by the ubiquitin-proteasome pathway. *Cell* **83,** 121–127.

27. Rock, K. L., Gramm, C., Rothstein, L., et al. (1994) Inhibitors of the proteasome block the degradation of most cell proteins and the generation of peptides presented on MHC class I molecules. *Cell* **78,** 761-771.

28. Fenteany, G., Standaert, R. F., Lane, W. S., Choi, S., Corey, E. J., and Schreiber, S. L. Inhibition of proteasome activities and subunit-specific amino-terminal threonine modification by lactacystin. *Science* **268,** 726-731.

4

Role of Tumor Suppressors in DNA Damage Response

Bin-Bing S. Zhou, Michael R. Mattern, and Kum Kum Khanna

1. Introduction

Studies of human cancer predisposition syndromes and mouse knockout models have revealed several connections between defects in DNA damage response and tumorigenesis. Several recently identified genes, including *ATM* (ataxia telangiectasia mutated), *BRCA1*, *BRCA2*, and *Nbs1* (Nijmegen breakage syndrome gene), act as tumor suppressors by regulating various aspects of the cellular damage response pathway. In all of these cases, inactivating mutations cause defects in DNA damage signaling, giving rise to some form of chromosomal instability and predisposition to cancer. This chapter is concerned with the relationship between these tumor suppressors and the cellular DNA damage response, and focuses primarily on how defects in the tumor suppressor-dependent DNA damage signaling pathway may cause genetic instability, thereby facilitating tumorigenesis, and, in some cases, providing an opportunity for selective therapy.

2. Tumor Suppressors Affecting Genetic Instability

Most cancer cells are genetically unstable, the instability existing at either the nucleotide or chromosome level *(1)*. In a small subset of tumors, instability is observed as base substitution or the deletion/insertion of a few nucleotides. In these tumors, "microsatellite" repeats [including poly(CA)] are similarly affected, giving rise to the term "microsatellite instability" (MIN). Although it is rare in sporadic colorectal cancers, MIN occurs in most cancers of patients with hereditary non-polyposis colorectal cancer (HNPCC) *(2)*. The identification of mismatch repair gene hMSH2 on human chromosome 2 (**ref.** *3*) and the demonstration of inactivating mutations in this gene in chromosome-2-linked HNPCC kindreds *(4)* suggest that the MIN phenotype is the result of defective mismatch repair. Mutation in another human mismatch repair gene, MLH1, also leads to a MIN phenotype in cancer patients *(5)*. This genetic evidence is strongly supported by the finding that tumors exhibiting MIN lack detectable mismatch repair activity in biochemical assays *(6,7)*. It is possible that these tumors are defective in certain aspect of DNA damage signaling as well.

From: *Methods in Molecular Biology, Vol. 223: Tumor Suppressor Genes: Regulation, Function, and Medicinal Applications.* Edited by: Wafik S. El-Deiry © Humana Press Inc., Totowa, NJ

In most other cancers, instability is observed at the chromosome level with changes in chromosome number and/or chromosome structure. This kind of chromosome instability is called CIN, presenting as wholesale changes in the genome of cancer cells. Chromosome number instability (aneuploidy) results in the loss or gain of entire chromosomes or large chromosome fragments *(1)* and most likely reflects a dysfunctional chromosome segregation apparatus (mitotic checkpoint) *(8)*. Alternatively, structural abnormalities (gross chromosomal abberations) involves the breakage and joining of DNA segments, and the underlying cause seems to be related to defects in sensing and repair of DNA strand breaks.

Several genes involved in the DNA damage response pathway have been implicated in human tumorigenesis. Well-documented examples include *p53, ATM, BRCA1, BRCA2, Nbs1,* and *Chk2.* The best characterized of these genes is *p53*, which encodes a tumor suppressor protein that, in response to DNA damage, elicits either cell cycle arrest or apoptosis. It is now clear that the functioning of the p53 DNA damage signaling pathway is lost in most, if not all, human cancers *(9)*. Ataxia-telangiectasia (AT), in which *ATM* is mutated, is associated with a high incidence of leukemia and lymphoma that develop in childhood. Some individuals carrying a mutation of ATM in one allele are at increased risk of developing breast cancer *(10)*. The breast cancer susceptibility genes *BRCA1* and *BRCA2* encode multifunctional proteins, the mutant phenotypes of which predispose to both breast and ovarian cancer *(11)*. Tumorigenesis in individuals with germline *BRCA* mutations requires somatic inactivation of the remaining wild-type allele. Because loss of heterozygosity is detected in the *BRCA1* and *BRCA2* genes in cancer cells from *BRCA* mutation carriers, BRCA1 and BRCA2 are considered to be tumor suppressors. The autosomal recessive genetic disorder, Nijmegen breakage syndrome (NBS, in which Nbs1 is mutated), is characterized by an abnormally high risk for development of lymphatic tumors *(12)*. Lymphocytes from NBS patients exhibit considerable overlap of cellular phenotype with lymphocytes derived from AT patients with respect to radiation sensitivity, defective repair of breaks in DNA, and chromosomal instability in the form of chromatid breaks and recombination figures *(13)*. Li-Fraumeni syndrome (LFS) is a hereditary cancer susceptibility syndrome characterized by the occurrence of sarcoma, breast cancer, and brain tumors in multiple relatives. Although germline *p53* mutations account for most LFS cases, *CHK2* mutations have recently been causally linked with a group of LFS patients with apparently normal p53 *(14)*. For this reason, Chk2 has been suggested as a tumor suppressor gene as well.

3. The DNA Damage Response Pathway

Many tumor suppressors discussed here are involved in discrete steps of DNA damage activated transduction pathway, which leads to checkpoint response, increased repair, or, in certain circumstances, apoptosis. Although these genes play unique roles in the DNA damage response pathway, loss of function of any of them leads to genetic instability or lack of cell death in extensively damaged cells, resulting in tumor formation (summarized in **Fig. 1**). The DNA damage response pathway is a network of interacting pathways consisting of sensors, transducers, and effectors (**Fig. 2**). These pathways are not only surveyors for occasional damage, but also firmly integrated components of cellular physiology. They are critical for the cell's response to DNA damage, and its progression through its cycle *(15)*.

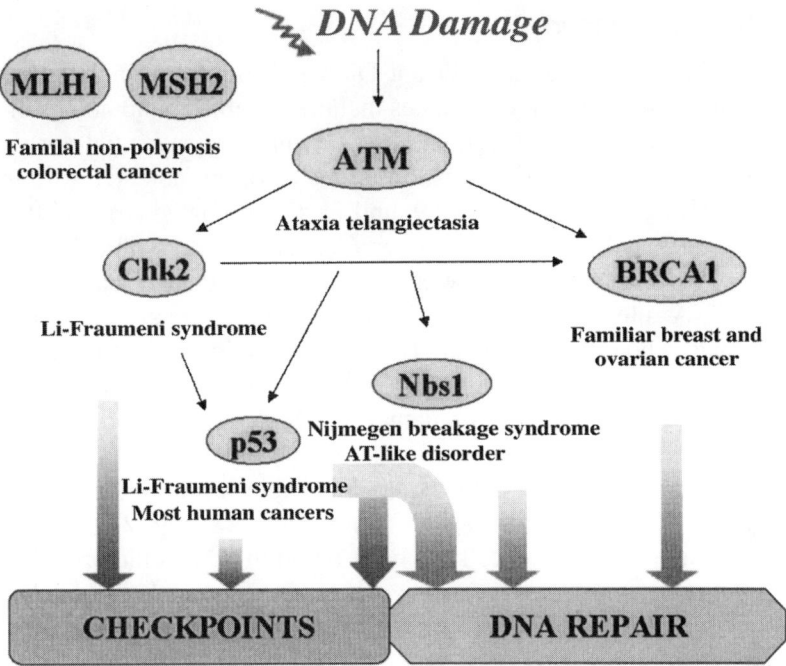

Fig. 1. A network of tumor suppressors that regulate cellular responses to DNA damage. Depicted are the tumor suppressors involved in DNA damage signaling and those human cancer predisposition syndromes associated with their deficiencies. Loss of function of any of these genes leads to defects in checkpoint and/or repair activity, resulting in genetic instability.

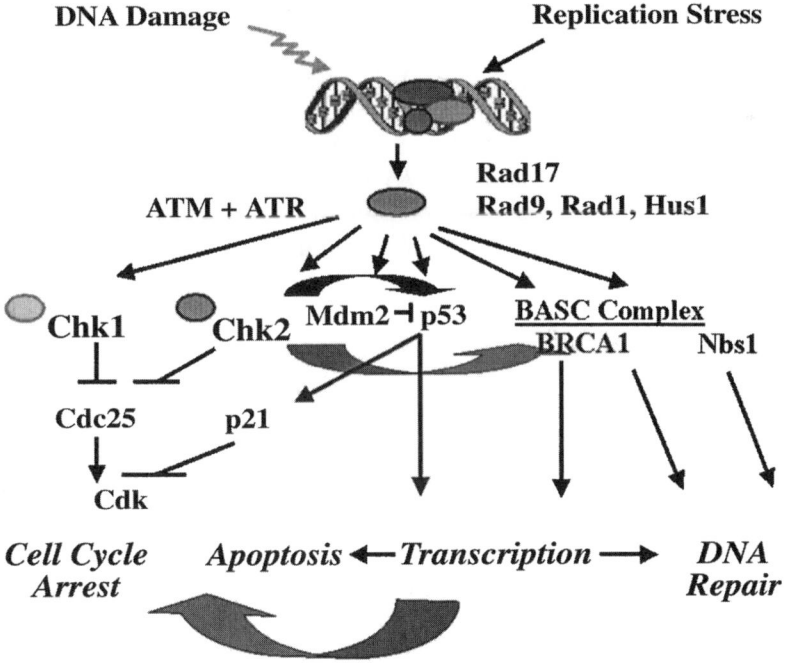

Fig. 2. Organization of the mammalian DNA damage response pathway. Depicted are the major signaling pathways and key proteins involved. Arrowheads represent positively acting steps, while perpendicular ends represent inhibitory steps. Adapted from *(15)* with permission from Nature publishing group.

3.1. DNA Damage Sensors

The proteins that initially sense aberrant DNA structures and initiate the signaling response are currently unknown. Candidates include mammalian homologs of four conserved yeast genes, *Rad1, Rad9, Hus1,* and *Rad17*, which play essential roles in the activation of the DNA damage response pathway and have the potential to interact with nucleic acids. Among them, Rad1, Rad 9, and Hus1 are related structurally to PCNA and form a DNA damage-responsive complex *(16)*. Thus it is possible that, like PCNA, these proteins form a doughnut-like hetero trimeric complex that loads onto damaged DNA just as PCNA is loaded onto primed DNA. The fourth conserved checkpoint protein, Rad17, shares homology with all five subunits of replication factor C (RFC) (reviewed in **ref. *17***). This homology suggests a model in which Rad17 either alone or with other RFC subunits interacts with damaged DNA and then loads the Rad1-Rad9-Hus1 complex onto DNA, which in turn leads to the transduction of DNA damage signal through activation of ATM and ATR (ataxia-telangiectasia Rad3-related), proximal serine-threonine kinases in the damage response pathway (**Fig. 2**). No mutations of *Rad1, Rad9 , Hus1*, or *Rad17* genes have been found in human tumors, and functions for the mammalian gene products have not yet been characterized completely. However, consistent with yeast studies, inactivation of mouse *Hus1* results in impaired response to cellular genotoxic stress *(18)*.

Alternative proteins have also been suggested as candidate sensors, for example, the tumor suppressor BRCA1. A defect in the ability of mouse cells lacking BRCA1 exon 11 to arrest in G2 has been reported, consistent with a defect in sensing or signaling *(19)*. BRCA1 is part of a large complex named BASC (BRCA1-associated genome surveillance complex) that contains multiple tumor suppressors, including ATM, Nbs1, and mismatch proteins (MSH2/6 and MLH2) (**ref. *20***). Such biochemical evidence of association suggests that these tumor suppressors function together to suppress genetic instability. Initial support for mismatch repair proteins as signaling molecule came from the observation that *MLH1*-defective cells fail to undergo ATM-dependent phosphorylation of c-Abl in response to cis-platinum treatment *(21)*. While the model that postulates BASC as a sensor complex is speculative, it is interesting that many of its component proteins have the ability to recognize aberrant DNA structures and transmit the information to ATM and BRCA1.

3.2. Proximal DNA Damage Response Kinases

According to the simplest model, putative DNA damage sensors bind to damaged DNA directly, leading to activation of proximal kinases such as ATM. ATM shares its C-terminal architecture with a large group of PI-3 kinase-related anzymes that includes DNA-PKcs (catalytic subunit of DNA-dependent protein kinase) and ATR. Although ATM kinase activity can be activated by DNA damage in vivo *(22,23)*, the precise mechanism of activation remains unknown. Direct activation by DNA has not been clearly established. Some studies report that there is no stimulation by DNA in vitro *(24,25)*, while others have found that small amounts of purified ATM bind to and are activated by DNA containing double-strand breaks (DSB) in vitro *(26)*. It is not clear, however, whether ATM has the ability to bind DNA breaks by itself or requires other, yet unidentified proteins complexed with it for efficient DNA binding. ATM may ultimately resemble DNA-PK, which possesses an intrinsic weak DNA-end binding activity that is greatly increased by addition of the Ku70/80 heterodimer *(27)*. Thus, any protein to which ATM binds is a candidate sensor, whose identification and characterization is essential for the molecular elucidation of ATM activation and DNA damage recognition.

The phenotypes of cells from AT patients have been particularly helpful in revealing ATM function. AT cells are defective in checkpoint arrests throughout the cell cycle, including G1 arrest *(28)*, reduction in DNA synthesis *(29)*, and G2 arrest *(30)*. Activated ATM catalyzes phosphorylation cascades to transmit signals to checkpoint and repair proteins (reviewed in **refs. 15** and *31*). A number of genuine tumor suppressors have been shown to lie downstream of ATM to trigger cell cycle arrest, DNA repair, or apoptosis after DNA damage, as discussed below in connection with effector pathways.

Similar to ATM, ATR is a proximal kinase that is a central component of the DNA damage response. Current data suggest that ATM primarily controls the response to ionizing radiation (IR), whereas ATR responds to other types of damage including ultraviolet radiation (UV), alkylating agents, and inhibitors of DNA replication. Unlike ATM, ATR is an essential kinase *(32,33)*, and its mutations have not been found in any diseases. Nevertheless, it is interesting to note that the incidence of large benign tumors has been found to be increased in *Atr* heterozygote mice compared with wild-type *(32)*, suggesting a connection between earlier-stage hyperplastic progression and ATR.

3.3. Downstream DNA Damage Response Kinases

Chk2, a serine-threonine kinase, is phosphorylated and activated in response to various DNA-damaging agents *(34–37)*. Its phosphorylation in response to IR is ATM-dependent. Chk2 contains a FHA domain and a SQ cluster domain (SCD) in its amino terminus, and a kinase domain at its carboxyl terminus. Thr-68 in the SCD is phosphorylated by ATM in vivo and in vitro in response to damage *(38–41)*. Chk2 phosphorylation is required for its activation; however, in-vitro phosphorylation of Chk2 by ATM does not seem to be sufficient to activate it (unpublished results). Chk2 was initially identified as a homolog of *Saccharomyces cerevisiae* Rad53 and *Saccharomyces pombe* Cds1. Mec1 (ATM/ATR homolog)-dependent phosphorylation and activation of Rad53 in budding yeast requires an association between the FHA domain of Rad53 and phosphorylated Rad9, subsequent to DNA damage *(42)*. Thus, it is likely that Chk2 activation involves both direct phosphorylation and protein–protein interaction. Consistent with the function of Chk2 as a damage response kinase, *Chk2*-null mutant mouse embryonic stem cells have been shown to be defective in maintaining G2 arrest in response to IR *(43)*.

Downstream of ATM and ATR, Chk1, another serine/threonine kinase in the DNA damage-response pathway, shares some substrate specificity with Chk2. Unlike Chk2, Chk1 does not appear to be activated significantly in response to DNA damage. Nevertheless, Chk1 is phosphorylated in response to IR in mammals *(44,45)* and in yeast *(46,47)*. Mice lacking *Chk1* die in early embryogenesis, similarly to *Atr−/−* mice *(44,48)*. Embryonic stem cells lacking *Chk1* are defective in their G2 checkpoint, as previously observed in other organisms. The facts that *Atr* and *Chk1* mutant mice have similar phenotypes and ATR controls Chk1 phosphorylation in vivo *(44)* suggest that Chk1 is a key effector of the ATR pathway, and may be responsible for ATR's essential function. To date, Chk1 mutations have not been found in human cancers, although it is known that mouse *Chk1* heterozygosity modestly enhances tumorigenesis *(44)*.

3.4. Phosphorylation-Dependent Activation of Effectors

Downstream of DNA damage response kinases are effectors that execute the various functions of the DNA damage responses. Many of these, such as p53, Cdc25C, BRCA1, and Nbs1, are substrates of ATM and Chk2, and are involved in cell cycle control, apoptosis, and DNA repair.

Both ATM and Chk2 are critical for the regulation of p53 stability after IR. IR-treated *Atm−/−* and *Chk2−/−* mouse thymocytes fail to stabilize and activate p53, induce p53-dependent genes, or trigger apoptosis *(43,49)*. Their inabilities to activate p53 and p21 indicate that *Atm* and *Chk2* mutants are not able to arrest in G1 in response to IR. Genetic evidence for a common pathway for Chk2 and p53 is also provided by the identification of mutations in the *CHK2* gene in a subset of LFS cancer families that lack p53 mutations *(14)*. ATM phosphorylates p53 at Ser 15 in vitro and in vivo *(22–24)*. Chk2 phosphorylation of p53 on Ser-20 *in vitro* *(43,50,51)* has been shown to interfere with binding to Mdm2, a ubiquitin ligase that targets p53 for proteosomal degradation *(52)*. Further, ATM phosphorylates Mdm2 *(53)*, although the significance of this activity is unknown. It is still unclear why ATM and Chk2, the latter of which depends on ATM for activation, both target p53, but it seems reasonable to postulate that they synergize to ensure that the p53 pathway is fully active after DNA damage.

AT cells have the property of "radioresistant DNA synthesis" *(29)*, a phenomenon caused by S-phase checkpoint defect. The recent demonstration that Chk2 links ATM to a S-phase checkpoint indicates that Chk2 might participate in multiple pathways of checkpoint control *(54)*. In this instance, Chk2 activation enhances the degradation of Cdc25A phosphatase, leading to a failure to activate Cdk2, resulting in the inhibition of S phase entry. Furthermore, the activation of Chk2, degradation of Cdc25A and inhibition of Cdk2 are all defective in AT cells *(54)*. It is conceivable that differences in G1 and S-phase checkpoint control lie downstream of ATM and Chk2.

Chk1 and Chk2 have roles in the G2/M checkpoint activated after DNA damage as well. Chk1 *(45)* and Chk2 (**refs. 34, 35,** and *37*) have been shown in vitro to phosphorylate Cdc25C at Ser-216. This modification is believed to be involved in the inactivation and translocation of Cdc25C into the cytoplasm mediated by 14-3-3 binding after DNA damage *(55–57)*. Biochemical confirmation of this model using cultured mammalian cells has not yet been achieved, possibly because Cdc25C is normally phosphorylated at Ser-216 all through the interphase and DNA damage merely prolongs this phosphorylation. Moreover, Chk2 regulation of p53 is also expected to be critical for the G2 DNA damage checkpoint because p53 targets such as p21 and 14-3-3σ have roles in the maintenance of G2 arrest in mammalian cells *(58,59)*.

Following their own activations, ATM and Chk2 signal to BRCA1 after IR treatment. Recent evidence suggests that BRCA1 is regulated by phosphorylation in response to DNA damage and complexes with Rad51 to perform homologous recombination (HR) repair of DSBs (reviewed in **ref. 60**). A portion of BRCA1 also colocalizes with another protein complex, Rad50/Mre11/Nbs1, in nuclear foci that are distinct from those of Rad51 (**ref. 61**). ATM, ATR, and Chk2 catalyze most of the damage-induced phosphorylation of BRCA1. ATM and ATR can phosphorylate BRCA1 at Ser-1423, -1524, and other sites *(62–65)*. BRCA1 phosphorylation in response to IR is largely ATM-dependent, while phosphorylation in response to UV and hydroxyurea requires ATR *(63,65)*. Chk2 can phosphorylate BRCA1 on Ser-988 in vitro *(66)*. The physiological relevance of ATM-, ATR-, and Chk2-dependent phosphorylation of BRCA1 is not yet clear, but it seems likely that these phosphorylations affect the repair function of BRCA1. In a recent report, Bonner and Colleagues *(67)* have identified a role of BRCA1 in the early response to DNA damage. Within minutes of irradiation, a histone H2A family member, H2Ax, becomes phosphorylated and forms foci with BRCA1 at sites of DSBs in cells *(68)*. These foci generally appear several hours before other events such as collocalization of Rad51 or Rad50 with BRCA1 (**ref. *67***). These results imply that BRCA1 links the detection of DNA damage to the actual repair of the lesion. Depending on the context, BRCA1 may have multiple functions, as it has also been shown to force the damaged cells to engage in cell cycle arrest or apoptosis *(60)*.

Another recently identified target of ATM is the Nbs1 protein. Cells from AT and NBS patients have strikingly similar responses to DNA damage, including increased chromosome breakage and radioresistant DNA synthesis, suggesting that ATM and Nbs1 function in the same pathway. Further evidence for this model has come from observations that ATM directly phosphorylates Nbs1 at several sites that are required for its function in vivo *(69–72)*. Nbs1 resides in a complex with Mre11 and Rad50, which redistributes from a diffuse to a focused staining pattern after γ-irradiation *(73)*. Recently, Mre11 was shown to be mutated in individuals with another AT-like disorder, AT-LD *(74)*, providing further evidence for a link between these proteins. The functions of the Rad50-Mre11-Nbs1 complex are not yet fully understood. The fact that it posseses exonuclease and helicase activities suggests a role in the processing of IR-induced DNA damage before it is repaired by the HR or nonhomologous end joining (NHEJ) pathways (reviewed in **ref. *75***). Recent work has indicated that the *S. pombe* counterparts of Rad50 and Mre11, while critical for HR, do not have an essential role for NHEJ *(76),* raising the possibility that their mammalian homologs have a more restricted function than those of *S. cerevisiae.*

3.5. Transcriptional Regulation

In addition to activate phosphorylation events, DNA damage induces transcription of a large number of genes involved in cell cycle regulation, DNA repair, and apoptosis. The tumor suppressor p53 is a transcription factor that is responsible for many of these changes. Cells defective in p53 are unable to arrest in G1 after IR, and show reduced repair or apoptosis under various circumstances. Part of the ability of p53 to arrest G1 cells results from activation of transcription of p21, a tight-binding inhibitor of cyclin-dependent kinases that control entry into S phase (reviewed in **ref. *77***). Several other cell cycle regulators are also induced by p53, including GADD45 and members of the 14-3-3 family. Fibroblasts lacking p53 have been shown to be defective in global excision repair of cyclobutane dimers. Consistent with this observation, the p48 gene, which is mutated in *Xeroderma pigmentosum* group E cells, is induced by the p53-dependent DNA damage response pathway *(78)*. Recently, a newly identified nuclear-localized subunit of ribonucleotide reductase, p53R2, was found to be induced by p53 in response to DNA damage. Blocking p53R2 expression increases cell killing by a variety of DNA-damaging agents *(79),* suggesting a role for p53R2 in DNA repair. Several p53 targets might be important for damage-induced apoptosis, although it is still unclear how p53 distinguishes between repair-related and apoptosis-related targets. Bax, one of the first apoptotic genes shown to be transcriptionally regulated by p53, is not induced by p53 under all circumstances, and is not absolutely required for p53-mediated death *(80)*. One interesting target of p53, which might transmit signals to apoptotic machinery, is PERP, a four-span plasma membrane protein with homology to the PMP-22/Gas3 family *(81)*. PERP transcript is associated exclusively with the apoptotic rather than cell-cycle arrest function of p53. Other possible p53 apoptotic targets include Noxa *(82)*, a protein containing a BH3 domain that interacts with the antiapoptotic Bcl2 protein, and p53 AIP1 *(83)*, a novel protein with no known homologs. An additional important p53-induced protein is Mdm2, which escorts p53 from the nucleus and targets it for proteasomal degradation. Mdm2 thus ensures that the p53 signal is transient and carefully regulated. All of these kinase cascades and transcription programs are likely utilized to determine how various damage responses are executed after DNA damage.

4. Implications of Checkpoint and Repair Defects for Cancer Therapy

During the past several years, considerable progress has been made in elucidating various components of the eukaryotic DNA damage response; many of these components were found to be tumor suppresor genes. Response pathways have been shown to control the activation of DNA repair pathways, transcriptional programs, and, in some cell types, induction of cell death by apoptosis, in addition to regulating cell cycle arrest (**Fig. 2**). Knowledge of molecular mechanisms of the DNA damage response and the genetic and biochemical distinctions between normal and cancer cells provide a useful conceptual framework for identifying novel strategies for cancer therapy *(84)*. Paradoxically, the same defects that lead to tumorgenesis could render cancer cells selectively vulnerable to therapy. AT cells, NBS cells, and BRCA1-deficient cells, characterized by increased susceptibility of donor patients to cancer, are all hypersensitive to IR. Likewise, cancer cell survival after chemotherapy or radiotherapy is diminished owing to specific checkpoints and/or repair pathways that have been lost. The G1 checkpoint is frequently inactivated in cancer by mutations in the p53 tumor suppressor pathway. On the other hand, mutations in the genes involved in the G2 checkpoint are extremely rare, and this checkpoint is functional in many cancer cells. Recent studies have provided evidence nonetheless that the cytotoxicity and selectivity of DNA-damaging agents can be enhanced by agents that disrupt the G2 checkpoint in cancer cells *(85)*. The observation that p53-deficient cells are sensitized preferentially to radiation-induced killing by caffeine and UCN-01 suggests that targeting the G2/M checkpoint will introduce an element of selectivity into current cancer therapy. Thus, drugs that inhibit specific checkpoints or repair pathways may enable the design of improved therapeutic strategies. In the foregoing, it is evident that a full understanding of the roles of tumor suppressors in the DNA damage response pathway will afford a spectrum of mechanism-based therapeutic treatments. It may be possible in the near future to evaluate patients at presentation for checkpoint and repair activities and apply this knowledge to the design of optimal treatment protocols.

References

1. Lengauer, C., Kinzler, K. W., and Vogelstein, B. (1998) Genetic instabilities in human cancers. *Nature* **396**, 643–649.
2. Aaltonen, L. A., Peltomaki, P., Leach, F. S., et al. (1993) Clues to the pathogenesis of familial colorectal cancer. *Science* **260**, 812–816.

3. Fishel, R., Lescoe, M. K., Rao, M. R., et al. (1993) The human mutator gene homolog MSH2 and its association with hereditary nonpolyposis colon cancer. *Cell* **75**, 1027–1038.
4. Leach, F. S., Nicolaides, N. C., Papadopoulos, N., et al. (1993) Mutations of a mutS homolog in hereditary nonpolyposis colorectal cancer. *Cell* **75**, 1215–1225.
5. Peltomaki, P. and de la Chapelle, A. (1997) Mutations predisposing to hereditary nonpolyposis colorectal cancer. *Adv Cancer Res.* **71**, 93–119.
6. Parsons, R., Li, G. M., Longley, M. J., et al. (1993) Hypermutability and mismatch repair deficiency in RER+ tumor cells. *Cell* **75**, 1227–1236.
7. Umar, A., Boyer, J. C., Thomas, D. C., et al. (1994) Defective mismatch repair in extracts of colorectal and endometrial cancer cell lines exhibiting microsatellite instability. *J. Biol. Chem.* **269**, 14367–14370.
8. Cahill, D. P., Lengauer, C., Yu, J., et al. (1998) Mutations of mitotic checkpoint genes in human cancers. *Nature* **392**, 300–303.
9. Levine, A. J. (1997) p53, the cellular gatekeeper for growth and division. *Cell* **88**, 323–331.
10. Khanna, K. K. (2000) Cancer risk and the ATM gene: a continuing debate. *J. Natl. Cancer Inst.* **92**, 795–802.
11. Welcsh, P. L., Owens, K. N., and King, M. C. (2000) Insights into the functions of BRCA1 and BRCA2. *Trends Genet.* **16**, 69–74.
12. Digweed, M., Reis, A., and Sperling, K. (1999) Nijmegen breakage syndrome: consequences of defective DNA double strand break repair. *Bioessays* **21**, 649–656.
13. Seemanova, E., Passarge, E., Beneskova, D., Houstek, J., Kasal, P., and Sevcikova, M. (1985) Familial microcephaly with normal intelligence, immunodeficiency, and risk for lymphoreticular malignancies: a new autosomal recessive disorder. *Am. J. Med. Genet.* **20**, 639–648.
14. Bell, D. W., Varley, J. M., Szydlo, T. E., et al. (1999) Heterozygous germ line hCHK2 mutations in Li-Fraumeni syndrome. *Science* **286**, 2528–2531.
15. Zhou, B.-B. and Elledge, S. J. (2000) The DNA damage response: putting checkpoints in perspective. *Nature* **408**, 433–439.
16. Volkmer, E. and Karnitz, L. M. (1999) Human homologs of *Schizosaccharomyces pombe* rad1, hus1, and rad9 form a DNA damage-responsive protein complex. *J. Biol. Chem.* **274**, 567–570.
17. O'Connell, M. J., Walworth, N. C., and Carr, A. M. (2000) The G2-phase DNA-damage checkpoint. *Trends Cell Biol.* **10**, 296–303.
18. Weiss, R. S., Enoch, T., and Leder, P. (2000) Inactivation of mouse hus1 results in genomic instability and impaired responses to genotoxic stress. *Genes Dev.* **14**, 1886–1898.
19. Xu, X., Weaver, Z., Linke, S. P., et al. (1999) Centrosome amplification and a defective G2-M cell cycle checkpoint induce genetic instability in BRCA1 exon 11 isoform-deficient cells. *Mol. Cell* **3**, 389–395.
20. Wang, Y., Cortez, D., Yazdi, P., Neff, N., Elledge, S. J., and Qin, J. (2000) BASC, a super complex of BRCA1-associated proteins involved in the recognition and repair of aberrant DNA structures. *Genes Dev.* **14,** 927–939.
21. Gong, J. G., Costanzo, A., Yang, H. Q., et al. (1999) The tyrosine kinase c-Abl regulates p73 in apoptotic response to cisplatin-induced DNA damage. *Nature* **399,** 806–809.
22. Canman, C. E., Lim, D. S., Cimprich, K. A., et al. (1998) Activation of the ATM kinase by ionizing radiation and phosphorylation of p53. *Science* **281,** 1677–1679.
23. Khanna, K. K., Keating, K. E., Kozlov, S., et al. (1998) ATM associates with and phosphorylates p53: mapping the region of interaction. *Nat. Genet.* **20,** 398–400.
24. Banin, S., Moyal, L., Shieh, S., et al. (1998) Enhanced phosphorylation of p53 by ATM in response to DNA damage. *Science* **281,** 1674–1677.
25. Chan, D. W., Son, S. C., Block, W., et al. (2000) Purification and characterization of ATM from human placenta. A manganese-dependent, wortmannin-sensitive serine/threonine protein kinase. *J. Biol. Chem.* **275**, 7803–7810.

26. Smith, G. C., Cary, R. B., Lakin, N. D., et al. (1999) Purification and DNA binding properties of the ataxia-telangiectasia gene product ATM. *Proc. Natl. Acad. Sci. USA* **96,** 11134–11139.

27. Gottlieb, T. M. and Jackson, S. P. (1993) The DNA-dependent protein kinase: requirement for DNA ends and association with Ku antigen. *Cell* **72,** 131–142.

28. Kastan, M. B., Zhan, Q., el-Deiry, W. S., et al. (1992) A mammalian cell cycle checkpoint pathway utilizing p53 and GADD45 is defective in ataxia-telangiectasia. *Cell* **71,** 587–597.

29. Painter, R. B. and Young, B. R. (1980) Radiosensitivity in ataxia-telangiectasia: a new explanation. *Proc. Natl. Acad. Sci. USA* **77,** 7315–7317.

30. Paules, R. S., Levedakou, E. N., Wilson, S. J., et al. (1995) Defective G2 checkpoint function in cells from individuals with familial cancer syndromes. *Cancer Res.* **55,** 1763–1773.

31. Khanna, K. K. and Jackson, S. P. (2001) DNA double-strand breaks: signaling, repair and the cancer connection. *Nat. Genet.* **27,** 247–254.

32. Brown, E. J. and Baltimore, D. (2000) ATR disruption leads to chromosomal fragmentation and early embryonic lethality. *Genes Dev.* **14,** 397–402.

33. de Klein, A., Muijtjens, M., van Os, R., et al. (2000) Targeted disruption of the cell-cycle checkpoint gene ATR leads to early embryonic lethality in mice. *Curr. Biol.* **10,** 479–482.

34. Matsuoka, S., Huang, M., and Elledge, S. J. (1998) Linkage of ATM to cell cycle regulation by the Chk2 protein kinase. *Science* **282,** 1893–1897.

35. Blasina, A., Van de Weyer, I., Laus, M. C., Luyten, W. H. M. L., Parker, A. E., and McGowan, C. H. (1999) A human homologue of the checkpoint kinase Cds1 directly inhibits Cdc25 phosphatase. *Curr. Biol.* **9,** 1–10.

36. Brown, A. L., Lee, C. H., Schwarz, J. K., Mitiku, N., Piwnica-Worms, H., and Chung, J. H. (1999) A human Cds1-related kinase that functions downstream of ATM protein in the cellular response to DNA damage. *Proc. Natl. Acad. Sci. USA* **96,** 3745–3750.

37. Chaturvedi, P., Eng, W.-K., Zhu, Y., et al. (1999) Mammalian Chk2 is a downstream effector of the ATM-dependent DNA damage checkpoint pathway. *Oncogene* **18,** 4047–4054.

38. Zhou, B.-B. S., Chaturvedi, P., Spring, K., et al. (2000) Caffeine abolishes the mammalian G2/M DNA damage checkpoint by inhibiting ataxia-telangiectasia-mutated kinase activity. *J. Biol. Chem.* **275,** 10342–10348.

39. Matsuoka, S., Rotman, G., Ogawa, A., Shiloh, Y., Tamai., K., and Elledge, S. J. (2000) ATM phosphorylates Chk2 in vivo and in vitro. *Proc. Natl. Acad. Sci. USA* **97,** 10389–10394.

40. Melchionna, R., Chen, X. B., Blasina, A., and McGowan, C. H. (2000) Threonine 68 is required for radiation-induced phosphorylation and activation of Cds1. *Nat. Cell Biol.* **2,** 762–765.

41. Ahn, J. Y., Schwarz, J. K., Piwnica-Worms, H., and Canman, C. E. (2001) Threonine 68 phosphorylation by ataxia telangiectasia mutated is required for efficient activation of Chk2 in response to ionizing radiation. *Cancer Res.* **60,** 5934–5936.

42. Sun, Z., Hsiao, J., Fay, D. S., and Stern, D. F. (1998) Rad53 FHA domain associated with phosphorylated Rad9 in the DNA damage checkpoint [see comments]. *Science* **281,** 272–274.

43. Hirao, A., Kong, Y. Y., Matsuoka, S., et al. (2000) DNA damage-induced activation of p53 by the checkpoint kinase Chk2. *Science* **287,** 1824–1827.

44. Liu, Q., Guntuku, S., Cui, X. S., et al. (2000) Chk1 is an essential kinase that is regulated by Atr and required for the G(2)/M DNA damage checkpoint. *Genes Dev.* **14,** 1448–1459.

45. Sanchez, Y., Wong, C., Thoma, R. S., et al. (1997) Conservation of the Chk1 checkpoint pathway in mammals: linkage of DNA damage to Cdk regulation through Cdc25. *Science* **277,** 1497–1501.

46. Sanchez, Y., Bachant, J., Wang, H., et al. (1999) Control of the DNA damage checkpoint by chk1 and rad53 protein kinases through distinct mechanisms. *Science* **286,** 1166–1171.

47. Walworth, N. C. and Bernards, R. (1996) rad-dependent response of the chk1-encoded protein kinase at the DNA damage checkpoint. *Science* **271,** 353–356.

48. Takai, H., Tominaga, K., Motoyama, N., et al. (2000) Aberrant cell cycle checkpoint function and early embryonic death in Chk1(−/−) mice. *Genes Dev.* **14**, 1439–1447.
49. Xu, Y. and Baltimore, D. (1996) Dual roles of ATM in the cellular response to radiation and in cell growth control. *Genes Dev.* **10**, 2401–2410.
50. Chehab, N. H., Malikzay, A., Appel, M., and Halazonetis, T. D. (2000) Chk2/hCds1 functions as a DNA damage checkpoint in G(1) by stabilizing p53. *Genes Dev.* **14**, 278–288.
51. Shieh, S. Y., Ahn, J., Tamai, K., Taya, Y., and Prives, C. (2000) The human homologs of checkpoint kinases Chk1 and Cds1 (Chk2) phosphorylate p53 at multiple DNA damage-inducible sites. *Genes Dev.* **14**, 289–300.
52. Chehab, N. H., Malikzay, A., Stavridi, E. S., and Halazonetis, T. D. (1999) Phosphorylation of Ser-20 mediates stabilization of human p53 in response to DNA damage. *Proc. Natl. Acad. Sci. USA* **96**, 13777–13782.
53. Khosravi, R., Maya, R., Gottlieb, T., Oren, M., Shiloh, Y., and Shkedy, D. (1999) Rapid ATM-dependent phosphorylation of MDM2 precedes p53 accumulation in response to DNA damage. *Proc. Natl. Acad. Sci. USA* **96**, 14973–14977.
54. Falck, J., Mailand, N., Syljuasen, R. G., Bartek, J., and Lukas, J. (2001) The ATM-Chk2-Cdc25A checkpoint pathway guards against radioresistant DNA synthesis. *Nature* **410**, 842–847.
55. Peng, C. Y., Graves, P. R., Thoma, R. S., Wu, Z., Shaw, A. S., and Piwnica-Worms, H. (1997) Mitotic and G2 checkpoint control: regulation of 14-3-3 protein binding by phosphorylation of Cdc25C on serine-216. *Science* **277**, 1501–1505.
56. Dalal, S. N., Schweitzer, C. M., Gan, J., and DeCaprio, J. A. (1999) Cytoplasmic localization of human cdc25C during interphase requires an intact 14-3-3 binding site. *Mol. Cell Biol.* **19**, 4465–4479.
57. Yang, J., Winkler, K., Yoshida, M., and Kornbluth, S. (1999) Maintenance of G2 arrest in the *Xenopus* oocyte: a role for 14-3-3-mediated inhibition of Cdc25 nuclear import. *EMBO J.* **18**, 2174–2183.
58. Bunz, F., Dutriaux, A., Lengauer, C., et al. (1998) Requirement for p53 and p21 to sustain G2 arrest after DNA damage. *Science* **282**, 1497–1501.
59. Chan, T. A., Hermeking, H., Lengauer, C., Kinzler, K. W., and Vogelstein, B. (1999) 14-3-3Sigma is required to prevent mitotic catastrophe after DNA damage. *Nature* **401**, 616–620.
60. Scully, R. and Livingston, D. M. (2000) In search of the tumour-suppressor functions of BRCA1 and BRCA2. *Nature* **408**, 429–432.
61. Zhong, Q., Chen, C. F., Li, S., et al. (1999) Association of BRCA1 with the hRad50-hMre11-p95 complex and the DNA damage response. *Science* **285**, 747–750.
62. Cortez, D., Wang, Y., Qin, J., and Elledge, S. J. (1999) Requirement of ATM-dependent phosphorylation of brca1 in the DNA damage response to double-strand breaks. *Science* **286**, 1162–1166.
63. Tibbetts, R. S., Cortez, D., Brumbaugh, K. M., et al. (2000) Functional interactions between BRCA1 and the checkpoint kinase ATR during genotoxic stress. *Genes Dev* **14**, 2989–3002.
64. Gatei, M., Scott, S. P., Filippovitch, I., et al. (2000a) Role for ATM in DNA damage-induced phosphorylation of BRCA1. *Cancer Res.* **60**, 3299–3304.
65. Gatei, M., Zhou, B.-B., Hobson, K., Scott, S., Young, D., and Khanna, K. K. (2001) ATM and ATR mediate phosphorylation of Brca1 at distinct and overlapping sites: in vivo assessment using phospho-specific antibodies. *J. Biol. Chem.* **276**, 17276–17280.
66. Lee, J. S., Collins, K. M., Brown, A. L., Lee, C. H., and Chung, J. H. (2000) hCds1-mediated phosphorylation of BRCA1 regulates the DNA damage response. *Nature* **404**, 201–204.
67. Paull, T. T., Rogakou, E. P., Yamazaki, V., Kirchgessner, C. U., Gellert, M., and Bonner, W. M. (2000) A critical role for histone H2AX in recruitment of repair factors to nuclear foci after DNA damage. *Curr. Biol.* **10**, 886–895.

68. Rogakou, E. P., Pilch, D. R., Orr, A. H., Ivanova, V. S., and Bonner, W. M. (1998) DNA double-stranded breaks induce histone H2AX phosphorylation on serine 139. *J. Biol. Chem.* **273**, 5858–5868.

69. Lim, D. S., Kim, S. T., Xu, B., et al. (2000) ATM phosphorylates p95/nbs1 in a S-phase checkpoint pathway. *Nature* **404**, 613–617.

70. Gatei, M., Young, D., Cerosaletti, K. M., et al. (2000) ATM-dependent phosphorylation of nibrin in response to radiation exposure. *Nat. Genet* **25**, 115–119.

71. Zhao, S., Weng, Y. C., Yuan, S. S., et al. (2000) Functional link between ataxia-telangiectasia and Nijmegen breakage syndrome gene products. *Nature* **405**, 473–477.

72. Wu, X., Ranganathan, V., Weisman, D. S., et al. (2000) ATM phosphorylation of Nijmegen breakage syndrome protein is required in a DNA damage response. *Nature* **405**, 477–482.

73. Carney, J. P., Maser, R. S., Olivares, H., et al. (1998) The hMre11/hRad50 protein complex and Nijmegen breakage syndrome: linkage of double-strand break repair to the cellular DNA damage response. *Cell* **93**, 477–486.

74. Stewart, G. S., Maser, R. S., Stankovic, T., et al. (1999) The DNA double-strand break repair gene hMRE11 is mutated in individuals with an ataxia-telangiectasia-like disorder. *Cell* **99**, 577–587.

75. Haber, J. E. (1999) DNA recombination: the replication connection. *Trends Biochem. Sci.* **24,** 271–275.

76. Manolis, K. G., Nimmo, E. R., Hartsuiker, E., Carr, A. M., Jeggo, P. A., and Allshire, R. C. (2001) Novel functional requirements for non-homologous DNA end joining in *Schizosaccharomyces pombe. EMBO J.* **20,** 210–221.

77. Elledge, S. J., Winston, J., and Harper, J. W. (1996) A question of balance: the role of cyclin-kinase inhibitors in development and tumorigenesis. *Trends Cell Biol.* **6**, 388–392.

78. Hwang, B. J., Ford, J. M., Hanawalt, P. C., and Chu, G. (1999) Expression of the p48 xeroderma pigmentosum gene is p53-dependent and is involved in global genomic repair. *Proc. Natl. Acad. Sci. USA* **96,** 424–428.

79. Tanaka, H., Arakawa, H., Yamaguchi, T., et al. (2000) A ribonucleotide reductase gene involved in a p53-dependent cell-cycle checkpoint for DNA damage. *Nature* 404, 42–49.

80. Vousden, K. H. (2000) p53: death star. *Cell* **103**, 691–694.

81. Attardi, L. D., Reczek, E. E., Cosmas, C., et al. (2000) PERP, an apoptosis-associated target of p53, is a novel member of the PMP-22/gas3 family. *Genes Dev.* **14**, 704–718.

82. Oda, E., Ohki, R., Murasawa, H., et al. (2000a) Noxa, a BH3-only member of the Bcl-2 family and candidate mediator of p53-induced apoptosis. *Science* **288**, 1053–1058.

83. Oda, K., Arakawa, H., Tanaka, T., et al. (2000) p53AIP1, a potential mediator of p53-dependent apoptosis, and its regulation by Ser-46-phosphorylated p53. *Cell* **102**, 849–862.

84. Hartwell, L. H., and Kastan, M. B. (1994) Cell cycle control and cancer. *Science* **266**, 1821–1828.

85. O'Connor, P. M. (1997) Mammalian G1 and G2 phase checkpoints. *Cancer Surv.* **29**, 151–182.

5

In Situ Hybridization in Cancer and Normal Tissue

Shrihari Kadkol, Jeremy Juang, and Tzyy-choou Wu

1. Introduction

With recent advances in the Human Genome Project and progress in microarray and gene chip technology, an explosion of genomic information is becoming available. One of the major ways to understand this vast amount of data is by bestowing topological information to help us identify the functional relevance of specific genes and to explain the pathogenesis of disease processes. *In situ* hybridization is a technique that will further our understanding of the function of genes within normal tissue along with the pathogenesis of a disease process within abnormal cells. This chapter provides an overview of *in situ* hybridization, followed by general principles and protocols.

1.1. Overview

In situ hybridization is a molecular technique to localize nucleic acid sequences in chromosomes, cell populations, or morphologically preserved tissue sections. The power of this technique lies in the fact that nucleic acid detection can be accomplished in the context of chromosomal, cellular, or tissue morphology and hence localization can be achieved to a specific chromosome, cellular population, or a specific location within a tissue. This localization is important to derive, for instance, the deletion or amplification of a particular portion of chromosome and the functional relevance of the expression of a specific gene or to identify target cells infected by a virus. Such spatial information may be critical to understand the function of genes within a normal tissue and the pathogenesis of a disease process. Other methods to identify nucleic acids include solution hybridization and general polymerase chain reaction (PCR)-based methods in which tissues or cells are homogenized before analysis. Although these methods identify the presence of a specific nucleic acid, they do not provide information about the localization of the signal to specific cell populations in relation to tissue morphology. Sensitivity of *in situ* hybridization has markedly improved in recent years because of better probe labeling methods and signal amplification detection systems, to the extent that very few copies of target nucleic acid can be routinely detected.

In situ hybridization can be used to detect either DNA or RNA in cells or tissue sections. Applications of this technique include analysis of gene expression, identification

From: *Methods in Molecular Biology, Vol. 223: Tumor Suppressor Genes: Regulation, Function, and Medicinal Applications.* Edited by: Wafik S. El-Deiry © Humana Press Inc., Totowa, NJ

of foreign DNA or RNA in cells infected by an infectious agent such as a virus, bacteria, or fungus, and the analysis of chromosomal DNA for gene localization, gene amplification, loss of genetic material, or translocation. *In situ* hybridization encompasses several protocols, depending on the analysis to be performed. Chromosomal aberrations are generally analyzed with long probes by fluorescence *in situ* hybridization on metaphase spreads or interphase nuclei of tissue sections. Other DNA targets such as infectious agents and cellular mRNA expression are analyzed by short probes or by oligonucleotide probes on frozen or formalin-fixed, paraffin-embedded tissue sections and visualized by either fluorescent or chromogenic detection methods.

In situ hybridization can also be done on tissue sections to analyze gene expression or identify infectious agents in diagnostic pathology or for research applications. An example of this is the diagnosis of pituitary adenomas by the *in situ* identification of prolactin mRNA expression *(1)*. Similarly, gastrin mRNA in tumors of patients with Zollinger-Ellison syndrome *(2)* and albumin mRNA in hepatocellular tumors *(3,4)* can be identified by *in situ* hybridization. Monoclonal analysis of lymphoid malignancies can be aided by *in situ* detection of light-chain mRNA restriction *(5)*. Viral infections that can be diagnosed with this technique include human papilloma virus (HPV) *(6–12)*, Epstein-Barr virus (EBV) *(13–20)*, cytomegalovirus (CMV) *(21–25)*, herpes simplex virus (HSV) *(26,27)*, hepatitis B and C *(28,29)*, and many others. Also, with HPV one can use type-specific probes to distinguish low-risk from high-risk oncogenic groups. Other infectious organisms that have been studied by *in situ* hybridization include *Pneumocystis carinii (30)*, *helicobacter pylori (31)*, sexually transmitted diseases *(32)*, *Aspergillus (33),* and mycobacteria *(34)*. A major advantage of *in situ* hybridization is that it provides results in a much shorter time when compared to culture methods.

1.2. Fluorescence In Situ Hybridization (FISH)

Fluorescence *in situ* hybridization (FISH) was introduced in 1986 *(35,36)*. This technology generally uses DNA probes that are directly labeled with fluorescence or with nonisotopic haptens (acetylaminoflurene, biotin, bromodeoxy uridine, digoxigenin, mercury tritophonol, rhodamine, and sulfonate) followed by fluorescent-labeled streptavidin or antibody against the ligands. Furthermore, in multicolor FISH analysis, simultaneous detection of several genetic abnormalities can be performed by co-hybridizing more than one fluorescent-labeled probe or label combinations to reveal multiple aberrations in a single hybridization. FISH is an important research tool in areas of gene mapping, exploration of chromosomal and gene abnormalities, gene expression, genome organization, identification of infectious diseases, and investigation of viral integration sites in mitotic and interphase cells (for review, see **ref.** *37*). Because of its ability to visualize DNA or RNA sequences of interest, FISH technology is also a powerful tool in diagnostic and prognostic pathology. It can be applied to clinical samples such as amniotic fluid cells, chorionic villus sampling, bone marrow, frozen specimens, touch preparations, cytological samples, products of conception, germ cells, and formalin-fixed paraffin-embedded tissue specimens. In cancer diagnostics, the primary application of FISH is to identify chromosomal and gene markers of disease *(38–44)*. More recently, FISH has been used to determine amplification status of the *Her 2/neu* gene on paraffin sections of breast and other tumor tissues *(45–50)*. Tumors with amplified *Her 2/neu* gene have been implicated to have a poor prognosis in breast cancer. The amplified gene results in overexpression of the protein and monoclonal antibody-based therapies targeting the Her 2/neu protein are undergoing clinical trials. Two techniques related to FISH, chromosome painting and comparative genomic hybridization, will be examined further.

Chromosome painting, or chromosome *in situ* suppression hybridization (CISS), refers to a modified form of traditional fluorescence *in situ* hybridization. This technique takes advantage of the available library of flow-sorted human chromosomes to generate probes that subsequently hybridize along the length of chromosome. The hybridization is attained by suppressing cross-hybridization to repetitive DNA in other chromosomes. This suppression is accomplished by blocking labeled, repetitive DNA with an excess of unlabeled DNA sequences. Chromosome painting permits the visualization of individual chromosomes in metaphase or interphase cells, and it is especially important in human pathology because it can identify chromosomal aberrations both numerically and structurally *(51–55)*.

Forward chromosome painting uses the chromosome painting probes generated from a normal chromosome, while reverse chromosome painting uses chromosome painting probes generated from aberrant chromosomes either flow-sorted or microdissected. The advantage of the flow-sorted chromosome painting probes is that they ensure high complexity of the painting probes due to the number of targets that can be amplified easily. On the other hand, microdissected chromosome painting probes have the advantage of generating region-specific probes for chromosomal bands in addition to entire chromosome painting probes. Additional protocols for the generation of chromosome painting probes have also appeared recently *(56–60)*. By improving signal-to-noise ratio and staining homogeneity, these probes provide a more efficient method of locating abnormalities in chromosomes.

Comparative genomic hybridization (CGH) is also a technique based on FISH methodology. In CGH, tumor DNA and the normal reference DNA are labeled by nick translation with different fluorochromes and hybridized simultaneously to a normal metaphase spread. Digital analysis is then employed to provide an over- or underrepresentation of DNA sequences in the tumor sample, a result of imbalanced fluorescence intensity ratio. An overrepresentation signifies amplification, while underrepresentation signifies deletion in the tumor genome. Although CGH is very powerful in identifying regions of loss of heterozygocity and oncogene amplification, it lacks the ability to identify structural rearrangements such as inversions and translocations *(61–64)*.

1.3. In Situ Hybridization with Amplification of Target Sequence

Several techniques to amplify the target sequence have been performed in tissue sections in conjunction with *in situ* hybridization. For example, *in situ* PCR and reverse transcriptase PCR (RT-PCR) are performed on tissue sections to detect targets that occur at very low copy numbers *(65–69)*. Such targets are generally below the detection limit of routine *in situ* hybridization. These methods result in target amplification on the slide, and this is followed by *in situ* hybridization with a specific probe. However, they require a thermocycler with a special slide block. Another method to amplify RNA is self-sustained replication or the 3SR method. This method has been used to successfully amplify dopamine receptor, CFTR, and measles virus mRNA on tissue sections. The advantages of this method are that amplification occurs isothermally and that the presence of DNA does not pose a problem since only RNA is specifically amplified without any denaturation step by a combination of reverse transcriptase, RNase H, and T7 RNA polymerase *(70–74)*.

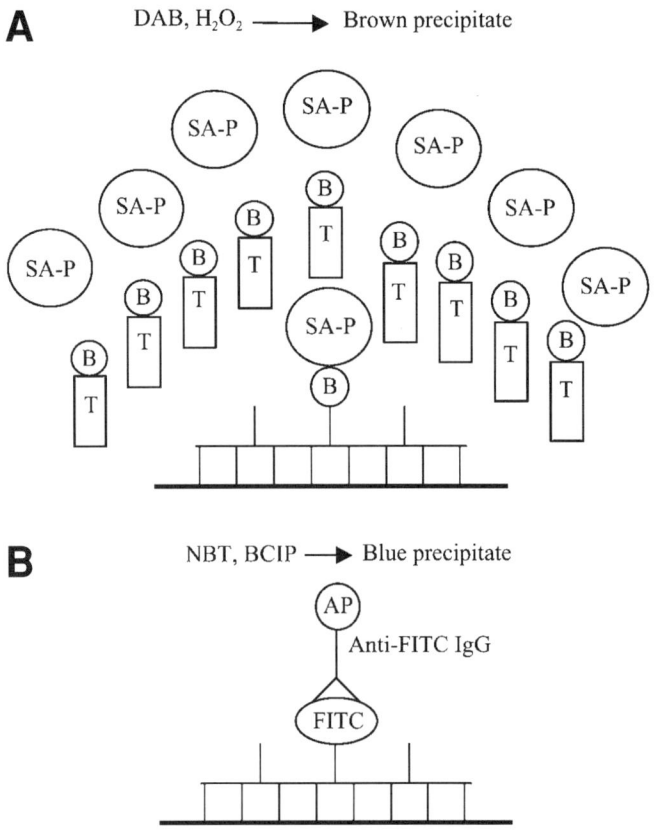

Fig. 1. **(A)** Schematic representation of catalyzed signal amplification detection of a hybridized probe labeled with biotic. B, biotin; T, tyramide; SA-P, streptavidin peroxidase; DAB, H_2O_2, chromogen, substrate. **(B)** Schematic representation of routine detection of a hybridized probe labeled with FITC. FITC, fluoresceirisothiocyanate; AP, alkaline phosphatase; NBT, BCIP, chromogen, substrate.

1.4. In Situ Hybridization with Signal Amplification

Recently, the *catalyzed signal amplification* (CSA) methodology has greatly improved the sensitivity of detection by routine *in situ* hybridization *(75–77)*. A schematic representation of the steps involved is shown in **Fig. 1A**. Results from *in situ* analysis of HPV infection utilizing type-specific biotinylated probes and catalyzed signal amplification are shown in **Fig. 2**. In this method, after hybridizing the probe, a peroxidase moiety is introduced on the slide: with biotin-labeled probes, one can use streptavidin-peroxidase; with digoxigenin- or FITC-labeled probes, one can use anti-digoxigenin or anti-FITC antibody conjugates linked to peroxidase. Then, tyramide linked to biotin or FITC or DNP is applied. The peroxidase from the previous step oxidizes the tyramide, which in turn binds to the aromatic groups of certain amino acids and results in additional deposition of label at the site of hybridization. After this, detection is performed with another round of streptavidin-peroxidase for biotin labels or with anti-FITC or anti-DNP antibodies conjugated with either peroxidase or alkaline phosphatase for FITC or DNP labels. This cycle can be repeated to amplify the signal. This method of catalyzed signal amplification results in intense signal amplification, to the extent that we are routinely able to detect a single copy of HPV-16 in

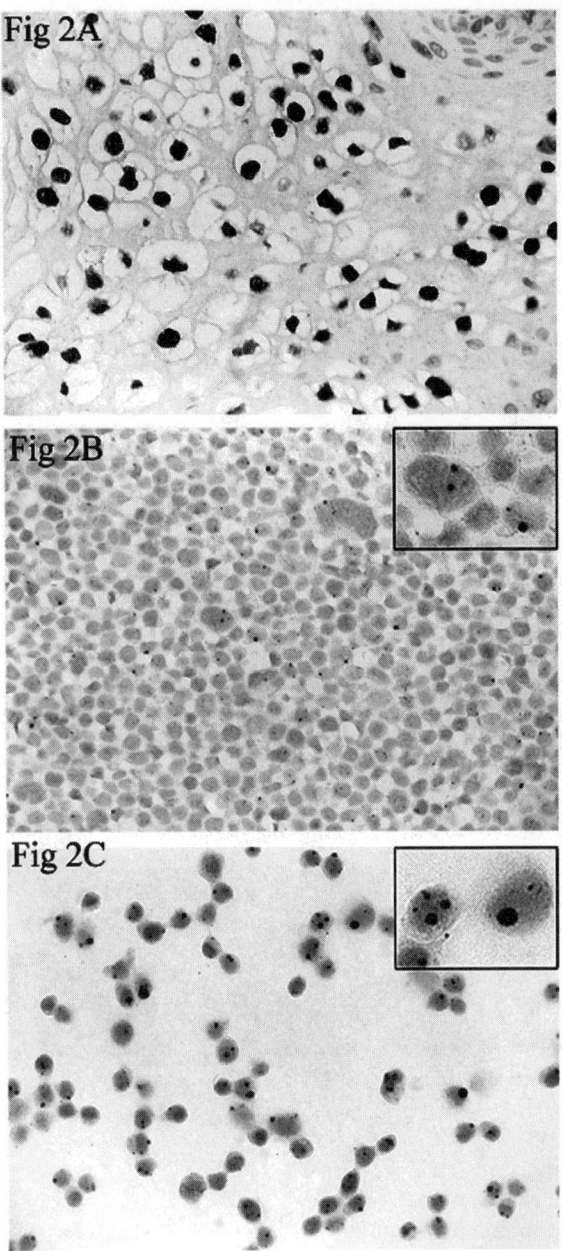

Fig. 2. *In situ* hybridization analysis of human papillomavirus infection in formalin-fixed and paraffin-embedded tissues with a type-specific biotinylated probe and catalyzed signal amplication detection. The dark areas are the sites of hybridization. **(A)** HPV 6/11 detection in a condyloma (episomal infection, several copies/nucleus). **(B)** HPV 16 detection in SiHa cells (1–2 integrated copies/nucleus, dotlike pattern). **(C)** HPV 18 detection in HeLa cells (20–50 integrated copies/nucleus, dotlike pattern).

SiHa cells (*see* **Fig 2**) by this method using a commercially available kit (Dako Genpoint kit, Dako, CA). If endogenous biotin in tissue sections results in background, biotin-based amplification systems will increase background. In these circumstances, going to a biotin-free system such as FITC or DNP labels will solve the problem. Between all steps of the amplification process, the sections must be washed with buffer thoroughly to avoid background. Several labels linked to tyramide are available commercially from NEN Life Science Products. Final chromogenic detection is performed with either DAB or alkaline phosphatase. The CSA system can also be used for fluorescent detection. One can substitute the second streptavidin-peroxidase step with streptavidin linked to a fluorochrome such as rhodamine or Texas Red to obtain a fluorescent signal.

The general principle of catalyzed signal amplification is to introduce a peroxidase moiety after the posthybridization stringency wash. This then forms the basis for additional tyramide-label deposition. Several labels and fluorochromes linked to tyramide are available commercially. The deposited label can then be detected by fluorescent or histochemical methods as described earlier. Since several label moieties are deposited around the site of initial hybridization, the result is an intense amplification of the signal at the hybridization site. The main drawback of the CSA system is that it can result in a higher level of background staining if appropriate precautions are not taken. They include adequate washes between steps with shaking, and prevention of slide drying at every point in time. Appropriate dilutions of the amplification reagents have to be adjusted rather empirically. Also, endogenous biotin activity in certain tissues could be a problem with streptavidin-based systems.

2. Materials

2.1. Reagents

1. PCR or RT-PCR reagents (Qiagen).
2. Expression vector (pCRII TOPO or pCR 4.1, Invitrogen).
3. Nucleic acid labeling kits.
4. 10% Neutral buffered formalin.
5. Xylene, ethanol.
6. Proteinase K, pepsin.
7. Hydrogen peroxide.
8. Tris-HCl-buffered saline, pH 7.5 (100 mM Tris-Cl, pH 7.5, 150 mM NaCl).
9. Tris-HCl-buffered saline, pH 9.5 (100 mM Tris-Cl, pH 9.5, 100 mM NaCl, 50 mM MgCl$_2$).
10. Wash buffer: 100 mM Tris-Cl, pH 7.5, 150 mM NaCl with 0.05% Tween-20.
11. 20× saline-sodiumatrate citrate (SSC), 10% sodium dodecyl sulfate (SDS), formamide, salmon sperm DNA, Denhardt's solution.
12. Detection reagents (anti-digoxigenein-alkaline phosphatase, streptavidin-peroxidase, or streptavidin alkaline phosphatase).
13. Chromogens: DAB, NBT/BCIP (Vector Laboratories, Roche Molecular Biochemicals).
14. Catalyzed signal amplification kits (Genpoint, Dako; TSA-Plus system, NEN Life Sciences).
15. Counterstains: hematoxylin, nuclear fast red.
16. Nuclease-free water.
17. Mounting media: Permanox, Glycergel (Dako).

2.2. Equipment

1. Tissue processor to embed tissues in paraffin.
2. Microtome.
3. Organosilane or poly-L-lysine-coated glass slides.
4. Glass cover slips.
5. Microscope.
6. Incubators.
7. Water baths.
8. Hybridization boxes.
9. Coplin jars, glass dishes.
10. Orbital shaker.

2.3. Computer Software

1. Primer and probe analysis: Oligo 5.0 (Molecular Biology Insights, Cascade, CO).
2. Sequence analysis: BLAST (www.ncbi.nlm.nih.gov).

3. Methods

3.1. General Principles

The following paragraphs discuss general principles related to *in situ* hybridization. Representative working protocols for basic *in situ* hybridization are provided at the end of this section. The steps of *in situ* hybridization can be divided into (1) specimen preparation, (2) nucleic acid retrieval, (3) probes, (4) hybridization conditions, (5) posthybridization stringency washs, (6) detection of the hybrid, (7) counterstains, and (8) general considerations. Each one of these steps will be explained in the following section.

3.1.1. Specimen Preparation

Frozen or fixed tissue specimens can be used for *in situ* hybridization. Immediate freezing or adequate fixation after tissue removal is extremely critical in achieving optimal results. This is especially important in the analysis of gene expression to preserve otherwise rapid degradation of RNA by ubiquitous ribonucleases. Upon removal, tissues should be sliced to 3–5 mm in thickness and either snap-frozen in liquid nitrogen for long-term storage or fixed in a suitable fixative such as 10% neutral buffered formalin, pH 7.4. Formalin is a cross-linking fixative and is optimal for most *in situ* hybridization applications. Other fixatives that heavily cross-link, such as gluteraldehyde or precipitant solutions such as Bouins solution, are generally not suitable *(78,79)*. Fixation should be performed in at least 10 vol of 10% neutral buffered formalin for 18–24 h. Overfixation for times greater than 24 h should be avoided to prevent excessive cross-linking. Fixed tissues should be embedded in paraffin and sectioned on a microtome at 4–5 µm thickness. Sections should be placed on poly-L-lysine- or organosilane-coated glass slides to prevent subsequent detachment as the sections go through several of the steps in the protocol. Paraffin-embedded sections are dewaxed by immersing slides for 5 min each in 100% xylene ×2, 100% ethanol ×2, 95%, 80%, and 50% ethanol. Finally, sections are washed twice in distilled water. Frozen specimens are sectioned at 4–5 μm thickness in a cryostat, placed on coated slides, and fixed in 100% acetone or anhydrous methanol for 30 min at 4°C.

3.1.2. Nucleic Acid Retrieval

DNA or RNA targets must be unmasked before hybridization of sections fixed with cross-linking fixatives such as formalin. This step enables penetration of the probes and detection reagents into the tissue. Unmasking is generally performed by treatment of sections with either (1) 0.2 N HCl, (2) protease digestion with pepsin or proteinase K, or (3) by heat-induced retrieval. Different tissues and targets may require any one or a combination of the conditions listed above to unmask the targets adequately. The success of *in situ* hybridization depends in large part on optimal target unmasking. The exact mechanism by which these reagents work to unmask targets remains unknown. It is believed that acid digestion removes proteins bound to nucleic acids and hence makes target sequences available for hybridization. Protease digestion is believed to remove cross-links created during fixation by formalin. Heat-induced retrieval is a relatively new method to unmask target sequences and is an adapatation of the protocol used for unmasking antigens in immunohistochemistry. Frozen sections generally do not require extensive target retrieval, or may require very mild acid or protease digestion.

For detection of genomic sequences by FISH, deparaffinized sections are dipped in 0.2 N HCl for 10–20 min and then washed extensively in buffer such as TBS, pH 7.5 (100 mM Tris-Cl, pH 7.5, 150 mM NaCl). They are further digested with protease. For mRNA or viral target detection in formalin-fixed sections, we have found that 10 μg/mL proteinase K in TBS, pH 7.5, at 37°C for 30 min or 0.1% pepsin in 0.01 N HCl (pH 2.0) for 15–30 min at 37°C is generally adequate for protease digestion of deparaffinized sections. Following digestion, sections are washed extensively in TBS. When establishing a protocol, it is advisable to perform protease digestion with different amounts of the enzyme for different periods of time to determine optimal conditions. Different tissues and probes behave differently, and this step has to be optimized empirically for each set of tissues and probes. Over- or underdigestion of tissue sections remarkably influences the end result, and in our experience protease digestion remains an extremely important step in achieving good results. Heat-induced retrieval is a target unmasking method that is performed on formalin-fixed sections to increase sensitivity or reduce background in certain circumstances. It is generally performed by immersing slides in 0.01 M citrate buffer, pH 6.0, at 95°C for 10–15 min in a Coplin jar in a water bath or by heating the solution with the slides in a microwave oven. Another technique for performing heat-induced retrieval is the capillary gap method. In this method, two slides with the tissue sections are opposed to each other to form a gap of about 150–200 μm between them. Such gap-forming slides are available commercially (Ventana Medical System, Tucson, AZ). The sandwich is dipped vertically into a small container with citrate buffer, which rises in the gap to bathe the sections. The container with the dipped sandwich is then placed in a steamer with boiling water. The citrate buffer that bathes the sections in the gap reaches 95°C and target retrieval is performed at this temperature for 10–15 min. After steaming, the slides are allowed to cool to room temperature for 15 min, then separated, washed in water, and then placed in TBS, pH 7.5. Sections can be further protease digested if necessary. Heat-induced retrieval can increase sensitivity or can reduce background in particular circumstances. For interphase FISH on formalin-fixed and paraffin-embedded tissue sections, heat-induced retrieval has also been performed with 50% glycerol/0.1× SSC for 3 min at 90°C *(80)*.

3.1.3. Probes

Hybridization probes are of different types and have distinct advantages depending on the intended application. One may wish to identify an entire family of closely related transcripts with a long probe and with low-stringency wash conditions, or identify individual members of a gene family by highly specific oligonucleotide probes and high-stringency wash conditions. Probes should be free of hairpin structures and should have a GC content of approximately 50%. Potential probes should be analyzed by computer software programs such as Oligo 5.0 (Molecular Biology Insights, Cascade, CO) to rule out hairpins and duplexes. They should be analyzed with genome databases such as BLAST (www.ncbi.nlm.nih.gov) to check for specificity and potential cross-hybridization with other known sequences. Oligonucleotide probes should not have more than three G's in a row, since such probes are "sticky" and can lead to nonspecific background *(81)*. General factors to consider while choosing a probe are probe length, probe type, and probe labeling—isotopic vs nonisotopic.

3.1.3.1. PROBE LENGTH

Identification of genomic sequences for gene localization, gene amplification, or translocations is generally performed with very long probes that are several kilobases in length. At the other end of the spectrum are oligonucleotide probes that range from 25 to 40 bases in length. In general, routine *in situ* hybridization is performed with probes ranging in length from 300 to 500 bp. These probes penetrate tissues well and have adequate sensitivity of detection due to sufficient probe label moieties. Penetration of probes into the tissue section is a very important aspect of *in situ* hybridization. Shorter probes penetrate better into tissue sections than longer probes. Hence oligonucleotide probes penetrate extremely well compared to longer probes. However, the apparently increased sensitivity of this approach is somewhat offset by the decreased ability to put sufficient label moieties on a shorter probe or on an oligonucleotide probe. Longer probes (300–500 bp) can be labeled with more label moieties than a shorter or oligonucleotide probe and hence provide good sensitivity of detection if tissue penetration is adequate. Longer probes also form more stable duplexes with their targets and hence will be more resistant to stringent wash conditions than an oligonucleotide probe.

3.1.3.2. PROBE TYPE

Probes can be double-stranded DNA (dsDNA), single-stranded RNA (ssRNA), or oligonucleotides. Double-stranded DNA probes need to be denatured before hybridization and can be generated by PCR or cloning. These probes can be internally labeled by nick translation, random priming, or PCR. However, dsDNA probes can self-anneal during hybridization and there by reduce hybridization efficiency to the target. Single-stranded RNA probes of defined lengths are more sensitive than double-stranded DNA probes of comparable lengths. Since they do not self-anneal, due to the presence of only one strand, all of the probe is available to hybridize with the target sequence. RNA probes yield higher specificity than DNA probes because the RNA:RNA and RNA:DNA hybrids tolerate higher-stringency wash conditions due to greater hybrid stability than

DNA:DNA hybrids. Hence tissues hybridized with RNA probes can be washed at higher stringencies to achieve greater specificity. To synthesize RNA probes, the first step is to subclone the desired DNA sequence into an expression vector. Subsequently, sense and antisense probes RNA probes are synthesized and simultaneously labeled internally by in-vitro transcription. Several commercially available kits exist for in-vitro transcription (Roche Molecular Biochemicals, Indianapolis, IN). Oligonucleotide probes ranging in length from 25 to 40 bases are the most specific and may be the only type of useful probe if one wishes to distinguish single base changes. However, oligonucleotide probes can be labeled only by end labeling or by tailing with label moieties by utilizing for example the 3-tailing kit from Roche Molecular Biochemicals. This limits the maximum number of label moieties that can be added to the probe and hence decreases overall sensitivity, although this decrease may be offset by increased probe penetration into tissue sections, which favors increased sensitivity. Furthermore, the sensitivity may be improved using the signal amplification strategy. Oligonucleotide probes have the distinct advantage that they can be easily synthesized commercially in large quantities at low cost.

Other, less frequently used probe types include peptide nucleic acid probes (PNAs) and single-stranded DNA probes. PNA probes have a peptide backbone attached to the nucleotide bases, and the hybridization still follows the Watson-Crick base-pairing rules *(82–91)*. PNA probes have a very high affinity for their target due to the replacement of the negatively charged phosphate backbone with a neutral peptide backbone and hence can be washed at very high stringencies if extreme specificity is desired. They are also resistant to enzymatic cleavage and hence are very stable. Single-stranded DNA probes can be generated and labeled by asymmetric PCR and have been used for *in situ* hybridization. Since they are single-stranded, the entire probe is available for hybridization. Also, RNase is not a major concern when using these probes *(92–94)*.

3.1.3.3. PROBE LABELING

In situ hybridization probes can be labeled isotopically or nonisotopically. Isotopic labels include ^{3}H, ^{35}S, and ^{33}P. Nonisotopic labels include biotin, FITC, digoxigenin, 2,4-dinitrophenyl, and other haptens. These labels are added to the probe internally by nick translation, random priming, or PCR for DNA probes and during in-vitro transcription for RNA probes. Oligonucleotide probes are labeled by end labeling or tailing with label moieties. After labeling, probes are purified by ethanol precipitation or column-based methods to remove unincorporated label. The choice of the label depends on the desired sensitivity. Isotopic probes offer the highest sensitivity and have excellent resolution. Such probes can be used for quantitative evaluation by grain counts. However, they are far less convenient to use, since isotopic probes have the disadvantages associated with radioactive use—relatively short half-lives of the label, radiation exposure, and strict adherence to disposal regulations. Also, the signal can be visualized only by an autoradiographic process with variably long exposure times. Nonisotopic labeling offers distinct advantages over isotopically labeled probes. Nonisotopically labeled probes are stable for 2–3 yr if stored properly. They are extremely convenient to use, without the hazards and risks of radioactive labels. No special precautions need to be taken. Final chromogenic detection requires very short times, and the color retention is permanent with excellent tissue morphology. Disadvantages of nonisotopic labels include background problems in some tissues and inability to detect rare sequences. Background problems arise due to the endogenous presence of the label moiety in some tissues. The kidney, brain, prostate, and liver are rich in endogenous biotin and one frequently encounters background problems when *in situ* hybridization is performed with biotin-labeled probes on these tissues. Heat-induced retrieval tends to reduce this background staining. Alternatively, FITC, digoxigenin, or 2,4-dinitrophenyl can be used to label the probes. These moieties do not occur endogenously and may be preferable to biotin if background problems persist. Estimation of probe labeling efficiency is a must before probes are used for *in situ* hybridization. For isotopically labeled probes, this may be as simple as measuring radioactive counts in a scintillation counter. For nonisotopically labeled probes, a series of dilutions is spotted in parallel with known standards on a membrane and incubated with the detection reagents in an enzyme-linked immunoassay or, in the case of biotin-labeled probes, with streptavidin-peroxidase. Color development from the dilution series is compared against the standards and the actual labeled probe concentration is calculated.

3.1.4. Hybridization

Hybridization should be performed at reasonably high stringency to ensure specificity. One important concept is generally to hybridize about 15–20°C below the calculated T_m of the hybrid. The T_m is defined as the emperature at which half of the probe-target molecules dissociate. Conditions that affect the T_m include the salt concentration, formamide content, GC content of the probe, and the probe length. Higher salt concentration, no formamide, and lower temperature favor hybridization and nonspecificity, whereas lower salt concentration, formamide, and higher temperature favor specificity. Each percent increase in formamide will reduce the T_m by approximately 0.6°C for a DNA:DNA hybrid. Several mathematical formula exist to calculate the T_m of a hybrid. These calculations should be used as a starting point and adjustments made by trial and error *(95)*.

DNA:DNA

$$T_m = 81.5 + 16.6 \log_{10} \frac{[Na^+]}{1 + 0.7[Na^+]} + 0.41\left(\frac{\%G + C}{D}\right) - 500 - 0.61(\% \text{ formamide}) - P$$

RNA:RNA

$$T_m = 78 + 16.6 \log_{10} \frac{[Na^+]}{1 + 0.7[Na^+]} + 0.7\left(\frac{\%G + C}{D}\right) - 500 - 0.61(\% \text{ formamide}) - P$$

RNA:DNA

$$T_m = 67 + 16.6 \log_{10} \frac{[Na^+]}{1 + 0.7[Na^+]} + 0.8\left(\frac{\%G + C}{D}\right) - 500 - 0.61(\% \text{ formamide}) - P$$

Oligonucleotides

$$T_d = 4(G + C) + 2(A + T)$$

$$(D = \text{duplex length}, \ P = \% \text{ mismatch}, \ \% \ G + C = \text{integer value})$$

Hybridization solutions frequently contain blocking reagents. These include 100–400 μg/mL denatured salmon sperm DNA, 1×–5× Denhardt's solution, 100–200-μg/mL yeast tRNA, 0.1–1% SDS, and 5–10% dextran sulfate. By sequestering water molecules, dextran sulfate raises the relative concentration of the probe in the hybridization solution. Also, dextran sulfate thickens the probe solution and hence facilitates application to tissue sections. For long probes approximately 1–10 ng/μL and for oligonucleotide probes 1–10 pmol/µL are generally used. A common starting hybridization solution for a probe about 500 bases and with a 50% GC content is 50% formamide, 2× SSC, 0.1% SDS, 2× Denhardt's solution, 5% dextran sulfate, and 400-µg/mL denatured salmon sperm DNA. Depending on the size of the tissue sections, 20–40 µL of hybridization solution is applied to the section. After applying the hybridization solution, the sections are cover-slipped. For dsDNA probes, the slides are denatured at 95°C for 5 min on a hot plate to denature the target and probe. For RNA probes, the slides are denatured at 65°C to release secondary structures in the probe or target even though the probe and target are single-stranded. Subsequently, the sections are hybridized overnight at a temperature 15–20°C below the calculated T_m of the hybrid in a humidified box. Hybridization times can be shortened empirically. If background is a problem, a prehybridization step for 1 h in the hybridization buffer without the probe is recommended at the same temperature at which hybridization will take place. For RNA probes, immersing the sections in 0.1 M triethanolamine buffer, pH 8.0, for 10 min before hybridization reduces nonspecific probe binding. Also, for RNA probes, a brief treatment with 40 µg/mL of RNase A in 2× SSC at 37°C for 15–20 min following hybridization will reduce nonspecific background.

3.1.5. Posthybridization Washes

After hybridization, the sections are immersed in 2× SSC/0.1% SDS to wash off unbound probe and to float away the cover slips. Slides are washed with shaking for at least 5 min × 3. Sections are then immersed in the high-stringency wash solution at the appropriate temperature and washed for a specified period of time. The high-stringency wash temperature should be 8–10°C below the calculated T_m of the hybrid. For example, this can be achieved using a solution containing 0.1× to 0.2× SSC/0.1% SDS with or without formamide. For oligonucleotide probes, the high-stringency wash temperature may be as close as 2–3°C below the calculated T_m.

3.1.6. Detection

Detection systems depend on the label incorporated into the probe. For biotin-labeled probes, sections are incubated with streptavidin-peroxidase conjugate and color development is performed with Diaminobenzadine (DAB). For digoxigenein-based probes, sections are incubated in 1.5 mU/mL of anti-digoxigenin antibody. This antibody can be obtained conjugated to alkaline phosphatase or peroxidase (Roche Molecular Biochemicals). For FITC- or DNP-labeled probes, anti-FITC or anti-DNP antibodies conjugated to peroxidase or alkaline phosphatase are commercially available (Roche, NEN Life Science Products, Boston, MA). When using an antibody conjugate following the posthybridization wash, the sections have to be blocked by incubating them in 5–10% normal serum from the same animal from which the antibody was made. Antibody conjugates can then be applied for about 30 min–1 h. Both the blocking serum and the antibody conjugate should be made up in a suitable buffer such as TBS, pH 7.5, with 0.05% Tween-20 or PBS, pH 7.4, with 0.05% Tween-20. After incubation in the antibody conjugate or streptavidin-peroxidase conjugate, sections are washed for 10 min × 3 in TBS, pH 7.5, with 0.05% Tween-20 or PBS, pH 7.4, with 0.05% Tween-20. NBT/BCIP is the chromogen/substrate combination of choice for alkaline phosphatase-based systems and results in a dark blue precipitate at the site of hybridization. Levamisole can be added

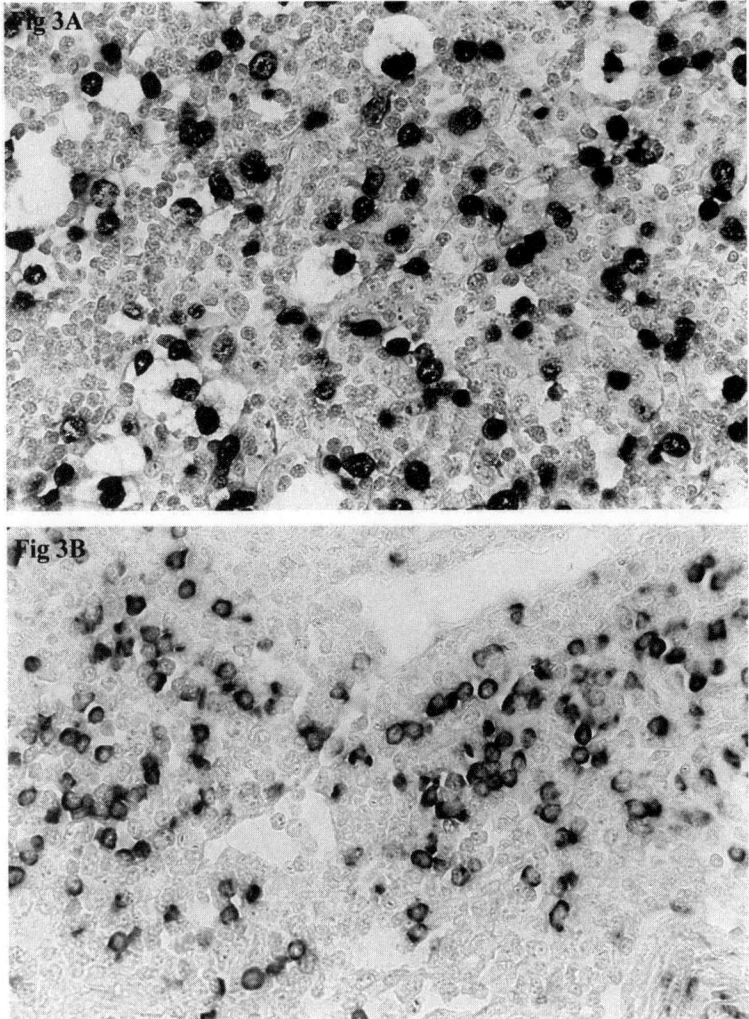

Fig. 3. *In situ* hybridization analysis in a formalin-fixed and paraffin-embedded tissue with a FITC-labeled oligonucleotide probe and routine detection with anti-FITC-AP and NBT, BCIP. The dark nuclei are the sites of hybridization. **(A)** EBER mRNA in latent EBV infection in a lymph node. **(B)** κ light-chain-bearing lymphocytes in a lymph node.

to the color substrate solution to inhibit endogenous alkaline phosphatase activity. The precipitate begins to form slowly, and it may take 2–3 h to develop fully. Sections can also be placed overnight in the NBT/BCIP color substrate solution. DAB/H_2O_2 is the chromogen/substrate combination for peroxidase-based systems and results in a brown precipitate at the hybridization site. Color development occurs very quickly, within minutes (*see* **Fig. 3**). Important sources of these detection systems include Roche Molecular Biochemicals, Vector Laboratories (Burlingame, CA), Dako (Carpinteria, CA), and many others. A schematic representation of routine detection is shown in **Fig. 1B**. Results from *in situ* analysis and routine detection of EBV infection and κ light-chain expression are shown in **Fig. 3**.

3.1.7. Counterstains

For fluorescent probes, slides can be counterstained with DAPI (Sigma Chemical Co., St. Louis, MO). For chromogenic detection, slides are counterstained with either dilute hematoxylin or neutral fast red for about 30 s–1min and then washed in water. Slides developed with DAB can be dehydrated in alcohol, cleared with xylene, and then mounted with Permount. Slides developed with NBT/BCIP should be mounted in aqueous mountant (Glycergel, Dako) to prevent the precipitate from crystallizing in organic mountant.

3.1.8. General Considerations

It is important to maintain the right temperatures during protease digestion, hybridization, and posthybridization wash steps. Placing too many slides in a single container will drop the temperature significantly. It is important not to dry the sections at any time, since drying results in background. Slides should always be washed with sufficient detergent containing buffer with shaking between steps. All steps should be performed with gloved hands, especially for RNA *in situ* hybridization, to avoid RNase contamination. Glassware and plasticware should be RNase-free. Water should be treated with 0.1% DEPC and autoclaved to eliminate RNase. Positive and negative controls should be used with every run. The ideal controls would be sections negative and positive for a particular target. A parallel slide without the probe should be run to identify nonspecific color development by the detection systems. This can occur if sections contain endogenous biotin or alkaline phosphatase. For all peroxidase-based systems, sections should be quenched for endogenous peroxidase by immersing them in aqueous 0.3% H_2O_2 for 10 min prior to hybridization. This is followed by several washes in TBS, pH 7.5. For RNA probes, a section should be hybridized with a sense probe as control. Since endogenous transcripts are of the sense orientation, one should not see a signal with the sense probe. However, sense probes can give rise to a signal if stretches of its sequence are complementary to some other sense mRNA in the tissue. Also, a parallel slide should be treated with either DNase or RNase prior to hybridization, to confirm the specificity of hybridization.

3.2. Conventional In Situ Hybridization (see Note 1)

For formalin-fixed, paraffin-embedded sections with a digoxigenin- or FITC-labeled probe of 500 bp and 50% GC content:

1. Cut 4–5-μm sections and place on coated slides (organosilane- or poly-L-lysine-coated).
2. Bake at 65°C for 1 h to melt the paraffin.
3. Deparaffinize: 100% xylene, 5 min ×2, 100% ethanol, 5 min ×2; 95% ethanol, 5 min, 70% ethanol, 5 min; 50% ethanol, 2 min, dH$_2$O, 3 min ×2 with shaking.
4. Immerse in 10 μg/mL proteinase K in TBS, pH 7.5, for 15–30 min at 37°C.
5. Wash in TBS, pH 7.5, with shaking, 10 min ×3.
6. Immerse in 0.1 *M* triethanolamine buffer, pH 8.0, for 10 min.
7. Wash in 2× SSC, 5 min ×3.
8. Apply 100 μL of hybridization solution (without the probe) and incubate at the hybridization temperature for 1 h in a humidified box.
9. Suction off the fluid and apply 20–40 μL of hybridization solution (with labeled probe) and cover-slip the slides. Use 1–3 ng/μL of probe for routine probes (300–500 bp in length), or for oligonucleotides use 1–10 pmol/mL of probe. Hybridization solution contains 50% for-

mamide, $2\times$ SSC, $1\times$ Denhardt's solution, 0.1% SDS, 5% dextran sulfate, and 400 µg/mL denatured salmon sperm DNA.

10. For dsDNA probes, denature at 95°C on a heating block for 5 min. For RNA probes, denature at 65°C for 5 min. Incubate the slides in a humidified box at the appropriate temperature for 16–18 h.
    a. For a DNA probe of 500 bp and 50% GC content, calculated T_m in 50% formamide, $2\times$ SSC = 61°C. Hybridize at 45°C ($20\times$ SSC = 3.3 M Na$^+$).
    b. For a 500-base RNA probe, calculated T_m in 50% formamide, $2\times$ SSC = 72°C. Hybridize at 55°C.
11. Wash the slides in $2\times$ SSC/0.1% SDS, 5 min $\times3$ at room temperature. Let the cover slips fall off into the first wash.
12. Wash the slides in the posthybridization high-stringency wash at an appropriate temperature for 15–30 min. For the 500-bp DNA probe of 50% GC content, calculated T_m in $0.1\times$ SSC/0.1% SDS is 71°C, hence wash at 60–63°C for 20 min. Calculated T_m in $0.1\times$ SSC/0.1% SDS/50% formamide is 40°C, hence wash at 30–33°C for 20 min. For an RNA probe of 500 bases and 50% GC content, calculated T_m in $0.1\times$ SSC/0.1% SDS is 82°C, hence wash at 72–74°C for 20 min. Calculated T_m in $0.1\times$ SSC/0.1% SDS/50% formamide is 51°C, hence wash at 40–43°C for 20 min. Make sure that the temperature of the wash solution does not drop significantly after addition of the slides.
13. Immerse the slides in TBS, pH 7.5, with 0.05% Tween-20 for 5 min (TBS-T).
14. Apply 100 µL of 10% normal sheep serum in TBS-T at room temperature for 30 min.
15. Suction off the fluid and apply 100 µL of 1.5 mU/mL of sheep anti-digoxigenin-alkaline phosphatase conjugate (Roche, if a digoxigenin-labeled probe is used) or anti-FITC- alkaline phosphatase conjugate (Roche, if a FITC-labeled probe is used) in TBS-T to each section. Incubate at room temperature for 30 min.
16. Wash slides in TBS-T for 5 min $\times3$.
17. Immerse slides in TBS, pH 9.5, for 3 min to activate the alkaline phosphatase,
18. Apply 100 µL of NBT/BCIP chromogen/substrate (Roche) solution appropriately diluted in TBS, pH 9.5, to each section. Incubate for 30 min–3 h. Follow color development under a microscope.
19. Wash slides under a stream of in dH$_2$O.
20. Counterstain for 30 s with nuclear fast red (Vector Laboratories), rinse with dH$_2$O.
21. Suction off water, air-dry sections, apply a drop of aqueous mountant (Glycergel, Dako), and cover slip.

3.3. Catalyzed Signal Amplification (CSA) Detection

Catalyzed signal amplification is performed after the posthybridization stringency wash. Two commercial sources of reagents used for CSA kits and reagents are NEN Life Science Products and Dako. All steps are performed at room temperature. For a biotin-labeled probe and with the Dako Genpoint kit:

1. Apply a 1:200–1:500 dilution of primary streptavidin-peroxidase for 15 min.
2. Wash in TBS-T, 5 mins $\times3$.
3. Apply 2–3 drops of biotinyl tyramide (supplied as working solution) for 15 min.
4. Wash in TBS-T, 5 min $\times3$.
5. Apply 2–3 drops of secondary streptavidin-peroxidase (supplied as working solution) for 15 min.
6. Wash in TBS-T, 5 min $\times3$.
7. Develop color with DAB/H$_2$O$_2$.
8. Follow color development under the microscope.
9. Wash sections in dH$_2$O.

10. Counterstain with dilute hematoxylin for 30 s, blue in tap water 1 min.
11. Dehydrate in graded series of ethanol, clear in xylene, and mount with organic mountant.

For a digoxigenin-labeled probe and with the TSA-AP kit from NEN Life Science Products:

1. Block sections with provided blocking solution for 30 min.
2. Incubate in 1:100 dilution of anti-digoxigenin-peroxidase (Roche) in TBS-T for 30 min.
3. Wash sections in TBS-T, 5 min ×3 with shaking.
4. Apply a 1:50 dilution of DNP-tyramide to the sections and incubate for 10 min.
5. Wash sections in TBS-T, 5 min ×3.
6. Incubate in 1:100 dilution of anti-DNP-AP conjugate for 30 min.
7. Wash sections in TBS-T, 5 min ×3.
8. Immerse slides in TBS, pH 9.5, for 3 min to activate the alkaline phosphatase.
9. Develop color with NBT/BCIP (Roche).
10. Wash slides in dH_2O.
11. Counterstain with nuclear fast red for 30 s, rinse in dH_2O.
12. Air-dry, mount with aqueous mountant (Glycergel).

3.4. Interphase FISH (see Note 2)

For formalin-fixed and paraffin-embedded sections, FISH protocol is similar to what has been described above with some modification. DNA probes can be labeled with fluorescence or with nonisotopic haptens such as biotin, digoxigenin, rhodamine, bromodeoxy uridine, sulfonate, acetylaminoflurene, and mercury tritophonol.

1. Cut 4–5-μm sections and place on coated slides (organosilane–or poly-L-lysine-coated).
2. Bake at 65°C for 1 h to melt the paraffin.
3. Deparaffinize: 100% xylene, 5 min ×2, 100% ethanol, 5 min ×2, air-dry slides.
4. Immerse the slides in 0.2 N HCl for 15 min (optional).
5. Wash in dH_2O, 5 min ×3.
6. Perform heat-induced retrieval by the capillary gap method in 1× Target Unmasking fluid (Vector Laboratories) for 12 min at 95°C and let slides cool down at room temperature for an additional 15 min before proceeding to the next step. Heat-induced retrieval can also be performed by immersing the slides in a tube containing 50% glycerol/0.1× SSC at 90°C for 3 min in a water bath. The slides are then allowed to cool to room temperature in 2× SSC *(80)*.
7. Wash in 2× SSC, 5 min ×2
8. Protease digest slides with 10–200 μg/mL of proteinase K in TBS, pH 7.5, for 5–15 min at 37°C. Concentration of enzyme and times must be adjusted empirically.
9. Wash in 2× SSC, 5 min ×3 with shaking.
10. Dehydrate slides in 70% ethanol, 95% ethanol, and 100% ethanol for 2 min each. Air-dry the slides.
11. Denature sections by immersing in 70% formamide/2× SSC at 72°C for 5 min.
12. Wash slides in 70% ethanol for 1 min and in 85% ethanol for 1 min with shaking to remove formamide.
13. Immerse in 100% ethanol for 1 min to dehydrate slides. Air-dry slides.
14. Denature the labeled dsDNA probe solution before use by heating to 95–100°C for 10 min. Apply denatured probe solution to the denatured sections—about 10–30 μL/specimen in hybridization solution. Coverslip sections and seal the edges of the coverslip with rubber cement to prevent evaporation. Incubate in a humidified box at 37°C for 16–18 h.
15. Remove the rubber cement and wash slides in 2× SSC/0.1% SDS at room temperature. Let the cover slips fall off into the wash buffer; do not peel them off.

16. Wash slides in the high-stringency posthybridization wash solution ($0.1–2\times$ SSC/0.1% SDS with or without formamide) at a predetermined temperature based on calculated T_m of the duplex. This step may have to be adjusted empirically.
17. The sections can be treated with fluorescent-labeled streptavidin or antibody against these ligands if the probe is not directly labeled with fluorescence.
18. Counterstain with 0.15 mM DAPI (Sigma, St. Louis, MO) in $2\times$ SSC for 5 min, destain for 5 min in $2\times$ SSC.
19. Dehydrate in 70%, 95%, and 100% ethanol for 2 min each, air-dry sections, and mount with antifade solution (Vector Laboratories, CA).
20. Slides are then visualized under an epifluorescence microscope with appropriate filters.

3.5. Combination of CSA and FISH

Catalyzed signal amplification has also been used to amplify signals for FISH after hybridization and high-stringency wash steps *(96)*.

1. Block sections with blocking solution for 30 min.
2. Apply 1:500 dilution of streptavidin-peroxidase for 30 min.
3. Wash in TBS-T, 5 min $\times$3.
4. Apply 1:50 dilution of biotinyl tyramide (diluent supplied) to the sections for 15 min.
5. Wash in TBS-T, 5 min $\times$3.
6. Apply streptavidin-fluorochrome conjugate for 30 min (follow manufacturer's recommendation).
7. Wash in TBS-T, 5 min $\times$3.
8. Counterstain and view under fluorescence microscope with appropriate filters.

4. Notes

1. If peroxidase-based systems are used, quench endogenous peroxidase activity by immersing sections in aqueous 0.3% H_2O_2 for 10 min followed by several washes in TBS, pH 7.5, after target retrieval and before hybridization. If using a biotin-labeled probe, substitute streptavidin-peroxidase or streptavidin-alkaline phosphatase complex for anti-digoxigenin-AP after the posthybridization wash. Dilutions to be used depend on the source, follow the manufacturer's recommendations (Dako, Vector Laboratories, Roche). After washing, develop color with DAB/H_2O_2 (Vector Laboratories) if peroxidase was used and NBT/BCIP (Roche, Vector Laboratories) if alkaline phosphatase was used. Follow manufacturer's recommendations for dilutions and develop color with appropriate chromogen/substrate.
2. If the probe was labeled with biotin, then after the posthybridization stringency wash, the sections are incubated with streptavidin linked to a fluorochrome (NEN Life Science products, Molecular Probes). Follow the manufacturer's recommendations.

References

1. McNicol, A. M., Walker, E., Farquharson, M. A., and Teasdale, G. M. (1991) Pituitary macroadenomas associated with hyperprolactinaemia: immunocytochemical and in-situ hybridization studies. *Clin. Endocrinol. (Oxf.).* **35,** 239–244.
2. Larsson, L. I. and Hougaard, D. M. (1992) Detection of gastrin and its messenger RNA in Zollinger-Ellison tumors by non-radioactive in situ hybridization and immunocytochemistry. *Histochemistry* **97,** 105–110.
3. Yamaguchi, K., Nalesnik, M. A., and Carr, B. I. (1993) In situ hybridization of albumin mRNA in normal liver and liver tumors: identification of hepatocellular origin. *Virchows Arch. B Cell. Pathol. Incl. Mol. Pathol.* **64,** 361–365.

4. Rajkumar, S. V., Richardson, R. L., and Goellner, J. R. (1998) Diagnostic value of albumin gene expression in liver tumors: case report and review of the literature. *Mayo Clin. Proc.* **73,** 533–536.
5. Inagaki, H., Nonaka, M., Nagaya, S., Tateyama, H., Sasaki, M., and Eimoto, T. (1995) Monoclonality in gastric lymphoma detected in formalin-fixed, paraffin- embedded endoscopic biopsy specimens using immunohistochemistry, in situ hybridization, and polymerase chain reaction. *Diagn. Mol. Pathol.* **4,** 32–38.
6. Nuovo, G. J. and Richart, R. M. (1989) A comparison of slot blot, Southern blot, and in situ hybridization analyses for human papillomavirus DNA in genital tract lesions. *Obstet. Gynecol.* **74,** 673–678.
7. Walboomers, J. M., Melchers, W. J., Mullink, H., et al. (1988) Sensitivity of in situ detection with biotinylated probes of human papilloma virus type 16 DNA in frozen tissue sections of squamous cell carcinomas of the cervix. *Am. J. Pathol.* **131,** 587–594.
8. Nagai, N., Nuovo, G., Friedman, D., and Crum, C. P. (1987) Detection of papillomavirus nucleic acids in genital precancers with the in situ hybridization technique. *Int. J. Gynecol. Pathol.* **6,** 366–679.
9. Shah, H. (1996) Papillomaviruses, in *Fields Virology*, 3rd ed. Lippincott Raven, Philadelphia, 2077–2109.
10. Lie, A. K., Skjeldestad, F. E., Hagen, B., Johannessen, E., Skarsvag, S., and Haugen, O. A. (1997) Comparison of light microscopy, in situ hybridization and polymerase chain reaction for detection of human papillomavirus in histological tissue of cervical intraepithelial neoplasia. *Apmis* **105,** 115–120.
11. Autillo-Touati, A., Joannes, M., d'Ercole, C., et al. (1998) HPV typing by in situ hybridization on cervical cytologic smears with ASCUS. *Acta. Cytol.* **42,** 631–638.
12. Unger, E. R., Vernon, S. D., Lee, D. R., Miller, D. L., and Reeves, W. C. (1998) Detection of human papillomavirus in archival tissues. Comparison of in situ hybridization and polymerase chain reaction. *J. Histochem. Cytochem.* **46,** 535–540.
13. Wu, T. C., Mann, R. B., Epstein, J. I., et al. (1991) Abundant expression of EBER-1 small nuclear RNA in nasopharyngeal carcinoma. *Am. J. Pathol.* 1461–1469.
14. Chang, K. L., Chen, Y. Y., Shibata, D., and Weiss, L. M. (1992) Description of an in situ hybridization methodology for detection of Epstein-Barr virus RNA in paraffin-embedded tissues, with a survey of normal and neoplastic tissues. *Diagn. Mol. Pathol.* **1,** 246–255.
15. d'Amore, F., Johansen, P., Houmand, A., Weisenburger, D. D., and Mortensen, L. S. (1996) Epstein-Barr virus genome in non-Hodgkin's lymphomas occurring in immunocompetent patients: highest prevalence in nonlymphoblastic T-cell lymphoma and correlation with a poor prognosis. Danish Lymphoma Study Group, LYFO. *Blood* **87,** 1045–1055.
16. Huh, J., Seoh, J. Y., and Kim, S. S. (1998) Cell block preparation of a Burkitt's lymphoma cell line as a positive control for in situ hybridization for Epstein-Barr virus. *Acta Cytol.* **42,** 1144–1148.
17. Mikata, A., Li, D. X., Kurosu, K., Oda, K., Yumoto, N., and Tamaru, J. I. (1997) Reappraisal of the relationship between immunoglobulin heavy chain gene rearrangement and Epstein-Barr virus infection in Reed-Sternberg cells of Hodgkin's disease. *Leuk. Lymphoma* **28,** 145–152.
18. Shimakage, M., Nakamine, H., Tamura, S., Takenaka, T., Yutsudo, M., and Hakura, A. (1997) Detection of Epstein-Barr virus transcripts in anaplastic large-cell lymphomas by mRNA in situ hybridization. *Hum. Pathol.* **28,** 1415–1419.
19. Khalidi, H. S., Lones, M. A., Zhou, Y., Weiss, L. M., and Medeiros, L. J. (1997) Detection of Epstein-Barr virus in the L & H cells of nodular lymphocyte predominance Hodgkin's disease: report of a case documented by immunohistochemical, in situ hybridization, and polymerase chain reaction methods. *Am. J. Clin. Pathol.* **108,** 687–692.
20. Li, D., Oda, K., Mikata, A., and Yumoto, N. (1995) Epstein-Barr virus genomes in Hodgkin's disease and non-Hodgkin's lymphomas. *Pathol. Int.* **45,** 735–741.

21. Sheehan, M. M., Coker, R., and Coleman, D. V. (1998) Detection of cytomegalovirus (CMV) in HIV+ patients: comparison of cytomorphology, immunocytochemistry and in situ hybridization. *Cytopathology* **9,** 29–37.
22. Ozono, K., Mushiake, S., Takeshima, T., and Nakayama, M. (1997) Diagnosis of congenital cytomegalovirus infection by examination of placenta: application of polymerase chain reaction and in situ hybridization. *Pediatr. Pathol. Lab. Med.* **17,** 249–258.
23. Musiani, M., Zerbini, M., Venturoli, S., et al. (1994) Rapid diagnosis of cytomegalovirus encephalitis in patients with AIDS using in situ hybridisation. *J. Clin. Pathol.* **47,** 886–891.
24. Nuovo, M. A., Nuovo, G. J., Becker, J., Gallery, F., Delvenne, P., and Kane, P. B. (1993) Correlation of viral infection, histology, and mortality in immunocompromised patients with pneumonia. Analysis by in situ hybridization and the polymerase chain reaction. *Diagn. Mol. Pathol.* **2,** 200–209.
25. Walts, A. E., Marchevsky, A. M., and Morgan, M. (1991) Pulmonary cytology in lung transplant recipients: recent trends in laboratory utilization. *Diagn. Cytopathol.* **7,** 353–358.
26. Wright, C. A., Haffajee, Z., van Iddekinge, B., and Cooper, K. (1995) Detection of herpes simplex virus DNA in spontaneous abortions from HIV- positive women using non-isotopic in situ hybridization. *J. Pathol.* **176,** 399–402.
27. Annunziato, P., Lungu, O., Gershon, A., Silvers, D. N., LaRussa, P., and Silverstein, S. J. (1996) In situ hybridization detection of varicella zoster virus in paraffin- embedded skin biopsy samples. *Clin. Diagn. Virol.* **7,** 69–76.
28. Ohsawa, M., Tomita, Y., Hashimoto, M., Kanno, H., and Aozasa, K. (1998) Hepatitis C viral genome in a subset of primary hepatic lymphomas. *Mod. Pathol.* **11,** 471–478.
29. Yamada, S., Koji, T., Nozawa, M., Kiyosawa, K., and Nakane, P. K. (1992) Detection of hepatitis C virus (HCV) RNA in paraffin embedded tissue sections of human liver of non-A, non-B hepatitis patients by in situ hybridization. *J. Clin. Lab. Anal.* **6,** 40–46.
30. Hayashi, Y., Watanabe, J., Nakata, K., Fukayama, M., and Ikeda, H. (1990) A novel diagnostic method of *Pneumocystis carinii*. In situ hybridization of ribosomal ribonucleic acid with biotinylated oligonucleotide probes. *Lab. Invest.* **63,** 576–580.
31. Barrett, D. M., Faigel, D. O., Metz, D. C., Montone, K., and Furth, E. E. (1997) In situ hybridization for *Heliobacter pylori* in gastric mucosal biopsy specimens: quantitative evaluation of test performance in comparison with the CLOtest and thiazine stain. *J. Clin. Lab. Anal.* **11,** 374–379.
32. Horn, J. E., Quinn, T., Hammer, M., Palmer, L., and Falkow, S. (1986) Use of nucleic acid probes for the detection of sexually transmitted infectious agents. *Diagn. Microbiol. Infect. Dis.* 101S–109S.
33. Montone, K. T. and Litzky, L. A. (1995) Rapid method for detection of *Aspergillus* 5S ribosomal RNA using a genus-specific oligonucleotide probe. *Am. J. Clin. Pathol.* **103,** 48–51.
34. Arnoldi, J., Schluter, C., Duchrow, M., et al. (1992) Species-specific assessment of *Mycobacterium leprae* in skin biopsies by in situ hybridization and polymerase chain reaction. *Lab. Invest.* **66,** 618–623.
35. Pinkel, D., Straume, T., and Gray, J. W. (1986) Cytogenetic analysis using quantitative, high-sensitivity, fluorescence hybridization. *Proc. Natl. Acad. Sci. USA* **83,** 2934–2938.
36. Pinkel, D., Gray, J. W., Trask, B., van den Engh, G., Fuscoe, J., and van Dekken, H. (1986) Cytogenetic analysis by in situ hybridization with fluorescently labeled nucleic acid probes. *Cold Spring Harb. Symp. Quant. Biol.* **51,** 151–157.
37. Luke, S. and Shepelsky, M. (1998) FISH: recent advances and diagnostic aspects. *Cell Vis.* **5,** 49–53.
38. Mark, H. F. (1999) Fluorescence in situ hybridization analysis of biomarkers in cancer. *Exp. Mol. Pathol.* **67,** 131–134.
39. Bench, A. J., Nacheva, E. P., Hood, T. L., et al. (2000) Chromosome 20 deletions in myeloid malignancies: reduction of the common deleted region, generation of a PAC/BAC contig and

identification of candidate genes. UK Cancer Cytogenetics Group (UKCCG). *Oncogene* **19,** 3902–3913.

40. Diebold, J., Mosinger, K., Peiro, G., et al. (2000) 20q13 and cyclin D1 in ovarian carcinomas. Analysis by fluorescence in situ hybridization. *J. Pathol.* **190,** 564–571.

41. Kaltz-Wittmer, C., Klenk, U., Glaessgen, A., et al. (2000) FISH analysis of gene aberrations (MYC, CCND1, ERBB2, RB, and AR) in advanced prostatic carcinomas before and after androgen deprivation therapy. *Lab. Invest.* **80,** 1455–1464.

42. Staff, S., Nupponen, N. N., Borg, A., Isola, J. J., and Tanner, M. M. (2000) Multiple copies of mutant BRCA1 and BRCA2 alleles in breast tumors from germ-line mutation carriers. *Genes Chromosomes Cancer* **28,** 432–442.

43. Oliver, R. T. (2000) Current opinion in germ cell cancer 2000. *Curr. Opin. Oncol.* **12,** 249–254.

44. Kytola, S., Rummukainen, J., Nordgren, A., et al. (2000) Chromosomal alterations in 15 breast cancer cell lines by comparative genomic hybridization and spectral karyotyping. *Genes Chromosomes Cancer* **28,** 308–317.

45. Pauletti, G., Godolphin, W., Press, M. F., and Slamon, D. J. (1996) Detection and quantitation of HER-2/neu gene amplification in human breast cancer archival material using fluorescence in situ hybridization. *Oncogene* **13,** 63–72.

46. Saffari, B., Jones, L. A., el-Naggar, A., Felix, J. C., George, J., and Press, M. F. (1995) Amplification and overexpression of HER-2/neu (c-erbB2) in endometrial cancers: correlation with overall survival. *Cancer Res.* **55,** 5693–5698.

47. Ooi, A., Kobayashi, M., Mai, M., and Nakanishi, I. (1998) Amplification of c-erbB-2 in gastric cancer: detection in formalin- fixed, paraffin-embedded tissue by fluorescence in situ hybridization. *Lab. Invest.* **78,** 345–351.

48. Xing, W. R., Gilchrist, K. W., Harris, C. P., Samson, W., and Meisner, L. F. (1996) FISH detection of HER-2/neu oncogene amplification in early onset breast cancer. *Breast Cancer Res. Treat.* **39,** 203–212.

49. Ross, J. S., Sheehan, C. E., Hayner-Buchan, A. M., et al. (1997) Prognostic significance of HER-2/neu gene amplification status by fluorescence in situ hybridization of prostate carcinoma. *Cancer* **79,** 2162–2170.

50. Szollosi, J., Balazs, M., Feuerstein, B. G., Benz, C. C., and Waldman, F. M. (1995) ERBB-2 (HER2/neu) gene copy number, p185HER-2 overexpression, and intratumor heterogeneity in human breast cancer. *Cancer Res.* **55,** 5400–5407.

51. Perissel, B., Coupier, I., De Latour, M., et al. (2000) Structural and numerical aberrations of chromosome 22 in a case of follicular variant of papillary thyroid carcinoma revealed by conventional and molecular cytogenetics. *Cancer Genet. Cytogenet.* **121,** 33–37.

52. Gribble, S., Andrews, K., Williams, D., et al. (2000) Fluorescence in situ hybridization detection of two telomeres on the short arm of a derived chromosome 16 in an infant with thrombocytopenia. *Cancer Genet. Cytogenet.* **120,** 99–104.

53. Reddy, K. S., Parsons, L., Wang, S., Mak, L., Dighe, P., and Yu, T. L. (2000) FISH analysis of an AML-M5a with segmental rearrangements involving 11q23-MLL region. *Cancer Genet. Cytogenet.* **118,** 48–51.

54. Brezinova, J., Zemanova, Z., Cermak, J., and Michalova, K. (2000) Fluorescence in situ hybridization confirmation of 5q deletions in patients with hematological malignancies. *Cancer Genet. Cytogenet.* **117,** 45–49.

55. Yehuda, O., Abeliovich, D., Ben-Neriah, S., et al. (1999) Clinical implications of fluorescence in situ hybridization analysis in 13 chronic myeloid leukemia cases: Ph-negative and variant Ph-positive. *Cancer Genet. Cytogenet.* **114,** 100–107.

56. Durm, M., Hausmann, M., Aldinger, K., Ludwig, H., and Cremer, C. (1996) Painting of human chromosome 8 in fifteen minutes. *Z. Naturforsch. [C]* **51,** 435–439.

57. Shi, L., Zhu, T., Morgante, M., Rafalski, J. A., and Keim, P. (1996) Soybean chromosome painting: a strategy for somatic cytogenetics. *J. Hered.* **87,** 308–313.

58. Durm, M., Schussler, L., Munch, H., et al. (1998) Fast-painting of human metaphase spreads using a chromosome-specific, repeat-depleted DNA library probe. *Biotechniques* **24,** 820–825.

59. Gerdes, A. M., Pandis, N., Bomme, L., et al. (1997) Fluorescence in situ hybridization of old G-banded and mounted chromosome preparations. *Cancer Genet. Cytogenet.* **98,** 9–15.

60. Roberts, I., Wienberg, J., Nacheva, E., Grace, C., Griffin, D., and Coleman, N. (1999) Novel method for the production of multiple colour chromosome paints for use in karyotyping by fluorescence in situ hybridisation. *Genes Chromosomes Cancer* **25,** 241–250.

61. Thompson, C. T. and Gray, J. W. (1993) Cytogenetic profiling using fluorescence in situ hybridization (FISH) and comparative genomic hybridization (CGH). *J. Cell. Biochem. Suppl.* 139–143.

62. Weiss, M. M., Hermsen, M. A., Meijer, G. A., et al. (1999) Comparative genomic hybridisation. *Mol. Pathol.* **52,** 243–251.

63. Wallrapp, C., Muller-Pillasch, F., Micha, A., et al. (1999) Strategies for the detection of disease genes in pancreatic cancer. *Ann NY Acad. Sci.* **880,** 122–146.

64. Jarosova, M., Holzerova, M., Jedlickova, K., et al. (2000) Importance of using comparative genomic hybridization to improve detection of chromosomal changes in childhood acute lymphoblastic leukemia. *Cancer Genet. Cytogenet.* **123,** 114–122.

65. Long, A. A. (1998) In-situ polymerase chain reaction: foundation of the technology and today's options. *Eur. J. Histochem.* **42,** 101–109.

66. Komminoth, P. and Long, A. A. (1993) In-situ polymerase chain reaction. An overview of methods, applications and limitations of a new molecular technique. *Virchows Arch. B Cell. Pathol. Incl. Mol. Pathol.* **64,** 67–73.

67. Mee, A. P., Denton, J., Hoyland, J. A., Davies, M., and Mawer, E. B. (1997) Quantification of vitamin D receptor mRNA in tissue sections demonstrates the relative limitations of in situ-reverse transcriptase-polymerase chain reaction. *J. Pathol.* **182,** 22–28.

68. Bates, P. J., Sanderson, G., Holgate, S. T., and Johnston, S. L. (1997) A comparison of RT-PCR, in-situ hybridisation and in-situ RT-PCR for the detection of rhinovirus infection in paraffin sections. *J. Virol. Meth.* **67,** 153–160.

69. Thaker, V. (1999) In situ RT-PCR and hybridization techniques. *Meth. Mol. Biol.* **115,** 379–402.

70. Vaughan, C. J., Aherne, A. M., Lane, E., Power, O., Carey, R. M., and O'Connell, D. P. (2000) Identification and regional distribution of the dopamine D(1A) receptor in the gastrointestinal tract. *Am. J. Physiol. Regul. Integr. Comp. Physiol.* **279,** R599–R609.

71. Sato, F. and Sato, K. (2000) cAMP-dependent Cl(−) channel protein (CFTR) and its mRNA are expressed in the secretory portion of human eccrine sweat gland. *J. Histochem. Cytochem.* **48,** 345–354.

72. O'Connell, D. P., Aherne, A. M., Lane, E., Felder, R. A., and Carey, R. M. (1998) Detection of dopamine receptor D1A subtype-specific mRNA in rat kidney by in situ amplification. *Am. J. Physiol.* **274,** F232–F241.

73. Mueller, J. D., Putz, B., and Hofler, H. (1997) Self-sustained sequence replication (3SR): an alternative to PCR. *Histochem. Cell Biol.* **108,** 431–437.

74. Hofler, H., Putz, B., Mueller, J. D., Neubert, W., Sutter, G., and Gais, P. (1995) In situ amplification of measles virus RNA by the self-sustained sequence replication reaction. *Lab. Invest.* **73,** 577–585.

75. Bobrow, M. N., Shaughnessy, K. J., and Litt, G. J. (1991) Catalyzed reporter deposition, a novel method of signal amplification. II. Application to membrane immunoassays. *J. Immunol. Meth.* **137,** 103–112.

76. Bobrow, M. N., Harris, T. D., Shaughnessy, K. J., and Litt, G. J. (1989) Catalyzed reporter deposition, a novel method of signal amplification. Application to immunoassays. *J. Immunol. Meth.* **125.** 279–285.
77. Komminoth, P. and Werner, M. (1997) Target and signal amplification: approaches to increase the sensitivity of in situ hybridization. *Histochem. Cell Biol.* **108,** 325–333.
78. McAllister, H. A. and Rock, D. L. (1985) Comparative usefulness of tissue fixatives for in situ viral nucleic acid hybridization. *J. Histochem. Cytochem.* **33,** 1026–1032.
79. Nuovo (1991) Comparison of Bouins solution and buffered formalin on the detection rate by in situ hybridization of human papillomavirus DNA in genital tract lesions. *J. Histotech.* 13–18.
80. Hyytinen, E., Visakorpi, T., Kallioniemi, A., Kallioniemi, O. P., and Isola, J. J. (1994) Improved technique for analysis of formalin-fixed paraffin embedded tumors by fluorescence in situ hybridization. *Cytometry* **16,** 93–99.
81. Hougaard, D. M., Hansen, H., and Larsson, L. I. (1997) Non-radioactive in situ hybridization for mRNA with emphasis on the use of oligodeoxynucleotide probes. *Histochem. Cell Biol.* **108,** 335–344.
82. Chen, C., Wu, B., Wei, T., Egholm, M., and Strauss, W. M. (2000) Unique chromosome identification and sequence-specific structural analysis with short PNA oligomers. *Mamm. Genome* **11,** 384–391.
83. Worden, A. Z., Chisholm, S. W., and Binder, B. J. (2000) In situ hybridization of *Prochlorococcus* and *Synechococcus* (marine cyanobacteria) spp. with RRNA-targeted peptide nucleic acid probes. *Appl. Environ. Microbiol.* **66,** 284–289.
84. Drobniewski, F. A., More, P. G., and Harris, G. S. (2000) Differentiation of *Mycobacterium tuberculosis* complex and nontuberculous mycobacterial liquid cultures by using peptide nucleic acid-fluorescence in situ hybridization probes. *J. Clin. Microbiol.* **38,** 444–447.
85. Stender, H., Lund, K., Petersen, K. H., et al. (1999) Fluorescence in situ hybridization assay using peptide nucleic acid probes for differentiation between tuberculous and nontuberculous mycobacterium species in smears of mycobacterium cultures. *J. Clin. Microbiol.* **37,** 2760–2765.
86. Prescott, A. M. and Fricker, C. R. (1999) Use of PNA oligonucleotides for the in situ detection of *Escherichia coli* in water. *Mol. Cell. Probes* **13,** 261–268.
87. Just, T., Burgwald, H., and Broe, M. K. (1998) Flow cytometric detection of EBV (EBER snRNA) using peptide nucleic acid probes. *J. Virol. Meth.* **73,** 163–174.
88. Taneja, K. L. (1998) Localization of trinucleotide repeat sequences in myotonic dystrophy cells using a single fluorochrome-labeled PNA probe. *Biotechniques* **24,** 472–476.
89. Taneja, K. L., Chavez, E. A., Coull, J., and Lansdorp, P. M. (2001) Multicolor fluorescence in situ hybridization with peptide nucleic acid probes for enumeration of specific chromosomes in human cells. *Genes Chromosomes Cancer* **30,** 57–63.
90. Zhang, X., Ishihara, T., and Corey, D. R. (2000) Strand invasion by mixed base PNAs and a PNA-peptide chimera. *Nucleic Acids Res.* **28,** 3332–3338.
91. Ray, A. and Norden, B. (2000) Peptide nucleic acid (PNA): its medical and biotechnical applications and promise for the future. *FASEB J.* **14,** 1041–1060.
92. Knuchel, M. C., Graf, B., Schlaepfer, E., et al. (2000) PCR-derived ssDNA probes for fluorescent in situ hybridization to HIV-1 RNA. *J. Histochem. Cytochem.* **48,** 285–294.
93. Konat, G. W. (1996) Generation of high efficiency ssDNA hybridization probes by linear polymerase chain reaction (LPCR). *Scanning Microsc. Suppl.* **10,** 57–60.
94. Hannon, K., Johnstone, E., Craft, L. S., et al. (1993) Synthesis of PCR-derived, single-stranded DNA probes suitable for in situ hybridization. *Anal. Biochem.* **212,** 421–427.
95. Wetmur, J. G. (1991) DNA probes: applications of the principles of nucleic acid hybridization. *Crit. Rev. Biochem. Mol. Biol.* **26,** 227–259.
96. Adler, K., Erickson, T., and Bobrow, M. (1997) High sensitivity detection of HPV-16 in SiHa and Caski cells utilizing FISH enhanced by TSA. *Histochem. Cell Biol.* **108,** 321–324.

6

Genetic Strategies in *Saccharomyces cerevisiae* to Study Human Tumor Suppressor Genes

Takahiko Kobayashi, Ting Wang, Hua Qian, and Rainer K. Brachmann

1. Introduction

Baker's yeast or *Saccharomyces cerevisiae* is the most intensely studied eukaryotic microorganism for very good reasons. Many fundamental processes such as cell cycle control and DNA repair show evolutionary conservation from yeast to human, thus making studies in yeast highly relevant to our understanding of mammalian cells. Yeast also allows for the efficient pursuit of questions that are often almost impossible to answer in mammalian assay systems.

The advantages of yeast are numerous, such as rapid growth, maintenance on plates, the ability to test numerous phenotypes at the same time, and the easy introduction and propagation of DNA as plasmids (**Fig. 1A**). Yeast can be maintained as haploid or diploid cells; haploid cells of opposite mating types are mated to form a diploid (**Fig. 1B**). Rapid mating assays take advantage of this and have been used extensively to perform two-hybrid assays on a genome-wide scale *(1,2)*.

Yeast is particularly unique in that homologous recombination can be efficiently mediated through short fragments of linear DNA. This allows for the targeted insertion of DNA into the yeast genome (**Fig. 1C**) *(3,4)*. Homologous recombination can also be used to repair a gapped plasmid with linear DNA that has homology to the ends of the gapped plasmid (**Fig. 1D**) *(5)*. Among other applications, this technique has been exploited for the rapid mutagenesis of a cDNA of interest *(6)*.

It comes as no surprise, then, that researchers have not only studied the biology of yeast, but have converted yeast into a tool that allows them to address some of their favorite questions. The most prominent example of such a yeast adaptation is the two-hybrid system that is discussed separately (*see* Chapter 14). This chapter addresses other interesting strategies that take advantage of yeast in order to gain new insights into tumor suppressor genes. Studies of true yeast homologs for human tumor suppressor genes, such as DNA repair enzymes, have not been included; instead, the emphasis is on the general concepts of how to adapt yeast for studies of human tumor suppressor proteins without yeast homologs. These strategies may be applied, in principle, to any other protein of interest. Standard protocols for yeast maintenance and manipulation

From: *Methods in Molecular Biology, Vol. 223: Tumor Suppressor Genes: Regulation, Function, and Medicinal Applications.* Edited by: Wafik S. El-Deiry © Humana Press Inc., Totowa, NJ

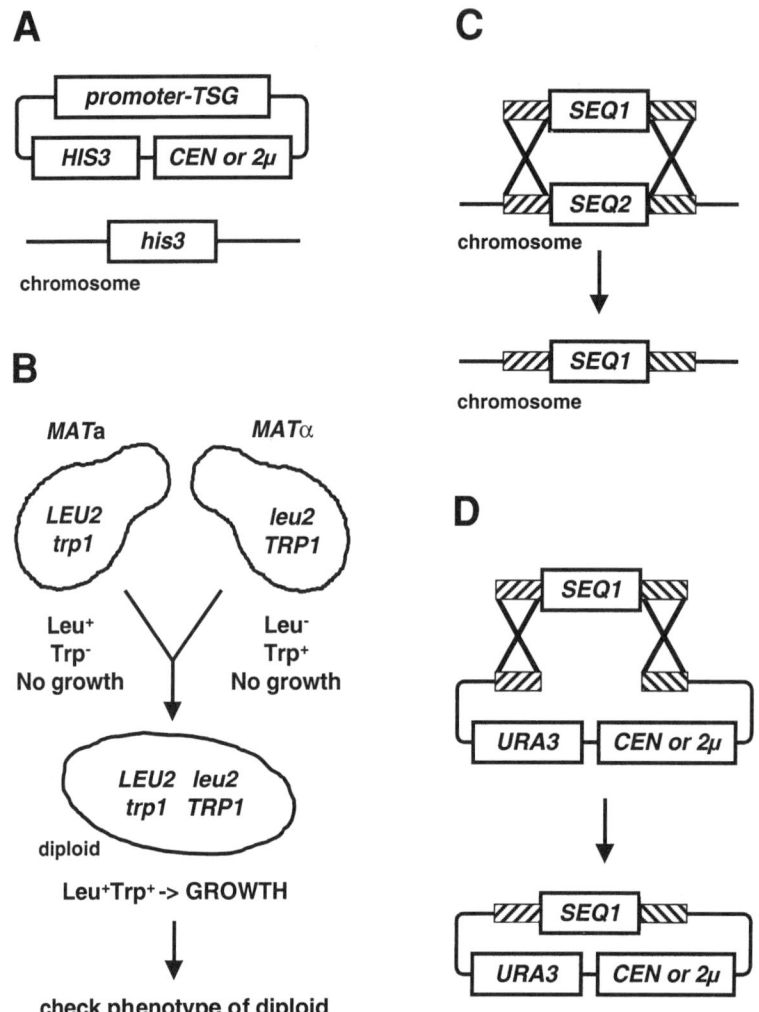

Fig. 1. Common genetic manipulations of and in *S. cerevisiae*. **(A)** Episomal DNA can be stably maintained in yeast through the use of auxotrophic markers. For example, expression of the *HIS3* gene enables yeast to grow on plates lacking histidine (Synthetic Complete or SC-His). If the *HIS3* gene of a yeast strain is inactivated (*his3*; note, all inactivating mutations in yeast are denoted by lowercase), propagation of plasmids in yeast cells can be ensured through the auxotrophic marker *HIS3*. Centromere-containing plasmids (*CEN*) are maintained at one or two copies per yeast cell, while numerous copies of *2μ*-based plasmids (typically 20–50) may be present. **(B)** Haploid yeast cells of opposite mating types (*MAT*a and *MAT*α) can be mated to form diploid cells. The diploid cells can be selected for by using auxotrophic markers. In the example, only diploid cells have one functional gene each of *LEU2* and *TRP1* that allows them to grow on plates lacking leucine and tryptophan (SC-Leu-Trp). Using mating assays, one can, for example, quickly combine numerous cancer mutants of the transcription factor p53 with numerous binding sites of downstream target genes of p53. **(C)** Fragments of linear DNA can mediate homologous recombination in yeast. This can be used to integrate any DNA sequence (*SEQ1*) into the yeast genome. The integration is mediated through flanking sequences of 30 bp or longer that have homology with sequences of the desired integration site (*SEQ2*). This strategy is frequently used to knock out a particular yeast gene. It also serves to integrate expression cassettes for proteins of interest. **(D)** Homologous recombination can also be exploited to repair a gapped plasmid in yeast using linear DNA with sequences at the ends that are homologous to the ends of the gapped plasmid. Besides simple construction of a plasmid in yeast, the process can be used to mutagenize a sequence of interest (*SEQ1*) prior to evaluation in yeast.

have been covered excellently and in great detail elsewhere *(7,8)*. The frequent focus on the p53 tumor suppressor protein is not so much a bias of the authors, but a reflection of the currently available literature.

2. Yeast Assays to Identify and Investigate Tumor Suppressor Gene Mutations Occurring in Human Cancers

2.1. Functional Assessment of Tumor Suppressor Gene Mutations

Using yeast assays to identify mutations in human tumor suppressor genes has several attractive aspects. The most obvious advantage is the immediate functional readout that can classify, for example, a missense mutation as a mutation deleterious for the activity of a tumor suppressor protein, as opposed to a mere polymorphism without impact on protein activity. Yeast assays provide instant information for both alleles of a tumor suppressor gene. Lastly, the diagnostic steps are straightforward and give answers quite rapidly. The challenge lies in engineering yeast strains that accurately reflect the biologic activity of a tumor suppressor protein. This has been achieved to a large extent for some tumor suppressor proteins. For others, strategies have been developed that circumvent these requirements.

2.1.1. Evaluating the Transcriptional Activity of Mutated Tumor Suppressor Proteins

A good candidate for a yeast functional assay is a transcription factor, since the extensive use of the yeast two-hybrid system (basically a transcription factor reconstitution assay) has led to the development of many useful yeast strains that can be easily converted into a one-hybrid assay (for example, a transcription factor fused to the GAL4 DNA-binding domain) or an assay for the native transcription factor, assuming specific DNA-binding sites for the transcription factor are known.

To date, the bulk of the literature on functional yeast assays is on the tumor suppressor protein p53. The p53 protein is a transcription factor that becomes activated in response to upstream signals, such as DNA damage, oncogene overexpression, or hypoxia, and protects humans from cancer by inducing DNA repair, cell cycle arrest, and/or apoptosis (see Volume I, Chapter 7). The important role of p53 is reflected in the high number of human cancers, approximately 50%, that carry *p53* gene mutations, a finding that has caused great interest in classifying cancers based on their p53 status.

One of the first two reports that determined the presence of a transcriptional activation domain in the amino-terminus of p53 fused the DNA-binding domain of the yeast transcription factor GAL4 with the first 73 amino acids of p53 and showed that this fusion protein could activate transcription in a p53-dependent manner in yeast and mammalian cells *(9)*. Once a DNA sequence had been identified to which p53 could specifically bind *(10)*, a yeast strain was engineered in which full-length p53 functioned as a transcription factor and drove the expression of *lacZ* *(11)*. Shortly thereafter, the assay was modified to allow for the rapid assessment of the p53 status of cancer cells, a technique called functional analysis of separated alleles in yeast, or FASAY *(12)*. mRNA from a cancer or any other tissue is reverse transcribed and the *p53* cDNA PCR-amplified with a high-fidelity polymerase. The PCR product for the *p53* cDNA, representing a mix of both alleles, and a gapped expression plasmid are co-transformed into a yeast reporter strain that will repair the gapped plasmid by invivo homologous recombination with the PCR product. The resulting yeast clones are then assessed for their ability to grow on plates lacking histidine. His$^+$ colonies represent clones expressing functional p53 able to drive the expression of the reporter gene *HIS3*, His$^-$ colonies represent clones with *p53* expression plasmids carrying a mutated *p53* cDNA (**Fig. 2A**). This assay was later modified to contain *ADE2* as the reporter gene that identifies *p53* mutations by the color of yeast clones on plates containing limiting adenine *(13)*. Wild-type p53 drives *ADE2* expression and leads to white colonies, mutant p53 results in Ade$^-$ cells that are red because an intermediate of the adenine metabolism accumulates.

These types of yeast functional assays for p53 have since been widely used to evaluate cancer cell lines and tissues and have been compared to other methods of *p53* mutation detection. In general, yeast assays assess p53 cancer mutants correctly, since transcriptional activity of p53 protein correlates well with its tumor suppressor activity. However, as with almost any other assay, there are also limitations. For example, since the p53 protein is analyzed separated from its normal cellular context, yeast assays do not address posttranscriptional inactivation of wild-type p53 protein that might be present in the cancer cell. Also, the p53 DNA-binding site used in the yeast assays may not be entirely representative of the ever-increasing number of p53 downstream target genes and their binding sites. This question has thus far been only partially addressed by analyzing p53 mutants with a few additional p53-binding sites of genes mediating cell cycle arrest or apoptosis *(14,15)*.

Yeast p53 assays have also been used to determine whether p53 cancer mutants have a dominant-negative effect on wild-type p53. The advantages of yeast are that they are genetically well defined and that use of centromere-containing plasmids with the same promoter for wild-type and mutant p53 results in approximately equal expression levels. Depending on the type of yeast assay, the dominant-negative effect of p53 cancer mutants on wild-type p53 was scored either by reduction of *URA3* expression *(16)*, resulting in colonies resistant to 5-fluoroorotic acid (5-Foa), or *ADE2* expression *(17,18),* resulting in pink or red colonies.

Using strategies similar to the initial characterization of p53 in yeast, a fusion protein between the GAL4 DNA-binding domain and the very carboxy-terminus of BRCA1 demonstrated a transactivation domain in BRCA1 whose activity is affected by *BRCA1* germline mutations *(19)*.

2.1.2. Evaluating the Effects of Mutated Tumor Suppressor Proteins on Yeast

While the assay of Monteiro and colleagues for the BRCA1 carboxy-terminus relies on a transcriptional readout *(19)*, Humphrey and colleagues took advantage of the fact that BRCA1 inhibits yeast growth *(20)*. This effect is mediated by the same carboxy-terminal region that has transcriptional activity, suggesting that both assays will identify a similar set of BRCA1 mutants. This illustrates the possibility to identify tumor suppressor mutants because they have lost an activity, as assayed in yeast, that the wild-type protein possesses. This concept is most elegantly realized in the case of the von Recklinghausen neurofibromatosis (*NF1*) gene. The NF1 protein was found to have a central region with similarity to the catalytic segments of ras GTPase-activating proteins

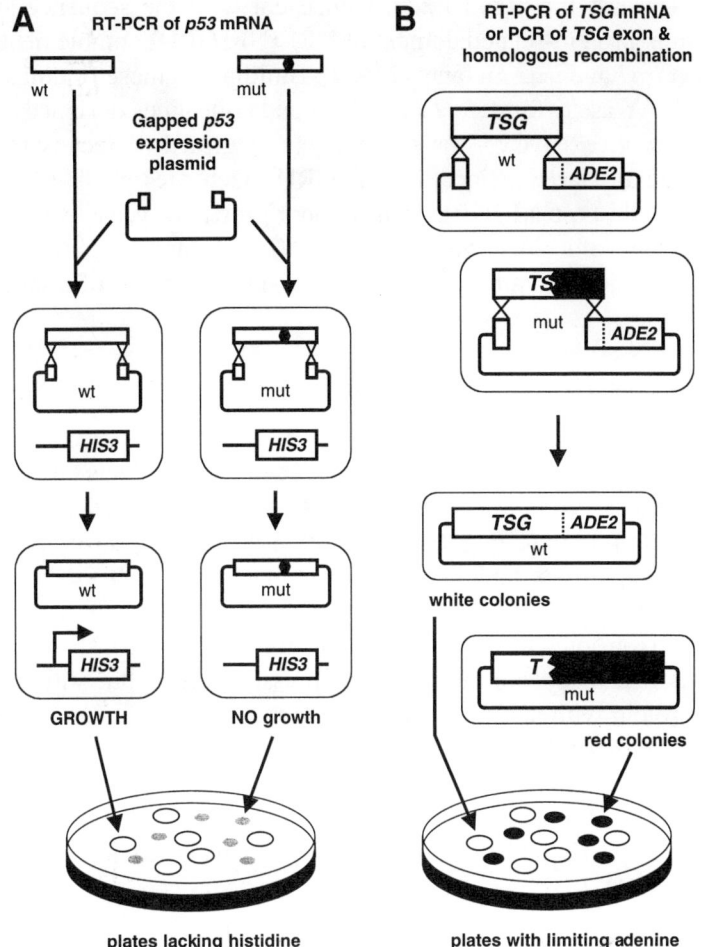

Fig. 2. Rapid evaluation of tumor suppressor mutants in *S. cerevisiae*. **(A)** Functional analysis of separated alleles in yeast or FASAY for p53. mRNA derived from a cancer or any other tissue is reverse-transcribed and the *p53* cDNA PCR-amplified with a high-fidelity polymerase. In the example, the left PCR product (wt) is derived from the wild-type *p53* allele, the right PCR product with the black triangle (mut) from the mutated *p53* allele. The PCR products for the *p53* cDNAs and gapped expression plasmid are cotransformed into a yeast reporter strain that will repair the gapped plasmid by in-vivo homologous recombination with the PCR product. The resulting yeast clones are selected for using the auxotrophic marker of the repaired plasmid and then replica-plated (not shown in the figure) to plates lacking histidine. His$^+$ colonies represent clones expressing functional p53 able to drive the expression of the reporter gene *HIS3*, His$^-$ colonies represent clones with *p53* expression plasmids carrying a mutated *p53* cDNA. Besides *HIS3*, *ADE2* is also frequently used as the reporter gene. **(B)** Identification of mutations by creating fusion genes between a tumor suppressor and an auxotrophic marker gene. This type of yeast assay does not require a known function of the wild-type tumor suppressor protein in yeast. If the tumor suppressor gene (*TSG*) contains a nonsense or frameshift mutation, a truncated protein will result that does not contain the auxotrophic marker (mut). Wild-type sequence will result in a functional fusion protein (wt). These assays do not detect missense mutations that inactivate the tumor suppressor protein but still result in a full-length fusion protein with an active auxotrophic marker. The source of DNA for the PCR amplification can be cDNA or genomic DNA for the evaluation of a particular exon. In-vivo homologous recombination between the PCR product and the gapped plasmid containing the auxotrophic marker (*ADE2* in the example) creates the fusion gene. The intact fusion protein will result in white colonies; truncated proteins without auxotrophic marker will lead to red (and slightly smaller) colonies.

(GAPs), such as yeast Ira1p and Ira2p. Consistent with the sequence similarity, the GTPase-activating protein-related domain of NF1 (NF1GRD) complemented most phenotypes of yeast *ira1* and *ira2* mutants *(21–23)*. Building on these findings and the principles of the FASAY assay for p53 *(12)*, Ishioka and colleagues designed an assay using a temperature-sensitive *CDC25* yeast strain whose growth defect is rescued by the expression of human Ha-*ras*. Since the NF1GRD inactivates p21^{ras} by stimulating its GTPase activity, wild-type NF1GRD leads to poor growth of yeast, while the NF1GRD of NF1 mutants does not affect yeast growth *(24)*. Similar to the BRCA1 assays, the assay for NF1 has the limitation of evaluating a single domain of a large tumor suppressor protein.

2.1.3. Yeast Assays That Do Not Rely on a Normal Function of the Tumor Suppressor Protein

It is often difficult to create a yeast assay that adequately reflects the normal function of a tumor suppressor protein, particularly if its function is not even fully understood. To circumvent this problem, investigators have used assays that create fusion genes between a tumor suppressor and an auxotrophic marker gene. If the tumor suppressor gene contains a nonsense or frameshift mutation, a truncated protein will result that does not contain the auxotrophic marker (**Fig. 2B**). These assays obviously do not detect missense mutations that inactivate the tumor suppressor protein but still result in a full-length fusion protein with an active auxotrophic marker. The fusion genes are created by PCR amplification of the tumor suppressor gene and homologous recombination of the PCR product with a gapped yeast expression plasmid in yeast. In this case, the source of DNA can be cDNA (as described for p53) or genomic DNA for the evaluation of a particular exon. Such screening assays have been developed for *BRCA1 (25)*, *APC (25,26)*, and *PTEN (27)*.

2.2. Studying the Effects of Carcinogens on Tumor Suppressor Genes

The p53 yeast assays described under **Subheading 2.1.1.** have also been used extensively to study the effects of specific mutagens on the *p53* gene. Excellent international databases with more than 14,000 entries of *p53* mutations are available to allow immediate comparison of laboratory results to data from human cancers *(28,29)*. The spectrum of *p53* mutations is quite unique and complex; approximately three-quarters of all mutations are missense mutations affecting the p53 core domain responsible for binding to p53 DNA-binding sites, and 926 amino acid changes in the p53 core domain have been reported *(29)*. This wealth of data has enabled researchers to make important links between mutagens and specific *p53* cancer mutations *(30)*. Functional yeast assays for p53 have contributed to the field of molecular epidemiology; and the important aspects of such usage have been reviewed recently in great detail *(31)*. In general, the yeast expression plasmid for p53 is exposed in vitro to known or suspected mutagens. The mutated *p53* expression plasmid is then directly transformed into yeast, or the *p53* ORF is PCR-amplified and cloned into a gapped expression plasmid using homologous recombination in yeast *(32–34)*.

p53 mutations that spontaneously occur within yeast cells have been studied as well *(16,35)*. The effects of mutagens on the *p53* cDNA can also be analyzed in live yeast cells; yeast may even be engineered to express human enzymes that are involved in metabolically activating a promutagen *(36)*.

2.3. Restoring Function to Tumor Suppressor Cancer Mutants

Yeast provides powerful strategies to perform structure–function analyses *(6)* that can be easily applied to the question of whether function can be restored to tumor suppressor cancer mutants. Because of excellent p53 yeast assays, the high frequency of *p53* gene mutations in human cancers, and the unique pattern of missense mutations for the *p53* gene that affect the core DNA-binding domain *(29)*, p53 is an ideal subject for such studies. p53 cancer missense mutants are full-length and, contrary to wild-type p53, have a long half-life and accumulate to high levels in cancer cells *(37,38)*, probably because of their inability to transactivate the *MDM2* gene that is part of a negative feedback loop for p53 *(39)*. If a small molecule could restore function to this large pool of inactive p53 protein, the impact on cancer therapy would be considerable *(40)*. As a proof of principle, small compounds were identified in a drug screen that restored function to at least some p53 cancer mutants *(41)*.

In an alternative strategy, common *p53* cancer mutations have been investigated for intragenic second-site suppressor mutations that can overcome the deleterious effects of the cancer mutations. PCR-mediated mutagenesis was combined with gap repair of a *p53* expression plasmid through in-vivo homologous recombination *(42)*. Such suppressor mutations were indeed identified, and a comprehensive study for the eight most common *p53* cancer mutations is currently underway (T. Baroni, T. Wang, H. Qian, and R. K. Brachmann, unpublished data). The long-term goal of these studies is the identification of a suppressor motif that suppresses a significant subset of *p53* cancer mutations, determination of the suppressor mechanism through crystallographic studies of p53 double mutants (one cancer and one suppressor mutation), and eventual re-creation of the suppressor mechanism through small molecules.

3. Yeast Assays to Investigate the Biology of Tumor Suppressor Proteins
3.1. Improving on Natural Tumor Suppressor Proteins

The idea of improving on the function of a tumor suppressor protein is an attractive one. It may be possible to exploit such an "optimized" tumor suppressor protein for therapeutic purposes. An "optimized" version may also turn into a great tool to better understand the biology of a tumor suppressor protein. For example, a p53 protein that is more efficient in inducing apoptosis may be of interest for gene therapy approaches, but may also lead to the identification of additional important apoptosis pathways that p53 triggers. In pursuit of p53 proteins with altered DNA-binding specificities, Freeman and colleagues used PCR mutagenesis of the *p53* cDNA followed by in-vivo gap repair and identified two interesting mutants, Ser121Phe and Thr123Ala *(43)*. Ser121Phe was later shown to induce apoptosis and activate certain genes more efficiently than wild-type p53 *(44)*. Inga and colleagues improved further on the basic selection scheme by expressing p53 under the control of the inducible *GAL* promoter, which results in very low to very high expression of p53 depending on the yeast growth conditions *(45)*.

3.2. Dissecting the Requirements for the Activity of Tumor Suppressor Proteins in Yeast

Yeast assays can be used to identify cofactors that a tumor suppressor protein requires to be active in yeast. These findings may then lead to the identification of the human homologs of such yeast cofactors.

Using the transcriptional activity of p53 as a readout, one study identified the p53-activating kinase PAK1 *(46)* and another characterized the dependence of p53 activity on the redox condition of yeast cells by studying yeast deletion mutants of the thiore-doxin reductase *TRR1* gene *(47,48)*. p53 transcriptional activity was also used to iden-tify human genomic fragments to which p53 binds by inserting them upstream of the *HIS3* reporter gene and selecting for His$^+$ colonies (yeast enhancer trap system) *(49)*. Further analysis of these genomic fragments led to the identification of several p53 downstream target genes, including *p53AIP1 (50)*.

Once an activity for a tumor suppressor protein is established in yeast, such as the rescue of *ira1* and *ira2* yeast mutants by the GAP-related domain of neurofibromin *(21–23)*, the assay can be used further for structure–function analyses *(51,52)*. This can even be extended to studying the effects of mutations on the yeast protein that were mod-eled after human tumor mutations *(53)*. And finally, this assay can be used to study the inhibitory effects of small molecules *(54)*.

3.3. Studying the Effects of Tumor Suppressor Proteins on Yeast

Often, the function of a tumor suppressor protein is initially only poorly or partially understood. Phenotypes in yeast may provide a first clue to some aspect of the function of a tumor suppressor protein or may explore other activities of a tumor suppressor pro-tein besides those already firmly established. The challenge, however, is to transfer these findings into the biology of mammalian cells.

One approach has been to overexpress a tumor suppressor protein and study its effects on yeast. For example, p53 was found to cause a significant G1 cell cycle arrest when coexpressed with the human cell cycle-regulated protein kinase CDC2Hs *(55)*. Simi-larly, overexpression of cyclin-dependent kinase 4 (Cdk4) in *S. cerevisiae* surprisingly led to a growth arrest that was overcome by the Cdk-inhibitor p16(INK4A) *(56)*. Alter-natively, yeast mutants can be studied whose phenotypes are rescued by a tumor sup-pressor protein. Koerte and colleagues isolated a yeast mutant (*rft1-1*) that required p53 for viability, and it was demonstrated that p53 rescued the *rft1-1* mutation by forming a protein–protein complex with the Rft1 protein *(57)*. The retinoblastoma protein (pRb) was found to interact with hsHec1p, a human protein essential for proper chromosome segregation, and pRb reduced chromosome segregation errors fivefold in yeast that were sustained by a temperature-sensitive allele of *hsHEC1 (58)*.

3.4. Yeast Dissociator Assays to Identify Inhibitors of Tumor Suppressor Proteins

Typically, the function of tumor suppressor proteins is lost in cancers due to gene muta-tions *(59,60)*. However, in any given type of cancer, mutations of a specific tumor suppres-sor gene rarely affect 100% of cases. This may indicate that loss of function of a particular tumor suppressor protein is not an obligatory event in the development of a certain cancer, or, conversely, that loss of function is absolutely essential, but achieved by other means.

For example, a tumor suppressor pathway rather than a tumor suppressor protein may be targeted, as appears to be the case for the retinoblastoma protein *(61)* (*see* Volume I, Chapter 1). Alternatively, a tumor suppressor may be targeted directly, but at the protein and not the DNA level. Examples of this have been described for p53: inactivation by overexpressed MDM2, a physiologic negative regulator of p53, or by the E6 protein of high-risk human papilloma viruses *(62)*. p53 may also be inappropriately exported from the nucleus and sequestered in the cytoplasm *(63)* or stabilized and inactivated in the nucleus *(64)*, two modes of inactivation that are currently not understood at a molecular level.

To find such inhibition pathways for tumor suppressor proteins, the concept of counterselection can be exploited in yeast *(65)*. The interaction of two hybrid proteins or the activity of a one-hybrid protein or a native transcription factor drives the expression of a reporter gene that results in a no-growth phenotype of yeast. Introduction of a protein that interferes with the assay results in decreased expression of the reporter gene and, thus, growth of the yeast cells. This selection scheme was first described as the "reverse two-hybrid and one-hybrid system" using the *URA3* reporter gene *(16,66,67)*, but the no-growth phenotype can also be achieved by using the *CYH2* gene or a cascade consisting of the *Escherichia coli* Tn*10*-encoded *tet*-repressor and the reporter gene *HIS3* *(68,69)*.

This counterselection strategy has been most extensively applied to questions regarding p53 *(16,70)*. In the "p53 dissociator assay," p53 drives the expression of the *URA3* gene, allowing yeast cells to grow on plates lacking uracil. On plates containing 5-Foa, the yeast cells do not grow, because the *URA3* gene product is involved in converting 5-Foa into a toxic substance *(71)*. Introduction of a p53 inhibitor, such as MDM2 or SV40 large T-antigen, into the p53 dissociator assay results in decreased *URA3* expression and, thus, survival on plates containing 5-Foa (**Fig. 3**) *(70)*.

Several successful screens have been performed, and the known negative regulator MDM2 was isolated, thus validating the approach. In addition, 53BP1 was identified, a protein that appears to have beneficial effects on p53 activity in mammalian cells *(72)*. The screens also isolated hADA3, a protein that is part of histone acetyltransferase complexes. Further characterization determined that hADA3 plays an important role in the cascade of posttranscriptional modifications that turn p53 into an active transcription factor *(70)*. The isolation of other proteins, in addition to p53 inhibitors, is likely due to overexpression of these, often truncated, proteins in an artificial assay and is an added benefit, since, for example, hADA3 would not have been isolated in a two-hybrid screen for p53 (T. Wang, T. Kobayashi, and R. K. Brachmann, unpublished results).

The principles of yeast counterselection systems can be applied to many other tumor suppressor-related questions. For example, proteins and small compounds can be identified that disrupt a protein–protein interaction important for the activity of a tumor suppressor protein, or that, on the contrary, inactivate a tumor suppressor protein. In the case of a transcription factor, several mechanisms of inhibition may be identified that might be missed in a two-hybrid screen, such as degradation or posttranslational modifications.

4. Conclusions

Saccharomyces cerevisiae is an extremely versatile organism that can be adapted in many ways to answer questions regarding human tumor suppressor proteins not conserved in evolution and, thus, without a homolog in yeast. Any researcher who is tempted to exploit yeast for these purposes will face three challenges that are sometimes easily

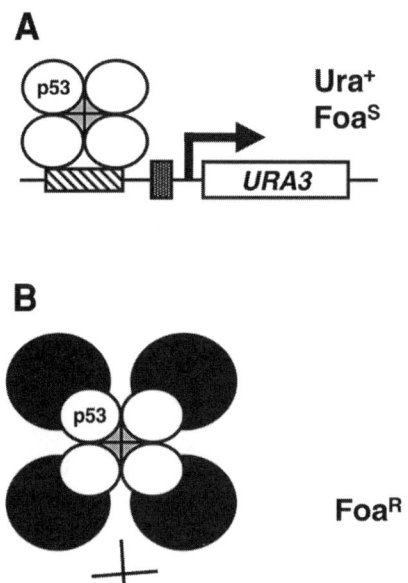

Fig. 3. The concept of counterselection in *S. cerevisiae*. **(A)** In the example of the "p53 dissociator assay," p53 (depicted as the transcriptionally active homo-tetramer) drives the expression of the *URA3* gene, allowing yeast cells to grow on plates lacking uracil (phenotype Ura$^+$). On plates containing 5-Foa, the yeast cells do not grow, because the *URA3* gene product is involved in converting 5-Foa into a toxic substance (phenotype Foa-sensitive or Foa^S). **(B)** Introduction of a p53 inhibitor into the p53 dissociator assay results in decreased *URA3* expression and, thus, survival on plates containing 5-Foa (phenotype Foa-resistant or Foa^R). Novel negative regulators of p53 can be identified in large-scale dissociator screens and may fall into various classes of inhibitors, such as proteins that induce p53 degradation, modify p53 posttranslationally, bind to its transactivation domain or bind to its DNA-binding domain, as shown in the example.

solved, other times never overcome. (a) The yeast assay for a human tumor suppressor protein should reflect, to some extent, the actual tumor suppressing function of the protein studied. (b) If the yeast adaptation for a human tumor suppressor protein is intended to be used for large-scale genetic screens, the problem of false positives in the assay system has to be appropriately addressed. (c) The intriguing findings obtained in a yeast assay have to be validated in mammalian assay systems and put into the context of the known biology of a tumor suppressor protein in a meaningful way. If all three challenges are met, the rewards are likely to be substantial.

References

1. Kolonin, M. G., Zhong, J., and Finley, R. L. (2000) Interaction mating methods in two-hybrid systems. *Meth. Enzymol.* **328**, 26–46.
2. Cagney, G., Uetz, P., and Fields, S. (2000) High-throughput screening for protein-protein interactions using two- hybrid assay. *Meth. Enzymol.* **328**, 3–14.
3. Baudin, A., Ozier-Kalogeropoulos, O., Denouel, A., Lacroute, F., and Cullin, C. (1993) A simple and efficient method for direct gene deletion in *Saccharomyces cerevisiae*. *Nucleic Acids Res.* **21**, 3329–3330.

4. Lorenz, M. C. Muir, R. S., Lim, E., McElver, J., Weber, S. C., and Heitman, J. (1995) Gene disruption with PCR products in Saccharomyces cerevisiae. *Gene* **158**, 113–117.

5. Ma, H., Kunes, S., Schatz, P. J., and Botstein, D. (1987) Plasmid construction by homologous recombination in yeast. *Gene* **58**, 201–216.

6. Muhlrad, D., Hunter, R., and Parker, R. (1992) A rapid method for localized mutagenesis of yeast genes. *Yeast* **8**, 79–82.

7. Burke, D., Dawson, D., and Stearns, T. (2000) *Methods in Yeast Genetics: A Cold Spring Harbor Laboratory Course Manual*, Cold Spring Harbor Laboratory Press, Plainview, NY, 205.

8. Lundblad, V. (1998) *Saccharomyces cerevisiae*, in *Current Protocols in Molecular Biology*, Vol. 2 (Ausubel, F. M., ed.) Wiley, New York, 13.10.11–13.14.17.

9. Fields, S. and Jang, S. K. (1990) Presence of a potent transcription activating sequence in the p53 protein. *Science* **249**, 1046–1049.

10. Kern, S. E., Kinzler, K. W., Bruskin, A., et al. (1991) Identification of p53 as a sequence-specific DNA-binding protein. *Science* **252**, 1708–1711.

11. Scharer, E. and Iggo, R. (1992) Mammalian p53 can function as a transcription factor in yeast. *Nucleic Acids Res.* **20**, 1539–1545.

12. Ishioka, C., Frebourg, T., Yan, Y. X., et al. (1993) Screening patients for heterozygous p53 mutations using a functional assay in yeast. *Nat. Genet.* **5**, 124–129.

13. Flaman, J. M., Frebourg, T., Moreau, V., et al. (1995) A simple p53 functional assay for screening cell lines, blood, and tumors. *Proc. Natl. Acad. Sci. USA* **92**, 3963–3967.

14. Robert, V., Michel, P., Flaman, J. M., et al. (2000) High frequency in esophageal cancers of p53 alterations inactivating the regulation of genes involved in cell cycle and apoptosis. *Carcinogenesis* **21**, 563–565.

15. Di Como, C. J. and Prives, C. (1998) Human tumor-derived p53 proteins exhibit binding site selectivity and temperature sensitivity for transactivation in a yeast-based assay. *Oncogene* **16**, 2527–2539.

16. Brachmann, R. K., Vidal, M., and Boeke, J. D. (1996) Dominant-negative p53 mutations selected in yeast hit cancer hot spots. *Proc. Natl. Acad. Sci. USA* **93**, 4091–4095.

17. Inga, A., Cresta, S., Monti, P., et al. (1997) Simple identification of dominant p53 mutants by a yeast functional assay. *Carcinogenesis* **18**, 2019–2021.

18. Marutani, M., Tonoki, H., Toda, M., et al. (1999) Dominant-negative mutations of the tumor suppressor p53 relating to early onset of glioblastoma multiforme. *Cancer Res.* **59**, 4765–4769.

19. Monteiro, A. N., August, A., and Hanafusa, H. (1996) Evidence for a transcriptional activation function of BRCA1 C-terminal region. *Proc. Natl. Acad. Sci. USA* **93**, 13595–13599.

20. Humphrey, J. S., Salim, A., Erdos, M. R., Collins, F. S., Brody, L. C., and Klausner, R. D. (1997) Human BRCA1 inhibits growth in yeast: potential use in diagnostic testing. *Proc. Natl. Acad. Sci. USA* **94**, 5820–5825.

21. Xu, G. F., Lin, B., Tanaka, K., et al. (1990) The catalytic domain of the neurofibromatosis type 1 gene product stimulates ras GTPase and complements ira mutants of *S. cerevisiae*. *Cell* **63**, 835–841.

22. Martin, G. A., Viskochil, D., Bollag, G., et al. (1990) The GAP-related domain of the neurofibromatosis type 1 gene product interacts with ras p21. *Cell* **63**, 843–849.

23. Ballester, R., Marchuk, D., Boguski, M., et al. (1990) The NF1 locus encodes a protein functionally related to mammalian GAP and yeast IRA proteins. *Cell* **63**, 851–859.

24. Ishioka, C., Ballester, R., Engelstein, M., et al. (1995) A functional assay for heterozygous mutations in the GTPase activating protein related domain of the neurofibromatosis type 1 gene. *Oncogene* **10**, 841–847.

25. Ishioka, C., Suzuki, T., FitzGerald, M., et al. (1997) Detection of heterozygous truncating mutations in the BRCA1 and APC genes by using a rapid screening assay in yeast. *Proc. Natl. Acad. Sci. USA* **94**, 2449–2453.

26. Furuuchi, K., Tada, M., Yamada, H., et al. (2000) Somatic mutations of the APC gene in primary breast cancers. *Am. J. Pathol.* **156**, 1997–2005.

27. Zhang, C. L., Tada, M., Kobayashi, H., Nozaki, M., Moriuchi, T., and Abe, H. (2000) Detection of PTEN nonsense mutation and psiPTEN expression in central nervous system high-grade astrocytic tumors by a yeast-based stop codon assay. *Oncogene* **19**, 4346–4353.

28. Beroud, C. and Soussi, T. (1998) p53 gene mutation: software and database. *Nucleic Acids Res* **26**, 200–204.

29. Hernandez-Boussard, T., Rodriguez-Tome, P., Montesano, R., and Hainaut, P. (1999) IARC p53 mutation database: a relational database to compile and analyze p53 mutations in human tumors and cell lines. International Agency for Research on Cancer. *Hum. Mutat.* **14**, 1–8.

30. Hussain, S. P. and Harris, C. C. (1998) Molecular epidemiology of human cancer: contribution of mutation spectra studies of tumor suppressor genes. *Cancer Res.* **58**, 4023–4037.

31. Fronza, G., Inga, A., Monti, P., et al. (2000) The yeast p53 functional assay: a new tool for molecular epidemiology. Hopes and facts. *Mutat. Res.* **462**, 293–301.

32. Inga, A., Iannone, R., Monti, P., et al. (1997) Determining mutational fingerprints at the human p53 locus with a yeast functional assay: a new tool for molecular epidemiology. *Oncogene* **14**, 1307–1313.

33. Moshinsky, D. J. and Wogan, G. N. (1997) UV-induced mutagenesis of human p53 in a vector replicated in *Saccharomyces cerevisiae*. *Proc. Natl. Acad. Sci. USA* **94**, 2266–2271.

34. Murata, J., Tada, M., Iggo, R. D., Sawamura, Y., Shinohe, Y., and Abe, H. (1997) Nitric oxide as a carcinogen: analysis by yeast functional assay of inactivating p53 mutations induced by nitric oxide. *Mutat. Res.* **379**, 211–218.

35. Epstein, C. B., Attiyeh, E. F., Hobson, D. A., Silver, A. L., Broach, J. R., and Levine, A. J. (1998) p53 mutations isolated in yeast based on loss of transcription factor activity: similarities and differences from p53 mutations detected in human tumors. *Oncogene* **16**, 2115–2122.

36. Sengstag, C., Morbe, J. L., and Weibel, B. (1999) Codon 249 of the human TP53 tumor suppressor gene is no hot spot for aflatoxin B1 in a heterologous background. *Mutat. Res.* **430**, 131–144.

37. Lowe, S. W. (1995) Cancer therapy and p53. *Curr. Opin. Oncol.* **7**, 547–553.

38. Donehower, L. A. and Bradley, A. (1993) The tumor suppressor p53. *Biochim. Biophys. Acta* **1155**, 181–205.

39. Lane, D. P. and Hall, P. A. (1997) MDM2—arbiter of p53's destruction. *Trends Biochem. Sci.* **22**, 372–374.

40. Gibbs, J. B. and Oliff, A. (1994) Pharmaceutical research in molecular oncology. *Cell* **79**, 193–198.

41. Foster, B. A., Coffey, H. A., Morin, M. J., and Rastinejad, F. (1999) Pharmacological rescue of mutant p53 conformation and function. *Science* **286**, 2507–2510.

42. Brachmann, R. K., Yu, K., Eby, Y., Pavletich, N. P., and Boeke, J. D. (1998) Genetic selection of intragenic suppressor mutations that reverse the effect of common p53 cancer mutations. *EMBO J.* **17**, 1847–1859.

43. Freeman, J., Schmidt, S., Scharer, E., and Iggo, R. (1994) Mutation of conserved domain II alters the sequence specificity of DNA binding by the p53 protein. *EMBO J.* **13**, 5393–5400.

44. Saller, E., Tom, E., Brunori, M., et al. (1999) Increased apoptosis induction by 121F mutant p53. *EMBO J.* **18**, 4424–4437.

45. Inga, A., Monti, P., Fronza, G., Darden, T., and Resnick, M. A. (2001) p53 mutants exhibiting enhanced transcriptional activation and altered promoter selectivity are revealed using a sensitive, yeast-based functional assay. *Oncogene* **20**, 501–513.

46. Thiagalingam, S., Kinzler, K. W., and Vogelstein, B. (1995) PAK1, a gene that can regulate p53 activity in yeast. *Proc. Natl. Acad. Sci. USA* **92**, 6062–6066.

47. Merrill, G. F., Dowell, P., and Pearson, G. D. (1999) The human p53 negative regulatory domain mediates inhibition of reporter gene transactivation in yeast lacking thioredoxin reductase. *Cancer Res.* **59**, 3175–3179.

48. Pearson, G. D. and Merrill, G. F. (1998) Deletion of the *Saccharomyces cerevisiae* TRR1 gene encoding thioredoxin reductase inhibits p53-dependent reporter gene expression. *J. Biol. Chem.* **273**, 5431–5434.

49. Tokino, T., Thiagalingam, S., El-Deiry, W. S., Waldman, T., Kinzler, K. W., and Vogelstein, B. (1994). p53 tagged sites from human genomic DNA. *Hum. Mol. Genet.* **3**, 1537–1542.

50. Oda, K., Arakawa, H., Tanaka, T., et al. (2000) p53AIP1, a potential mediator of p53-dependent apoptosis, and its regulation by Ser-46-phosphorylated p53. *Cell* **102**, 849–862.

51. Gutmann, D. H., Boguski, M., Marchuk, D., Wigler, M., Collins, F. S., and Ballester, R. (1993) Analysis of the neurofibromatosis type 1 (NF1) GAP-related domain by site-directed mutagenesis. *Oncogene* **8**, 761–769.

52. Mori, S., Satoh, T., Koide, H., Nakafuku, M., Villafranca, E., and Kaziro, Y. (1995) Inhibition of Ras/Raf interaction by anti-oncogenic mutants of neurofibromin, the neurofibromatosis type 1 (NF1) gene product, in cell- free systems. *J. Biol. Chem.* **270**, 28834–28838.

53. Gil, R. and Seeling, J. M. (1999) Characterization of *Saccharomyces cerevisiae* strains expressing ira1 mutant alleles modeled after disease-causing mutations in NF1. *Mol. Cell Biochem.* **202**, 109–118.

54. Golubic, M., Tanaka, K., Dobrowolski, S., et al. (1991) The GTPase stimulatory activities of the neurofibromatosis type 1 and the yeast IRA2 proteins are inhibited by arachidonic acid. *EMBO J.* **10**, 2897–2903.

55. Nigro, J. M., Sikorski, R., Reed, S. I., and Vogelstein, B. (1992) Human p53 and CDC2Hs genes combine to inhibit the proliferation of Saccharomyces cerevisiae. *Mol. Cell. Biol.* **12**, 1357–1365.

56. Moorthamer, M., Panchal, M., Greenhalf, W., and Chaudhuri, B. (1998) The p16(INK4A) protein and flavopiridol restore yeast cell growth inhibited by Cdk4. *Biochem. Biophys. Res. Commun.* **250**, 791–797.

57. Koerte, A., Chong, T., Li, X., Wahane, K., and Cai, M. (1995) Suppression of the yeast mutation rft1-1 by human p53. *J. Biol. Chem.* **270**, 22556–22564.

58. Zheng, L., Chen, Y., Riley, D. J., Chen, P. L., and Lee, W. H. (2000) Retinoblastoma protein enhances the fidelity of chromosome segregation mediated by hsHec1p. *Mol. Cell. Biol.* **20**, 3529–3537.

59. Macleod, K. (2000) Tumor suppressor genes. *Curr. Opin. Genet. Dev.* **10**, 81–93.

60. Haber, D. and Harlow, E. (1997) Tumour-suppressor genes: evolving definitions in the genomic age. *Nat. Genet.* **16**, 320–322.

61. Sellers, W. R. and Kaelin, W. G., Jr. (1997) Role of the retinoblastoma protein in the pathogenesis of human cancer. *J. Clin. Oncol.* **15**, 3301–3312.

62. Vogelstein, B., Lane, D., and Levine, A. J. (2000) Surfing the p53 network. *Nature* **408**, 307–310.

63. Stommel, J. M., Marchenko, N. D., Jimenez, G. S., Moll, U. M., Hope, T. J., and Wahl, G. M. (1999) A leucine-rich nuclear export signal in the p53 tetramerization domain: regulation of subcellular localization and p53 activity by NES masking. *EMBO J.* **18**, 1660–1672.

64. Lutzker, S. G. and Levine, A. J. (1996) A functionally inactive p53 protein in teratocarcinoma cells is activated by either DNA damage or cellular differentiation. *Nat. Med.* **2**, 804–810.

65. Sikorski, R. S. and Boeke, J. D. (1991) In vitro mutagenesis and plasmid shuffling: from cloned gene to mutant yeast. *Meth. Enzymol.* **194**, 302–318.

66. White, M. A. (1996) The yeast two-hybrid system: forward and reverse. *Proc. Natl. Acad. Sci. USA* **93**, 10001–10003.

67. Vidal, M., Brachmann, R. K., Fattaey, A., Harlow, E., and Boeke, J. D. (1996) Reverse two-hybrid and one-hybrid systems to detect dissociation of protein-protein and DNA-protein interactions. *Proc. Natl. Acad. Sci. USA* **93**, 10315–10320.

68. Shih, H. M., Goldman, P. S., De Maggio, A. J., Hollenberg, S. M., Goodman, R. H., and Hoekstra, M. F. (1996) A positive genetic selection for disrupting protein-protein interactions: identification of CREB mutations that prevent association with the coactivator CBP. *Proc. Natl. Acad. Sci. USA* **93**, 13896–13901.

69. Leanna, C. A. and Hannink, M. (1996) The reverse two-hybrid system: a genetic scheme for selection against specific protein/protein interactions. *Nucleic Acids Res.* **24**, 3341–3347.

70. Wang, T., Kobayashi, T., Takimoto, R., et al. (2001) hADA3 is required for p53 activity. *EMBO J.* **20**, 6404–6413.

71. Boeke, J. D., LaCroute, F., and Fink, G. R. (1984) A positive selection for mutants lacking orotidine-5'-phosphate decarboxylase activity in yeast: 5-fluoro-orotic acid resistance. *Mol. Gen. Genet.* **197**, 345–346.

72. Iwabuchi, K., Li, B., Massa, H. F., Trask, B. J., Date, T., and Fields, S. (1998) Stimulation of p53-mediated transcriptional activation by the p53-binding proteins, 53BP1 and 53BP2. *J. Biol. Chem.* **273**, 26061–26068.

7

Electrophoretic Mobility Shift Analysis of the DNA Binding of Tumor Suppressor Gene Products

Edward C. Thornborrow, Matthew Maurer, and James J. Manfredi

1. Introduction

Central to the tumor suppressor activity of certain proteins is the ability to interact physically with DNA. A well-studied example of this is the tumor suppressor p53 *(1,2)*. The p53 protein has been implicated in several diverse growth-related pathways, including apoptosis and cell cycle arrest *(3,4)* and the *p53* gene is mutated in the majority of human cancers *(5,6)*. At its amino-terminus, the protein contains a potent transcriptional activation domain *(7)* which is linked to a central core domain that mediates sequence-specific DNA binding *(8–10)*. Both of these domains have been shown to be important for p53-mediated growth suppression *(11)*. The relevance of the DNA-binding domain to the tumor suppressor activity of p53 is further highlighted by the fact that the major sites of mutation found in human cancers are localized to this region *(12)*. Several of these mutations have been shown to abolish the ability of p53 to function as a transcriptional activator *(13–15)*. A DNA consensus sequence through which p53 binds and activates transcription has been identified. This sequence consists of two palindromic decamers of 5'-RRRCWWGYYY-3' (where R is a purine, Y is a pyrimidine, and W is an adenine or thymine) separated by 0–13 bp *(16–18)*. Through sequences similar to this consensus, p53 has been shown to activate the transcription of many genes, including *bax, p21, mdm2, gadd45, IGF-BP3,* and *cyclin G (19–26)*, which leads to the various physiologic outcomes that contribute to the ability of p53 to function as a tumor suppressor.

Several experimental methods have been developed to investigate the DNA-binding properties of proteins, including electrophoretic mobility shift assays (EMSA) *(27,28)*, DNA footprinting (this volume), and chromatin immunoprecipitation (this volume). EMSA is popular as a technically easy, rapid, and inexpensive in-vitro assay that can provide several lines of information concerning the DNA-binding properties of a protein. Using this method, one can address questions of binding affinity and specificity, as well as analyze the DNA sequence determinants for protein binding. In fact, EMSA was an integral technique used in identifying the DNA-binding domain of the tumor suppressor p53 *(8–10)*.

In EMSA, a radiolabeled oligonucleotide (probe) of interest, typically less than 200 bp, is incubated with the protein under investigation. Native polyacrylamide gel elec-

From: *Methods in Molecular Biology, Vol. 223: Tumor Suppressor Genes: Regulation, Function, and Medicinal Applications.* Edited by: Wafik S. El-Deiry © Humana Press Inc., Totowa, NJ

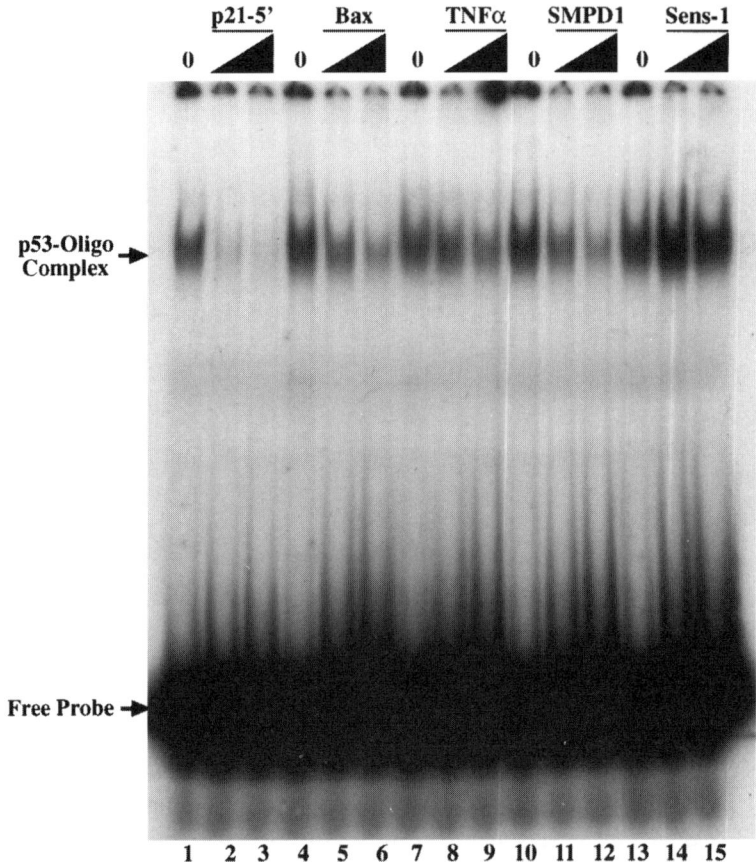

Fig. 1. p53 demonstrates sequence-specific binding to putative response elements in the *TNFα* and *SMPD1* promoters. A radiolabeled probe corresponding to the upstream p53 response element in the human *p21* gene (p21-5′) was incubated with purified p53 protein in the absence of competitor (lanes 1, 4, 7, 10, and 13), in the presence of a 17- or 42-fold molar excess of unlabeled probe (lanes 2 and 3, respectively), or in the presence of a 200- (lanes 5, 8, 11, and 14) or 500-fold (lanes 6, 9, 12, and 15) molar excess of unlabeled oligonucleotide corresponding to the p53 response element of the human *Bax* promoter (lanes 5 and 6) or to putative p53 response elements in the *TNFα* (lanes 8 and 9) and *SMPD1* (lanes 11 and 12) promoters. The Sens-1 oligonucleotide (lanes 13–15) represents an unrelated nonspecific competitor. The p53–DNA complex and free probe are shown.

trophoresis (PAGE) is used subsequently to separate probe bound to protein from free probe. Due to its relatively small size and high negative charge, free probe migrates rapidly through the gel. Probe bound by protein, however, is retarded in the gel due to the larger size of the complex. When the gel is exposed to autoradiography film, the free probe is visualized as a band at the bottom of the gel, while protein-bound probe produces a "band shift" of slower mobility (**Figs. 1** and **2**).

Issues of specificity and affinity in DNA binding are easily addressed with this technique using competition assays. For this purpose, EMSA reactions are conducted in the presence of an excess of a second, nonradiolabeled oligonucleotide (competitor). If the competitor also is capable of binding the protein in question, it will compete the radiolabeled probe. Less radiolabeled probe will be complexed by protein and retarded in the gel, producing a band shift of lighter intensity on the autoradiography film (**Fig. 1,** com-

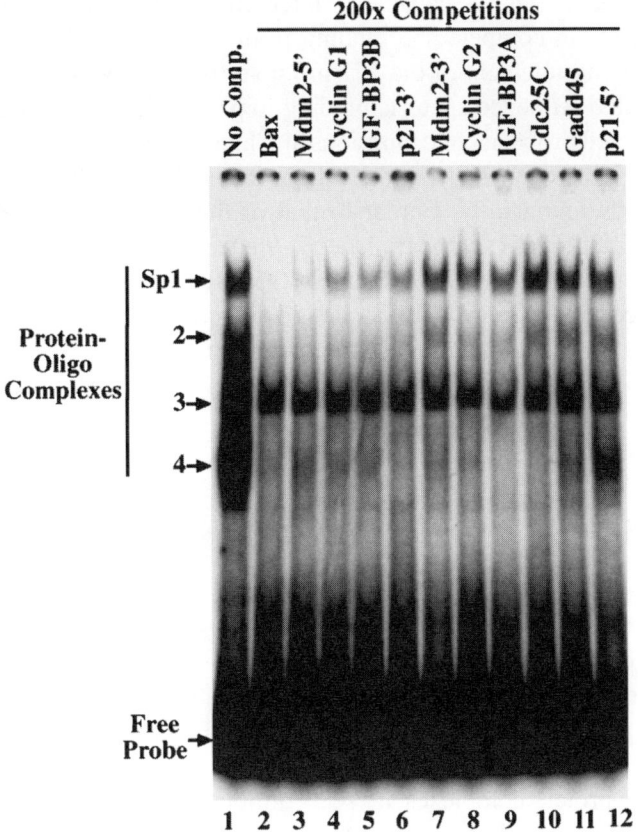

Fig. 2. Sp1 demonstrates marked specificity for the p53 response element of the human *Bax* promoter. HeLa cell nuclear extract was incubated with a radiolabeled probe corresponding to the p53 response element of the human *bax* promoter in the absence of competitor (lane 1) or the presence of a 200-fold molar excess of oligonucleotide corresponding to the p53 response elements of the indicated genes (lanes 2–12). Four distinct complexes are observed, indicated by the arrows. The top complex was shown to contain Sp1 by antibody super-shift *(29)*.

pare lane 1 with lanes 2–3). In this way, multiple oligonucleotides can be compared for their affinity to a particular protein (**Fig. 1**). The higher the affinity of the oligonucleotide for the protein under study, the fewer molar equivalents will be needed to achieve complete competition of the radiolabeled probe (**Fig. 1,** compare lanes 2–3 to lanes 5–6, 8–9, and 11–12). The DNA sequence determinants for protein binding can be determined this way by comparing the affinities of variously mutated oligonucleotides. The specificity of a DNA–protein interaction also can be demonstrated in a similar fashion. For the latter, a competing oligonucleotide is used that is a mutated version of the probe or a completely unrelated oligonucleotide. If the original protein–probe band-shift complex represents a specific interaction then neither the mutated nor the unrelated oligonucleotide will compete the radiolabeled probe, and no decrease in the band-shift intensity is observed (**Fig. 1,** compare lane 13 to lanes 14–15).

An example of a competition assay is shown in **Fig. 1**. Sequence analysis of the promoters for the *TNFα* and acid sphingomyelinase (*SMPD1*) genes demonstrated two regions that closely resemble the DNA-binding consensus sequence for p53. To determine if these two elements are capable of interacting with p53 in a sequence-specific manner, their ability to compete p53 binding from a radiolabeled probe corresponding to the 5′ p53 response element of the *p21* promoter (p21-5′) was examined. These elements demonstrated sequence-specific binding to p53 as shown by their ability to compete the radiolabeled probe at concentrations at which a nonspecific competitor, Sens-1, did not (**Fig. 1,** compare lanes 7–9 and 10–12 to lanes 13–15). The affinity of these elements for p53 was found to be similar to that of the p53 response element from the human *bax* promoter (**Fig. 1,** compare lanes 7–9 and 10–12 to lanes 4–6), which is significantly less than that of the p21-5′ element (**Fig. 1,** compare lanes 4–6 to lanes 1–3). It should be noted that in-vitro DNA binding does not necessarily correlate with functionality in cells. As an example of this, neither of these elements was capable of conferring p53 responsiveness on a heterologous promoter when examined in transient transfection assays (data not shown).

In addition to studies with purified protein, EMSA can be used to analyze the DNA-binding properties of proteins present in a crude extract. In this instance, it is not uncommon to observe multiple band shifts for a given oligonucleotide probe. This is demonstrated in **Fig. 2.** Previously we identified an Sp1-binding site contained within the p53 response element of the human *bax* promoter *(29)*. When HeLa nuclear extract is analyzed by EMSA with a radiolabeled probe corresponding to this *bax* element, four protein complexes are observed, including the previously identified Sp1 complex (**Fig. 2,** lane 1). Demonstrating an overlap between p53- and Sp1-binding sites, a subset of p53 response elements displayed various degrees of affinity for Sp1 as seen by their ability to compete the radiolabeled probe for binding to p53 (**Fig. 2**).

When multiple protein complexes are seen, antibody specific for the protein of interest can be added to the EMSA reaction to identify which band shift corresponds to a probe complex containing that particular protein. The addition of the antibody will further retard the mobility of the protein–DNA complex in the gel, producing a "super-shifted" complex (**Fig. 3**), if binding of the antibody does not inhibit the DNA-binding activity of the protein. Should the binding of the antibody block the interaction of the protein with DNA, inclusion of the antibody in the EMSA reaction will cause the original band shift to decrease in intensity or to disappear. This typically occurs when the epitope for the antibody resides within the DNA-binding region of the protein. In addition to aiding in the identification of a particular protein complex, antibodies also have been used to investigate conformational variations of protein bound to DNA. An example of this is the use of mAb421, a C-terminal specific monoclonal antibody to p53, to distinguish between two subsets of p53 response elements. This antibody has been shown to enhance the binding of p53 to one class of response elements, while inhibiting or having no effect on the binding of p53 to another class of elements *(30,31)*. Such a result is demonstrated in **Fig. 3.** We recently identified two p53 response elements in the murine *bax* gene, one in the promoter and one in the first intron (unpublished data). To determine the effects of mAb421 on the ability of p53 to bind these two elements, a super-shift EMSA with purified p53 was performed using either the intronic element (**Fig. 3,** lanes 1 and 2) or the promoter element (**Fig. 3,** lanes 3 and 4) as a radiolabeled probe. The 421 antibody effectively super-shifted and enhanced the binding of p53 to the intronic element (**Fig. 3,** compare lanes 1 and 2). This antibody, however, failed to either super-shift or enhance the binding to the promoter element (**Fig. 3,** compare lanes 3 and 4), suggesting that the p53 bound to

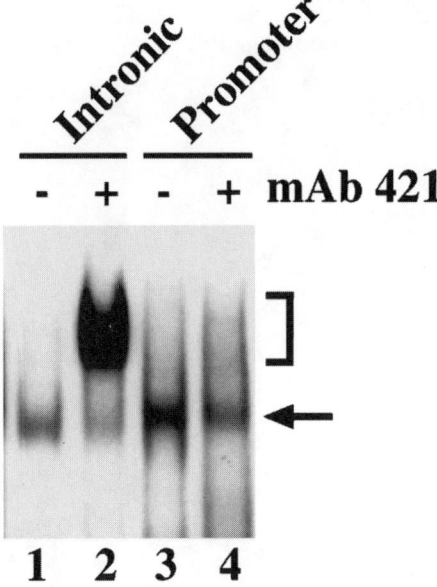

Intronic *Promoter*

- + - + **mAb 421**

1 2 3 4

Fig. 3. Monoclonal antibody 421 enhances the binding of p53 to the intronic response element but inhibits the binding of p53 to the promoter element from the murine *bax* gene. An electrophoretic mobility shift assay was performed, incubating 50 ng of purified p53 with 3 ng of radiolabeled probe corresponding to either the intronic (lanes 1 and 2) or promoter (lanes 3 and 4) p53 response elements of the murine *bax* gene in the absence (lanes 1 and 3) or presence of monoclonal antibody 421 (lanes 2 and 4). The arrow indicates the position of the p53–DNA complex and the bracket indicates the position of the super-shifted antibody–p53–DNA complex.

this element is in a distinct conformation in which the 421 epitope is not accessible to the antibody.

Use of the EMSA technique in conjunction with proteolytic digestion can be used to examine the contributions of various domains within a protein to its overall DNA-binding properties. An example of this approach is shown in **Fig. 4**. The p53 protein was subjected to partial proteolysis by thermolysin and its DNA-binding activity to its two response elements within the *p21* promoter (p21 5′ and p21 3′) was examined (**Fig. 4A**). The p53–DNA complex without digestion showed similar binding to both sites (**Fig. 4A,** compare lanes 1 and 9). Digestion with thermolysin, however, produced species of p53 that bound significantly better to the p21 5′ element as compared to the 3′ element (**Fig. 4A,** compare lanes 5–8 and 13–16). This suggests that thermolysin digestion is removing a portion of p53 that contributes to the ability of the protein to bind the p21 3′ response element, and that distinct regions of the protein contribute to the ability of p53 to distinguish between different response elements.

Here is presented a protocol for electrophoretic mobility shift assay that was derived from the methods originally presented by Lasser et al. *(32)* and Funk et al. *(17)*. This protocol has been optimized for recombinant baculovirus expressing His-tagged human p53 (**Fig. 1**), but has been used effectively with other recombinant proteins as well as with crude nuclear extracts (**Fig. 2**). As such, this protocol represents a good starting point, but as with all new applications of a technique, certain parameters may require optimization (*see* **Note 1**).

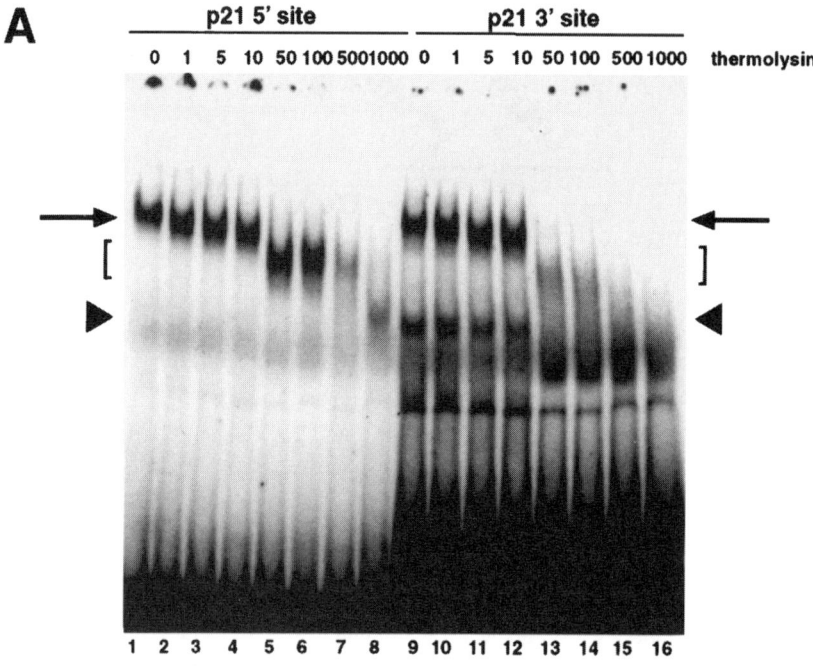

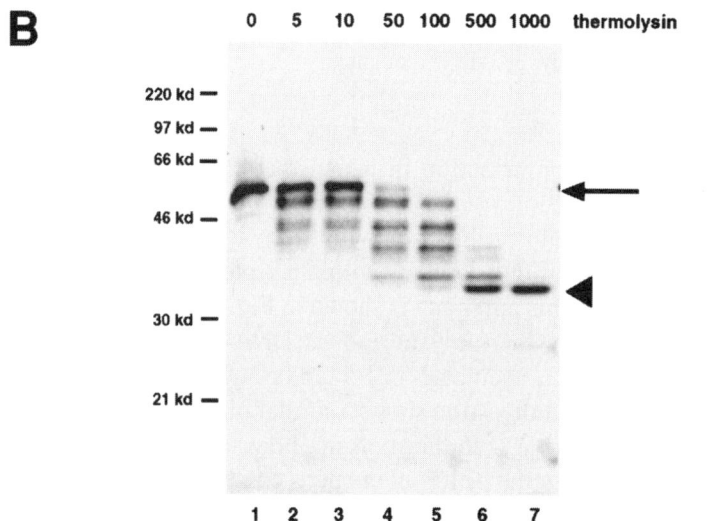

Fig. 4. A thermolysin-resistant domain of p53 has higher affinity for the 5' response element from the *p21* promoter as compared to the 3' element. **(A)** Three nanograms of either p21 5' (lanes 1–8) or p21 3'-labeled probe (lanes 9–16) was incubated with 10 ng of purified p53. After 10 min, 0–1000 ng of thermolysin was added, and samples were incubated for an additional 10 min at room temperature. The digestion was stopped with the thermolysin inhibitor phosphoramidon. Samples were electrophoresed as normal. The long arrow indicates the position of the nondigested p53-DNA complex, whereas the arrowhead indicates the position of the thermolysin-digested p53-DNA complex. **(B)** The EMSA reactions described in **(A)** were also analyzed by SDS-PAGE and immunoblot for p53, using the core domain-specific antibody mAb 240. Lanes 1–7 show the progressive digestion of p53 by thermolysin from the intact protein (lane 1, long arrow) to a domain that is apparently resistant to further digestion (lane 7, arrowhead). Molecular weights are noted to the left.

2. Materials

2.1. Nuclear Extraction

1. Lysis buffer: 20 mM HEPES (pH 7.5), 20% (v/v) glycerol, 10 mM NaCl, 1.5 mM MgCl$_2$, 0.2 mM EDTA, 0.1% (v/v) Triton X-100, 1 mM dithiothreitol (DTT), 1 mM phenylmethylsulfonyl fluoride (PMSF), 50 μM leupeptin, 50 μg/mL aprotinin. Make as needed and keep at ≤4°C.
2. Nuclear extraction buffer (NEB): Lysis buffer with 500 mM NaCl.
3. 1× Phosphate-buffered saline (PBS).

2.2. Oligonucleotide Probe Preparation

1. 10× Klenow buffer.
2. DNA Polymerase I (Klenow).
3. [a-^{32}P] TTP (10 mCi/mL).
4. [a-^{32}P] ATP (10 mCi/mL).
5. 3 M sodium acetate (pH 5.2).
6. 100% ethanol.
7. 70% ethanol.

2.3. Quantification of Radiolabeled Probe by TCA Precipitation

1. 5% Trichloroacetic acid (TCA).
2. 0.1 M tetrasodium pyrophosphate.
3. Salmon sperm DNA (1 mg/mL): Dissolve 100 mg salmon sperm DNA in approximately 80 mL of 20 mM Tris-HCl (pH 7.5), 10 mM EDTA. Boil for 10 min with stirring and cool to room temperature. Bring up to a final volume of 100 mL and store at 4°C.
4. GF/B glass microfiber filters (Whatman).
5. 1% TCA.
6. ScintiVerse (Fisher Scientific).

2.4. Electrophoretic Mobility Shift Assay

1. 23% Acrylamide/0.5% bis-acrylamide (electrophoresis grade). Filter through 0.2-μm filter and store in a dark glass bottle at 4°C. Unpolymerized acrylamide is neurotoxic. When working with acrylamide crystals the use of a face mask is highly recommended, and when working with acrylamide in any form one should always wear gloves.
2. 10× Tris/boric acid/EDTA (TBE) buffer.
3. 10% (w/v) ammonium persulfate in water. Store at 4°C for <1 mo.
4. N, N, N′, N′-tetramethylethylenediamine (TEMED).
5. 5× EMSA buffer: 100 mM HEPES (pH 7.5), 0.5 mM EDTA, 50% (v/v) glycerol, 10 mM MgCl$_2$. Store at 4°C.
6. 40 mM spermidine: Dissolve 0.1 g spermidine trihydrochloride in 10 mL water. Aliquot as 1 mL and store at −20°C.
7. 10 mM dithiothreitol (DTT). Store at −20°C.
8. 500 μg/mL poly(dI-dC): Dissolve 10 A$_{260}$ units of poly(dI-dC) · poly(dI-dC) (Pharmacia, approximately 500 μg) in 1 mL water. Store at −20°C.
9. 30 mg/mL Bovine serum albumin (BSA). Store at 4°C.
10. 1× Sample buffer: 0.04% (w/v) bromophenol blue, 0.04% (w/v) xylene cyanol FF, 5% (v/v) glycerol.
11. 0.5× TBE
12. 3 MM Whatman chromatography paper.
13. Autoradiography film, cassette, and intensifying screen.

3. Methods

3.1. Preparation of Nuclear Extracts

The protocol listed here is a high-salt extraction of isolated nuclei that has been used successfully to prepare nuclear extracts from both eukaryotic tissue culture cells as well as from Sf9 insect cells. Following nuclear extraction, a protein of interest can be purified further using various chromatographic techniques. For EMSA analysis, however, the protein sample ultimately should be in nuclear extraction buffer (NEB). If additional purification steps are taken, therefore, the final protein sample should be dialyzed against NEB. Unless otherwise indicated, all reagents and samples should be maintained on ice or at 4°C.

1. Remove the medium from an almost-confluent 100-mm tissue culture dish of adherent cells.
2. Wash dish with 5 mL of 1× PBS (for Sf9 or other suspension culture cells, wash cell pellet with 5× pellet volume of 1× PBS).
3. Repeat PBS wash two additional times, removing as much PBS as possible after the final wash.
4. Add 1 mL Lysis buffer to the plate (for suspension cells resuspend pellet with 5× original cell pellet volume of lysis buffer).
5. With a cell lifter, detach the cells from the bottom of the dish by scraping.
6. Transfer the cell slurry to a tube and centrifuge at 500 × *g* for 5 min.
7. Aspirate supernatant and resuspend pellet in 200 μL of nuclear extraction buffer (for suspension cells use 3× original cell pellet volume).
8. Rock sample end over end for 1 h.
9. Centrifuge for 10 min at 18,300 × *g*.
10. Collect supernatant as nuclear extract and freeze in liquid nitrogen. Store at −70°C.

3.2. Oligonucleotide Probe Preparation

Due to their rapid and inexpensive production, synthetic oligonucleotides have become a popular source of EMSA probes. Complimentary single-stranded oligonucleotides are produced and annealed to form double-stranded oligonucleotides for use as EMSA probes and competitors. Oligonucleotides for use as probes should be constructed with 5′-overhangs as shown below:

5′-AATTNNNNNNNNNNNNN
3′-NNNNNNNNNNNNNNTTAA-5′

Klenow enzyme then is used to fill in the overhangs with radiolabeled nucleotides. Alternatively, blunt-ended oligonucleotide probes can be 5′-end-labeled with polynucleotide kinase. Probes generated by this latter method, however, are significantly reduced in their cpm. Below is a protocol for Klenow labeling of double-stranded oligonucleotides with 5′-overhangs. This protocol typically yields probes of ~1,000,000 cpm/μL. As with all procedures involving radioactivity, appropriate safety precautions and disposal protocols should be followed.

1. Mix equal amounts of "top" and "bottom" single-stranded oligonucleotides in water to produce a stock solution of 20 pmol double-stranded oligonucleotide per microliter (*see* **Note 2**).
2. Heat the solution at 95°C for 4 min, 65°C for 10 min, and then slowly cool by turning the heat block off and allowing it to cool to room temperature. (Store at −20°C.)

3. Prepare the following mixture: 2 μL 10× Klenow buffer, 2 μL annealed double-stranded oligonucleotide (20 pmol/μL), 5 μL $[a\text{-}^{32}P]$ TTP (10 mCi/mL), 5 μL $[a\text{-}^{32}P]$ ATP (10 mCi/mL), 5 μL H_2O, 1 μL Klenow enzyme (1–5 U/μL).

4. Incubate at room temperature for 15 min. Due to the 3′-to-5′ exonuclease activity of Klenow enzyme, extended incubation times can lead to the removal of incorporated radiolabel.

5. Stop the labeling reaction with the addition of 20 μL H_2O, 4 μL of 3 M sodium acetate (pH 5.2), and 80 μL 100% ethanol.

6. Freeze in a dry ice–ethanol bath for 20 min or overnight at −70°C.

7. Centrifuge for 10 min at 18,300 × g at 4°C.

8. Remove the supernatant, and carefully wash the pellet with 1 mL ice-cold 70% ethanol.

9. Remove the supernatant and dry the pellet in a Speed Vac or comparable vacuum centrifuge for 10 min or until all the ethanol has evaporated. Avoid overdrying the pellet, as this will make subsequent resuspension difficult.

10. Add 100 μL H_2O to the pellet and let stand at room temperature for ≥ 15 min.

11. Vortex for 15 s and then briefly centrifuge the sample to collect all liquid at the bottom of the tube. The final concentration of the probe is 0.4 pmol/μL. (Store in an appropriate vessel at −20°C.)

3.3. Quantification of Radiolabeled Probe by TCA Precipitation

To determine the efficiency of Klenow radiolabeling of the double-stranded oligonucleotide probe, the probe is acid-precipitated onto a glass filter, which is then quantitated in a scintilation counter.

1. Prepare the following mixture in a 1.5-mL Eppendorf tube: 500 μL 5% TCA, 50 μL of 0.1 M tetrasodium pyrophosphate, 50 μL salmon sperm DNA (1 mg/mL), 1 μL radiolabeled probe (0.4 pmol/μL).

2. Incubate on ice for 15 min.

3. Vacuum-filter through a Whatman GF/B glass microfiber filter.

4. Wash with 1 mL 1% TCA by adding the TCA to the Eppendorf tube and then pouring it through the filter. Repeat this wash two additional times.

5. Wash with 1 mL of 70% ethanol by adding the ethanol to the Eppendorf tube and then pouring it through the filter. Repeat this wash two additional times.

6. Transfer the glass microfiber filter to a scintillation vial, add an appropriate volume of ScintiVerse, and quantitate the $[^{32}P]$ in a scintillation counter. Klenow labeling typically yields ~1,000,000 cpm/μL probe. Probes should be ≥ 200,000 cpm/μL for subsequent use in EMSA analysis.

3.4. Electrophoretic Mobility Shift Assay

Once an appropriate source of protein has been obtained and an oligonucleotide probe has been annealed and labeled, one can proceed with the electrophoretic mobility shift assay. If a competition assay or super-shift/blocking assay is to be performed, then competitor oligonucleotides need to be annealed (**Subheading 3.2., steps 1** and **2**), and the appropriate antibodies should be obtained. The following EMSA sample conditions represent an optimization for binding of baculovirus expressing recombinant His-tagged human p53 purified from Sf9 nuclear extracts to a synthetic oligonucleotide probe containing the p53 response element from the human p21 promoter. Other applications of this protocol may require further optimization of binding conditions (*see* **Note 1**).

1. Clean a set of electrophoresis gel plates, spacers, and a well-forming comb.

2. Assemble the electrophoresis plates, ensuring a tight seal between the spacers and plates. The use of Vaseline on the spacers can facilitate this process. The absence of leaks can be confirmed by pipetting water between the glass plates.

3. Prepare a 4% polyacrylamide gel as follows: Combine 7.8 mL 23% acrylamide/0.5% bis-acrylamide, 1.25 mL 10× TBE, 36 mL H_2O, 350 µL 10% APS, 35 µL TEMED in a 50-mL tube. Unpolymerized acrylamide is neurotoxic and should be handled with care. Gloves should be worn whenever working with solutions containing acrylamide.

4. Gently mix the solution be inverting the tube 4–5 times.

5. Pour the solution between the electrophoresis gel plates, filling to just below the top of the plates.

6. Gently insert the well-forming comb, being careful not to trap air bubbles beneath the comb. The gel should polymerize in approximately 30 min. If the gel is not to be used immediately, cover in plastic wrap and store overnight at 4°C.

7. Prepare a "master mix" that contains the following for each sample: 6 µL 5× EMSA buffer, 1.5 µL of 40 mM spermidine, 1.5 µL of 10 mM DTT, 1 µL of 0.5-µg/µL poly(dI/dC), 1 µL of 30-mg/mL BSA, 0.3 µL of 0.4-pmol/µL [^{32}P] probe ($\geq$ 200,000 cpm/µL).

8. Vortex the master mix and briefly centrifuge to collect the liquid at the bottom of the tube (*see* **Note 3**).

9. For each sample, prepare a tube on ice containing 11 µL master mix, 5 µL of protein sample in nuclear extraction buffer, and enough water to bring the final volume to 30 µL (final concentrations per sample: 23.3 mM HEPES (pH 7.5), 83.3 mM NaCl, 13.3% glycerol, 0.7 mM DTT, 0.13 mM EDTA, 2.2 mM MgCl$_2$, 0.02% Triton X-100, 2 mM spermidine, 16.7 µg/mL poly(dI/dC), 4 pmol/ml [^{32}P] probe). Competitor oligonucleotides and antibody should be added as desired (*see* **Note 4**), adjusting the water volume to maintain a total sample volume of 30 µL.

10. Vortex each sample and briefly centrifuge to collect the liquid at the bottom of the tube.

11. Incubate samples at room temperature for 20 min (*see* **Note 5**).

12. Attach the gel to the electrophoresis unit and fill the buffer reservoirs with 0.5× TBE. Be sure to remove any air bubbles trapped at the bottom of the gel.

13. Flush wells with 0.5× TBE and load each sample onto the gel. Load 30 µL of 1× sample buffer into an empty well.

14. Electrophorese the gel at 4°C with a constant 200 V until the bromophenol blue dye is approximately 6 cm from the bottom of the gel (~2 h) (*see* **Note 6**).

15. Following electrophoresis, separate the glass plates and notch a corner of the gel to help with orientation. If upon separation of the glass plates air bubbles become trapped beneath the gel, gently run water between the gel and the glass plate to remove the bubbles.

16. Transfer the gel to a piece of 3 MM Whatman paper by placing the paper onto the exposed gel and then slowly lifting it from the glass plate.

17. Cover the gel with plastic wrap and dry the gel on a vacuum drier for 1 h at 80°C. The use of a second piece of 3 MM Whatman paper between the gel and the dryer will help prevent radioactive contamination of the drying equipment.

18. Place the gel in a film cassette with an intensifying screen and expose to autoradiography film overnight at −70°C.

4. Notes

1. The binding of DNA by protein can be effected by NaCl, MgCl$_2$, EDTA, DTT, and poly(dI/dC) concentrations (**Fig. 5**). As each protein–DNA interaction is unique, it may be necessary to optimize the DNA-binding conditions for a particular electrophoretic mobility shift assay. If modifications are necessary, the reagents should be varied one at a time. The

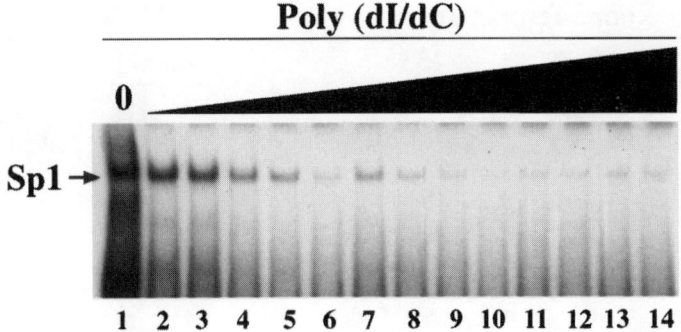

Fig. 5. Optimization of poly(dI/dC) concentration. HeLa cell nuclear extract was incubated with radiolabeled probe corresponding to the p53 response element of the human *Bax* promoter in the absence of poly(dI/dC) (lane 1) and presence of increasing amounts of poly(dI/dC), ranging from 17 to 217 µg/mL (lanes 2–14). In the absence of poly(dI/dC), a number of complexes are observed (lane 1). As the concentration of poly(dI/dC) is increased, many of these complexes are competed away, increasing the specificity of the assay (compare lane 1 to lanes 2 and 3). At higher concentrations, however, the nonspecific DNA begins to compete the probe for binding to Sp1 (lanes 3–14), which is known to be a specific interaction *(29)*. The Sp1-oligonucleotide complex is indicated by the arrow.

use of a nonspecific competitor such as poly(dI/dC) is essential to the sensitivity of the assay. An excessive amount of nonspecific competitor, however, can decrease the sensitivity of the assay by competing the protein from the probe (**Fig. 5**).

2. The following equation may be beneficial in converting the units of a single-stranded oligonucleotide from µg to pmol:

Oligonucleotide in pmole = (weight in µg × 1,000,000)/(length in bases × 327)

Example: 10 µg of a 38-mer oligonucleotide:

$$(10 \times 1,000,000)/(38 \times 327) = 805 \text{ pmol}$$

3. It is important to vortex the master mix sufficiently before adding it to each EMSA sample. The reagents contained in the master mix vary in density, and will separate into distinct layers if not properly mixed. Vortexing ensures that every sample receives the appropriate proportion of each reagent. Unexpected variations in sample lanes observed by autoradiography, such as "empty lanes," are often due to a failure to vortex the master mix solution appropriately.

4. In competition assays, the appropriate amount of competitor to use is determined empirically. Typically, this is in the range of 10- to 1000-fold molar excess. Since each EMSA sample contains 0.12 pmol of [^{32}P]-labeled probe, samples usually contain 1.2–120 pmol of competitor (**Fig. 1**). For super-shift/blocking assays, the appropriate volume of antibody to use also is determined empirically by testing several amounts. To control for nonspecific antibody effects, an unrelated antibody or preimmune serum should be examined in parallel.

5. Some nuclear extracts contain a high level of nuclease activity that can degrade the radiolabeled probe during the 20-min sample incubation period (**Subheading 3.4.**, **step 11**). This is demonstrated on the autoradiograph as a signal migrating below the free probe, representing free nucleotides (**Fig. 6,** lanes 1–5). Nuclease activity that requires divalent cations can be inhibited by increasing the level of EDTA in the EMSA sample buffer. EDTA, however, may also adversely effect the protein–DNA interaction being assayed (*see* **Note 1**). An alternative approach, therefore, is to conduct the 20-min sample incubation on ice (**Subheading 3.4.**, **step 11**), which will slow the rate of nuclease activity (**Fig. 6,** compare lanes 1–5 to lanes 12–14).

6. During polymerization of the polyacrylamide gel, APS, catalyzed by TEMED, produces free radicals that cause the cross-linking of acrylamide and bis-acrylamide. These free radicals, however, may also interfere with certain protein–DNA interactions. To avoid this complica-

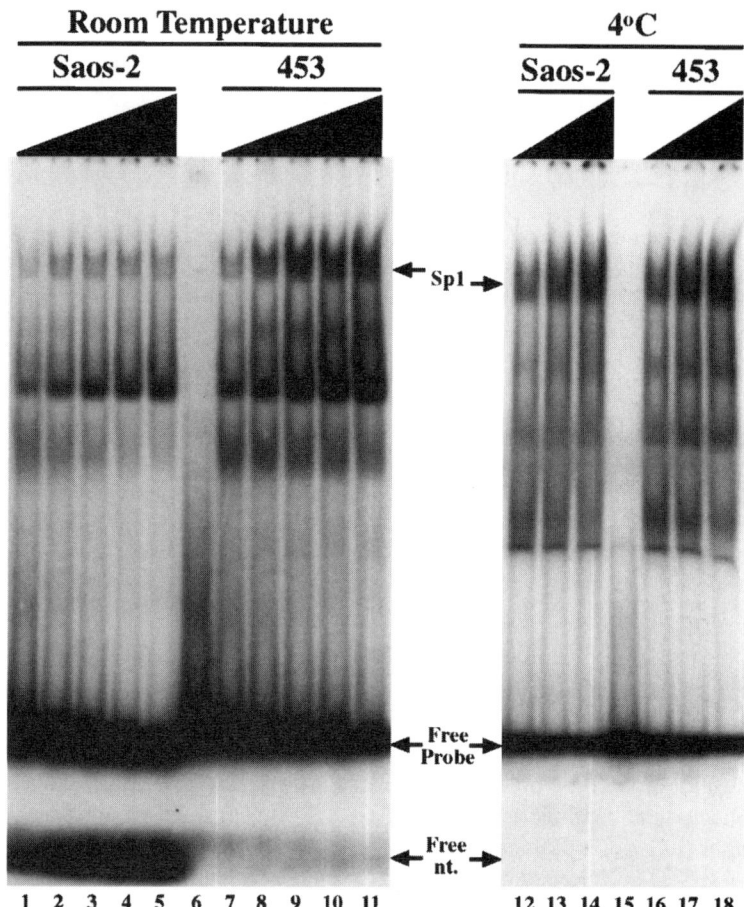

Fig. 6. Saos-2 nuclear extracts contain a nuclease activity that is inhibited at 4°C. Increasing amounts of Saos-2 (lanes 1–5 and 12–14) or MDA-MB-453 (lanes 7–11 and 16–18) nuclear extract were incubated with a radiolabeled probe corresponding to the p53 response element of the human *Bax* promoter at room temperature (lanes 1–11) or at 4°C (lanes 12–18). When the DNA-binding reaction is incubated at room temperature, a significant amount of free nucleotide is observed in the Saos-2 samples (lane 1–5), representing nuclease activity. When the DNA-binding reaction is performed at 4°C, however, this nuclease activity is not present, as demonstrated by the absence of free nucleotides (lanes 12–14).

tion, one can prerun the polyacrylamide gel to dissipate the free radicals. Prior to loading EMSA samples (**Subheading 3.4.**, **step 13**), the gel is electrophoresed at 200 V for approximately 2 h at 4°C. This also ensures that the gel is at a uniform temperature before separating EMSA samples.

References

1. Bargonetti, J., Friedman, P. N., Kern, S. E., Vogelstein, B., and Prives, C. (1991) Wild-type but not mutant p53 immunopurified proteins bind to sequences adjacent to the SV40 origin of replication. *Cell* **65**, 1083–1091.
2. Kern, S. E., Kinzler, K. W., Bruskin, A., et al. (1991) Identification of p53 as a sequence-specific DNA-binding protein. *Science* **252**, 1708–1711.

3. Ko, L. J. and Prives, C. (1996) p53: puzzle and paradigm. *Genes Dev.* **10**, 1054–1072.
4. Levine, A. J. (1997) p53, the cellular gatekeeper for growth and division. *Cell* **88**, 323–331.
5. Greenblatt, M. S., Bennett, W. P., Hollstein, M., and Harris, C. C. (1994) Mutations in the p53 tumor suppressor gene: clues to cancer etiology and molecular pathogenesis. *Cancer Res.* **54**, 4855–4878.
6. Hollstein, M., Rice, K., Greenblatt, M. S., et al. (1994) Database of p53 gene somatic mutations in human tumors and cell lines. *Nucleic Acids Res.* **22**, 3551–3555.
7. Fields, S. and Jang, S. K. (1990) Presence of a potent transcription activating sequence in the p53 protein. *Science* **249**, 1046–1049.
8. Bargonetti, J., Manfredi, J. J., Chen, X., Marshak, D. R., and Prives, C. (1993) A proteolytic fragment from the central region of p53 has marked sequence-specific DNA-binding activity when generated from wild-type but not from oncogenic mutant p53 protein. *Genes Dev.* **7**, 2565–2574.
9. Pavletich, N. P., Chambers, K. A., and Pabo, C. O. (1993) The DNA-binding domain of p53 contains the four conserved regions and the major mutation hot spots. *Genes Dev.* **7**, 2556–2564.
10. Wang, Y., Reed, M., Wang, P., et al. (1993) p53 domains: identification and characterization of two autonomous DNA- binding regions. *Genes Dev.* **7**, 2575–2586.
11. Pietenpol, J. A., Tokino, T., Thiagalingam, S., el-Deiry, W. S., Kinzler, K. W., and Vogelstein, B. (1994) Sequence-specific transcriptional activation is essential for growth suppression by p53. *Proc. Natl. Acad Sci. USA* **91**, 1998–2002.
12. Prives, C. (1994) How loops, beta sheets, and alpha helices help us to understand p53. *Cell* **78**, 543–546.
13. Kern, S. E., Pietenpol, J. A., Thiagalingam, S., Seymour, A., Kinzler, K. W., and Vogelstein, B. (1992) Oncogenic forms of p53 inhibit p53-regulated gene expression. *Science* **256**, 827–830.
14. Raycroft, L., Wu, H. Y., and Lozano, G. (1990) Transcriptional activation by wild-type but not transforming mutants of the p53 anti-oncogene. *Science* **249**, 1049–1051.
15. Unger, T., Nau, M. M., Segal, S., and Minna, J. D. (1992) p53: a transdominant regulator of transcription whose function is ablated by mutations occurring in human cancer. *EMBO J.* **11**, 1383–1390.
16. El-Deiry, W. S., Kern, S. E., Pietenpol, J. A., Kinzler, K. W., and Vogelstein, B. (1992) Definition of a consensus binding site for p53. *Nat. Genet.* **1**, 45–49.
17. Funk, W. D., Pak, D. T., Karas, R. H., Wright, W. E., and Shay, J. W. (1992) A transcriptionally active DNA-binding site for human p53 protein complexes. *Mol. Cell. Biol.* **12**, 2866–2871.
18. Halazonetis, T. D., Davis, L. J., and Kandil, A. N. (1993) Wild-type p53 adopts a "mutant"-like conformation when bound to DNA. *EMBO J.* **12**, 1021–1028.
19. Buckbinder, L., Talbott, R., Velasco-Miguel, S., et al. (1995) Induction of the growth inhibitor IGF-binding protein 3 by p53. *Nature* **377**, 646–649.
20. El-Deiry, W. S., Tokino, T., Velculescu, V. E., et al. (1993) WAF1, a potential mediator of p53 tumor suppression. *Cell* **75**, 817–825.
21. El-Deiry, W. S., Tokino, T., Waldman, T., et al. (1995) Topological control of p21WAF1/CIP1 expression in normal and neoplastic tissues. *Cancer Res.* **55**, 2910–2919.
22. Hollander, M. C., Alamo, I., Jackman, J., Wang, M. G., McBride, O. W., and Fornace, A. J., Jr. (1993) Analysis of the mammalian gadd45 gene and its response to DNA damage. *J. Biol. Chem.* **268**, 24385–24393.
23. Macleod, K. F., Sherry, N., Hannon, G., et al. (1995) p53-dependent and independent expression of p21 during cell growth, differentiation, and DNA damage. *Genes Dev.* **9**, 935–944.
24. Miyashita, T. and Reed, J. C. (1995) Tumor suppressor p53 is a direct transcriptional activator of the human bax gene. *Cell* **80**, 293–299.
25. Perry, M. E., Piette, J., Zawadzki, J. A., Harvey, D., and Levine, A. J. (1993) The mdm-2 gene is induced in response to UV light in a p53-dependent manner. *Proc. Natl. Acad. Sci. USA* **90**, 11623–11627.

26. Zauberman, A., Lupo, A., and Oren, M. (1995) Identification of p53 target genes through immune selection of genomic DNA: the cyclin G gene contains two distinct p53 binding sites. *Oncogene* **10**, 2361–2366.
27. Ceglarek, J. A. and Revzin, A. (1989) Studies of DNA-protein interactions by gel electrophoresis. *Electrophoresis* **10**, 360–365.
28. Revzin, A. (1989) Gel electrophoresis assays for DNA-protein interactions. *Biotechniques* **7**, 346–355.
29. Thornborrow, E. C. and Manfredi, J. J. (2001) The tumor suppressor protein p53 requires a cofactor to transcriptionally activate the human bax promoter. *J. Biol. Chem.* **276,** 15,598–15,608.
30. Resnick-Silverman, L., St Clair, S., Maurer, M., Zhao, K., and Manfredi, J. J. (1998) Identification of a novel class of genomic DNA-binding sites suggests a mechanism for selectivity in target gene activation by the tumor suppressor protein p53. *Genes Dev.* **12**, 2102–2107.
31. Thornborrow, E. C., and Manfredi, J. J. (1999) One mechanism for cell type-specific regulation of the bax promoter by the tumor suppressor p53 is dictated by the p53 response element. *J. Biol. Chem.* **274**, 33747–33756.
32. Lassar, A. B., Davis, R. L., Wright, W. E., et al. (1991) Functional activity of myogenic HLH proteins requires hetero-oligomerization with E12/E47-like proteins in vivo. *Cell* **66**, 305–315.

8

Analysis of Gene Promoter Regulation by Tumor Suppressor Genes

Kumaravel Somasundaram, Sanjeev Das, Smita Lakhotia, and Narendra Wajapeyee

1. Introduction

Gene expression is controlled at several steps. Transcription is one of the important steps at which regulation occurs. Tumor suppressors such as p53, which are basically transcription factors, carry out their function primarily by the transcriptional activation of the target genes. The promoter region of a gene plays a major role in its transcriptional regulation, so it would be very useful for laboratories involved in the study of the tumor suppressor function to know the techniques used to analyze the regulation of gene promoters, particularly if they encode transcription factors.

This chapter is restricted to techniques that make use of various reporter genes to study the regulation of promoters by transient transfections (in vitro). It gives the various options available to a scientist who wants to study the regulation of a promoter by a tumor suppressor. It includes various reporter systems available, methods to assay the reporter enzymes, choice of reporter plasmids available from different sources, various methods of transfection into eukaryotic cells, and analysis of promoter–reporter constructs.

1.1. Overview

The most common method of analysis of promoter regulation is the use of promoter–reporter fusions. Typically, one transfects a promoter–reporter fusion construct, wherein a promoter is fused to a reporter gene so that by assaying the reporter gene product one can study the activity of promoter. The basic requirements are as follows: an easily assayable reporter protein that does not affect any physiologic process inside the cells, an efficient transfection method, and a suitable cell line. Reporter assays could be done both in vitro and in vivo. In-vitro assays refer to procedures in which the reporter protein is quantified using the cell lysate or the cultured medium for the secreted reporter proteins. In-vivo reporter assays are procedures in which the reporter protein is detected in either live or fixed cells or tissues. This method provides information regarding the cell-type specificity of promoters/enhancers as well as the tissue distribution of transcription factors, but it is less quantitative than in-vitro reporter systems.

From: *Methods in Molecular Biology, Vol. 223: Tumor Suppressor Genes: Regulation, Function, and Medicinal Applications.* Edited by: Wafik S. El-Deiry © Humana Press Inc., Totowa, NJ

1.2. Reporter Genes

Several reporter genes are available for promoter analysis. An ideal reporter should have the following features: the reporter protein should be absent from the host; a simple, rapid, sensitive, and cheap assay method should be available to detect the reporter protein; the assay method should have a broad linear range so that different levels of activation of a promoter can be studied; and the reporter gene expression should not affect the normal physiology of the cells.

1.2.1. Chloramphenicol Acetyltransferase (CAT)

Chloramphenicol acetyltransferase (CAT) is one of the most frequently used reporter proteins. CAT is an enzyme that catalyzes the transfer of acetyl group from acetyl coenzyme A to chloramphenicol. The most common CAT assay is based on the incubation of cell lysates with ^{14}C-chloramphenicol and acetyl CoA. Thin-layer chromatography is used to separate the acetylated and unacetylated forms. The nonisotopic method (CAT ELISA) directly quantifies CAT using an anti-CAT antibody, thus measuring the CAT protein directly instead of measuring CAT activity.

Being a prokaryotic enzyme, CAT does not face any competing activities in most eukaryotic cells, thus high signal-to-background ratios are obtained in these assays. The half-life of CAT enzyme is about 50 h in mammalian cells, which means that it is a very stable enzyme (*1*). The sensitivity of the CAT assay is increased with the use of nonradioactive substrate. There are also disadvantages in using CAT as a reporter. The radioactive CAT assay requires expensive radioactive substrate. The assays are time-consuming to perform. The sensitivity of the CAT assay is less as compared to that of other nonisotopic reporter assays, and the linear range is narrow. The stability of the CAT can result in masking of the induction due to addition of an effector in inducible systems.

1.2.2. Firefly Luciferase

Luciferase is encoded by the *luc* gene of the firefly *Photonis pyralis* (*2*). It provides a nonisotopic reporter system that is widely used in mammalian cells. Luciferase catalyzes a bioluminescent reaction that requires luciferin (the substrate), coenzyme A, ATP, Mg^{2+}, and molecular O_2 (*3*). Mixing of luciferin reagent with the cell lysate results in a flash of light, which can be detected using either a luminometer or a liquid scintillation counter. Light intensity is proportional to luciferase concentration in the range 10^{-16}–10^{-8} *M*.

The half-life of luciferase enzyme is only 3 h, so it acts as a good candidate for studying inducible systems (*1*). Luciferase assay is about 10–1000 times more sensitive than CAT assay and has a broad linear range. The assay system is nonradioactive, relatively inexpensive, and rapid compared to CAT assay. Minimal endogenous luciferase activity is seen in mammalian cells. Some of the disadvantages are that luciferase lacks "reproducibility" between samples due to rapid emission of light, the half-life of the luciferase in the sample is very short, and the assay requires a luminometer or scintillation counter, which can be a limiting factor.

1.2.3. β-Galactosidase

The lacZ gene from *Escherichia coli* that encodes β-galactosidase is one of the versatile reporters available. This reporter can be used for both in-vivo and in-vitro assay formats with different substrates. The enzyme catalyzes the hydrolysis of various β-galactosides. Apart from its use as a reporter, it is also used as an internal control to normalize the transfection efficiency variability between different samples. Two commonly available assays are the colorimetric assay using *o*-nitrophenyl β-D-galactopyranoside *(4)* and the fluorometric assay using 4-methylumbelliferyl-β-D-galactosidase *(5)*. However, these two methods are less sensitive as compared to the assay using chemiluminescent 1,2-dioxetane, which is 50,000-fold more sensitive *(6)*.

1.2.4. Secreted Alkaline Phosphatase (SEAP)

Secreted alkaline phosphatase (SEAP) differs from the above-mentioned reporters in that it is secreted from transfected cells and can thus be assayed using a small aliquot of the culture medium *(7)*. The SEAP is a truncated protein that lacks the membrane-anchoring domain, so is secreted in the medium. Hence, there is no need to prepare cell lysate. The kinetics of the gene expression can be studied by repeated collection of the medium from the same cultures. Since the transfected cells are not disturbed, they can be used for further investigation. The endogenous background activity for SEAP is almost zero. The SEAP assay can be automated by using 96-well microliter plates. Currently available assays use a nonisotopic method. This reporter may not be suitable for cells derived from lung, testes, and cervix, as these have low levels of placental-type alkaline phosphatase.

1.2.5. Human Growth Hormone (hGH)

The secretary property of hGH makes this protein an attractive choice as a genetic reporter for most mammalian cell types *(8)*. It has the same advantages as SEAP. However, it is less preferred due to the relatively low sensitivity of the procedure and the need for a hazardous radioimmunoassay to quantitate hGH. It is also used as an internal control in transfection assays.

1.2.6. Green Fluorescent Protein (GFP)

Light is produced in several bioluminescent jellyfishes as a result of energy transfer to green fluorescent proteins (GFPs). One such GFP, from *Aequorea victoria*, fluoresces in vivo after receiving energy from the Ca^{2+}-activated photoprotein aequorin *(9)*. The purified protein also absorbs blue light and emits green light (absorption and emission similar to that of fluorescein), which is detected using a fluorescent microscope or by fluorescent-activated cell sorting (FACS). Expression of cloned GFP gives a bright green fluorescence when blue or UV light excites the cells (prokaryotic or eukaryotic). The expression of the protein is species independent and requires no other product from the jellyfish. One of the main advantages is that the GFP reporter expression can be visualized in live cells. The GFP expression has no apparent toxicity to the cells. One disadvantage is that signal intensity may be too low for certain applications.

2. Materials

2.1. Assay for Chloramphenicol Acetyl Transferase Acivity

1. Cells transfected with CAT expression plasmid.
2. Phosphate-buffered saline (PBS): 137 mM NaCl, 2.7 mM KCl, 4.3 mM Na_2HPO_4, 1.4 mM KH_2PO_4.
3. Trypsin (0.5%)/EDTA, pH 8.0 (0.2%).
4. 0.25 M Tris-Cl (pH 7.8).

5. ^{14}C-chloramphenicol (50 µCi/mL; NEN cat. no. NEC 408A).
6. 10 mM acetyl CoA (Sigma cat. no. C3144; to prepare 10 mM acetyl CoA (FW 809.6), dissolve 25 mg in 3.1 mL of water. Store as aliqouts at –70°C).
7. Thin-layer chromatography (TLC) tank.
8. Thin-layer chromatography plate (Sigma cat. no. 212, 278-5).
9. Ethyl acetate.
10. Chloroform/methanol (19:1).

2.2. Assay for β-Galactosidase Activity

1. Cells transfected with β-galactosidase expression plasmid.
2. Phosphate-buffered saline (PBS): 137 mM NaCl, 2.7 mM KCl, 4.3 mM Na$_2$HPO$_4$, 1.4 mM KH$_2$PO$_4$.
3. 0. 25 M Tris-Cl (pH 7.8).
4. 100× Mg solution: 0.1 M MgCl$_2$, 4.5 M β-mercaptoethanol.
5. 1× ONPG solution: 4-mg/mL solution of ONPG in 0.1 M sodium phosphate (pH 7.5).
6. 0.1 M sodium phosphate (pH 7.5).

2.3. Assay for Luciferase Activity

1. Cells transfected with luciferase expression plasmid.
2. Phosphate-buffered saline (PBS).
3. Trypsin-EDTA.
4. Triton/glycylglycine lysis buffer: 1% (v/v) Triton X-100, 25 mM glycylglycine, pH 7.8, 15 mM MgSO$_4$, 4 mM EGTA, 1 mM DTT,
15. Luciferase assay reagent (20 mM Tricine, 1.07 mM Mg(CO$_3$)$_4$ Mg(OH)$_2$ · 5H$_2$O, 2.67 mM MgSO$_4$, 0.1 mM EDTA, 33.3 mM DTT, 270 µM coenzyme A, 470 µM luciferin, 530 µM ATP). To make luciferase assay reagent, put 400 µL of 0.5 M tricine (Sigma cat. no. T9784), 0.0052 g of Mg(CO$_3$)$_4$ Mg(OH)$_2$ · 5H$_2$O (Sigma cat. no. M-5671), 0.0066 g of MgSO$_4$, and 2 µL of 0.5 M EDTA in a test tube. Add water to make up the volume to 8.5 mL. Put the tube in a boiling water bath for a few minutes until all the contents dissolve. Cool it and adjust the pH to 7.8. Now add 333 µL of 1 M DTT, 0.00207 g of coenzyme A (FW. 767.5; Sigma cat. no. C3144), 0.00142 g of luciferin (FW.302.3; Sigma cat. no. L6882), and 0.0032 g of ATP (FW. 602.8), and make up the volume to 10 mL. Store the reagent in small aliquots at –70°C.

2.4. Calcium Phosphate-Mediated Transfection

1. 2× HEPES-buffered saline (280 mM NaCl, 10 mM KCl, 1.5 mM Na$_2$HPO$_4$ · 2H$_2$O, 12 mM dextrose, 50 mM HEPES). To make 2× HEPES-buffered saline (100 mL), take 1.6 g of NaCl, 0.074 g of KCl, 0.027 g of Na$_2$HPO$_4$ · 2H$_2$O, 0.2 g of dextrose, and 1 g of HEPES and dissolve in 80 mL of water. Adjust the pH to 7.05, make up the volume to 100 mL and filter-sterilize. Make several batches of the reagent and test for transfection efficiency. The batch that gives highest transfection efficiency can be used for regular experiments. Store it in 5-ml aliqouts at –20°C.
2. 1 M calcium chloride.
3. Sterile water.
4. Phosphate-buffered saline (PBS).

3. Methods
3.1. Reporter Gene Products

This section describes in detail the commonly used method for assaying three different reporter enzymes: chloramphenicol acetyl transferase (CAT), β-galactosidase, and luciferase.

3.1.1. Chloramphenicol Acetyl Transferase (CAT) Activity

The most commonly used radioactive method, which is the measure of acetylation of ^{14}C-labeled chloramphenicol, is described below *(10)*.

3.1.1.1. MAKING THE CELL EXTRACT

1. Wash the cells transfected with CAT reporter plasmid twice with ice-cold PBS.
2. Add trypsin-EDTA solution and incubate the plate at 37°C. A 35-mm Petri dish requires 200 μL of trypsin-EDTA solution.
3. After 5–10 min, tap the dish so that the cells come off. Add 1 mL PBS to the dish and bring the cells into suspension by gentle splashing with a 1000-μL tip. Collect the cells in a micocentrifuge tube kept on ice.
4. Spin at 960*g* for 5 min.
5. Resuspend the pellet in 100 μL of the ice-cold 0. 25 *M* Tris-Cl (pH 7.8) (*see* **Note 1**) and vortex for about 30 s.
6. Disrupt the cells by three cycles of freezing in dry ice and ethanol and thawing at 37°C.
7. Spin at 7840*g* for 5 min and collect the supernatant in a fresh microcentrifuge tube.

3.1.1.2. CAT Assay

1. Quantify the protein by any of the standard methods, such as the Bradford method *(11)* or Lowry method *(12)*.
2. Set up the CAT assay reaction as follows:
   a. Cell extract equivalent to 20 μg total protein.
   b. 2 μL (0.1 μC) ^{14}C-chloramphenicol.
   c. 0.25 *M* Tris-HCl, make the volume to 90 μL.
3. Incubate at 37°C for 5 min.
4. Add 10 μL of 10 m*M* acetyl CoA to each of the tubes, mix well, and incubate at 37°C for 1 h.
5. While the reaction is proceeding, prepare the thin-layer chromatography (TLC) chamber and plate as follows:
   a. TLC chamber: Add 200 mL of TLC solvent by mixing chloroform and methanol (95/5), close the chamber with a lid, and leave it undisturbed for at least 1 h for the solvent to equilibrate.
   b. TLC plate: Cut a 20 × 20-cm TLC plate into two halves. Use one half for an experiment. Using a soft pencil, draw a line 1.5 cm from the edge along the length of the cut plate and mark dots at 1- to 1.5-cm intervals along this line.
6. Add 500 μL of ice-cold ethyl acetate at the end of reaction and extract the acetylated chloramphenicol into the upper organic phase by vigorous vortexing for 30 s.
7. Centrifuge the tubes at 7840*g* for 5 min and carefully transfer exactly 450 μL of the top layer into a fresh tube.
8. Evaporate the ethyl acetate completely under vacuum. We do it by placing the tubes in a rotating evaporator (Savant SpeedVac) for about 30 min.
9. Redissolve the reaction products in 25 μL of ethyl acetate. Make sure to rinse down the sides of the tubes. Mix well and centrifuge briefly to bring all the liquid to the bottom of the tube.
10. Using yellow tips, spot the dissolved reaction products from **step 9** onto the dots marked on the TLC plate. Apply 3–4 μL at a time, and allow sufficient time to allow the spots to dry completely between applications.
11. Place the spotted TLC plate in the TLC chamber, close the lid, and allow the solvent front to move up to about 5 mm from the top.
12. Take out the TLC plate, air-dry it, and expose the plate to a Phosphoimager screen or to an X-ray film. The autoradiogram will have up to 5 spots for each sample, which in ascending

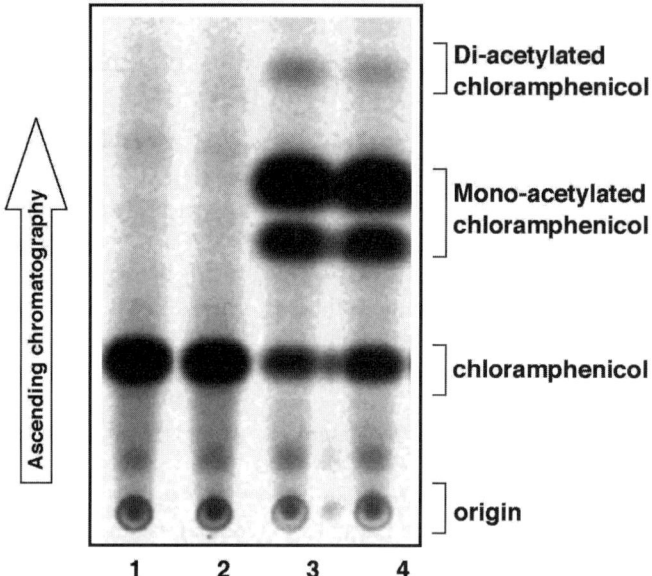

Fig. 1. Results of a typical CAT assay experiment. SW480 cells were transfected with G5E1BCAT (reporter plasmid) either alone (lanes 1 and 2) or with Gal4:p73αAD (transcription plasmid; lanes 3 and 4). The cells were harvested 48 h later and CAT assay was performed. G5E1BCAT contains five copies of Gal4-binding sites fused and the basal promoter of E1B gene of adenovirus cloned upstream of CAT gene. Gal4:p73αAD contains the DNA-binding domain of Gal4 fused to the N-terminus transcription activation domain of p73α.

order are: a faint spot at the origin, unacetylated chloramphenicol, two forms of monoacetylated chloramphenicol (the higher spot being more intense than the lower one), and diacetylated chloramphenicol (**Fig. 1**).

13. If exposure is done using a Phosphoimager, then quantification of spots can be done using different software available. For quantification of spots on an X-ray film, identify and cut out the spots, add to scintillation fluid in a scintillation vial, and count in a scintillation counter. Calculate the CAT activity as follows:

$$\% \text{ acetylated} = \frac{\text{counts in acetylated species}}{\text{counts in acetylated species} + \text{counts in nonacetylated species}}$$

14. The assay for CAT activity is not in linear range when the conversion of chloramphenicol to acetyl chloramphenicol is above 30%. If a sample gives a conversion above 30%, it has to be diluted or assayed for shorter period to get an accurate measure of CAT activity.

3.1.2. Assay for β-Galactosidase Activity

The colorimetric assay using *o*-nitrophenyl-β−D-galactopyranoside (ONPG) to measure β-galactosidase activity is described below.

3.1.2.1. MAKING THE CELL EXTRACT

Follow the procedure described under **Subheading 3.1.1.1.**

3.1.2.2. β-GALACTOSIDASE ASSAY

1. Quantify the protein by any of the standard methods, such as the Bradford method *(11)* or the Lowry method *(12)*.

2. Set up the assay reaction as follows:
   a. Cell extract equivalent to 20 µg total protein.
   b. 66 µL of 1× ONPG solution.
   c. 3 µL of 100× Mg solution.
   d. 0.1 M sodium phosphate, make the volume to 300 µL.
3. Incubate the reaction at 37°C for 30–60 min or until a faint yellow color develops.
4. Stop the reaction by adding 500 µL of 1 M Na_2CO_3.
5. Take OD at 420 nm. The linear range is 0.2µ–0.8 OD.

3.1.3. Assay for Luciferase Activity

The following protocol describes the assay for measuring luciferase activity. Luciferase reporter has become more popular because it is more sensitive than CAT reporter. The individual assays are much simpler, quicker, and easier to carry out.

3.1.3.1. MAKING THE CELL EXTRACT

1. Wash the cells transfected with luciferase reporter plasmid twice with ice-cold PBS.
2. Add trypsin-EDTA solution and incubate the plate at 37°C. A 35-mm Petri dish requires 200 µL of trypsin-EDTA solution.
3. After 5–10 min, tap the dish so that the cells come off. Add 1 mL PBS to the dish and bring the cells into suspension by gentle splashing with a 1000-µL tip. Collect the cells in a micocentrifuge tube kept on ice.
4. Spin at 960g for 5 min. Throw the supernatant and loosen the pellet by tapping or vortexing.
5. Add 100 µL of the ice-cold triton/glycylglycine lysis buffer and vortex for about 30 s. Incubate at 4°C for 30 min.
6. Spin at 7840g for 5 min and collect the supernatant in a fresh microcentrifuge tube (*see* **Note 2**).

3.1.3.2. LUCIFERASE ASSAY

1. Quantify the protein by any of the standard methods, such as the Bradford method *(11)* or the Lowry method *(12)*.
2. Transfer the cell extract equivalent to 20 µg of total protein to a 0.5-mL Eppendorf tube containing 100 µL of luciferase assay reagent and mix well. Place the tube in a glass scintillation vial and immediately measure the luciferase acitivity in a scintillation counter.
3. To use a luminometer, follow the instructions of the manufacturer.

3.2. Reporter Vectors Available

In this section, we attempt to provide information on various vectors available for promoter analysis from both academic and commercial sources. Typically, a reporter plasmid contains a multiple cloning site (MCS) upstream of the reporter gene. MCS allows for the insertion of foreign DNA containing putative promoter sequence. The presence of a polyadenylation site downstream of the reporter gene ensures proper and efficient processing of the reporter transcript. In addition, the vector also contains a bacterial origin of replication for propagation of the vector in *E. coli* and an antibiotic resistance marker such as β-lactamase, which encodes resistance to ampicillin for the specific selection of bacteria containing the plasmid. **Table 1** gives the list of various vectors available for promoter regulation studies.

Vectors are also available in which a basal promoter is present upstream of the reporter gene (**Table 2**). These vectors provide a constitutive level of expression of the

Table 1
Promoterless Reporter Vectors

S. no.	Reporter gene	Name of the vector	Properties	Unique sites*	Source/ reference
1	CAT	pCAT3—basic	4.0 kb/Amp^r	Many sites[a]	Promega
		pCAT3—enhancer	4.2 kb/Amp^r	Many sites[a]	Promega
		pA$_{10}$cat$_2$	NA/Amp^r	*Sph*I/*Bg*/II	(13)
		p300—CAT	4.8 kb/Amp^r	Many sites[b]	(14)
2	Luciferase	pGL3—basic	4.8 kb/Amp^r	Many sites[c]	Promega
		pGL3—enhancer	5 kb/Amp^r	Many sites[c]	Promega
		pOLUC	4.07 kb/Amp^r	Many sites[d]	(15)
		pRL—null	3.3 kb/Amp^r	Many sites[e]	Promega
3	β-Galactosidase	pβgal—basic	7.5 kb/Amp^r	Many sites[f]	Clontech
		pβgal—enhancer	7.7 kb/Amp^r	Many sites[f]	Clontech
4	EGFP	pEGFP-1	4.2 kb/kan^r/Neo$^{r\#}$	Many sites[g]	Clontech
		pd2EGFP-1	4.3 kb/kan^r/Neo$^{r\#}$	Many sites[h]	Clontech
5	EYFP	pEYFP-1	4.2 kb/kan^r/Neo$^{r\#}$	Many sites[g]	Clontech
6	SEAP	pSEAP	5.8 kb/Amp^r	Many sites[i]	Applied Biosystems
		pSEAP2—basic	4.7 kb/Amp^r	Many sites[j]	Clontech
		pSEAP2—enhancer	4.9 kb/Amp^r	Many sites[j]	Clontech
7	Human growth hormone (LGH)	pφGH	4.8 kb/N.A.	Many sites[k]	(8)

*a. Kpn I *Sac* I *Mlu* I *Nhe* I *Xma* I *Xho* I *Bgl* II; **b.** *Hind* III *Sph* I *Pst* I *Sal* I *Xba* I *Bam*H I *Sma* I *Kpn* I *Sac* I c. *Kpn* I *Sac* I *Mlu* I *Nhe* I *Xma* I *Xho* I *Bgl* II *Hind* III; **d.** *Sph* I *Sal* I *Xba* I *Bam*H I *Sma* I *Hind* III; **e.** *Bgl* II *Xho* I *Sac* I *EcolC*R I *Hind* III *Nde* I Nsi I *Sph* I Spe I *Nar* I *Sal* I *Mlu* I *EcoR* I *Xma* I *Sma* I *Pst* I; **f.** *Sma* I *Xma* I *Asp*718 I *Kpn* I *Nhe* I *Xho* I *Bgl* II *Hind* III; **g.** *Eco*47 III *Bgl* II *Xho* I *Sac* I *Ecl*136 II *Hind* III *Eco*R I *Pst* I *Sal* I *Acc* I *Kpn* I *Asp*718 I *Sac* II *Apa* I *Apa* I *Bsp*120 I *Xma* I *Sma* I *Bam*H I *Age* I; **h.** *Eco*47 III *Bgl* II *Xho* I *Sac* I *Ecl*136 II *Hind* III *Eco*R I *Sal* I *Acc* I *Kpn* I *Asp*718 I *Sac* II *Apa* I *Bsp*120 I *Xma* I *Sma* I *Bam*H I; **i.** *Asp*718 I *Kpn* I *Mlu* I *Nhe* I *Xho* I *Bgl* II *Hind* III *Sph* I; **j.** *Asp*718 I *Kpn* I *Mlu* I *Nhe* I *Srf* I *Xho* I *Bgl* II *Hind* III *Bst*B I *Nru* I *Eco*RI; **k.** *Hind* III *Sal* I *Hind* II *Xba* I.

N.A.: information not available.
Kan^r, prokaryotic selection marker.
Neo^r, eukaryotic selection marker.

reporter gene in a wide variety of cells and are used to identify and characterize activator-binding sites and enhancer sequences.

Vectors with a complete promoter upstream of a reporter gene are also available (**Table 3**). These vectors as well as those carry basal promoters upstream of the reporter gene may also be used as positive control to normalize transfection efficiency of experimental reporter constructs.

3.3. Transfection Methods

In order to study the regulation of promoters in the cellular context, it is necessary to have suitable methods to introduce DNA into tissue culture cells. DNA transfection is the process whereby the nucleic acid sequences are introduced into the cells. Biochemical methods of transfection include DEAE-dextran *(17),* calcium phospohate precipitation *(18),* and liposome-mediated transfection methods *(19).*

Table 2
Reporter Vector Available with Basal Promoters

S. no.	Reporter gene	Name of the vector	Properties	Promoter[a]	Unique sites	Source/ reference
1	CAT	pBLCAT2	4.5 kb/Amp^r	Thymidine kinase	Many sites[b]	*(16)*
		pCAT3—promoter	4.2 kb/Amp^r	SV40	Many sites[c]	Promega
2	Luciferase	pGL2—promoter	5.7 kb/Amp^r	SV40	Many sites[d]	Promega
		pGL3— promoter vector	5 kb/Amp^r	SV40	Many sites[e]	Promega
3	β-Galacto- sidase	pβgal—promoter	7.7 kb/Amp^r	SV40	Many sites[f]	Clontech

a. SV40, early promoter of SV40 (without enhancer); thymidine kinase, promoter of thymidine kinase gene.
b. Hind III *Sph* I *Pst* I *Sal* I *Xba* I *Bam*H I; *c. Kpn* I *Sac* I *Mlu* I *Nhe* I *Xma* I *Xho* I *Bgl* II.
d. Sma I *Kpn* I *Sac* I *Mlu* I *Nhe* I *Xho* I *Bgl* II; *e. Kpn* I *Sac* I *Mlu* I *Nhe* I *Xma* I *Xho* I *Bgl* II.
f. Sma I *Xma* I *Asp*718 I *Kpn* I *Nhe* I *Xho* I *Bgl* II.

These methods depend on the ability of the formed DNA complex to bind to the negatively charged plasma membrane and subsequent intake by endocytosis. Physical transfection methods include direct microinjection and electroporation. While microinjection involves mechanical perforation of the cell membrane, electroporation exposes target cells to brief, defined electrical pulses to create transient pores that allow nucleic acids to cross the cell membrane. The cell type used for the transfection is the major factor, which determines the efficiency of transfection. A suitable method of transfection for a cell type has to be identified by trying different methods (*see* **Note 3**).

Biochemical methods have been used very commonly for the last three decades. The DEAE method works on the principle that the binding of positively charged DEAE–dextran to the negatively charged phosphate groups of the DNA results in the formation of a complex, which when applied to the cells binds to the negatively charged plasma membrane. Calcium phosphate-mediated transfection is one very commonly used method of transfection. This method requires the formation of insoluble calcium phosphate–DNA precipitate. The size of these precipitate complexes affects the efficiency of transfection. This method works with wide variety of cell types with varying efficiency. A detailed protocol for the calcium phosphate method of transfection is given below. The liposome method of transfection has become the most versatile biochemical-based transfection method because of its high efficiency, lower toxicity, and broad host cell range. This method relies on the principle that the positively charged small lipids bind to both phosphate groups of DNA and the negatively charged plasma membrane of the cells. Polycationic lipid reagents yield higher transfection efficiency than monocationic lipids because of the presence of highly positively charged amine head-groups. Transfection reagents available as kits from various manufacturers are given in **Table 4**. Details of other methods of transfection are available elsewhere *(20,21)*.

Table 3
Control Reporter Expression Vectors

S. no.	Reporter gene	Name of the vector	Properties	Promoter[a]	Source
1	CAT	pCAT3—control	4.4 kb/Amp^r	SV40	Promega
		pSV2CAT	N.A.	SV40	*(10)*
2	Luciferase	pGL2—control	6.0 kb/Amp^r	SV40	Promega
		pRL-SV40	3.7 kb/Amp^r	SV40	Promega
		pRL-CMV	4.0 kb/Amp^r	CMV	Promega
3	β-Galactosidase	pβgal—control	7.9 kb/Amp^r	SV40	Clontech
		pSV-β-gal—control	6.82 kb/Amp^r	SV40	Promega
		pCMVβ	7.2 kb/Amp^r	CMV	Clontech
		pCMV.Sport-β-gal	7.8 kb/Amp^r	CMV	Life Technologies
4	EGFP	pEGFP-N1,N2,N3	4.7 kb/Kan^r/Neo$^{r\#}$	CMV	Clontech
		pEGFP-C1,C2,C3	4.7 kb/Kan^r/Neo$^{r\#}$	CMV	Clontech
		pd2EGFP-N1	4.8 kb/Kan^r/Neo$^{r\#}$	CMV	Clontech
5	SEAP	pCMV/SEAP	7.58 kb/Amp^r	CMV	Applied Biosystems
		pSV40/SEAP	6.2 kb/Amp^r	SV40	Applied Biosystems
6	Human growth hormone (hGH)	pXGH5	6.67 kb/N.A.	mMT-I	*(8)*

Kan^r, prokaryotic selection marker; # Neo^r, eukaryotic selection marker.

a. CMV, immediate early promoter of human cytomegalovirus (CMV); SV40, early promoter and enhancer of SV40; mMT-1, mouse metallothionein promoter.

3.3.1. Calcium Phosphate–Mediated Transfection

1. Plate cells at a density of 0.5×10^6 per 35-mm dish or 1.5×10^6 per 60-mm dish 1 d before transfection. It is important that cells are harvested from an exponentially growing culture. Incubate the cells for 20–24 h at 37°C in a serum-containing medium in an atmosphere of 5% CO_2.
2. Next day, change the medium a few hours before the transfection.
3. Prepare the DNA–calcium phosphate precipitate: For 35-mm dish, place the DNA sample(s) to be transfected in an Eppendorf tube and make up the volume to 94.5 μL using sterile water. Add 125 μL 2× HEPES-buffered saline to the DNA solution and mix the contents well. Add 31 μL of 1 *M* calcium chloride solution dropwise to the DNA–HEPES solution over a period of 30 s with gentle mixing. We usually keep the Eppendorf tube containing DNA–HEPES solution under constant mixing with the use of a vortex mixer while adding the $CaCl_2$ in drops.
4. Keep the tube at room temperature for 45 min.
5. At the end of incubation, mix and add the contents of the tube into the medium above the cells in drops. Swirl the dish gently to mix the medium, which will become yellow-orange and turbid. An alternative method is to remove the medium and add the precipitate directly to the cells. Incubate the cells for 15 min at room temperature, and then add medium to the dish.
6. Remove the medium from the dish 16–24 h after adding the DNA, wash twice with PBS, add fresh medium containing 10% serum, and continue incubation.
7. Harvest the cells after 1–2 d after medium change and carry out the reporter assay.

3.4. Co-transfection with Control Plasmid

The transfection efficiency between samples can sometimes vary as many as two- to threefold, which if not taken into consideration can easily mislead. Co-transfecting an internal control plasmid carrying a reporter other than the one used for promoter analysis experiments can monitor the difference in the transfection efficiency between samples. Normalizing the data for transfection efficiency with the use of internal control makes the results reproducible. Internal controls are particularly important if the activation or repression is of the order of two- to threefold. Basically, one can choose any reporter expression vector (**Tables 2** and **3**). β-Galactosidase reporter is very often used as an internal control reporter. CMV or RSV promoter-based reporter expression plasmids are very commonly used. It is important to note that the genes used in the transfection experiments should not affect the promoter of the internal control plasmid. For example, SV40 promoter-based reporter vectors are not suitable for internal control in experiments involving adenovirus oncogene E1A, because E1A has been shown to inhibit transcription from SV40-promoter. The inclusion of internal control reporter also warrants the right choice of the lysis buffer (*see* **Note 4**). Triton X-100-based lysis protocol is not suitable from CAT enzyme as the sensitivity of the CAT assay can be reduced 10-fold by the detergent.

3.5. Analysis of cis-Acting Elements in a Promoter

Analysis of the promoter of a target gene will give a lot of valuable information on the regulation of the target gene by the tumor suppressor gene. The regulation can be studied by directly quantifying the levels of specific mRNAs using the techniques such as Northern hybridization or nuclease protection assays. However, these procedures are time-consuming and are not always practical to analyze many different promoter gene constructs. An alternative method is to fuse the promoter sequences of the target gene with a reporter gene. To analyze a complete promoter, the DNA fragment carrying the promoter sequence can be cloned upstream of the reporter gene in a vector lacking a promoter (**Table 1**). If the promoter under study is a weak one, one can use vectors, which carry an enhancer element (**Table 1**). Following transfection of the promoter/reporter construct, one measures the reporter protein level or enzyme activity as a measure of promoter activity.

Once the assay system for the complete promoter is established, one can characterize the promoter for the *cis*-regulatory elements by several methods. For example, *cis*-regulatory elements can be characterized by constructing successive deletion mutants (**Fig. 2**), by altering the sequence orientation, or by site-directed mutagenesis. Activation of the promoter by a transcription factor can be studied by co-transfecting the cDNA, which encodes the transcription factor along with the promoter–reporter construct. For example, the response element for activation by p53 has been mapped in the promoter of p53 target genes p21^{WAF1} *(22,23)*, GADD45 *(24)*, Bax *(25)*, c-Met *(26)*, PCNA *(27)*, IGF-BP3 *(28)*, and KILLER/DR5 *(29)*. The p21^{WAF1} promoter is one of the very well characterized promoters for its activation by several transcription factors or pathways. Activation of p21^{WAF1} promoter by p53 *(22,23)*, STAT1 *(30)*, AP2 *(31)*, c-Jun *(32)*, Sp1 and Smad1 *(33)*, BRCA1 *(34)*, and TGF-β *(35)* are well documented. The requirement of transcription factor can also be studied by the expression of the reporter construct in different cell types and organisms or by external manipulation of the source of the factor. For example, involvement of a signaling pathway in the activation of a transcription factor can be studied by activating the pathway by treating the cells with activating drug **(Fig. 2)**.

Table 4
Commercially Available Transfection Reagents

Company name	Product name	Type
Life Technologies (www.lifetech.com)	Calcium Phosphate Transfection System	Calcium phosphate
	DMRIE-C Reagent	Monocationic lipid (DMRIE) and cholesterol
	Lipofectin® Reagent	Monocationic lipid (DOTMA:DPOE)
	LipofectAMINE™ Reagent	Polycationic lipid (DOSPA:DOPE)
	LipofectAMINE™ Plus Reagent	Polycationic lipid (DOSPA:DOPE)
	LipofectACE™ Reagent	Monocationic lipid (DDAB:DOPE)
	CellFECTIN® Reagent	Cationic lipo-polyamine (spermine groups covalently attached; TM-TPS-DOPE)
Promega (www.promega.com)	Tfx™ Reagents	Three different formulations of cationic lipid (DOPE)
	Profection®	Calcium phosphate/DEAE dextran
	Transfectam®	Cationic lipo-polyamine (spermine groups covalently attached)
Qiagen (www.qiagen.com)	SuperFect™ Transfection Reagent	Activated dendrimer
	Effectene™ Transfection Reagent	Cationic lipid
Clontech (www.clontech.com)	CLONfectin™ Transfection Reagent	Cationic lipid
Stratagene (www.stratagene.com)	LipoTAXI™ Transfection Reagent	Novel liposome reagent
	Transfection MBS Mammalian Transfection Kit	Calcium phosphate together with specially modified bovine serum
	Mammalian Transfection Kit	Calcium phosphate/DEAE dextran
Boehringer Mannheim/ Roche Biochemicals (www.biochem.boehringer-manheim.com)	FUGENE™ 6 Transfection Reagent	Blend of lipids and proprietary compounds.
	DOSPER Liposomal Transfection Reagent	Polycationic lipid
	DOTAP Liposomal Transfection Reagent	Monocationic lipid
	X-tremeGENE Q2 Transfection Reagent	Proprietary reagent
Invitrogen (www. invitrogen. com)	PerFect Transfection Kit	Eight different lipids, monocationic and polycationic
	Calcium Phosphate Transfection Kit	Calcium phosphate

Table 4
Commercially Available Transfection Reagents *(continued)*

Pharmacia Biotech q (www.biotech.pharmacia.se)	CellPhect Transfection Kit	Calcium phosphate/DEAE dextran
Wako Bioproducts (www.panvera.com)	Gene transfer HMG-1, HMG-2 mixture	Cationic lipid Nuclear protein
Molecular Research Laboratories (www.molecula.com)	Maxfect™ Genfect™	Liposomal polycationic formulation Liposomal polycationic formulation
Speciality Media (www. specialitymedia.com)	Mammalian Cell Transfection Kit Transient Transfection Expression Kit	Calcium phosphate DEAE/dextran
Biontex (www.biontex.com)	Insectogene™ (specific for insect cells) Metafectene	Polycationic lipid Polycationic lipid transfection reagent based on RMA Technology (Repulsive Membrane Acidosis)
5 Primefi 3 Prime (www.5prime.com)	Calcium Phosphate Transfection Kit 20X DEAE Dextran Transfection solution	Calcium phosphate DEAE dextran
TaKaRa (distributed by PanVera) (www.panvera.com)	RetroNectin™	Recombinant human fibronectin fragment CH-296
Mirus Corporation (distributed by PanVera)	TransIT™ Polyamine Transfection Reagents (LT1, LT2, 100) TransIT™ PanPak	Cationic lipo-polyamine Three cationic Lipids

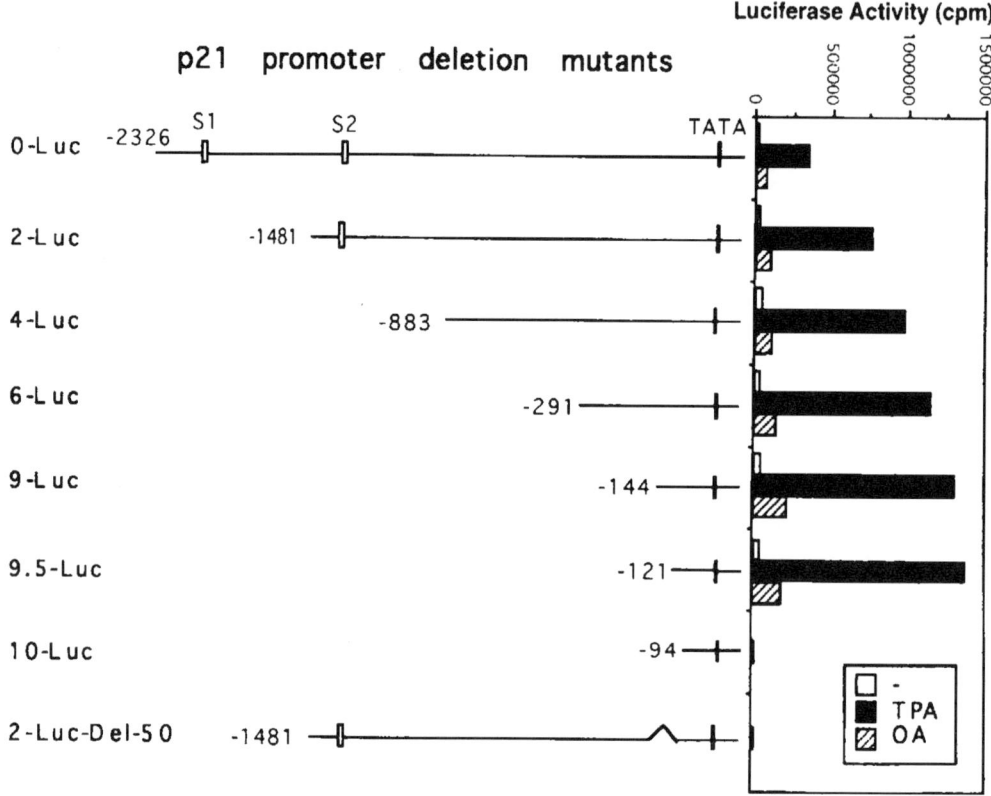

Fig. 2. Mapping of TPA response element to an AP2-binding region in the p21^{WAF1} promoter. The luciferase activity as a measure of activity of various deletion constructs of promoter of p21^{WAF1} in the absence and presence of TPA is studied. (From Zeng et al., 1997, with permission *(31)*)].

4. Notes

1. The pellet should be loosened before the addition of lysis buffer. This can be done by striking the tube few times against an Eppendorf tube rack. If the pellet is not resuspended properly, it will result in loss of total protein.
2. Repeated freeze-thawing of the cell lysate results in loss of luciferase activity.
3. We routinely use GFP expression vector for standardizing the transfection method. The advantage is that the efficiency of transfection can be monitored by looking directly at the live cells under a fluorescent microscope for the green fluorescence. Alternatively, one can use β-galactosidase expression vector, in which case the cells need to be fixed and stained using X-gal.
4. The reporter lysis buffer developed by Promega (cat. no. E397A) is suitable for luciferase, β-galactosidase, and CAT enzymes.

References

1. Thompson, J. F., Hayes, L. S., and Lloyd, D. B (1993) Modulation of firefly luciferase stability and impact on studies of gene regulation. *Gene* **103**, 171–177.
2. De Wet, J. R., Wood, K. V., DeLuca, M., Helinski, D. R., and Subramani, S. (1987) Firefly luciferase gene: structure and expression in mammalian cells. *Mol. Cell. Biol.* **7**, 725–737.
3. Bronstein, I., Fortin, J., Stanley, P. E.,. Stewart, G. S., and Kricka, L. J. (1994) Chemiluminescent and bioluminescent reporter gene assays. *Anal. Biochem.* **219**, 169–181.

4. Miller, J. (1972) *Experiments in Molecular Genetics*. Cold Spring Harbor Laboratory, Cold Spring Harbor, NY.

5. Young, D. C., Kingsley, S. D., Ryan, K. A., and Dutko F. J.(1993) Selective inactivation of eukaryotic B-galactosidase in assays for inhibitors of HIV-1 TAT using bacterial B-galactosidase as a reporter enzyme. *Anal. Biochem.* **215**, 24–30.

6. Jain, V. and Magrath, I. (1991) A chemiluminescent assay for quantitation of β-galactosidase in the femtogram range: application to quantitation of β-galactosidase in lacZ transfected cells. *Anal. Biochem.* **199**, 119–124.

7. Berger, J., Hauber, J., Hauber, R., Geiger, R., and Cullen B. R. (1988) Secreted placental alkaline phosphatase: a powerful new quantitative indicator of gene expression in eukaryotic cells. *Gene* **66**, 1–10.

8. Selden, R. F., Howie, K. B., Rowe, M. E., Goodman, H. M., and Moore, D. D. (1986) Human growth hormone as a reporter gene in regulation studies employing transient gene expression. *Mol. Cell. Biol.* **6**, 3173–3179.

9. Inouye, S. and Tsuji, F. I. (1994) *Aequorea* green fluorescent protein: expression of the gene and fluorescence characteristics of the recombinant protein. *FEBS Lett.* **341**, 277–280.

10. Gorman, C. M., Moffat, L. F., and Howard, B. H.(1982) Recombinant genomes which express chloramphenicol acetyltransferase in mammalian cells. *Mol. Cell Biol.* **2**, 1044–1051.

11. Bradford, M. M. (1976) A rapid and sensitive method for the quantitation of microgram quantities of protein utilizing the principle of protein dye binding. *Anal. Biochem.* **72**, 248–254.

12. Lowry, O. H., Rosebrough, N. J., Farr, A. L., and Randall, R. J. (1951) Protein measurement with the Folin phenol reagent. *J. Biol. Chem.* **193**, 265–275.

13. Laiminis, L. A., Gruss, P., Pozzatti, R., and Khoury, G. (1984) Characterization of enhancer elements in the long terminal repeat of Moloney murine sarcoma virus. *J. Virol.* **49**, 183–189.

14. Gilman, M. Z., Wilson, R. N., and Weinberg, R. A (1986) Multiple protein binding sites in 5′-flanking region regulate c-fos expression. *Mol. Cell. Biol.* **6**, 4305–4316.

15. Nordeen, S. K. (1988) Luciferase reporter gene vector for analysis of promoters and enhancers. *Biotechniques* **6**, 454–457.

16. Luckow, B. and Schütz, G. (1987) CAT constructions with multiple unique sites for functional analysis of eukaryotic promoters and regulatory elements. *Nucleic Acids Res.* **15**, 5490.

17. Vaheri, A. and Pagano, J. S. (1965) Infectious poliovirus RNA: a sensitive method of assay. *Virology* **27**, 434–436.

18. Graham, F. L. and van der Eb, A. J. (1973) A new technique for the assay of infectivity of human adenovirus 5 DNA. *Virology* **52**, 456-467.

19. Felgner, P. L., Gadek, T. R., Holm, M., et al. (1987) Lipofectin: a highly efficient, lipid-mediated DNA-transfection procedure. *Proc. Natl. Acad. Sci USA* **84**, 7413–7417.

20. Sambrook, J., Fritsch E. F., and Maniatis, T. (eds.) (1989) *Molecular Cloning: A Laboratory Manual*. Cold Spring Harbor Laboratory Press, Cold Spring Harbor, NY.

21. Ausubel, F. M., Brent, R., Kingston, R. E., et al. (eds.) (1994) *Current Protocols in Molecular Biology*. Wiley, New York.

22. El-Deiry, W. S., Tokino, T., Velculescu, V. E., et al. (1993) *WAF1*, a potential mediator of p53 tumor suppression. *Cell* **75**, 817–825.

23. El-Diery, W. S., Tokino, T., Waldman, T., et al. (1995) Topological control of p21$^{WAF1/CIP1}$ expression in normal and neoplastic tissues. *Cancer Res.* **55**, 2910–2919.

24. Hollander, M. C., Alamo, I., Jackman, J., Wang, M. G., McBride, O. W. and Fornace, A. C., Jr. (1993) Analysis of the mammalian *gadd45* gene and its response to DNA damage. *J. Biol. Chem.* **268**, 24385–24393.

25. Miyashita, T. and Reed, J. C. (1995) Tumor suppressor p53 is a direct transcriptional activator of human *bax* gene. *Cell* 80, 293–299.

26. Seol, D., Chen, Q., Smith, M. L., and Zarnegar, R. (1999) Regulation of the *c-met* proto-oncogene promoter by p53. *J. Biol. Chem.* **274**, 3565–3572.

27. Morris, G. F., Bischoff, J. R., and Mathews, M. B. (1996) Transcriptional activation of the human proliferating-cell nuclear antigen promoter by p53. *Proc. Natl. Acad. Sci. USA* **93**, 895–899.

28. Buckbinder, L., Talbott, R., Velasco-Miguel, S., et al. (1995) Induction of the growth inhibitor IGF-binding protein 3 by p53. *Nature* **377**, 646–649.

29. Takimoto, R. and El-Deiry, W. S. (2000) Wild-type p53 transactivates the KILLER/DR5 gene through an intronic sequence-specific DNA-binding site. *Oncogene* **19**, 1735–1743.

30. Chin, Y. E., Kitagawa, M., Su, W. C., You, Z. H., Iwamoto, Y., and and Fu, X. Y. (1996) Cell growth arrest and induction of cyclin-dependent kinase inhibitor p21 WAF1/CIP1 mediated by STAT1. *Science* **272**, 719–722.

31. Zeng, Y. X., Somasundaram, K., and El-Diery, W. S.(1997) AP2 inhibits cancer growth and activates p21WAF1/CIP1 expression. *Nat. Genet.* **15**, 78–82.

32. Kardassis, D., Papakosta, P., Pardali, K., and Moustakas, A. (1999) c-Jun transactivates the promoter of the human p21$^{WAF1/Cip1}$ gene by acting as a superactivator of the ubiquitous transcription factor Sp1. *J. Biol. Chem.* **274**, 29572–29581.

33. Moustakas, A. and Kardassis, D. (1998) Regulation of the human p21/WAF1/Cip1 promoter in hepatic cells by functional interactions between Sp1 and Smad family members. *Proc. Natl. Acad. Sci. USA* **95**, 6733–6738.

34. Somasundaram, K., Zhang, H., Zeng, Y., et al. (1997) Arrest of the cell cycle by the tumor suppressor BRCA1 requires the CDK-inhibitor p21$^{WAF1/Cip1}$. *Nature* **389**, 187–190.

35. Datto, M. B., Yu, Y., and Wang, X. (1995) Fuctional analysis of the transforming growth factor β responsive elements in the WAF1/Cip1/p21 promoter. *J. Biol. Chem.* **270**, 28623–28628.

9

DNA Footprinting

M. Stacey Ricci and Wafik S. El-Deiry

1. Introduction

The classic paper by Schmitz and Galas *(1)* established the usefulness of footprinting analysis for identifying protein-bound sites on DNA. The basis of the footprinting technique is that DNA-bound proteins protect the phosphodiester backbone of DNA from modification or cleavage by external agents, such as deoxyribonuclease. The technique is used most commonly to identify protected DNA sequences that are binding sites for transcription factors or other proteins involved in transcription initiation. When used in combination with other methods that confirm high-affinity protein–DNA interactions, such as electrophoretic mobility shift assay or chromatin immunoprecipitation, footprinting adds important evidence that delineates the residues critical for protein binding.

Variations of the footprinting method include both in-vitro and in-vivo assays. In-vitro studies are relatively straightforward, requiring a protein source, the DNA fragment to be footprinted, and the use of sequencing gel apparatus. The in-vivo technique requires more reagents because it couples the footprinting assay with ligation-mediated polymerase chain reaction (PCR) (*see* **Note 1**). Here we present two protocols for alternative methods of in-vitro DNA protection assay, DNase I protection and methylation interference. Both methodologies share several steps in common. A known DNA fragment containing the protein-binding recognition sequence is radiolabeled at one end and mixed with the protein(s) of interest. These protein-bound DNA fragments are isolated by immunoprecipitation and the bound DNA is then cleaved with attacking agents and separated on a sequencing gel. The fragments absent from the DNA ladder, when compared to that of cleaved nonfootprinted DNA, enables the identification of residues important for protein binding. Schematics of the methods are presented in **Figs. 1** and **2**.

The major difference between the two assays is not simply the way in which DNA is cleaved. In the case of Dnase I protection, the immunoprecipitated protein-bound DNA is incubated with deoxyribonuclease that cleaves the unprotected DNA, i.e., the DNA not bound by protein. Therefore, protein binding itself prevents cleavage of the DNA at those residues. The DNA sequences that are protected from cleavage will be absent in the DNA ladder run on the gel (**Fig. 3**).

From: *Methods in Molecular Biology, Vol. 223: Tumor Suppressor Genes: Regulation, Function, and Medicinal Applications.* Edited by: Wafik S. El-Deiry © Humana Press Inc., Totowa, NJ

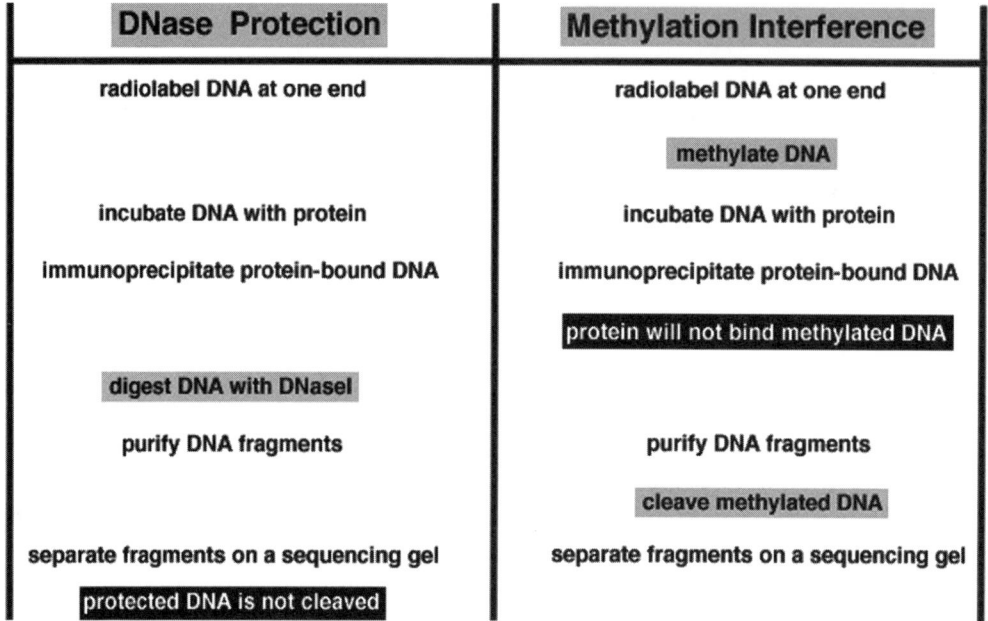

Fig. 1. Procedures used in two common in-vitro DNA footprinting techniques.

Methylation interference does not rely on protein protection of DNA for selecting protein-bound sequences. Rather, the DNA fragments are first modified by the addition of methyl groups to residues at a frequency of approximately one residue per DNA fragment. If the modification is present in the DNA-binding region of the protein, then this modification will interfere with protein recognition of DNA-binding sequences. Therefore, fragments containing modifications at these residues will be absent from the immunoprecipitated DNA. The protein-bound DNA sequences are then identified by their absence in the DNA ladder on the sequencing gel.

Both techniques offer the ability to identify specific nucleotides that are protein-bound. However, the different methods of DNA attack will provide unique information. DNase I protection offers the advantage of permitting the identification of multiple binding sites within one strand of DNA, whereas methylation interference does not. This was particularly useful when examining the human *bax* gene promoter, a sequence that contains four potential c-Myc-binding sites, known as E-boxes *(2)*. DNase I footprinting analysis showed that only one of the four binding sites was a high-affinity binding site for c-Myc. Methylation interference involves first methylating guanines and, to a lesser degree, adenines, using dimethyl sulfate. These methylated residues are later cleaved specifically by piperidine. In our experience, this method generally produces a clearer banding pattern than Dnase I protection. We have used both DNase I protection and methylation interference assays to better define the consensus p53 tumor suppressor DNA-binding site *(3)*.

The following protocol describes how to prepare end-labeled DNA to be footprinted, whether it is to be PCR-amplified or removed from a plasmid. The user must have a protein source and an antibody specific to the protein of interest. Sequencing gel apparatus and knowledge of its use is also required. All other reagents are commercially available, and most laboratories are equipped with the necessary instruments.

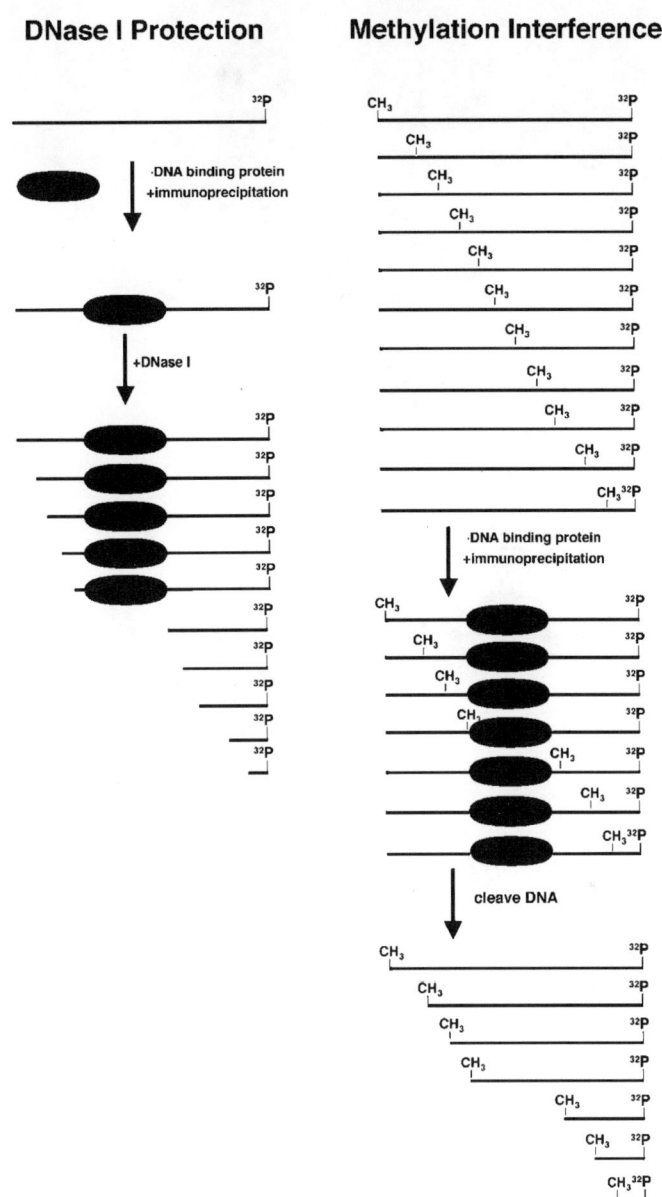

Fig. 2. Generation of DNA footprints.

2. Materials

2.1. DNA Labeling

Radioactively labeling one end of a DNA fragment can be done using a cloned insert removed from a plasmid or by using PCR. If the DNA is to be removed from a plasmid, then the enzymes used to release the fragments must generate 5′-overhangs (recessed 3′-ends). This is necessary because end-labeling is done using DNA polymerase to fill in the recessed ends.

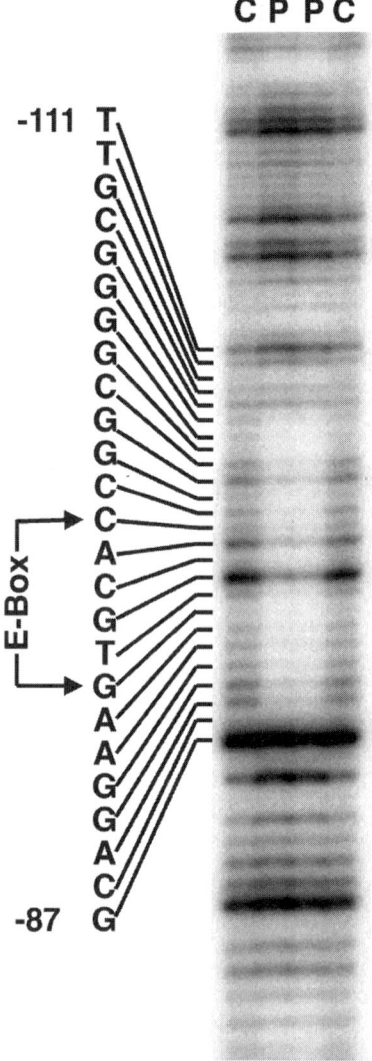

Fig. 3. Deoxyribonuclease protection. Shown is a footprint using in-vitro-translated c-Myc and Max proteins that bind to the region of the human *bax* gene (numbered relative to the translational start site). The footprint shows that residues flanking the E-box are also protected from DNase I attack. Lanes marked with "P" refer to protected DNA, and those marked "C" are the unprotected controls.

2.1.1. Using PCR

If the DNA sequence will be removed from a plasmid by enzymatic digestion, you will need the following reagents:

1. Plasmid containing the DNA sequence.
2. Enzyme that will produce a 5′-overhang (e.g., HindIII), which can be subsequently filled in with DNA polymerase (*see* **Subheading 3.1.2.**) plus the accompanying 10× buffer.
3. One [α-^{32}P]-labeled dNTP to be used in the fill-in reaction. It is preferable to use either [α^{32}P]-dCTP or -dATP, since these are the most stable of the radiolabeled dNTPs (New England Nuclear; 10 Ci/mmol).

4. 5 m*M* solutions of the remaining dNTPs to be used in the fill-in reaction (dATP, dGTP, dCTP, or dTTP).

2.1.2. Using a Plasmid

If the DNA sequence to be used will be PCR-amplified, then each primer first must be radioactively labeled for use in the amplification reaction:

1. 350 ng/μL solution of each primer.
2. [γ^{32}P]-ATP (New England Nuclear BLU-502A; 10 Ci/mmol).
3. T4 polynucleotide kinase and accompanying 10× buffer.
4. 10 m*M* solution of each dNTP for PCR amplification.
5. *PfuTurbo*® DNA polymerase plus accompanying 10× buffer (Stratagene cat. no. 600250) or another suitable high-fidelity DNA polymerase used in polymerase chain reactions.

2.2. Probe Purification

1. 100% Ethanol.
2. 70% Ethanol.
3. Glycogen (Roche Molecular Biochemicals 901393, 0.02 mg/mL).
4. 10 m*M* ammonium acetate.
5. TE buffer: 10 m*M* Tris-Cl, pH 7.5, 1 m*M* EDTA. Add 10 mL 1 *M* Tris-Cl, pH 7.5, 2 mL 0.5 *M* EDTA, pH 8.0, to a final volume of 1 L).
6. Microspin™ G-50 spin column (Amersham Pharmacia Biotech).

2.3. Nondenaturing Polyacrylamide Gel Electrophoresis

1. Glass plates, spacers, and combs for casting vertical gels and accompanying acrylamide gel electrophoresis apparatus.
2. 29/1 (w/w) acrylamide/bisacrylamide solution (29 g acrylamide, 1 g *N-N'*-bismethylene acrylamide H$_2$O to 100 mL; store at 4°C) (*see* **Note 2**).
3. Tris-acetate-EDTA electrophoresis buffer (50× stock solution): 242 g Tris base, 57.1 mL glacial acetic acid, 37.2 g Na$_2$EDTA·2H$_2$O, H$_2$O to 1 L—the working solution will be 40 m*M* Tris·acetate, 2 m*M* EDTA·2H$_2$O, pH ~8.5).
4. TEMED (*N,N,N',N'*-tetramethylethylenediamine; store at 4°C).
5. 10% (w/v) Ammonium persulfate (APS) in water; store at 4°C.
6. 5× DNA loading buffer: 50 m*M* Tris·Cl, pH 8.0, 50 m*M* EDTA, pH 8.0, 50% glycerol, 0.25% bromphenol blue, 0.25% xylene cyanol.
7. Razor blade.
8. Plastic wrap.
9. X-ray film.
10. Reagents for probe purification (*see* **Subheading 2.2.**).

2.4. Immunoprecipitation

1. End-over-end rotation device.
2. Protein A/G-linked agarose beads for use with most monoclonal antibodies and goat or rabbit polyclonal antibodies. Suspend beads to a concentration of 1.5 mg/mL for use in each immunoprecipitation reaction.
3. 5× DNA-binding buffer: 500 m*M* NaCl, 100 m*M* Tris-Cl, pH 7.0, 50% glycerol, 5% NP40, 25 m*M* dithiothreitol (DTT). Add 2.92 g NaCl, 10 mL 1 *M* Tris-Cl, pH 7.0, 50 mL glycerol, 5 mL NP-40, 2.5 mL 1 *M* DTT, bring to 100 mL.
4. 100-ng/μL poly(dI-dC)/poly(dI-dC) or 10-μg/μL sheared salmon sperm DNA (*see* **Subheading 3.2.2.**).

5. Antibody to the DNA-binding protein.

2.5. DNase I Protection

1. DNA-binding buffer (*see* **Subheading 2.4.**).
2. Deoxyribonuclease I (DNase I) (GibcoBRL 150 U/mL cat. no.18047-019).
3. 1 M MgCl$_2$.
4. TE buffer: 10 mM Tris-Hcl, pH 7.5, 1 mM EDTA.
5. 25/24/1 (v/v/v) phenol/chloroform/isoamyl alcohol (made with buffered phenol). For purification of DNA, it is good practice to use phenol that has been redistilled before use, since oxidation products of phenol can damage and introduce breaks into nucleic acid chains. Redistilled phenol for use in nucleic acid purification is commercially available, but must be buffered before use.
6. Proteinase K/SDS solution: Add 5 mg proteinase K/mL of 10% sodium dodecyl sulfate (SDS).

2.6. Methylation Interference

1. DMS buffer: 50 mM cacodylic acid-Cl, pH 7.0, 1 mM EDTA.
2. 10% Dimethyl sulfoxide (DMS): 100 µL dimethyl sulfate, 900 µL ethanol (*see* **Note 3**).
3. DMS stop buffer: For 20 reactions use 500 µL 3 M sodium acetate, 71 µL 1 M β-mercaptoethanol, 60 µL glycogen (Roche Molecular Biochemicals 901393, 0.02 mg/mL), 370 µL H$_2$O).
4. 1 M piperidine: 200 µL 10 M piperidine, 1800 µL H$_2$O—make fresh.

2.7. Sequencing Gel

1. 10× TBE buffer: 890 mM Tris-borate, pH 8.3–8.9, 20 mM EDTA. Add 108 g Tris base, 55 g boric acid, 40 mL 0.5 M EDTA, pH 8.0.
2. 38% Acrylamide/2% bisacrylamide solution: mix 38 g acrylamide (ultrapure) and 2 g N-N'-bismethylene acrylamide (ultrapure) in about 60 mL H$_2$O. Heat if necessary to dissolve, but do not heat above 55°C. Adjust volume to 100 mL. Store protected from light at 4°C.
3. Urea (ultrapure).
4. Formamide loading buffer: 0.05% (w/v) bromphenol blue, 0.05% (w/v) xylene cyanol, 20 mM EDTA prepared in deionized formamide.

3. Methods
3.1. Labeling the DNA

The DNA fragments to be footprinted are labeled at one end only. It is a good idea to perform the footprint twice, using the same DNA fragment labeled at each of the two ends. This helps to obtain good resolution of the entire sequence and adds reassurance that the same pattern is reproducible. Two methods for end-labeling DNA fragments are provided (*see* **Note 4**).

3.1.1. End-Labeling and Amplifying the DNA Using PCR

This procedure involves end-labeling one of the primers to be used to amplify the DNA fragment that will be footprinted. First the primer is labeled at the 5'-end using T4 polynucleotide kinase. This radiolabeled primer is used in the amplification reaction.

1. Add 3.5 µg primer, plus 50 µCi [γ^{32}P]-ATP to a 50-µL reaction containing 30 Richardson units of polynucleotide T4 kinase. Incubate the reaction for 30 min at 37°C. To inactivate

the kinase incubate further for 10 min at 70°C. This entire volume can be used in five 100-µL PCR reactions.

2. Perform PCR using the entire kinase reaction, plus 3.5 µg of the unlabeled primer. To the 50-µL reaction, add 50 of µL 10× *PfuTurbo*® buffer, 315 µL H_2O, 50 µL 10 mM dNTP mix, 1 µg DNA template, and 10 µL *PfuTurbo*® DNA polymerase. Perform 35 cycles of PCR at the appropriate annealing temperature. Pellet the labeled DNA fragment from the unincorporated radionucleotides (*see* **Subheading 3.1.3.**) and quantify by placing the microcentrifuge tube containing the DNA pellet into a scintillation vial for use in a scintillation counter.

3.1.2. End-Labeling Using Cloned Inserts

This procedure is used if the DNA fragment can be released from a plasmid by restriction enzymes that produce 5′-overhangs. The plasmid is cut with this enzyme, purified, and then the ends are filled in using a DNA polymerase plus the appropriate radiolabeled dNTP. After purification of the linear radiolabeled plasmid, the fragment is released by a second restriction enzyme digest.

1. Digest up to 10 µg DNA, in separate reactions, either 3′ or 5′ of the insert using an enzyme that produces 5′-overhangs;
2. Ethanol-precipitate the DNA: Bring the enzyme reaction volume to 200 µL, add 3 µL glycogen, 100 µL 10 M ammonium acetate, and 700 µL ethanol and chill by placing at −70°C for 10 min, or at –20°C for 30 min. Spin at top speed in a microcentrifuge for 15 min at 4°C. Wash the pellet once in 70% ethanol and dry under vacuum.
3. Suspend the pellet in a 25-µL DNA Polymerase I, Large (Klenow) fragment reaction containing a final concentration of 0.2 mM of each dNTP needed and 0.4 µCi of [α^{32}P]-dNTP. As an example, if HindIII was used, it leaves a 5′-TCGA-3′ overhang. Therefore, include 1 µL of 5 mM dATP, dTTP, and dGTP and use [α^{32}P]-dCTP. Incubate the reaction for 30 min at room temperature. Heat-inactivate the enzyme at 80°C for 5 min. To release the labeled insert from the plasmid, add 65 µL water, 10 µL of 10× reaction buffer, and the restriction enzyme that will release the insert.

3.1.3. Probe Purification

To reduce the potential for radioactive contamination prior to gel electrophoresis, remove the unincorporated radioactivity by passing the labeled DNA fragment over a G50 spin column or by ethanol precipitation. When ethanol-precipitating DNA, it is advantageous for improved recovery to add a carrier agent, such as glycogen.

1. Bring the DNA sample volume to 200 µL with water. Add 3 µL of glycogen to the DNA sample and 100 µL 10 M ammonium acetate and mix. Add 700 µL 100% ethanol and chill by placing it at –70°C for 10 min, or at –20°C for 30 min. Spin at 4°C for 15 min. Remove the radioactive supernatent carefully and place it into the radioactive liquid waste. Wash the pellet with 1 mL 70% ethanol. Remove any residual liquid and dispose in the radioactive waste. Allow the pellet to air dry at room temperature. Suspend in 40 µL TE buffer.
2. It is now advantageous to further purify your probe by native polyacrylamide gel electrophoresis. This will eliminate any radiolabeled primers or undigested plasmids from your preparation. Choose a percentage gel that best suits your fragment size (**Table 1**).
3. Acrylamide percentages refers to amount of acrylamide using a w/w stock solution of 29/1 acrylamide/bisacrylamide. The buffering system used is 40 mM Tris-acetate (~pH 8.5), 2 mM EDTA. Pour the gel and allow time for polymerization.
4. Add 10 µL of 5× DNA loading buffer to your sample(s), load onto the gel, and run at 5 V/cm until the desired resolution is obtained.

Table 1
Concentrations of Acrylamide Giving Maximum Resolution of DNA Fragments[a]

Acrylamide (%)	Size of fragments separated (bp)	Migration of bromphenol blue marker (bp)	Migration of xylene cyanol marker (bp)
3.5	100–1000	100	460
5.0	100–500	65	260
8.0	60–400	45	160
12.0	50–200	20	70
20.0	5–100	12	45

[a]Data taken from **ref. 4**.

5. Disassemble the apparatus and remove one plate. Carefully wrap the gel and attached glass plate with plastic wrap and expose the gel to film for 1–10 s. Using the processed film as a guide, cut out the desired fragment(s) from the gel and place into microcentrifuge tubes. Crush the gel fragment into smaller pieces. Add enough TE buffer to cover the gel fragments, using no more than 200 µL if possible. Incubate at room temperature with slow agitation or end-over-end rotation for 3 h or more (overnight will produce the best results).
6. Pellet the gel fragments by spinning for 1 min in a microcentrifuge. Pipet off the supernatent and place into another microcentrifuge tube. Bring the volume to 200 µL and ethanol precipitate as described above. Suspend the pellet in 50 µL TE and quantify the radioactivity by scintillation counting. A useful working concentration is 50,000 cpm/µL.

3.2. DNase I Protection Assay

In this protocol, DNA is incubated with the protein(s) of interest and then subjected to immunoprecipitation using antibody-conjugated agarose beads. These protein–DNA–bead complexes are incubated with DNase I, which cleaves the unprotected DNA. It is recommended to test a range of DNase I concentrations using the naked DNA fragment to find the optimum DNase I concentration necessary. This reaction is also useful because digestion of the unprotected DNA is necessary for comparison with the footprinted DNA.

3.2.1. Protecting the DNA

Using purified proteins will provide the cleanest results when footprinting. Given the cost and the myriad of possible technical difficulties involved in generating purified proteins, using in-vitro translated proteins, with rabbit reticulocyte lysates for example, has provided good results *(2)*. Such lysates are commercially available. Producing in-vitro translated proteins requires plasmids containing the coding sequences for the protein(s) of interest, whose transcription can be driven by one of several bacterial RNA polymerases.

The amount of protein necessary to produce good results needs to be derived empirically for each protein source. However, a good starting point for purified proteins is 100 ng purified protein or 5 µL of in-vitro-translated lysate.

3.2.2. Immunoprecipitating the DNA

1. For the immunoprecipitation all incubations are at 4°C and use gentle mixing in the end-over-end rotation device or a suitable substitute.
2. Add 100,000 cpm DNA with protein in a total volume of 25 µL of 1× DNA-binding buffer in a microcentrifuge tube and incubate for 30 min.
3. Add 1 µg of the appropriate antibody and incubate for 30 min.

4. Spin down 1.5 mg protein A/G agarose beads for 15 s in a microcentrifuge, wash once by removing the supernatent, suspend the beads in 1 mL DNA binding buffer, and spin again for 15 s. Remove the supernatent. Add the protein–DNA complex to the pelleted beads. Also add 10 μg sheared salmon sperm if using a crude protein mixture, such as rabbit reticulocyte lysate, or 100 ng poly dI-dC/poly dI-dC if using purified proteins, as nonspecific competitor, and incubate for a final 30 min.
5. Wash the beads 3 times as described above.

3.2.3. Digesting Unprotected DNA

The amount of DNase I to add to the reactions needs to be determined empirically. A good working range for optimization are dilutions from 1:10 to 1:1000.

1. Deoxyribonuclease digestion is performed directly with the DNA–protein-bound agarose beads. Suspend the beads in 150 μL DNA-binding buffer containing the DNase I plus 5 m*M* magnesium chloride. *It is essential to add this cation, otherwise the enzyme will have no activity.* Incubate at room temperature for 2 min.
2. Stop the reaction by adding 100 μL stop buffer (80 μL TE plus 20 μL Proteinase K/SDS solution) and incubate at 48°C for 30 min.
3. Raise the volume of the reaction to 200 μL by adding 100 μL TE. Extract the DNA using phenol/chloroform. Ethanol precipitate the aqueous layer (being sure to add a DNA carrier) and wash the pellet once with 70% ethanol.

3.2.4. Running the Sequencing Gel

1. Prepare a 34 × 42-cm denaturing polyacrylamide/urea/TBE (Tris-borate-EDTA) sequencing gel following a standard protocol *(5)*. Use a 6% gel when the first binding site is approximately 100 bp from the [^{32}P] label. Use a higher percentage gel if greater resolution is needed.
2. Count each of the pellets to be loaded using a scintillation counter. This is important because the loading will be equalized between the protected and unprotected DNA. A good technique for helping to visualize differences between protected and unprotected DNA is to flank two lanes of the protected ladders on either side with unprotected DNase I treated DNA (*see* **Fig. 3**).
3. Suspend the pellets in formamide loading buffer to a concentration of 1000 cpm/μL. Denature the DNA samples for 10 min at 90°C. Place on ice.
4. Load 2000 cpm DNA per lane. This will give sharp clear bands after overnight exposure to film. Try to load equivalent volumes on the gel.
5. It is important to run a labeled known sequence on the gel for sizing purposes. Performing a sequencing reaction and loading this known sequence will provide a good reference, or labeling a commercially available low-molecular-weight DNA ladder or a plasmid digest is also an option.
6. Dry the gel and expose to film.

3.3. Methylation Interference

The basis of this assay is that the DNA fragment to be footprinted is first methylated prior to protein binding. Methylated residues in the protein binding region of the fragment should interfere with protein binding. Dimethyl sulfate methylates DNA at the N-7 position of guanine residues, and at the N-3 position of adenines. Adenines can be detected in this assay, but these reactions are much weaker in comparison to guanines. The methylation reaction usually will methylate DNA on one residue per strand of DNA. Therefore, if the DNA-binding protein can not bind methylated residues, as is the case for transcription factors, then sequences with methylated residues in the DNA binding site will not immunoprecipitate and will not be present on the sequencing gel following chemical cleavage.

3.3.1. Methylate the DNA

1. Perform the methylation reactions in duplicate for each condition to be tested. In a micro-centrifuge tube, combine 100,000 cpm of the DNA (2 μL), 200 μL DMS buffer, and 5 μL 10% dimethyl sulfate.
2. Incubate for 5 min at room temperature, then place immediately on ice. Add 50 μL chilled DMS stop solution.
3. Ethanol-precipitate the DNA, wash two times with 70% ethanol, and quantify the pellets. Recovery should be 70–90% of input.

3.3.2. Protecting the DNA

1. Perform the binding assay and immunoprecipitate as described under **Subheading 3.2.2.** Treat both protected and unprotected DNA to the same purification steps.
2. Quantify the DNA pellets.

3.3.3. Cleave Methylated DNA

1. Add 100 μL 1 *M* piperidine to the pellets.
2. Incubate for 30 min at 90°C.
3. Dry the pellet under vacuum.
4. Quantify the pellet and suspend in DNA loading buffer to at a concentration of 1000 cpm/μL.
5. Prepare a sequencing gel and the samples as described above and analyze the samples.

4. Notes

1. A criticism of in-vitro footprinting experiments is that an accurate cellular context of the protein-bound chromatin structure is not reproduced. The in-vivo footprinting technique was developed to address this issue *(6)*. This assay involves two steps (for a detailed protocol, also *see* **ref. 7**). First, DNA is modified, in the cellular context (cell cultures, whole sections, etc.), using a chemical, physical or enzymatic agent. If DNA is protected by bound proteins, then the modification will not occur. Second, the DNA sequence of interest is visualized using a ligation-mediated PCR technique. The theory here is that the genomic DNA is cleaved at many positions in the sequence of interest, except those regions protected by proteins. If a single strand of DNA is synthesized using a primer that anchors the sequence at one end, and a double-stranded DNA linker is annealed to each of the newly generated blunt ends that are made at each of the cleavage sites, then all the cleaved sequences can be amplified using a primer matching the linker and one gene-specific primer. Where the DNA was protected, there is no cleavage and therefore no ends able to be ligated. These absent sequences can be visualized as footprints on a sequencing gel.
2. Caution: Acrylamide is a neurotoxin. Always wear gloves when working with the unpolymerized monomer. Prepared acrylamide solutions are also commercially available.
3. Caution: DMS is toxic and must be used in a fume hood with careful handling. Liquids containing DMS should be disposed of in a designated DMS waste bottle, and pipet tips that come into contact with DMS should be placed in a separate DMS solid-waste bottle for disposal by institutional safety officers.
4. Resolution decreases if the DNA fragment is larger than 0.2 kb, but if the footprint is near either end of the DNA fragment, fragments as large as 0.5 kb will work.

References

1. Schmitz, A. and Galas, D. J. (1978) The interaction of RNA polymerase and lac repressor with the lac control region. *Nucleic Acids Res.* **6**, 111–137.
2. Mitchell, K. O., Ricci, M. S., Miyashita, T., et al. (2000) Bax is a transcriptional target and mediator of c-myc-induced apoptosis. *Cancer Res.* **60**, 6318–6325.
3. El-Deiry, W. S., Kern, S. E., Pietenpol, J. A., Kinzler, K. W., and Vogelstein, B. (1992) Definition of a consensus binding site for p53. *Nat. Genet.* **1**, 45–49.
4. Chory, J. and Pollard, J. D. (1999) Resolution and recovery of small DNA fragments, in *Current Protocols in Molecular Biology* (Moore, D., ed.) Wiley, New York, vol. 1, pp. 2.7.1–2.7.8.
5. Slatko, B. E. and Albright, L. M. (1992) Denaturing gel electrophoresis for sequencing, in *Current Protocols in Molecular Biology* (Ausubel, F. M. and Albright, L. M., eds.) Wiley, New York, vol. 1, pp. 7.6.1–7.6.13.
6. Garrity, P. and Wold, B. (1992) Effects of different DNA polymerases in ligation-mediated PCR: enhanced genomic sequencing and in vivo footprinting. *Proc. Natl. Acad. Sci. USA* **89**, 1021–1025.
7. Mueller, P. R., Wold, B., and Garrity, P. A. (1992) Ligation-mediated PCR for genomic sequencing and footprinting, in *Current Protocols in Molecular Biology* (Coen, D. M., ed.) Wiley, New York, vol. 2, pp. 15.15.11–15.15.26.

10

Identification of DNA-Binding of Tumor Suppressor Genes by Chromatin Immunoprecipitation

Timothy K. MacLachlan and Wafik S. El-Deiry

1. Introduction

The processes of DNA replication and transcription are intimately affected by the status of chromatin structure. Proteins bind and dissociate at rapid rates in order to control the levels of activity at the genome. Much work has been done over the years to determine the function of transcription factors and proteins that control the speed and fidelity of DNA replication. However, the bulk of these studies have attempted simply to correlate the expression levels of a DNA-binding protein with the function attributed to it, and not necessarily the dynamics of actual association with DNA. Indeed, in the field of transcriptional control, proteins have been shown to bind a consensus sequence in the promoter region of a particular gene, but these experiments are performed entirely in vitro, out of context with the genomic structure of the actual promoter region of the gene *(1)*. In the last few years, with the advent of highly specific antibodies, the development of the chromatin immunoprecipitation assay has been a major step forward in understanding the true in-vivo nature of and dynamics of DNA protein binding *(2–4)*. This technique allows the covalent cross-linking of DNA-binding proteins to chromatin and coincident precipitation of the DNA fragment with the protein upon antibody binding. Here we describe a protocol for the immunoprecipitation of transcription factors bound to promoter sequences in cells, and use the p53 tumor suppressor as an example (**Fig. 1**).

2. Materials

1. 3×10^6 cells treated appropriately to induce DNA binding of the transcription factor of interest.
2. Formaldehyde (Sigma cat. no. F8775).
3. Salmon sperm DNA/protein A agarose: Protein A agarose (Sigma cat. no. P3391) swelled in ChIP-protein A solution (10 mM Tris-HCl pH 8.0, 1 mM EDTA, 200 µg/mL sonicated salmon sperm DNA (Sigma cat. no. D1626), 500 µg/mL bovine serum albumin, 0.05% sodium azide) as a 50% slurry.

From: *Methods in Molecular Biology, Vol. 223: Tumor Suppressor Genes: Regulation, Function, and Medicinal Applications.* Edited by: Wafik S. El-Deiry © Humana Press Inc., Totowa, NJ

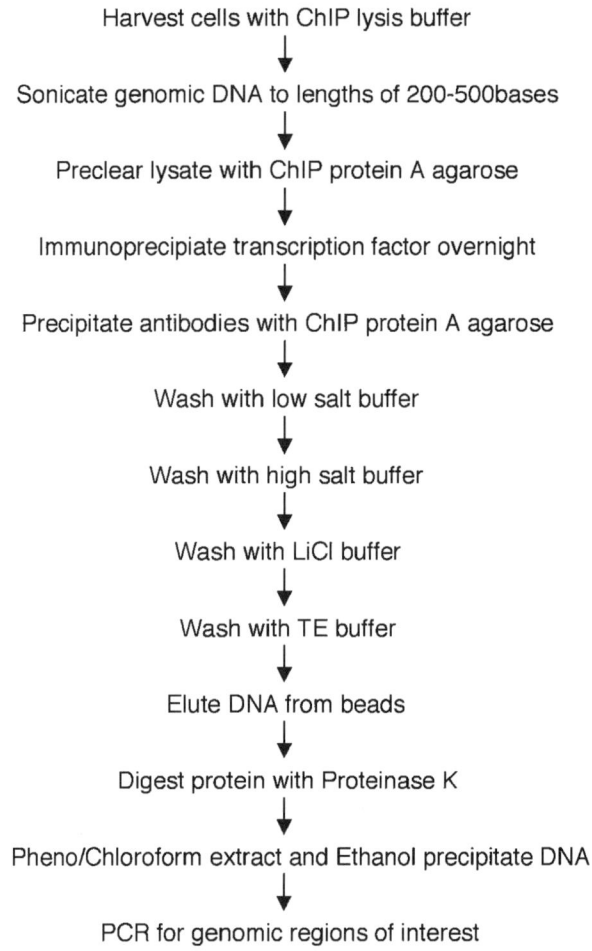

Harvest cells with ChIP lysis buffer

Sonicate genomic DNA to lengths of 200-500bases

Preclear lysate with ChIP protein A agarose

Immunoprecipiate transcription factor overnight

Precipitate antibodies with ChIP protein A agarose

Wash with low salt buffer

Wash with high salt buffer

Wash with LiCl buffer

Wash with TE buffer

Elute DNA from beads

Digest protein with Proteinase K

Pheno/Chloroform extract and Ethanol precipitate DNA

PCR for genomic regions of interest

Fig. 1. Flowchart of the chromatin immunoprecipitation assay.

 4. Phosphate-buffered saline (PBS).
 5. Lysis buffer: 50 mM Tris-HCl pH 8.0, 10 mM EDTA, 1% sodium dodecyl sulfate (SDS), 1/100 dilution of protease inhibitor cocktail (Sigma cat. no. P8340).
 6. Sonic dismembranator: For example, a Fisher 60 sonic dismembranator equipped with a 2-mm tip.
 7. Dilution buffer: 16.7 mM Tris-HCl pH 8.0, 1.2 mM EDTA, 167 mM NaCl, 1.1% Triton X-100, 0.01% SDS, 1/100 dilution of Sigma protease inhibitor cocktail.
 8. Low-salt wash: 20 mM Tris-HCl pH 8.0, 2 mM EDTA, 150 mM NaCl, 1% Triton X-100, 0.1% SDS.
 9. High-salt wash: 20 mM Tris-HCl pH 8.0, 2 mM EDTA, 500 mM NaCl, 1% Triton X-100, 0.1% SDS.
10. LiCl wash: 10 mM Tris-HCl pH 8.0, 1 mM EDTA, 1% deoxycholate, 1% NP-40, 0.25 M LiCl.
11. TE wash: 10 mM Tris-HCl pH 8.0, 1 mM EDTA.
12. Elution buffer: 1% SDS, 0.1 M NaHCO$_3$ (this must be prepared fresh each time).
13. 5 M NaCl.

14. Proteinase K solution: 1 part 0.5 *M* EDTA to 2 parts 1 *M* Tris-HCl pH6.5. To this, add 2 μL of 10-mg/mL proteinase K.

3. Methods (*see* Note 1)

3.1. Setting up Conditions for Sonication

1. Plate 10 T25 flasks of the cell line to be used.
2. Add formaldehyde directly to the medium to obtain a final concentration of 1%.
3. Incubate the cells for 10 min at 37°C.
4. Harvest each in 200 μL of lysis buffer (no protease inhibitors needed at this step), scrape cells into a 1.7-mL tube.
5. Sonicate each tube on ice using the following combinations:
    a. 10% power for 1, 2, or 3 10-s pulses.
    b. 20% power for 1, 2, or 3 10-s pulses.
    c. 30% power for 1, 2, or 3 10-s pulses.
6. Run a 20-μL aliquot of each sample on a 1% agarose gel and stain with ethidium bromide. Choose the condition that provided the bulk of DNA sheared to 200–1000 bases in length (*see* **Note 2**) (*see* **Fig. 2A**).

3.2. Immunoprecipitation

1. Treat 3×10^6 cells as desired to allow transcription factor binding to DNA. Cross-link proteins to DNA by adding formaldehyde to the medium to obtain a final concentration of 1%. This step may have to be extensively modified depending on the transcription or replication factor being studied (*see* **Notes 3** and **4**).
2. Wash the cells thoroughly with PBS containing protease inhibitors.
3. Scrape cells and pellet in a microcentrifuge at $700 \times g$ at 4°C.
4. Lyse cells for 10 min on ice with 200 μL lysis buffer and scrape lysate into 1.7-mL tube. All steps from here on in should be performed on ice.
5. Sonicate the DNA as determined in **step 4**.
6. Pellet cell debris in a microcentrifuge at $700 \times g$ at 4°C.
7. Dilute the resulting supernatant 10-fold with dilution buffer. Set aside 20 μL for quantitation of DNA used for each immunoprecipitation.
8. Add 80 μL of salmon sperm DNA/protein A to lysate and rock end over end for 30 min at 4°C to preclear the lysate.
9. Pellet the beads by centrifugation.
10. Add an appropriate amount of antibody (for p53 Pab421 antibody, previous experiments determined that 10 μL was sufficient to immunoprecipitate p53 protein from 100 mg of total protein. Therefore, 10 μL was used here for the chromatin immunoprecipitation) to 1 mL of lysate. Rock this and the other 1 mL with no antibody (negative control) added end over end overnight at 4°C.
11. Add 80 μL of salmon sperm DNA/protein A to all tubes and rock end over end for 1 h at 4°C.
12. Spin beads and remove supernatant. The supernatant can be saved to represent DNA unbound by protein recognized by antibody.
13. Perform the following washes:
    a. Low-salt wash
    b. High-salt wash
    c. LiCl wash
    d. TE wash
    e. TE wash
    in 1 mL with 3–5 min rocking end over end for each wash.

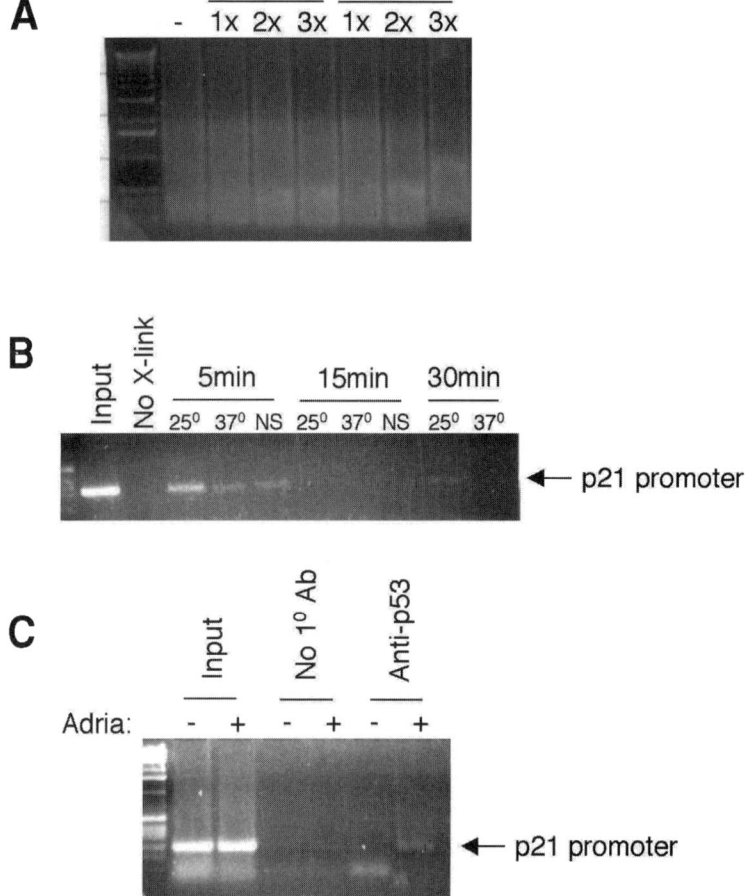

Fig. 2. Preparation of DNA–protein lysates for chromatin immunoprecipitation. **(A)** H460 cells were lysed and sonicated at 10% or 20% maximum power for 1, 2, or 3 10-s pulses. 10-mL aliquots were run on a 1% agarose gel. Note the progressive decrease in DNA smear at 3000 bases and increase at 500 and 200 bases. **(B)** H460 cells were treated with adrimycin and then exposed to 1% formaldehyde at 250, 370, or 370°C in the absence of serum in the medium. The cells were then incubated for 5, 15, or 30 min. A ChIp was then performed on p53 binding to the p21^{WAF1} promoter. **(C)** Using the predetermined conditions described for the previous two figures, H460 cells were treated with adriamycin or left untreated, and analyzed for binding of p53 to the promoter of p21 ^{WAF1}.

14. Add 250 µL of elution buffer to pelleted beads and rock on a shaker at room temperature for 15 min. Pellet beads and transfer supernatant to a separate tube and repeat with an additional 250 µL.
15. Add 20 µL of 5 M NaCl to the 500 µL of eluted protein/DNA and heat at 65°C for 4 h to reverse the cross-links.
16. Add 32 µL of Proteinase K solution and heat at 45°C for 1 h to degrade protein.
17. Extract protein with phenol/chloroform and precipitate by ethanol in the presence of glycogen (20 µg/precipitation).
18. Resuspend pellet in water and proceed to polymerase chain reaction (PCR) genomic regions thought to be bound by the protein of interest. The use of ultrapure water is essential at this point, as contamination by even the slightest bit of genomic DNA will result in the appearance of positive bands in the negative control lanes.

4. Notes

1. The methods described in this chapter were culled from several protocols and arranged as necessitated for analyzing binding of p53 to the p21^{WAF1} promoter. These protocols are described in **refs. *2–6***.
2. Depending on the sonicator used, the types of conditions tested may need to be modified than what is described under **Subheading 3.1.**
3. The extent to which proteins are cross-linked will have an effect on how the protein is recognized by the antibody, and how strong the covalent bond is between the protein and DNA. The factors to take into consideration are (a) adding formaldehyde to medium containing 10%, 1%, or no serum; (b) incubation of the cells at 370°C or 250°C; (c) incubation with formaldehyde for 5, 10, 15, or 30 min.
4. It is best to do a test experiment on a gene that is known to be regulated by a particular protein through a known sequence (*see* **Fig. 2B**, using p53 binding to the p21^{WAF1} promoter).

References

1. Futcher, B. (2000). Microarrays and cell cycle transcription in yeast. *Curr. Opin. Cell Biol.* **12**, 710–715.
2. Kuo, M. H. and Allis, C. D. (1999). In vivo cross-linking and immunoprecipitation for studying dynamic Protein:DNA associations in a chromatin environment. *Methods* **19**, 425–433.
3. Orlando, V. (2000). Mapping chromosomal proteins in vivo by formaldehyde-crosslinked-chromatin immunoprecipitation. *Trends Biochem. Sci.* **25**, 99–104.
4. Orlando, V., Strutt, H., and Paro, R. (1997). Analysis of chromatin structure by in vivo formaldehyde cross-linking. *Methods* **11**, 205–214.
5. Murphy, M., Ahn, J., Walker, K. K., et al. (1999). Transcriptional repression by wild-type p53 utilizes histone deacetylases, mediated by interaction with mSin3a. *Genes Dev.* **13**, 2490–2501.
6. Szak, S. T., Mays, D., and Pietenpol, J. A. (2001). Kinetics of p53 binding to promoter sites in vivo. *Mol. Cell. Biol.* **21**, 3375–3386.

11

Co-immunoprecipitation of Tumor Suppressor Protein-Interacting Proteins

Li-Kuo Su

1. Introduction

Tumor suppressor proteins, like most proteins, require interacting with other proteins in the cell to function properly. These interacting proteins may be regulators or substrates of this tumor suppressor protein. A tumor suppressor protein may also associate with its interacting proteins to form a functional complex to perform an activity. Identification and characterization of proteins that associate with a tumor suppressor protein can provide important information for the tumor suppressor's function and how this function is regulated.

Co-immunoprecipitation is the gold standard for demonstrating a protein–protein interaction. Co-immunoprecipitation has been used to investigate the function and the regulation of tumor suppressor genes since the first tumor suppressor gene was identified. Soon after the cloning of the retinoblastoma (*RB1*) tumor suppressor gene, co-immunoprecipitation was used to show that pRB associated with oncoproteins encoded by DNA tumor virus, including E1A of adenovirus, large T-antigen of SV40 virus, and E7 of human papilloma virus *(1–4)*. These findings allowed the formulation of the hypothesis that one major function of DNA tumor viral oncoproteins was to interact physically with pRB and block the normal function of pRB. The suggestion that the tumor suppressor function of pRB was regulated by phosphorylation and that an under-phosphorylated form of pRB was the active form while the hyperphosphorylated species were inactive ones originated from the observation that T-antigen preferentially associated with underphosphorylated pRB *(5)*. Co-immunoprecipitation also was used to reveal proteins that associated with the APC tumor suppressor protein and led to the identification of β-catenin as an APC-associated protein *(6,7)*. It is clear now that APC regulates β-catenin's activity and plays an important role in the WNT signal transduction pathway *(8,9)*. More recently, co-immunoprecipitation has been used to isolate a BRCA1-associated protein complex. The analysis of these BRCA1-associated proteins suggests that BRCA1 is important for DNA damage recognition and repair *(10)*.

One main proposes of the co-immunoprecipitation is to detect and identify proteins in addition to the tumor suppressor protein recognized directly by the antibody used in

From: *Methods in Molecular Biology, Vol. 223: Tumor Suppressor Genes: Regulation, Function, and Medicinal Applications.* Edited by: Wafik S. El-Deiry © Humana Press Inc., Totowa, NJ

the immunoprecipitates. The major challenge in carrying out co-immunoprecipitation and in interpreting the result of co-immunoprecipitation is to determine whether these additional proteins are really proteins that associate with the tumor suppressor protein. These additional proteins could be immunoprecipitated by an antibody against the tumor suppressor protein because they contain amino acid sequences similar to that of the tumor suppressor protein recognized by the antibody. BRCA1 was reported by a group of investigators to be a secreted protein because a protein detected by a presumably BRCA1-specific antibody had many characteristics of a secreted protein *(11)*. It was later realized that the BRCA1 antibody used in this study cross-reacted with epidermal growth factor receptor (EGFR), and the protein being studied in the earlier study was EGFR *(12)*. This cross-reaction probably resulted because a region of EGFR had an amino acid sequence very similar to that of the BRCA1 peptide used to raise this antibody *(12)*. Proteins can also be "immunoprecipitated" by an antibody because they are "sticky" and associate nonspecifically with the antibody.

In regular immunoprecipitation, one can increase the specificity of the immunoprecipitation and achieve the immunoprecipitation of only the specific antigen by using different concentrations and kinds of salt and/or detergent. However, these conditions may disrupt the protein–protein interaction. Because the purpose of co-immunoprecipitation usually is to reveal proteins that associate with the tumor suppressor protein being immunoprecipitated, these stringent conditions may not be appropriate. Several experiments may be performed to determine whether a co-immunoprecipitated protein is a specific associated protein. The first is to perform the co-immunoprecipitation experiments using antibodies that recognize different regions of the tumor suppressor protein of interest. Proteins that associate with the tumor suppressor protein specifically should be co-immunoprecipitated with the tumor suppressor protein by these different antibodies. However, occasionally, an antibody may interfere with the protein–protein interaction. Harlow et al. *(13)* showed that different monoclonal antibodies against E1A co-immunoprecipitated different sets of cellular proteins with E1A. This was probably because a monoclonal antibody competed with E1A-associated proteins for binding to the same region of E1A. One should always perform negative control immunoprecipitations. One negative control is to perform immunoprecipitations using antibodies that do not detect the tumor suppressor protein being studied. Any protein that is "co-immunoprecipitated" by antibodies that cannot recognize the tumor suppressor protein is likely to be a protein that interacts nonspecifically with antibodies. However, one should keep in mind that some proteins may nonspecifically bind only some antibodies and not others. An additional control experiment is to perform immunoprecipitation using cell lines that do not express the tumor suppressor proteins being studied or that express only mutants that cannot be immunoprecipitated by the antibody used. A true associated protein should be co-immunoprecipited from cell lines that express the tumor suppressor proteins that can be immunoprecipitated by the antibodies used. A true associated protein should not be co-immunoprecipitated from cell lines that either do not express the tumor suppressor protein or that express only mutants that cannot be immunoprecipitated by the antibodies used. Although such cell lines may not be always available, this control experiment is the best way to reveal proteins that interact nonspecifically with an antibody, and this control immunoprecipitation should be performed whenever it is possible.

There are several methods to detect proteins present in the immunoprecipitate. Methods commonly used are immunoblot analysis and detection of radiolabeled proteins by autoradiography. Immunoblot analysis is the preferred method if one wants to test whether a specific protein interacts with the tumor suppressor protein being studied and antibodies to this protein are available. When the associated proteins are unknown and the identification of all associated proteins is intended, the metabolic labeling method is preferred. In this latter approach, cells are usually metabolically labeled with ^{35}S-methionine by incubating in a methionine-free medium with added ^{35}S-methionine for a period of time to allow ^{35}S-methionine to incorporate into proteins. Cells are then lysed, the tumor suppressor protein is immunoprecipitated, and then all proteins in the immunoprecipitate are resolved by sodium dodecyl sulfate (SDS)-polyacrylamide gel electrophoresis (PAGE). The radiolabeled proteins on the gel are then detected by autoradiography. Cells labeled metabolically with other radioisotopes also have been used to identify proteins that associated with tumor suppressor proteins. The interaction between MDM2 and p53 was revealed by the identification of a cellular phosphoprotein that co-immunoprecipitated with p53 using lysates prepared from cells metabolically labeled with ^{32}P *(14)*.

Like most techniques used in studying proteins, there is no universal co-immunoprecipitation condition that is appropriate for every tumor suppressor protein. The major challenge for the identification of proteins that associate with a tumor suppressor protein by co-immunoprecipitattion is to identify a condition for immunoprecipitation that minimizes the nonspecific immunoprecipitation but does not disrupt the interaction between the tumor suppressor protein and its associated proteins. I will therefore use the co-immunoprecipitation of catenins with APC (**Fig. 1**) only as an example in this chapter.

2. Materials

1. Hank's balanced salt solution (HBSS) (Life Technologies, Rockville, MD).
2. Dulbecco's modified Eagle's medium (DMEM) without methionine (Life Technologies).
3. ^{35}S-methionine/cysteine (Tran^{35}S-Label, ICN, Costa Mesa, CA).
4. Lysis buffer: 50 mM Tris-HCl, pH 7.5; 100 mM NaCl, and 0.5% Nonidet P-40, protease inhibitors [0.2% 4-(2-aminoethyl)benzenesulfonylfluoride (Calbiochem, La Jolla, CA), and 0.01 mg/mL each of chymostatin, leupeptin, antipain, and pepstatin A (all from Sigma, St. Louis, MO), and phosphatase inhibitors (0.1 mM Na$_3$VO$_4$ and 50 mM NaF [both from Sigma]) (*see* **Note 1**).
3. Normal mouse immunoglobulin G (IgG) (Sigma).
4. Monoclonal antibody against APC, Ab5 (Oncogene Research, Boston, MA).
5. Protein A agarose (Sigma) suspension: 50 mg/mL in lysis buffer.
6. Laemmli sample buffer: 60 mM Tris-HCl, pH 6.8, 2% SDS, 100 mM dithiothreitol (DTT), 10% glycerol, and 0.01% bromophenol blue.
7. Fix solution: 5% methanol and 5% acetic acid in water.

3. Methods

1. Plate cells in a 25-cm^2 cell culture flask so that the density of the cells is about 70% confluent at the day of experiment (*see* **Note 2**).
2. Remove the medium, wash cells once with DMEM without methionine, add 1.5 mL DMEM without methionine (*see* **Note 3**).
3. Add ^{35}S-labeled methionine to cells at the final specific activity of 200 μCi/mL, incubate 4 h in an appropriate cell culture incubator (*see* **Note 4**).
4. Remove the medium, wash cells with HBSS (*see* **Note 5**).
5. Place the dish on ice and add 0.3 mL ice-cold lysis buffer to cells.

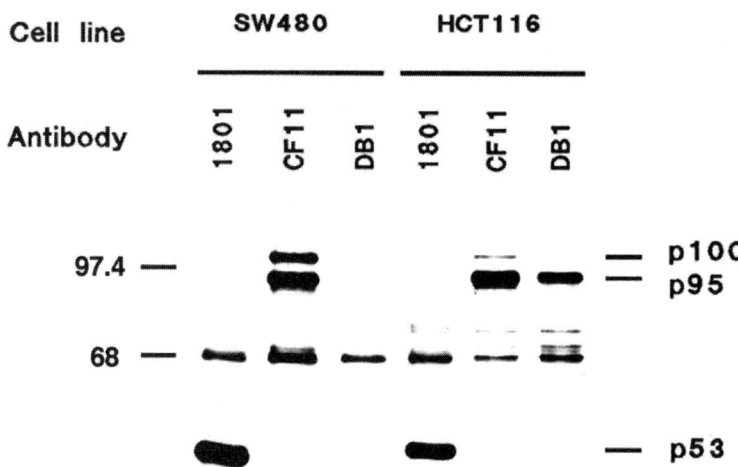

Fig. 1. Identification of APC associated proteins by co-immunoprecipitation. SW480 and HCT116 are human colon cancer cell lines, SW480 expresses only a truncated APC protein, while HCT116 expresses the wild-type APC. Lysates were prepared from cells metabolically labeled with ^{35}S-methionine, and immunoprecipitation was performed using the indicated antibodies. Immunoprecipitated proteins were resolved by SDS-PAGE and detected by fluorography using Kodak X-Omat AR film. CF11 is a monoclonal antibody against the amino terminus of APC, therefore could immunoprecipitate both the truncated APC in SW480 cells and the wild-type APC in HCT116 cells. DB1 is a monoclonal antibody against the carboxyl terminus of APC, therefore could only immunoprecipitate the wild-type APC in HCT116 cells, not the truncated APC in SW480 cells. A monoclonal antibody against p53, 1801, was used as a negative control. Positions of molecular weight standards (in kilodaltons) are shown on the left. The APC protein itself is not visible, probably because of its slow turnover rate resulting in its low radiolabeling. The presence of the 68-kDa protein in the immunoprecipitates of APC and non-APC antibodies strongly suggested that it nonspecifically associated with antibodies and was not an APC-associated protein. In contrast, p95 and p100 were likely to be APC-associated proteins because they were co-immunoprecipitated with APC in HCT116 cell when two antibodies that bound different regions of APC were used and they were not co-immunoprecipitated with p53. This suggestion was supported by the immunoprecipitation of APC using SW480 cells, because p95 and p100 presented in the immunoprecipitate only when the antibody that could detect the truncated APC was used, not in the immunoprecipitate only when the antibody that could not detect the truncated APC was used. (Reprinted with permission from Su et al., *Science*, vol. 262, p. 1734, copyright 1993 American Association for the Advancement of Science.)

6. Lyse cells for 30 min, rock the cell culture dish a few times to make lysis buffer cover all cells during this time.
7. Transfer lysate, and cell debris if there is any, to a 1.5-mL tube.
8. Centrifuge at 4°C at full speed in a microcentrifuge for 10 min.
9. Transfer the supernatant to a new tube.
10. Add 1–2 μg of normal mouse IgG to the supernatant, mix well and incubate on ice for 45 min (*see* **Note 6**).
11. Add 50 μL of protein A agarose suspension to the tube, incubate at 4°C with rocking for 45 min (*see* **Note 7**).
12. Spin down the protein A agarose, transfer the supernatant to a new 1.5-mL tube (*see* **Note 8**).
13. Add 1 μg APC monoclonal antibody to the precleared lysate, mix well, and incubate on ice for 1 h.
14. Add 50 μL of protein A agarose suspension to the tube, incubate at 4°C with rocking for 1 h.
15. Spin down the protein A agarose, remove the supernatant.
16. Add 1 mL lysis buffer to the protein A agarose, mix well by vortexing for a few seconds.

17. Repeat **steps 15** and **16** two more times.
18. Spin down the protein A agarose, remove as much supernatant as possible.
19. Add Laemmli sample buffer, mix well, and boil for 10 min in a boiling-water bath.
20. Load sample on to an 8% SDS-polyacrylamide gel (*see* **Note 9**).
21. Electrophoresis until the bromophenol blue dye reaches the bottom of the gel.
22. Remove the gel and place it in the fix solution.
23. Rock the gel in the fix solution at room temperature for at least 30 min, change the fix solution at least once during this period of time.
24. Dry the gel on a gel dryer.
25. Expose the dried gel to Kodak BioMax MR film (*see* **Note 10**).

4. Notes

1. For each specific tumor suppressor protein, one should test different lysis buffers to identify the one that is the best for the tumor suppressor protein being studied. One should try various concentration of salt, usually NaCl. One should also try different kinds and concentrations of detergent. To preserve protein–protein interaction, nonionic detergents such as Nonidet P-40 and Triton X-100 are preferred. Ionic detergents such as SDS and sodium deoxycholate are more likely to disrupt protein–protein interaction.
2. Cells can be grown in either dish or flask. However, ^{35}S is volatile and can contaminate the incubator. The contamination is less severe when ^{35}S-methionine is added to cells in flask than to cells in dish. Check the incubator carefully and clean it up after the incubation.
3. Dialyzed fetal bovine serum may be added to the labeling media!
4. To increase the labeling efficiency of proteins, cells can be methionine-starved by incubating in labeling media for 30 min to 4 h before adding ^{35}S-methionine. Cells also can be incubated with a higher final specific activity of ^{35}S and/or be labeled for a longer period of time.
5. Phosphate-buffered saline (PBS) can be used in the place of HBSS.
6. Normal IgG or normal serum from the species that used to generate the antibody used in **step 13** should be used in this step to preclear the cell lysate.
7. Use protein A or protein G according to the one used in **step 14**. Protein A should be used for rabbit, human, and mouse polyclonal antibodies. For mouse monoclonal antibody with IgG$_1$ isotype, protein G may be better than protein A. Protein G should be used for polyclonal antibodies prepared from goat, rat, donkey, and horse.
8. The centrifugation can be done at full speed using a nanofuge for 10–20 s or at about 2000 $\times$ g in a microcentrifuge for 10–20 s.
9. If the molecular weights of the proteins in the immunoprecipitate are not known, several gels with different concentrations of acrylamide should be used. Alternatively, a gradient gel with 5–15% acrylamide may be used.
10. BioMax MR film is a very sensitive film for detecting medium energy isotopes such as ^{35}S, and makes the use of fluorography unnecessary. Because of its sensitivity, be sure to use the recommended safety lights. Safety lights that do not cause problem for less sensitive films, such as Kodak X-Omat AR films, may expose BioMax MR films.

Acknowledgment

I thank Dr. Sue-Hwa Lin for critical reading of this manuscript.

References

1. Whyte, P., Buchkovich, K. J., Horowitz, J. M., et al. (1988) Association between an oncogene and an anti-oncogene: the adenovirus E1A proteins bind to the retinoblastoma gene product. *Nature* **334,** 124–129.

2. DeCaprio, J. A., Ludlow, J. W., Figge, J., et al. (1988) SV40 large tumor antigen forms a specific complex with the product of the retinoblastoma susceptibility gene. *Cell* **54,** 275–283.
3. Dyson, N., Howley, P. M., Munger, K., and Harlow, E. (1989) The human papilloma virus-16 E7 oncoprotein is able to bind to the retinoblastoma gene product. *Science* **243,** 934–937.
4. Munger, K., Werness, B. A., Dyson, N., Phelps, W. C., Harlow, E., and Howley, P. M. (1989) Complex formation of human papillomavirus E7 proteins with the retinoblastoma tumor suppressor gene product. *EMBO J.* **8,** 4099–4105.
5. Ludlow, J. W., DeCaprio, J. A., Huang, C. M., Lee, W. H., Paucha, E., and Livingston, D. M. (1989) SV40 large T antigen binds preferentially to an underphosphorylated member of the retinoblastoma susceptibility gene product family. *Cell* **56,** 57–65.
6. Su, L.-K., Vogelstein, B., and Kinzler, K. W. (1993) Association of the APC tumor suppressor protein with catenins. *Science* **262,** 1734–1737.
7. Rubinfeld, B., Souza, B., Albert, I., et al. (1993) Association of the APC gene product with beta-catenin. *Science* **262,** 1731–1734.
8. Morin, P. J. (1999) Beta-catenin signaling and cancer. *Bioessays* **21,** 1021–1030.
9. Polakis, P. (2000) Wnt signaling and cancer. *Genes Dev.* **14,** 1837–1851.
10. Wang, Y., Cortez, D., Yazdi, P., Neff, N., Elledge, S. J., and Qin, J. (2000) BASC, a super complex of BRCA1-associated proteins involved in the recognition and repair of aberrant DNA structures. *Genes Dev.* **14,** 927–939.
11. Jensen, R. A., Thompson, M. E., Jetton, T. L., et al. (1996) BRCA1 is secreted and exhibits properties of a granin. *Nat. Genet.* **12,** 303–308.
12. Wilson, C. A., Payton, M. N., Pekar, S. K., et al. (1996) BRCA1 protein products: antibody specificity. *Nat. Genet.* **13,** 264–265.
13. Harlow, E., Whyte, P., Franza, B. R., and Schley, C. (1986) Association of adenovirus early-region 1A proteins with cellular polypeptides. *Mol. Cell. Biol.* **6,** 1579–1589.
14. Momand, J., Zambetti, G. P., Olson, D. C., George, D., and Levine, A. J. (1992) The mdm-2 oncogene product forms a complex with the p53 protein and inhibits p53-mediated transactivation. *Cell* **69,** 1237–1245.

12

Microarray Approaches for Analysis of Tumor Suppressor Gene Function

Sally A. Amundson and Albert J. Fornace, Jr.

1. Introduction

Many tumor suppressor genes are known to function at least in part through regulation of the transcription of downstream effector genes (**Table 1**). A major example of such a transcriptional regulator is p53, one of the most commonly mutated tumor suppressor genes in human cancer *(1)* and hence one of the most exhaustively studied. Estimates based on a survey of p53 binding sites in the genome put the number of p53-regulated genes at several hundred *(2)*, while the finding that p53 can affect the expression of some genes in the absence of direct DNA binding may increase this number *(3)*. Genes known to be regulated by p53 play roles in many important cellular processes, including cell cycle progression, DNA repair, and apoptosis (**Table 2**). Loss of such a tumor suppressor gene can disrupt the regulation of multiple genes, affecting numerous cellular pathways and leading to a variety of phenotypic changes. Comparative analysis of complex patterns of gene expression can therefore provide a powerful tool to develop insight into mechanisms of tumor suppressor gene function involving transcriptional regulation.

Techniques for simultaneously comparing expression levels of thousands of genes, or even the entire expressed genome, have become widely accessible in recent years. Sequencing-based methods of large-scale highly parallel gene expression studies, including serial analysis of gene expression (SAGE) *(4)*, can be biased toward detection of highly expressed or strongly induced transcripts. Multigene nylon filter arrays hybridized to radioactive probes provide a useful method of screening for genes with alterations in expression levels, although such differential hybridization screening has its own limitations *(5)*. The use of fluorescent probes labeled with different fluorochromes and co-hybridized to the same microarray can circumvent some of these problems.

Microarrays can be constructed either by direct printing of cDNAs on a glass surface *(6)*, or photolithographic synthesis of oligonucleotides *in situ (7)*. Both types of microarray are currently in wide use. Newer methods of array production, involving the use of bubble jet *(8)* and inkjet *(9,10)* printing technology to produce either oligonucleotide or cDNA microarrays, are also under development. Such refinements may increase the flexibility of the technology while making microarrays even more widely accessible.

From: *Methods in Molecular Biology, Vol. 223: Tumor Suppressor Genes: Regulation, Function, and Medicinal Applications.* Edited by: Wafik S. El-Deiry © Humana Press Inc., Totowa, NJ

Table 1
Tumor Suppressor Genes That Can Act as Transcription Factors

TP53	Tumor protein p53
WT1	Wilm's tumor 1
VHL	Von Hippel-Lindau syndrome
MEN1	Multiple endocrine neoplasia, type 1
TCF7	Transcription factor 7
MXI1	MAX-interacting protein 1
BRCA1	Breast cancer 1, early onset
BRCA2	Breast cancer 2, early onset
ATM	Ataxia telangiectasia
NBL1	Neuroblastoma candidate region, suppression of tumorigenicity 1
NME1	Nonmetastatic cells 1, protein expressed in
EGR1	Early growth response 1

Table 2
Examples of Tp53 Effector Genes with Roles in Cellular Stress Response Processes

Cell Cycle Control	Apoptosis	DNA Repair	Other
CIP1/WAF1	BAX	XPC	MDM2
ClnG	BCL-X	DDB2	FRA1
ClnD1	PAG608	GADD45A	ATF3
WIP1	FAS/APO1	PCNA	14-3-3σ
EGF-R	KILLER/DR5		Rb
TGF-α	TRUNDD		c-MYC
Rb	TRID		MMP2
PCNA	seven in absentia		MAP4
GADD45A	IGF-BP3		TSP1 & 2
14-3-3σ	PIG1 to PIG14		BAI-1
BTG2			WIG1
seven in absentia			amyloid
IGF-BP3			GML
PIG1 to PIG14			bFGF
inhibin-β			PIR121

The protocols in this chapter are for use with cDNA microarrays for fluorescent hybridization, and require access to such arrays, either from a core printing facility, or from a commercial source, and to an appropriate scanning device and data analysis software. The specific methods covered, therefore, focus on preparation and labeling of the RNA samples to be compared, and hybridization of the labeled samples to the arrays. A general discussion of array preparation is included to provide context, and references to detailed methodology are given to assist readers interested in construction of their own printed arrays. Scanning and data extraction steps are discussed generally, but due to the wide variety of platforms available for these tasks, specific protocols will depend on the system used. Finally, some examples of informatic approaches that have been applied to analysis of gene expression data related to tumor suppressor genes are discussed. Again, this section does not detail specific methods, but is intended to serve as an introduction to the type of data analyses that are being actively developed in tandem with our ability to obtain vast gene expression data sets.

1.1. Microarray Production

The cDNAs used in the printing of microarrays are generally prepared by polymerase chain reaction (PCR) amplification from purified plasmid DNA. A common strategy used for human and mouse sequences is amplification of cloned expressed sequence tags (ESTs) using primers to the vector sequence adjacent to the cloning site. Selection of ESTs representing the 3' ends of the genes to be screened on the array enhances hybridization when an oligo d(T) primer is later used for reverse transcription of the cellular mRNA pools to be compared.

The purified PCR products are resuspended in 3× saline-sodium citrate (SSC) at 100–500 µg/mL and specialized pens moved by a highly accurate industrial robot are used to deposit several nanoliters of this solution onto poly-L-lysine-coated glass microscope slides. The use of a poly-L-lysine coating on the slides increases their hydrophobicity, reduces the spreading out of the printed DNA spots, and allows UV cross-linking of the printed DNA to the slide. After the slides have been cross-linked, the poly-L-lysine coating leaves charged amines on the slide surface. These can cause nonspecific electrostatic binding of the labeled cDNA during the hybridization step, resulting in high fluorescent background. The charged amines can be reacted with succinic anhydride in a buffer with a high organic solvent content *(11)* in a chemical passivation step that circumvents this problem. Detailed protocols for the handling and amplification of EST clones and the printing and preparation of slides can be found at http://www.nhgri.nih.gov/DIR/Microarray/main.html.

2. Materials

2.1. Preparation of RNA Samples for Microarray Analysis

1. Liquid nitrogen.
2. Trizol reagent (Life Technologies, Rockville, MD).
3. Chloroform.
4. Dulbecco's Phosphate Buffered Saline (DPBS).
5. GTC lysis buffer: (4 M guanidine thiocyanate, 10 mg/mL sarkosyl, 50 mM Tris-HCl (pH 7.5). Immediately before use, add 10 µL β-mercaptoethanol/mL lysis solution.
6. 2 M sodium acetate (pH 4.0).
7. Phenol, water-saturated.
8. Chloroform/isoamyl alcohol (24/1).
9. Ethanol (200 proof, USP ethyl alcohol).
10. Rneasy Maxi Kit (Qiagen, Valencia, CA).
11. Tissue homogenizer.
12. Sonicator.

2.2. Preparation of Labeled cDNA: Direct Incorporation of Fluorochrome

1. Whole-cell RNA.
2. Rnase-free water.
3. MicroCon 30 (Millipore, Bedford, MA).
4. 2.0 µg/µL Oligo dT$_{20}$ (Life Technologies, Rockville, MD).

5. 5× First-strand synthesis buffer for Superscript II (Life Technologies, Rockville, MD).
6. 0.1 M dithiothreitol (DTT).
7. 10× Low-T NTP mix: 5 mM dGTP, 5 mM dATP, 5 mM dCTP, 2 mM dTTP (Pharmacia).
8. RNasin RNase inhibitor (Promega).
9. Superscript II Reverse Transcriptase (Life Technologies, Rockville, MD).
10. Cy3-dUTP (1 mM) and Cy5-dUTP (1 mM) (NEN Life Sciences, Boston, MA).
11. 0.5 M EDTA (pH 8.0).
12. 1 N NaOH.
13. 1 M Tris-HCl (pH 7.5).
14. MicroCon 100 (Millipore, Bedford, MA).
15. TE (pH 7.4).
16. 1 mg/mL C_0t-1 DNA (Life Technologies, Rockville, MD).
17. Agarose.
18. Tris-acetate buffer.
19. PCR thermal cycler.
20. Fluorescence scanner (i.e., Molecular Dynamics Storm).

2.3. Alternative Labeling Protocol: Amino-Allyl Coupling of Fluorochromes

1. Whole-cell RNA.
2. 5.0 µg/µL Oligo dT$_{20}$ (Life Technologies, Rockville, MD).
3. RNase-free water.
4. 5× First-strand synthesis buffer for Superscript II (Life Technologies, Rockville, MD).
5. 0.1 M DTT.
6. Superscript II Reverse Transcriptase (Life Technologies, Rockville, MD).
7. 40× amino-allyl dNTP mix (20 mM dATP; 20 mM dCTP; 20 mM dGTP; 12 mM dTTP (100 mM stocks from Pharmacia); 8 mM amino-allyl-dUTP (Sigma).
8. 0.5 M EDTA (pH 8.0).
9. 1 N NaOH.
10. 1 M Tris-HCl (pH 7.5).
11. TE (pH 7.4).
12. MicroCon 30 (Millipore, Bedford, MA).
13. Monofunctional Cy3 and Cy5 dye (Amersham Pharmacia Biotech, Piscataway, NJ).
14. 0.1 M sodium bicarbonate (pH 9.0).
15. 4 M hydroxylamine.
16. Speed-Vac.
17. PCR thermal cycler.

2.4. Microarray Hybridization

1. cDNA microarrays.
2. Fluorescently labeled cDNA probes.
3. 20× SSC: 3 M NaCl, 0.3 M sodium citrate, pH 7.4.
4. 8-mg/mL poly d(A).
5. 4-mg/mL yeast tRNA.
6. 10-mg/mL C_0t-1 DNA, human or mouse (Life Technologies, Rockville, MD).
7. 50× Denhardt's blocking solution: 10 mg/mL Ficoll, 10 mg/mL bovine serum albumin (BSA) (Pentax Fraction V), 10 mg/mL polyvinylpyrrolidone.
8. 10% sodium dodecyl sulfate (SDS).
9. Microarray hybridization chambers.

10. Water bath.
11. Wash buffer: 0.5× SSC, 0.01% SDS.
12. 0.06× SSC.
13. Clinical centrifuge.
14. Fluorescent microarray imaging device and data extraction software.

3. Methods

3.1. Preparation of RNA Samples for Microarray Analysis

1. For RNA extraction from tissues, fresh tissue samples should be flash-frozen in liquid nitrogen, or immediately homogenized in Trizol solution (*see* **Note 1**). To a 15-mL polypropylene centrifuge tube containing 4 mL Trizol, add 100 mg frozen or fresh tissue. Dissociate thoroughly using a rotating-blade tissue homogenizer.
2. Add 800 µL chloroform and shake vigorously for 15–20 s. Incubate at room temperature for 3 min.
3. Centrifuge at 12,000 × g for 15 min at 4°C. Transfer the aqueous (top) phase into a clean polypropylene tube.
4. Cells from tissue culture can be prepared similarly up to this point, first washing cells or cell pellets twice with DPBS, and adding 1 mL Trizol per 2×10^7 cells. Alternatively, lysis in GTC and subsequent phenol extraction can be used. GTC lysates may be stored conveniently at –80°C in order to collect samples throughout the course of an experiment.
5. Cells growing in suspension must be pelleted by centrifugation, washed in DPBS, then up to 10^8 lymphoid cells can be resuspended in 5 mL GTC lysis buffer. (Add 10 µL/mL β-mercaptoethanol immediately before use.) Cells growing as monolayers may be rinsed in DPBS, then lysed *in situ*. For larger cells containing greater amounts of cytoplasm, it is advisable to use a higher proportion of GTC lysis buffer: up to 5×10^7 cells per 5 mL lysis buffer.
6. Tissue cultures lysates, whether in GTC or trizol, should be disrupted using several 5–10-s bursts of sonication to reduce the viscosity of the solution.
7. To each 5 mL GTC lysate, add 0.5 mL 2 M NaOAc. Invert to mix.
8. Add 1 vol water-saturated phenol. Invert to mix, then add 2 mL chloroform/isoamyl alcohol 24/1. Shake vigorously for at least 20 s, and incubate on ice for at least 20 min. Centrifuge and reserve the aqueous phase as in **step 3**.
9. To the recovered aqueous phase, add 0.53 vol of 200-proof ethanol dropwise while vortexing the tube (*see* **Note 2**).
10. Immediately add this mixture to an Rneasy maxicolumn and centrifuge in a clinical centrifuge at 3000 × g for 5 min at room temperature.
11. To increase recovery of RNA, pour the flow-through back over the column and repeat the centrifugation.
12. Discard the flow-through, wash with 15 mL of RW1 buffer (from Qiagen Rneasy kit) and centrifuge again at 3000 × g for 5 min.
13. Discard the flow-through, wash with 10 mL of RPE buffer (from Qiagen Rneasy kit), centrifuging for 2 min at 3000 ×g.
14. Repeat with a fresh 10 mL of RPE buffer (Qiagen), this time centrifuging for 10 min to dry the column.
15. Place the column in a clean 50-mL tube, and elute RNA with 1 mL Rnase-free water (included in Rneasy kit). Let stand for 1 min, then centrifuge at 3000 × g for 3 min.
16. Repeat elution **step 15** with another 1 mL water. Quantitate the RNA and store at –80°C. The ratio of absorbance at 260/280 nm should be very near to 2.0 (*see* **Note 3**).

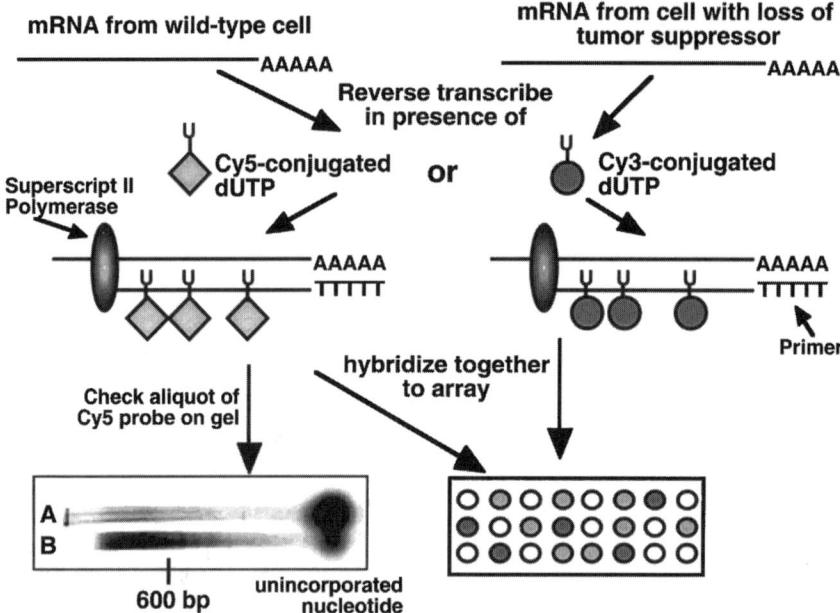

Fig. 1. Making fluorescent labeled cDNA by direct incorporation of fluorochrome (*see* **Subheading 3.2.**). A single round of reverse transcription is used to make cDNA from the mRNA samples to be compared. Cy5 (diamonds) or Cy3 (circles) conjugated to dUTP will be incorporated by the Superscript reverse transcriptase. The probes are then hybridized together to the microarray, washed, and scanned using a confocal laser scanning microscope. Transcripts present in greater proportion in the wild-type sample, here labeled in green, will produce a green spot on the final microarray image, while transcripts increased with the loss of a tumor suppressor, here the sample labeled in red, will yield a red spot. Equal representation in both original mRNA pools results in a yellow spot. If a sample of the labeled probe is run on an agarose gel (*see* **Subheading 3.2.**) incorporated Cy5 can be easily detected as in the example in the lower left of this figure. Lane A is the result of poor incorporation with most of the Cy5 being present as unincorporated nucleotide, while lane B shows good incorporation, and a strong signal in the region of high-molecular-weight transcripts.

3.2. Preparation of Labeled cDNA: Direct Incorporation of Fluorchrome

In this approach, the cDNA to be hybridized to the microarray is synthesized from the RNA isolated under **Subheading 3.1.** Reverse transcription is carried out in the presence of dUTP conjugated to either Cy3 or Cy5 fluorescent dyes (**Fig. 1**). Despite a relatively low incorporation rate of dUTP containing these bulky fluorescent groups, consistently good results have been obtained using such protocols.

1. Aliquot the RNA to be labeled, usually between 40 and 100 µg per sample. Concentrate the RNA by adding each sample to 400 µL water in a MicroCon 30 cartridge. Centrifuge at 16 000 × g for approximately 4.5 min (*see* **Note 4**).
2. Add an additional 200 µL water to each cartridge, pipetting to mix the samples. Centrifuge at 16,000 × g for approximately 2.5 min. Further centrifugation may be needed to reduce the volume of RNA to 16 µL. Monitor volume carefully to avoid drying completely.
3. Recover the RNA by inverting the cartridge over a clean collection tube and centrifuging at 500 × g for 3 min. Bring the volume to 16 µL total in a 0.2-mL thin-wall PCR tube. Add 1 µL oligo-dT primer (2 µg/µL), and preanneal for 10 min at 70°C in a PCR thermal cycler block.
4. While the primer is annealing, make a reaction mix containing 8 µL 5× first-strand buffer, 4 µL 0.1 *M* DTT, 4 µL 10× low-T NTP mix, 1 µL Rnasin (30 U/µL), 2 µL Superscript II

enzyme (200 U/µL) per reaction. (It is generally advisable to make enough reaction mix for an extra reaction, to ensure sufficient volume for all reactions.)

5. After preannealing, cool the reactions on ice for 2 min. Then add 4 µL either Cy5-dUTP or Cy3-dUTP, and 19 µL of the reaction mix from **step 4** to each tube. Mix well, and incubate at 42°C (*see* **Note 5**).

6. After 30 min, add another 2 µL Superscript II, mix well, and continue incubating the reaction at 42°C for an additional 30–60 min.

7. Remove from the heat block and stop the reaction by adding 5 µL of 0.5 M EDTA.

8. In order to hydrolyze the remaining RNA, add 10 µL 1 N NaOH and incubate at 65°C for 30–60 min, then cool to room temperature (*see* **Note 6**).

9. Add 25 µL 1 M Tris-HCL (pH 7.5) to neutralize the solution.

10. Prepare a MicroCon 100 cartridge with 400 µL TE (pH 7.5) and 20 µg of C_0t-1 DNA. Add the neutralized reaction and pipette to mix. Centrifuge 12 min at $500 \times g$.

11. Wash the samples with an additional 200 µL TE, mix by pipetting, and centrifuge again for about 10 min. Reduce the volume to around 15–18 µL by continuing to centrifuge as needed, and recover the labeled cDNA by inverting the cartridge over a clean collection tube and spinning at $500 \times g$ for 3 min.

12. Incorporation of Cy5 may be checked by running a 2–3-µL aliquot of the Cy5-labeled cDNA on a 2% agarose (TAE) gel without ethidium bromide. Scan the gel on a Molecular Dynamics Storm fluorescence scanner set to detect red fluorescence, 200-µm resolution and 1000 V on the PMT. Successful labeling should yield a dense smear of probe in the range of 400–1000 bp. An excess of low-molecular-weight transcripts and unincorporated nucleotides indicates weak labeling (*see* **Fig. 1**).

3.3. Alternative Labeling Protocol: Amino-Allyl Coupling of Fluorochromes

Incorporation of an amino-allyl-modified dUTP in the cDNA synthesis step, with subsequent coupling of monofunctional Cy3 or Cy5 dyes may circumvent problems associated with unequal incorporation rates of dUTP coupled to the two different fluorochromes (**Fig. 2**).

1. Prepare RNA samples for labeling as in **steps 1–3** of the preceding protocol, but adjust the final volume to 14.5 µL in a 0.2-mL thin-wall PCR tube. Add 1 µL of oligo-dT primer (5 µL/µL). Incubate for 10 min at 70°C to preanneal, then put on ice for 10 min.

2. Prepare a reaction mix containing 6 µL 5× first-strand buffer, 1 µL 40× amino-allyl NTP mix, 3 µL 0.1 M DTT, 3 µL water, and 2 µL Superscript II per reaction.

3. Add 15 µL of reaction mix from **step 2** to each reaction. Mix and incubate at 42°C for 2 h. A second aliquot of Superscript II enzyme may be added midway through the reaction.

4. Stop the reaction with 10 µL 0.5 M EDTA. Add 10 µL 1 N NaOH, mix well, and hydrolyze remaining RNA for 20–30 min at 65°C.

5. Cool to room temperature and add 25 µL 1 M Tris-HCl (pH 7.5) to neutralize.

6. Add each reaction to a MicroCon 30 cartridge containing 450 µL water. Centrifuge at $16,000 \times g$ for 6–8 min. Wash two more times with 450 µL water followed by centrifugation and collect RNA in a clean tube by centrifuging for 3 min at $500 \times g$. The target volume at this step is 4.5 µL. The samples can be further concentrated using a Speed-Vac if necessary.

7. Resuspend one dye vial each of monofunctional Cy3 and Cy5 in 72 µL RNase-free water. Use a Speed-Vac to dry down individual aliquots of 4.5 µL. Store these at 4°C until needed.

8. Prepare a fresh aliquot of monofunctional Cy3 and Cy5 for each reaction by resuspending in 4.5 µL 0.1 M Na-bicarbonate (pH 9.0). Mix the appropriate fluorochrome with each RNA sample, and incubate at room temperature for 1 h in the dark.

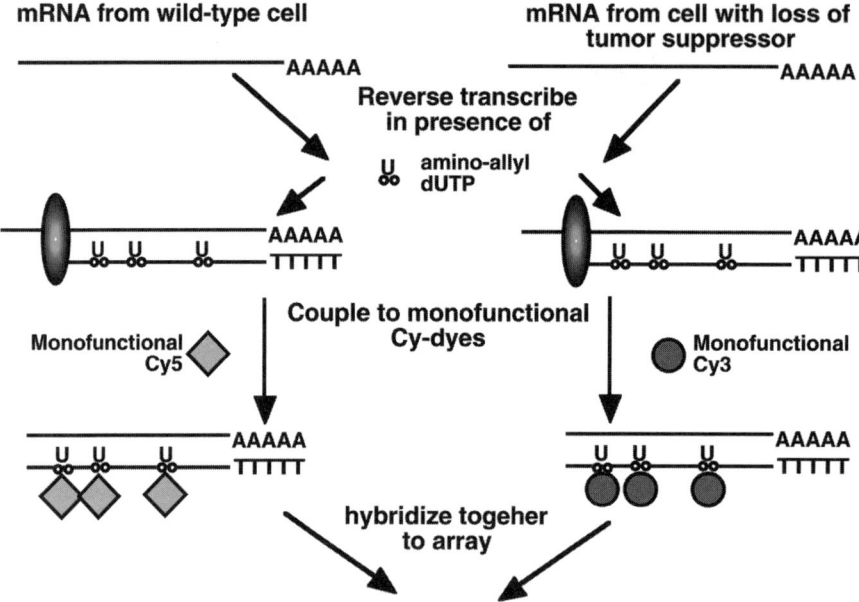

mRNA from wild-type cell   mRNA from cell with loss of tumor suppressor

Fig. 2. Making fluorescent labeled cDNA by amino-allyl coupling of fluorochromes (*see* **Subheading 3.3.**). In this case, the reverse transcription reaction is carried out in the presence of dUTP modified only with an amino-allyl site. This creates much less steric hinderance than the presence of the bulky cyanine dyes during this step. Chemical coupling is then used to label the amino-allyl sites within the newly synthesized cDNA with the appropriate reactive cyanine fluorescent molecule. Hybridization and scanning are then carried out as with the standard labeling technique.

   9. Add 4.5 µL 4 *M* hydroxylamine to each reaction. Incubate for 15 min at room temperature in the dark.

10. Pool the Cy3 and Cy5 reactions for each set of samples being compared, and clean on a MicroCon 30 cartridge as in **step 6**. The total volume should be concentrated to around 30–34 µL. Incorporation of Cy5 may be checked by running 4–5 µL on a gel as described in **step 12** of the previous protocol.

3.4. Microarray Hybridization

Blocking conditions may need to be individually adjusted, as arrays and samples will vary somewhat. The conditions given below for a 40-µL hybridization may be used as a starting point for hybridization under a 24 × 50-mm cover slip. If a smaller area is to be covered by the hybridization, the volume should be adjusted proportionately.

   1. Combine the two labeled samples to be compared in a thin-wall PCR tube in a total volume of 30 µL. To this, add 6 µL 20× SSC, 1 µL poly d(A) (8 mg/mL), 1 µL yeast tRNA (4 mg/mL), 1 µL human C_0t-1 DNA (10 mg/mL) (mouse C_0t-1 DNA should be used for blocking of mouse arrays), and 1 µL 50× Denhardt's blocking solution. Mix well and incubate at 98°C for 2 min.

   2. Cool rapidly to 25°C, then add 0.6 µL 10% SDS. Mix well and centrifuge for 5 min at 14,000 × *g* (*see* **Note 7**).

   3. Apply the hybridization mixture to a glass cover slip. Apply the inverted microarray and allow surface tension to draw the cover slip gently onto the array. A very delicate touch is required in this step to prevent the formation of bubbles under the cover slip.

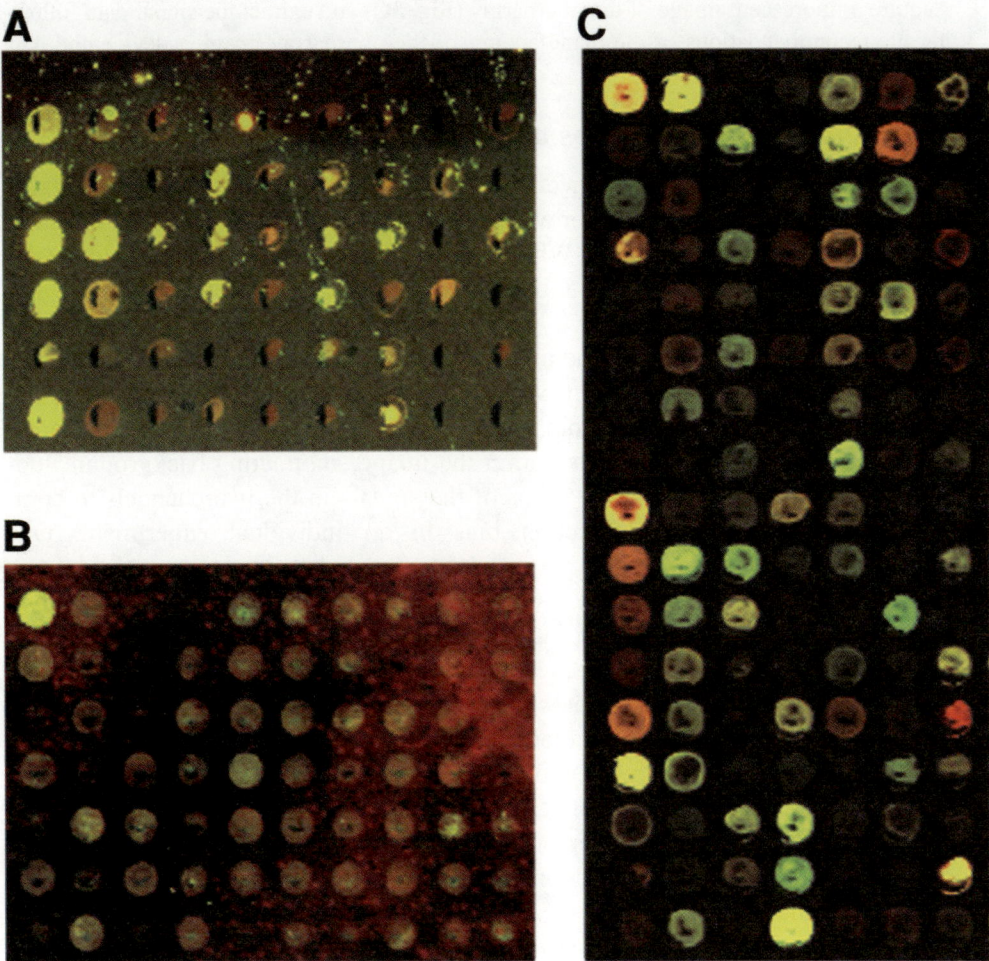

Fig. 3. The good, the bad, the ugly: examples of microarray hybridization outcomes. **(A)** typical example of a microarray dehydrating during hybridization or subsequent processing. Note the high degree of overall fluorescent haze and poor signal quality. **(B)** Example of chamber leakage during hybridization. This tends to degrade portions of the image, creating variable signal quality and areas of high local background. **(C)** Hybridization showing uniformly low background and good signal quality. Some features will show low or absent signal with even the best hybridization, as not all genes are expressed in all samples. This should not be considered a sign of failure.

4. Transfer the slide to a microarray hybridization chamber. Add 20 μL 3× SSC to the chamber, to prevent drying of the hybridization solution (**Fig. 3A**), seal the chamber, and submerge in a 65°C water bath for 16–20 h.
5. Remove the chamber from the water bath and dry carefully, especially around the seals, before unsealing it and removing the slide. Leaking of the chamber, or introducing water when unsealing the chamber can spoil the hybridization (**Fig. 3B**).
6. Place the slide in a Coplin jar filled with 0.5× SSC/ 0.01% SDS wash buffer. Allow the cover slip to float off, remove it from the jar, and allow the slide to wash for 3–5 min.
7. Transfer the slide to fresh wash buffer for another 3–5 min.
8. Wash the slide in 0.06 × SSC for 3–5 min. Immediately spin-dry by centrifuging in a slide rack in a clinical centrifuge for 3 min at 200*g* (*see* **Note 8**).

9. Acquire a fluorescent image of the microarray (**Fig. 3C**) and extract the signal data following the recommendations of the manufacturer of the microarray imaging device and data extraction software used.

3.5. Data Analysis and Examples of Informatic Approaches

Although the specific protocols involved in adjusting a microarray scanning device to obtain the best possible image differ widely depending on the manufacturer, the basic principles are more or less the same. Whatever the device used, the scan should ideally make use of as much of the scanner's linear detection range as possible, and both signals should be scaled to occupy close to the same range. The laser power and voltage of the photomultiplier detectors can generally be adjusted to achieve this goal. Once a satisfactory image has been captured, it can be loaded into a variety of data-extraction software packages for analysis. Custom analysis tools are available, and have different algorithms for identifying the signals within the image, subtracting background fluorescence levels, and normalizing the ratios of the signals in the two channels to compensate for any labeling or detection bias in an individual experiment. One straightforward method of testing the performance of array hybridizations is to compare samples from the same source in the two channels. A scatterplot of the signal intensities in the two channels will give an idea of the amount of error in the system when measuring identical signals. Similarly, repeating hybridization of two different samples after switching the fluorochrome with which each is labeled and comparing the resulting ratios will give an idea of the accuracy of detection of different expression levels.

It is also advisable to verify a selection of the results obtained by microarray analysis. Quantitative hybridization *(12)* or real-time PCR are the preferable methods, but a general confirmation of relative differences between two samples may also be obtained from Northern blot analysis. It is important for such verification purposes to use the same EST sequence that served as the detector of interest on the array, and to test the same RNA samples that were used for the array. Such testing should be applied to ESTs showing a broad range of signal intensities and relative expression ratios.

The extremely complex nature of the large data sets obtained from microarray experiments demands advanced approaches for meaningful analysis. Numerous groups are devising relational databases and search tools tailored specifically for the management of microarray expression data, and many are already available commercially. One of the most widely used tools for organizing and sorting microarray data is cluster analysis. This includes a number of approaches for arranging expression patterns so that genes behaving most similarly in a series of experiments are ordered closest together. Clustering algorithms applied successfully to gene expression data include agglomerative hierarchical clustering *(13)*, and divisive clustering with two-dimensional sorting to order both genes and experiments *(14)*. Self organizing maps and k-means clustering have also been applied to microarray data. For a recent review and discussion of these approaches, *see* **ref. 15**.

Such cluster analysis of microarray gene expression patterns has already been used to reveal signatures indicative of different stages of immune cell function and physiology *(15)*, and to identify subclassifications of diffuse large B-cell lymphoma correlating with patient survival *(15)*. A similar approach with clustering of expression data from melanomas has revealed a set of genes correlating with aggressively metastatic tumors *(16)*. Such use of expression profiling to define molecular phenotypes of cancer correlating with clinical outcome has great potential for improvements of therapy. Distinct expression profiles have also been defined recently in human breast cancers with mutation in either the *BRCA1* or *BRCA2* tumor suppressor gene *(17)*. Not only does this allow more precise classification of nonhereditary breast cancers, but the availability of information on differential regulation of a large number of genes upon inactivation of these tumor suppressors may provide insight into their mechanisms of action.

In a slightly different approach, gene expression levels in the cell lines of the National Cancer Institute's Anti-Cancer Drug Screen (NCI-ACDS) have also been compared using microarray analysis. This panel *(18,19)* consists of 60 human tumor cell lines of different tissues of origin (breast, central nervous system, colon, bone marrow, lung, skin, ovary, prostate, and kidney). Clustering of the gene expression measurements revealed a clear tissue of origin signature in these tumor cell lines, with cell lines from the same tissues showing expression patterns most similar to each other, with only a few exceptions *(20)*. One great strength of the NCI-ACDS system is the availability of results of cytotoxicity assays in all 60 cell lines for nearly 100,000 compounds *(21)*. Clustering analysis could therefore be applied to this data to identify relationships between gene expression and drug activity patterns to aid both the drug discovery process and help provide a molecular basis for the selection of cancer therapy *(22)*. These sorts of cell line studies also have great potential for the study of tumor suppressor gene function.

The characterization of the status of the p53 pathway in the NCI-ACDS cell lines provides an example of a very simple form of this type of relational analysis in characterization of tumor suppressor genes. For each cell line in the panel, the p53 pathway was assessed by complete sequencing of the open reading frame, measurement of levels of p53 protein, functional evaluation based on a yeast transcription assay, measurement of γ-ray induction of G1 arrest, and on measurement of γ-ray induction of mRNA for *MDM2*, *CIP1/WAF1*, and *GADD45A* *(23)*. In this case, the COMPARE program (see http://dtp.nci.nih.gov) was used to search for correlations between these measurements of p53 function and the activity patterns of a set of 123 standard chemotherapy agents that included the majority of clinically approved anticancer drugs. A strong correlation was found between an intact p53 pathway and sensitivity to most classes of chemotherapy agents. An exception to this trend was the class of antimitotic agents, which include such drugs as vincristine, vinblastine, and paclitaxel (taxol). No correlation with p53 status was found for the antimitotic class of chemotherapy agents, perhaps suggesting an advantage to these drugs in the treatment of p53-mutant tumors. This finding provides insight into the physiologic and pharmacologic effects of inactivation of the p53 tumor suppressor pathway, one of the most common alterations in human cancers. As the availability of large gene expression data sets increases and informatic tools are refined, studies such as these are likely to have a profound impact on the study of tumor suppressor gene function.

4. Notes

1. Any delay in the initial lysis of samples can result in gene expression changes not associated with the parameters under study, thus adding large amounts of noise to the final data. Extremely high-quality RNA is a prime requirement for successful microarray analysis. Impurities in the RNA can inhibit fluorochrome incorporation in subsequent steps, and cellular contaminants, such as carbohydrates, may produce high background fluorescence on the hybridized arrays. Experience has shown that including precipitation steps at an early stage of extraction tends to

exacerbate these problems. Degradation of the RNA will obviously also lead to decreased hybridization specificity, and all standard RNA handling procedures should be observed.

2. The slow addition of ethanol at this step is necessary to prevent precipitation of the RNA at areas of high local ethanol concentration. Such local precipitation can reduce RNA yields.

3. While microarray analysis is extremely sensitive to many factors, and can often be difficult to troubleshoot, poor-quality RNA is a major cause of poor results. Even RNA with a good $A_{260/280}$ ratio that appears intact by Northern analysis can label poorly, and in cases of poor hybridization results, starting over with a fresh preparation of RNA is often advisable. Care should be taken not to exceed the binding capacity of the columns. For some samples, improved results can be obtained by doing a second round of trizol extraction before proceeding to the Qiagen columns, or by adding a DNase digestion step.

4. As an alternative to concentration by Centricon, the RNA can be precipitated by addition of 2 vol of 3.3% sodium acetate (3 *M* solution, pH 5.2) in absolute ethanol. Incubate on ice for 20 min, then microfuge at 4°C for 20 min. Wash twice with 70% ethanol and allow to dry. Precipitation does not seem to be harmful at this stage, as carbohydrates and other contaminants that can pose problems at earlier stages in the isolation should no longer be present.

5. Although relatively stable, the cyanine dyes used in this protocol are photosensitive, and once they have been added to the reactions, it is advisable to protect them from light exposure as much as possible. Care should also be taken in the storage of these dyes to avoid photobleaching, especially if they are not used very rapidly.

6. This is a relatively sensitive step, and care should be taken to avoid any impurities in the NaOH used. Make fresh NaOH if there is any sign of discoloration, and avoid storage in glass. If a color change is visible in the Cy5 reactions when the NaOH is added it will likely result in weak signal, and fresh NaOH should be made. Hydrolyzing for excessive time can result in weakened signal. Insufficient hydrolysis, however, will also produce weak signal due to hybridization with unlabeled RNA when this is not removed from the reaction.

7. Excessive cooling at this step will cause the SDS to precipitate when it is added, contributing to hybridization background. Occasionally, when the reactions have been microfuged, some debris will be visible at the bottom of the tube. Care should be taken to avoid pipetting this onto the array.

8. If a rotor for microtiter plates is not available, good results can also be obtained by centrifuging individual slides in uncapped 50-mL polypropylene tubes in a standard rotor, or by washing for 2 min in 100% isopropanol and then air-drying. If the slides are allowed to air-dry slowly, this frequently will contribute to high background.

Acknowledgment

The authors would like to thank the Cancer Genetics Branch of the National Human Genome Research Institute for their implementation and many refinements of these techniques.

References

1. Levine, A. J., Momand, J., and Finlay, C. A. (1991) The p53 tumor suppressor gene. *Nature* **351**, 453–456.

2. Hartwell, L. H. and Kastan, M. B. (1994) Cell cycle control and cancer. *Science* **266**, 1821–1828.

3. Zhan, Q., Chen, I. T., Antinore, M. J., and Fornace, A. J., Jr. (1998) Tumor suppressor p53 can participate in transcriptional induction of the *GADD45* promoter in the absence of direct DNA binding. *Mol. Cell. Biol.* **18**, 2768–2778.

4. Velculescu, V. E., Zhang, L., Vogelstein, B., and Kinzler, K. W. (1995) Serial analysis of gene expression. *Science* **270**, 484–487.

5. Fargnoli, J., Holbrook, N. J., and Fornace, A. J., Jr. (1990) Low-ratio hybridization subtraction. *Anal. Biochem.* **187**, 364–373.

6. Schena, M., Shalon, D., Davis, R. W., and Brown, P. O. (1995) Quantitative monitoring of gene expression patterns with a complementary DNA microarray. *Science* **270**, 467–470.

7. Lockhart, D. J., Dong, H., Byrne, M. C., et al. (1996) Expression monitoring by hybridization to high-density oligonucleotide arrays. *Nat. Biotechnol.* **14**, 1675–1680.

8. Okamoto, T., Suzuki, T., and Yamamoto, N. (2000) Microarray fabrication with covalent attachment of DNA using bubble jet technology. *Nat. Biotechnol.* **18**, 438–441.

9. Medlin, J. (2001) Array of hope for gene technology. *Environ. Health Perspect.* **109**, A34–A37.

10. Shoemaker, D. D., Schadt, E. E., Armour, C. D., et al. (2001) Experimental annotation of the human genome using microarray technology. *Nature* **409**, 922–927.

11. DeRisi, J., Penland, L., Brown, P. O., et al. (1996) Use of a cDNA microarray to analyse gene expression patterns in human cancer. *Nat. Genet.* **14**, 457–460.

12. Koch-Paiz, C. A., Momenan, R., Amundson, S. A., Lamoreaux, E., and Fornace, A. J., Jr. (2000) Estimation of relative mRNA content by filter hybridization to a polyuridylic probe. *Biotechniques* **29**, 706–714.

13. Eisen, M. B., Spellman, P. T., Brown, P. O., and Botstein, D. (1998) Cluster analysis and display of genome-wide expression patterns. *Proc. Natl. Acad. Sci. USA* **95**, 14863–14868.

14. Alon, U., Barkai, N., Notterman, D. A., et al. (1999) Broad patterns of gene expression revealed by clustering analysis of tumor and normal colon tissues probed by oligonucleotide arrays. *Proc. Natl. Acad. Sci. USA* **96**, 6745–6750.

15. Alizadeh, A. A., Eisen, M. B., Davis, R. E., et al. (2000) Distinct types of diffuse large B-cell lymphoma identified by gene expression profiling [see comments]. *Nature* **403**, 503–511.

16. Bittner, M., Meltzer, P., Chen, Y., et al. (2000) Molecular classification of cutaneous malignant melanoma by gene expression profiling. *Nature* **406**, 536–540.

17. Hedenfalk, I., Duggan, D., Chen, Y., et al. (2001) Gene-expression profiles in hereditary breast cancer. *N. Engl. J. Med.* **344**, 539–548.

18. Monks, A., Scudiero, D., Skehan, P., et al. (1991) Feasibility of a high-flux anticancer drug screen using a diverse panel of cultured human tumor cell lines. *J. Natl. Cancer Inst.* **83**, 757–766.

19. Grever, M. R., Schepartz, S. A., and Chabner, B. A. (1992) The National Cancer Institute: cancer drug discovery and development program. *Semin. Oncol.* **19**, 622–638.

20. Ross, D. T., Scherf, U., Eisen, M. B., et al. (2000) Systematic variation in gene expression patterns in human cancer cell lines. *Nat. Genet.* **24**, 227–235.

21. Mai, M., Yokomizo, A., Qian, C., et al. (1998) Activation of p73 silent allele in lung cancer. *Cancer Res.* **58**, 2347–2349.

22. Scherf, U., Ross, D. T., Waltham, M., et al. (2000) A gene expression database for the molecular pharmacology of cancer. *Nat. Genet.* **24**, 236–244.

23. O'Connor, P. M., Jackman, J., Bae, I., et al. (1997) Characterization of the p53 tumor suppressor pathway in cell lines of the National Cancer Institute anticancer drug screen and correlations with the growth-inhibitory potency of 123 anticancer agents. *Cancer Res.* **57**, 4285–4300.

13

Analysis of Tumor Suppressor Gene-Induced Senescence

Judith Campisi

1. Introduction

Cellular senescence refers to the response of cells to a variety of stimuli, many of which have the potential to induce preneoplastic or neoplastic phenotypes. These stimuli include dysfunctional telomeres, DNA damage, disrupted chromatin structures, the expression of certain oncogenes, and supraphysiologic mitogenic signals. The most prominent outcome of the senescence response is an essentially irreversible arrest of cell proliferation. In addition, senescent cells acquire marked changes in cell morphology and function. In some cases, senescent cells also acquire resistance to certain signals that induce apoptotic cell death *(1–4)*.

Several lines of evidence support the idea that, at least in mammals, cellular senescence evolved as a fail-safe mechanism to prevent the growth of cells at risk for neoplastic transformation *(2,3)*. In this regard, cellular senescence resembles apoptosis. However, whereas apoptosis kills and eliminates potential cancer cells, cellular senescence irreversibly arrests their growth. Recent findings raise the possibility that cellular senescence is an antagonistically pleiotropic trait—that is, while it protects young organisms from developing cancer prior to and during the period of reproduction, it can have unselected deleterious effects late in life *(3,5,6)*. The senescence response is critically dependent on the activity of several tumor suppressor genes. Most notable among these are those that participate in the p53 and pRB tumor suppressor pathways *(4,7,8)*.

Perhaps the most critical evidence that cellular senescence suppresses tumorigenesis derives from studies of intact organisms. Mice carrying germline deletions in genes that regulate the p53 and/or pRB pathway have been especially useful. Cells derived from such mice generally resist or fail to respond to senescence-inducing signals. Invariably, the animals are cancer-prone and typically die of cancer at an early age *(4,7,9,10)*. Similarly, in humans, the vast majority of malignant tumors harbor mutations in the p53, pRB, or both pathway, and germline mutations in p53 or pRB cause cancer-prone syndromes in humans *(11–13)*. However, although in-vivo studies are used to critically test the biologic consequences of cellular senescence, cell culture studies are generally needed to determine whether and how cells respond to senescence-inducing signals. Cell

From: *Methods in Molecular Biology, Vol. 223: Tumor Suppressor Genes: Regulation, Function, and Medicinal Applications.* Edited by: Wafik S. El-Deiry © Humana Press Inc., Totowa, NJ

culture studies have provided much of our understanding about the nature of the senescent phenotype, the stimuli that induce the response, and the mechanisms responsible for the senescence growth arrest. They have also shed considerable light on the types of mutations that allow cells to overcome the growth arrest. Interestingly, despite four decades of research, there is still debate as to whether cellular senescence occurs in vivo, and whether the senescent phenotype observed in cell cultures actually occurs in vivo. At present, we can only say that mutations that inactivate either the p53 or pRB tumor suppressor pathway increase the incidence of cancer in intact organisms, and confer upon cells from those organisms resistance to senescence-inducing stimuli. In addition, cells with at least some features associated with the senescent phenotype have been detected in cells in intact or freshly isolated tissues *(14–17)*.

2. Materials

1. Standard reagents for mammalian cell culture.
2. ^{3}H-thymidine or bromo-deoxyuridine (BrdU).
3. Phase-contrast microscope.
4. Bright-field microscope.
5. Cell fixation buffer: 2% formaldehyde + 0.2% glutaraldehyde in phosphate-buffered saline (PBS).
6. X-gal (5-bromo-4-chloro-3-indolyl-β-D-galactoside).
7. Dimethylformamide.
8. Citric acid/Na phosphate buffer, 0.2 *M* (100 mL): 36.85 mL 0.1 *M* citric acid solution, 63.15 mL 0.2 *M* sodium phosphate (dibasic) solution; verify that pH is 6.0.
9. Citric acid solution (0.1 *M*): 2.1 g citric acid monohydrate ($C_6H_8O_7 \cdot H_2O$)/100 mL water.
10. Sodium phosphate solution (0.2 *M*): 2.84 g sodium dibasic phosphate (Na_2HPO_4) or 3.56 g sodium dibasic phosphate dihydrate ($Na_2HPO_4 \cdot\cdot H_2O$)/100 mL water.
11. Potassium ferrocyanide.
10. Potassium ferricyanide.
11. Sodium and magnesium chloride.
12. Photographic emulsion and fixer (Kodak).
13. Standard reagents for mRNA reverse transcription and polymerase chain reaction (RT-PCR).
14. Oligonucleotide primers.
15. Cell staining solution (*see* **Table 1**).

3. Methods

3.1. Replicative Senescence

Cellular senescence was first formally described as the process that limits the proliferation of human cells in culture *(18–19)*. This process—referred to here as replicative senescence—is now known to be driven by the progressive shortening of telomeres *(20–24)*. Telomere shortening occurs as a consequence of DNA replication and an absence or insufficient activity of the enzyme telomerase *(20)*. Typically, human cells senesce when the average telomere length (or, more accurately, the average length of the terminal restriction fragment, or TRF) reaches 4–6 kb (reduced from approximately 15 kb in the germline) *(21,23)*.

It is now widely believed that only one or very few telomeres need reach a critically short length in order to induce the senescence response. It has also been cogently argued that cells sense and respond to disruption of the telomeric structure, rather than telom-

Table 1
Cell Staining Solution

Component	Volume	Final concentration
20-mg/mL X-gal in dimethylformamide	1 mL (*see* **Note 1**)	1 mg/mL
0.2 *M* citric acid/Na phosphate buffer,		
pH = 6.0 (pH is important!)	4 mL	40 m*M*
100 m*M* potassium ferrocyanide	1 mL	5 m*M*
100 m*M* potassium ferricyanide	1 mL	5 m*M*
5 *M* sodium chloride	0.6 mL	150 m*M*
1 *M* magnesium chloride	0.04 mL	2 m*M*
Water	12.4 mL	Total volume = 20 mL

ere length *per se* (25,26). According to this idea, the terminal telomeric structure can vacillate between closed and open forms. The closed form is the capped or protected structure, whereas cells sense the open form as a disrupted or dysfunctional telomere, and respond by arresting growth with a senescent phenotype. The structure of short telomeres is thought to be less stable, and hence more likely to experience structural disruption (26). An interesting correlate of this idea is that any perturbation that disrupts a telomere, whether or not it causes shortening, can induce a senescence response. Whatever the etiology of the disruption, the p53 and pRB tumor suppressor pathways are essential for the senescence response to dysfunctional telomeres (21,27,28).

Replicative senescence has been variably termed "natural" or "true" senescence, whereas the induction of senescence by nontelomeric stimuli is frequently referred to as "premature" senescence. There is little basis for these qualitative terms. The phenotype of cells induced to senesce by short telomeres, oxidative damage, or oncogenic RAS (which do not shorten the telomeres) is very similar (29,30). Moreover, there is increasing evidence that a common set of regulatory tumor suppressor genes lies at the heart of the senescence response, regardless of the inducing stimulus (4,7,8). Thus, replicative senescence is "natural" or "true" only by virtue of its history. Here, I refer to the senescent phenotype induced by cell division, telomere erosion, and subsequent telomere dysfunction as replicative senescence. I refer to the senescent phenotype induced by any stimulus, regardless of its origin, as cellular senescence.

Replicative senescence (dependent on telomere length) is not unique to human cells (31). However, cells from some vertebrates proliferate for only a limited number of divisions despite very long telomeres and the presence of telomerase activity (32–34). This behavior is especially common among cells from rodent species. Although the limited growth potential of cultured rodent cells is often termed replicative senescence, it is clear that telomere length does not limit the proliferation of these cells (35). For this reason, I will first discuss the replicative senescence of human cells, and then discuss the process that limits the proliferation of rodent cells.

By way of definition, the number of population doublings (PD) a cell culture can achieve is termed its replicative life span. Cultures that undergo senescence are said to have a finite or mortal replicative life span, and are termed cell strains. By contrast, cultures that proliferate indefinitely are said to have an indefinite or immortal replicative life span, and are referred to as cell lines.

In general, the replicative life span of a cell culture depends on the cell type and the donor species, genotype, and, to some extent, age *(1,36)*. Fibroblasts are the most widely used cell type for studying cellular senescence because they were among the earliest cells to be cultured, and they remain relatively easy to propagate in culture. For general cell culture methods, the reader is referred to any of several handbooks written on this topic (e.g., **ref. 37**). Unless noted otherwise, the methods described below are optimized for fibroblasts, although they often can be used without modification for other cell types.

The replicative life span of a culture is one of the most sensitive means by which to assess the action of oncogenes or tumor suppressor genes that regulate cellular senescence.

3.1.1. Human Cells

Human fibroblasts can have replicative life spans that range from <10 PD to >80 PD. In general, fetal and neonatal cells proliferate for 50–80 PD, whereas fibroblasts from adult donors typically have shorter replicative life spans (<40 PD). There are, however, exceptional strains in both categories, and the replicative life span of fibroblasts can vary considerably depending on the donor and tissue *(1,36,38)*.

The fibroblast strains that were used in early studies of cellular senescence were derived from fetal lung (e.g., WI-38, IMR-90) *(18,19)*. These strains are still in frequent use. However, fibroblast strains have also been developed from a variety of fetal and adult tissues, and used with success and similar results *(36)*. Although there is no question that fibroblasts from diverse tissues and developmental stages differ in many ways, it is remarkable how similarly they behave with regard to the environmental and genetic control of replicative life span. For example, low oxygen tension (2–5%) modestly extends the replicative life span of human fibroblasts from different origins *(39,40)*. Likewise, inactivation of p53 extends the replicative life span of fibroblasts from diverse tissues and donors, as well as nonfibroblastic cells from different tissues and donors *(41–46)*. Thus, there are many options for choosing a cell strain to study the control of cellular senescence. It is, of course, wise when working with uncharacterized strains to be sure it has the general characteristics of the many normal human fibroblast strains that have been studied.

The replicative senescence of human fibroblasts has the following general features:

1. The telomeres progressively shorten with increasing PDs *(47,48)*.
2. The cells arrest growth with a G1 DNA content and senescent phenotype (discussed below) *(1,36,49)*.
3. Manipulations that inactivate critical components of either the p53 or pRB pathways extend the replicative life span, although the cells eventually arrest growth (with subsenescent telomere lengths) *(21,44,50)*.
4. Inactivation of both the p53 and pRB pathways synergistically extend the replicative life span, but the population enters an unstable state termed crisis, from which rare immortal variants may arise *(21,44,51)*.
5. The cells can acquire an indefinite replicative life span when they express hTERT, the catalytic component of human telomerase *(24,52)*.

To the extent it has been studied, these features also hold true for a number of other human cell types (for example, pigmented retinal epithelial cells and endothelial cells) *(24,53)*. There is some debate, however, as to whether at least some human epithelial cells can be immortalized by hTERT alone. In some cases, hTERT failed to immortalize human keratinocytes or mammary epithelial cells unless p16, a critical component of the pRB pathway, was inactivated *(54)*. On the other hand, propagating keratinocytes or mammary epithelial cells on a feeder layer appeared to abrogate the need for p16 inactivation *(55)*. Moreover, under some culture conditions, mammary epithelial cells appear to silence p16 (or the culture conditions favor the growth of cells with silenced p16), and these cells eventually arrest growth with characteristics that differ in several ways from those of senescent fibroblasts *(56)*. Clearly, culture conditions need to be optimized and the cultures need to be characterized, especially for nonfibroblastic cell types, before one can determine the replicative life span and understand the senescent phenotype.

3.1.2. Replicative Life-Span Determinations

Determining the replicative life span of a culture is quite simple, requiring little more than the ability to serially subcultivate (passage) the cells and determine the cell number at each passage. There are, however, some pitfalls to be avoided and additional manipulations that can make the determination more accurate and valuable.

1. At the outset, it helps to know the fraction of cells in the culture that are capable of proliferation (presenescent), which can be determined by measuring the fraction of cells capable of DNA synthesis (maximum labeling index). To determine the maximum labeling index, seed the cells in small culture dishes or wells at relatively low density ($0.5–1 \times 10^3/cm^2$), incubate with ^{3}H-thymidine (or BrdU) for an interval equivalent to 2–3 doublings (generally 3–4 d), and determine the percentage of cells with radiolabeled nuclei after autoradiography (or BrdU-labeled nuclei after immunostaining) (*see* **ref. *57*** for detailed methods).

2. During serial subcultivation, avoid seeding the cells at densities that are too low, and subculture the cells before they reach confluence. We generally seed human fibroblasts at $1–3 \times 10^{-3}/cm^2$, and subculture them when they reach 80% confluence.

3. We recommend determining the maximum labeling index at each passage, or at least every second or third passage for cultures with long replicative life spans. This determination can be particularly informative when oncogenes or tumor suppressor genes are manipulated and alter the balance between entry into S phase, cell division, and cell death.

4. Some laboratories declare a culture senescent if the cells fail to double after an arbitrary interval, generally 2–4 wk. Other laboratories declare a culture senescent when the maximum labeling index falls below an arbitrary value, generally <5% or <10%. Both criteria are valid, but the maximum labeling index provides a more quantitative measure of proliferative capacity. In addition, some features of senescent human fibroblasts—for example, the increase in p16—occurs only several weeks after a culture has reached senescence *(50)*. Thus, cultures with a labeling index of 10% or 1% would fail to double in 2 wk, but the 10% cultures would have relatively low p16, compared to cultures with a labeling index of 1%.

5. The maximum labeling index is generally a more meaningful measure of proliferative capacity than the calculated number of completed PDs. The PD number assumes 100%, or at least an invariant, plating efficiency. However, variability in culture media and serum lots, trypsinization conditions and intervals, surface of the culture vessels, and tissue culture technique can affect the plating efficiency and viability of cells, which in turn can affect the PD calculation.

6. Replicative life span (PDs to senescence) is not an absolute number. When a culture is established from freshly explanted tissue, the starting population is generally considered the first confluent culture that grows from the explant. But the number of cells and PDs they completed depends, of course, on the size of the explant and culture vessel, and is another reason why maximum labeling index is a more meaningful measure of proliferative capacity.

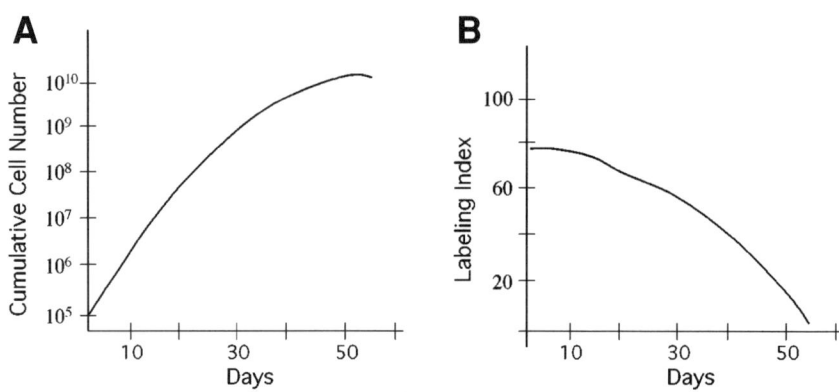

Fig. 1. Replicative life-span determination. Normal human fibroblasts (1×10^5 per 75-cm^2 flask, and having a maximum labeling index of approximately 78%) were serially subcultivated for approximately 50 d. Each time the cultures reached 80% confluence, they were trypsinized and $1–2 \times 10^5$ were seeded into a new flask. At each passage, the cell number and maximum labeling index was determined. Cell proliferation ceased after approximately 17 PD, when the cumulative cell number reached approximately 2×10^{10} and the labeling index had declined <2%. **(A)** Semilog plot showing the cumulative cell number vs time in culture. **(B)** Linear plot showing the maximum labeling index vs time in culture.

Replicative life span is generally displayed on a semilog plot of cumulative cell number vs time in continuous culture (**Fig. 1A**), cumulative cell number vs PDs, or PDs vs time in culture—all with similarly shaped curves. Less conventionally, replicative life span can be displayed on a linear plot of the maximum labeling index vs time in culture (**Fig. 1B**).

3.1.3. Manipulating Replicative Life Span

A number of manipulations have been shown to alter the replicative life span of human fibroblasts, most notably genetic interventions that alter the activity of certain tumor suppressor genes *(4,7,24,51,52,58,59)*. Relatively few manipulations have been shown to shorten replicative life span, as opposed to arresting cell proliferation within a relatively short interval. For example, certain cyclin-dependent kinases cause cells to arrest growth with a senescent phenotype when overexpressed, but this occurs within one or two cell cycles *(60)*. By contrast, a number of manipulations have been shown to extend replicative life span, and studies of this sort have been paramount in identifying the tumor suppressor genes and pathways that regulate cellular senescence. For example, the expression of viral oncogenes such as HPV16-E6, HPV16-E7, or SV40 T antigen extends the replicative life span of human fibroblasts, and mutational analyses of these genes established the importance of the p53 and pRB pathways for allowing cells to sense and respond to telomere shortening *(51)*.

In general, two methods have been used to express genes in human cells in order to determine their effect on replicative life span.

First, expression vectors can be introduced into the cells by DNA-mediated transfection. Single cell clones (stable transfectants) can then be isolate, expanded, and followed during serially subcultivated until senescence. This method requires that a relatively long-lived strain be used as the starting population (e.g., BJ neonatal foreskin fibroblasts, which generally proliferate for 80–90 PD) *(24)* because many clones will senesce before expanding enough to determine the cell number accurately. In addition, because the replicative life spans of single cell clones from a given culture are broadly distributed about a characteristic mean, this method also requires that multiple control and experimental clones be examined in order for the results to achieve statistical significance.

More recently, amphotropic retroviruses have been used to express genes in human cell strains. High-titer retroviruses have the advantage of infecting and integrating into most (>90%) of the cells in the starting population, eliminating the necessity to isolate single cell clones. In addition, most retroviral vectors carry antibiotic resistance genes, which permits rapid elimination of any uninfected cells in the culture. Advances in retroviral production, such as transient transfection protocols for virus production *(61)*, make the use of retroviruses the preferred method for expressing genes that affect replicative life span.

1. Because retroviruses integrate into dividing cells, and are relatively short-lived at 37°C, multiple infections may be needed to obtain a culture in which most of the cells have integrated viral genomes. We recommend beginning with a relatively sparse, proliferating cells (20–30% confluent), and carrying out infections for 5–7 h each on 2–3 successive days (or more for very slow-growing cultures), with medium changes in between infections. After infection, the cultures are trypsinized and diluted for antibiotic selection.
2. Antibiotic selection can begin 12–24 h after the last infection. A pilot experiment should be done first to determine the antibiotic sensitivity of the culture. Ideally, uninfected cells should die within 3–7 d.
3. For most experiments, it is important to know the infection efficiency. A culture that is mock-infected (treated identically to infected cultures) but spared antibiotic selection can serve this purpose. At the end of selection, the number of cells in the infected cultures divided by the number of cells in the mock-infected culture provides an estimate of the infection efficiency. The infection efficiency of control and experimental viruses should be similar.

Figure 2 shows a typical result when human fibroblasts are infected with a retrovirus carrying the HPV16-E6 oncogene, which inactivates p53, or a retrovirus carrying hTERT. Inactivation of p53 typically extends the replicative life span by 5–8 PD, whereas hTERT typically renders the cells of essentially indefinite division.

3.1.4. Rodent Cells

As noted earlier, in general, rodent cells have replicative life spans that are substantially shorter than those of human cells. Most studies on cellular senescence in rodent cells have used cells from mice and, to a lesser extent, rats. Mouse and rat embryo fibroblasts typically cease division after 5–10 PD in culture. In our experience, mouse embryo fibroblasts (MEFs) are much more sensitive than human fibroblasts to the 20% oxygen in which cells are usually cultured, and their replicative life span can be at least doubled by culturing the cells in lower than ambient oxygen concentrations (3–5%) (Parrinello and Campisi, unpublished). However, most studies of cellular senescence using mouse or rat fibroblasts have used standard 20% oxygen culture conditions.

Replicative senescent rodent cells share a number of features with replicatively senescent human cells, but there are also a number of dissimilarities.

1. In contrast to human cells, there is no evidence that telomere erosion limits the replicative life span of rodent cells in culture *(34,35,62)*.
2. Rodent cells arrest growth with a typical senescent morphology (discussed below), but very little has been done to characterize the senescent phenotype of rodent fibroblasts, much less

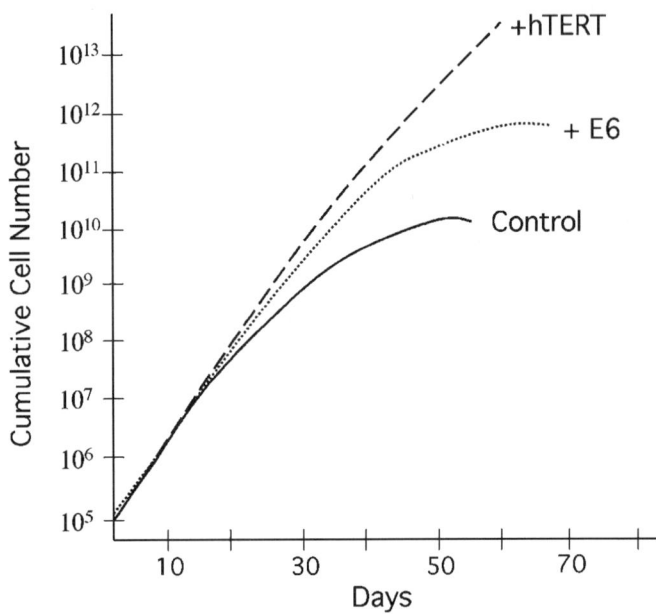

Fig. 2. Extension of replicative life span. Normal human fibroblasts were allowed to proliferate in culture as described in **Fig. 1**. At the start of the experiment, parallel cultures were infected with retroviruses carrying either no insert (Control), the HPV-E6 oncogene (+ E6), or a cDNA encoding the catalytic subunit of telomerase (hTERT). Shown is a semilog plot of the cumulative cell numbers vs time in culture.

determine whether or to what extent it resembles the senescent phenotype of human fibroblasts.

3. Manipulations that inactivate critical components of either the p53 or pRB pathway extend the replicative life span of mouse and rat fibroblasts, and in most cases fully immortalizes them *(4,7,28,58,59,63,64)*.

4. In general, rodent fibroblasts spontaneously immortalize at frequencies that range from 1 in 10^{-4} to 1 in 10^{-7}. This is in sharp contrast to human fibroblasts, which rarely, if ever, immortalize spontaneously. Immortal rodent fibroblasts frequently have multiple chromosomal abnormalities, and immortalization appears to be preceded by a state that resembles that of crisis in human cells *(65,66)*.

5. Rodent cells frequently express telomerase in culture and in vivo, yet the cells still have only a finite replicative life span in culture *(33,34,67–69)*.

3.2. Cellular Senescence

As noted earlier, cells undergo a senescence growth arrest in response to a number of stimuli, most, if not all, of which have in common the potential to cause neoplastic transformation *(3)*. These findings support the idea that replicative senescence is a specific example of a more generally responsive process, termed here cellular senescence.

Consistent with the idea the malignant transformation can be induced by multiple pathways, the number of stimuli and agents that have been reported to elicit a senescence response in mammalian cells has been growing steadily over the past decade. These stimuli are too numerous to discuss comprehensively and in detail here. However, some of the major types of stimuli/agents that can induce the senescence response, and

Table 2
Inducers of Cellular Senescence

Class	Specific agent	Reference(s)
DNA-damaging agents	Ionizing radiation	*(70–72)*
	Hydrogen peroxide	*(40)*
	Chemotherapeutic drugs	*(73) (74)*
Chromatin-remodeling agents	Histone deacetylase inhibitors	*(75)*
	Methyltransferase inhibitors	*(76)*
Strong mitogenic signals	Oncogenic RAS	*(29)*
	Activated MEK	*(77)*
	Oncogenic RAF	*(78)*
	Overexpressed E2F1	*(79)*
Tumor suppressor genes/	p16	*(60,80,81)*
growth inhibitors	p14^{ARF}	*(79)*
	PML	*(82,83)*
	p21	*(60,84,85)*
	PI3 kinase inhibitors	*(86)*

a few examples of each, are given in **Table 2**. For detailed protocols on how to induce senescence by one or more of these specific agents, the reader should consult the references provided in **Table 2**.

In general, agents that induce a senescence response can be classified into one of two groups. The first group includes those that cause DNA damage or impose a similar cellular stress *(40,70–74)*, those that disrupt chromatin normal organization *(75,76)*, and those that deliver strong mitogenic signals *(29,77–79)*. All these agents have the potential to induce or promote neoplastic transformation. The second group is likely to be the effectors or mimics of the senescence response. This group includes the overexpression of certain tumor suppressor genes *(60,79,80–83)*, and the overexpression or treatment with certain growth inhibitors *(60,84–86)*.

To the extent it has been studied, inactivation of either the p53 or pRB tumor suppressor pathway abrogates the senescence response to those agents that act upstream of p53 or pRB. Moreover, human cells that have been immortalized by ectopic expression of hTERT (which abrogates replicative senescence) nonetheless senesce in response to nontelomeric senescence inducers such as oncogenic RAS *(87)*. This finding suggests that telomerase expression can protect cells only from replicative senescence, but not against other inducers of senescence. This finding may also explain why rodent cells senescence under standard culture conditions (20% oxygen), despite telomerase expression.

3.3. The Senescent Phenotype

The senescence response entails many changes in gene expression *(1,3,38,49)*. Some of these changes relate directly to the irreversible growth arrest that is the hallmark of cellular senescence. Other changes, however, appear to be unrelated to the growth arrest but relate closely to cellular function. The functional changes associated with cellular senescence have been most thoroughly studied in human fibroblasts, where it appears that the senescence response "locks" the cells in an activated state that they only transiently experience during their physiologic function in extracellular matrix remodeling and wound healing *(88,89)*. The characteristics of the senescent phenotype that are discussed below pertain largely to fibroblasts. However, where these characteristics are pertinent to the senescent phenotype of other cell types, or cells from other species, is noted and discussed.

In general, multiple measures of the senescent phenotype should be used in determining whether and to what extent a gene or treatment causes a senescent response. The most commonly used criteria are (a) growth arrest, (b) senescent morphology, (c) expression of the senescence-associated β-galactosidase, (d) increased expression of one or more gene that is upregulated by replicatively senescent cells.

3.3.1. Growth Arrest

Senescent human fibroblasts arrest growth with a G1 DNA content *(1,36,38)*. This is also true for the majority of other normal human cell types in which the senescent phenotype has been studied. One possible exception are the long-lived human mammary epithelial cells that eventually overgrow the primary cultures from explants of normal breast tissue. These long-lived cells have methylated and silenced the p16 locus, and arrest growth in culture with a substantial fraction of cells in G2 *(56)*. These cells may be preneoplastic, as preneoplastic cells have been shown to exist in even apparently normal tissue *(90–92)*. In all cell types and species in which it has been studied, the senescence growth arrest appears to be essentially irreversible, in that no physiologic mitogens or stimuli have been identified that can reverse it. Nonphysiologic mitogens, such as viral oncogenes, can abrogate the senescence response and do so largely by inactivating the p53 and/or pRB tumor suppressor pathways *(50,93–98)*.

3.3.2. Morphology

In general, senescent cells acquire an enlarged morphology relative to their presenescent counterparts *(1,36,38)*. In the case of adherent cells, senescence is also accompanied by an increase in intracellular actin stress fibers, easily seen using a phase-contrast microscope, and increased adherence to the substratum. An enlarged, flattened morphology appears to be a general characteristic of virtually all senescent cell types that have been studied. It has been noted in many nonfibroblastic human cell types, and in cells from a number of nonhuman species, including rodent cells. **Figure 3** shows the morphology of early passage (presenescent) and senescent human fibroblasts, which is typical of the senescent phenotype of many mammalian cells.

3.3.3. Senescence-Associated β-Galactosidase

A common feature of senescent cells is the expression of a neutral (pH 6) β-galactosidase, termed the senescence-associated β-galactosidase (SA-βgal) *(14)*. This enzyme marker is strongly associated with the senescent phenotype, although it is not necessarily senescence-specific (which is to be expected if cellular senescence is indeed antagonistically pleiotropic). SA-βgal is expressed in a variety of replicatively senescent human cell types, including senescent fibroblasts, adult melanocytes, endothelial cells, keratinocytes, and mammary epithelial cells. It is also induced by diverse senescence-inducing stimuli, including short telomeres, agents that cause DNA damage or perturb chromatin structure, and strong mitogenic signals *(29,60,73–75,78,79)*. SA-βgal expression varies among species. It is weaker in senescent rodent cells, compared to senescent

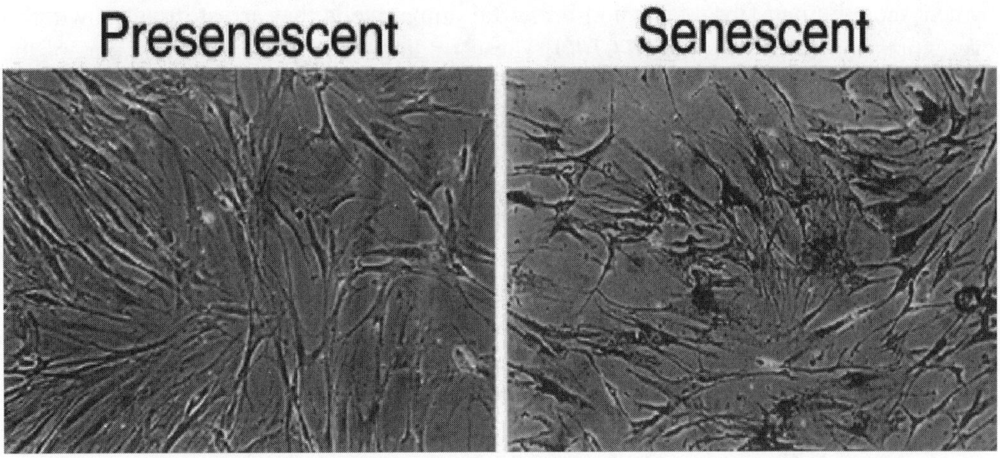

Presenescent   Senescent

Fig. 3. Normal human fibroblasts were stained for senescence-associated β-galactosidase activity SA-βgal, according the protocol given in the text. The cells were viewed and photographed under phase-contrast microscopy.

human cells, but relatively strong in nonhuman primate cells. Because the assay is simple, SA-βgal activity has become a widely used marker for the senescent phenotype. It is best viewed using a bright-field microscope. **Figure 3** shows presenescent and senescent human fibroblasts stained for SA-βgal.

3.3.3.1. SA-βGAL STAINING

For cells on tissue culture dishes or slides:

1. Wash cells twice with PBS.
2. Fix 3–5 min at room temperature with 2% formaldehyde + 0.2% glutaraldehyde in PBS (*see* **Note 2**).
3. Wash cells 2–3 times with PBS.
4. Add staining solution (1–2 mL per 35-mm dish, or immerse the slide) (**Table 1**).
5. Incubate at 37°C (*not* in a CO_2 incubator).

Blue color is detectable in some cells within 2 h, but staining is maximal in 12–16 h.

3.3.3.2. COMBINED ^{3}H-THYMIDINE AUTORADIOGRAPHY AND SA-βGAL STAINING

1. Label cells with ^{3}H-thymidine. We use 10 μCi/mL (50-80 Ci/mmol) for 48–72 h.
2. Wash cells 2× with PBS, fix and stain for SA-βgal activity, as described above.
3. After the color develops, wash cells twice with PBS, twice with methanol, and air-dry.
4. Coat the cells with photographic emulsion (such as Kodak, NTB2), store in the dark 12–48 h, develop, fix, and rinse well with water.

3.3.4. Senescent Gene Expression Pattern

Microarray and similar analyses have revealed many changes in gene expression that accompany the senescent phenotype of human fibroblasts *(30,99)*. Interestingly, although there are some differences, the phenotype of human fibroblasts induced to senesce by diverse stimuli (short telomere, oncogenic RAS, hydrogen peroxide) are quite similar. In addition, there is overlap in the patterns of gene expression displayed by senescent human fibroblasts and immortal human cells that arrest growth owing to overexpression of p53 or p21 *(84,100)*. These findings suggest that at least part of the senescent pattern of gene expression results from the activation of tumor suppressor genes that establish and maintain the senescent growth arrest.

One of the striking features of the senescent phenotype of human fibroblasts is an increase in the expression of genes that encode secreted proteins. These proteins have been reviewed elsewhere, but include matrix metalloproteinases (such as MMP-1 and stromelysin-1), inflammatory cytokines (such as interleukin-1), and growth factors (such as heregulin and other EGF-like growth factors) *(1,88)*. Ideally, in assessing whether or to what extent a gene or treatment induces a senescent phenotype, microarray analyses provide the most comprehensive assessment of the senescent pattern of gene expression. However, as this is not always feasible, the most commonly assayed senescence-associated genes in human fibroblasts are MMP-1, stromelysin-1, and plasminogen activator inhibitor-1 (PAI-1) *(29,77,79)*. Most often, these genes are assayed by RT-PCR. PCR primers that have been used successfully for this purpose on total RNA isolated from human cells *(79)* are

PAI-1:
Forward—5′ GTCATAGTCTCAGCCCGCATG
Reverse—5′ TTTCCTTCAGAAAGAGTCATAAC

MMP-1:
Forward—5′ CCAGTGACTGCACATGAGGTTC
Reverse—5′ CCTCTAGAGTCACTGATACACA

Stromelysin-1:
Forward—5′ GACACACACTTTGAAGAGTAACAG
Reverse—5′ GTCTGTTGCACACGAGTGCTTCC

4. Notes

1. Store X-gal stock solution at –20°C. Buffer and salts may be prepared several days or weeks ahead of time. However, X-gal is not stable in aqueous solution. It should be added the day of the assay, no earlier.
2. 3% Formaldehyde also works, but 2% formaldehyde + 0.2% glutaraldehyde preserves morphology somewhat better.

References

1. Campisi, J., Dimri, G. P., and Hara, E. (1996) Control of replicative senescence, in *Handbook of the Biology of Aging* (Schneider, E., and Rowe, J., eds.) Academic Press, New York, pp. 121–149.
2. Smith, J. R. and Pereira-Smith, O. M. (1996) Replicative senescence: implications for in vivo aging and tumor suppression. *Science* **273**, 63–67.
3. Campisi, J. (2000) Cancer, aging and cellular senescence. *In Vivo* **14**, 183–188.
4. Bringold, F. and Serrano, M. (2000) Tumor suppressors and oncogenes in cellular senescence. *Exp. Gerontol.* **35**, 317–329.
5. Campisi, J. (1996) Replicative senescence: an old lives tale? *Cell* **84**, 497–500.
6. Campisi, J. (1997) Aging and cancer: the double-edged sword of replicative senescence. *J. Am. Geriatr. Soc.* **45**, 1–6.
7. Lundberg, A. S., Hahn, W. C., Gupta, P., and Weinberg, R. A. (2000) Genes involved in senescence and immortalization. *Curr. Opin. Cell Biol.* **12**, 705–709.

8. Itahana, K., Dimri, G., and Campisi, J. (2001) Regulation of cellular senescence by p53. *Eur. J. Biochem.* **268**, 2784–2791.

9. Ghebranious, N. and Donehower, L. A. (1998) Mouse models in tumor suppression. *Oncogene* **17**, 3385–3400.

10. Sherr, C. J. (1998) Tumor surveillance via the ARF-p53 pathway. *Genes Dev.* **12**, 2984–2891.

11. Hollstein, M., Sidransky, D., Vogelstein, B., and Harris, C. C. (1991) p53 mutation in human cancer. *Science* **253**, 49–53.

12. Paggi, M. G., Baldi, A., Bonetto, F., and Giordano, A. (1996) Retinoblastoma protein family in cell cycle and cancer: a review. *J. Cell. Biochem.* **62**, 418–430.

13. Mcleod, K. (2000) Tumor suppressor genes. *Curr. Opin. Genet. Dev.* **10**, 81–93.

14. Dimri, G. P., Lee, X., Basile, G., et al. (1995) A novel biomarker identifies senescent human cells in culture and in aging skin in vivo. *Proc. Natl. Acad. Sci. USA* **92**, 9363–9367.

15. Mishima, K., Handa, J. T., Aotaki-Keen, A., Lutty, G. A., Morse, L. S., and Hjelmeland, L. M. (1999) Senescence-associated beta-galactosidase histochemistry for the primate eye. *Invest. Ophthalmol. Vis. Sci.* **40**, 1590–1593.

16. Pendergrass, W. R., Lane, M. A., Bodkin, N. L., et al. (1999) Cellular proliferation potential during aging and caloric restriction in rhesus monkeys (*Macaca mulatta*). *J. Cell. Physiol.* **180**, 123–130.

17. Paradis, V., Youssef, N., Dargere, D., Ba, N., Bonvoust, F., and Bedossa, P. (2001) Replicative senescence in normal liver, chronic hepatitis C, and hepatocellular carcinomas. *Hum. Pathol.* **32**, 327–332.

18. Hayflick, L. and Moorhead, P. S. (1961) The serial cultivation of human diploid cell strains. *Exp. Cell Res.* **25**, 585–621.

19. Hayflick, L. (1965) The limited in vitro lifetime of human diploid cell strains. *Exp. Cell Res.* **37**, 614–636.

20. Levy, M. Z., Allsopp, R. C., Futcher, A. B., Greider, C. W., and Harley, C. B. (1992) Telomere end-replication problem and cell aging. *J. Mol. Biol.* **225**, 951–960.

21. Wright, W. E. and Shay, J. W. (1996) Mechanism of escaping senescence in human diploid cells, in *Modern Cell Biology Series—Cellular Aging and Cell Death* (Holbrook, N. J., Martin, G. R., and Lockshin, R. A., eds.). Wiley & Sons, Inc., New York, pp. 153–167.

22. Campisi, J. (1997) The biology of replicative senescence. *Eur. J. Cancer* **33**, 703–709.

23. Chiu, C. P. and Harley, C. B. (1997) Replicative senescence and cell immortality: the role of telomeres and telomerase. *Proc. Soc. Exp. Biol. Med.* **214**, 99–106.

24. Bodnar, A. G., Ouellette, M., Frolkis, M., et al. (1998) Extension of life span by introduction of telomerase into normal human cells. *Science* **279**, 349–352.

25. Ouelette, M. M., Liao, M., Herbert, B. S., et al. (2000) Subsenescent telomere lengths in fibroblasts immortalized by limiting amounts of telomerase. *J. Biol. Chem.* **275**, 10072–10076.

26. Blackburn, E. H. (2000) Telomere states and cell fates. *Nature* **408**, 53–56.

27. Chin, L., Artandi, S. E., Shen, Q., et al. (1999) p53 deficiency rescues the adverse effects of telomere loss and cooperates with telomere dysfunction to accelerate carcinogenesis. *Cell* **97**, 527–538.

28. Serrano, M., Lee, H., Chin, L., Cordon-Cardo, C., Beach, D., and DePinho, R. A. (1996) Role of the INK4A locus in tumor suppression and cell mortality. *Cell* **85**, 27–37.

29. Serrano, M., Lin, A. W., McCurrach, M. E., Beach, D., and Lowe, S. W. (1997) Oncogenic ras provokes premature cell senescence associated with accumulation of p53 and p16INK4a. *Cell* **88**, 593–602.

30. Shelton, D. N., Chang, E., Whittier, P. S., Choi, D., and Funk, W. D. (1999) Microarray analysis of replicative senescence. *Curr. Biol.* **9**, 939–945.

31. Rohme, D. (1981) Evidence for a relationship between longevity of mammalian species and life spans of normal fibroblasts in vitro and erythrocytes in vivo. *Proc. Natl. Acad. Sci. USA* **78**, 5009–5013.

32. Kakuo, S., Asaoka, K., and Ide, T. (1999) Human is a unique species among primates in terms of telomere length. *Biochem. Biophys. Res. Commun.* **263**, 308–314.

33. Sherr, C. J. and DePinho, R. A. (2000) Cellular senescence: mitotic clock or culture shock? *Cell* **102**, 407–410.

34. Wright, W. E. and Shay, J. W. (2000) Telomere dynamics in cancer progression and prevention: fundamental differences in human and mouse telomere biology. *Nat. Med.* **6**, 849–851.

35. Campisi, J. E. G.-P. (2001) From cells to organisms: can we learn about aging from cells in culture? *Exp. Gerontol.* **36**, 607–618.

36. Stanulis-Praeger, B. (1987) Cellular senescence revisited: a review. *Mech. Aging Dev.* **38**, 1–48.

37. Freshney, R. I. (1987) *Animal Cell Culture: A Practical Approach*, IRL Press, Oxford, UK.

38. Cristofalo, V. J. and Pignolo, R. J. (1993) Replicative senescence of human fibroblast-like cells in culture. *Physiol. Rev.* **73**, 617–638.

39. Balin, A. K., Fisher, A. J., and Carter, D. M. (1984) Oxygen modulates growth of human cells at physiological partial pressures. *J. Exp. Med.* **160**, 152–166.

40. Chen, Q., Fischer, A., Reagan, J. D., Yan, L. J., and Ames, B. N. (1995) Oxidative DNA damage and senescence of human diploid fibroblast cells. *Proc. Natl Acad. Sci. USA* **92**, 4337–4341.

41. Shay, J. W., Pereira-Smith, O. M., and Wright, W. E. (1991) A role for both Rb and p53 in the regulation of human cellular senescence. *Exp. Cell Res.* **196**, 33–39.

42. Band, V. (1995) Preneoplastic transformation of human mammary epithelial cells. *Semin. Cancer Biol.* **6**, 185–192.

43. Bond, J. A., Wyllie, F. S., and Wynford-Thomas, D. (1994) Escape from senescence in human diploid fibroblasts induced directly by mutant p53. *Oncogene* **9**, 1885–1889.

44. Hara, E., Tsuri, H., Shinozaki, S., and Oda, K. (1991) Cooperative effect of antisense-Rb and antisense-p53 oligomers on the extension of lifespan in human diploid fibroblasts, TIG-1. *Biochem. Biophys. Res. Commun.* **179**, 528–534.

45. Klingelhutz, A., Foster, S. A., and McDougall, J. K. (1996) Telomerase activation by the E6 gene product of human papillomavirus type 16. *Nature* **380**, 79–83.

46. Whikehart, D. R., Register, S. J., Chang, Q., and Montgomery, B. (2000) Relationship of telomeres and p53 in aging bovine corneal endothelial cell cultures. *Invest. Ophthalmol. Vis. Sci.* **41**, 1070–1075.

47. Harley, C. B., Futcher, A. B., and Greider, C. W. (1990) Telomeres shorten during aging of human fibroblasts. *Nature* **345**, 458–460.

48. Allsopp, R. C., Vaziri, H., Patterson, C., et al. (1992) Telomere length predicts replicative capacity of human fibroblasts. *Proc. Natl. Acad. Sci. USA* **89**, 10114–10118.

49. Goldstein, S. (1990) Replicative senescence: the human fibroblast comes of age. *Science* **249**, 1129–1133.

50. Dulic, V., Beney, G. E., Frebourg, G., Drullinger, L. F., and Stein, G. H. (2000) Uncoupling between phenotypic senescence and cell cycle arrest in aging p21-deficient fibroblasts. *Mol. Cell. Biol.* **20**, 6741–6754.

51. Campisi, J. (1999) Replicative senescence and immortalization, in *The Molecular Basis of Cell Cycle and Growth Control* (Stein, G., Baserga, R., Giordano, A., and Denhardt, D., eds.). Wiley-Liss, New York, pp. 348–373.

52. Vaziri, H. and Benchimol, S. (1998) Reconstitution of telomerase activity in normal human cells leads to elongation of telomeres and extended replicative life span. *Curr. Biol.* **8**, 279–282.

53. Yang, J., Chang, E., Cherry, A. M., et al. (1999) Human endothelial cell life extension by telomerase expression. *J. Biol. Chem.* **274**, 26141–26148.

54. Kiyono, T., Foster, S. A., Koop, J. I., McDougall, J. K., Galloway, D. A., and Klingelhutz, A. J. (1998) Both Rb/p16INK4a inactivation and telomerase activity are required to immortalize human epithelial cells. *Nature* **396**, 84–88.

55. Ramirez, R. D., Morales, C. P., Herbert, B. S., et al. (2001) Putative telomere-independent mechanisms of replicative aging reflect inadequate growth conditions. *Genes Dev.* **15**, 398–403.

56. Romanov, S. R., Kozakiewicz, B. K., Holst, C. R., Stampfer, M. R., Haupt, L. M., and Tlsty, T. D. (2001) Normal human mammary epithelial cells spontaneously escape senescence and acquire genomic changes. *Nature* **409**, 633–637.

57. Campisi, J. (1999) Cellular aging/Replicative senescence, in *Studies of Aging: Springer Lab Manual* (Sternberg, H., and Timiras, P. S., eds.). Springer-Verlag, Berlin, pp. 35–45.

58. Jacobs, J. J., Kieboom, K., Marino, S., DePinho, R. A., and van Lohuizen, M. (1999) The oncogene and Polycomb-group gene bmi-1 regulates cell proliferation and senescence through the ink4a locus. *Nature* **397**, 164–168.

59. Jacobs, J. J., Keblusek, P., Robanus-Maandag, E., et al. (2000) Senescence bypass screen identifies TBX2, which represses Cdkn2a (p19/ARF) and is amplified in a subset of human breast cancers. *Nat. Genet.* **26**, 291–299.

60. McConnell, B. B., Starborg, M., Brookes, S., and Peters, G. (1998) Inhibitors of cyclin-dependent kinases induce features of replicative senescence in early passage human diploid fibroblasts. *Curr. Biol.* **8**, 351–354.

61. Yang, S., Delgado, R., King, S. R., et al. (1999) Generation of retroviral vector for clinical studies using transient transfection. *Hum. Gene Ther.* **10**, 123–132.

62. Weng, N. P. and Hodes, R. J. (2000) The role of telomerase expression and telomere length maintenance in human and mouse. *J. Clin. Immunol.* **20**, 257–267.

63. Harvey, M., Sands, A. T., Weiss, R. S., et al. (1993) In vitro growth characteristics of embryo fibroblasts isolated from p53-deficient mice. *Oncogene* **8**, 2457–2467.

64. Mcleod, K. (1999) pRb and E2F-1 in mouse development and tumorigenesis. *Curr. Opin. Genet. Dev.* **9**, 31–39.

65. Ponten, J. (1976) The relationship between in vitro transformation and tumor formation in vivo. *Biochim. Biophys. Acta* **458**, 397–422.

66. Sager, R. (1984) Resistance of human cells to oncogenic transformation. *Cancer Cells* **2**, 487–493.

67. Hodes, R. J. (1999) Telomere length, aging and somatic cell turnover. *J. Exp. Med.* **190**, 153–156.

68. Prowse, K. R. and Greider, C. W. (1995) Developmental and tissue-specific regulation of mouse telomerase and telomere length. *Proc. Natl. Acad. Sci. USA* **92**, 4818–4822.

69. DePinho, R. A. (2000) The age of cancer. *Nature* **408**, 248–254.

70. DiLeonardo, A., Linke, S. P., Clarkin, K., and Wahl, G. M. (1994) DNA damage triggers a prolonged p53-dependent G1 arrest and long-term induction of Cip1 in normal human fibroblasts. *Genes Dev.* **8**, 2540–2551.

71. Robles, S. J. and Adami, G. R. (1998) Agents that cause DNA double strand breaks lead to p16INK4a enrichment and the premature senescence of normal fibroblasts. *Oncogene* **16**, 1113–1123.

72. Suzuki, K., Mori, I., Nakayama, Y., Miyakoda, M., Kodama, S., and Watanabe, M. (2001) Radiation-induced senescence-like growth arrest requires TP53 function but not telomere shortening. *Radiat. Res.* **155**, 248–253.

73. Robles, S. J., Buchler, P. W., Negrusz, A., and Adami, G. R. (1999) Permanent cell cycle arrest in asynchronously proliferating normal human fibroblasts treated with doxorubicin or etoposide but not camptothecin. *Biochem. Pharmacol.* **58**, 675–685.

74. Chang, B. D., Broude, E. V., Dokmanovic, M., et al. (1999) A senescence-like phenotype distinguishes tumor cells that undergo terminal proliferation arrest after exposure to anticancer agents. *Cancer Res.* **59**, 3761–3767.

75. Ogryzko, V. V., Hirai, T. H., Russanova, V. R., Barbie, D. A., and Howard, B. H. (1996) Human fibroblast commitment to a senescence-like state in response to histone deacetylase inhibitors is cell cycle dependent. *Mol. Cell. Biol.* **16**, 5210–5218.

76. Young, J. I. and Smith, J. R. (2001) DNA methyltransferase inhibition in normal human fibroblasts induces a p21-dependent cell cycle withdrawal. *J. Biol. Chem.* **276**, 19610–19616.

77. Lin, A. W., Barradas, M., Stone, J. C., van Aelst, L., Serrano, M., and Lowe, S. W. (1998) Premature senescence involving p53 and p16 is activated in response to constitutive MEK/MAPK mitogenic signaling. *Genes Dev.* **12**, 3008–3019.

78. Zhu, J., Woods, D., McMahon, M., and Bishop, J. M. (1998) Senescence of human fibroblasts induced by oncogenic raf. *Genes Dev.* **12**, 2997–3007.

79. Dimri, G. P., Itahana, K., Acosta, M., and Campisi, J. (2000) Regulation of a senescence checkpoint response by the E2F1 transcription factor and p14/ARF tumor suppressor. *Mol. Cell. Biol.* **20**, 273–285.

80. Dai, C. Y. and Enders, G. H. (2000) p16 INK4a can initiate and autonomous senescence program. *Oncogene* **19**, 1613–1622.

81. Uhrbom, L., Nister, M., and Westermark, B. (1997) Induction of senescence in human malignant glioma cells by p16INK4a. *Oncogene* **15**, 505–514.

82. Ferbeyre, G., de Stanchina, E., Querido, E., Baptiste, N., Prives, C., and Lowe, S. W. (2000) PML is induced by oncogenic ras and promotes premature senescence. *Genes Dev.* **14**, 2015–2027.

83. Pearson, M., Carbone, R., Sebastiani, C., et al. (2000) PML regulates p53 acetylation and premature senescence induced by oncogenic RAS. *Nature* **406**, 207–210.

84. Chang, B. D., Watanabe, K., Broude, E. V., et al. (2000) Effects of p21Waf1/Cip1/Sdi1 on cellular gene expression: implications for carcinogenesis, senescence, and age-related diseases. *Proc. Natl. Acad. Sci. USA* **97**, 4291–4296.

85. Kagawa, S., Fujiwara, T., Kadowaki, Y., et al. (1999) Overexpression of the p21/sdi1 gene induces senescence-like state in human cancer cells: implications for senescence-directed molecular therapy for cancer. *Cell Death Differ.* **6**, 765–772.

86. Collado, M., Medema, R. H., Garcia-Cao, I., et al. (2000) Inhibition of the phosphoinositide 3-kinase pathway induces a senescence-like arrest mediated by p27 Kip1. *J. Biol. Chem.* **275**, 21960–21968.

87. Wei, S., Wei, S., and Sedivy, J. M. (1999) Expression of catalytically active telomerase does not prevent premature senescence caused by overexpression of oncogenic Ha-Ras in normal human fibroblasts. *Cancer Res.* **59**, 1539–1543.

88. Campisi, J. (1998) The role of cellular senescence in skin aging. *J. Invest. Dermatol.* **3**, 1–5.

89. Krtolica, A., Parrinello, S., Lockett, S., Desprez, P., and Campisi, J. (2001) Senescent fibroblasts promote epithelial cell growth and tumorigenesis: A link between cancer and aging. *Proc. Natl. Acad. Sci. USA* **98**, 12,072–12,077.

90. Deng, G., Lu, Y., Zlotnikov, G., Thor, A. D., and Smith, H. S. (1996) Loss of heterozygosity in normal tissue adjacent to breast carcinomas. *Science* **274**, 2057–2059.

91. Jonason, A. S., Kunala, S., Price, G. T., et al. (1996) Frequent clones of p53-mutated keratinocytes in normal human skin. *Proc. Natl. Acad. Sci. USA* **93**, 14025–14029.

92. Cha, R. S., Thilly, W. G., and Zarbl, H. (1994) N-nitroso-N-methylurea-induced rat mammary tumors arise from cells with preexisting oncogenic Hras1 gene mutations. *Proc. Natl. Acad. Sci. USA* **91**, 3749–3753.

93. Ide, T., Tsuji, Y., Ishibashi, S., and Mitsui, Y. (1983) Reinitiation of host DNA synthesis in senescent human diploid cells by infection with simian virus 40. *Exp. Cell Res.* **143**, 343–349.

94. Gorman, S. D. and Cristofalo, V. J. (1985) Reinitiation of cellular DNA synthesis in BrdU-selected nondividing senescent WI38 cells by simian virus 40 infection. *J. Cell. Physiol.* **125**, 122–126.

95. Galloway, D. A. and McDougall, J. K. (1989) Human papillomaviruses and carcinomas. *Adv. Virus Res.* **37**, 125–171.

96. Shay, J. W., Wright, W. R., and Werbin, H. (1991) Defining the molecular mechanisms of human cell immortalization. *Biochim. Biophys. Acta* **1071**, 1–7.

97. Kierstead, T. D. and Tevethia, M. J. (1993) Association of p53 binding and immortalization of primary C57BL/6 mouse embryo fibroblasts by using simian virus 40 T-antigen mutants bearing internal overlapping deletion mutations. *J. Virol.* **67**, 1817–1829.

98. Sakamoto, K., Howard, T., Ogryzko, V., et al. (1993) Relative mitogenic activities of wild-type and retinoblastoma binding defective SV40 T antigens in serum deprived and senescent human fibroblasts. *Oncogene* **8**, 1887–1893.

99. Linskens, M. H. K., Feng, J., Andrews, W. H., et al. (1995) Cataloging altered gene expression in young and senescent cells using enhanced differential display. *Nucleic Acids Res.* **23**, 3244–3251.

100. Komarova, E. A., Diatchenko, L., Rokhlin, O. W., et al. (1998) Stress-induced secretion of growth inhibitors: a novel tumor suppressor function of p53. *Oncogene* **17**, 1089–1096.

14

Yeast Two-Hybrid Screening as a Means of Deciphering Tumor Suppressor Pathways

E. Robert McDonald III and Wafik S. El-Deiry

1. Introduction

With the identification and functional characterization of various tumor suppressor genes, the multiprotein pathways in which they function are beginning to be delineated. Mutations in critical pathways, such as the basal cell cycle machinery, DNA surveillance and repair, apoptosis, and checkpoint control ultimately provide a selective growth advantage leading to unfettered cell proliferation and tumor formation. Understanding the cross-talk between these pathways as well as protein–protein interactions within the pathways themselves has become increasingly important in order to design better therapeutic strategies. Therefore methods involved in identifying novel protein–protein interactions in mammalian cells have become just as valuable as those that served initially to identify the tumor supresssor genes. Large-scale biochemical purification of interacting proteins is a viable strategy that was, in one case, used to identify a number of novel proteins (i.e., TRAF1, TRAF2, c-IAP1, and c-IAP2) *(1,2)* that interact with the tumor necrosis factor receptor 2. However, due to the difficulty of reconstituting in-vivo interactions in a large-scale purification scheme as well as still having to clone your gene of interest, this method can prove to be difficult as well as time-consuming.

In order to identify protein–protein interactions more rapidly and circumvent the problems associated with biochemical purification, the idea of yeast two-hybrid screening was born following significant advances in the understanding of yeast transcriptional activators. Seminal work from the Ptashne and Struhl laboratories demonstrated that a classical yeast transcriptional activator (GAL4) contained two separable domains: a DNA-binding domain (DBD) and a transcriptional activation domain (AD) *(3,4)*. Further experiments demonstrated that "hybrid proteins" that contained a LexA DNA-binding domain along with a heterologous acidic activation domain such as GAL4 AD were still able to activate transcription of reporter genes with LexA upstream activating sequences (UASs) *(5)*. These concepts were further extended in the laboratory of Dr. Stanley Fields in order to lay the foundation for the transcriptionally based yeast two-hybrid assays so commonly used today (**Fig. 1**). The tech-

From: *Methods in Molecular Biology, Vol. 223: Tumor Suppressor Genes: Regulation, Function, and Medicinal Applications.* Edited by: Wafik S. El-Deiry © Humana Press Inc., Totowa, NJ

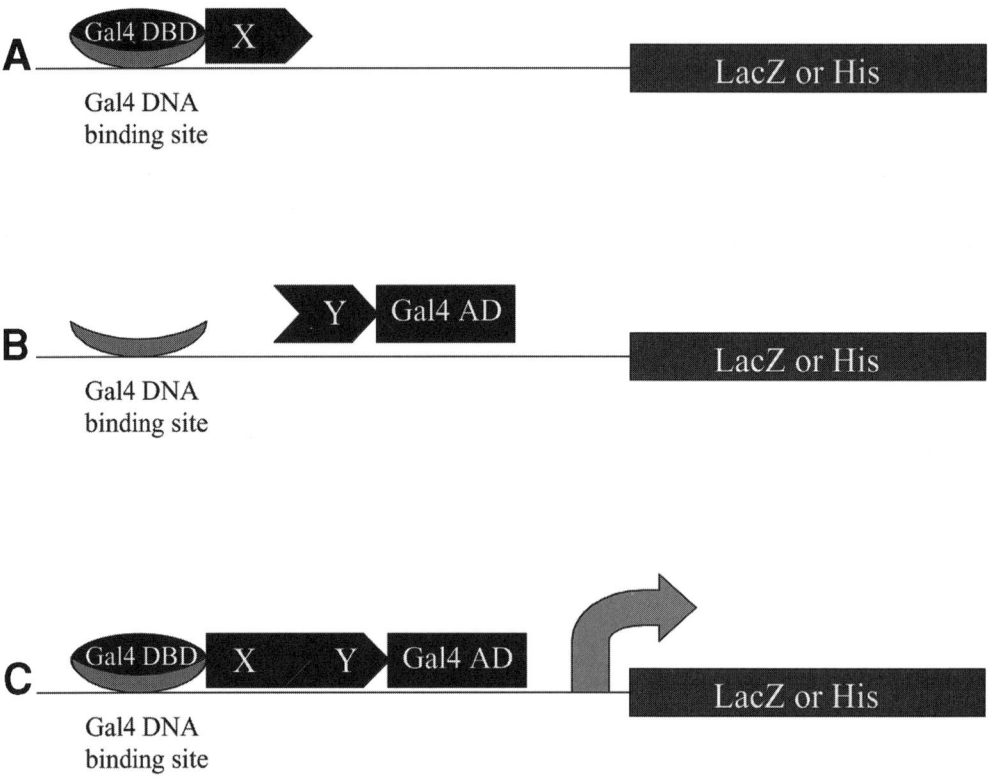

Fig. 1. Transcriptional-based GAL4 yeast two-hybrid screening methodology. **(A)** Fusion between protein X and the GAL4 DNA-binding domain (DBD) creates hybrid protein that is able to bind the GAL4 upstream activating sequence (UAS) within the promoter of the reporter gene(s) of the engineered yeast strain. However, no transcription of the reporter gene occurs, due to the absence of a transcriptional activation domain. **(B)** Fusion between protein Y and GAL4 activation domain (AD) creates hybrid with transcriptional activation potential but lacking the ability to bind DNA. **(C)** Introduction of both GAL4DBD-X and GAL4AD-Y allows for interaction between X and Y within yeast nucleus. Close juxtaposition of GAL4DBD and AD via X/Y interaction creates a DNA-binding, transcription-competent complex that activates transcription of the reporter genes.

nique takes advantage of the aforementioned modular composition of the yeast GAL4 gene so that when the two domains are brought into close proximity in the presence of GAL4 UASs, transcription of a reporter gene results *(6)*. This allows for the creation of a fusion of a known protein or protein domain (X) to the GAL4 DNA-binding domain (GAL4DBD-X) in conjunction with a putative X-interacting protein (Y) fused to the GAL4 activation domain (GAL4AD-Y), so that interaction of X and Y results in the close juxtaposition of the GAL4 domains and leads to the activation of reporter genes **(Fig. 1)**. Activation of reporter genes such as LacZ or a nutritional reporter allows for a measurable yeast phenotype that serves as a readout of protein–protein interactions *(6)*. This technique can be extended even further to search for novel X-interacting proteins through the use of cDNA libraries fused to the GAL4AD in order to identify networks of proteins that cooperate in a linear fashion in pathways such as tumor suppression.

1.1. The Original: Transcriptional-Based Yeast Two-Hybrid Screening

The GAL4- and LexA-derived systems that rely on transcriptional activation of reporter genes have been used successfully for over 10 years to identify novel proteins involved in a host of cellular activities. The impact of this system was immediately felt in the field of cell cycle control. Screens with the positive cell cycle regulator cdk2 led to the discovery of a naturally occurring *c*dk *i*nhibitor *p*rotein CIP1 or p21 *(7)* as well as the Rb-related protein, p130 *(8)*. An *in*hibitor of cd*k4* (INK4) or p16 was also discovered via yeast two-hybrid with cdk4 as bait *(9)*, and subsequently p16's role in tumor suppression has been well documented *(10)*. Yeast two-hybrid methods also aided our understanding of the regulation of the tumor suppressor, p53. Mdm2, which is responsible for regulating the rapid turnover of p53 and by extension its activity, was shown via yeast two-hybrid to be a target of ARF (alternative reading frame of the p16 cdk inhibitor) *(11)*. In addition, the minimal regions and critical residues important for the p53/SV40 large T-antigen interaction were delineated in a directed yeast two-hybrid setting *(12)*. Finally, the most recent and dramatic example of the power of this system stems from the study of the regulation of apoptosis. Ligand/receptor-mediated apoptosis such as the Fas/FasL pathway is mediated through a series of protein–protein interactions between conserved motifs in order to generate a death-inducing signaling complex (DISC) at the cell membrane *(13)*. All the members of the Fas DISC were discovered through yeast two-hybrid screens. The receptor was shown to interact with a novel molecule, FADD, through a death domain (DD) motif *(14,15)*. Subsequently, FADD via its death effector domain (DED) was used to identify caspase-8 *(16)*, a cellular cysteine protease which, upon its activation, leads to the dismantling of the cell. FLASH, a positive regulator of caspase-8 activation *(17)*, as well as the negative regulator of DISC-induced apoptosis, c-FLIP or CASH *(18)*, were also discovered via yeast two-hybrid. Because of the modular nature of the protein motifs in this pathway, this area was particularily amenable to two-hybrid screening and helped to reveal an entire signaling pathway.

The GAL4-based system will be the focus of this chapter because of its longevity and success in diverse areas of biology; however, problems with this system do exist, which has led to variations on the original two-hybrid scheme. One of the most commonly encountered problems with a transcriptional-based reporter readout system is that construction of a GAL4DBD fusion protein with your gene of interest can lead to activation of the yeast reporters, so called "self-activation." Your gene may have no role in transcriptional activation but when fused to a GAL4DBD can result in reporter activation, thereby making it useless for screening purposes. Acidic regions are thought to be responsible for activation, therefore deletion of these may eliminate self-activation; however, this may ultimately affect proper protein folding. There is no way to predict how your protein will behave until it is tested. The second major downfall to this system is due to the fact that the proteins must be able to reside and interact within the yeast nucleus to initiate transcriptional activation. A strong intrinsic nuclear export signal on either protein or an inability to interact in a nuclear environment may result in "false negatives." Protein–protein interactions that require specific (i.e., phosphorylation), non-nuclear (i.e., glycosylation), or nonyeast posttranslational modifications are also not well suited for study in this type of system. The utility of this system for receptor–ligand interactions is probably limited as well. Nevertheless, both the GAL4 and LexA transcriptional-based systems have proved invaluable in delineating tumor suppressor pathways.

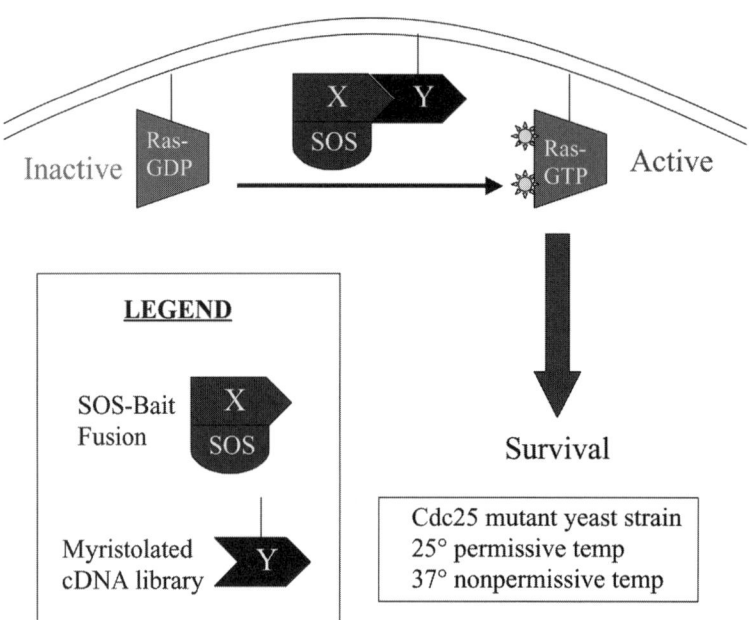

Fig. 2. Nontranscriptional-based yeast two-hybrid screening method that involves reconstitution of the yeast Ras pathway. This two-hybrid screening method takes place within the yeast cytoplasm and does not rely on transcriptional activation for reporter readout. Instead, this assay is carried out in a temperature-sensitive yeast strain that carries a mutation in the Ras guanyl nucleotide exchange factor (GEF), cdc25, and requires reconstitution of the Ras pathway by introduction of the human GEF, SOS. The yeast strain is viable at 25°C but not at 37°C. Introduction of a myristolated cDNA library targets proteins to the cytoplasmic side of the yeast cell in close proximity to inactive Ras molecules. Introduction of a SOS-protein X fusion that is able to interact with library protein Y reconstitutes the activity of the Ras pathway at the nonpermissive temperature of 37°C. This allows for survival of the yeast colony and the opportunity to identify the library protein that interacts with protein X.

1.2. The Alternative: Nontranscriptional, Cytoplasmic-Based Yeast Two-Hybrid

A non-transcription-based system involving reconstitution of the Ras pathway within the cytoplasm of yeast was designed in an attempt to circumvent the two main problems of a nuclear, transcription-based yeast two-hybrid system *(19)* (**Fig. 2**). This system uses a temperature-sensitive yeast strain that carries a mutation in the yeast Ras guanyl nucleotide exchange factor (GEF), cdc25, which is viable at 25°C but not at 37°C. Introduction of the human Ras GEF, Sos, is able to rescue the lethality at 37°C when targeted to the plasma membrane where Ras normally resides. By creating and introducing a SOS bait fusion protein (SOS-X) and a myristolated cDNA library into this cdc25 mutant yeast strain, proteins that interact with your gene X will recruit the SOS-X fusion to the membrane due to the myristolation signal on the library constructs. Once at the membrane, the SOS-X fusion will reconstitute the Ras signaling pathway and allow these clones to grow at the restrictive temperature of 37°C. The addition of galactose-inducible promoters on the vectors allows for plating of the initial screen on glucose, followed by replica plating to galactose and incubation at 37°C to yield putative interacting clones. This technique was first described by the Karin laboratory *(19)*. By using this method, they were able to use a known transcription factor, c-Jun, as bait in order to search for novel protein-binding partners. Because of its strong transactivation ability, a transcription-based system such as GAL4 was not an option. In addition, they were able to demonstrate not only an interaction between two cytoplasmic mammalian proteins but also between two nuclear proteins, c-Jun and c-Fos, within the yeast cytoplasm using this system. This system has since been commercialized by Stratagene under the name CytoTrap Two-Hybrid System. This system alleviates the two major pitfalls associated with the transcription-based systems, self-activation of bait constructs and the requirement for nuclear localization for interaction. Concerns involving the use of a yeast environment to study mammalian protein interactions still persist in this system, such as toxicity of certain mammalian proteins to yeast as well as the absence of certain posttranslational modifications in yeast. An additional concern with this system appears to be the instability or rate of reversion of the mutant cdc25 yeast strain back to wild-type, so extreme care must be taken and proper controls maintained to ensure that the phenotype of the yeast is as expected. The power of this system has not been realized but should be instrumental to laboratories interested in studying protein–protein interactions involving transcription factors or highly acidic proteins.

1.3. Adaptations on the Original: Yeast One-Hybrid, Yeast Three-Hybrid, and Nonyeast Two-Hybrid

The yeast one-hybrid system provides a tool to identify novel proteins that bind a specific DNA sequence (20). The same activation-domain fusion libraries can be used in this type of screen, but the yeast reporter strain must be engineered to contain a reporter gene that is responsive to your particular DNA-binding sequence. This technique was first used successfully to identify the OLF1 transcription factor (21) and subsequently helped to identify proteins important for the recognition of the origins of DNA replication (22). Because of the frequency of multiprotein complexes within the cell, a yeast three-hybrid system was described to demonstrate ternary interactions. The third protein is normally expressed from a third vector whose expression is negatively regulated by the presence of methioine in the medium (23). Deletion of methionine along with the other nutritional selections associated with the two-hybrid screen can reveal the importance of the third protein. The ternary complex of cyclinH/cdk7/MAT1, which acts to activate cyclin/cdk complexes and promote cell cycle progression, was shown using this method (23). Screens of this nature can yield proteins that either enhance or inhibit the interaction of the original two proteins. Again, this technology has been commercialized by Clontech, which features a pBridge vector that expresses both the GAL4DBD fusion as well as the methionine-controlled third protein partner. Lastly, a three-hybrid system used to identify RNA–protein interactions has also been described (24) and is available from Invitrogen.

A mammalian two-hybrid system was described (25) in the hopes of creating the optimal system to study mammalian protein–protein interactions. This system, however, is not suited to screening cDNA libraries for novel interacting proteins, but rather is more instrumental when confirming the interaction between two known proteins. Stratagene created a system in which luciferase is the readout for interaction, whereas Clontech offers a CAT-based system. Isolation of novel protein interactions from a yeast screen can be verified in this mammalian setting in addition to other, more biochemical approaches such as co-immunoprecipitation techniques. Recently, Stratagene has announced the availability of a new bacterial two-hybrid system. Because of the faster growth rate and higher transformation efficiency of bacteria as well as the ability to propagate putative positive clones directly, this system could drastically reduce the time involved in carrying out screens. Similar concerns exist in bacteria as in yeast when trying to study mammalian protein interactions, such as toxicity to the host from foreign proteins as well as the lack of proper posttranslational modifications. Time will tell if the bacterial system provides a true alternative to the yeast-based transcriptional system that has garnered so much success.

2. Materials

2.1. Handling Yeast and Performing Lithium Acetate Transformation

1. YPD media: 20 g peptone +10 g yeast extract + 950 mL dH$_2$O (pH to 5.8). Autoclave and, after cooling, add 50 mL sterile 40% dextrose (2%). If making plates, add agar to 2% prior to autoclaving.
2. SD media: 6.7 g yeast nitrogen base without amino acids + 950 mL dH$_2$O + appropriate amino acid dropout powder from Clontech (alternatively, add sterile 100× amino acids individually after autoclaving), pH to 5.8. Autoclave and, after cooling, add 50 mL sterile 40% dextrose (2%). Add necessary concentration of 3-AT after cooling (heat-labile). If making plates, add agar to 2% prior to autoclaving.

    100× amino acid stocks:
    L-Isoleucine, 3 g/L
    L-Valine, 15 g/L
    L-Adenine hemisulfate salt, 2 g/L
    L-Arginine HCl, 2 g/L
    L-Histidine HCL monohydrate, 2 g/L
    L-Leucine, 10 g/L
    L-Lysine HCl, 3 g/L
    L-Methionine, 2 g/L
    L-Phenylalanine, 5 g/L
    L-Threonine, 20 mg/L
    L-Tryptophan, 2 g/L
    L-Tyrosine, 3 g/L
    L-Uracil, 2 g/L
3. 10× lithium acetate (LiAc): 1 *M*, pH 7.5, and autoclave to sterilize.
4. 10× TE: 0.1 *M* Tris-HCl, 10 m*M* EDTA, pH 7.5, and autoclave to sterilize.
5. 50% PEG 3350: Polyethylene glycol, average MW 3350, filter-sterilize.
6. Salmon sperm carrier DNA: Make 10 mg/mL with 1× TE; dissolve overnight with shaking at 4°C; sonicate 2–3 times on 3/4 power with probe sonicator followed by boiling for 20 min; check that average size is between 2 and 15 kb by gel electrophoresis.
7. PEG/LiAc (make fresh each time): 8 mL 50% PEG + 1 mL 10× TE + 1 mL 10× LiAc.
8. Dimethyl sulfoxide (DMSO), 3-AT: Available from Sigma.

2.2. β-Gal Filter Lift Assay

1. Z buffer: 16.1 g Na$_2$HPO$_4$·7H$_2$O, 5.5 g NaH$_2$PO$_4$·H$_2$O, 0.75 g KCl, 0.246 g MgSO$_4$·7H$_2$O, pH to 7.0, adjust volume to 1 L and autoclave.
2. X-gal: 20 mg/mL in N,N-dimethylformamide (DMF).

3. Z-buffer/X-gal solution (make fresh before using): 100 mL Z-buffer, 0.27 mL β-mercaptoethanol, 1.67 mL X-gal.

2.3. Yeast Miniprep

1. Lyticase (available as powder from Sigma): make 1 U/μL stock in 1× TE.
2. ChromaSpin-1000 columns (available from Clontech).
3. 20% sodium dedecyl sulfate (SDS)

3. Methods

Yeast two-hybrid techniques have become extremely accessible to researchers in all areas of science due to their commercialization by a variety of companies. This has allowed nonyeast-based labs to use the system successfully without having an extensive knowledge of yeast genetics. Clontech has developed a large product section devoted to detecting protein–protein interactions through the use of the GAL4 transcriptional yeast two-hybrid system developed in the laboratory of Dr. Fields. The original technology has been improved upon enormously with the advent of their MATCHMAKER Two-Hybrid System 3. In addition, they offer reagents to help in yeast transformation and minipreps as well as premade GAL4AD cDNA libraries from a variety of species and tissues. These products made available by Clontech allow a scientist inexperienced in yeast methods the opportunity to conduct a screen successfully in the least amount of time possible. Stratagene also offers a transcriptional-based GAL4 system as well as offering the only nontranscriptional yeast system (CytoTrap) and the only bacterial two-hybrid system (BacterioMatch) available. Invitrogen offers a transcriptional-based LexA system (Hybrid Hunter) originally described by Gyuris et al. *(26)*, as well as an RNA/protein two-hybrid system. The LexA-based system is able to detect some interactions that the GAL4 system does not and vice versa, but the reason for these differences is not understood. The protocols discussed here are based on the MATCHMAKER system from Clontech, and helpful hints are provided based on extensive experience with this system. **Figure 3** outlines the steps involved in carrying out a two-hybrid screen with this system. For more in-depth general yeast protocols or more information on a LexA-based system, the reader is referred to excellent references in *Current Protocols in Molecular Biology (27)*. For online yeast two-hybrid systems and protocols, the reader is referred to a site maintained by the Golemis laboratory at the Fox Chase Cancer Center and links therein, http://www.fccc.edu:80/research/labs/golemis/interactiontrapinwork.

3.1. Handling of Yeast and Transformations

1. Restreak frozen yeast strain on YPD plate and incubate at 30°C until colonies form (3–5 d).
2. Take isolated colony from YPD plate and restreak on various SD plates (i.e., SD-Leu, SD-Trp, etc.) to verify that the strain has the proper phenotype (yeast generally take longer to grow on SD media; 4–6 d).

Once the behavior of the strain is verified, use these yeast to conduct a pilot yeast transformation with positive and negative controls supplied with the kit. In the case of the MATCHMAKER system a p53/SV40 large T-antigen positive interaction control is supplied along with empty GAL4DBD and GAL4AD vectors and a GAL4DBD-LaminC for

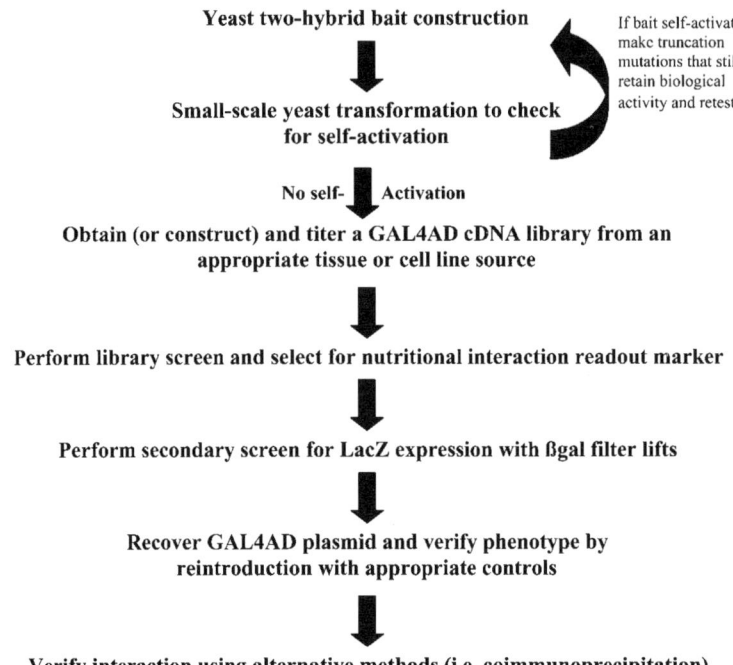

Fig. 3. Procedure to carry out a yeast two-hybrid screen using the GAL4 transcriptional-based system.

a negative control. The GAL4DBD vectors have a Trp-selectable marker and the GAL4AD vectors have a Leu-selectable marker. In order to maintain yeast that have successfully retained the plasmid(s) of interest, simply delete that amino acid(s) from the SD plate and grow at 30°C following tranformation and plating. In order to select for interaction, determine which reporter construct(s) your yeast strain contains and omit that amino acid as well (e.g., His). MATCHMAKER systems have a variety of strains available for library screening (HF7c, AH109, CG-1945, Y190) and for direct protein–protein interaction studies (SFY526, Y187). The newer system offers a yeast strain (AH109) with three reporters for interaction readout (His, Ade, and LacZ), while the other strains utilize His and LacZ. Testing for interaction between two specific proteins can be carried out in these strains as well or in the single reporter strains (SFY526 and Y187) by measuring for LacZ expression alone. 3-Amino-1,2,4-triazole (3-AT) inhibits any residual His protein that may exist in the aforementioned strains in order to suppress any background growth when conducting screens. Yeast are more susceptible to the toxicity of 3-AT after transformation, therefore titration of 3-AT for your specific strain *after transformation* is necessary. A range of 0.1–0.5 m*M* for HF7c is good for library screening purposes. Unlike bacteria, for good transformation efficiencies in yeast, fresh competent cells must be prepared every time. The following small-scale lithium acetate yeast transformation protocol can be used for testing the behavior of your strain and positive controls as well as for testing your bait for self-activation prior to library screening.

1. Inoculate 50 mL of YPD with fresh colony of your yeast strain and grow overnight at 30°C shaking at 250 rpm (create single yeast cells from colony through vortexing to insure proper growth).
2. Innoculate 300 mL fresh YPD with enough overnight culture to get $OD_{600} \approx 0.2$.

3. Incubate at 30°C with shaking until $OD_{600} = 0.5 \pm 0.1$.
4. Spin down yeast at $1000g$ at room temperature for 10–15 min.
5. Pool cells in one 50-mL tube by resuspending pellets with a total of 50 mL sterile water and spin down again at $1000g$ at room temperature
6. Resuspend pellet in 1.5 mL 1× TE/LiAc.
7. The cells are now ready to be used for transformation; it is best to use them immediately, but if necessary, storage at room temperature for a few hours slightly reduces transformation efficiency.
8. For each transformation, add 250 ng of DNABD and 250 ng AD plasmid; 100 µg salmon sperm carrier DNA; 100 µL competent yeast; and 600 µL freshly made PEG/LiAc.
9. Vortex at high speed to mix and shake at 30°C at 250 rpm for 30 min.
10. Add 70 µL DMSO and mix by inverting gently (no vortex).
11. Incubate at 42°C for 15 min, followed by 2 min on ice.
12. Microfuge at MAX for 1 min and dump supernatent.
13. Resuspend in 400 µL 1× TE and plate 200 µL on SD-Leu/-Trp to select for both plasmids and 200 µL SD-Leu/-Trp/-His + 3-AT to select for interaction if using strain with His nutritional marker.
14. Incubate at 30°C until colonies appear; can also test for LacZ readout at this time with protocol listed below

Once you have completed these initial transformations and are comfortable with the techniques, proceed to bait construction and testing followed by library screenings.

3.2. Constructing and Testing a Bait Vector for Library Screening Purposes

Constructing a bait protein that does not self-activate but retains proper folding in the yeast environment is the single most important step when preparing to screen for interacting proteins. Many proteins with no known role in transcriptional activation when fused to either the GAL4 or LexA DNA-binding domain can activate transcription of the yeast reporter genes. If this is the case for your protein of interest, try to identify a domain or region within the protein that retains biologic activity in mammalian cells and use that as bait. The modular death domains and death effector domains have been used successfully in two-hybrid screens because they retain activity and structure when expressed. If you are unable to create such a bait, you can try a LexA-based system, knowing that the result may be the same. Alternatively, you can try the non-transcription-based system from Stratagene (CytoTrap) to circumvent the problem. It is also worth examining the biology of your pathway to determine whether another downstream protein could be used to isolate the protein of interest. Once a bait vector is constructed, generating a fusion between the GAL4DBD and your sequence through standard cloning techniques, test it for self-activation by using the small-scale yeast transformation protocol in a co-transformation with an empty AD plasmid. As long as neither reporter is activated, the bait can be used for library screening purposes.

3.3. Library Selection and Screening Procedure

Knowing the expression pattern of your bait protein or a cell line in which functional studies have been performed is crucial information when choosing the appropriate cDNA source for your library screen. High-quality GAL4AD-cDNA libraries are commercially available from a variety of cell lines and tissues, making it unneccessary to construct one *per se*. Clontech even offers libraries pretransformed into yeast, thereby making it necessary only to introduce the bait plasmid and start selection. If, however, a specialized tissue or treatment to a particular cell line may enhance expression of the putative target protein, construction of the library would be worthwhile. Kits are now available to aid in cDNA library construction.

Screening multiple libraries also enhances the possibility of identifying novel interacting proteins. One such example includes screens conducted with the prodomain of caspase-10. Screening of over 10 million independent clones in a fetal brain library yielded 3 clones of the interacting protein, FLASH. The adapter molecule, FADD, was isolated twice in an independent screen with a HeLa cell library, and a novel interacting protein, CARP was discovered in a third screen with a kidney library. Even though all three are *bona fide* interactors and are expressed in each tissue, they were isolated from different library screens. It is important to screen at least 1 million independent clones in order to ensure that the screen covered the complexity of the cDNA library. Determining how many clones were screened will be discussed following the library screen transformation protocol.

1. Innoculate 150 mL of YPD with fresh colony of your yeast strain and grow overnight at 30°C, shaking at 250 rpm (create single yeast cells from colony through vortexing to ensure proper growth).
2. Innoculate 1 L fresh YPD with enough overnight culture to get $OD_{600} \approx 0.2$.
3. Incubate at 30°C with shaking until $OD_{600} = 0.5 \pm 0.1$.
4. Spin down yeast at 1000*g* at room temperature for 10–15 min.
5. Pool cells in one 50-mL tube by resuspending pellets with a total of 50-mL sterile water and spin down again at 1000*g* at room temperature.
6. Resuspend pellet in 8 mL 1× TE/LiAc.
7. The cells are now ready to be used for transformation; it is best to use them immediately to ensure a high transformation efficiency.
8. For library transformation, add 1 mg of GAL4-DNABD-bait; 0.5 mg of GAL4AD-cDNA library; 20 mg salmon sperm carrier DNA; 8 mL competent yeast; and 60 mL freshly made PEG/LiAc (*see* **Note 1**).
9. Vortex at high speed to mix and shake at 30°C at 250 rpm for 30 min.
10. Add 7 mL DMSO and mix by swirling gently (no vortex).
11. Incubate at 42°C for 15 min (swirling once or twice), followed by 2 min on ice.
12. Spin at 1000 rpm for 15 min and dump supernatent.
13. Resuspend in 5 mL of 1× TE and plate 200 µL on 150-mm SD-Leu/-Trp/-His+3-AT to select for interaction if using strain with His nutritional marker; note the total volume of the resuspended yeast in order to calculate number of clones screened.
14. Incubate at 30°C until colonies appear (at least 1 wk; allow longer incubation to select for weak interactions).
15. Restreak colonies on 150-mm SD-Leu/-Trp/-His+3-AT plates along with p53/SV40 LT antigen positive control in a grid to facilitate β-gal filter lifts (**Subheading 3.4.**) (*see* **Note 2**).

In order to assess co-transformation efficiency and the number of clones screened, plate 1:10, 1:100, and 1:1000 dilutions of the library screen on SD-Leu/-Trp plates. Count colonies on plate with 30–300 clonies and multiply by total resuspension volume from **step 13**. Divide this number by (volume plated on dilution plate × dilution factor × µg of limiting plasmid [AD]). This determines the cotransformation efficiency in cfu/µg. To determine number of clones screened, just multipy cotransformation efficiency by amount of library plasmid used.

3.4. β-Gal Filter Lifts to Assess Readout of LacZ Reporter Gene

Once colonies appear after performing the library screen, streak each positive onto a fresh 150-mm SD-Leu/-Trp/-His+3-AT plate in a gridlike fashion. This alleviates the need to perform a filter lift on every plate used in the screen and will reduce the number to a few plates. In addition, occasionally some of the restreaked colonies will not regrow, reducing the number of positives to screen. It is important also to include a positive interaction control on the grid to ensure that the assay is performing as expected. When directly testing two proteins for interaction, a liquid β-gal assay using either Y187 or Y190 can be used to assess strength of interaction more quantitatively. For the purposes of conducting a library screen, only the filter lift assay will be discussed below.

1. Place Whatman filter paper on tray and presoak in Z-buffer/X-gal solution.
2. Take 150-mm yeast plates with fresh-grown grid colonies (3 d of growth) and place a filter paper on plate; avoid bubbles or creases when laying down filter.
3. Allow filter to sit on plates while you mark the orientation of the filter with a needle containing India ink or some type of dye.
4. Remove filter from plate and submerge in liquid nitrogen for at least 10 s to ensure lysis of the yeast; this step is crucial to allow for substrate–enzyme interaction.
5. Remove filter and allow it to thaw at room temperature for a few minutes.
6. Place thawed filter with yeast facing up on the presoaked filter paper on the tray.
7. Allow incubation at 30°C or room temperature and monitor for β-gal activity; use of HF7c in this type of assay along with p53/SV40 positive control yields a blue color in ~30 min; allow for continued incubation for library clones, but overnight incubation should be avoided as it tends to yield false positives.
8. Restreak and repeat the lifts once or twice to convince yourself that these colonies are true positives at the level of yeast before attempting to isolate the AD plasmids.

If the initial library screen yielded a large number of His+ colonies (hundreds) that all eventually turned blue after performing the β-gal lifts, retest the bait for self-activation. The prolonged incubation period of the library screen procedure may have uncovered a residual self-activation phenotype that was not detected when first testing your bait. If this is the case, modify your bait and repeat the procedure.

3.5. Recovering AD Plasmid via Yeast Miniprep for Identification of Interactor

A variety of yeast miniprep protocols to recover the AD plasmid have been published. After trying a variety of techniques including detergent steps and glass beads to generate DNA useful for PCR or *Escherichia coli* transformation, our experience with Clontech ChromaSpin-1000 columns in conjunction with SDS and lyticase treatments turned out to be the most reproducible. This protocol is listed below, but any method of recovery that results in isolation of the plasmid is acceptable. One of the simplest but most time-saving changes made to Clontech's MATCHMAKER system 3 is the generation of DBD and AD plasmids with different bacterial selectable markers (Amp and Kan). Previously, both plasmids were ampicillin-resistant, which required the use of a special *E. coli* strain that utilized the nutritional selection marker (Leu; KC8 or HB101 strain) to specifically recover the AD plasmid. Making a kanamycin-resistant bait plasmid allows for the use of general *E. coli* cloning strains that have a high transformation efficiency and are EndA−.

1. Innoculate 2 mL SD-Leu/-Trp with single His+/β-gal + colony from screen.
2. Vortex to disperse colony and incubate at 30°C at 250 rpm overnight or until medium becomes sufficiently turbid.
3. Spin down yeast in Eppendorf tube at Max for a few minutes.
4. Dump supernatent and resuspend pellet with 50 μL of lyticase (1 U/μL)
5. Incubate at 37°C for 1 h to weaken cell wall.
6. During incubation, prepare Clontech ChromaSpin-1000 columns per manufacturer's instructions.
7. Add 10 μL of 20% SDS to lyticase-treated yeast and vortex for least 1 min.
8. Subject yeast to one freeze/thaw cycle using liquid nitrogen.
9. Add sample to middle of prepared column and spin in microfuge at recommended speed.
10. Plasmid DNA is in the flow-through; ethanol-precipitate, wash thoroughly, and resuspend in small-volume (30–50 μL) to use for *E. coli* transformation or polymerase chain reaction (PCR)

Once the plasmid has been recovered, it is imperative to reintroduce it back into yeast via small-scale transformation with the original bait to recapitulate the phenotype. The use of loss of function deletion or point mutants of the bait as well as nonspecific controls such as Lamin C, p53, or empty DBD vector reveal the specificity of the interaction. Recovery of multiple clones of the same gene within the context of one screen provides further evidence for specificity. If the novel interacting protein fulfills the yeast specificity criteria, then co-immunoprecipitation experiments in mammalian cells will begin to address the question in a more physiological setting. Yeast two-hybrid techniques can also be used further to map the domains necessary for the interaction. These systems provide a powerful tool to help delineate pathways that transduce signals using protein–protein interactions that ultimately affect cell division and/or survival.

4. Notes

1. Perform small-scale negative and positive controls as outlined under **Subheading 3.1.**
2. It is not recommended to add X-gal directly to the plates in order to assess β-gal activity, because of the relative insensitivity of this approach.

References

1. Rothe, M., Wong, S. C., Henzel, W. J., and Goeddel, D. V. (1994) A novel family of putative signal transducers associated with the cytoplasmic domain of the 75 kDa tumor necrosis factor receptor. *Cell* **78**, 681–692.
2. Rothe, M., Pan, M. G., Henzel, W. J., Ayres, T. M., and Goeddel, D. V. (1995) The TNFR2-TRAF signaling complex contains two novel proteins related to baculoviral inhibitor of apoptosis proteins. *Cell* **83**, 1243–1252.
3. Keegan, L., Gill, G., and Ptashne, M. (1986) Separation of DNA binding from the transcription-activating function of a eukaryotic regulatory protein. *Science* **231**, 699–704.
4. Hope, I. A. and Struhl, K. (1986) Functional dissection of a eukaryotic transcriptional activator protein, GCN4 of yeast. *Cell* **46**, 885–894.
5. Brent, R. and Ptashne, M. (1985) A eukaryotic transcriptional activator bearing the DNA specificity of a prokaryotic repressor. *Cell* **43**, 729–736.
6. Fields, S. and Song, O. (1989) A novel genetic system to detect protein-protein interactions. *Nature* **340**, 245–246.
7. Harper, J. W., Adami, G. R., Wei, N., Keyomarsi, K., and Elledge, S. J. (1993) The p21 Cdk-interacting protein Cip1 is a potent inhibitor of G1 cyclin-dependent kinases. *Cell* **75**, 805–816.

8. Hannon, G. J., Demetrick, D., and Beach, D. (1993) Isolation of the Rb-related p130 through its interaction with CDK2 and cyclins. *Genes Dev.* **7**, 2378–2391.

9. Serrano, M., Hannon, G. J., and Beach, D. (1993) A new regulatory motif in cell-cycle control causing specific inhibition of cyclin D/CDK4. *Nature* **366**, 704–707.

10. Kamb, A. (1995) Cell-cycle regulators and cancer. *Trends Genet.* **11**, 136–140.

11. Zhang, Y., Xiong, Y., and Yarbrough, W. G. (1998) ARF promotes MDM2 degradation and stabilizes p53: ARF-INK4a locus deletion impairs both the Rb and p53 tumor suppression pathways. *Cell* **92**, 725–734.

12. Li, B. and Fields, S. (1993) Identification of mutations in p53 that affect its binding to SV40 large T antigen by using the yeast two-hybrid system. *FASEB J.* **7**, 957–963.

13. Ashkenazi, A. and Dixit, V. M. (1998) Death receptors: signaling and modulation. *Science* **281**, 1305–1308.

14. Chinnaiyan, A. M., O'Rourke, K., Tewari, M., and Dixit, V. M. (1995) FADD, a novel death domain-containing protein, interacts with the death domain of Fas and initiates apoptosis. *Cell* **81**, 505–512.

15. Boldin, M. P., Varfolomeev, E. E., Pancer, Z., Mett, I. L., Camonis, J. H., and Wallach, D. (1995) A novel protein that interacts with the death domain of Fas/APO1 contains a sequence motif related to the death domain. *J. Biol. Chem.* **270**, 7795–7798.

16. Boldin, M. P., Goncharov, T. M., Goltsev, Y. V., and Wallach, D. (1996) Involvement of MACH, a novel MORT1/FADD-interacting protease, in Fas/APO-1- and TNF receptor-induced cell death. *Cell* **85**, 803–815.

17. Imai, Y., Kimura, T., Murakami, A., Yajima, N., Sakamaki, K., and Yonehara, S. (1999) The CED-4-homologous protein FLASH is involved in Fas-mediated activation of caspase-8 during apoptosis. *Nature* **398**, 777–785.

18. Goltsev, Y. V., Kovalenko, A. V., Arnold, E., Varfolomeev, E. E., Brodianskii, V. M., and Wallach, D. (1997) CASH, a novel caspase homologue with death effector domains. *J. Biol. Chem.* **272**, 19641–19644.

19. Aronheim, A., Zandi, E., Hennemann, H., Elledge, S. J., and Karin, M. (1997) Isolation of an AP-1 repressor by a novel method for detecting protein-protein interactions. *Mol. Cell. Biol.* **17**, 3094–3102.

20. Fields, S. and Sternglanz, R. (1994) The two-hybrid system: an assay for protein-protein interactions. *Trends Genet.* **10**, 286–292.

21. Wang, M. M. and Reed, R. R. (1993) Molecular cloning of the olfactory neuronal transcription factor Olf-1 by genetic selection in yeast. *Nature* **364**, 121–126.

22. Li, J. J. and Herskowitz, I. (1993) Isolation of ORC6, a component of the yeast origin recognition complex by a one-hybrid system. *Science* **262**, 1870–1874.

23. Tirode, F., Malaguti, C., Romero, F., Attar, R., Camonis, J., and Egly, J. M. (1997) A conditionally expressed third partner stabilizes or prevents the formation of a transcriptional activator in a three-hybrid system. *J. Biol. Chem.* **272**, 22995–22999.

24. SenGupta, D. J., Zhang, B., Kraemer, B., Pochart, P., Fields, S., and Wickens, M. (1996) A three-hybrid system to detect RNA-protein interactions in vivo. *Proc. Natl. Acad. Sci. USA* **93**, 8496–8501.

25. Dang, C. V., Barrett, J., Villa-Garcia, M., Resar, L. M., Kato, G. J., and Fearon, E. R. (1991) Intracellular leucine zipper interactions suggest c-Myc hetero-oligomerization. *Mol. Cell. Biol.* **11**, 954–962.

26. Gyuris, J., Golemis, E., Chertkov, H., and Brent, R. (1993) Cdi1, a human G1 and S phase protein phosphatase that associates with Cdk2. *Cell* **75**, 791–803.

27. Ausubel, F. M., Brent, R., Kingston, et al. (1996) *Current Protocols in Molecular Biology.* Wiley, Boston.

15

Somatic Cell Knockouts of Tumor Suppressor Genes

Ruth Hemmer, Wenyi Wei, Annie Dutriaux, and John M. Sedivy

1. Introduction

Gene targeting is the modification of specific DNA sequences in a living organism. Three requirements must be met in order for gene targeting to be successful *(1)*. The process must be directed, so that it affects only the locus of choice. The targeting procedure requires specificity, such that a predetermined sequence can be inserted or substituted at the target locus. Finally, the process should be efficient.

Targeted homologous recombination is the most direct and unambiguous way to eliminate gene function and to establish the null phenotype. Homologous recombination can also be exploited to engineer other genetic changes, for example, substitution of coding regions (knock-ins), conditional knockouts, alterations of noncoding regulatory sequences, and so forth. These methods have created a spectacular boom in mouse genetics, in which approximately 2000 genetically altered mouse strains have been created by gene targeting in embryonic stem (ES) cells.

Studies in cell culture have also benefited significantly from gene targeting. Cells with genetic modifications engineered by targeted homologous recombination can be obtained by two distinct routes. First, gene targeting can be performed in ES cells, and cells or cell lines can be derived from the resultant mutant animals or embryos. Second, gene targeting can be performed directly in somatic cells. The ES cell–transgenic mouse route has the clear advantage of a well-developed methodology and an extensive supporting infrastructure. Many institutions have core facilities that will create transgenic mice and even derive cell lines so that individual investigators do not have to acquire expertise in these procedures and can focus on utilizing the genetically altered cells in their research programs. The disadvantage of this approach is that it is currently limited to mouse and rat cells.

1.1. Gene Targeting in Somatic Cells

The aim of cell culture is to develop well-defined experimental systems that offer the advantages of clonal homogeneity, an easily regulated external environment, and the ability to perform extensive biochemical characterization. A large number of cell lines derived from a variety of somatic cell types and normal as well as pathologic states are readily available. One obvious advantage of cell culture is the ability to address tissue-

From: *Methods in Molecular Biology, Vol. 223: Tumor Suppressor Genes: Regulation, Function, and Medicinal Applications.* Edited by: Wafik S. El-Deiry © Humana Press Inc., Totowa, NJ

or cell type-specific gene functions without having to contend with the complexities of animal models, for example, embryonic lethality. Cell culture models have thus been used extensively in studies of basic or ubiquitous cellular processes, such as DNA replication or transcriptional mechanisms, cell cycle regulation, and so on.

Gene targeting directly in somatic cells has the advantage of being applicable to all species. Gene targeting in human somatic cells is expected to have important applications in areas in which rodent model systems do not accurately represent human biology or disease processes. The recent discovery that isogenic DNA *(2)* appears not to be required for efficient gene targeting in human cells *(3)* has opened up exciting new possibilities (*see* **Note 1**). Furthermore, gene targeting is feasible not only in established cell lines but also in nonimmortalized or primary human cells *(12)*.

Gene targeting in somatic cells has also recently opened a new avenue for creating genetically modified transgenic animals in species such as sheep, cows, or pigs, in which the ES cell technology has been difficult or impossible to translate into practice *(13)*. For example, it has been possible to perform gene targeting in sheep fibroblasts and then create animals from the modified cells by nuclear transplantation *(14)*. The disadvantage of gene targeting in somatic cells is that a minimum of two sequential rounds of targeting must be performed to achieve a homozygous state. Furthermore, as will be explored in detail in the following sections, the methodology has been more difficult to develop *(4)*.

1.2. Principles of Gene Targeting

1.2.1. Efficiency of Homologous Recombination

Gene targeting exploits endogenous cellular pathways of homologous recombination. These processes appear to be intrinsically inefficient, and gene targeting is made even more difficult by interference from nonhomologous recombination, which is typically orders of magnitude more frequent in mammalian cells. The efficiency of recovering homologously targeted clones is the ratio of two absolute frequencies: homologous recombination at the target locus and the total number of nonhomologous recombination events. For reasons that are currently not understood, the absolute frequency of homologous recombination in somatic cells is some two order of magnitude lower than in murine ES cells *(15,16)*. It should not come as a surprise that this fact alone has been the chief obstacle to the widespread use of gene targeting in somatic cells.

1.2.2. Genetic Enrichment Procedures

A typical gene targeting experiment in any mammalian cell involves introduction of vector DNA into the target cell (typically by electroporation or lipofection), selection of drug-resistant colonies, cloning and establishment of cell lines, and finally screening for homologously recombined clones by polymerase chain reaction (PCR) or Southern blotting. The rate-limiting step is the screening of drug-resistant colonies, and if the frequency of the sought targeted events in this pool falls significantly below 1%, the entire procedure rapidly becomes prohibitively labor-intensive.

Gene targeting was made feasible by the development of genetic enrichment procedures designed to prevent nonhomologous recombinants from scoring as colonies. Two fundamentally different enrichment methods have been developed. The first one is, in genetic terms, negative, since it selects against recombination events at the incorrect

PNS Vector

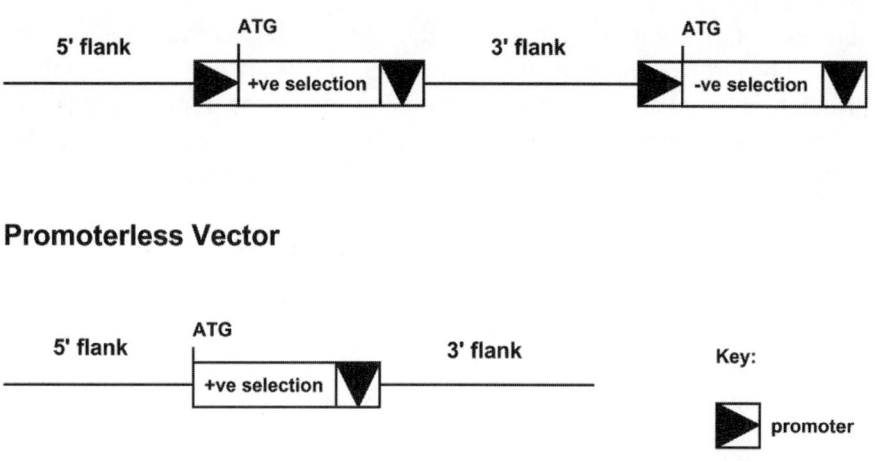

Promoterless Vector

Fig 1. PNS and promoterless targeting vectors.

(nonhomologous) loci *(17)*. The method relies on the use of a negatively selectable gene that is placed on the flanks of the targeting vector (**Fig. 1**), and is commonly referred to as the "positive-negative selection" (PNS). The second method is, in genetic terms, positive, since it selects for recombination at the correct (homologous) locus *(18,19)*. This method relies on the use of a positively selectable gene whose expression is made conditional on recombination at the homologous target site (**Fig. 1**), and is referred to as the "promoterless" selection.

In a PNS vector, the positively and negatively selectable genes are functionally independent expression cassettes, and each contains its own promoter and polyadenylation signals. Nonhomologous events are recovered because of damage to the negatively selectable gene, which allows the expression of the positively selectable gene from many chromosomal locations. Empirical evidence in many laboratories has shown this to be a frequent event; consequently, PNS vectors typically achieve enrichments of only 2–5-fold. PNS vectors are typically used for gene targeting in ES cells, in which high enrichments are not necessary because of the relatively high frequencies of homologous recombination. In practical terms, gene targeting experiments in ES cells usually produce several homologously recombined cell lines after screening 50–100 drug-resistant clones.

In order for gene targeting in somatic cells to become feasible, the problem of low efficiency had to be resolved. In a promoterless vector the positively selectable gene lacks a promoter, and is expressed from the promoter of the target gene following homologous recombination. Nonhomologous events are recovered when the vector adventitiously integrates close to some chromosomal promoter that allows sufficient expression to generate a drug-resistant clone. These events are relatively rare, and promoterless vectors typically achieve enrichments of 100–500-fold *(16)*. Furthermore, the selection parameters can be easily manipulated to achieve enrichments of 1000–5000-fold. Such powerful selections appear to be absolutely essential for efficient gene targeting in somatic cells. In practical terms, the development of highly efficient promoterless vectors has increased the efficiency of gene targeting in somatic cells to the level observed in ES cells.

A number of limitations are inherent in the promoterless vector design. First, silent genes cannot be targeted. Second, promoters cannot be easily targeted, since the enrichment principle is based on the absence of promoter sequences from the targeting vector. Finally, intergenic regions are difficult to target because of the absence of easily definable promoter elements that can be used to select for integration.

1.2.3. Targeting of Human Cells Without Isogenic DNA

Soon after gene targeting in ES cells became established it was demonstrated that the frequency of homologous recombination was adversely affected by the presence of vector–target mismatches (2). This observation led to the practice of constructing targeting vectors from isogenic DNA. For gene targeting in murine ES cells the scientific community has adopted a single inbred genetic background. In contrast, the number of commonly used human cell lines is very large. A requirement for isogenic DNA would thus present a significant impediment, because it would necessitate the customization of vectors for each cell line. Furthermore, due to the outbreeding of most populations and the consequent heterozygosity of human cells, a strict requirement for isogenic DNA would necessitate the construction of allele-specific vectors.

A preliminary study did not find a significant requirement for isogenic DNA in human cells (3). The data showed that using nonisogenic DNA resulted in gene targeting frequencies in the 5–30% range. It was also encouraging that in several cases in which homozygous knockouts were generated, the two gene copies were targeted with similar frequencies. The practical consequence of this observation is that one vector can be used for widespread genetic analyses in a large number of human experimental cell culture systems.

1.2.4. Choosing a Cell Line

There is no evidence to date suggesting that any one human somatic cell line is more or less recombinogenic than another (4). Thus, the most important criterion for choosing a cell line should be that it represents an adequate model system for the topic under investigation. Second, the cloning efficiency of the selected cell line needs to be high, as the drug selection process requires that single cells have the ability to expand into clonal cultures. The ploidy is also important, because in most cases each gene copy has to be targeted independently. This means that a diploid cell requires two successive rounds of gene targeting, while a triploid cell line would require three, and so on. With new methods of recycling vectors (see **Subheading 3.8.5.**) (20,21), it is no longer prohibitive to contemplate multiple rounds of gene targeting; however, targeting three copies of a gene would clearly be more time-consuming than targeting only two. With aneuploid cell lines it is most constructive to assess the ploidy of the target gene directly by fluorescence in situ hybridization (FISH). Lastly, if use of nonimmortalized cell lines is contemplated, the proliferative capacity of the cultures must be considered to ensure that the required number of gene targeting events can be achieved before the onset of senescence. It has been estimated that 30–35 generations are required for one round of gene targeting (12).

2. Materials

2.1. Targeting Vector Preparation

1. Qiagen plasmid maxi kit (Qiagen, Valencia, CA, http://www.qiagen.com, cat. no. 12362).
2. Appropriate restriction enzymes and buffers.
3. 1 M EDTA.
4. 5 M potassium acetate.
5. TE (10 mM Tris-HCl, pH 8.0, 1 mM EDTA).
6. Buffer-equilibrated phenol.
7. Chloroform/isoamyl alcohol (24/1).
8. Absolute ethanol.
9. 70% ethanol.
10. 0.1 ×TE (1 mM Tris-HCl, pH 8.0, 0.1 mM EDTA).
11. Agarose gel electrophoresis apparatus.
12. F10 nutrient mixture (Life Technologies, Gaithersburg, MD, http://www.lifetech.com, cat. no. 81200-040).

2.2. Culture of LF1 Cells (see ref. 12)

1. F10 complete medium: F10 nutrient mixture supplemented with 15% fetal bovine serum (Hyclone Laboratories, Logan, UT, http://www.hyclone.com, cat. no. SH30071.03), 2 mM L-glutamine (Life Technologies, cat. no. 25030-081), and 50 U/mL penicillin-streptomycin (Life Technologies, cat. no. 15070-063).
2. Trypsin (Life Technologies, cat. no. 15050-057).
3. Phosphate-buffered saline (PBS), pH 7.4.

2.3. Electroporation

1. F10 nutrient mixture.
2. F10 complete medium.
3. Linearized vector DNA.
4. Gene Pulser II and 0.4-cm cuvette (Bio-Rad, Hercules, CA, http://www.bio-rad.com, cat. no. 165-2088).

2.4. Drug Selection

1. F10 complete medium.
2. Trypsin.
3. Selection drugs: histidinol·2HCl (Sigma, St. Louis, MO, http://www.sigma-aldrich.com, cat. no. H6647), hygromycin B (Calbiochem, La Jolla, CA, http://www.calbiochem.com, cat. no. 400051), Geneticin (Life Technologies, cat. no. 11811-031).

2.5. Colony Screening

1. F10 complete medium.
2. Alamar Blue (Trek Diagnostic Systems, Westlake, OH, http://www.treks.com, cat. no. 00-100).
3. 24-well cell culture dish (Corning, Corning, NY, http://www.corning.com, cat. no. 3524).
4. PBS.
5. Sterile cloning rings.
6. Sterile forceps.
7. Sterile silicone vacuum grease (Corning, Corning, NY, http://www.corning.com).

3. Methods

In the following sections are presented detailed protocols developed in our laboratory for gene targeting of nonimmortalized human fibroblasts. These methods are quite general, however, and have been adapted to several established cell lines in our as well as in other laboratories. Indications are given throughout the protocol where significant changes may have to be made for use with other cell lines.

3.1. Preparation of the Targeting Vector

The targeting vector plasmid DNA is purified using the Qiagen maxiprep procedure according to the manufacturer's instructions. The DNA should be free of contaminating RNA. The DNA concentration is measured by optical absorbance (260 nm) in a spectrophotometer, and 100 ng of uncut DNA is run on an agarose gel to verify the optical density measurement and to ascertain that the DNA is not degraded or otherwise contaminated.

The targeting vector plasmid DNA is linearized with an appropriate restriction enzyme. The restriction enzyme should cut one (or both) of the targeting vector–plasmid backbone junctions. Ten micrograms of plasmid DNA is used in one electroporation procedure (a typical targeting experiment utilizes two side-by-side electroporation procedures, for a total 20 µg of vector DNA). 10 µg of vector DNA is digested in a single 50 µL reaction with 10–20 units of the appropriate restriction enzyme. Following incubation for 2−3 h, 1 µl (approximately 200 ng) of the digest is run on an agarose gel to ascertain that the cutting was complete. The reaction is terminated by adding EDTA to a final concentration of 15 mM. Potassium acetate is added to a final concentration of 0.25 M and the total volume is increased to 100 µL with TE. The DNA solution is extracted once with 100 µL of equilibrated phenol and twice with 100 µL of chloroform/isoamyl alcohol. The DNA is precipitated by addition of 225 µL of absolute ethanol. The precipitation tube is placed for a minimum of 1 h in a –20°C freezer. If the linearized vector needs to be stored for longer periods of time, it should be stored as an ethanol precipitate.

Approximately 2 h before electroporation, the DNA is recovered by centrifugation in a microcentrifuge (13,200g, 30 min, 4°C). The DNA pellet should be small but clearly visible. No carrier should be necessary. The DNA pellet is washed carefully with 600 µL of 70% ethanol (room temperature). After removing the ethanol, the tube is briefly centrifuged (5 s) to spin down any remaining ethanol on the tube walls, and the last droplet of ethanol is carefully removed. The pellet is air dried for 10–20 min, or lightly dried in a Speed-Vac (5 min). Since the DNA may be difficult to resuspend, it is important to make sure that all traces of ethanol are evaporated but that the pellet is not over-dried. The DNA pellet is resuspended in 50 µL of 0.1 × TE by repeated vortexing and pipeting. One common cause for failure of the electroporation procedure is not to resuspend the targeting vector DNA adequately. It is thus important to run 1 µL of the final preparation on an agarose gel as a final check of recovery (including adequate resuspension) and integrity of the targeting vector DNA. The targeting DNA can be stored at 4°C while the cells are being harvested for the electroporation.

Immediately prior to electroporation the 50 µL of the targeting DNA solution is mixed with 750 µL of room-temperature F10 nutrient mixture (without serum, L-glutamine, or penicillin-streptomycin). This solution is referred to as the electroporation mixture. The electroporation mixture should not be kept for extended periods of time, as F10 contains magnesium, and even a trace of nuclease can damage the targeting DNA.

3.2. Culture of LF1 Cells Prior to Electroporation

One targeting experiment typically involves the electroporation of approximately 5×10^7 cells in two cuvettes (2.5×10^7 cells per cuvette, use 10 μg of DNA for each cuvette). For young to middle-aged human fibroblasts, this number of cells can be harvested from 35 to 40 10-cm tissue culture dishes in which the cells are approximately 40–50% confluent. To achieve maximum efficiency it is important that the cells are in the exponential phase of growth and proliferating vigorously. In order to minimize the number of cell divisions used up during the targeting procedure, it is essential to build up the cell numbers as rapidly and efficiently as possible. Therefore, close attention should be paid to the passaging regimen. Cells must not be allowed to become more than 80% confluent at any time, and the trypsinization procedure must be performed as gently and quickly as possible to ensure good recovery and viability of cells. It is important that only freshly made complete culture medium is used during the targeting procedure (complete medium with serum should not be stored, as this will reduce single cell cloning efficiency).

To thaw a vial of cells from liquid nitrogen (LN2) storage, first add 20 mL of F10 complete medium (warmed to 37°C) to a 10-cm culture dish, which is then placed in a CO_2 incubator to equilibrate the pH. The vial is thawed by transferring directly from the LN2 freezer into a 37°C water bath. When the last chunk of ice is melted, the vial is immediately sprayed with 70% ethanol, placed in a laminar-flow hood, and the entire contents are quickly transferred into the 10-cm culture dish. The dish is swirled gently to disperse the cells and replaced in the 37°C incubator. The dish is observed under a microscope after 2–3 h to check that the cells are beginning to adhere. Attachment should be complete after 4–6 h. At this point, the medium is gently removed by aspiration, and replaced with 10 mL of fresh complete medium.

The dish is incubated until the cells are no more than 80% confluent. At this point the cells are trypsinized and diluted in the 1:4 to 1:6 range. Higher dilutions should be avoided. This passaging regimen is repeated until the necessary number of dishes is obtained. The cells should be well dispersed at all times; if cells appear to grow in "clumps," "islands," or "rafts," the stringency of the trypsinization should be increased. The dishes harvested for the actual electroporation should have the cells evenly dispersed and no more than 40–50% confluent.

3.3. Harvest of LF1 Cells for Electroporation

The cells are washed twice with 10 mL of room-temperature PBS and drained well. Each 10-cm dish is overlaid with 0.5 mL of trypsin. The dishes are kept at room temperature with occasional tilting to spread the thin film of trypsin evenly over their surface. The cells need to be sufficiently trypsinized to get good recovery, but gently and rapidly to avoid loss of viability. Progress of trypsinization should be monitored microscopically. Trypsinization should be continued until the majority (>90%) of the cells can be dislodged by tapping the dish. After the cells are tapped loose, they are resuspend in 4.5 mL of F10 complete medium and transferred to a 50-mL sterile centrifuge tube. Using this procedure, 10 dishes can be collected in one tube. The harvested cells are kept at room temperature as additional dishes are trypsinized and pooled. Speed is essential in order to reduce the time of exposure to trypsin as well as the overall harvest time. Trypsinizing 4–5 dishes at a time is a good compromise.

Once all the cells are harvested, the tubes are centrifuged at 500g for 5 min at room temperature. Each pellet (from ten 10-cm dishes) is resuspended in 2 mL of F10 nutrient mixture (without serum, L-glutamine, or penicillin-streptomycin) by gently tapping the tube. Five milliliters of F10 nutrient mixture is added to each tube, and the contents are mixed by rocking the tube back and forth. The contents of 2 tubes (14 mL total) are transferred into one 15-ml centrifuge tube and centrifuged at 500g for 5 min at room temperature. The supernatant is carefully aspirated, and the last remnants are removed with a pipetman. It is important to remove as much of the liquid as possible without disturbing the pellet.

3.4. Electroporation Procedure

One electroporation mixture (0.8 mL total volume containing 10 μg of DNA (*see* **Subheading 3.1.**) is added to one cell pellet (approx. 2.5 × 10^7 cells). The cells are loosened gently by tapping the tube, and the resuspension process is completed by pipetting slowly up and down three to five times with a P1000 pipetman. The mixture of cells and DNA is transferred to a 0.4-cm-gap electroporation cuvette and immediately electroporated. The internal self-calibration check should be done immediately prior to electroporation. Settings of 260 V and 975 μF have been found to be optimal for LF1 cells. The time constant should be recorded; under the conditions described here the time constant should be in the range of 10–13 ms. Optimal electroporation conditions should be determined empirically for each new cell line. The cuvette is allowed to sit undisturbed for 5 min after electroporation. The entire electroporation procedure is performed at room temperature.

During the 5-min recovery period a viscous plug of lysed cells and DNA will form at the top of the cuvette, and should be scooped and dragged out with a 200-μL pipet tip. The remainder of the cell suspension should be removed with a clean 200-μL tip and transferred to a 50-mL centrifuge tube containing 40 mL of F10 complete medium. (Pipetting with a 200-μL tip should be done very slowly, as the cells are sensitive to shear forces. A useful precaution is to cut off the very end of the pipet tip to increase its bore.) The contents of the tube are mixed thoroughly by gently rocking and inverting the tube. The resulting cell suspension is divided among eight 10-cm dishes (5 mL per dish). The procedure is completed by adding 5 mL of F10 complete medium to each dish (to bring the total volume up to 10 mL), swirling each dish to thoroughly disperse the cells, and placing all the dishes in a 37°C incubator.

3.5. Postelectroporation Handling of Cells and Selection Regimens

Following the electroporation procedure, the cells are incubated for 36–48 h without any drugs; this incubation is commonly referred to as the "expression period." The dishes should be examined microscopically 8–12 h after electroporation, at which point the cells should have attached and flattened out to assume a typical fibroblastic morphology. There should be a significant amount of floating (dead) cells. The medium should be replaced with fresh F10 complete medium to remove the dead cells and debris. At the end of the 36–48 h expression period the dishes should be 80–100% confluent, at which point they should be trypsinized, diluted, and plated directly into appropriate selective medium. The dilution ratio depends on the total number of cells being plated and the type and concentration of the drug being used for selection.

Most drugs are maximally effective on actively growing cells, and some drugs do not work at all on confluent or near-confluent cultures. For example, for good killing with G418, in the range of 500–800 μg/mL LF1 cells should be plated at no more than 20–30% confluence. Thus, dishes that are 80–100% confluent at the time of trypsinization should be split 1:4. High concentrations of drug allow for lower split ratios because cells are killed more rapidly. The split ratios are always calculated on the basis of surface area, not volume, in cases where different culture vessels may be used at different points in the procedure. It should be kept in mind that some proliferation does occur during the expression period, so colonies recovered from cells descended from the same expression dish may contain sibling clones. All selection dishes should be marked with a code that identifies the expression dishes from which they were derived. This is important, because any homologous hits ultimately recovered from different expression dishes can be assumed to be independent.

The selection process is most conveniently done in 10-cm dishes, if the colonies will be cloned with cloning rings, and in 24-well microtiter plates in cases where screening with Alamar blue is desired (see below). One 24-well microtiter plate is approximately equivalent to one 10-cm dish in terms of total surface area. For experiments using 10-cm dishes the medium should be replaced once after 4–6 d of selection to remove the dead cells. If 24-well microtiter plates are used, no medium change is necessary. Colonies will become visible in the microscope after 6–8 d; however, colonies will not be visible to the naked eye for 12–14 d.

3.6. Colony Screening

3.6.1. Alamar Blue Method

As outlined above, a typical gene targeting experiment starts with 5×10^7 cells, which are electroporated side by side in two cuvettes and subsequently plated into sixteen 10-cm dishes. The total number of dishes (either 10-cm or 24-well microtiter) under selection will grow to 64 if a 1:4 split is used. This number of dishes cannot be significantly reduced without compromising the effectiveness of the drug selection. Since the total number of recovered drug-resistant clones is typically in the 100–300 range, each dish will contain one to three colonies, or, if 24-well microtiter plates are used, only 1–3 of 24-wells will be positive. Therefore, the probability that each of the positive wells is clonal is almost 100%, thus obviating the need for any further subcloning or single cell cloning. The problem with using 24-well microtiter plates is the difficulty of identifying wells with colonies: 64 microtiter plates contain a total of 1536 wells, making manual microscopic observation prohibitively time-consuming as the primary means of detection.

A convenient method of rapidly identifying positive wells takes advantage of Alamar blue, a vital redox dye that can be used in a colorimetric assay to detect actively metabolizing cells. The dye is added directly to the growth medium, and if living cells are present the medium changes color from light blue to pink. Typically, Alamar blue is added after 8–12 d of selection using a multichannel pipettor. The amount of Alamar Blue added depends on the type of culture medium being used. F10 medium is relatively light in color, and 10 μL of Alamar blue added to 1 mL of medium is sufficient to elicit a detectable color change. A medium containing higher concentrations of phenol red, such as DMEM, requires 20 µL or more of Alamar blue per 1 mL of medium to provide reliable detection. Since Alamar blue is costly, it makes good sense to switch to a colorless (phenol red-free) medium for the selection step, in which case 5 µL of Alamar blue per 1 mL of medium will be sufficient. The time required for the color change will depend on the size of the colony. For larger colonies the color change may be detected as soon as 24 h after addition of the dye, but a reliable color change may not be detected for 72 h or even more if the colonies are small or slowly growing. There have been no indications of toxicity even during extended periods of incubation in the presence of Alamar blue. It is helpful to place the 24-well plates on a pure white background to visualize the color. Once colonies have been identified, they are trypsinized and transferred to duplicate 24-well plates to begin the process of expansion for screening as well as freezing down.

3.6.2. Ring Cloning Method

In some instances a more traditional approach of selection in 10-cm dishes and establishment of clonal cell lines by isolating individual colonies with cloning rings may be preferable. One method of spotting colonies is to observe the dish with the naked eye from below against a dark background. A more rigorous method is exhaustive microscopic examination under low magnification. A systematic approach to scanning the entire surface of the dish is required. Spotted colonies can be conveniently marked with a felt-tip pen on the bottom of the plate.

In preparation for ring cloning, the silicone vacuum grease is spread in a uniform layer on the bottom of a 10-cm Pyrex petri dish, wrapped in foil, and sterilized by autoclaving. Cloning rings are autoclaved in Pyrex beakers covered with foil. Subsequently, the cloning rings are manipulated with ethanol sterilized forceps. By pressing or dabbing each cloning ring into the vacuum grease, one surface can be lightly but uniformly coated. The dishes should be well marked in advance to indicate the location of the colonies.

A dish is rinsed twice with 5 mL of room-temperature PBS and drained very well. Using the sterile forceps, a grease-coated ring is placed firmly over a mark indicating a colony. The process is repeated in rapid succession for all colonies on a plate. It is essential to avoid desiccation of the cells during this procedure. Then 100 µL of room-temperature trypsin is pipetted into each ring and allowed to act for several minutes at room temperature. The progress of trypsinization can be followed by microscopic observation. The trypsin is pipetted up and down 5–6 times to loosen the cells from the plate, and the entire cell suspension is then transferred into a well of a 24-well microtiter plate containing 1 mL of prewarmed F10 complete medium.

3.7. Screening for Recombinants

Screening strategies do not differ significantly from those used in murine ES cells and will thus not be described in detail. Two basic methods are used: PCR and Southern hybridization. Since it is prudent to perform targeting experiments on a scale that will yield 200–400 drug-resistant colonies, a PCR strategy is generally preferable as the primary screen. The main danger of PCR strategies is the possibility of false negative results, and thus, of missing genuinely targeted clones. It is therefore important to choose the primers carefully and to optimize the reaction empirically to ensure that the detection limit is well within the required sensitivity range. This can be easily performed on an in-vitro-constructed template of the same structure as the desired in-vivo recombination event. The inclusion of primers of known sensitivity to an endogenous target is also useful as a positive control for the robustness of each individual PCR reaction.

The general strategy is to PCR across the short arm of the targeting vector by placing one primer in the selectable gene and the other in chromosomal sequences outside of those contained in the targeting vector. Ideally, the amplification product should be less than 2 kb, since amplification of longer fragments is less reliable and the risk of false negatives can be significantly increased. If screening of large numbers of clones is necessary, it is feasible to perform PCR on pools of clones. The one aspect where gene targeting in somatic cells differs from that in ES cells is the recovery of DNA for PCR. In general, a rapid, "single-tube" method is preferred, but the method of harvesting the cells and the quality of the resulting DNA need to be addressed for each cell line. In most cases sufficient DNA for several PCR reactions can be obtained from cells grown in a single well of a 24-well plate. Before DNA can be extracted, a duplicate set of all clones must be generated and set aside. Ideally, the PCR analysis should be performed rapidly so that nontargeted clones can be efficiently discarded; if this is not feasible, the entire duplicate set of clones should be cryopreserved. Since methods of subculture and cryopreservation can vary significantly from cell line to cell line, they will not be discussed in detail.

After the primary screen, the fidelity of the recombination events must be verified by Southern hybridization analysis. In some cases it is possible to skip the primary PCR screen and proceed directly with the Southern hybridization analysis. It is highly recommended that probes for this analysis are chosen on both the 5'and 3' sides of the targeted locus, and that at least one of the probes falls outside the sequences contained in the targeting vector. The former is recommended because it is sometimes observed that a single targeting event can arise by a combination of both homologous and illegitimate events, and that the illegitimate events can include large deletions of neighboring sequences. Use of outside probes is prudent because they detect exclusively changes in the chromosomal target locus; probes homologous to any part of the targeting vector will recognize many nonhomologous insertion events and may lead to ambiguity in the interpretation of results.

3.8. Design of Promoterless Targeting Vectors

3.8.1. Basic Principles

A promoterless vector is constructed such that, following homologous recombination with the chromosome, the selectable gene in the vector is expressed from the target gene promoter (**Subheading 3.8.2.**). A number of points should be considered in the design process. The principal design feature is that the vector must not contain any promoter sequences. Thus, the location of the target gene promoter should be known to ensure that it is excluded from the vector. Targeting vectors typically contain a 5' arm, a 3' arm, and a central region of nonhomology that contains the selectable gene (**Fig. 1**) (*see* **Note 2**). The 5' and 3' arms are the regions of homology with the chromosomal target site through which homologous recombination takes place. The 5' and 3' arms can contain exons, introns, or intergenic regions including repetitive elements.

The total homology with the target locus (5' arm and 3' arm combined) should be in the range of 6–10 kb. If the screening of drug-resistant cell lines will be done by PCR, the vector should have one arm in the range 0.8–2 kb, as fragments that are longer than 2 kb can be difficult to amplify by PCR. Arms shorter than 0.8 kb may result in decreased frequencies of recombination. Either the 5′ or the 3′ arm can be the short arm as long as the total length of homology adds up to 6 kb or more. It is generally counterproductive to increase total homology beyond 10 kb, since such vectors are often more difficult to construct by common recombinant DNA methods, and large DNA fragments may be difficult to propagate even in medium-copy-number plasmids. The pUC backbone can be recommended for vectors up to 18–20 kb, but it should be kept in mind that instability is sequence-dependent and has been observed with significantly shorter fragments. Very-high-copy plasmids (such as Bluescript) are not recommended because of stability problems.

3.8.2. Strategies

Since in promoterless vectors the positively selectable gene is not a self-contained expression module, careful design is important to ensure adequate expression of the drug-resistant phenotype following homologous recombination *(4)*. Several strategies have been developed to accommodate different target gene architectures. In the following discussion the neomycin phosphotransferase gene *(Neo)* will be used as the drug-resistance gene, but the principles apply to any selectable gene.

A simple and effective strategy is to replace the coding region of the target gene with the coding region of the drug-resistance gene (**Fig. 2**). In this case, the ATG codon of *Neo* will become synonymous with the ATG codon of the target gene. The advantage of this design is that no modifications are made to *cis*-acting elements of the target gene, such as splice consensus sequences or translation initiation signals (Kozak box). The rationale is straightforward: if these regulatory sequences provide for the expression of the target gene product in the cell under investigation, the same sequences should also direct adequate expression of the Neo protein (*see* **Note 3**).

In most cases the coding region of the target gene will be divided into several exons, but it would be impractical to do likewise with the Neo open reading frame. Therefore, the Neo open reading frame is used as a single exon, resulting in an exon of some 900 bp. The Neo open reading frame is followed with a strong polyadenylation signal. It is important to note that this strategy is usually limited to genes whose translation is initiated in exon 2 or downstream exons. The reason for this is that the length of the 5′ arm is limited by the promoter on the 5′ end, which must not be included in the vector, and the ATG start codon on the 3′ end, which is where the Neo coding region is inserted. Most genes that initiate translation in exon 2 have a first intron long enough to use as a 5′ arm. Any 5′untranslated region could also be used to extend the length of the 5′ arm.

A straightforward application of this strategy is to insert the Neo-poly A cassette into the first coding exon of the target gene (**Fig. 2**). With this method, no sequences are deleted from the target gene, Neo protein is expressed from the native ATG codon of the target gene, and the Neo open reading frame is immediately followed by a polyadenylation site. There is one potential problem that should be considered with this approach. Since the target gene is not deleted and even strong SV40 polyadenylation signals can give some readthrough, there is a finite probability that some RNA will be synthesized downstream of Neo. Whether any of this RNA could actually code for protein will depend on many factors, such as cryptic splice sites and internal initiations of translation. Expression of protein fragments from downstream exons could elicit unwanted phenotypes. However, there are many examples in which this simple strategy gave clean knockouts.

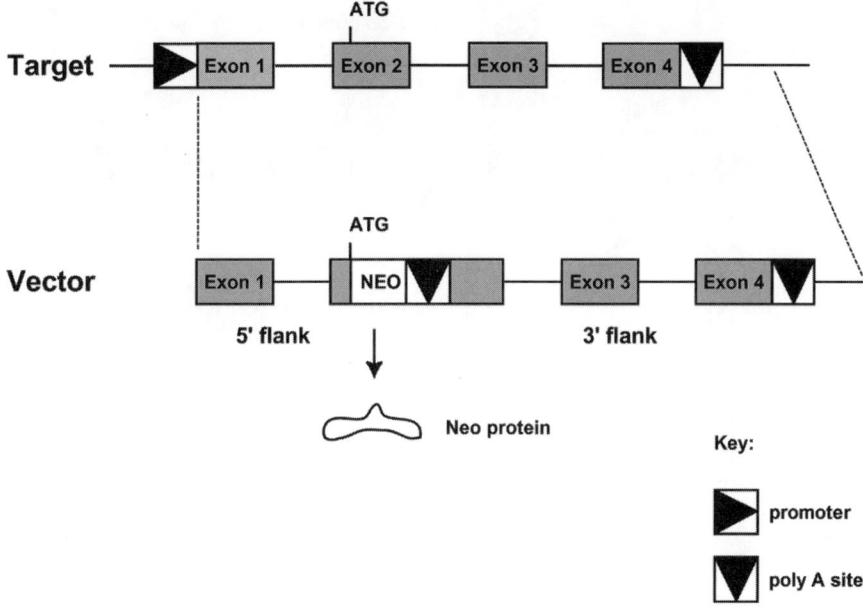

Fig. 2. Promoterless strategy for targets with translational starts in exon 2 or downstream exons.

To avoid these problems it is possible to remove some or all of the coding exons downstream of Neo. This results in an insertion-deletion type of substitution being introduced into the chromosome. The 3′ arm used to direct this type of recombination would therefore consist of intergenic regions located 3′ of the target gene. This may be the preferred approach in cases where the target gene is compact, and exons are spread over a region of less than 10 kb. However, as the size of the target gene increases, so does the size of the desired chromosomal deletion, and it is uncertain how large a deletion can be sustained without compromising the ability of the vector to undergo efficient homologous recombination. There are examples of deletions up to 90 kb having been successfully made in ES cells *(22,23)*, but experience in somatic cells is very limited.

There are two strategies that can be used to target genes whose translation starts in exon 1 (**Fig. 3**). In the first method the Neo open reading frame is inserted in frame into a downstream coding exon (**Fig. 3A**). This will result in the synthesis of a fusion polypeptide whose N-terminal sequences are derived from the target gene. In most cases the insertion point can be placed relatively close to the 5′ end to minimize the length of protein sequences translated from the target gene. This will reduce the possibility of dominant-defective phenotypes being elicited by the fusion polypeptide. The disadvantage of this approach is that the Neo enzymatic activity of such fusion polypeptides can be variable. To be prudent the Neo activity should be tested empirically, and several different fusions may need to be constructed.

In the second strategy the picornavirus IRES (internal ribosome entry site sequence) is used to construct a bi-cistronic message (**Fig. 3B**). Such vectors contain a full-length Neo open reading frame (with an optimized Kozak box) preceded by an IRES and followed by a polyadenylation signal. The IRES is additionally preceded by translational stop codons in all frames to terminate translation of the target gene initiated in exon 1.

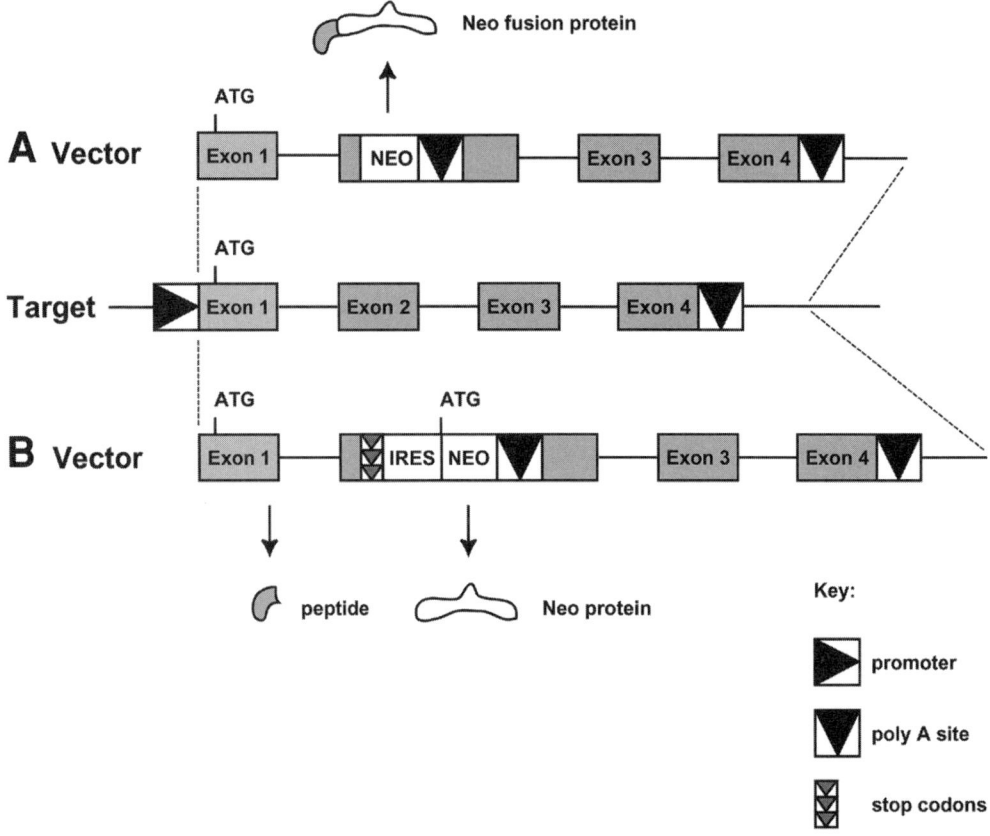

Fig. 3. Promoterless strategies for targets with translational starts in exon 1.

As a result, only a short peptide is typically synthesized from the target gene itself. This strategy can also be combined with the deletion of downstream target gene exons.

3.8.3. Drug-Resistance Genes

Three drug-resistance genes have been used for the construction of promoterless vectors: neomycin phosphotransferase (Neo), hygromycin phosphotransferase (Hyg), and histidinol dehydrogenase (His). Two parameters should be considered when evaluating any positively selectable gene: the level of expression necessary for cell survival in the presence of sufficient drug to kill sensitive cells, and the strength of the target gene promoter. Additional selectable genes, such as puromycin acetyltransferase, blasticidin deaminase, and the gene encoding a bleomycin-binding protein, may also be suitable for gene targeting, but their use has not been documented to date.

Evidence from several laboratories indicates that *Neo* is a powerful selectable gene; this means that even very low levels of expression can generate a drug-resistant clone. The minimum effective level of Neo expression has been estimated to be in the range of 5000 molecules of Neo protein per cell. A side-by-side comparison of *Neo*, *Hyg*, and *His (16)* showed that *His* was a weaker selectable gene than *Neo*, and that *Hyg* was the weakest of all three. The minimum effective expression levels for *Hyg* and *His* in terms of molecules per cell have not been reported to date. *c-Myc*, a gene with a very weak promoter, could not be targeted with *Hyg (16)*, whereas the Cdk inhibitor p21 could be targeted *(12)*. *c-Myc* could, however, be targeted with *His (24)*, and both could be targeted with *Neo (12,16)*.

The disadvantage of using a powerful selectable gene is that nonhomologous events are recovered more frequently, since many nonhomologously recombined clones exceed the low expression threshold required to generate a drug-resistant colony. However, a powerful selectable gene is required to target genes expressed at low levels. It is important to note that the efficacies of selective genes can be significantly adjusted by varying the concentration of the drug added to the medium *(16)*.

3.8.4. Knock-in Strategies

A knock-in is a substitution or insertion of a foreign coding sequence into a chromosomal gene, usually engineered such that the knocked-in gene is placed under control of the target promoter. It should be noted that the promoterless vector design dictates that each knockout is also automatically a knock-in of the selectable gene. Expression of *Neo* (or another selectable gene) can thus be monitored to study the expression patterns of the targeted gene. It is often desirable to knock-in a reporter gene, such as ß-galactosidase or GFP, whose activity can be easily monitored (**Fig. 4**). Two alternatives can be pursued to integrate the additional open reading frame into the targeting vector: the creation of bicistronic expression units using the IRES sequence (**Fig. 4A**), or in-frame fusion to create bifunctional hybrid polypeptides, such as GFP-Neo (**Fig. 4B**). In addition to reporter genes, an often encountered knock-in application is the substitution of mutated or tagged versions of the target gene. Finally, knock-ins can be made with completely unrelated genes, for example, to direct the synthesis of a therapeutic protein a specifically desired cell type or temporal pattern of expression.

The creation of bicistronic expression units using the IRES sequence is a frequently used strategy. Most applications to date have exploited the IRES signal from picornaviruses, which is approximately 400 bp in length. It should be noted that the expression level of the second coding sequence is 4- to 5-fold lower relative to the first coding sequence *(25)*. This is due to a lower efficiency of initiation of translation at the internal, IRES directed site. Thus, if one of the genes in a bicistronic unit is to be expressed at a higher level, it should be placed first. It should also be noted that some commercially available IRES sequences have been purposefully modified to further reduce the expression level of the second gene, and should not be used for the construction of targeting vectors unless low expression of the downstream gene is specifically desired.

Several IRES sequences unrelated to the picornavirus IRES have recently been discovered *(26)* in cellular genes, but have not been exploited to date for the construction of targeting vectors. These additional IRES sequences, some of which are considerably shorter than the picornavirus IRES, can be exploited for the construction of multicistronic expression units. Attempts to construct such units using multiple picornavirus IRES sequences have been compromised by instability problems due to homologous recombination between the tandemly repeated sequence elements.

3.8.5. Recyclable Targeting Vectors

"Recycling" a targeting vector means to remove the selectable gene from an already targeted gene such that the same drug selection can be used to target the second copy of the gene, or a different gene altogether. Typically, the coding sequence of the drug resist-

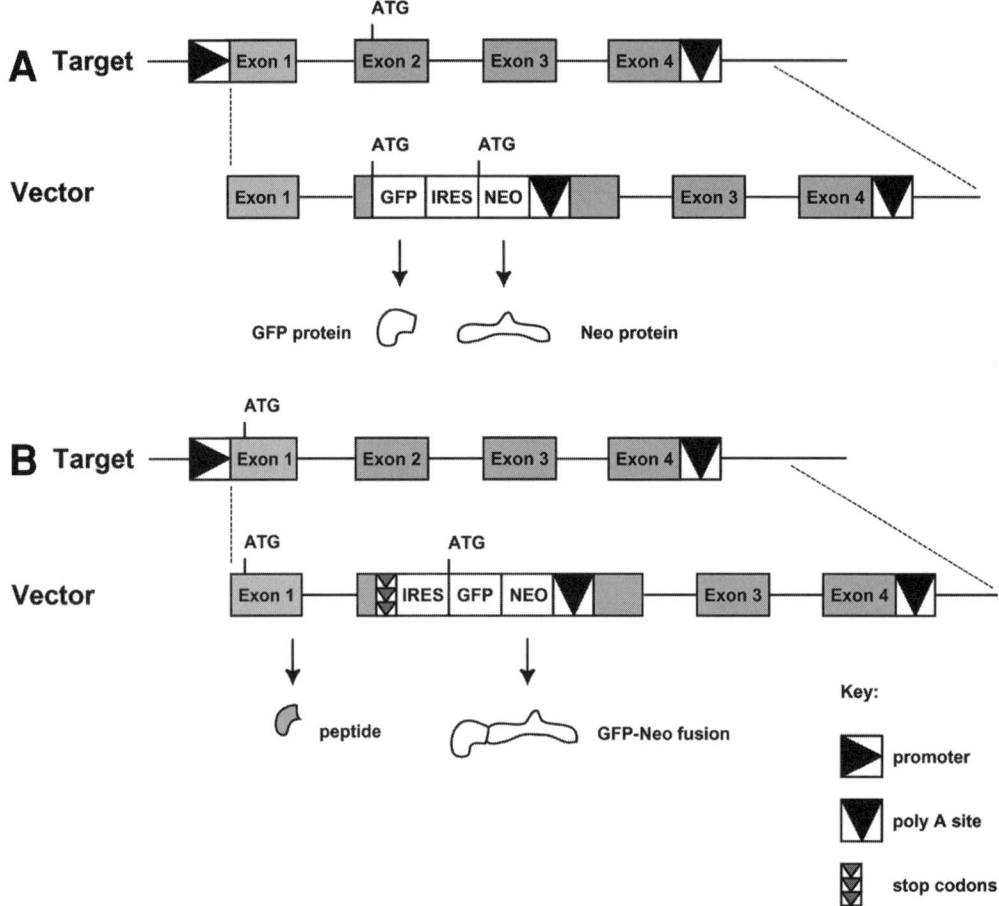

Fig. 4. Promoterless strategies for knock-in vectors.

ance gene (**Fig. 5**) is flanked with *Lox* sites *(20)*; after successful gene targeting, Cre protein is transiently expressed in the cells, inducing the looping out and segregation of the drug resistance gene. This strategy can be especially effective when used with a vector designed to delete a substantial portion of the target gene. The chief reason for adopting this strategy is to obviate the need to construct two different vectors to generate a homozygous knockout, and can be a considerable saving of time and effort. It should be noted that in the second round of gene targeting the vector has an equal probability of recombining with either the previously targeted chromosome or the untargeted chromosome; however, it is usually straightforward to design PCR or Southern hybridization strategies to distinguish all possible recombination events.

In earlier experiments using this strategy, Cre was expressed using transient transfection of a plasmid vector. Since the fraction of productively transfected cells may be quite low, the recovery of the desired looped-out events can be made significantly easier by introducing a negatively selectable gene into the vector (**Fig. 5**). More recently, adenovirus vectors expressing Cre have become available. Using this method, expression of Cre is typically so efficient that no negative selection is needed, and the sought events can be recovered with good efficiency by simple dilution and single-cell cloning of the infected pool of cells.

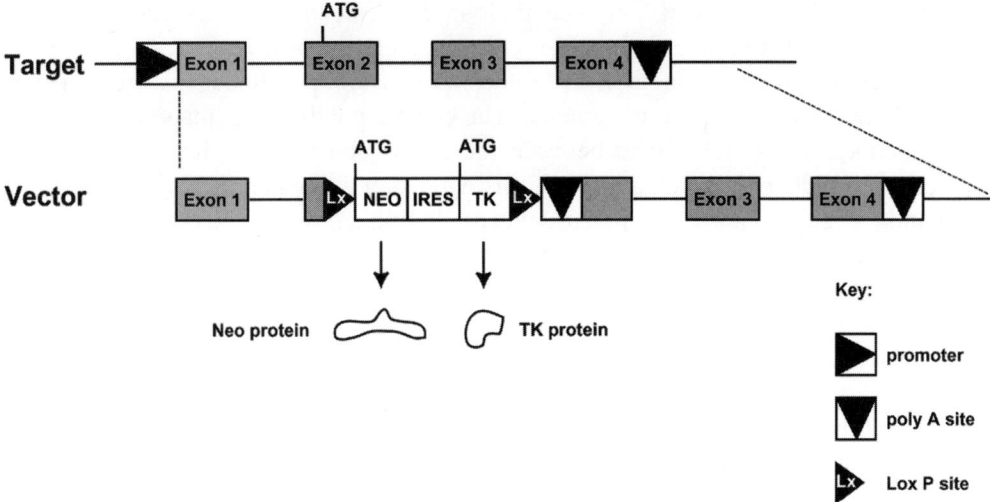

Fig. 5. Recyclable targeting vector based on Cre-Lox site-specific recombination.

It should be noted that the Cre-Lox recycling strategy should not be used for gene targeting of nonimmortalized (presenescent) cells. This is because the process of segregating the drug resistance gene requires additional cell division, and presenescent cells may not have sufficient life span to complete the subsequent second targeting event. Therefore, in presenescent cells it is necessary, at this time, to use two vectors with distinct selectable genes.

3.8.6. Vector Assembly

There are two fundamentally different construction methods to link up the vector arms with the internal selectable cassette: one employs conventional recombinant DNA technologies, while the other employs in-vivo homologous recombination in *Saccharomyces cerevisiae* or *Eshcerichia coli*. The latter method holds the promise of very significantly reducing the time and effort required to assemble even complicated vectors. However, these methods are new and still under development; consequently, it is not yet clear which strategies will become widely used. A comprehensive review of this topic is beyond the scope of this chapter; however, the relevant primary references are provided for those interested in this approach *(27–31)*.

The traditional approach exploiting conventional molecular cloning and PCR needs little comment; however, a few issues of specific interest for the construction of promoterless vectors will be mentioned. In the postgenomic era it is generally assumed that the entire genomic sequence of the target region of interest will be known during the vector design process. While this is by no means essential, especially with respect to the long vector arm, it does aid considerably in vector design. In general, it is most convenient to assemble the targeting vector in modular fashion, involving a minimum of three discrete cassettes: the 5' arm, the central portion containing the selectable gene(s), and the 3' arm. The critical features of several selectable cassettes have been discussed above, and a number of such cassettes are now available. The design process thus involves, in stepwise fashion, (a) choice of selection strategy to match the target gene architecture, (b) choice of selectable cassette to accomplish the desired selection strategy, and (c) design of a specific molecular cloning scheme to link the selectable cassette with the vector arms to provide homology for the targeting process.

A convenient cloning strategy is to synthesize a custom polylinker to accept in step-wise fashion the desired vector segments. The custom polylinker is made by annealing two oligonucleotides and cloned between the EcoRI and HindIII sites of pUC18/19, replacing the pUC polylinker. Although this entails an extra step, it is easy to accomplish and can significantly aid the subsequent cloning steps. It should be remembered that the extensive regions of genomic sequence typically present in promoterless vectors (8–12 kb) are likely to contain a significant number of sites for restriction enzymes with 6 base recognition sequences. Thus, complex polylinkers containing numerous sites can actually hinder the construction process, since preferably the junctions between the different vector segments should contain uniquely cutting restriction sites. This not only aids the assembly process, but also allows subsequent modification of any functional component of the targeting vector.

PCR can be very useful in the construction of targeting vectors. It can be used to amplify chromosomal sequences for the vector arms either from BAC clones or even from total genomic DNA. PCR can be used to place desired restriction sites on the ends of vector segments, or even to link vector segments together directly. If PCR is used for any steps involving the selectable cassette, at a minimum the entire open reading frame of the drug-resistance gene should be sequenced in the final vector, since even a single nucleotide change can introduce a mutation with debilitating effects on the drug-resistance phenotype.

4. Notes

1. References for gene targeting in somatic cells prior to 1999 can be found in reviews *(3,4)*. Recent references are *(5–10)*. Gene targeting in the chicken DT40 cell line is not covered in this chapter, but see recent **ref. 11.**
2. The exception would be vectors designed to integrate into the chromosome by a single crossover event *(1)*, but this design is not in common use.
3. One exception would be a situation in which the target gene contained a weak Kozak box, and it was desirable to increase the expression o the inserted Neo protein. In this case an optional Kozak box could be substituted

5. References

1. Sedivy, J. M. and Joyner, A. (1992) *Gene Targeting.* Freeman, San Francisco.
2. te Riele, H., Maandag, E. R., and Berns, A. (1992) Highly efficient gene targeting in embryonic stem cells through homologous recombination with isogenic DNA constructs. *Proc. Natl. Acad. Sci. USA* **89,** 5128–5132.
3. Sedivy, J. M., Vogelstein, B., Liber, H. L., Hendrickson, E. A., and Rosmarin, A. (1999) Gene targeting in human cells without isogenic DNA. *Science* **283,** 9a.
4. Sedivy, J. M. and Dutriaux, A. (1999) Gene targeting and somatic cell genetics: a rebirth or a coming of age? *Trends Genet.* **15,** 88–90.
5. Park, B. H., Vogelstein, B., and Kinzler, K.W. (2001) Genetic disruption of *PPARδ* decreases the tumorigenicity of human colon cancer cells. *Proc. Natl. Acad. Sci. USA* **98,** 2598–2603.
6. Rhee, I., Jair, K. W., Yen, R. W., et al. (2000) CpG methylation is maintained in human cancer cells lacking DNMT1. *Nature* **404,** 1003–1007.
7. Chan, T. A., Hwang, P. M., Hermeking, H., Kinzler, K. W., and Vogelstein, B. (2000) Cooperative effects of genes controlling the G2/M checkpoint. *Genes Dev.* **14,** 1584–1588.

8. Zhang, L., Yu, J., Park, B. H., Kinzler, K. W., and Vogelstein, B. (2000) Role of BAX in the apoptotic response to anticancer agents. *Science* **290,** 989–992.

9. Chan, T. Y., Hermeking, H., Lengauer, C., Kinzler, K. W., and Vogelstein, B. (1999) 14-3-3σ is required to prevent mitotic catastrophe after DNA damage. *Nature* **401,** 616–620.

10. Faria, T. N., Mendelsohn, C., Chambon, P., and Gudas, L. J. (1999) The targeted disruption of both alleles of RARß2 in F9 cells results in the loss of retinoic acid-associated growth arrest. *J. Biol. Chem.* **274,** 26783–26788.

11. Lahti, J. M. (1999) Use of gene knockouts in cultured cells to study apoptosis. *Methods* **17,** 305–312.

12. Brown, J., Wei, W., and Sedivy, J. M. (1997) Bypass of senescence after disruption of p21*CIP1/WAF1* gene in normal diploid human fibroblasts. *Science* **277,** 831–834.

13. Polejaeva, I. A., Chen, S. H., Vaught, T. D., et al. (2000) Cloned pigs produced by nuclear transfer from adult somatic cells. *Nature* **407,** 86–90.

14. McCreath, K. J., Howcroft, J., Campbell, K. H., Colman, A., Schnieke, A. E., and Kind, A. J. (2000) Production of gene-targeted sheep by nuclear transfer from cultured somatic cells. *Nature* **405,** 1066–1069.

15. Arbones, M. L., Austin, H. A., Capon, D. J., and Greenburg, G. (1994) Gene targeting in normal somatic cells: inactivation of the interferon-gamma receptor in myoblasts. *Nat. Genet.* **6,** 90–97.

16. Hanson, K. D. and Sedivy, J. M. (1995) Analysis of biological selections for high-efficiency gene targeting. *Mol. Cell. Biol.* **15,** 45–51.

17. Mansour, S. L., Thomas, K. R., and Capecchi, M. R. (1988) Disruption of the proto-oncogene *int-2* in mouse embryo-derived stem cells: a general strategy for targeting mutations to non-selectable genes. *Nature* **336,** 348–352.

18. Jasin, M. and Berg, P. (1988) Homologous integration in mammalian cells without target gene selection. *Genes Dev.* **2,** 1353–1363.

19. Sedivy, J. M. and Sharp, P. A. (1989) Positive genetic selection for gene disruption in mammalian cells by homologous recombination. *Proc. Natl. Acad. Sci. USA* **86,** 227–231.

20. Abuin, A. and Bradley, A. (1996) Recycling selectable markers in mouse embryonic stem cells. *Mol. Cell. Biol.* **16,** 1851–1856.

21. Ray, M. K., Fagan, S. P., and Brunicardi, F. C. (2000) The Cre-loxP system: a versatile tool for targeting genes in a cell- and stage-specific manner. *Cell Transplant.* **9,** 805–815.

22. Zhang, H., Hasty, P., and Bradley, A. (1994) Targeting frequency for deletion vectors in embryonic stem cells. *Mol. Cell. Biol.* **14,** 2404–2410.

23. Medina-Martinez, O., Bradley, A., and Ramirez-Solis, R. (2000) A large targeted deletion of Hoxb1-Hoxb9 produces a series of single-segment anterior homeotic transformations. *Dev. Biol.* **222,** 71–83.

24. Mateyak, M. K., Obaya, A. J., Adachi, S., and Sedivy, J. M. (1997) Phenotypes of c-Myc-deficient rat fibroblasts isolated by targeted homologous recombination. *Cell Growth Differ.* **8,** 1039–1048.

25. Davies, M. V. and Kaufman, R. J. (1992) The sequence context of the initiation codon in the encephalomyocarditis virus leader modulates efficiency of internal translation initiation. *J. Virol.* **66,** 1924–1932.

26. Chappell, S. A., Edelman, G. M., and Mauro, V. P. (2000) A 9-nt segment of a cellular mRNA can function as an internal ribosome entry site (IRES) and when present in linked multiple copies greatly enhances IRES activity. *Proc. Natl. Acad. Sci. USA* **97,** 1536–1541.

27. Baudin, A., Ozier-Kalogeropoulos, O., Denouel, A., Lacroute, F., and Cullin, C. (1993) A simple and efficient method for direct gene deletion in *Saccharomyces cerevisiae*. *Nucleic Acids Res.* **21,** 3329–3330.

28. Zhang, Y. M., Buchholz, F., Muyrers, J. P., and Stewart, A. F. (1998) A new logic for DNA engineering using recombination in *Escherichia coli*. *Nat. Genet.* **20,** 123–128.

29. Liu, Q., Li, M. Z., Leibham, D., Cortez, D., and Elledge, S. J. (1998) The univector plasmid-fusion system, a method for rapid construction of recombinant DNA without restriction enzymes. *Curr. Biol.* **8,** 1300–1309.
30. Yu, D., et al. (2000) An efficient recombination system for chromosome engineering in *Escherichia coli. Proc. Natl. Acad. Sci. USA* **97,** 5978–5983.
31. Datsenko, K. A. and Wanner, B. L. (2000) One-step inactivation of chromosomal genes in *Escherichia coli* K-12 using PCR products. *Proc. Natl. Acad. Sci. USA* **97,** 6640–6645.

16

Colony Growth Suppression by Tumor Suppressor Genes

Gen Sheng Wu

1. Introduction

The ability of a tumor suppressor gene to inhibit cell growth is critical for its tumor suppression. Such activity is carried out largely by inducing cell cycle arrest and apoptosis. For example, when the *p53* tumor suppressor gene is activated by DNA damage *(1)*, it inhibits the growth of tumor cells by inducing cell cycle arrest or apoptosis *(2)*. Despite a temporal and reversible event, cell cycle arrest allows damaged DNA to be repaired and thereby prevents abnormal DNA propagation and the emergence of cancer cells. On the other hand, p53-induced apoptosis is powerful and irreversible *(3)*. When p53 is introduced into tumor cells, for example via p53 adenovirus infection, the cells undergo rapidly apoptosis *(4)*. Conversely, loss of p53-mediated apoptosis has been implicated not only in tumor progression, but also in the drug-resistant phenotype of tumor cells *(5)*. Thus, the tumor suppressor gene often possesses these two features. Accordingly, several techniques have been designed to analyze cell cycle arrest and apoptosis induction to characterize the growth inhibition activity of tumor suppressor genes, including flow cytometry assay *(6)*, MTT assay *(7),* and clonogenic survival assay *(8)*. Among these, the clonogenic survival assay is the most commonly used assay to assess growth inhibition in vitro. Basically, this method examines how the gene of interest affects cell growth in cultured cells. It includes three steps. The first step is to construct a mammalian expression vector that can express the gene of interest. This involves standard molecular cloning techniques. The second step introduces the expression vector into human cancer cell lines via a carrier such as lipofectin. The last step is to select the cells with appropriate antibiotics in order to kill untransfected cells and allow transfected cells to expand and can form colonies if the gene of interest has no effect on cell growth (*see* **Note 1**). Conversely, there are few or no surviving colonies left in the culture if the gene of interest has an ability to inhibit cell growth.

2. Materials

1. pcDNA3 (Invitrogen).
2. 5× Ligase reaction buffer.
3. T4 DNA ligase (5 U/µL).

From: *Methods in Molecular Biology, Vol. 223: Tumor Suppressor Genes: Regulation, Function, and Medicinal Applications.* Edited by: Wafik S. El-Deiry © Humana Press Inc., Totowa, NJ

207

4. DH10B bacteria strain.
5. LB-agar plates with ampicillin (50 µg/mL).
6. SOC medium (Invitrogen).
7. LB broth with ampicillin (50 µg/mL).
8. Cell culture medium RPMI 1640.
9. Lipofectin (Gibco/BRL).
10. Opti-MEM medium.
11. Phosphate-buffered saline (PBS).
12. Fetal bovine serum (FBS).
13. G418.
14. 0.1% Crystal violet.

3. Methods

3.1. Construction of TSG Expression Vector

A number of vectors can be used for this assay (*see* **Note 2**). Here the vector pcDNA3 is used to construct TSG expression vector. Before starting ligation reaction, completely digest pcDNA3 with endonclease (S), and purify it. The gene of interest is also digested with same endonclease (S), if the ligation reaction is carried out with cohesive ends.

3.1.1. Ligation

1. Add the following components in a sterile 1.5-mL tube: 50 ng pcDNA3, 0.15 µg Insert, 4 µL 5× ligase reaction buffer, water to 9 µL, 1 µL T4 DNA ligase (5U).
2. Mix gently and spin briefly. Meanwhile, set up two control reactions; one with the pcDNA3 vector alone and another with the gene of interest alone.
3. Incubate the tubes at 16°C overnight.

3.1.2. Transformation

1. Preparation of LB agar plates with amplicillin (50 µg/mL).
2. Add 2 µL of the ligation reaction to 50 µL of competent DH10B cells (Invitorgen) in a 1.5-mL tube and mix by tapping gently.
3. Incubate the tube on ice for 30 min.
4. Transfer the tube in a 42°C water bath and incubate for exactly 30 s. Do not mix and shake.
5. Place the tube on ice for 2 min.
6. Add 250 µL of prewarmed SOC medium to the tube.
7. Place the tube in a rack and shake it at 225 rpm at 37°C for 1 h.
8. Plate 20–200 µL of the transformation reaction onto LB agar plates.
9. Invert the plates and incubate at 37°C overnight (*see* **Note 3**).

3.1.3. Preparation of Plasmids

1. Pick up 10 individually isolated colonies to ten 50-mL centrifuge tubes, with each tube containing 10 mL LB with ampicillin (50 µg/mL).
2. Place the tubes in a 37°C incubator and shake it at 225 rpm overnight.
3. Miniprep plasmids using the method described by Sambrook et al. *(9)* or using a commercial plasmid purification kit.

4. Digest the plasmids with corresponding enzyme (S) to verify the insert.
5. Purify the plasmid if necessary and determine its concentration.
6. Proceed to the next step.

3.2. Transfection

Transfection efficiency is dependent on the cell types and the methods used (*see* **Note 4**). Here the human lung cancer cell line H460 and Lipofectin reagent (Gibco/BRL) are used to perform the transfection procedure.

1. Split H460 cells into 6 T25 flasks at a density of 1 million per flask.
2. Incubate the cells at 37°C in a CO_2 incubator for 18–24 h.
3. Prepare eight 12 × 75-mm sterile tubes (in duplicate for each plasmid) and add the following components in the tubes:

Name	Dc1/Dc2	De1/De2	Lc1/Lc2	Le1/Le2
Opti-MEM medium	100 μL	100 μL	100 μL	100 μL
pcDNA3	2 μg	—	—	—
pcDNA3 with insert	—	2 μg	—	—
Lipofectin	—	—	20 μL	20 μL

4. Mix well and allow to stand at room temperature for 30–45 min.
5. Add Dc1 to Lc1 tube, Dc2 to Lc2, De1 to Le1, and De2 to Le2, in drops with gentle shaking. The combinations of De with Le are the experiment group, while the combinations of Dc with Lc serve as a vector control. Incubate them at room temperature for 15 min, which allows them to form DNA–lipofectin complexes.
6. Wash cells twice with prewarmed PBS.
7. Add 1.8 mL of Opti-MEM into the tubes with the DNA–lipofectin complexes. Mix gently and then overlay onto cells (*see* **Note 5**).
8. Incubate the cells at 37°C in an incubator with 5% CO_2 and 95% air for 6–10 h.
9. Replace Opti-MEM with normal RPMI 1640 with 10% FBS and antibiotics, leave the cells in a incubator at 37°C overnight, and proceed to the next step for selection.

3.3. Selection

1. Split cells into two T25 flasks and add G418 (Geneticin) at 500 μg/mL in the medium to select resistant colonies.
2. Change medium every 2 d with RPMI 1640 medium in the presence of 500 μg/mL of G418.
3. For H460 cells, evident resistant colonies appear in about 2–3 wk.

3.4. Staining

1. Discard the medium.
2. Add 2 mL of 0.1% crystal violet and incubate at room temperature for 10 min.
3. Wash the cells four times with PBS.
4. Count colonies that consist of ≥ 50 cells (*see* **Note 6**). A few or no surviving colonies are expected in the flasks in which the cells are transfected with pcDNA3-insert, if this gene has an ability to inhibit cell growth, and many colonies are expected in cells transfected with pcDNA empty vector.

4. Notes

1. It is important to use the appropriate concentration of antibiotics to select the cells. If the concentration of antibiotics used is too high, it can kill cells in which the gene of interest is expressed, although this gene has no effect on cell growth. On the other hand, if the concentration of antibiotics used is too low, it is unable to kill all untransfected cells, which can give you a high background. Therefore, we recommend that, before starting the clonogenic assay, plate cells in a 24-well plate and select them with antibiotics, such as G418, at 100, 200, 300, 400, 500, 600, 700, 800, and 1000 µg/mL for 2 wk. The lowest concentration that completely kills cells is optimal concentration for your selection.
2. Many expression vectors can be used for constructing the expression of the gene of interest; even the episomal expression vector, such as pCEP4 vector, can be used for in this assay.
3. Many colonies can be seen in the plates that are plated with the ligation reaction of pcDNA3-insert, whereas there are no colonies in the plates that are with either insert or vector alone.
4. Several methods can be used to transfect cells including electroporation, calcium chloride, Lipofectamine (Gibco/BRL) and Lipofectin. Because the transfection efficiency varies depending on cell types and transfected methods used, it is recommended that before staring the experiment, search the transfection literature from Gibco/BRL to determine which method is optimal for the cells you plan to use, since the transfection protocol have been documented in many cell lines.
5. Cells can be transfected in the presence of serum.
6. To count colonies, it is important to count the colonies that contain $\geq$ 50 cells, because colonies with less 50 cells may not be truly resistant colonies.

References

1. Kastan, M. B., Onyekwere, O., Sidransky, D., Vogelstein, B., and Craig, R. W. (1991) Participation of p53 protein in the cellular response to DNA damage. *Cancer Res.* **51**, 6304–6311.
2. Levine, A. J. (1997) p53, the cellular gatekeeper for growth and division. *Cell* **88**, 323–331.
3. El-Deiry, W. S. (1998) Regulation of p53 downstream genes. *Semin. Cancer Biol.* **8**, 345–357.
4. Blagosklonny, M. V. and El-Deiry, W. S. (1996) In vitro evaluation of a p53-expressing adenovirus as an anti-cancer drug. *Int. J. Cancer* **67**, 386–392.
5. Lowe, S. W. and Lin, A. W. (2000) Apoptosis in cancer. *Carcinogenesis* **21**, 485–495.
6. Macey, M. G. (1988) Flow cytometry: principles and clinical applications. *Med. Lab. Sci.* **45**, 165–173.
7. Carmichael, J., DeGraff, W. G., Gazdar, A. F., Minna, J. D., and Mitchell, J. B. (1987) Evaluation of a tetrazolium-based semiautomated colorimetric assay: assessment of chemosensitivity testing. *Cancer Res.* **47**, 936–942.
8. El-Deiry, W. S., Tokino, T., Velculesue, V. E., et al. (1993) WAF1, a potential mediator of p53 tumor suppression. *Cell* **75**, 817–825.
9. Sambrook, J., Fritsch, E. F., and Maniatis, T. (1989) *Molecular Cloning: A Laboratory Manual*, 3rd ed. Cold Spring Harbor Laboratory, Cold Spring Harbor, NY.

17

Flow Cytometric Analysis of Cell Cycle Control by Tumor Suppressor Genes

David T. Dicker and Wafik S. El-Deiry

1. Introduction

Determination of cell cycle distribution is an important tool in cell biology. Flow cytometric analysis of cellular DNA can provide rapid, quantitative information about DNA ploidy and cell cycle distribution. Presented here are three basic protocols for analysis of DNA by flow cytometry. There are many variations of these protocols that may be more appropriate for particular cell types or applications *(1)*.

2. Materials

2.1. Cell Cycle Analysis: PI Staining with or Without GFP Reporter Gene

1. Fetal bovine serum (FBS).
2. Phosphate-buffered saline (PBS).
3. Propidium iodide (PI) stock solution: 500 µg/mL propidium iodide, in H_2O.
4. RNAse stock solution: 10 mg/mL RNAse A, in H_2O, heat to 100°C for 15 min.
5. PI/RNAse staining solution: 300 µL PBS with 1% FBS + 30 µL PI stock solution + 8 µL RNAse solution.
6. Phosphate–citric acid buffer: 192 mL 0.2 M Na_2HPO_4 + 8 mL 0.1 M citric acid, pH 7.8.
7. Ethanol.

2.2. Hoechst 33342 Staining of Live Cells

1. Hoechst 33342 staining solution: Hoechst 33342, 1.0 mg/mL in dH_2O. Stock dye solution may be frozen and is stable for at least 1 mo at 4°C in the dark.
2. 3,3-Dipentyloxacarbocyanine iodide ($DiOC_5$ *[3]*): 1.0 mg/mL in dimethyl sulfoxide or verapamil 100 mM in dH_2O (*see* **Note 1**).

2.3. Bromodeoxyuridine Staining

1. Phosphate-buffered saline (PBS).
2. BrdU solution: 1 mM bromodeoxyuridine (BrdU) in PBS. BrdU is light-sensitive; store solution in the dark at –20°C.

From: *Methods in Molecular Biology, Vol. 223: Tumor Suppressor Genes: Regulation, Function, and Medicinal Applications.* Edited by: Wafik S. El-Deiry © Humana Press Inc., Totowa, NJ

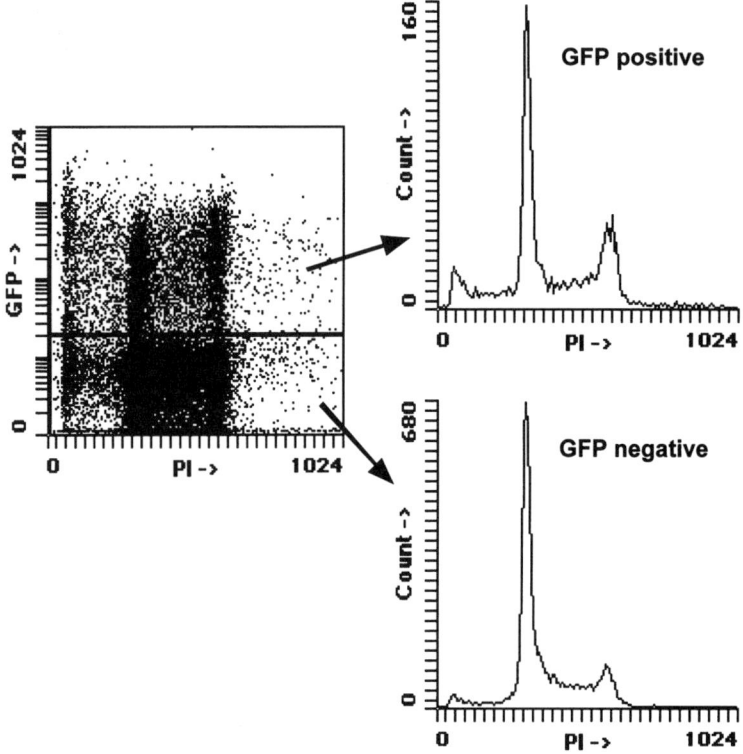

Fig. 1. SW480 cells co-tranfected with EGFP and pCEP4-FLIP.

3. PBS/BSA solution: 1% bovine serum albumin (w/v) in PBS.
4. Tween/PBS/BSA: 0.5% Tween-20 (v/v) in PBS/BSA.
5. HCl/Triton: 2 N HCl with 0.5% Triton X-100 (v/v).
6. Anti-BrdU from Becton-Dickinson (cat. no. 347580).
7. Goat F(ab)$_2$ anti-mouse IgG (H+L)-FITC: Dilute 1:5 in PBS, final concentration = 140 μg/mL.
8. PI solution: 5 μg/mL propidium iodide (PI) in H$_2$O.

3. Methods

3.1. Cell Cycle Analysis: PI Staining with or Without GFP Reporter Gene

Propidium iodide has long been used to analyze DNA content in cells (2). Quantitative analysis of DNA content in cells can be used to calculate cell cycle distributions in populations of cells. The addition of a GFP reporter gene allows the researcher to discriminate between transfected and nontransfected cells.

If using EGFP reporter gene, start with **step 1**. Otherwise, skip this step.

1. When transfecting your cells using your favorite method, co-transfect pEGFPN-Spectrin (3) as a 1/10 fraction of the total DNA. Allow at least 8 h before analysis, so sufficient EGFP can be synthesized.
2. Harvest ~10^6 cells; if cells are adherent, be sure to collect cells that have become detached.
3. Spin down cells and wash in 2 mL PBS/1% FBS.
4. Spin down and resuspend in 1 mL PBS.

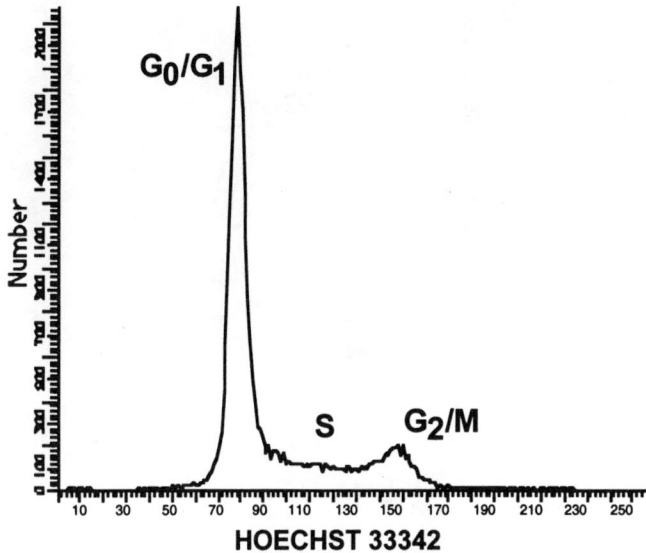

Fig. 2. SW480 cells stained with Hoechst 33342/ DiOC$_5$.

5. Add 9 mL cold (4°C) EtOH while vortexing; let stand at 4°C at least 20 min, protected from light (cells can be stored indefinitely at this stage at 4°C).
6. Spin down cells and wash in 4 mL PBS/1% FBS.
7. Spin down and resuspend in 1 mL PBS.
8. Add 0.5 mL phosphate–citric acid buffer, let stand at room temperature for 5 min.
9. Spin down and resuspend in 300 μL PI/RNase staining solution.
10. Incubate at 37°C for 15 min or at room temperature for 30 min.
11. Analyze by flow cytometry.

Figure 1 shows SW480 cells co-transfected with EGFP and pCEP4-FLIP

3.2. Hoechst 33342 Staining of Live Cells

The use of the permeant DNA dye Hoechst 33342 allows analysis of live cells. Some cells require the drug efflux pump blocked (using verapamil or DiOC$_5$) to obtain stoichiometric staining *(4)*. Hoechst 33342 is a UV-excited fluorochrome and therefore requires a cytometer with a UV source, such as a UV laser or arc lamp.

1. Hoechst 33342 staining solution is added to ~10^6 cells in culture medium at 1.0–5.0 μg/mL.
2. DiOC$_5$ is added at the same time at 0.1–0.3 μg/mL, or verapamil at 100 μ*M* (*see* **Note 2**).
3. Adherent cells should be stained while attached.
4. Incubate cells at 37°C for 30–60 min.
5. Harvest cells if adherent, trypsin and trypsin-neutralizing solutions used should contain the same dye concentration as above.
6. Cells are analyzed without washing, while in the media containing the Hoechst 33342 staining solution.
7. Analyze by flow cytometry using UV excitation.

Figure 2 shows SW480 cells stained with Hoechst 33342/DiOC$_5$.

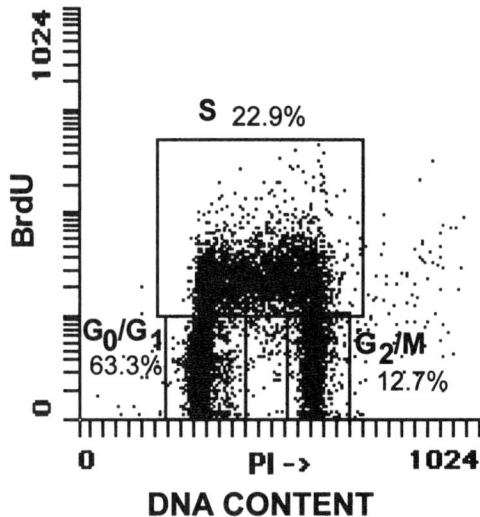

Fig. 3. SW480 cells assayed using the BrdU protocol.

3.3. Bromodeoxyuridine Staining (see Note 3)

Detection of replicating cells can be achieved by incubating cells with the thymidine analog 5-bromo-2-deoxyuridine (BrdU), and subsequent immunostaining with antibodies developed against BrdU *(5)*. Simultaneous measurement of DNA content can provide additional information to the investigator *(6)*.

1. Add BrdU solution directly to the culture medium at 10 μL/mL of medium (final concentration 10 μ*M*).
2. Incubate cells for 30 min in 37°C incubator.
3. Harvest ~10^6 cells and wash cells twice with PBS/BSA solution.
4. Resuspend the cell pellet in 200 μL cold PBS.
5. Add cells slowly, while vortexing, to 5 mL ice-cold 70% ethanol.
6. Incubate cells for at least 30 min on ice or at −20°C. (The fixed cells are now stable and can be stored in the dark at –20°C.)
7. Spin down cells, remove ethanol, and loosen cell pellet.
8. Add 1 mL HCl/Triton solution slowly, while vortexing.
9. Incubate at room temperature for 30 min.
10. Spin down cells and wash in PBS/FBS solution.
11. Resuspend cells in 1 mL Tween/PBS/BSA solution.
12. Add 20 μL of anti-BrdU antibody to each tube mix and incubate at room temperature for 30 min.
13. Spin down cells and resuspend in 50 mL Tween/PBS/BSA solution.
14. Add 7 μL of the diluted goat F(ab)$_2$ anti-mouse IgG-FITC, mix, and incubate 30 min at room temperature.
15. Spin down and resuspend in 0.5–1.0 mL PI solution.
16. Analyze by flow cytometry or store in the dark at 4°C until analyzed (stable for up to 1 wk). Figure 3 shows SW480 cells assayed using the BrdU protocol.

4. Notes

1. DiOC$_5$ *(3)* or verapamil is added to prevent cells from actively pumping out the Hoechst; it may not be necessary in all cell types.
2. DiOC$_5$ fluoresces green and therefore cannot be used in conjunction with other green fluorescent dyes.
3. This procedure should be done in minimal light.

References

1. Darzynkiewicz, Z. and Juan, G. (1997) Nucleic acid analysis, in *Current Protocols in Cytometry* (Robinson, J. P., Darzynkiewicz, Z., Dean, P. N., et al., eds.). Wiley, New York, pp. 7.0.1–7.13.24.
2. Crissman, H. A. and Steinkamp J.A. (1973) Rapid simultaneous measurement of DNA, protein and cell volume in single cells from large mammalian cell populations. *J. Cell Biol.* **59**, 766.
3. Kalejta, R. F., Shenl, T., and Beavis, A. S. (1997) Use of a membrane-localized green fluorescent protein allows simultaneous identification of transfected cells and cell cycle analysis by flow cytometry. *Cytometry* **29**, 286–291.
4. Krishan, A. (1987) Effect of drug efflux blockers on vital staining of cellular DNA with Hoechst 33342. *Cytometry* **8**, 642.
5. Gratzner, H. G. (1982) Monoclonal antibody to 5-bromo-2-deoxyuridine: a new reagent for the detection of DNA replication *Science* **218**, 747–748.
6. Dolbeare, F., Gratzner, H. G., Pallvicini, M., and Gray, J. W. (1983) Flow cytometric measurement of total DNA content and incorporated bromodeoxyuridine. *Proc. Natl. Acad. Sci. USA* **80**, 5573–5577.

18

Analysis of Cyclin-Dependent Kinase Activity

Timothy K. MacLachlan and Wafik S. El-Deiry

1. Introduction

The number of kinases, enzymes whose purpose is to phosphorylate other proteins on serine, threonine, or tyrosine residues, in mammalian cells is presently known to number in the several hundreds or even thousands *(1,2)*. This particular posttranslational modification can dramatically affect the overall function of the affected protein, and is a very tightly regulated process. Therefore, it is no surprise that many avenues of research focus at least partially on the alterations in activity of certain kinases. The cell cycle kinases in particular fluctuate in maximal phosphotransfer ability with respect to the cell cycle. The five cyclin-dependent kinases (CDKs) that possess the major cell cycle phase-transition activities have specific areas of the cell cycle that require their kinase abilities—CDKs 3, 4, and 6 are active during G1 phase, CDK2 functions during the G1-to-S transition and S phase, and cdc2 (also known as cdk1) is essential for the G2-to-mitosis transition *(3)*. In vivo, each kinases activity is affected by the binding of its activating subunit, the cyclin, but also by modifying phosphorylation on the kinase itself. Several lines of evidence have suggested that the specific inhibition of individual CDKs by chemical agents is effective in arresting the cell cycle—a line of research that has spawned the design of such chemicals for therapy in diseases with abnormal cell proliferation such as cancer *(4–6)*. Assays that detect the activity of endogenous CDK activity are therefore needed to test chemical candidates for such therapy. Here we present a protocol for analysis of kinase activity for cyclin-dependent kinases using general substrates that will detect the broad activity of cyclin-dependent kinases (**Fig. 1**).

2. Materials

1. 3×10^6 cells treated (or untreated) as necessary
2. Lysis buffer: 50 m*M* Tris-HCl, pH 7.4, 0.1% Triton X-100, 5 m*M* EDTA, 250 m*M* NaCl, 50 m*M* NaF, 0.1 m*M* Na$_3$VO$_4$, Protease inhibitors (1 m*M* phenyl-methylsulfonyl fluoride [PMSF] and 10 µg/mL leupeptin, or a 1:100 dilution of protease inhibitor cocktail [Sigma cat. no. P8340]).

From: *Methods in Molecular Biology, Vol. 223: Tumor Suppressor Genes: Regulation, Function, and Medicinal Applications.* Edited by: Wafik S. El-Deiry © Humana Press Inc., Totowa, NJ

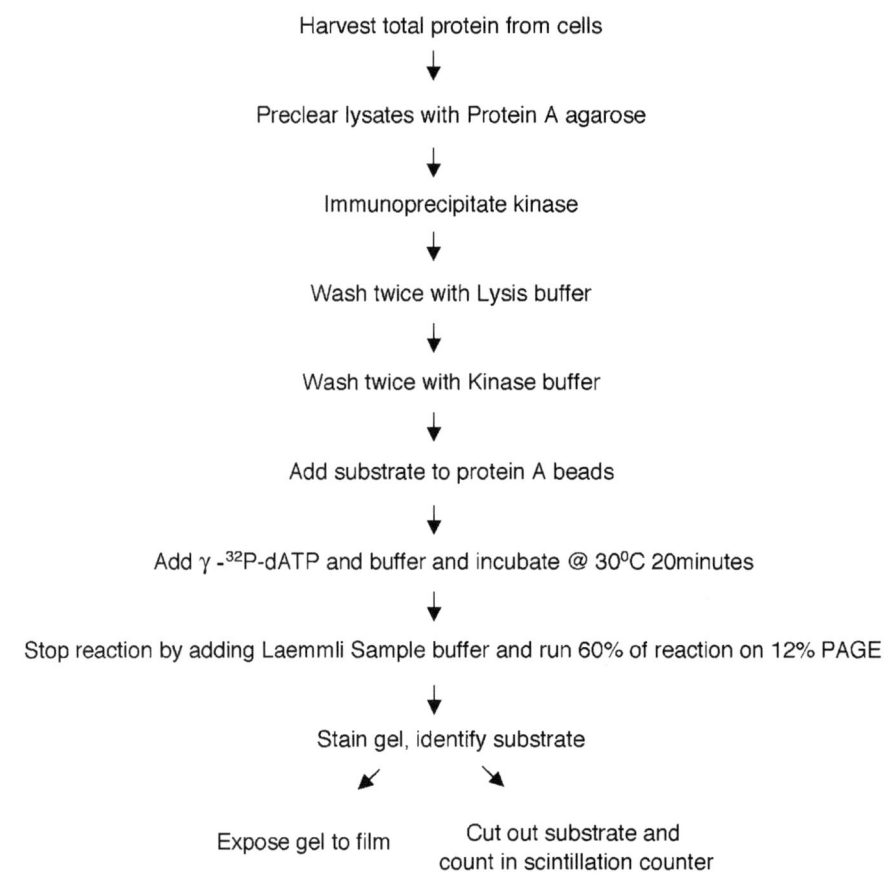

Fig. 1. Flowchart of analysis of cyclin-dependent kinase activity.

3. Bradford reagent for protein quantification (Sigma cat. no. 6916).
4. Kinase buffer: 20 mM HEPES, pH 7.4, 10 mM magnesium acetate.
5. 25 μCi γ-^{32}P-dATP per reaction.
6. 10× Assay buffer: 100 mM magnesium acetate, 200 μM adenosine triphosphate, 10 mM dithiothreitol (stored in 500-mL aliquots at –80°C).
7. Reaction mix: (2.5 × a) μL 10× assay buffer [(17.5 × a] – a) μL dH$_2$O (25 μCi × a) γ-^{32}P-dATP, where a is the number of samples.
8. Phosphate-buffered saline (PBS).
9. Appropriate antibody (for cdk2 assays, Santa Cruz anti-cdk2 [cat. no. M2-G]; for cyclinB1-cdc2 assays, Santa Cruz anti-cyclin B1 [cat. no. GNS1]; for cdk4 assays, Santa Cruz anti-cdk4 [cat. no. H-22]).
10. Protein A agarose (Sigma cat. no. P3391) in lysis buffer as a 50% solution once completely swelled.
11. Substrate (for cdk2 and cyclin B1 assays, Histone H1 Roche Molecular Biochemicals cat. no. 0223549; for cdk4 assays) (Rb Santa Cruz cat. no. sc-4112). Resuspend Histone H1 in kinase buffer to 1 mg/mL, prepare 50-mL aliquots in 1.5-mL tubes and store at −80°C
12. Laemmli sample buffer: 125 mM Tris-HCl (pH 6.8), 1.43 M β-mercaptoethanol, 4% sodium dodecyl sulfate (SDS), 20% glycerol, 0.05% bromophenol blue.
13. 12% SDS polyacrylamide gel.

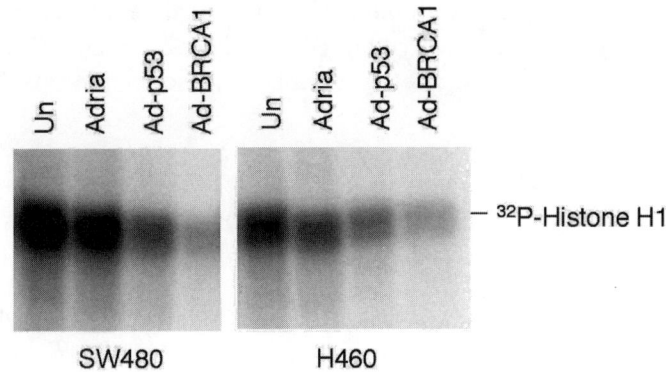

Fig. 2. SW480 (mutant p53) and H460 (wild-type p53) cells were treated with adriamycin (0.2 µg/mL) or infected with 50MOI of the indicated adenoviruses or left untreated. Twenty-four hours later, protein lysates were immunoprecipitated with anti-cdk2 antibodies. Immunoprecipitates were run through a histone H1 kinase assay and electrophoresed through a 12% polyacrylamide gel.

3. Methods

1. Start with one T25 flask of cells (~3 × 10⁶ cells)—this should be enough for one IP/kinase assay. One T75 can be good for 3–4 assays.
2. Aspirate medium, wash once with PBS.
3. Add 300 µL of lysis buffer, shake flask to distribute equally, place flask on ice for 20 min.
4. Scrape cells from flask and place lysate in 1.7-mL tube.
5. Spin lysate for 10 min at 4°C.
6. Transfer cleared lysate to new 1.7-mL tube and quantitate protein by Bradford analysis. All steps from here on should be on ice.
7. Add 100 µg of protein to new 1.7-mL tube and bring up the volume to 300 µL with lysis buffer.
8. Preclear lysate by adding 30 µL of protein A sepharose and rock end over end at 40°C for 1 h.
9. Spin lysate and transfer supernatant to a new 1.7-mL tube.
10. Add appropriate amount of antibody (for Santa Cruz anti-cdk2 [M2-G]), –cyclin B1 [GNS1], or anti-cdk4 [#H-22], 5 mL is enough) and rock end over end at 4°C for 2 h or overnight. Include a negative control—normal mouse/rabbit serum or no antibody.
11. Add 30 µL of protein A sepharose and rock end over end for no more than 1 h.
12. Spin down agarose and aspirate the supernatant. Wash agarose beads twice with lysis buffer.
13. Wash twice with kinase buffer.
14. Add 1–2 µg of substrate and mix with beads with pipet tip.
15. Add 20 µL reaction mix, gently mix with pipet tip, and place immediately at 30°C for 20 min.
16. Stop the reaction by addition of 30 µL Laemmli sample buffer and vortex.
17. Spin down the Sepharose and add 30 µL of reaction to a 12% SDS-PAGE gel.
18. Run gel until the front is about an inch from the bottom. Cut off the front and the gel below it and throw it away—this contains all the unused free ³²P and is *very* radioactive.
19. Stain the gel in Coomassie blue and destain to identify where the substrate is. Dry the gel on blotting paper and expose to film (start with an hour exposure, then go from there).
20. The band intensity can be quantified by phosphorimaging, densitometry from film (*see* **Fig. 2**), or by cutting out the substrate band (as identified from staining) and counting in a scintillation counter.

The methods described here were culled from several protocols and arranged as necessitated for analyzing kinase activity of cdk2 and cdc2 and cdk4/6 directed toward histone H1 and pRb, respectively. These protocols are described in **refs. 7–11.**

References

1. Baltimore, D. (2001) Our genome unveiled. *Nature* **409,** 814–816.
2. Venter, J. C., Adams, M. D., Myers, E. W., et al. (2001) The sequence of the human genome. *Science* **291,** 1304–1351.
3. MacLachlan, T. K., Sang, N., and Giordano, A. (1995) Cyclins, cyclin-dependent kinases and cdk inhibitors: implications in cell cycle control and cancer. *Crit. Rev. Eukaryot. Gene Expr.* **5,** 127–156.
4. Gray, N., Detivaud, L., Doerig, C., and Meijer, L. (1999) ATP-site directed inhibitors of cyclin-dependent kinases. *Curr. Med. Chem.* **6,** 859–875.
5. Meijer, L., Leclerc, S., and Leost, M. (1999) Properties and potential-applications of chemical inhibitors of cyclin-dependent kinases. *Pharmacol. Ther.* **82,** 279–284.
6. Noble, M. E. and Endicott, J. A. (1999) Chemical inhibitors of cyclin-dependent kinases: insights into design from X-ray crystallographic studies. *Pharmacol. Ther.* **82,** 269–278.
7. Giordano, A., Lee, J. H., Scheppler, J. A., et al. (1991) Cell cycle regulation of histone H1 kinase activity associated with the adenoviral protein E1A. *Science* **253,** 1271–1275.
8. Giordano, A., Whyte, P., Harlow, E., Franza, B. R., Jr., Beach, D., and Draetta, G. (1989) A 60 kd cdc2-associated polypeptide complexes with the E1A proteins in adenovirus-infected cells. *Cell* **58,** 981–990.
9. Grana, X., De Luca, A., Sang, N., et al. (1994) PITALRE, a nuclear CDC2-related protein kinase that phosphorylates the retinoblastoma protein in vitro. *Proc. Natl. Acad. Sci. USA* **91,** 3834–3838.
10. Koff, A., Giordano, A., Desai, D., et al. (1992) Formation and activation of a cyclin E-cdk2 complex during the G1 phase of the human cell cycle. *Science* **257,** 1689–1694.
11. Stewart, N. T., Byrne, K. M., Hosick, H. L., Vierck, J. L., and Dodson, M. V. (2000) Traditional and emerging methods for analyzing cell activity in cell culture. *Meth. Cell Sci.* **22,** 67–78.

19

Tumor Suppressor Gene-Inducible Cell Lines

Michael Dohn, Susan Nozell, Amy Willis, and Xinbin Chen

1. Introduction

In a noncancerous mammalian cell, the growth-promoting effects of proto-oncogenes are counterbalanced by the growth-constraining effects of tumor suppressor genes (TSGs). The net result is a masterfully orchestrated display of cell proliferation in the absence of tumorigenesis. The significance of this relationship is most evident in tumor cells, where the accumulation of genetic mutations has upset this critical balance of power.

While it is clear how oncogenes contribute to the process of tumorigenesis, it is less well understood how TSGs function to inhibit tumorigenesis. Therefore, for those who study TSGs, two long-standing goals have been to understand the normal physiologic function of TSGs and to understand how the loss of TSGs contributes to the development of tumors. Historically, the function of a gene has been studied in cells or animals that lack the functional protein encoded by that gene. However, because many human tumors have already accomplished this, one must approach the investigation from a different perspective.

To fully elucidate the roles of TSGs in the control of proliferation and transformation in normal cells, we must first understand the effects of TSG expression in cancerous cells. Typically, TSGs are activated in response to cellular stresses, including hypoxia, nutrient depletion, UV, ionizing radiation, and DNA damage. TSG activation often results in cell cycle arrest or apoptosis, which prevents the passage of stress-induced genetic lesions to future generations. Therefore, these conditions can be utilized to activate TSGs in vivo and observe their effects on cell cycle arrest, DNA repair, differentiation, and apoptosis. However, stimuli that commonly activate TSGs can have broader side effects on other proteins and cellular processes. For example, the culture of cells under hypoxic conditions increases the activity of the p53 tumor suppressor protein; however, hypoxic conditions can also activate vascular endothelial growth factor (VEGF), a potent mitogen known to initiate angiogenesis *(1).* In cases such as these, the effects of TSG expression become muddled with pleiotropic effects of the activation conditions. Additionally, it is not yet clear how some tumor suppressors, such as PTEN, become activated in vivo. Ideally, we need to be able to activate TSGs artificially, without also affecting other cellular events. One approach to avoiding pleiotropic effects is

From: *Methods in Molecular Biology, Vol. 223: Tumor Suppressor Genes: Regulation, Function, and Medicinal Applications.* Edited by: Wafik S. El-Deiry © Humana Press Inc., Totowa, NJ

the use of inducible cell lines. Cells in culture can be stably transfected with inducible TSG expression vectors such that the addition or removal of a drug or hormone from the culture medium induces expression of the gene of interest. In this manner, TSGs can be turned on and off at will. In addition, the level of gene expression can be varied by altering the concentration of the inducing agent. Inducible expression systems also allow for the regulation of potentially toxic gene products that could be detrimental to cell growth if expressed constitutively. Furthermore, with dual expression systems, the cooperative effects of simultaneous expression of two TSGs can be examined. Inducible expression vectors under control of different inducing agents can be stably co-transfected into cells, and the addition or removal of the two inducing agents can produce conditions in which one, both, or neither TSG is expressed. This chapter focuses on three different inducible expression systems: tetracycline-off and -on systems, ecdysone-inducible systems, and estrogen receptor fusion protein systems. Additional inducible systems will be discussed in lesser detail.

1.1. Tetracycline (Tet)-off and Tet-on Systems

The first system for controlling gene expression, and one that is still widely used today, is the Tet-off system. After observing that the tetracycline resistance operon of the bacterial transposon *Tn-10* was capable of regulating eukaryotic gene expression, a modified version soon became available for use in mammalian cells *(2)*. By fusing the DNA-binding domain of the *Escherichia coli* tetracycline repressor (tetR) protein to the activation domain of the herpesvirus VP16 protein, the tetracycline transactivator (tTA) was born. Therefore, by positioning a target gene downstream of tet operator (tetO) sequences, one can control target gene expression through the addition or removal of tetracycline (**Fig. 1A**). In the presence of tetracycline (Tet) or the tetracycline-derivative doxycycline (Dox), the ability of tTA to bind DNA is abolished and the target gene is not expressed. However, in the absence of Tet or Dox, the tTA chimera can activate transcription of target genes by binding to the tetO sequences *(3–5)*.

The Tet-off system is advantageous for several reasons. Perhaps most appealing is the low level of background expression while in the "off" state. This is due to the use of tetO sequences, which are not native to mammalian cells and are therefore not targets of endogenous mammalian transcription factors. Additionally, one can control the level of gene expression by varying the amount of Tet or Dox present in the medium (*see* **Notes 1** and **2**). This allows the user to express the gene of interest at extremely low to extremely high levels in order to investigate the dose-dependent effects of TSG function. This is particularly helpful when studying genes whose expression may be lethal or cytotoxic, and hence must be studied using lower expression levels (*see* **Note 3**). Moreover, this system is advantageous in that neither Tet nor Dox interfere with mammalian physiology *(3,5)*.

However, there are some particulars that may be unfavorable to the user. First, some may find the persistent application of Tet or Dox to maintain silenced gene expression inconvenient. Second, induction of the target gene is determined by how efficient Tet or Dox is cleared from the environment (*see* **Notes 2** and **4**). Therefore, a second tetracycline-regulated system has been created to counter these points.

The Tet-on system was created through chemical mutagenesis and selection of a mutant tetracycline repressor protein that would bind to tetO sequences only in the pres-

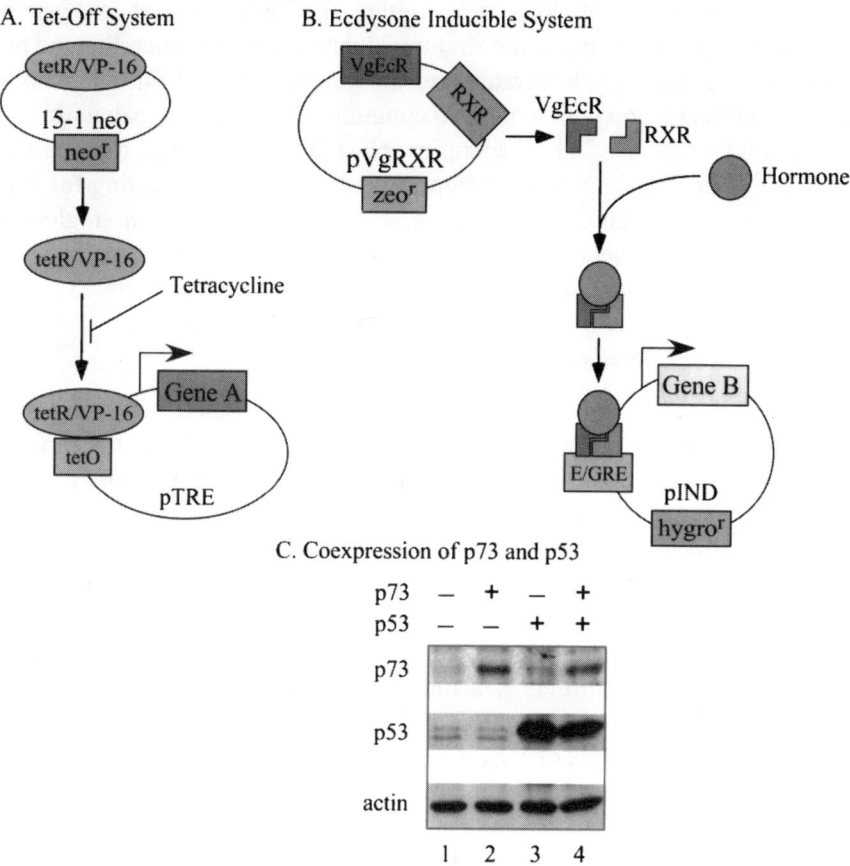

Fig. 1. Dual-inducible expression system. (**A**) Tetracycline-repressible system. (**B**) Ecdysone-inducible system. (**C**) Coexpression of p73 and p53. Cell extracts were purified from HTE-20-p53-p73-1 cells, which were uninduced (lane 1), induced to express p73 by adding hormone ponasterone in the culture medium (lane 2), induced to express p53 by withdrawal of tetracycline from the culture medium (lane 3), and induced to express both p73 and p53 (lane 4). The levels of p73, p53, and actin were assayed by Western blot analysis.

ence of Tet or Dox. When fused to the VP16 protein, a functional "reverse transactivator," or rtTA, was generated. Like its predecessor, this system offers the advantage of using the non-native operator sequence to keep background expression levels low. However, in this system induction of the target gene is only accomplished when Tet or Dox is present. For some, this system is ideal because it requires administration of Tet or Dox only when gene expression is desirable *(3–5)*.

Perhaps the simplest of the Tet-on systems is one that employs the native *E. coli* tetracycline repressor (TetR) protein (Invitrogen). In this system, TetR represses transcription through binding to strategically placed TetO sites upstream of the gene of interest. In the presence of Tet or Dox, the TetR binds Tet and undergoes a conformational change rendering it unable to bind the TetO sequences. This effectively reverses the transcriptional repression and allows transactivation of the gene of interest. While similar to the above Tet-on system that utilizes VP16 fusion proteins, this model does not use viral transactivators or components.

Typically, the generation of cell lines using either the Tet-off or Tet-on systems requires two stable transfection steps. In the first transfection, the gene encoding the relevant regulatory protein randomly integrates into the genome of the host cell. This is followed by a second transfection with a plasmid containing the target gene downstream of the regulatory tetO sequences. The generation of stable cell lines is a rewarding process but one that does take time, as each step can require weeks of waiting for cells to grow. As such, some have reported generating stable cell lines using a single-step transfection. However, many users of these systems advise performing the two-step selection process. The experimental design for this process is described below.

1.2. Ecdysone-Inducible System

Another system that allows for the control of gene expression is the ecdysone-inducible system (**Fig. 1B**). Ecdysone is an insect hormone that mediates the molting process through interactions with the ecdysone receptor and the USP protein *(6,7)*. When bound by ecdysone, the ecdysone receptor/USP heterodimer transactivates genes through ecdysone responsive elements *(6,7)*. This system has been modified for ecdysone-inducible gene expression in mammalian cells *(8)*. To establish cell lines that inducibly express a gene of interest, cells are co-transfected with two vectors. One vector is an inducible vector, pIND, that contains the gene to be inducibly expressed. The other vector, pVgRXR, expresses two proteins, a modified *Drosophila* ecdysone receptor (VgEcR) *(9–13)* and the mammalian homolog of the USP gene product, the retinoid X receptor (RXR) *(6,7)*. In the presence of ecdysone or one of its analogs (ponasterone A or muristerone A), VgEcR and RXR interact to form an active heterodimer that binds to the hybrid ecdysone/glucocorticoid-responsive element (E/GRE) in the inducible pIND vector *(8)*. The ecdysone receptor recognizes one half-site of the glucocorticoid-responsive element, while the retinoid X receptor recognizes one half-site of the ecdysone-responsive element. Because the responsive element is a hybrid, it prevents activation of transcription by any endogenous gene product that would bind the natural response element *(8)*, such as the farnesoid X receptor *(14)*. Therefore, expression is prevented in the absence of ecdysone, and basal levels of expression are very low *(15,16)* (*see* **Notes 5** and **6**). In addition, ecdysone has no known physiologic effects on mammalian cells *(8)*.

1.3. Estrogen Receptor Fusion Protein Systems

While the tetracycline- and ecdysone-inducible systems regulate gene activity at the level of transcription, the estrogen receptor (ER) fusion protein system involves the post-translational regulation of constitutively expressed genes. ERs are nuclear transcription factors that dimerize upon binding of estrogens or synthetic mimetics (reviewed in **ref. 13**). ER dimerization allows the transcription factors to interact directly with specific DNA sequences within the promoters of estrogen-responsive genes *(17)*. In the absence of estrogen, ERs are complexed with a variety of proteins, including Hsp90, that appear to inhibit ER activity *(18,19)*. Estrogen binding releases ER proteins from these complexes, thus reversing the steric inhibition and activating the ER dimers.

The ER-inducible expression system takes advantage of the ER–multiprotein complex that forms in the absence of estrogen. This system utilizes a fusion protein comprised of the hormone-binding domain of the ER fused to the protein of interest. In the absence of estrogens, the sequestration of these fusion proteins into multiprotein complexes inhibits any activity normally exhibited by the fused gene product. Upon treatment with estrogens, ER fusion proteins are released from this complex, and the activity of the fused gene product is then restored. In this manner, activation of a particular protein can be hormonally dependent.

Hormone-binding domains of steroid receptors have been used routinely to regulate heterologous proteins *(20)*, but modifications to and the wide use of the ER fusion protein system have rendered it most valuable. The most apparent obstacle to the original ER-inducible system was the presence of estrogens and ER agonists in serum and conventional culture media *(21)*. This drawback was overcome with the use of a mutant ER that is unable to bind estradiol but is responsive to the synthetic steroids tamoxifen and 4-hydroxytamoxifen *(22,23)*. This modification allowed for more stringent regulation of the ER fusion protein. Additionally, the lack of endogenous estrogen receptors in the majority of cell types makes the use of ER fusion proteins an attractive system for inducibly regulating protein activity. Consequently, the activities of numerous transcription factors, including p53 *(24,25)* and c-myc *(22,26)*, have been regulated successfully as ER fusion proteins.

1.4. Additional Inducible Systems

In addition to the tetracycline-, ecdysone-, and ER-inducible expression systems, other available systems using inducible promoter elements have been shown to be effective. The metallothionein (MT) promoter has been used quite extensively in inducible expression systems. The MT promoter contains metal-responsive elements *(27)*, and transcription of MT genes is regulated by the metal ions cadmium and zinc *(28)*. In an MT-inducible system, a gene placed downstream of the MT promoter is expressed when cadmium or zinc chlorides are added to the culture medium. Expression from this promoter can be regulated in a temporal and dose-dependent manner, and the absence of metal ions in most culture medium maintains low basal activity. A disadvantage of the metal ion-inducible system is that in addition to metal ions, certain hormones and cytokines are capable of regulating transcription from MT promoter elements *(29–31)*, including the transcription factor Sp1 *(32)*. Furthermore, MT genes have been found to be activated in response to various forms of oxidative stress, which indicates that induction of the MT promoter may be more complex than originally thought *(29)*. Nevertheless, inducible expression systems employing MT promoter elements can be effective under the proper culture conditions.

Another available inducible expression system utilizes a dexamethasone-inducible promoter present in the mouse mammary tumor virus long terminal repeat (MMTV LTR). This system generally uses the pMSG plasmid (Amersham Pharmacia), which confers resistance to mycophenolic acid by expressing the *E. coli* xanthine-guanine phosphoribosyltransferase *(gpt)* gene *(34)*. Expression from the MMTV promoter requires that cells are hormonally responsive, and internal homologies within the pMSG vector require that the bacterial strains used with this plasmid are recombination-negative *(recA13)*. These limitations on compatible cell lines and bacterial strains make the dexamethasone-inducible system less attractive than others, but it is still a valuable method of inducible expression.

A less commonly used inducible expression system involves isopropyl β-D-thiogalactopyranoside (IPTG). IPTG is a synthetic inducer of the *lac* operon and functions by inhibiting the Lac repressor. Although it was originally designed for inducible expression in bacteria, this system has been adapted for use in mammalian cells *(35,36)*. The IPTG-inducible system includes a eukaryotic Lac repressor-expressing vector, as well as a eukaryotic *lac* operator-containing vector into which the gene of interest is inserted. In the absence of IPTG, the constitutively expressed Lac repressor binds the *lac* operator elements and inhibits transcription of the downstream gene of interest. With the addition of IPTG, the Lac repressor loses its ability to inhibit the *lac* operator, thus allowing transcription of the downstream gene of interest. This expression system exhibits low basal levels of expression, and IPTG is nontoxic to mammalian cells and permits induction within 4–8 h.

1.5. Dual Inducible Systems

To examine potential cross-talk between TSGs, it is necessary to analyze the effects of direct and indirect interactions between two proteins. In the past, in order to study these types of interactions, a cell line that endogenously expresses one gene was transiently transfected with a vector that expressed a second gene; alternatively, two vectors were transiently co-transfected. However, transient transfections do not provide consistent levels of gene expression, which could generate misleading results regarding interactions between, or cooperative effects of, the two gene products.

To overcome this obstacle, a system that allows for stable, inducible expression of two genes within the same cell line is needed. This requires the generation of a stable cell line that constitutively expresses both a tTA-expressing vector of the tetracycline-regulated system and the pVgRXR vector of the ecdysone-inducible system (**Fig. 1**). This cell line is then stably transfected with the inducible vectors for these two systems, each constructed to express the genes to be studied. Therefore, removing or adding tetracycline or ponasterone A can induce expression of one, both, or neither gene (*see* **Note 7**). This dual-expression system is advantageous in studying interplay between gene products such as dominant negative effects, activation or repression of one protein by another, or physical interaction between two proteins.

2. Materials
2.1. Plasmids for Tetracycline-Regulated Systems
2.1.1. ClonTech System (www.clontech.com or 800-662-2566)

1. Plasmids encoding the tetracycline regulatory protein
   a. pTet-off Vector System (cat. no. K1620-A): This vector encodes the tetracycline activator that stimulates transactivation from TetO-regulated sites in the absence of Tet or Dox. This vector contains the neomycin cassette for use as a selectable marker.
   b. pTet-on Vector System (cat. no. K1621-A): This vector encodes the tetracycline repressor (rtTA) that inhibits transactivation from TetO-regulated sites in the absence of Tet or Dox. This vector contains the neomycin cassette for use as a selectable marker.
2. Plasmids for tetracycline-regulated target gene expression
   a. pTRE Vector (sold with Tet-off and Tet-on systems): This vector is included in both the Tet-off and Tet-on systems and does not contain a selectable marker. As such, a second plasmid containing a selectable marker may be co-transfected with pTRE.
   b. pBI Vector (cat. no. 6152-1): This vector is suitable for the expression of two independent genes through the use of independent multiple cloning sites, each downstream of a common tetO promoter sequence.

c. pBI Vector derivatives
  i. pBI-EGFP Tet Vector (cat. no. 6154-1): This vector contains the enhanced green fluorescent protein (EGFP) as a reporter gene.
  ii. pBI-G Tet Vector (cat. no. 6150-1): This vector contains the β-galactosidase gene as a reporter gene.
  iii. pBI-L Tet Vector (cat. no. 6151-1): This vector contains the firefly luciferase gene as a reporter gene.

2.1.2. Invitrogen System (www.invitrogen.com or 800-955-6288)

1. A plasmid encoding the tetracycline regulatory protein
   a. pcDNA6/TR (cat. no. V1025-20): This vector encodes the tetracycline repressor (TetR) that inhibits transactivation from TetO-regulated sites in the absence of Tet or Dox. This vector contains the blasticidin cassette for use as a selectable marker.
2. Plasmids for tetracycline-regulated target gene expression. Each of these vectors offers the following advantages: large multiple cloning site for cloning ease; complete CMV promoter-enhancer sequence containing two copies of the TetO sequence for maximal levels of expression; drug-resistance cassettes for use as selectable markers (either zeocin or hygromycin).
   a. pcDNA4/TO (cat. no. V1020-20) (zeocin)
   b. pcDNA4/TO/myc-His A, B, & C (cat. no. V1025-20) (zeocin): This vector is a derivative of the pcDNA4/TO vector and contains a copy of the c-myc epitope that allows the user to detect protein expression using an anti-Myc antibody. Additionally, the inclusion of a polyhistidine (6xHis) sequence allows the user to purify proteins using Invitrogen ProBond resin.
   c. pcDNA5/TO (cat. no. V1033-20) (hygromycin)

2.2. Plasmids for the Ecdysone-Regulated System from Invitrogen (www.invitrogen.com or 800-955-6288)

1. Plasmid encoding the ecdysone regulatory proteins
   a. pVgRXR: This vector expresses both the modified ecdysone receptor (VgEcR) and the retinoid X receptor (RXR), which, upon hormone binding, interact to form the heterodimeric nuclear receptor capable of binding to the hybrid ecdysone/glucocorticoid responsive element (E/GRE). The cytomegalovirus (CMV) enhancer-promoter drives expression of VgEcR, and the Rous sarcoma virus (RSV) promoter drives expression of RXR. This vector contains the zeocin cassette for use as a selectable marker.
2. Plasmids for ecdysone-inducible target gene expression. Each of these vectors offers the following advantages: multiple cloning sites for cloning ease; five hybrid E/GRE-binding sites for the modified ecdysone receptor; minimal heat-shock promoter for inducible transcription of the recombinant gene; drug-resistance cassettes for use as selectable markers.
   a. pIND (cat. no. V705-20): This vector contains a neomycin cassette for use as a selectable marker.
   b. pIND(SP1) (cat. no. V700-20): This vector is a derivative of the pIND vector and contains 3 *cis*-acting SP1 elements. SP1 transcription factors that are endogenous to mammalian cells interact with the SP1 elements and increase inducible expression by fivefold. However, basal expression levels are also greater with pIND(SP1) than with pIND.
   c. pIND/Hygro (cat. no. V725-20): This vector is identical to the pIND but contains a hygromycin cassette for use as a selectable marker.
   d. pIND(SP1)/Hygro (cat. no. V715-20): This vector is identical to the pIND(SP1) but contains a hygromycin cassette for use as a selectable marker.
   e. pIND/V5-His A, B, C (cat. no. V720-20) (neomycin): This pIND derivative encodes the V5 epitope and polyhistidine (6xHis) tag downstream from the multiple cloning site for

expression of fusion proteins. The C-terminal V5 epitope allows for simple detection and analysis of your fusion protein with anti-V5 antibody, and the 6xHis sequence allows rapid purification using Invitrogen ProBond resin.

   f. pIND(SP1)/V5-His A, B, C (cat. no. V710-20) (neomycin).

   g. pIND/V5-His TOPO TA Cloning Kit (cat. no. K1010-01) (neomycin): This vector is a derivative of pIND/V5-His and is designed for use with the Invitrogen TOPO TA Cloning Kit. This vector contains 3' T-overhangs for direct ligation of Taq-amplified PCR products.

2.3. Choice of Cell Line

2.3.1. Commercially Available Cell Lines

A number of commercially available cell lines are ready for use with either ClonTech or Invitrogen systems. These offer the user the advantage of beginning this process at the second stable transfection step.

1. Tetracycline-regulated system: ClonTech (www.clontech.com or 800-662-2566) has available a variety of Tet-On and Tet-Off cell lines that stably express rtTA or tTA, respectively. A few selected cell lines are shown below.
   a. CHO-AA8 Tet-Off (cat. no. C3004-1).
   b. HeLa Tet-Off1 (cat. no. C3005-1).
   c. MCF-7 Tet-Off (cat. no. C3007-1).
   d. MDCK Tet-Off7 (cat. no. C3017-1).
   e. Saos-2 Tet-Off6 (cat. no. C3013-1).
   f. 293 Tet-On (cat. no. C3003-1).
2. Ecdysone-inducible system: Invitrogen (www.invitrogen.com or 800-955-6288) has available three mammalian cell lines into which the pVgRXR vector has been stably integrated.
   a. EcR-CV-1 (cat. no. R640-07).
   b. EcR-293 (cat. no. R650-07).
   c. EcR-CHO (cat. no. R660-07).

2.3.2. Custom Cell Lines

Although the use of commercially available cell lines will spare time and expense, not all varieties of cell lines that express Tet-on, Tet-off, or pVgRXR are available. Therefore, custom cell lines may have to be generated to express the regulatory components for these inducible systems.

2.4. Transfections

2.4.1. Modified Calcium Phosphate Transfection

1. N,N-bis(2-hydroxyethyl)-2-aminoethanesulfonic acid (BES): Prepare 2× BES buffered saline (2× BBS) containing 50 mM BES (pH 6.95), 280 mM NaCl, and 1.5 mM Na$_2$HPO$_4$. Adjust pH with HCl and store at $-20°C$.
2. CaCl$_2$: Prepare 0.25 M solution for use in transfection and store at 4°C.

2.4.2. Lipsome-Mediated Transfection

For many, this method of transfection is ideal due to its simplicity, efficiency, and low interference with background effects. First, through the use of a cationic lipid such as DOTMA (N[1-(2,3-dioleyloxy)propyl]-N,N,N,-triethylammonium), the synthetic liposomes internalize nearly 100% of the target DNA. Likewise, the efficiency at which cells take up the liposome/DNA appears to be just as high. Additionally, unlike the calcium phosphate method, liposomal delivery appears to be unaffected by changes in pH levels. However, this method is not without some drawbacks. First, the time to transfect cells efficiently may require 2–24 h. Second, because the lipid is cationic, it is critical to establish a lipid:DNA ratio that favors the cationic lipid and not the anionic DNA. This is required to facilitate interaction with the negatively charged cell surface. Third, transfections performed using the liposome method must be done in the absence of serum, as there appears to be an inhibitory effect present in this component. Lastly, this method costs more per transfection than any other method described herein. Should users choose the liposome method for transfecting mammalian cells with DNA, they should follow the individual manufacturer's instructions. Listed below are suppliers who provide analogous liposomal delivery kits.

1. ClonTech (www.clontech.com or 800-662-2566): CLONfectin Transfection Reagent (cat. no. 8020-1).
2. Life Technologies (www.lifetech.com or 800-828-6686): Lipofectamine (cat. no. 8324SA).
3. Promega (www.promega.com or 800-356-9526): Tfx Reagents (cat. no. E2400, E2381, E2391, E1811).

2.4.3. Electroporation

1. 1× Trypsin/EDTA.
2. Dulbecco's modified Eagle's medium supplemented with 10% fetal bovine serum.
3. Electroporation cuvettes.
4. Electroporation instrument.

2.5. Colony Selection

1. Picking colonies:
   a. Cloning cylinders (Bellco Glass, 800-257-7043).
      10 × 10-mm Cloning cylinders (cat. no. 2090-01010).
   b. Forceps.
   c. Vacuum grease: This is used to create a tight seal between the cloning cylinder and cell culture dish during colony selection. Choose a vacuum grease that is autoclavable.
   d. 1× Trypsin/EDTA.
   e. Dulbecco's modified Eagle's medium supplemented with 10% fetal bovine serum.

3. Methods
3.1. Generation of Cell Lines Expressing Regulatory Proteins

The first transfection is performed with the plasmid containing the regulatory gene. Several widely accepted methods are available for transfecting plasmid DNA into mammalian cells. Ultimately, the user will decide which method is right for his or her project. The three most commonly used techniques are discussed below.

3.1.1. Modified Calcium Phosphate Transfection Method

This method is based on the premise that, when mixed, DNA and calcium phosphate will precipitate onto cells and be incorporated into cells by endocytosis *(37)*. Unlike other methods, several critical factors must be met to ensure success of this method.

First, this method is very sensitive to changes in pH, working optimally at pH 6.95. Second, the amount of DNA that can be transfected is limited to 30 μg but reportedly works best with around 10 μg. Lastly, this protocol requires nonlinearized circular plasmid DNA that is free of any contamination. Perhaps the greatest advantage of this method is its low cost of per transfection use.

1. Use cells on a 10-cm plate that is 70% confluent.
2. Mix 10 μg DNA with 0.5 mL of 2× BBS.
3. Mix DNA and BBS with 0.5 mL of 0.25 M $CaCl_2$ and incubate at room temperature for 30 min.
4. Remove media from cells.
5. Add the mixture of DNA, BBS, and $CaCl_2$ to cells and incubate at room temperature for 30 min.
6. Add 5 mL DMEM/FBS and grow cells at 37°C, 5% CO_2 overnight.
7. After 24 h, wash cells and replace medium.

3.1.2. Liposome Transfection Method

Follow manufacturer's guidelines.

3.1.3. Electroporation Method

This method is based on the idea that the use of a pulse electric field permeabilizes the cell membrane and therefore can be used to introduce DNA into target cells *(38)*.

1. Use cells that are 70% confluent.
2. Harvest cells using 1× trypsin/EDTA.
3. Pellet cells by centrifugation at 1500× g for 3 min.
4. Discard the supernatant and resuspend cells in 400 μL Dulbecco's modified Eagle's medium with 10% fetal bovine serum (DMEM/FBS).
5. Place 5 μL sheared salmon sperm DNA and 1–10 μg DNA to be transfected into electroporation cuvette.
6. Pipet cell suspension into cuvette and allow to stand for 10 min.
7. Electroporate cells at room temperature with a charge (Q) of 0.15–0.20 Co (*see* **Note 8**).
8. Allow cells to sit in cuvette for 10 min following electroporation.
9. Remove cells from cuvette with pipet and plate in 7 mL DMEM/FBS in 10-cm dish.
10. Allow cells to recover overnight at 37°C, 5% CO_2.
11. After 24 h, change medium to remove dead cells.

3.1.4. Transfection of Regulatory Plasmid

1. Using the method of choice, transfect one 10-cm plate of cells (approximately 70% confluent) with 10 μg of regulatory plasmid, i.e., a tTA-expressing vector for the tetracycline-regulated system or the pVgRXR vector for the ecdysone-inducible system.
2. Allow the cells to recover for 18 h at 37°C, 5% CO_2.
3. Trypsinize and resuspend the cells in DMEM/FBS.
4. Plate the cells in a series of dilutions in DMEM/FBS containing the drug to be used for selection. For H1299 cells, plate the cells in dilutions of 1:20, 1:40, and 1:60. For SAOS-2 cells, plate the cells in dilutions of 1:4 and 1:6. For MCF-7 cells, plate the cells in dilutions of 1:5 and 1:10. Dilutions of other cell types may need to be determined.
5. Replace medium every 2–3 d to remove debris and dead, untransfected cells.

3.1.5. Selection of Stably Transfected Clones

The dilutions cited in **step 4** of the above transfection are designed to subclone the population of stably transfected cells such that a maximum number of individual colonies are present on each dilution plate. If cells are not adequately dispersed to support the growth of individual colonies, further dilutions may be necessary. Because cells are grown in the presence of selection drug, only those in which the regulatory plasmid has been integrated into the genome will survive the selection period. However, it may be necessary to determine the minimal antibiotic concentration needed to prevent growth of the parental cell line. Additionally, for co-transfection of more than one vector for dual expression systems, the presence of multiple selection drugs may increase the sensitivity of parental cells to the selection drugs. Therefore, lower concentrations of each selection drug may be required.

3.1.5.1. SELECTIONS OF INDIVIDUAL CLONES

1. Check cells frequently for growth of colonies derived from single cells. This can easily be assessed by viewing the plate from underneath and in front of a light source, as colonies representing a single clone will appear as white spots on the plate.
2. When each clone is approximately 100 cells in number, use cloning cylinders and trypsin/EDTA to transfer each colony to individual wells of a 48-well plate. Typically, 30–50 colonies are selected for analysis.
3. Continue to grow cells in the appropriate drug-containing media.
4. Continue transferring clones to larger wells until they are transferred to 6-cm plates.

3.1.5.2. SCREENING CLONES

The ability of a clone to proliferate in the presence of selection drug implies that integration of the transfected vector into the host genome has occurred. However, clones should be screened to isolate the cell lines that are capable of inducibly expressing a protein from the transfected DNA.

1. At the 6-cm plate stage, the clones should be split 1:2. One plate (test plate) will be assayed for vector integration, while the other plate will be saved as a seed plate.
2. Transfer the cells from the 6-cm test plate into a 10-cm test plate and grow until they reach 70% confluency.
3. Transiently transfect the cells on the test plate with a reliably tested reporter vector. This plasmid should contain a reporter gene or a target gene whose expression is regulated by the gene product of the stably transfected vector. The firefly luciferase gene cloned in the pTRE or pIND vectors can be used to detect induction in tetracycline- or ecdysone-regulated systems, respectively. Alternatively, known genes can be cloned into the pTRE or pIND vectors for detection of induction by Western blot analysis.
4. Allow the transfected cells to recover for 18 h at 37°C, 5% CO_2.
5. Split the cells 1:2 and grow overnight in either the presence or absence of inducer, i.e., tetracycline or ponasterone A. Tetracycline should be at a final concentration of 2.0 µg/mL, and ponasterone A should be at a final concentration of 5 µM.
6. After 24 h, collect the cells and perform Western blot analyses to determine if gene expression is regulated by the inducer.
7. Positive clones should be immediately frozen down in several vials in an appropriate manner.

3.2. Generation of Cell Lines Expressing Inducible Proteins

The second transfection is performed with the plasmid containing the inducible gene. This transfection is similar to that described above, although here the cells used are those into which a regulatory plasmid has been stably integrated. Additionally, the transfection efficiency is determined prior to clonal selection to optimize proficiency and prevent the unnecessary loss of time and effort.

3.2.1. Transfection of Inducible Plasmid

By method of choice, transfection of the inducible plasmid containing the desired gene into cells expressing the regulatory gene is performed as above. For the Tet-off system, add tetracycline to the media to suppress expression.

1. Following transfection, wash the cells extensively if tetracycline is present in the medium, then trypsinize and resuspend the cells in DMEM/FBS.
2. Plate the cells in an appropriate series of dilutions in DMEM/FBS. Dilutions for different cell types may need to be determined. Add tetracycline to the medium if the Tet-off system is used.
3. Plate the remaining cells in two 6 cm plates. If using the Tet-off system, add tetracycline to the medium in one plate to inhibit expression. If using the ecdysone system, add ponasterone A to the medium in one plate to induce expression. Grow for 24 h and assay the protein expression by Western blot analysis.
4. If Western blot analysis shows proper induction of the inducible gene, add selection drug to the medium in the dilution plates, and replace medium every 2–3 d to remove debris and dead, untransfected cells. Otherwise, repeat the above transfection steps.

3.2.2. Selection of Stably Transfected Clones

3.2.2.1. SELECTION MARKERS

1. If using the pTRE vector for the tetracycline-regulated system, no selection marker is present for selecting positive clones. As mentioned above, a second plasmid containing a selectable marker may be co-transfected with pTRE. Co-transfection of an additional plasmid should be performed at a ratio of 1:10 in favor of pTRE. In theory, since pTRE is 10 times as abundant as the co-transfected plasmid, the selected cells should also contain pTRE.
2. If using the pIND vector for the ecdysone-inducible system, this vector will contain either the hygromycin or neomycin resistance cassette. Here, positive clones can be selected for by adding hygromycin or neomycin, respectively, to the culture medium.

3.2.2.2. SELECTION AND SCREENING OF INDIVIDUAL CLONES

As above, 30–50 colonies should be selected for analysis. Again, Western blot analysis with cell extracts from each clone grown in both the presence and absence of tetracycline or ponasterone A can be used to identify positive clones.

3.3. Generation of Cell Lines for Dual Inducible Systems

Since cell lines that constitutively express regulatory proteins for multiple inducible systems are not yet available commercially, cell lines capable of dual inducible expression need to be custom generated in the laboratory.

3.3.1. Transfection of Regulatory Plasmids into Tet-on, Tet-off, or pVgRXR-Expressing Cell Lines

As discussed above, cell lines that constitutively express regulatory proteins for either the tetracycline or ecdysone systems are available commercially. Cell lines such as these can be transfected with additional regulatory plasmids to generate cell lines with dual expression capabilities. Cell lines containing Tet-on or Tet-off components can be transfected with pVgRXR of the ecdysone system, and transfectants can be selected with the drug zeocin. Alternatively, cell lines containing pVgRXR can be transfected with Tet-on or Tet-off vectors, and transfectants can be selected with the drug neomycin.

3.3.2. Co-transfection of Two Regulatory Plasmids

If a specific cell line is not commercially available with tetracycline or ecdysone system components, a Tet-on or Tet-off vector and the pVgRXR vector can be co-transfected to generate cell lines with dual inducible expression capabilities.

3.3.3. Selection of Stably Dual-Transfected Clones

Selection of clones expressing components of both the tetracycline and ecdysone systems is similar to the selection of single-inducible clones. However, when examining potential positive clones, integration of both tetracycline and ecdysone regulatory vectors must be determined. Reporter vectors, such as pTRE-luciferase and pIND-luciferase, can be used to detect induction in tetracycline- and ecdysone-regulated systems, respectively. Additionally, selection of clones expressing inducible proteins from both tetracycline- and ecdysone-inducible vectors is similar to the selection of single inducible clones.

4. Notes

1. Stock solutions of tetracycline should be made in 70% ethanol, stored at $-20°C$ in the dark, and used within 2 mo. Because direct application of tetracycline in ethanol is cytotoxic, it is recommended that tetracycline in ethanol be mixed in the medium first and then applied to cells. Alternatively, doxycycline can be used. Stock solution of doxycycline is prepared in water, stored at 4°C in the dark, and used within 4 wk.
2. ClonTech makes a tetracycline-free fetal bovine serum that is appropriate for use with the tetracycline-off system.
3. Constitutive expression of most TSGs will induce cell cycle arrest or apoptosis. Therefore, it should be noted that while working with inducible TSGs in the Tet-off system, tetracycline should remain in the culture medium until expression is desired.
4. To increase the inducibility of protein expression in the Tet-off system, the plates could be washed a second time with serum-free medium and replaced with fresh medium 2–3 h after the initial wash. This will ensure that all of the tetracycline is removed from the system.
5. To enhance expression of an ecdysone-inducible gene, increase the induction time or increase the concentration of ponasterone A up to 10 μM.
6. Saez et al. *(39)* found that induction with ponasterone A or muristerone A in the presence of an RXR ligand (LG268) increases the level of induction 3–5-fold.
7. The Tet-off system has significantly higher levels of induction than the ecdysone-inducible system. Therefore, when expressing protein in the dual inducible system, longer induction time may be needed for the protein in the ecdysone-inducible system than for the protein in

the Tet-off system. However, the ecdysone-inducible system does have lower levels of basal expression than does the Tet-off system.

8. Using a higher charge will increase cell death, but in creating stable cell lines, a higher proportion of surviving cells will be transfected.

References

1. McMahon, G. (2000) VEGF receptor signaling in tumor angiogenesis. *Oncologist*, **5**, 3–10.
2. Gossen, M. and Bujard, H. (1992) Tight control of gene expression in mammalian cells by tetracycline- responsive promoters. *Proc. Natl. Acad. Sci. USA* **89**, 5547–5551.
3. Clackson, T. (1997) Controlling mammalian gene expression with small molecules. *Curr. Opin. Chem. Biol.* **1**, 210–218.
4. Rossi, F. M. and Blau, H. M. (1998) Recent advances in inducible gene expression systems. *Curr. Opin. Biotechnol.* **9**, 451–456.
5. Saez, E., No, D., West, A., and Evans, R. M. (1997) Inducible gene expression in mammalian cells and transgenic mice. *Curr. Opin. Biotechnol.* **8**, 608–616.
6. Yao, T. P., Segraves, W. A., Oro, A. E., McKeown, M., and Evans, R. M. (1992) *Drosophila* ultraspiracle modulates ecdysone receptor function via heterodimer formation. *Cell* **71**, 63–72.
7. Yao, T. P., Forman, B. M., Jiang, Z., et al. (1993) Functional ecdysone receptor is the product of EcR and Ultraspiracle genes. *Nature* **366**, 476–479.
8. No, D., Yao, T. P., and Evans, R. M. (1996) Ecdysone-inducible gene expression in mammalian cells and transgenic mice. *Proc. Natl. Acad. Sci. USA* **93**, 3346–3351.
9. Triezenberg, S. J., Kingsbury, R. C., and McKnight, S. L. (1988) Functional dissection of VP16, the trans-activator of herpes simplex virus immediate early gene expression. *Genes Dev.* **2**, 718–729.
10. Triezenberg, S. J., LaMarco, K. L., and McKnight, S. L. (1988) Evidence of DNA: protein interactions that mediate HSV-1 immediate early gene activation by VP16. *Genes Dev.* **2**, 730–742.
11. Sadowski, I., Ma, J., Triezenberg, S., and Ptashne, M. (1988) GAL4-VP16 is an unusually potent transcriptional activator. *Nature* **335**, 563–564.
12. Umesono, K. and Evans, R. M. (1989) Determinants of target gene specificity for steroid/thyroid hormone receptors. *Cell* **57**, 1139–1146.
13. Cress, W. D. and Triezenberg, S. J. (1991) Critical structural elements of the VP16 transcriptional activation domain. *Science* **251**, 87–90.
14. Forman, B. M., Goode, E., Chen, J., et al. (1995) Identification of a nuclear receptor that is activated by farnesol metabolites. *Cell* **81**, 687–693.
15. Underhill, T. M., Cash, D. E., and Linney, E. (1994) Constitutively active retinoid receptors exhibit interfamily and intrafamily promoter specificity. *Mol. Endocrinol.* **8**, 274–285.
16. Perlmann, T., Rangarajan, P. N., Umesono, K., and Evans, R. M. (1993) Determinants for selective RAR and TR recognition of direct repeat HREs. *Genes Dev.* **7**, 1411–1422.
17. Osborne, C. K., Zhao, H., and Fuqua, S. A. (2000) Selective estrogen receptor modulators: structure, function, and clinical use. *J. Clin. Oncol.* **18**, 3172–3186.
18. Pratt, W. B. (1990) Interaction of hsp90 with steroid receptors: organizing some diverse observations and presenting the newest concepts. *Mol. Cell. Endocrinol.* **74**, C69–C76.
19. Smith, D. F. and Toft, D. O. (1993) Steroid receptors and their associated proteins. *Mol. Endocrinol.* **7**, 4–11.
20. Picard, D. (1993) Steroid-binding domains for regulating the functions of heterologous proteins in cis. *Trends Cell Biol.* **3**, 278–280.
21. Berthois, Y., Katzenellenbogen, J. A., and Katzenellenbogen, B. S. (1986) Phenol red in tissue culture media is a weak estrogen: implications concerning the study of estrogen-responsive cells in culture. *Proc. Natl. Acad. Sci. USA* **83**, 2496–2500.

22. Littlewood, T. D., Hancock, D. C., Danielian, P. S., Parker, M. G., and Evan, G. I. (1995) A modified oestrogen receptor ligand-binding domain as an improved switch for the regulation of heterologous proteins. *Nucleic Acids Res.* **23**, 1686–1690.

23. Danielian, P. S., White, R., Hoare, S. A., Fawell, S. E. and Parker, M. G. (1993) Identification of residues in the estrogen receptor that confer differential sensitivity to estrogen and hydroxytamoxifen. *Mol. Endocrinol.* **7**, 232–240.

24. Vater, C. A., Bartle, L. M., Dionne, C. A., Littlewood, T. D., and Goldmacher, V. S. (1996) Induction of apoptosis by tamoxifen-activation of a p53-estrogen receptor fusion protein expressed in E1A and T24 H-ras transformed p53−/− mouse embryo fibroblasts. *Oncogene* **13**, 739–748.

25. Roemer, K. and Friedmann, T. (1993) Modulation of cell proliferation and gene expression by a p53-estrogen receptor hybrid protein. *Proc. Natl. Acad. Sci. USA* **90**, 9252–9256.

26. Eilers, M., Picard, D., Yamamoto, K. R., and Bishop, J. M. (1989) Chimaeras of myc oncoprotein and steroid receptors cause hormone- dependent transformation of cells. *Nature* **340**, 66–68.

27. Karin, M., Haslinger, A., Holtgreve, H., et al. (1984) Characterization of DNA sequences through which cadmium and glucocorticoid hormones induce human metallothionein-IIA gene. *Nature* **308**, 513–519.

28. Durnam, D. M. and Palmiter, R. D. (1981) Transcriptional regulation of the mouse metallothionein-I gene by heavy metals. *J. Biol. Chem.* **256**, 5712–5716.

29. Friedman, R. L. and Stark, G. R. (1985) alpha-Interferon-induced transcription of HLA and metallothionein genes containing homologous upstream sequences. *Nature* **314**, 637–639.

30. Hager, L. J. and Palmiter, R. D. (1981) Transcriptional regulation of mouse liver metallothionein-I gene by glucocorticoids. *Nature* **291**, 340–342.

31. Karin, M., Imbra, R. J., Heguy, A., and Wong, G. (1985) Interleukin 1 regulates human metallothionein gene expression. *Mol. Cell. Biol.* **5**, 2866–2869.

32. Westin, G. and Schaffner, W. (1988) A zinc-responsive factor interacts with a metal-regulated enhancer element (MRE) of the mouse metallothionein-I gene. *EMBO J.* **7**, 3763–3770.

33. Palmiter, R. D. (1998) The elusive function of metallothioneins. *Proc. Natl. Acad. Sci. USA* **95**, 8428–8430.

34. Mulligan, R. C. and Berg, P. (1981) Selection for animal cells that express the *Escherichia coli* gene coding for xanthine-guanine phosphoribosyltransferase. *Proc. Natl. Acad. Sci. USA* **78**, 2072–2076.

35. Wyborski, D. L. and Short, J. M. (1991) Analysis of inducers of the *E. coli* lac repressor system in mammalian cells and whole animals. *Nucleic Acids Res.* **19**, 4647–4653.

36. Fieck, A., Wyborski, D. L., and Short, J. M. (1992) Modifications of the *E. coli* Lac repressor for expression in eukaryotic cells: effects of nuclear signal sequences on protein activity and nuclear accumulation. *Nucleic Acids Res.* **20**, 1785–1791.

37. Chen, C. and Okayama, H. (1987) High-efficiency transformation of mammalian cells by plasmid DNA. *Mol. Cell Biol.* **7**, 2745–2752.

38. Chu, G., Hayakawa, H., and Berg, P. (1987) Electroporation for the efficient transfection of mammalian cells with DNA. *Nucleic Acids Res.* **15**, 1311–1326.

39. Saez, E., Nelson, M. C., Eshelman, B., et al. (2000) Identification of ligands and coligands for the ecdysone-regulated gene switch. *Proc. Natl. Acad. Sci. USA* **97**, 14512–14517.

20

In Vitro Models of Early Neoplastic Transformation of Human Mammary Epithelial Cells

Vimla Band

1. Introduction

About 80% of all human cancers are carcinomas, representing oncogenic transformation of epithelial cells. In-vitro models of epithelial cell transformation are therefore of great interest to elucidate the molecular mechanisms of human cancer. Malignant transformation is a multistep process in which genetic changes and environmental factors, including viruses, carcinogens, radiation, and dietary factors, impinge on common cellular pathways resulting in uncontrolled proliferation, a hallmark of tumorigenic process *(1,2)*. Understanding the nature of these cellular pathways is a central goal in cancer biology. A critical event in oncogenesis is the conversion of normal epithelial cells with a finite proliferative potential into cells endowed with an ability to multiply continuously, a trait that allows the accumulation of further genetic alterations resulting in full malignancy. In vitro, this behavior manifests as continuous proliferation of cells beyond their limited life span, a process referred to as immortalization. Understanding the biochemical basis of immortalization is therefore likely to point to crucial tumor suppressor pathways that ensure the untransformed state of normal epithelial cells.

Reduction mammoplasty specimens represent by far the best source of normal human mammary epithelial cells (MECs) for in-vitro studies. A second source of MECs is human milk from normal lactating women. Our laboratory as well as others has developed culture conditions for establishment and growth of MECs from reduction mammoplasty-derived tissue specimens and from milk *(3–6)*. Media that are commonly used to grow MECs from reduction mammoplasty specimens are either serum-free, such as the MCDB 170 medium *(3),* or contain relatively low concentrations of serum, as in the DFCI-1 medium *(4,5)* (**Table 1**). When mammary tissue is explanted in the DFCI-1 medium, a morphologically heterogeneous epithelial cell population emerges within a week (**Fig. 1**). These cell populations proliferate for 3–5 passages, at which time the majority of cells undergo senescence *(5)*. Recent studies have shown that these senescing epithelial cells show a senescent phenotype similar to that well documented for fibroblasts *(7)*. However, regular feeding of cell cultures during this stage leads to emergence of a homogenous population of cells. These cells continue to proliferate for 10–20

From: *Methods in Molecular Biology, Vol. 223: Tumor Suppressor Genes: Regulation, Function, and Medicinal Applications.* Edited by: Wafik S. El-Deiry © Humana Press Inc., Totowa, NJ

Table 1
Composition of DFCI-1 Medium

Components	500-mL bottle	Stock solution	Storage conditions
Alpha-MEM	250 mL	—	4°C
F12	250 mL	—	4°C
HEPES	5.0 mL	1 M	4°C
Insulin	0.5 mL	1 mg/mL in water (pH 2.0)	−20°C
Hydrocortisone	0.5 mL	0.1 mg/mL in ethanol	−20°C
EGF	0.25 mL	25 µg/mL in PBS	−20°C
Ascorbic acid	0.5 mL	10 mg/mL made in PBS	Fresh
Transferrin	0.5 mL	10 mg/mL in PBS	−20°C
Phosphoethanolamine	0.5 mL	14.1 mg/mL in PBS	−20°C
Estradiol	50 µL	$2 \times 10^{-5} M$ in ethanol	RT
Sodium selenite	50 µL	26 µg/mL in PBS	RT
Glutamine	5.0 mL	200 mM	4°C
Cholera toxin	50 µL	10 µg/mL in water	4°C
Triiodothyronine	25 µL	130 µg/mL in water	4°C
Ethanolamine	3 µL		RT
Bovine pituitary extract	1.25 mL	14 mg protein/mL[a]	−20°C
Gentamycin	2.5 mL	10 mg/mL	4°C
Fetal calf serum	5.0 mL		−20°C

[a]Prepared by first centrifuging at 10,000 rpm for 10 min and then sequential filtration through 0.8-, 0.4-, and 0.2-µm filters. RT, room temperature.

passages (30–60 population doublings) and then senesce *(5)* (**Fig. 1**). Interestingly, senescence at this time is characterized by genomic instability *(7)*. Over the years, we have made the observations that: (a) a homogenous population of cells does not emerge from each reduction mammoplasty sample; and (b) a homogenous cell population is not seen each time a culture is initiated from a single specimen. These observations suggest the presence of different types of epithelial cells in early cultures that may not be equally represented in different reduction mammoplasty samples. We and others have demonstrated that heterogeneous population of cells at early or preselection passages express a number of keratins characteristic of luminal (K8, K18, and K19) and basal (K5, K14, K17) cells, whereas the late passage/postselection homogenous population of cells expresses predominantly the basal cell markers *(8–10)*. These results indicate that preselection cells are a mixture of epithelial cell types, whereas postselection cells represent a particular type of cells (**Fig. 1**). However, at present, there are no definitive molecular markers to directly assess the relationship of the different cell types that emerge in vitro to cell types present in the mammary gland.

Multiple cell types are also observed when human milk-derived epithelial cells are cultured in DFCI-1 medium *(10)*. Similar to preselection mammoplasty-derived MECs, the milk-derived MECs also senesce by 3–5 passages *(6,10)*. Interestingly, the emergence of homogenous population of cells resembling those seen with reduction mammoplasty samples has not been reported with milk samples. Due to their extremely short life span, few laboratories have reported the use of milk specimens as a source of MECs *(6,10)*.

Given the ease with which MECs can be obtained from healthy subjects, the finite life span of emerging MECs with eventual senescence, and lack of spontaneous immor-

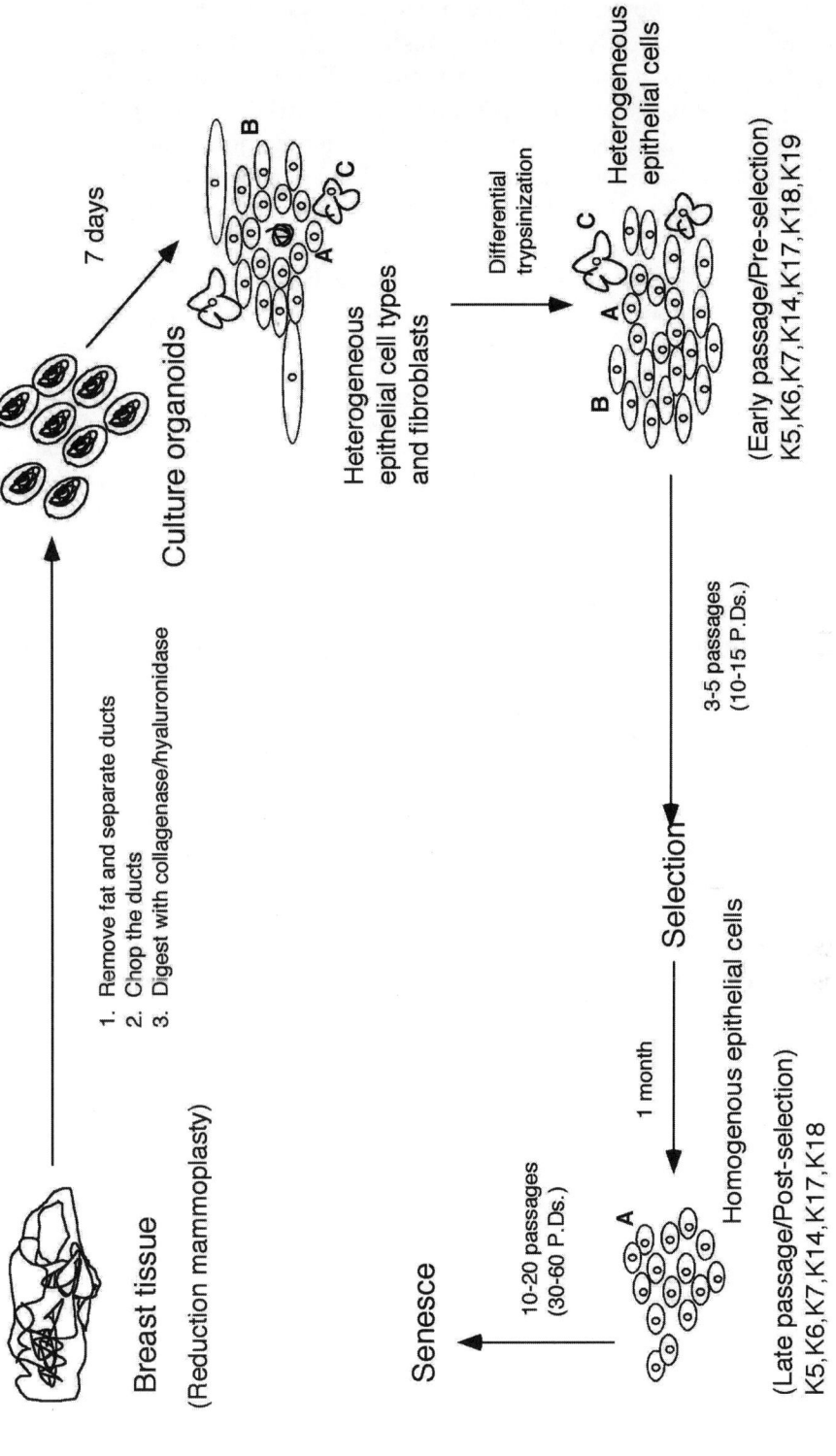

Fig. 1. Schematic representation of establishment of mammary epithelial cells from breast tissue specimens in DFCI-1 medium. A, B, and C denote three morphologically different types of epithelial cells present in tissue. P.D., population doubling, K, keratins.

talization in culture, models of MEC transformation have begun to provide a wealth of knowledge about tumor suppressor pathways relevant to human carcinomas *(11)*. Over the years, several models of early oncogenic transformation of MECs have been established *(11,12)*. The initial evidence for in-vitro immortalization of MECs was provided by the observations that treatment of MECs with benzo(a)pyrene led to an extension of life span followed by crisis and immortalization, although immortalization was infrequent *(13)*. The benzo(a)pyrene-immortalized cells did not exhibit significant anchorage independence or growth in nude mice, suggesting an early neoplastic transformation *(13)*. In our laboratory, we were unable to immortalize human MECs using carcinogens known to transform rodent mammary cells, including N-nitrosomethylurea (NMU), 7,12-dimethylbenz[a]anthracene, and benzo(a)pyrene (unpublished data). Similar to carcinogen-induced immortalization, we found that γ-radiation could induce the transformation of mammary epithelial cells, but relatively infrequently *(14)*. While we were able to achieve transformation of MECs using 2-Gy/d fractionated γ-irradiation for a total of 30 Gy in one study *(14)*, we failed to achieve immortalization or transformation of MECs in several other attempts, indicating that radiation-induced transformation is a relatively rare event (our unpublished data). Nonetheless, the availability of isogenic normal and radiation-transformed cells allowed the use of subtractive hybridization to identify a novel gene that is transcriptionally downregulated during breast cell transformation *(15)*. This gene, called *NES1* (Normal Epithelial Specific 1), is expressed in normal but not in radiation-transformed MECs, and its mRNA and protein expression is drastically downregulated or completely lost in a majority of breast cancer cell lines *(15)*. NES1 expression is also downregulated in prostate cancer cell lines as compared to nontumorigenic prostate epithelial cells *(16)*. Importantly, NES1 expression is lost in a majority of infiltrating ductal carcinomas and about half of *in situ* carcinomas *(17)*. Our recent studies have shown that tumor-specific loss of NES1 expression in cell lines and primary breast tumors is due to methylation of the NES1 gene *(18)*, indicating that loss of NES1 expression/hypermethylation could serve as a novel marker for early diagnosis of breast and other carcinomas. These findings suggest that, in addition to providing a possible tumor marker, inactivation of the NES1 gene expression may be linked to oncogenesis. Consistent with this possibility, transfection of NES1 into a NES1-negative breast cancer cell line MDA-MB-231 dramatically reduced its aggressive tumorigenic phenotype, as evidenced by decreased anchorage independence and inhibition of tumor formation in nude mice *(16)*.

In addition to chemical carcinogens and radiation, viral oncogenes have proven very useful to define cellular tumor suppressor pathways, which are among the crucial targets of transforming viral genes. Thus, a number of laboratories have attempted to transform MECs with viral oncogenes. Remarkably, however, a number of viral oncogenes, including the SV40 T-antigen, adenovirus E1A and E1B, and polyoma T-antigen, have proven to be inefficient at immortalizing human cells *(19,20)*. While SV40 T-antigen-induced MEC immortalization has been reported, it resembles the carcinogen- and radiation-induced transformation in that cells go through a long crisis period and the emergence of immortal cells is rare *(21)*. Over the last several years, we have carried out extensive analyses using human papilloma virus (HPV) oncogenes that have resulted in an efficient and reproducible system to immortalize human MECs.

HPVs have been associated with cervical carcinomas and benign genital warts *(22)* and transfection of cancer-associated HPV genome into keratinocytes, a natural host epithelial cell for these viruses, could induce their in-vitro immortalization *(23)*. Notably, we observed that transfection of HPV type 16 or 18 genome into MECs led to efficient immortalization *(24)*. The HPV-immortalized MECs did not exhibit anchorage independence and were unable to form tumors when implanted in nude mice, indicating an early neoplastic transformation *(24)*. The HPV genome contains seven early genes (designated E1–E7), of which E6 and E7 have been associated with transformation of human keratinocytes and fibroblasts *(25,26)*. Both E6 and E7 are required for efficient immortalization of human keratinocytes *(26)*. To examine the role of E6 and E7 in MEC immortalization, we used retroviral infection of early (passages 1–4, preselection) or late (passages 7–12, postselection) passages of MECs with HPV-16 E6 alone, E7 alone, or E6 + E7 together, and serially passaged the cultures to assess if immortalization had occurred *(10)*. Comparison of pre- and postselection cultures revealed that different MEC subtypes exhibit a remarkably different susceptibility to E6, E7, or their combination *(10)*. One MEC subtype was exclusively immortalized by E6 but not by E7; such cells predominated the postselection cultures but were rare in preselection cultures. A second cell type, present only at early passages of tissue-derived cultures, showed extension of life span and infrequent immortalization with E7 alone. Finally, E6 and E7 together were required to fully immortalize a large proportion of MECs in early-passage/preselection cultures. All of the immortalized cell types failed to exhibit anchorage-independent growth and did not grow as tumors in nude mice *(10)*.

The MECs present in human milk are thought to be differentiated luminal cells, which can be cultured in vitro for only 2–3 passage *(6)*. It has been shown that these milk cells can be immortalized by SV40 T-antigen *(27)*. We observed that neither E6 nor E7 induced the immortalization of milk-derived MECs. However, infection with E6 and E7 retroviruses together induced the complete immortalization of these cells *(10)*. Unlike E6 or E6/E7-immortalized tissue-derived cells, milk-derived immortal cells exhibited anchorage-independent growth. However, when injected into nude mice these cells failed to produce tumor *(10)*.

The ability of E6, E7, and or their combination to induce early transformation of MECs has provided practical models to understand the mechanism of MEC immortalization. One successful approach that other investigators and we have used is to identify cellular targets of E6 and E7 oncogenes and to characterize these targets for their role in MEC immortalization. Studies of HPV16 E6 targets have led these efforts. The HPV E6 protein has no known intrinsic enzymatic activity. Therefore, its dominant oncogenic function is thought to reflect its ability to interact with, and alter the function of, a number of cellular proteins that together constitute a cell's tumor suppressor machinery. Indeed, a number of E6-binding proteins have been identified recently. These include E6AP, a ubiquitin ligase which allows E6 to induce p53 degradation, E6BP (E6-binding protein or ERC55, a putative calcium-binding protein), paxillin (a focal adhesion protein involved in cell migration and regulation of actin cytoskeleton via integrins), hDlg (the human homolog of *Drosophila* disk large tumor suppressor protein), interferon regulatory factor-3 (a component of virus-activated transcription factor complex), multicopy maintenance protein 7 (a subunit of the replication licensing factor-M), Bak (bcl-2 homologous antagonist/killer, a pro-apoptotic protein), p300 (a transcriptional co-activator), E6TP1 (E6-targeted protein 1, a novel RapGAP), PKN (a Rho-regulated serine threonine kinase), Vartul (human homolog of the Drosophila Scribble tumor suppressor protein), and Gps2 (G protein suppressor of unknown function) *(28–39)*. The relative contribution of these various E6 targets in cellular transformation is under investigation in our laboratory and in other laboratories. It is important to note that the majority of cellular targets of E6 are well-known proteins implicated in a number of carcinomas. Therefore, the strategy to identify E6-binding proteins should help identify novel cellular proteins whose function is lost in cancer cells.

The E6–p53 interaction has received most attention thus far. E6 interacts with p53 indirectly through E6AP, which binds to E6 and p53 concurrently *(28)*. E6 targets p53 for E6AP-mediated ubiquitination and proteasome-mediated degradation *(40)*. Although the mutational analysis of E6 showed an important role of p53 degradation in E6-induced immortalization *(41)*, recent studies have identified three MEC-immortalizing E6 mutants (8S9A10T, F2V, and Y54H) that do not target p53 for degradation *(11,42)*. These results strongly support the critical roles of non-p53 targets in E6-induced epithelial cell transformation. Indeed, we found that binding to, and subsequent degradation of, RAP-GAP E6TP1 correlated strongly with immortalizing ability of E6 mutants *(36,43)*. Consistent with an important role of E6-mediated perturbation of the Rap pathway in immortalization, recent studies from our laboratory has shown that ectopic expression of an activated Rap-G12Vcan itself induce MEC immortalization (unpublished data).

The HPV16 E7-induced extension of life span of MECs present in early-passage cultures is relatively efficient and reproducible, whereas immortalization is rare and occurs only after a "crisis period" *(10)*. E7 is known to bind to Rb and its relatives p107, and p130, and inactivates their function *(44–46)*, implicating these tumor suppressor proteins in E7-induced immortalization. Similar to E6, a number of additional targets of E7 have been identified in recent years (summarized in **ref. *11***). Future studies should clarify which of these target(s) are necessary for E7-induced immortalization of epithelial cells.

Thus, these in-vitro models have helped us identify a number of novel cellular pathways that contribute to MEC transformation. Several of these pathways have a well-documented role in human cancer, such as p53 and Rb, while the role of the novel pathways identified through these model systems will require further direct studies.

2. Materials

1. Media: DFCI-1 medium consists of α-minimum essential medium/Ham's nutrient mixture F-12 (Gibco, BRL) supplemented with epidermal growth factor, triiodothyronine, HEPES, ascorbic acid, estradiol, insulin, hydrocortisone, ethanolamine, phosphoethanolamine, transferrin, L-glutamine, gentamycin (all from Sigma); sodium selenite (Amend Drugs and Chemical, New York); cholera toxin (Schwartz/Mann); fetal calf serum (Hyclone), and bovine pituitary extract (Hammond, Cell/Tech, Alameda, CA). The pH is 7.4 at 6.5% CO_2. The detailed protocol with stock solutions is presented in **Table 1**.
2. Phosphate-buffered saline with glucose (PBS): For 1 L PBS use 7.15 g HEPES, 0.23 g KCl, 7.7 g NaCl, 0.27 g $Na_2HPO_4 \cdot 7H_2O$, and 1.8 g of glucose. Make $10\times$ solution and refrigerate. Make $1\times$ solution in sterilized, deionized, double-distilled water.
3. Trypsin/EDTA solution: 0.025 g and 0.01 g, respectively, per 100 mL of PBS.
4. Freezing medium: DFCI-1 medium containing 10% dimethylsulfoxide (DMS) and 20% fetal calf serum (FCS).
5. $2\times$ HBS: 16.4 g NaCl, 11.9 g HEPES acid, 0.21 g Na_2HPO_4 in 800 mL of double-distilled, deionized water. Adjust pH to 7.05 with 5 *N* NaOH. Add water to 1000 mL. Filter through 0.2-μm filter and store at 4°C.

>AU: Sp out HBS first use.

3. Methods

3.1. Establishment and Culture of Normal Mammary Epithelial Cells

1. Normal human mammary tissue specimens from reduction mammoplasties are received in sterile tubes containing DFCI-1 medium with 10% FCS and gentamycin.
2. Mammary ducts are separated from fatty tissues using scissors. The ductal tissues are minced using tissue chopper (Mickle Laboratory Engineering, Gomshall, Surrey, U.K.).
3. The ductal tissues are digested with 100 U/mL each of collagenase and hylauronidase (Sigma) in DFCI-1 medium for 4 h at 37°C.
4. Minced and digested ductal tissue is either frozen (see below) or is cultured in flasks in DFCI-1 medium at 37°C in a humidified atmosphere of 6.5% CO_2.
5. Within 7 d, a morphologically heterogenous population of epithelial cells and fibroblasts emerges (**Fig. 1**).
6. Remove fibroblasts using trypsinization of cultures at room temperature for 3 min (*see* **Note 1**).
7. Remaining heterogeneous population of epithelial cells (usually referred to as early or pre-selection) proliferate for 3–5 passages (10–15 population doublings), followed by selection when majority of cells senesce.
8. Regular continuous feeding of cells at this time may give rise to a homogenous population of cells (usually referred as late or postselection) that proliferate further for 10–20 passages (30–60 population doublings), followed by senescence (**Fig. 1**).

3.2. Markers to Distinguish Pre- and Postselection Mammary Epithelial Cells

At present, immunocytochemistry with antibodies specific for various keratins provides the most reliable method to characterize pre- and postselection mammary epithelial cells. The following procedure using Vectastain kit (Vector Laboratories, CA) can be used.

1. Plate 2×10^4 cells per well in 8-well chamber slides (Marsh Biomedical Products, New York).
2. After 48 h, wash cells in PBS.
3. Fix in methanol and acetone (1/1) at room temperature for 10 min.
4. Wash with PBS twice (5 min each).
5. Add blocking serum (provided by supplier) for 20 min.
6. Add antikeratin antibodies (ICN Biomedicals) diluted in PBS for 30 min.
7. Wash with PBS twice (5 min each).
8. Add biotinylated antibody (provided by supplier) for 30 min.
9. Wash with PBS twice (5 min each).
10. Add Vectastain (provided by supplier) for 30 min.
11. Wash with PBS twice (5 min each).
12. Incubate in peroxidase substrate solution for 2 min (.01 g of 3,3′-diaminobenzidine (Sigma)/0.1 M Tris-HCl + equal volume of 0.02% H_2O_2).
13. Wash 5 min in tap water.
14. Clean and mount in Permount (Fisher Scientific).

3.3. Keratin Analysis by One-Dimensional or Two-Dimensional Electrophoresis

1. Plate MECs at 5×10^5 cells per 60-mm culture dish and grow until approximately 80% confluent.

2. Wash cells with PBS and label with 50 µCi/mL of [^{35}S] methionine for 2 h in methionine-free medium.

3. Wash cells with a buffered solution (137 m*M* NaCl, 3 m*M* KCl, 1.5 m*M* KH$_2$PO$_4$, 8 m*M* Na$_2$HPO$_4$ [pH 7.2] and 1 m*M* phenylmethyl-sulfonyl fluoride).

4. Lyse cells in 1 mL of extraction buffer (20 m*M* Tris-HCl [pH 7.5], 0.6 *M* KCl, 5 m*M* EDTA and 1.0% Triton X-100).

5. Pellet the insoluble material in an Eppendorf microcentrifuge for 4 min.

6. Resuspend the pellet in 1 mL of extraction buffer, pellet again, rinsed twice with PBS and 1 m*M* PMSF, and lyophilize the cell pellet.

7. Resuspend the pellet in loading sample buffer.

8. Remove an aliquot to determine the amount of radioactivity in the sample.

9. Load 100,000 cpm onto a denaturing 10% polyacrylamide gel or analyze by two-dimensional gel electrophoresis *(47)*.

3.4. Culture of Cells from Milk Specimens

1. Collect milk under sterile conditions in DFCI-1 medium using milk pump from lactating women volunteers after informed consent.

2. Mix milk sample with DFCI-1 medium at 1/5 ratio and spin at 2000 rpm for 15 min.

3. Resuspend cell pellet in DFCI-1 medium containing 10% FCS and culture in the same medium.

4. These cells proliferate for 3–5 passages similar to preselection tissue-derived cells and invariably senesce.

3.5. Introducing Genes into MECs

Due to the heterogeneous population of preselection cells, retroviral infection rather than transfection is the method of choice.

1. Introduce DNA constructs into retroviral vectors such as pLXSN (neo resistance marker) or pBabe-puro (puromycin resistance).

2. Production of retroviruses is done by transient transfection of retrovirus vector together with a packaging plasmid pIK into a packaging cell line tsa54 using the calcium phosphate method, as described previously *(48)*.

3. Collect supernatant containing the virus particles after 24 h of transfection. Centrifuge to remove cell debris, and store supernatants frozen at −70°C.

4. Infect logarithmically growing mammary epithelial cells with retrovirus after quick thaw of virus at 37°C. Infect cells three times, each time for 4 h, in the presence of polybrene (4 µg/mL), and select in 0.5 µg/mL puromycin or 100 µg/mL of G418 for 5–10 d.

5. This protocol typically results in 60–90% infection (antibiotic-resistant) rates.

6. Subculture cells at 2×10^5/T25 during regular passaging.

The other methods of choice for postselection mammary epithelial cells are stable transfection using the calcium phosphate method or Fugene-6 reagent (Roche). For transfection of postselection cells, Fugene reagent is the method of choice.

1. Plate 5×10^5 cells into 100-mm dishes for 48 h.

2. Mix 5 µg of DNA with 10 µL of Fugene reagent in optiMEM medium (Gibco, BRL).

3. Incubate for 15 min at room temperature.

4. Add to cells and incubate cells for 6 h at 37°C.

5. Cells are either harvested after 48 h (for transient transfections) or selected in an appropriate selection medium in order to obtain stable clones expressing the desired protein.

For the calcium phosphate method for stable transfections:

1. Plate 5×10^5 cells into 100-mm dishes for 24 h.
2. Take 8 μg of DNA in 420 μL of Tris/EDTA in a tube.
3. Add 60 μL of 2 *M* $CaCl_2$ with a pipetman with gentle vortexing.
4. Add this slowly to 480 μL of 2 × HBS (placed in a separate tube) while blowing air (slow to moderate bubbling) into the tube to ensure quick mixing.
5. Mix gently by tapping tube with finger.
6. Keep at room temperature for 30 min.
7. Add to cells and incubate cells at 37°C for 6 h.
8. Take out medium (remove each drop) and add 2 mL of 15% glycerol in PBS for 4 min.
9. Wash twice with PBS.
10. Add medium. Start selection after 48 h.

3.6. Immortalization Assays

1. Immortalization of cells is assessed by either continuous passaging of 2×10^5 cells per 25-cm^2 flask in selection medium containing DFCI-1 with either puromycin (0.5 μg/mL or G418 (100 μg/mL) at 1/10 ratio. Alternatively, cells can be cultured in a defined medium called DFCI-2. DFCI-2 medium contains all factors present in DFCI-1 medium (**Table 1**) except that it lacks fetal calf serum and bovine pituitary extract. In addition, it contains 0.05% bovine serum albumin (Sigma). DFCI-2 medium only allows the growth of immortal cells. Normal cells do not proliferate in DFCI-1 medium beyond three passage *(10)*.
2. Growth of cells beyond their life span indicates extention of life span (more than 10 passages or 30 population doublings after transfection).
3. Cells are considered immortal if cells fail to exhibit crisis and continue to proliferate beyond 30 passages (about 100 population doublings).
4. Immortal cells are routinely cultured at 1/10 split ratio in DFCI-2 medium.

3.7. Assessment of Transformation Phenotype of Immortal Cells

1. Add 2 mL of 0.5% agar (1/1 in 2× DFCI-1 medium) in a 60-mm dish.
2. Let the agar solidify at room temperature. These plates can be used after 1 h or stored at 4°C for up to a month. If stored, warm plates at 37°C before use.
3. Mix 5×10^4 cells with 0.3% agar and overlay this on top of 0.5% agar. The 0.3% agar can solidify quickly. To avoid this problem, keep the tubes containing 2 mL of 0.3% agar in a beaker containing water at 40°C. Add cells to agar, mix quickly, and overlay on top of the bottom agar.
4. Cells are fed weekly with 1× DFCI-1 medium and are examined microscopically for clonal growth after 2 wk.

3.8. Growth of Implanted Cells in Nude Mice

1. Grow cells to 70% confluence (logarithmically growing).
2. Trypsinize cells, count and resuspend at a density of 5×10^6 cells per 200 μL of saline. Keep cell suspension on ice until injection.
3. Inject 5×10^6 cells per 200 μL of saline subcutaneously in the mammary gland area (below the nipple) of BALB/c nude mice.
4. Examine mice daily for the appearance of palpable tumors.
5. Mice are monitored for up to 3 mo for the appearance of visible tumors.
6. Excised tumors are used for histopathology and to assess the ability of tumor cells to grow in culture.

4. Notes

1. Such treatment preferably removes fibroblasts but not epithelial cells, which are more tightly attached to plastic and require trypsinization for 5–10 min at 37°C.

5. References

1. Farber, E. (1984) The multistep nature of cancer development. *Cancer Res.* **44**, 4217–4223.
2. Klein, G. and Klein, E. (1985) Evolution of tumors and the impact of molecular oncology. *Nature* **315**, 190–195.
3. Hammond, S. L., Ham, R. G., and Stampfer, M. R. (1984) Serum-free growth of human mammary epithelial cells: rapid clonal growth in defined medium and extended serial passage with pituitary extract. *Proc. Natl. Acad. Sci. USA* **81**, 5435–5439.
4. Petersen, O. W. and van Deurs, B. (1987) Preservation of defined phenotypic traits in short-term cultured human breast carcinoma derived epithelial cells. *Cancer Res.* **47**, 856–866.
5. Band, V. and Sager, R. (1989) Distinctive traits of normal and tumor-derived human mammary epithelial cells expressed in a medium that supports long-term growth of both cell types. *Proc. Natl. Acad. Sci. USA* **86**, 1249–1253.
6. Taylor-Papadimitriou, J., Shearer, M., and Stoker, M. G. P. (1977) Growth requirements of human mammary epithelial cells in culture. *Int. J. Cancer* **20**, 903–908.
7. Romanov, S. R., Kozakiewicz, B. K., Holst, C. R., Stampfer, M. R., Haupt, L. M., and Tisty, T. D. (2001) Normal human mammary epithelial cells spontaneously escape senescence and acquire genomic changes. *Nature* **409**, 633–637.
8. Taylor-Papadimitriou, J., Stampfer, M., Bartek, J., et al. (1989) Keratin expression in human mammary epithelial cells cultured from normal and malignant tissue: relation to in vivo phenotypes and influence of medium. *J. Cell Sci.* **94**, 403–413.
9. Trask, D. K., Band, V., Zajchowski, D., Yaswen, P., Suh, T., and Sager, R. (1990) Keratin proteins as markers that distinguish normal and tumor-derived mammary epithelial cells. *Proc. Natl. Acad. Sci. USA* **87**, 2319–2323.
10. Wazer, D. E., Liu, X.-L., Chu, Q., Gao, Q., and Band, V. (1995) Immortalization of distinct human mammary epithelial cell types by human papilloma virus-16 E6 or E7. *Proc. Natl. Acad. Sci. USA* **92**, 3687–3691.
11. Ratsch, S. B., Gao, Q., Srinivasan, S., Wazer, D. E., and Band, V. (2001) Multiple genetic changes are required for efficient immortalization of different subtypes of normal human mammary epithelial cells. *Radiation Res.* **155,** 143–150.
12. Band, V. (1995) Preneoplastic transformation of human mammary epithelial cells. *Semin. Cancer Biol.* **6**, 185–192.
13. Stampfer, M. R. and Bartley, J. C. (1985) Induction of transformation and continuous cell lines from normal human mammary epithelial cells after exposure to benzo(a)pyrene. *Proc. Natl. Acad. Sci. USA* **82**, 2394–2398.
14. Wazer, D. E., Chu, Q., Liu, X-L., Gao, Q., Safaii, H., and Band, V. (1994) Loss of p53 protein during radiation transformation of primary human mammary epithelial cells. *Mol. Cell. Biol.* **14**, 2468–2478.
15. Liu, X.-L., Wazer, D. E., Watanabe, K., and Band V. (1996) Identification of a novel serine protease-like gene, the expression of which is down-regulated during breast cancer progression. *Cancer Res.* **56**, 3371–3379.
16. Goyal, J., Smith, K. M., Cowan, J. M., Wazer, D. E., Lee, S. W., and Band, V. (1998) The role for NES1 serine protease as a novel tumor suppressor. *Cancer Res.* **58**, 4782–4786.
17. Dhar, S., Bhargava, R., Yunes, M., et al. (2001) Analysis of normal epithelial cell-specific 1 (NES1)/Kallikrein 10 mRNA expression by in situ-hybridization, a novel marker for breast cancer. *Clin. Cancer Res.* **7,** 3393–3398.

18. Li, B., Goyal, J., Dhar, S., et al. CpG Methylation as a basis for breast tumor-specific loss of NES1/Kallikrein 10 expression. *Cancer Res.* **61,** 8014–8021.

19. Sager, R. (1984) Resistance of human cells to oncogenic transformation. *Cancer Cells* **2,** 487–494.

20. DiPaolo, J. A. (1983) Relative difficulties in transforming human and animal cells in vitro. *J. Natl. Cancer Inst.* **70,** 3–8.

21. Shay, J. W., Van Der Haegen, B. A., Ying, Y., and Wright, W. E. (1993) The frequency of immortalization of human fibroblasts and mammary epithelial cells transfected with SV40 large T-antigen. *Exp. Cell Res.* **209,** 45–52.

22. Zur Hausen, H. (1994) Molecular pathogenesis of cancer of the cervix and its causation by specific human papillomavirus types. *Curr. Topics Micro. Immunol.* **186,** 131–156.

23. Kaur, P. and McDougall, J. K. (1989) HPV-18 immortalization of human keratinocytes. *Virology* **173,** 302–310.

24. Band, V., Zajchowski, D., Kulesa, V., and Sager, R. (1990) Human papilloma virus DNAs immortalize normal human mammary epithelial cells and reduce their growth factor requirements. *Proc. Natl. Acad. Sci. USA* **87,** 463–467.

25. Munger, K., Phelps, W. C., Bubb, V., Howley, P. M., and Schlegel, R. (1989) The E6 and E7 genes of the human papilloma virus type 16 together are necessary and sufficient for the transformation of primary human keratinocytes. *J. Virol.* **63,** 4417–4421.

26. Watanabe, S., Kanda, T., and Yoshiike, K. (1989) Human papillomavirus type 16 transformation of primary human embryonic fibroblasts requires expression of open reading frames E6 and E7. *J. Virol.* **63,** 965–969.

27. Bartek, J., Bartkova, J., Kyprianou, N., et al. (1991) Efficient immortalization of luminal epithelial cells from human mammary gland by introduction of simian virus 40 large tumor antigen with a recombinant retrovirus. *Proc. Natl. Acad. Sci. USA* **88,** 3520–3524.

28. Huibregtse, J. M., Scheffner, M., and Howley, P. M. (1991) A cellular protein mediates association of p53 with the E6 oncoprotein of human papillomavirus types 16 or 18. *EMBO J.* **10,** 4129–4135.

29. Chen, J. J., Reid, C. E., Band, V., and Androphy, E. J. (1995) Interaction of papillomavirus E6 oncoproteins with a putative calcium-binding protein. *Science* **269,** 529–531.

30. Tong, X. and Howley, P. M. (1997) The bovine papillomavirus E6 oncoprotein interacts with paxillin and disrupts the actin cytoskeleton. *Proc. Natl. Acad. Sci. USA* **94,** 4412–4417.

31. Kiyono, T., Hiraiwa, A., Fujita, M., Hayashi, Y., Akiyama, T., and Ishibashi, M. (1997) Binding of high-risk human papillomavirus E6 oncoproteins to the human homologue of the Drosophila discs large tumor suppressor protein. *Proc. Natl. Acad. Sci. USA* **94,** 11612–11616.

32. Ronco, L. V., Karpova, A. Y., Vidal, M., and Howley, P. M. (1998) Human papillomavirus 16 E6 oncoprotein binds to interferon regulatory factor-3 and inhibits its transcriptional activity. *Genes Dev.* **12,** 2061–2072.

33. Kuhne, C. and Banks, L. (1998) E3-ubiquitin ligase/E6-AP links multicopy maintenance protein 7 to the ubiquitination pathway by a novel motif, the L2G box. *J. Biol. Chem.* **18,** 34302–34309.

34. Thomas, M. and Banks, L. (1999) Human papillomavirus (HPV) E6 interactions with Bak are conserved amongst E6 proteins from high and low risk HPV types. *J. Gen. Virol.* **80,** 1513–1517.

35. Zimmermann, H., Degenkolbe, R., Bernard, H. U., and O'Connor, M. J. (1999) The human papillomavirus type 16 E6 oncoprotein can down-regulate p53 activity by targeting the transcriptional coactivator CBP/p300. *J. Virol.* **73,** 6209–6219.

36. Gao, Q., Srinivasan, S., Boyer, S. N., Wazer, D. E., and Band, V. (1999) The E6 oncoproteins of high-risk papillomaviruses bind to a novel putative GAP protein, E6TP1, and target it for degradation. *Mol. Cell. Biol.* **19,** 733–744.

37. Gao, Q., Kumar, A., Srinivasan, S., et al. (2000) PKN binds and phosphorylates human papillomavirus E6 oncoprotein. *J. Biol. Chem.* **275,** 14824–14830.
38. Nakagawa, S. and Huibregtse, J. M. (2000) Human scribble (Vartul) is targeted for ubiquitin-mediated degradation by the high-risk papillomavirus E6 proteins and the E6AP ubiquitin-protein ligase *Mol. Cell. Biol.* **20,** 8244–8253.
39. Degenhardt, Y. Y. and Silverstein, S. J. (2001) Gps2, a protein partner for human papillomavirus E6 proteins. *J. Virol.* **75,**151–160.
40. Huibregtse, J. M., Scheffner, M., and Howley, P. M. (1993) Cloning and expression of the cDNA for E6-AP, a protein that mediates the interaction of the human papillomavirus E6 oncoprotein with p53. *Mol. Cell. Biol.* **13,** 775–784.
41. Dalal, S., Gao, Q., Androphy, E. J., and Band, V. (1996) Mutational analysis of human papillomavirus type 16 E6 demonstrates that p53 degradation is necessary for immortalization of mammary epithelial cells. *J. Virol.* **70,** 683–688.
42. Kiyono, T., Foster, S. A., Koop, J. I., McDougall, J. K., Galloway, D. A., and Klingelhutz, A. J. (1998) Both Rb/p16INK4a inactivation and telomerase activity are required to immortalize human epithelial cells. *Nature* **396,** 84–88.
43. Gao, Q., Singh, L., Kumar, A., Srinivasan, S., Wazer, D. E., and Band V. (2001) HPV 16 E6 induced degradation of E6TP1 correlates with its ability to immortalize human mammary epithelial cells. *J. Virol.* **75,** 4459–4466.
44. Nevins, J. R. (1992) E2F: a link between the Rb tumor suppressor protein and viral oncoproteins. *Science* **258,** 424–429.
45. Davies, R., Hicks, R., Crook, T., Morris, J., and Vousden K. (1993) Human papillomavirus type 16 E7 associates with histone H1 kinase and with p107 through sequences necessary for transformation. *J. Virol.* **67,** 2521–2528.
46. Boyer, S. N., Wazer, D. E., and Band, V. (1996) E7 protein of human papilloma virus-16 induces degradation of retinoblastoma protein through the ubiquitin-proteasome pathway. *Cancer Res.* **56,** 4620–4624.
47. O'Farrell, P. Z., Goodman, H. M., and O'Farrell, P. H. (1977) High resolution two-dimensional electrophoresis of basic as well as acidic proteins. *Cell* **12,** 1133–1141.
48. Dimri, G. P., Itahana, K., Acosta, M., and Campisi, J. (2000) Regulation of a senescence checkpoint response by the E2F1 transcription factor and p14ARF tumor suppressor. *Mol. Cell. Biol.* **20,** 273–285.

21

Tumor Suppression Through Angiogenesis Inhibition

Andrew M. Arsham and M. Celeste Simon

1. Background: Tumor Angiogenesis
1.1. Introduction

In the early 1970s, Judah Folkman first proposed the hypothesis that tumor growth was dependent on the recruitment of blood vessels, a process called angiogenesis, and that angiogenesis was both an important stage in tumor development and a promising therapeutic target. Based on his observations as a surgeon, he also postulated the existence of circulating molecules that could inhibit this process, so-called angiogenesis inhibitors. The ideas of Folkman and other pioneers (reviewed in **ref. 1**), received skeptically by the scientific community at the time, have become one of the central dogmas of cancer research. A PubMed search for the terms "tumor" and "angiogenesis" turns up 508 papers between 1981 and 1991, and 4604 between 1991 and 2001, a 9-fold increase (compared to a 1.3-fold increase in the word "tumor" alone over the same period). Nearly a thousand such papers have been published in the past year. In addition to the angiogenesis inhibitors that Folkman's own group isolated, upwards of 40 biological inhibitors of angiogenesis have been discovered in recent years, as well as numerous pharmacologic compounds with similar activities. Many of these molecules are currently undergoing clinical trials as anticancer drugs (reviewed in **ref. 2**).

This chapter consists of a brief discussion of the process of angiogenesis and its promise as a therapeutic target; in vivo animal models for studying tumor angiogenesis; laboratory methods for its quantitation and comparison; and a summary of the literature to date on some of the major genes involved. It will focus principally on in vivo genetic strategies for understanding angiogenesis as opposed to more clinically or therapeutically oriented studies. Our intent is to provide the reader with a strong understanding of the complex and interrelated processes on which angiogenesis depends, and on the genes that control them.

1.2. Angiogenesis as a Therapeutic Target

Three major drawbacks of current cancer therapy are toxicity, resistance, and lack of specificity. Most (if not all) chemotherapy drugs have toxic side effects, and tumor cells are able to develop chemoresistance over the course of treatment. In addition, both

From: *Methods in Molecular Biology, Vol. 223: Tumor Suppressor Genes: Regulation, Function, and Medicinal Applications.* Edited by: Wafik S. El-Deiry © Humana Press Inc., Totowa, NJ

chemotherapy and radiation therapy target all dividing cells, as opposed to just dividing tumor cells, leaving much room for improvement in the specificity of these treatments. Antiangiogenic therapy addresses these concerns in a number of ways. While angiogenesis is of critical importance in mammalian development and growth, its only role in adults is pathophysiologic, excepting the menstrual cycle. So inhibition of angiogenesis is, in principle, unlikely to have toxic side effects and is also targeted more directly at rapidly growing tumors, which depend on new vessels, as opposed to normal dividing cells, which should already have an adequate blood supply. In addition, cells cannot ultimately survive without O_2 and thus will not develop resistance to antiangiogenic therapy *(1)*. Angiogenesis inhibition is not a silver bullet: it is unknown what effects long-term inhibition of angiogenesis might have on patients, or on common pathophysiologic processes such as wound healing, tissue ischemia/hypoxia, and tissue repair. Like any therapy that has been extensively tested in rodent models but has yet to be fully explored in humans, safety and efficacy cannot be assumed; they must be demonstrated in well-controlled clinical trials. Antiangiogenic cancer therapy remains, however, a highly promising avenue of exploration.

1.3. Tumor Angiogenesis: A Synopsis

1.3.1. Growth

Initially, the development of cancer is driven by repeated rounds of stochastic genetic mutations and positive selection for cells with advantageous mutations. Oncogenesis is a multistep process of mutation and clonal selection that results in a highly proliferative cell mass that resists instructions to stop proliferating or to die *(3)*. One class of mutations that can contribute to tumor progression is inactivating mutations of tumor suppressor genes including cell cycle regulators, apoptosis effectors, death receptors and components of their signaling pathways, and genes that inhibit growth factor signaling. Another class of mutations is those that hyperactivate protooncogenes such as growth factors and their receptors, inhibitors of apoptosis, and positive regulators of the cell cycle.

1.3.2. Crisis

As tumor cells proliferate, the tumor will outstrip its resources and reach a metabolic crisis point. Two models exist to explain how this crisis point is reached. The "avascular growth" model posits an initial tumor growth phase that takes place in the absence of blood vessels. The alternative is the "vessel cooption" model, which posits that the tumor grows around, or "coopts" existing vessels *(4)*, which are then thought to regress and essentially withdraw support from the tumor. It is probable that both the avascular growth and vessel cooption models are correct depending on tumor types and model systems, and both are instructive. In either case, the tumor's huge potential for expansion is curtailed at a certain point by its inability to obtain the necessary nutrients. Since the diffusion limit of O_2 is 100–200 µm, the cells in the center of the proliferating tumor mass will no longer be able to obtain O_2 and other nutrients via simple diffusion from existing or regressed vessels. The propensity of tumor cells to have highly glycolytic metabolisms (known as the Warburg effect) also results in high lactic acid production. The tumor microenvironment is therefore hypoxic, hypoglycemic, and acidotic, resulting in central areas of necrosis and a block in tumor growth *(5–7)*.

1.3.3. Resolution: The Angiogenic Switch

The metabolic crisis is resolved when the tumor undergoes the "angiogenic switch," loosely defined as the point at which the ratio of angiogenic factors to antiangiogenic factors becomes greater than 1. Many different models can be invoked to explain the angiogenic switch, but one molecule that is common among them is vascular endothelial growth factor (VEGF), a highly potent and endothelial cell-specific cytokine. VEGF is thought to be the most important of many positive effectors of the process of angiogenesis. It can be upregulated by the hypoxia responsive transcription factor hypoxia inducible factor 1 (HIF-1, *see* **Subheading 4.3.**), either through hypoxia or HIF-enhancing mutations *(8–10)* (reviewed in **ref. *11***), as well as by many other oncogenic mutations *(12)*.

Once VEGF is secreted by tumor cells, it activates quiescent endothelial cells on nearby blood vessels. One of the first observable markers for such activated endothelium is integrin αvβ3, a receptor for a wide array of extracellular matrix (ECM) components. Integrins in general are thought to mediate cell adhesion and migration, and αvβ3 appears to be an important mediator of these processes during angiogenesis *(13)*.

Once the endothelial cells are activated and proliferating, the final barrier to O_2 delivery is the basement membrane, if present, separating the mature vessel from the tumor. The sprouting vessel must cross the basement membrane and migrate through the extracellular matrix into the tumor proper. These processes are just beginning to be understood in the context of tumor angiogenesis, and require the cooperation of the matrix metalloproteinases (MMPs), integrins and other adhesion molecules, and the components of the ECM. In contrast, the tissue inhibitors of metalloproteinases (TIMPs) and the thrombospondin (TSP) family of ECM proteins, among other factors, inhibit vessel migration and invasion. Negative feedback loops tightly regulate the progress of vessel invasion, for example when proinvasion proteases liberate inhibitors of endothelial cell growth like endostatin and angiostatin from the ECM. Thus, there is a fine balance between pro- and antiangiogenic stimuli, and this balance inherently favors stasis.

2. In Vivo Strategies for Genetic Analysis of Tumor Angiogenesis

2.1. Tumor Xenograft Experiments

In tumor xenograft experiments, transformed cells, established tumor cell lines, or embryonic stem (ES) cells are injected into mice, where they proliferate to form solid tumors. Cells are usually injected subcutaneously, though they can also be injected intracerebrally, or within other organs such as the pancreas and gut wall. For solid tumor formation, cells must be injected into an enclosed space where they can form a solid mass. These experiments are somewhat empirical, in that not all tumor lines will establish tumors when injected. Most experiments utilize 10^5–10^7 cells. The host mice must be either immunocompromised (e.g., nude, SCID, or $RAG^{-/-}$) or syngeneic to the tumor line. There are several established syngeneic tumor models such as the B16 melanoma line, which is derived from BL6 mice and will not be rejected by these mice.

Some important caveats of tumor xenograft experiments include: (a) the tumor cells are not growing in an environment like that in which they arose and therefore might respond differently to this milieu than to a more physiologically relevant location; (b) subcutaneous injection puts the tumor cells in a well-enclosed space that lacks much of an established vasculature—angiogenesis must then occur through a basement membrane, a potentially nonphysiologic situation *(14)*. Cranial and subcutaneous tumors can also be established using the dorsal window technique, which implants a clear glass window over the xenograft site, enabling intravital microscopy and observation and quantitation of angiogenesis during tumor growth. Intravital microscopy can be used to measure vessel quantity and morphology, hemodynamics, extravasation of labeled particles, and many other parameters (for example, *see* **ref. *15***). This provides an efficient way to do quantitative and qualitative time points, but only one face of the tumor is in contact with the host, the other being in contact with glass. Direct measurements have shown that this creates an O_2 gradient within the tumor that is qualitatively different than that in a standard xenograft model *(16)*.

2.1.1. ES Cell-Derived Teratomas

ES cells, when injected into immunocompromised mice, form teratomas consisting of all three primordial germ layers, reflecting the ability of ES cells in vivo to differentiate into all cell lineages. The ES-derived teratoma system has some unique advantages over more commonly used tumor models, namely, that genes can be ablated in ES cells via homologous recombination, and that ES cells are a nontransformed primary cell line. An investigator can thus test the importance of a gene in the absence of transforming oncogenes or the many unknown genetic lesions in cancer cells. Homozygous null ES cells can be derived by a number of methods, providing a way to study effects of gene ablation even if it confers an early embryonic lethality or the gene is haploinsufficient *(17)*.

The ES/teratoma system has its drawbacks. Since ES cells are totipotent, the development of the tumor in some ways recapitulates embryonic development. Thus, if the gene of interest has a developmental phenotype, it may be recapitulated in the teratoma, and careful histologic examination of the tumor will be necessary to determine whether the tissue composition and architecture of the tumors are different from relevant controls. Differentiation also presents a problem in that differentiated cells may produce compensatory molecules not expressed in ES cells themselves. Thus tumor growth may select for cells that express such molecules or family members redundant to the gene of interest, masking possible phenotypes.

2.1.2. Transformed Cell Lines

Many cells, such as transformed mouse embryo fibroblasts (MEFs) or tumor-derived cell lines, can also be xenografted into immunocompromised or syngeneic mice. The benefits of this system are as numerous as its caveats. On one hand, it provides an excellent way to study many different tumor types, and the opportunity to modify their gene expression in vitro by transfection or viral infection before assaying their in vivo properties in a well-controlled and flexible experimental system. In addition, MEFs nullizygous for a gene of interest can often be derived from mutant embryos, providing a way to study nullizygous cells that are not pluripotent, one way to address or circumvent the problems with using ES cells. On the other hand, few tumor lines are very well characterized with respect to the number and type of genetic mutations they possess, and that will necessarily affect the experimental outcome. Known metastatic cell lines, or nonmalignant cell lines that have been modified to express putative metastasis-inducing genes, can be injected intravenously to study the number and size of metastases, most often in lung or liver. Many of the benefits and caveats of solid tumor xenografts apply here as well.

2.2. Mutant or Transgenic Cells, Wild-Type Host

Gene expression within tumors is thought to be the primary driver of the angiogenic switch. One important test of the function of a gene is whether its presence, absence, modification, or overexpression within the tumor affects tumor growth and angiogenesis. Likely candidates include genes for secreted growth factors such as VEGF; signal transducers or transcription factors that may transactivate such growth factors; and receptors for proliferation or survival signals or their effectors. One type of experiment that this system enables are competition experiments that can distinguish between cell-intrinsic or autocrine, and cell extrinsic or paracrine defects.

2.3. Wild-Type Cells, Mutant or Transgenic Host

Host tissues make important contributions to tumor vasculature and architecture *(18,19)*. Thus, another test of a gene's function in tumor angiogenesis is whether its absence, modification, or overexpression within host tissues, especially the tumor stroma or surrounding tissues, can affect tumor angiogenesis. Among the candidate classes of genes are endothelial cell-specific receptors or their effectors; adhesion molecules and proteases or their inhibitors; components of the host ECM; and circulating antiangiogenic molecules. Animals with modifications in any of these candidates can be tested for their ability to support tumor xenografts.

2.4. Crosses Between Mutant or Transgenic Animals

Animals that are transgenic for a dominant oncogene or mutant for tumor suppressors exist that will form spontaneous tumors at known frequencies and of known morphologies and locations. These models can be used to test a gene product's role in tumorigenesis or angiogenesis in a more physiologic setting than the others discussed here. Mice transgenic or mutant for a gene of interest can be crossed with the oncogenic mice to see how tumor growth and angiogenesis are modified (for example, *see* **Subheading 4.6.**).

3. Techniques for Visualizing and Measuring Angiogenesis

3.1. Two-Dimensional

Tumor tissues can be frozen or fixed in paraformaldehyde or formalin, and sectioned using standard histologic techniques for two-dimensional analysis of vascular phenotype. This is performed by endothelial cell staining and microscopic analysis. This can be done with antibody to CD31/PECAM or von Willebrand factor (vWF), or by using an endothelial cell-specific lectin (Griffonia Simplicifolia Lectin I isolectin B4, Vector Labs). CD31/PECAM staining is generally not recommended for paraformaldehyde– or formalin-fixed sections, for which the lectin staining protocols work quite well.

Various quantitative methods can then be used to assess vascularity. Care should be taken to choose regions of high vascularity, as tumors, especially teratomas, can be highly heterogeneous. All solid tumors will have avascular or necrotic regions, and these regions should not be used to determine vascular density. Since, experimentally, one is really interested in assessing angiogenic potential, it is important to choose the most vascular region within the tumor, avoiding the tumor margins. It is most useful to select a highly vascular region at a relatively low magnification, and then move to higher magnification to evaluate multiple fields (3–5) within the vascular "hot spot" to obtain a statistically useful average.

Counting methods vary. One can simply count the number of vessels in a given field, or classify them according to size or morphology (for example **ref. 10**). Another, more formal approach, employs a Chalkley graticule, an eyepiece with a number of randomly distributed points on it, which overlays the microscope eyepiece. The graticule is rotated so that the highest number of points overlay areas of endothelial staining. These points are then counted and the highest 3 or 4 of 5 counts are averaged and expressed as a histogram.

3.2. Three-Dimensional

Three-dimensional imaging of mammalian vasculature requires that the animal be alive when perfused with the imaging agent. Fluorescent (FITC-labeled) tomato lectin (Vector Labs) or high-molecular-weight dextran (Mr 500,000, Molecular Probes, Inc.) can both be injected intravenously (usually via tail vein) and allowed to circulate through the animal for approximately 20 min. If a dorsal window model is being used, fluorescent intravital microscopy can then be performed (for example, see **ref. 15**) If not, the tumor can be removed, sectioned, and observed on a fluorescent confocal microscope. Common imaging software can also be used to quantitate the vasculature of the tumor, either by total fluorescence, average vessel diameter, total vessel length per unit area of observation field, number of branch points, or other morphologic criteria of interest to the investigator

Another class of reagents are curable latexes, such as Microfil MV-122 (Microfil, Inc.), which perfuse the animal and are allowed to polymerize before the surrounding tissue is cleared. This results in solid three-dimensional casts of the tumor vasculature. Quantitative methods similar to those mentioned above can be used (for example, *see* **ref. 20**).

4. Genes That Modulate Tumor Angiogenesis

4.1. VEGF

VEGF, a potent endothelial cell-specific cytokine, is widely acknowledged to be the *sine qua non* of angiogenesis and neovascularization. Even slight perturbations in the levels of VEGF during embryonic development are lethal: the *Vegf* gene is haploinsufficient *(21,22)*, the absence of two of the three splice variants confers postnatal lethality *(23)*, and even overexpression of VEGF protein causes developmental lethality *(24)*. VEGF appears to activate a number of signal transduction pathways by binding to FLK-1, one of its high-affinity, endothelial cell-specific, receptor tyrosine kinases. Downstream effectors include the MAPK, PI-3 kinase, src family kinase, ras, and PKC pathways. VEGF also appears to activate $\alpha v \beta 3$ integrin *(25)* (*see* **Subheading 4.7.**), which can colocalize with FLK-1 on the surface of endothelial cells *(13)*. The precise roles these different pathways play in endothelial cell activation, proliferation, survival, and morphology remain to be fully elucidated *(26)*.

In addition to being induced by hypoxia, VEGF has also been shown to be induced by hypoglycemia and by a wide array of oncogenes and repressed by a variety of tumor suppressors *(12)* Murine VEGF protein exists as three different isoforms of 120, 164, and 188 amino acids (121, 165, and 189 in humans), which are translated from three distinct mRNA splice variants and appear to have both overlapping and unique functions. The existing data on the relative importance and roles of the different isoforms, some of which are reviewed below, lacks consensus.

Vegf$^{-/-}$ ES cells form poor teratomas when injected into nude mice, due to a serious defect in angiogenesis *(15,22)*. Even in *Vegf*$^{-/-}$ tumors, VEGF protein levels within tumor tissues was still 50% of that in wild-type. This high level of host-derived VEGF demonstrates an important role for the surrounding host stromal tissues in VEGF production and tumor angiogenesis *(15)*.

Transformed *Vegf*$^{-/-}$ MEFs form fibrosarcomas on average 10% the size of *Vegf*$^{+/+}$ controls, with similar rates of proliferation within the tumor but with increased hypoxia and apoptosis, and marked decreases in vascular permeability and vascular surface area *(27)*. When these *Vegf*$^{-/-}$ MEFs were stably transfected to express only one of the major isoforms of VEGF, VEGF164 was found to almost completely rescue both tumor size and vascular density, demonstrating that it is necessary and sufficient for tumor-driven angiogenesis, while VEGF120 was only able to partly rescue the size and vascular density. Paradoxically, VEGF188, while actually increasing the vascular density of the tumors, was nonetheless unable to rescue the size defect, suggesting that the other splice variants are performing a nonangiogenesis-related function that is necessary for tumor growth *(28)*. This is especially interesting in light of findings that, in human lung cancers, VEGF188 was the isoform whose expression most strongly correlated with angiogenesis, relapse, and mortality *(29)*. Studies of the different isoforms have also been carried out in other model systems, yielding somewhat muddled results. One study revealed that MCF-7 breast carcinoma clones overexpressing VEGF121, but not those expressing the 165 or 189 isoform, formed significantly faster-growing and more vascular tumors when subcutaneously xenografted into nude mice *(30,31)*, though other studies showed that VEGF165 strongly and significantly increased vascular density in a similar system *(32)*. VEGF165 also seems to play a critical role in endothelial cell survival after angiogenesis itself, since repression of VEGF165 expression in established glioma xenografts caused extensive regression of newly formed vessels *(33)* VEGF165 or 121, when overexpressed in glioblastoma cells, cause severe intracerebral bleeding just 60–90 h after being implanted in the brains of nude mice. VEGF189-expressing tumors, though larger and more vascular than wild-type, lacked this extreme phenotype *(34)*. VEGF121 can also cause a human colon cancer line to become more tumorigenic, angiogenic, and metastatic *(35)* and can accelerate tumor growth in cancer-prone animals *(36)*.

Many groups have used sense and antisense expression constructs to confirm the importance of VEGF in tumor angiogenesis. Exogenously expressed VEGF increases the tumorigenicity and/or vascularity of tumor lines that expressed little endogenous VEGF or formed poorly vascularized tumors. Conversely, in highly angiogenic lines, expression of antisense VEGF constructs inhibits tumor growth and angiogenesis. These trends are observable over a wide variety of tumor models, such as melanoma *(37,38)*, glioblastoma *(39,40)*, squamous cell carcinoma *(41)*, hepatocellular carcinoma *(42)*, and thyroid carcinoma *(43)*.

Though the role and importance of VEGF in tumor angiogenesis is well established, some fundamental questions remain unanswered. The functions and roles that the different isoforms play remains unclear, especially given the somewhat contradictory data from different experimental systems. Also unclear is the role of VEGF in endothelial proliferation vs survival. It is clear, however, that VEGF is the central effector of the angiogenic switch and, as such, is the subject of intense research into mechanisms and signaling pathways as the search for new targets for antiangiogenic therapy continues. Drugs, antibodies, or blocking peptides that inhibit VEGF signaling are one promising area of inquiry and clinical investigation.

4.2. VEGF Receptors

VEGF binds to numerous receptors, including FLT-1 (VEGFR1), FLK-1 (KDR/ VEGFR2), and neuropilin-1 (NRP1). Since VEGF receptors sense and respond to VEGF signals coming from the tumor, expression of these molecules is clearly more important in host tissues than in the tumor itself. In both development and tumor growth, a consensus exists that FLK-1 is the most important receptor in terms of positive VEGF signals, and that FLT-1, at least in this physiologic context, exists mainly as an antagonist or sink for VEGF signaling (reviewed in **ref. 26**). Since hemizygosity for VEGF confers embryonic lethality, it is not surprising that posttranslational mechanisms exist to tightly regulate its signaling in vivo. NRP1 is a receptor for the semaphorin family of ligands, and is involved in both neural and cardiovascular development *(44)*. It has a broad expression profile, including endothelial cells and tumor cells, and in addition to its high affinity for semaphorin family members, has a high affinity for VEGF165 *(45)*. NRP1 lacks tyrosine kinase activity and appears to enhance VEGF/FLK1 interactions.

The most widespread method for ablating the activity of VEGF receptors in the context of adult mice is by overexpressing dominant interfering truncations or mutants, creating a ligand sink. Overexpression of either truncated or soluble FLT-1 in tumor cells interferes with tumor growth and vascularization *(46,47),* as does expression of soluble FLT-1 by muscle tissue distant from the xenograft site *(47)* An interesting model that further explores the role of FLT-1 in host tissue utilizes transgenic mice expressing a kinase dead, and thus dominant interfering, form of the receptor. Lewis lung carcinoma cells overexpressing PlGF-2, a specific ligand for FLT-1, grew much faster and formed more vascular tumors when injected into wild-type mice compared to transgenic mice, showing that FLT-1, in the presence of an appropriate ligand, can play a positive role in tumor growth and angiogenesis. Cells that overexpressed VEGF showed no differences in growth between the two types of mice, demonstrating the specificity of this interaction, and reinforcing the idea that FLT-1 has little or no positive signaling role with respect to VEGF *(48)*.

Studies of interfering mutations have also been carried out on FLK-1, to similar outcomes. In vivo retroviral infection of host endothelial cells with a dominant negative form of FLK1 prevented the subcutaneous growth of several different tumor types in nude mice *(49,50)*. A similar trend was seen in mouse mammary carcinoma lines *(51)* and rat gliomas *(52)*.

Neuropilin's role in tumor angiogenesis is unclear. It retains function as a VEGF receptor in the context of a tumor, as overexpression of NRP1 by rat prostate carcinoma cells increased tumor size by 2.5–7-fold, causing markedly increased microvessel density, increased endothelial proliferation, and less tumor cell apoptosis *(53)* This is counterintuitive, since VEGF receptors are thought to be required primarily on host endothelium. One possible model for NRP1's proangiogenic function in this context is that it binds VEGF to the surface of tumor cells, preventing VEGF diffusion and contributing to neovessel recruitment. Interestingly, a soluble splice variant of NRP1 has recently been identified that maintains its specificity for VEGF165 and causes extensive tumor hemorrhage, blood vessel damage, and apoptosis when expressed in rat prostate carcinoma cells, suggesting that it can behave as a VEGF165 antagonist *(54)* and nicely demonstrating that both the membrane-bound and soluble forms of neuropilin can interact with VEGF outside the context of the nervous system, though the significance of this interaction is unclear.

The interfering receptor studies outlined above must be interpreted cautiously—since the receptors used will bind VEGF and prevent VEGF signaling from occurring, many are more easily interpreted as VEGF neutralization experiments than as rigorous tests of the functions of the various receptors themselves. It is notable, however, that FLT-1, often thought of as an inhibitory molecule or a "sink" for VEGF, can have a positive effect on angiogenesis when an appropriate ligand (i.e., PlGF-2) is produced by the tumor. It is also interesting that NRP1 can, on the surface of tumor cells, contribute to angiogenesis, though it is thus far unclear whether this is physiologically important or not. Caveats aside, blocking VEGF signaling at the level of receptor/ligand interactions is a promising therapeutic approach, and various techniques for administering truncated or soluble VEGF receptors are currently being investigated, as are drugs that inhibit the signal transduction cascades downstream of the VEGF receptors.

4.3. Hypoxia-Inducible Factors

One important regulator of VEGF levels is hypoxia. The hypoxia-inducible factors (HIFs) are a family of heterodimeric transcription factors that are critical mediators of the cellular and organismal response to hypoxia. HIF-1, the best studied of these factors, consists of a constitutive subunit, ARNT (HIF-1β), and an oxygen-regulated subunit, HIF-1α. HIF-1α is posttranslationally upregulated at low O_2 concentrations and subsequently forms transcriptionally active heterodimers with ARNT. HIF-1 is known to transactivate a wide array of genes (over 30) that increase cellular glucose transport and glycolysis, increase the O_2 carrying capacity of existing vasculature, and activate angiogenesis via increased production of VEGF (*see* **Subheading 4.4.**) *(55)*. Mice that lack either HIF-1 subunit die *in utero* between E8.5 and 10 with a variety of cardiovascular, hematopoietic, and placental defects *(56–59)*. One interpretation of these studies is that O_2 levels can act as a developmental morphogen, and that, as in tumors, hypoxia can activate aspects of angiogenesis *(60)*. The roles of the more tissue-restricted ARNT2 and HIF-2α subunits have been less well studied, but both are essential *(61–63)*, and HIF-2α is regulated by O_2 in a nearly identical fashion to HIF-1α. In addition to being upregulated in vitro and in vivo by hypoxia, HIF activity also appears to be modulated by commonly mutated tumor pathways including the PI-3 kinase/Akt, MAPK, and ras pathways in certain tumor cell lines (reviewed in **ref. *11***). HIF-α subunits are also rendered constitutively stable by homozygous loss of the pVHL tumor suppressor gene, which produces the E3 ubiquitin ligase responsible for the ubiquitination and degradation of HIF-αs under normoxia *(64)*. In addition, several groups have found that HIF-1α can stabilize p53 in hypoxic cells, and that this interaction may play a role in hypoxia-mediated cell cycle arrest *(65–68),* though this interaction has not been universally observed *(69)*. p53 can repress HIF-1α mediated transcription in vitro *(70),* and can, through its negative regulator MDM2, downregulate HIF-1 activity via degradation of HIF-1α *(68)*.

The effect of HIF-1 on tumor angiogenesis has been studied in ES cells lacking either subunit. Teratomas derived from *Hif-1α*-null ES cells have yielded surprisingly varied and difficult to reconcile results. One group found that cells lacking HIF-1α formed tumors in $RAG^{-/-}$ mice that were roughly 25% the size of wild-type controls after 21 d, with a 40% decrease in vascular density as measured by CD31 immunohistochemistry *(57)*. In separate experiments, our group has shown a similar growth defect in tumors lacking HIF-1α, demonstrating in addition that genetic rescue of the cells with stably transfected HIF-1α cDNA rescued the size and vascular defects of the mutant tumors (A. M. Arsham and M. C. Simon, unpublished data). Another group found that loss of HIF-1α caused tumors that, after 8 wk of growth, were several times bigger than wild-type-derived tumors. These mutant tumors had increased proliferation and decreased apoptosis, despite having fewer large vessels and more hypoxia. Based on these and other observations, the authors postulated an important proapoptotic role for HIF-1α in addition to its known metabolic and VEGF-inducing roles *(10)*. Ablation of the *Vegf* gene's HIF-1-binding site caused greater reductions in tumor VEGF levels, and consequently in vascular density and permeability, than the loss of HIF-1α itself. This suggests either that other HIF family members are able to compensate for the loss of HIF-1α in vivo, or that other signal transduction pathways are able to sufficiently upregulate VEGF in the absence of HIF. Alternatively, the loss of the HRE may have ablated binding sites for other factors as well as for HIF, thus increasing the severity of the VEGF deficit *(15)*.

MEFs lacking HIF-1α form tumors much more poorly than wild-type cells, but with no apparent vascular defect, suggesting a role for HIF-1 in addition to its clear function in hypoxia-induced VEGF expression, and suggesting as well that other factors are competent to enact the angiogenic switch in the absence of HIF-1 *(71)*. However, some reports clearly demonstrate a pivotal role for HIF-1 in tumor angiogenesis. Tumors derived from a hepatocyte line mutated in the *Arnt* gene were much smaller and less vascular than those derived from the parental, i.e., *Arnt*-expressing line, displaying lower and aberrantly localized *Vegf* mRNA *(72)*. When HIF-1 activity is disrupted by overexpression of a dominant interfering HIF-1α truncation, much smaller, less vascular tumors are formed by two different tumor cell lines. In an in vivo competition experiment, cells with disrupted HIF-1 activity were effectively selected against in favor of control cells that retained HIF-1 activity, suggesting, in agreement with Ryan et al., that HIF transcriptional activity plays an important positive role in tumor cell survival in addition to its paracrine role in recruiting blood vessels *(73)*.

HIF-1's role in tumors is highly complex, and the literature reflects this complexity. HIF levels have been both positively *(74–76)* and negatively *(77)* correlated with tumor grade and progression, and various in vivo systems have shown both positive and negative roles for HIF in tumor growth. It positively affects tumor development through cell survival and VEGF production, and inhibits tumor growth by inducing apoptosis, perhaps through p53. It is difficult to envision a general strategy for disruption of angiogenesis through HIF inhibition, both because it has so many other functions and because the balance of its tumor-enhancing and -inhibiting functions varies so greatly between experimental systems—HIF's role in specific tumor types is likely to be highly empirical. While the study of HIFs provides crucial insight into tumor development, metabolism, and the effects of tumor microenvironment on growth, HIF-1 is problematic as a therapeutic target.

4.4. Angiopoietins and Ties

The angiopoietin (ang)/tie ligand/receptor system is known to play an important role in developmental angiogenesis, and is thought to be important for pathophysiologic angiogenesis as well. Tie1 and tie2, receptor tyrosine kinases, are expressed on vascular and hematopoietic cells. The function of tie1 remains elusive, and tie2 appears to be the primary receptor for the ligands ang1 and ang2. Ang1 appears to be an important signal for the maturation of developing vessels, and some data suggest that ang2 can function as an antagonist of this process. Thus, the balance of ang1 and ang2 seems important in the maturation vs regression of neovessels. Genetic data are scarce as to the actual role this system plays in tumor angiogenesis *(14)*.

Ang1/tie2 can represent an important angiogenic signal in tumors, as expression of a dominant negative tie2 can inhibit tumor growth and angiogenesis in breast carcinoma lines that are known to express high endogenous ang1. In contrast, cells that do not express high levels of ang1 (i.e., a line in which ang/tie signaling has not played a major role in tumor progression) were not appreciably inhibited by the dominant negative tie2 *(51)*. Interfering with tie2 can also block growth and angiogenesis of melanoma cells *(46)*. Soluble tie2 slows the growth of established tumors, and decreases the frequency and vascularity of metastases *(78)*.

It is clear enough that ang/tie signaling can modulate tumor angiogenesis, but it is also clear that this is not always the case, and may even vary on a tumor-to-tumor basis. Tumor vasculature displays many of the hallmarks of immature vessels, and it may be that ang/tie signaling does not play a broad role in this process. From a therapeutic standpoint, it is then necessary to figure out not just how ang/tie signaling affects tumor vasculature, but also in which tumors this is true, in order to target therapy appropriately. Ang/tie inhibition, while not applicable in all cases, may become an important backup therapy for tumors that do not respond to more general (i.e., VEGF-neutralizing) therapy.

4.5. Other Growth Factors and Cytokines

Several other growth factors and cytokines are clearly important for tumor growth and angiogenesis, among them members of the fibroblast growth factor (FGF) and interleukin (IL) families, although none is as central as VEGF, and none has been studied as extensively.

FGFs-1, -3, and -4 have each been implicated in the positive regulation of tumor angiogenesis in different tumor models *(79–84)*. FGF-2 (bFGF) can also increase tumor angiogenesis, but is not always able to concomitantly increase tumor size, suggesting that there are functionally overlapping, as well as nonoverlapping activities within this family *(20,84,85)*. FGF-2 also appears to be more important in the initial stages of tumor growth and angiogenesis, since repression of FGF-2 at the time of xenograft strongly inhibits tumor formation and growth, while repression in established tumors has little effect on their further progression *(86)*.

Interleukins (IL) are a family of small soluble cytokines that mediate inflammation and immunity. They are produced at the site of injury or infection by immune cells *(87)*. Stable overexpression of IL8 increases gastric carcinoma growth and vascularity *(88)*, and attenuation of IL-8 expression via an antisense construct retards intrapancreatic tumor angiogenesis and growth *(89)*. IL-8 appears, at least in part, to function through the upregulation of matrix metalloproteinases (*see* **Subheading 4.8.**), as exogenous expression in nonmetastatic melanoma *(90)*, prostate cancer *(91)*, or bladder cancer *(92)* increases their tumorigenic, angiogenic, and metastatic potential, which correlates with increased MMP activity. Conversely, antisense constructs for IL-8 inhibit both MMP activity and tumor aggressiveness and angiogenesis in highly metastatic prostate *(91)* or bladder *(92)* cancers.

Some members of the interleukin family, namely, IL-10, IL-12, and IL-18, have been shown to have tumor suppressive properties. Exogenous IL-12 and IL-18 each separately inhibit tumor growth in a mammary cell line—together they suppress tumorigenesis at distant sites, apparently through angiogenesis inhibition *(93)*. IL-12 also inhibits tumor angiogenesis in pancreatic and colon carcinomas, independently of the host's immune response, and apparently by induction of interferon-γ and repression of VEGF and MMP activity *(94,95)*. Stable expression of human or viral IL-10 also strongly suppresses tumorigenicity and angiogenicity of several tumor lines, apparently through a number of mechanisms including inhibition of macrophage secretion of VEGF and MMP-9 *(96,97)*, activation of NK-cell activity *(98)*, and induction of TIMP-1 *(99)*.

The FGFs represent a relatively minor target for antiangiogenic therapy. While they are able to induce or enhance various aspects of angiogenesis, they are not central players in the way that VEGF is. They may be valuable as complementary targets, but are not likely to be useful as primary therapeutic targets. The interleukins hold promise as tumor suppressive or antiangiogenic agents, but more needs to be known about their mechanism of action, and possible side effects, as they will have effects on the immune system aside from their intended therapeutic purposes. They will also be difficult to use in conjunction with radio- or chemotherapy, as they seem to require, at least in part, an intact immune system.

4.6. Thrombospondins

Thrombospondins are ECM proteins that exhibit strong antiangiogenic properties in vivo and in vitro. They are involved in a broad array of cell/cell and cell/ECM interactions, and seem to function as mediators of interactions between ECM components, cell surface proteins, and cytokines. Of the 5 known TSPs, only TSP-1 and -2 have been extensively studied with regard to tumor angiogenesis. This may be related to their structural similarity to each other, and dissimilarity to the other members of their family (reviewed in **ref. *100***). While the mechanism of TSP inhibition of angiogenesis activity is not yet clear, it may involve binding to its known receptor, CD36, negative modulation of FGF-2 signaling or MMP-9 activity, or positive modulation of TGF-β signaling. Transgenic expression of thrombospondin can negatively affect both normal developmental and tumor-associated angiogenesis *(101)*.

Overexpression of TSP-1 negatively affects growth and angiogenesis in numerous tumor models, including mammary *(102,103)*, prostate *(104)*, glioblastoma *(105)*, transformed 3T3 fibroblasts *(106)*, and squamous cell carcinoma *(107)*. In one squamous cell line, TSP-2 was shown to be an even more potent inhibitor of tumor growth and angiogenesis than TSP-1, independent of VEGF levels, and the two TSPs together completely abolished tumor growth *(108)*. In skin cancer cells that lack a copy of chromosome 15 (on which TSP-1 resides), tumor growth and angiogenesis were equally well suppressed by replacement of the chromosome as by transfection with TSP-1 cDNA, and the TSP-1 tumor suppression was relieved by injection of antisense oligonucleotides to TSP-1, suggesting that TSP itself was the major source of chromosome 15 tumor suppression *(109)*.

In transgenic mice that express the *neu* oncogene in breast tissues and develop spontaneous mammary tumors, the absence of endogenous TSP-1 accelerates tumorigenesis and growth, and leads to increased vascular density and surface area within tumors when compared to wild-type. In contrast, mice that overexpressed TSP-1 in mammary tissue produced fewer and smaller tumors, displaying decreased vascular density and surface area. In this study, both active MMP9 and receptor-bound VEGF inversely correlated with TSP-1 levels, strongly suggesting a mechanism of action for angiogenesis inhibition by TSP-1 *(110)*. This model represents an important advance in the study of angiogenesis, as it avoids most of the caveats associated with xenograft experiments. The location, etiology, and development of tumors much more closely approximates "normal" tumor pathology. Further, the modulation of tumorigenesis and angiogenesis by TSP-1 occurs in the context of tissue with a single, defined genetic lesion, as opposed to the background of multiple unknown mutations, aneuploidy, and genomic instability present, axiomatically, in all commonly used tumor cell lines.

The thrombospondins, as well as other proteins that contain the thrombospondin repeats, are potently antiangiogenic, and suggest new ways in which to target the process of angiogenesis. The TSPs and other endogenous inhibitors of angiogenesis, in addition to providing insight into the physiologic processes involved in vessel growth, are excellent candidates as angiogenesis inhibitors, especially since they are less likely to have toxic side effects than pharmacologic agents.

4.7. Integrins

Integrins are among the adhesion molecules that play critical roles in the migration and invasion of vessels. They are heterodimeric transmembrane proteins comprised of an α- and a β-subunit, and mediate adhesion interactions between cell surfaces and the ECM. Fifteen known α- and 8 β-subunits can dimerize in at least 20 combinations, providing for many different cell-type and substrate specificities, and integrin interaction with the ECM can initiate intracellular signaling events affecting cell survival, proliferation, and migration. Integrin αvβ3 mediates endothelial cell interactions with other cells and with the ECM, and initiates intracellular signaling when engaged by extracellular ligands. In addition, integrins seem to play a role in the activation and membrane localization of MMP2 *(111)* (reviewed in **ref. 13**).

Integrin α1-knockout mice have elevated MMP2 and MMP9 production, and are strikingly impaired in their ability to support xenograft growth—tumors in these animals were smaller and less vascular than those in wild-type hosts. Plasma from tumor-bearing mutant animals had elevated levels of the powerful ECM-derived antiangiogenic molecule angiostatin, suggesting that the reduced tumor growth was due to increased production of angiostatin by elevated MMP activity in the ECM *(112)*. This is somewhat counterintuitive, in that MMPs are generally thought of as promoting angiogenesis and invasion, and demonstrates the physiologic complexity of the balance between pro- and antiangiogenic factors. Interestingly, it has recently been shown that basic helix–loop–helix transcriptional inhibitors Id1 and Id3 are also required for host vasculature to invade growing tumors. Mice deficient for these two proteins cannot support tumor xenografts due to a failure of αvβ3 and MMP2 localization to the surface of invading endothelial cells *(113)*.

Integrin β1 also has a role in tumor angiogenesis. Unlike αv integrin, β1 appears to be necessary in tumor cells to mediate recruitment of host stromal tissues, and differentiation of tumor endothelium. Loss of β1 renders ES cells much less tumorigenic, and the tumors that do form are less than 5% the size of wild-type. Tumor vessels were small, morphologically defective, and composed exclusively of host-derived cells, as opposed to the mix of host- and tumor-derived endothelial cells seen in wild-type teratomas *(114)*.

Genetic data regarding the role of integrins in tumor angiogenesis are scarce, and do not address the integrin that seems to be the most deeply involved, αvβ3. While many studies have shown the effectiveness of antibody, peptide, and small molecule inhibitors of integrin function in inhibiting angiogenesis, they are beyond the scope of this chapter. What the genetic and inhibitor studies combine to show is that integrins are not only critical mediators of tumor angiogenesis, they are also promising targets, as the cell-type and substrate specificity of the integrins means that their activities can be exquisitely targeted, e.g., disruption of αvβ3 can be accomplished without simultaneously blocking other integrins that are fulfilling important roles in normal physiology. Integrins are thus one of the most promising antiangiogenic therapeutic targets currently under investigation.

4.8. Matrix Metalloproteinases and Their Inhibitors

MMPs are Zn^{2+}-dependent endopeptidases central to the regulation of ECM homeostasis. The MMP family is, taken together, able to degrade all known components of the ECM. MMPs are particularly important in developmental and reproductive processes, and in disease states such as wound healing, all processes that require active tissue remodeling. MMP activity is thought to play an important role in both the invasion of neovessels into tumors, and the invasion and intravasation of metastatic tumors into the vascular system. MMPs are secreted with autoinhibitory pro-domains, and the cleavage of these domains is one major regulatory mechanism of MMP activity. MMPs appear to promote tumor formation and progression in a variety of ways and at multiple stages of tumor development (reviewed in **refs. *115* and *116***). As was noted under **Subheading 4.7.**, however, MMPs, while clearing the way for blood vessel and/or tumor invasion of the ECM, also cause antiangiogenic molecules to be released from the ECM in the process. Thus MMPs can both promote angiogenesis and, indirectly, result in its inhibition. The tissue inhibitors of metalloproteinases (TIMPs) are a family of four highly related proteins that block MMP activity by interacting with the Zn^{2+}-binding pocket and active site of the proteinases, and also in some cases by binding to the MMP pro-domains. TIMPs are known to have tumor suppressor activity, though the mechanism(s) of such activity are unclear. Presumably, a large part of their antitumor activity is related to their MMP-inhibitory function, but they may also have a direct effect on endothelial cells. As with the MMPs, evidence exists that TIMPs may also be able to play against type, displaying tumor-promoting properties, either through the enhancement of ECM degradation or through a direct effect on cellular proliferation or survival (reviewed in **ref. *117***).

Inhibition of MMP2 activity by expression of antisense oligos can slow tumor progression by 70% and strongly inhibits angiogenesis in in vitro and in vivo assays, suggesting that angiogenesis inhibition is at least part of its mode of action *(118)*. Loss of MMP2 activity by host tissues is also antiangiogenic—xenograft size and rate of metastasis are much lower in MMP2-knockout mice than in wild-type *(119)*. MMP9, which enhances angiogenesis by localizing to the surface of activated endothelial cells *(120)*, also causes the release of VEGF from pancreatic islet cells *(121)*. Mice that lack MMP9 show much less in vivo tumorigenesis in a transgenic oncogene model *(121)*. Metalloelastase (MMP12), a protease that is associated in the generation of angiostatin from plasminogen, strongly inhibits tumor growth, angiogenesis, and metastasis, presumably though the generation of angiostatin *(122,123)*.

Data concerning the TIMPs are also somewhat counterintuitive and contradictory. TIMP1 actually enhances tumor angiogenesis in rat mammary carcinoma cells, concomitant with an upregulation of VEGF production in these and other mammary lines *(124)*. This may be cell-type-specific, or may be a function of the cytokine-like activity of TIMP-1 *(117)*. However, liver overexpression of TIMP1 strongly inhibited T-antigen (Tag)-induced tumorigenesis in transgenic mice. Expression of antisense TIMP1 enhanced Tag-induced tumorigenesis *(125)*. Thus, TIMP1 appears able to enhance tumor growth through VEGF induction, as well as to inhibit tumor growth through its MMP-inhibitory activity. The data concerning TIMP2 are more straightforward. Overexpression of TIMP2 inhibits the ability of a tumor to recruit host endothelium *(126)*, reverses the growth of established hepatomas and the formation of metastases *(127)*, and similarly inhibits the subcutaneous growth of a melanoma line *(128)*.

MMPs and TIMPs are indicative of the kinds of feedback loops that exist during the dynamic process of angiogenesis. In general, MMP activity tends to be proangiogenic and TIMP activity antiangiogenic, but the existing data show that the physiologic reality is not so clear-cut. Thus, these molecules do not seem to be likely therapeutic targets or effectors.

5. Conclusions

Inhibition of angiogenesis, whatever its caveats, promises to be not only an important therapeutic strategy for many types of cancer, but also an important basic scientific tool for understanding cancer. Many critical questions remain. While VEGF is clearly one of the most important effectors of tumor angiogenesis, the many ways in which it can be induced by mutation and microenvironment, and the precise functions of its isoforms and receptors requires further exploration. The role of complementary pathways such as ang/tie signaling and the numerous other growth factors and cytokines implicated in angiogenesis will be critical to determine whether these are valid therapeutic targets. The complex balance between pro- and antiangiogenic factors at the interface of tumor, host tissues, and ECM represents another major challenge. This interface is extremely dynamic, and the study of the interactions between proteinases, their inhibitors, adhesion molecules, growth factors and receptors, and the pro- and antiangiogenic molecules embedded in the ECM will be one of the most exciting areas of research in the near future. It will also be important to further develop current in vivo tumor models to more accurately reflect the physiologic milieus in which therapies will ultimately be deployed. One example is the *neu*/TSP-1 model described under **Subheading 4.6.**, in which the incidence and pathology of spontaneous tumors can be attributed to specific genetic alterations. The genetic precision of this type of model is complementary to the much faster but less controlled xenograft models in wide use, and the two together will be a powerful tool in the understanding of tumor angiogenesis from both basic science and clinical perspectives.

References

1. Kerbel, R. S. (2000) Tumor angiogenesis: past, present and the near future. *Carcinogenesis* **21**, 505–515.

2. Feldman, A. L. and Libutti, S. K. (2000) Progress in antiangiogenic gene therapy of cancer. *Cancer* **89**, 1181–1194.

3. Hanahan, D., and Weinberg, R. A. (2000) The hallmarks of cancer. *Cell* **100**, 57–70.

4. Holash, J., Wiegand, S. J., and Yancopoulos, G. D. (1999) New model of tumor angiogenesis: dynamic balance between vessel regression and growth mediated by angiopoietins and VEGF. *Oncogene* **18**, 5356–5362.

5. Helmlinger, G., Yuan, F., Dellian, M., and Jain, R. K. (1997) Interstitial pH and pO_2 gradients in solid tumors in vivo: high-resolution measurements reveal a lack of correlation. *Nat. Med.* **3**, 177–182.

6. Brown, J. M. and Giaccia, A. J. (1998) The unique physiology of solid tumors: opportunities (and problems) for cancer therapy. *Cancer Res.* **58**, 1408–1416.

7. Dang, C. V. and Semenza, G. L. (1999) Oncogenic alterations of metabolism. *Trends Biochem. Sci.* **24**, 68–72.

8. Shweiki, D., Itin, A., Soffer, D., and Keshet, E. (1992) Vascular endothelial growth factor induced by hypoxia may mediate hypoxia-initiated angiogenesis. *Nature* **359**, 843–845.

9. Shweiki, D., Neeman, M., Itin, A., and Keshet, E. (1995) Induction of vascular endothelial growth factor expression by hypoxia and by glucose deficiency in multicell spheroids: implications for tumor angiogenesis. *Proc. Natl. Acad. Sci. USA* **92**, 768–772.

10. Carmeliet, P., Dor, Y., Herbert, J. M., et al. (1998) Role of HIF-1alpha in hypoxia-mediated apoptosis, cell proliferation and tumour angiogenesis. *Nature* **394**, 485–490.

11. Semenza, G. L. (2000) Hypoxia, clonal selection, and the role of HIF-1 in tumor progression. *Crit. Rev. Biochem. Mol. Biol.* **35**, 71–103.

12. Claffey, K. P. and Robinson, G. S. (1996) Regulation of VEGF/VPF expression in tumor cells: consequences for tumor growth and metastasis. *Cancer Metastasis Rev.* **15**, 165–176.

13. Eliceiri, B. P. and Cheresh, D. A. (1999) The role of alphav integrins during angiogenesis: insights into potential mechanisms of action and clinical development. *J. Clin. Invest.* **103**, 1227–1230.

14. Yancopoulos, G. D., Davis, S., Gale, N. W., Rudge, J. S., Wiegand, S. J., and Holash, J. (2000) Vascular-specific growth factors and blood vessel formation. *Nature* **407**, 242–248.

15. Tsuzuki, Y., Fukumura, D., Oosthuyse, B., Koike, C., Carmeliet, P., and Jain, R. K. (2000) Vascular endothelial growth factor (VEGF) modulation by targeting hypoxia-inducible factor-1alpha→ hypoxia response element→ VEGF cascade differentially regulates vascular response and growth rate in tumors. *Cancer Res.* **60**, 6248–6252.

16. Dewhirst, M. W., Ong, E. T., Braun, R. D., et al. (1999) Quantification of longitudinal tissue pO_2 gradients in window chamber tumours: impact on tumour hypoxia. *Br. J. Cancer* **79**, 1717–1722.

17. Robertson, E. J. (1987) *Teratocarcinomas and Embryonic Stem Cells: A Practical Approach*. Practical Approach Series (Rickwood, D. and Hames, B. D. eds.) IRL Press, Oxford, UK.

18. Fukumura, D., Xavier, R., Sugiura, T., et al. (1998) Tumor induction of VEGF promoter activity in stromal cells. *Cell* **94**, 715–725.

19. Gerber, H. P., Kowalski, J., Sherman, D., Eberhard, D. A., and Ferrara, N. (2000) Complete inhibition of rhabdomyosarcoma xenograft growth and neovascularization requires blockade of both tumor and host vascular endothelial growth factor. *Cancer Res.* **60**, 6253–6258.

20. Konerding, M. A., Fait, E., Dimitropoulou, C., et al. (1998) Impact of fibroblast growth factor-2 on tumor microvascular architecture. A tridimensional morphometric study. *Am. J. Pathol.* **152**, 1607–1616.

21. Carmeliet, P., Ferreira, V., Breier, G., et al. (1996) Abnormal blood vessel development and lethality in embryos lacking a single VEGF allele. *Nature* **380**, 435–439.

22. Ferrara, N., Carver-Moore, K., Chen, H., et al. (1996) Heterozygous embryonic lethality induced by targeted inactivation of the VEGF gene. *Nature* **380**, 439–442.

23. Carmeliet, P., Ng, Y. S., Nuyens, D., et al. (1999) Impaired myocardial angiogenesis and ischemic cardiomyopathy in mice lacking the vascular endothelial growth factor isoforms VEGF164 and VEGF188. *Nat. Med.* **5**, 495–502.

24. Miquerol, L., Langille, B. L., and Nagy, A. (2000) Embryonic development is disrupted by modest increases in vascular endothelial growth factor gene expression. *Development* **127**, 3941–3946.

25. Byzova, V. T., Goldman, K. C., Pampori, N., et al. (2000) A mechanism for modulation of cellular responses to VEGF: activation of the integrins. *Mol. Cell.* **6**, 851–860.

26. Veikkola, T., Karkkainen, M., Claesson-Welsh, L., and Alitalo, K. (2000) Regulation of angiogenesis via vascular endothelial growth factor receptors. *Cancer Res.* **60**, 203–212.

27. Grunstein, J., Roberts, W. G., Mathieu-Costello, O., Hanahan, D., and Johnson, R. S. (1999) Tumor-derived expression of vascular endothelial growth factor is a critical factor in tumor expansion and vascular function. *Cancer Res.* **59**, 1592–1598.

28. Grunstein, J., Masbad, J. J., Hickey, R., Giordano, F., and Johnson, R. S. (2000) Isoforms of vascular endothelial growth factor act in a coordinate fashion To recruit and expand tumor vasculature. *Mol. Cell. Biol.* **20**, 7282–7291.

29. Yuan, A., Yu, C. J., Kuo, S. H.,et al. (2001) Vascular endothelial growth factor 189 mRNA isoform expression specifically correlates with tumor angiogenesis, patient survival, and postoperative relapse in non-small-cell lung cancer. *J. Clin. Oncol.* **19**, 432–441.

30. Zhang, H. T., Craft, P., Scott, P. A., et al. (1995) Enhancement of tumor growth and vascular density by transfection of vascular endothelial cell growth factor into MCF-7 human breast carcinoma cells. *J. Natl. Cancer Inst.* **87**, 213–219.

31. Zhang, H. T., Scott, P. A., Morbidelli, L., et al. (2000) The 121 amino acid isoform of vascular endothelial growth factor is more strongly tumorigenic than other splice variants in vivo. *Br. J. Cancer* **83**, 63–68.

32. Lewin, M., Bredow, S., Sergeyev, N., Marecos, E., Bogdanov, A., and Weissleder, R. (1999) In vivo assessment of vascular endothelial growth factor-induced angiogenesis. *Int. J. Cancer* **83**, 798–802.

33. Benjamin, L. E. and Keshet, E. (1997) Conditional switching of vascular endothelial growth factor (VEGF) expression in tumors: induction of endothelial cell shedding and regression of hemangioblastoma-like vessels by VEGF withdrawal. *Proc. Natl. Acad. Sci. USA* **94**, 8761–8766.

34. Cheng, S. Y., Nagane, M., Huang, H. S., and Cavenee, W. K. (1997) Intracerebral tumor-associated hemorrhage caused by overexpression of the vascular endothelial growth factor isoforms VEGF121 and VEGF165 but not VEGF189. *Proc. Natl. Acad. Sci. USA* **94**, 12081–12087.

35. Kondo, Y., Arii, S., Mori, A., Furutani, M., Chiba, T., and Imamura, M. (2000) Enhancement of angiogenesis, tumor growth, and metastasis by transfection of vascular endothelial growth factor into LoVo human colon cancer cell line. *Clin. Cancer Res.* **6**, 622–630.

36. Larcher, F., Murillas, R., Bolontrade, M., Conti, C. J., and Jorcano, J. L. (1998) VEGF/VPF overexpression in skin of transgenic mice induces angiogenesis, vascular hyperpermeability and accelerated tumor development. *Oncogene* **17**, 303–311.

37. Claffey, K. P., Brown, L. F., del Aguila, L. F., et al. (1996) Expression of vascular permeability factor/vascular endothelial growth factor by melanoma cells increases tumor growth, angiogenesis, and experimental metastasis. *Cancer Res.* **56**, 172–181.

38. Oku, T., Tjuvajev, J. G., Miyagawa, T., et al. (1998) Tumor growth modulation by sense and antisense vascular endothelial growth factor gene expression: effects on angiogenesis, vascular permeability, blood volume, blood flow, fluorodeoxyglucose uptake, and proliferation of human melanoma intracerebral xenografts. *Cancer Res.* **58**, 4185–4192.

39. Cheng, S. Y., Huang, H. J., Nagane, M., et al. (1996) Suppression of glioblastoma angiogenicity and tumorigenicity by inhibition of endogenous expression of vascular endothelial growth factor. *Proc. Natl. Acad. Sci. USA* **93**, 8502–8507.

40. Ma, J., Fei, Z. L., Klein-Szanto, A., and Gallo, J. M. (1998) Modulation of angiogenesis by human glioma xenograft models that differentially express vascular endothelial growth factor. *Clin. Exp. Metastasis* **16**, 559–568.

41. Detmar, M., Velasco, P., Richard, L., et al. (2000) Expression of vascular endothelial growth factor induces an invasive phenotype in human squamous cell carcinomas. *Am. J. Pathol.* **156**, 159–167.

42. Yoshiji, H., Kuriyama, S., Yoshii, J., et al. (1998) Vascular endothelial growth factor tightly regulates in vivo development of murine hepatocellular carcinoma cells. *Hepatology* **28**, 1489–1496.

43. Belletti, B., Ferraro, P., Arra, C., et al. (1999) Modulation of in vivo growth of thyroid tumor-derived cell lines by sense and antisense vascular endothelial growth factor gene. *Oncogene* **18**, 4860–4869.

44. Miao, H. Q. and Klagsbrun, M. (2000) Neuropilin is a mediator of angiogenesis. *Cancer Metastasis Rev.* **19**, 29–37.

45. Soker, S., Takashima, S., Miao, H. Q., Neufeld, G., and Klagsbrun, M. (1998) Neuropilin-1 is expressed by endothelial and tumor cells as an isoform-specific receptor for vascular endothelial growth factor. *Cell* **92**, 735–745.

46. Siemeister, G., Schirner, M., Weindel, K., et al. (1999) Two independent mechanisms essential for tumor angiogenesis: inhibition of human melanoma xenograft growth by interfering with either the vascular endothelial growth factor receptor pathway or the Tie-2 pathway. *Cancer Res.* **59**, 3185–3191.

47. Takayama, K., Ueno, H., Nakanishi, Y., et al. (2000) Suppression of tumor angiogenesis and growth by gene transfer of a soluble form of vascular endothelial growth factor receptor into a remote organ. *Cancer Res.* **60**, 2169–2177.

48. Hiratsuka, S., Maru, Y., Okada, A., Seiki, M., Noda, T., and Shibuya, M. (2001) Involvement of Flt-1 tyrosine kinase (vascular endothelial growth factor receptor-1) in pathological angiogenesis. *Cancer Res.* **61**, 1207–1213.

49. Millauer, B., Shawver, L. K., Plate, K. H., Risau, W., and Ullrich, A. (1994) Glioblastoma growth inhibited in vivo by a dominant-negative Flk-1 mutant. *Nature* **367**, 576–579.

50. Millauer, B., Longhi, M. P., Plate, K. H., et al. (1996) Dominant-negative inhibition of Flk-1 suppresses the growth of many tumor types in vivo. *Cancer Res.* **56**, 1615–1620.

51. Stratmann, A., Acker, T., Burger, A. M., Amann, K., Risau, W., and Plate, K. H. (2001) Differential inhibition of tumor angiogenesis by tie2 and vascular endothelial growth factor receptor-2 dominant-negative receptor mutants. *Int. J. Cancer* **91**, 273–282.

52. Machein, M. R., Risau, W., and Plate, K. H. (1999) Antiangiogenic gene therapy in a rat glioma model using a dominant-negative vascular endothelial growth factor receptor 2. *Hum. Gene Ther.* **10**, 1117–1128.

53. Miao, H. Q., Lee, P., Lin, H., Soker, S., and Klagsbrun, M. (2000) Neuropilin-1 expression by tumor cells promotes tumor angiogenesis and progression. *FASEB J.* **14**, 2532–2539.

54. Gagnon, M. L., Bielenberg, D. R., Gechtman, Z., et al. (2000) Identification of a natural soluble neuropilin-1 that binds vascular endothelial growth factor: In vivo expression and antitumor activity. *Proc. Natl. Acad. Sci. USA* **97**, 2573–2578.

55. Wenger, R. H. (2000) Mammalian oxygen sensing, signalling and gene regulation. *J. Exp. Biol.* **203 Pt 8**, 1253–1263.

56. Iyer, N. V., Kotch, L. E., Agani, F., et al. (1998) Cellular and developmental control of O_2 homeostasis by hypoxia- inducible factor 1 alpha. *Genes Dev.* **12**, 149–162.

57. Ryan, H. E., Lo, J., and Johnson, R. S. (1998) HIF-1a is required for solid tumor formation and embryonic vascularization. *EMBO J.* **17**, 3005–3015.

58. Maltepe, E., Schmidt, J. V., Baunoch, D., Bradfield, C. A., and Simon, M. C. (1997) Abnormal angiogenesis and responses to glucose and oxygen deprivation in mice lacking the protein ARNT. *Nature* **386**, 403–407.

59. Kozak, K. R., Abbott, B., and Hankinson, O. (1997) ARNT-deficient mice and placental differentiation. *Dev. Biol.* **191**, 297–305.

60. Maltepe, E. and Simon, M. C. (1998) Oxygen, genes, and development: an analysis of the role of hypoxic gene regulation during murine vascular development [In Process Citation]. *J. Mol. Med.* **76**, 391–401.

61. Tian, H., Hammer, R. E., Matsumoto, A. M., Russell, D. W., and McKnight, S. L. (1998) The hypoxia-responsive transcription factor EPAS1 is essential for catecholamine homeostasis and protection against heart failure during embryonic development. *Genes Dev.* **12**, 3320–3324.

62. Peng, J., Zhang, L., Drysdale, L., and Fong, G. H. (2000) The transcription factor EPAS-1/hypoxia-inducible factor 2alpha plays an important role in vascular remodeling. *Proc. Natl. Acad. Sci. USA* **97**, 8386–8391.

63. Keith, B., Adelman, D. M., and Simon, M. C. (2001) Targeted mutation of the murine Arylhydrocarbon receptor nuclear translocator-2 (*Arnt2*) gene reveals partial redundancy with *Arnt*. *Proc. Natl. Acad. Sci. USA* **98,** 6692–6697.

64. Ivan, M. and Kaelin, W. G. (2001) The von Hippel-Lindau tumor suppressor protein. *Curr. Opin. Genet. Dev.* **11**, 27–34.

65. Graeber, T. G., Osmanian, C., Jacks, T., et al. (1996) Hypoxia-mediated selection of cells with diminished apoptotic potential in solid tumours [see comments]. *Nature* **379**, 88–91.

66. An, W. G., Kanekal, M., Simon, M. C., Maltepe, E., Blagosklonny, M. V., and Neckers, L. M. (1998) Stabilization of wild-type p53 by hypoxia-inducible factor 1alpha. *Nature* **392**, 405–408.

67. Carmeliet, P., Dor, Y., Herbert, J. M., et al. (1998) Role of HIF-1alpha in hypoxia-mediated apoptosis, cell proliferation and tumour angiogenesis. *Nature* **394**, 485–490.

68. Ravi, R., Mookerjee, B., Bhujwalla, Z. M., et al. (2000) Regulation of tumor angiogenesis by p53-induced degradation of hypoxia-inducible factor 1alpha. *Genes Dev.* **14**, 34–44.

69. Wenger, R. H., Camenisch, G., Desbaillets, I., Chilov, D., and Gassmann, M. (1998) Up-regulation of hypoxia-inducible factor-1alpha is not sufficient for hypoxic/anoxic p53 induction. *Cancer Res.* **58**, 5678–5680.

70. Blagosklonny, M. V., An, W. G., Romanova, L. Y., Trepel, J., Fojo, T., and Neckers, L. (1998) p53 inhibits hypoxia-inducible factor-stimulated transcription. *J. Biol. Chem.* **273**, 11995–11998.

71. Ryan, H. E., Poloni, M., McNulty, W., et al. (2000) Hypoxia-inducible factor-1alpha is a positive factor in solid tumor growth. *Cancer Res.* **60**, 4010–4015.

72. Maxwell, P. H., Dachs, G. U., Gleadle, J. M., et al. (1997) Hypoxia-inducible factor-1 modulates gene expression in solid tumors and influences both angiogenesis and tumor growth. *Proc. Natl. Acad. Sci. USA* **94**, 8104–8109.

73. Kung, A. L., Wang, S., Klco, J. M., Kaelin, W. G., and Livingston, D. M. (2000) Suppression of tumor growth through disruption of hypoxia-inducible transcription. *Nat. Med.* **6**, 1335–1340.

74. Zhong, H., De Marzo, A. M., Laughner, E., et al. (1999) Overexpression of hypoxia-inducible factor 1alpha in common human cancers and their metastases. *Cancer Res.* **59**, 5830–5835.

75. Bos, R., Zhong, H., Hanrahan, C. F., et al. (2001) Levels of hypoxia-inducible factor-1alpha during breast carcinogenesis. *J. Natl. Cancer Inst.* **93**, 309–314.

76. Zagzag, D., Zhong, H., Scalzitti, J. M., Laughner, E., Simons, J. W., and Semenza, G. L. (2000) Expression of hypoxia-inducible factor 1alpha in brain tumors: association with angiogenesis, invasion, and progression. *Cancer* **88**, 2606–2618.

77. Blancher, C., Moore, J. W., Talks, K. L., Houlbrook, S., and Harris, A. L. (2000) Relationship of hypoxia-inducible factor (HIF)-1alpha and HIF-2alpha expression to vascular

endothelial growth factor induction and hypoxia survival in human breast cancer cell lines. *Cancer Res.* **60**, 7106–7113.

78. Lin, P., Buxton, J. A., Acheson, A., et al. (1998) Antiangiogenic gene therapy targeting the endothelium-specific receptor tyrosine kinase Tie2. *Proc. Natl. Acad. Sci. USA* **95**, 8829–8834.

79. McLeskey, S. W., Tobias, C. A., Vezza, P. R., Filie, A. C., Kern, F. G., and Hanfelt, J. (1998) Tumor growth of FGF or VEGF transfected MCF-7 breast carcinoma cells correlates with density of specific microvessels independent of the transfected angiogenic factor. *Am. J. Pathol.* **153**, 1993–2006.

80. Zhang, L., Kharbanda, S., Chen, D., et al. (1997) MCF-7 breast carcinoma cells overexpressing FGF-1 form vascularized, metastatic tumors in ovariectomized or tamoxifen-treated nude mice. *Oncogene* **15**, 2093–2108.

81. Pili, R., Chang, J., Muhlhauser, J., Crystal, R. G., Capogrossi, M. C., and Passaniti, A. (1997) Adenovirus-mediated gene transfer of fibroblast growth factor-1: angiogenesis and tumorigenicity in nude mice. *Int. J. Cancer.* **73**, 258–263.

82. Li, J. J., Friedman-Kien, A. E., Cockerell, C., Nicolaides, A., Liang, S. L., and Huang, Y. Q. (1998) Evaluation of the tumorigenic and angiogenic potential of human fibroblast growth factor FGF3 in nude mice. *J. Cancer Res. Clin. Oncol.* **124**, 259–264.

83. Jouanneau, J., Moens, G., Montesano, R., and Thiery, J. P. (1995) FGF-1 but not FGF-4 secreted by carcinoma cells promotes in vitro and in vivo angiogenesis and rapid tumor proliferation. *Growth Factors* **12**, 37–47.

84. Jouanneau, J., Plouet, J., Moens, G., and Thiery, J. P. (1997) FGF-2 and FGF-1 expressed in rat bladder carcinoma cells have similar angiogenic potential but different tumorigenic properties in vivo. *Oncogene* **14**, 671–676.

85. Coltrini, D., Gualandris, A., Nelli, E. E., et al. (1995) Growth advantage and vascularization induced by basic fibroblast growth factor overexpression in endometrial HEC-1-B cells: an export-dependent mechanism of action. *Cancer Res.* **55**, 4729–4738.

86. Giavazzi, R., Giuliani, R., Coltrini, D., et al. (2001) Modulation of tumor angiogenesis by conditional expression of fibroblast growth factor-2 affects early but not established tumors. *Cancer Res.* **61**, 309–317.

87. Abbas, A. K., Lichtman, A. H., and Pober, J. S. (1994) *Cellular and Molecular Immunology.* Saunders, Philadelphia.

88. Kitadai, Y., Takahashi, Y., Haruma, K., et al. (1999) Transfection of interleukin-8 increases angiogenesis and tumorigenesis of human gastric carcinoma cells in nude mice. *Br. J. Cancer* **81**, 647–653.

89. Shi, Q., Abbruzzese, J. L., Huang, S., Fidler, I. J., Xiong, Q., and Xie, K. (1999) Constitutive and inducible interleukin 8 expression by hypoxia and acidosis renders human pancreatic cancer cells more tumorigenic and metastatic. *Clin. Cancer Res.* **5**, 3711–3721.

90. Luca, M., Huang, S., Gershenwald, J. E., Singh, R. K., Reich, R., and Bar-Eli, M. (1997) Expression of interleukin-8 by human melanoma cells up-regulates MMP-2 activity and increases tumor growth and metastasis. *Am. J. Pathol.* **151**, 1105–1113.

91. Inoue, K., Slaton, J. W., Eve, B. Y., et al (2000) Interleukin 8 expression regulates tumorigenicity and metastases in androgen-independent prostate cancer. *Clin. Cancer Res.* **6**, 2104–2119.

92. Inoue, K., Slaton, J. W., Kim, S. J., et al. (2000) Interleukin 8 expression regulates tumorigenicity and metastasis in human bladder cancer. *Cancer Res.* **60**, 2290–2299.

93. Coughlin, C. M., Salhany, K. E., Wysocka, M., et al. (1998) Interleukin-12 and interleukin-18 synergistically induce murine tumor regression which involves inhibition of angiogenesis. *J. Clin. Invest.* **101**, 1441–1452.

94. Sunamura, M., Sun, L., Lozonschi, L., et al. (2000) The antiangiogenesis effect of interleukin 12 during early growth of human pancreatic cancer in SCID mice. *Pancreas* **20**, 227–233.

95. Duda, D. G., Sunamura, M., Lozonschi, L., et al. (2000) Direct in vitro evidence and in vivo analysis of the antiangiogenesis effects of interleukin 12. *Cancer Res.* **60**, 1111–1116.
96. Huang, S., Xie, K., Bucana, C. D., Ullrich, S. E., and Bar-Eli, M. (1996) Interleukin 10 suppresses tumor growth and metastasis of human melanoma cells: potential inhibition of angiogenesis. *Clin. Cancer Res.* **2**, 1969–1979.
97. Huang, S., Ullrich, S. E., and Bar-Eli, M. (1999) Regulation of tumor growth and metastasis by interleukin-10: the melanoma experience. *J. Interferon Cytokine Res.* **19**, 697–703.
98. Cervenak, L., Morbidelli, L., Donati, D., et al. (2000) Abolished angiogenicity and tumorigenicity of Burkitt lymphoma by interleukin-10. *Blood* **96**, 2568–2573.
99. Stearns, M. E., Garcia, F. U., Fudge, K., Rhim, J., and Wang, M. (1999) Role of interleukin 10 and transforming growth factor beta1 in the angiogenesis and metastasis of human prostate primary tumor lines from orthotopic implants in severe combined immunodeficiency mice. *Clin. Cancer Res.* **5**, 711–720.
100. Lawler, J. (2000) The functions of thrombospondin-1 and-2. *Curr. Opin. Cell Biol.* **12**, 634–640.
101. Iruela-Arispe, M. L., Vazquez, F., and Ortega, M. A. (1999) Antiangiogenic domains shared by thrombospondins and metallospondins, a new family of angiogenic inhibitors. *Ann. N.Y. Acad. Sci.* **886**, 58–66.
102. Weinstat-Saslow, D. L., Zabrenetzky, V. S., VanHoutte, K., Frazier, W. A., Roberts, D. D., and Steeg, P. S. (1994) Transfection of thrombospondin 1 complementary DNA into a human breast carcinoma cell line reduces primary tumor growth, metastatic potential, and angiogenesis. *Cancer Res.* **54**, 6504–6511.
103. Iruela-Arispe, M. L., Lombardo, M., Krutzsch, H. C., Lawler, J., and Roberts, D. D. (1999) Inhibition of angiogenesis by thrombospondin-1 is mediated by 2 independent regions within the type 1 repeats. *Circulation* **100**, 1423–1431.
104. Jin, R. J., Kwak, C., Lee, S. G., et al. (2000) The application of an antiangiogenic gene (thrombospondin-1) in the treatment of human prostate cancer xenografts. *Cancer Gene Ther.* **7**, 1537–1542.
105. Tenan, M., Fulci, G., Albertoni, M., et al. (2000) Thrombospondin-1 is downregulated by anoxia and suppresses tumorigenicity of human glioblastoma cells. *J. Exp. Med.* **191**, 1789–1798.
106. Castle, V. P., Dixit, V. M., and Polverini, P. J. (1997) Thrombospondin-1 suppresses tumorigenesis and angiogenesis in serum- and anchorage-independent NIH 3T3 cells. *Lab. Invest.* **77**, 51–61.
107. Streit, M., Velasco, P., Brown, L. F., et al. (1999) Overexpression of thrombospondin-1 decreases angiogenesis and inhibits the growth of human cutaneous squamous cell carcinomas. *Am. J. Pathol.* **155**, 441–452.
108. Streit, M., Riccardi, L., Velasco, P., et al. (1999) Thrombospondin-2: a potent endogenous inhibitor of tumor growth and angiogenesis. *Proc. Natl. Acad. Sci. USA* **96**, 14888–14893.
109. Bleuel, K., Popp, S., Fusenig, N. E., Stanbridge, E. J., and Boukamp, P. (1999) Tumor suppression in human skin carcinoma cells by chromosome 15 transfer or thrombospondin-1 overexpression through halted tumor vascularization. *Proc. Natl. Acad. Sci. USA* **98**, 12,485–12,490.
110. Rodriguez-Manzaneque, J. C., Lane, T. F., Ortega, M. A., Hynes, R. O., Lawler, J., and Iruela-Arispe, M. L. (2001) Thrombospondin-1 suppresses tumor growth by a novel mechanism that includes blockade of matrix metalloproteinase-9 activation. *Cell* **85**, 683–693.
111. Brooks, P. C., Stromblad, S., Sanders, L. C., et al. (1996) Localization of matrix metalloproteinase MMP-2 to the surface of invasive cells by interaction with integrin alpha v beta 3. *Cell* **85**, 683–693.
112. Pozzi, A., Moberg, P. E., Miles, L. A., Wagner, S., Soloway, P., and Gardner, H. A. (2000) Elevated matrix metalloprotease and angiostatin levels in integrin alpha 1 knockout mice cause reduced tumor vascularization. *Proc. Natl. Acad. Sci. USA* **97**, 2202–2207.

113. Lyden, D., Young, A. Z., Zagzag, D., et al. (1999) Id1 and Id3 are required for neurogenesis, angiogenesis and vascularization of tumour xenografts. *Nature* **401**, 670–677.
114. Bloch, W., Forsberg, E., Lentini, S., et al. (1997) Beta 1 integrin is essential for teratoma growth and angiogenesis. *J. Cell Biol.* **139**, 265–278.
115. Stetler-Stevenson, W. G. (1999) Matrix metalloproteinases in angiogenesis: a moving target for therapeutic intervention. *J. Clin. Invest.* **103**, 1237–1241.
116. McCawley, L. J. and Matrisian, L. M. (2000) Matrix metalloproteinases: multifunctional contributors to tumor progression. *Mol. Med. Today.* **6**, 149–156.
117. Blavier, L., Henriet, P., Imren, S., and Declerck, Y. A. (1999) Tissue inhibitors of matrix metalloproteinases in cancer. *Ann. N.Y. Acad. Sci.* **878**, 108–119.
118. Fang, J., Shing, Y., Wiederschain, D., et al. (2000) Matrix metalloproteinase-2 is required for the switch to the angiogenic phenotype in a tumor model. *Proc. Natl. Acad. Sci. USA* **97**, 3884–3889.
119. Itoh, T., Tanioka, M., Yoshida, H., Yoshioka, T., Nishimoto, H., and Itohara, S. (1998) Reduced angiogenesis and tumor progression in gelatinase A-deficient mice. *Cancer Res.* **58**, 1048–1051.
120. Yu, Q. and Stamenkovic, I. (2000) Cell surface-localized matrix metalloproteinase-9 proteolytically activates TGF-beta and promotes tumor invasion and angiogenesis. *Genes Dev.* **14**, 163–176.
121. Bergers, G., Brekken, R., McMahon, G., et al. (2000) Matrix metalloproteinase-9 triggers the angiogenic switch during carcinogenesis. *Nat. Cell. Biol.* **2**, 737–744.
122. Gorrin-Rivas, M. J., Arii, S., Furutani, M., et al. (2000) Mouse macrophage metalloelastase gene transfer into a murine melanoma suppresses primary tumor growth by halting angiogenesis. *Clin. Cancer Res.* **6**, 1647–1654.
123. Matsuda, K. M., Madoiwa, S., Hasumi, Y., et al. (2000) A novel strategy for the tumor angiogenesis-targeted gene therapy: generation of angiostatin from endogenous plasminogen by protease gene transfer. *Cancer Gene Ther.* **7**, 589–596.
124. Yoshiji, H., Harris, S. R., Raso, E., et al. (1998) Mammary carcinoma cells over-expressing tissue inhibitor of metalloproteinases-1 show enhanced vascular endothelial growth factor expression. *Int. J. Cancer.* **75**, 81–87.
125. Martin, D. C., Ruther, U., Sanchez-Sweatman, O. H., Orr, F. W., and Khokha, R. (1996) Inhibition of SV40 T antigen-induced hepatocellular carcinoma in TIMP-1 transgenic mice. *Oncogene* **13**, 569–576.
126. Vergani, V., Garofalo, A., Bani, M. R., et al. (2001) Inhibition of matrix metalloproteinases by over-expression of tissue inhibitor of metalloproteinase-2 inhibits the growth of experimental hemangiomas. *Int. J. Cancer.* **91**, 241–247.
127. Brand, K., Baker, A. H., Perez-Canto, A., et al. (2000) Treatment of colorectal liver metastases by adenoviral transfer of tissue inhibitor of metalloproteinases-2 into the liver tissue. *Cancer Res.* **60**, 5723–5730.
128. Valente, P., Fassina, G., Melchiori, A., et al. (1998) TIMP-2 over-expression reduces invasion and angiogenesis and protects B16F10 melanoma cells from apoptosis. *Int. J. Cancer.* **75**, 246–253.

22

Flow Cytometric Analysis of Tumor Suppressor Gene-Induced Apoptosis

David T. Dicker and Wafik S. El-Deiry

1. Introduction

As interest in cell death has recently increased, many new techniques have been developed to detect and quantitate apoptosis in cells. Multiparameter flow cytometric-based assays provide statistical information on measured parameters at the single-cell level, and are therefore well suited for measuring apoptosis *(1)*. Since apoptosis has many functional expressions, it is often desirable for investigators to demonstrate cell death by several different methods. A sampling of flow cytometric assays used in our laboratory is presented here.

2. Materials
2.1. Sub-G_1 Fraction: PI Staining with or Without GFP Reporter Gene

1. Fetal bovine serum (FBS).
2. Phosphate-buffered saline (PBS).
3. Propidium iodide (PI) stock solution: 500 μg/mL propidium iodide, in H_2O.
4. RNAse stock solution: 10 mg/mL RNAse A, in H_2O, heat to 100°C for 15 min.
5. PI/RNAse staining solution: 300 μL PBS with 1% FBS + 30 μL PI stock solution + 8 μL RNAse solution.
6. Phosphate–citric acid buffer: 192 mL 0.2 M Na_2HPO_4 + 8 mL 0.1 M citric acid, pH 7.8.
7. Ethanol

2.2. Terminal Deoxynucleotidyl Transferase-Mediated dUTP Nick End Labeling (TUNEL)

1. APO-BRDU kit (Phoenix Flow Systems, cat. no. AU1001).
2. Phosphate-buffered saline (PBS).
3. Formaldehyde solution: 1% prepared freshly from paraformaldehyde in PBS.
4. Ethanol.

2.3. Caspase Activity: Fluorogenic Substrate

1. PhiPhiLux-G_1D_2 kit (OncoImmunin, cat. no. 260-066).

From: *Methods in Molecular Biology, Vol. 223: Tumor Suppressor Genes: Regulation, Function, and Medicinal Applications.* Edited by: Wafik S. El-Deiry © Humana Press Inc., Totowa, NJ

2. Fetal bovine serum.
3. Growth media appropriate for cells to be assayed.

2.4. Caspase Activity: Fluorescent Inhibitor

1. Growth media appropriate for cells to be assayed.
2. CaspACE reagent (Promega, cat. no. G7461).
3. Phosphate-buffered saline (PBS).

2.5. Antibody to Active Caspase-3 with or Without Reporter Gene

1. Fetal bovine serum (FBS).
2. Phosphate-buffered saline (PBS).
3. Cytofix/Cytoperm solution, Pharmigen kit, cat. no. 2075KK.
4. Perm/Wash solution, Pharmigen kit, cat. no. 2075KK.
5. Anti-active caspase-3 monoclonal antibody.
6. Goat $F(ab')_2$ anti-rabbit IgG-PE.

2.6. Plasma Membrane Changes: Annexin-V

1. Annexin V-EGFP kit, BioVision, cat. no. K104-100
2. Media appropriate for cells being assayed.

2.7. Mitochondrial Membrane Changes: Apo2.7 Assay

1. Fetal bovine serum (FBS).
2. Phosphate-buffered saline (PBS).
3. Digitonin stock: 25 mg/mL digitonin, in PBS. Boil until completely disolved, store at −20°C.
4. APO2.7-PE, Immunotech, cat. no. IM2088

2.8. Mitochondrial Membrane Changes: $DiOC_6(3)$

1. Hanks' buffered saline.
2. 3,3'-dihexyloxacarbocyanine iodide, $DiOC_6(3)$.
3. $DiOC_6(3)$ stock solution: 1 mM in dimethyl sulfoxide.
4. Media appropriate for cells to be assayed.

3. Methods

3.1. DNA Fragmentation Assays

One of the classic features of apoptosis is the activation of the apoptosis-associated endonuclease(s) that degrades nuclear DNA *(2–4)*. Apoptotic cells can be identified within a population as cells with fractional DNA content following extraction of the degraded DNA. DNA cleavage generates a large number of strand breaks. These strand breaks can also serve as a marker for apoptosis *(5)*.

3.1.1. Sub-G_1 Fraction: PI Staining with or Without GFP Reporter Gene (see Fig. 1)

If using EGFP reporter gene, start with **step 1.** Otherwise, skip this step.

1. When transfecting your cells using your favorite method, co-transfect pEGFPN-Spectrin *(6)* as a 1/10 fraction of the total DNA. Allow at least 8 h before analysis, so sufficient EGFP can be synthesized.

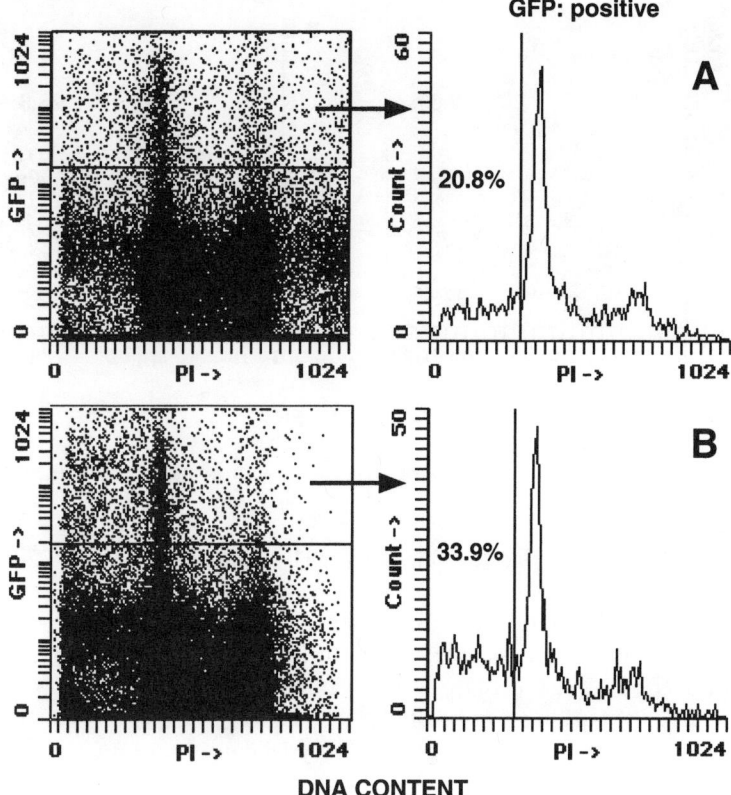

Fig. 1. FaDu cells transfected with DR4 and GFP reporter gene (**A**) control or (**B**) treated with 20 ng/mL TRAIL for 10 h.

2. Harvest ~10^6 cells; if cells are adherent, be sure to collect cells that have become detached.
3. Spin down cells and wash in 2 mL PBS/1% FBS.
4. Spin down and resuspend in 1 mL PBS.
5. Add 9 mL cold (4°C) EtOH, while vortexing, let stand at 4°C > 20 min, protected from light (cells can be stored indefinitely at this stage at 4°C).
6. Spin down cells and wash in 4 mL PBS/1% FBS.
7. Spin down and resuspend in 1 mL PBS.
8. Add 0.5 mL phosphate–citric acid buffer, let stand at room temperature for 5 min.
9. Spin down and resuspend in 300 µL PI/RNase staining solution.
10. Incubate at 37°C for 15 min or at room temperature for 30 min.
11. Analyze by flow cytometry.

3.1.2. Terminal Deoxynucleotidyl Transferase-Mediated dUTP Nick End Labeling (TUNEL) (see **Fig. 2**)

1. Treat cells with your favorite apoptosis inducer.
2. Harvest ~10^6 cells; if cells are adherent, be sure to collect cells that have detached.
3. Spin down and resuspend in 0.5 mL PBS at 2–4 × 10^6 cells/mL.
4. Add cells to 5 mL 1% formaldehyde solution, 4°C.
5. Fix on ice 15 min.
6. Wash 2× in cold PBS, resuspend in 0.5 mL cold PBS.

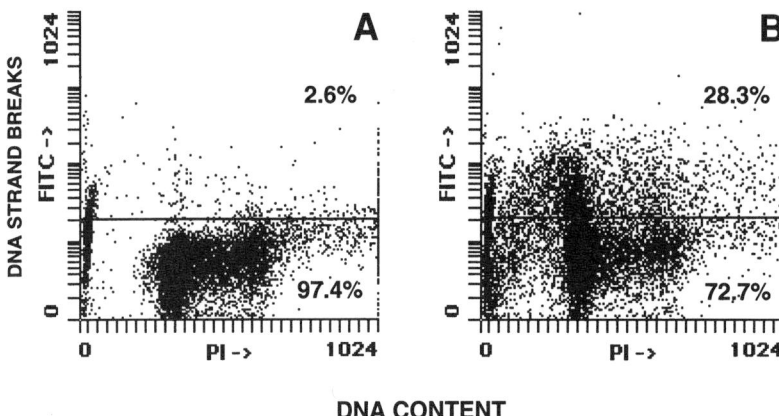

DNA CONTENT

Fig. 2. SW480 cells (**A**) control or (**B**) treated with 20 ng/mL TRAIL for 6 h.

7. Add cells slowly, while mixing to cold 70% EtOH. Maintain on ice or at –20°C at least 30 min. Cells can be stored up to 1 wk.
8. Spin down cells and wash in 1 mL wash buffer.
9. Prepare reaction solution (per sample): 10.00 μL TdT reaction buffer, 0.75 μL TdT enzyme, 8.00 μL Br-dUTP, 32.25 μL dH$_2$O.
10. Resuspend cells in 50 μL reaction solution.
11. Incubate at 37°C for 60 min, agitating tubes every 15 min.
12. Add 1 mL rinse buffer, spin down, and wash cells 1× in rinse buffer.
13. Prepare antibody solution (per sample): 5.00 μL fluorescein–PRB-1, 95.00 μL rinse buffer.
14. Resuspend cells in 100 μL antibody solution.
15. Incubate 30 min at room temperature in the dark.
16. Add 300 μL PI/RNase A solution to cells.
17. Incubate at room temperature for 30 min protected from light.
18. Analyze by flow cytometry within 3 h.

3.2. Caspase Activity

The caspase family of cyteine proteases plays a key role in apoptosis *(7)*. The caspases are synthesized in an inactive pro-caspase form. During apoptosis these pro-caspases are processed by proteolytic cleavage to form active enzymes. The presence of these active forms in cells can serve as a marker of apoptosis *(8)*.

3.2.1. Caspase Activity: Fluorogenic Substrate (see **Fig. 3**)

PhiPhiLux-G$_1$D$_2$ is a caspase-3 substrate that, when cleaved, has the following fluorescent characteristics: λ_{ex} = 505 nm, λ_{em} = 530 nm.

1. Treat cells with your favorite apoptosis inducer in 6-well plate.
2. Harvest ~10^6 cells; if cells are adherent, be sure to collect cells that have detached.
3. Spin down and resuspend in 1.5 mL medium, transfer to 1.7-mL microfuge tube.
4. Spin down and carefully remove all media. Loosen pellet by flicking with finger.
5. Add 50 μL PhiPhiLux substrate* + 5 μL fetal bovine serum.
6. Incubate at 37°C in a CO$_2$ incubator for 60 min with caps off or loosened.
7. Add 500 μL ice-cold flow cytometry dilution buffer (FCDB).
8. Spin down and resuspend in 500 μL cold FCDB; maintain samples on ice.

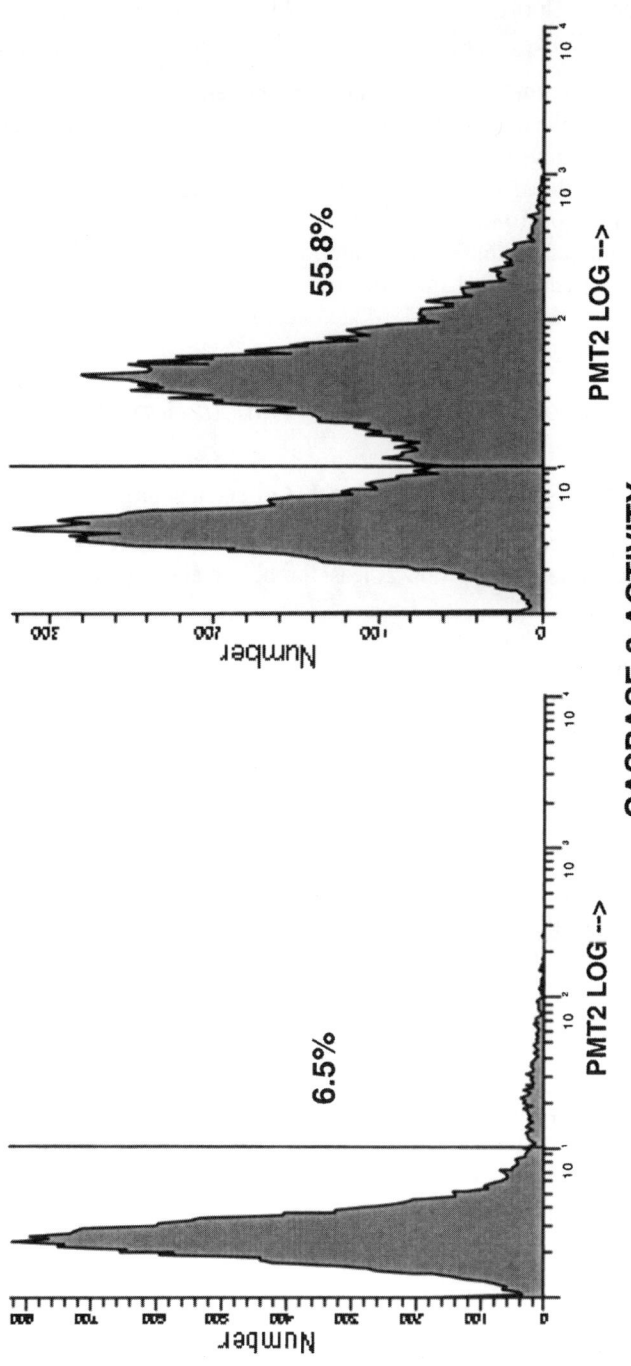

Fig. 3. SW480 cells (**A**) control or (**B**) treated with 10 ng/mL TRAIL for 17 h and assayed with the fluorogenic substrate PhiPhiLux.

9. Analyze by flow cytometry within 60 min.

3.2.2. Caspase Activity: Fluorescent Inhibitor (see **Fig. 4**)

FITC-VAD-FMK is a fluoroisothiocyanate (FITC) conjugate of the cell permeant caspase inhibitor VAD-FMK. This inhibitor binds irreversibly to activated caspases.

1. Treat cells with your favorite apoptosis inducer in a 6-well plate.
2. Harvest ~10^6 cells; if cells are adherent, be sure to collect cells that have detached.
3. Spin down cells and remove supernatant.
4. Resuspend in 75 µL 10 µM FITC-VAD-FMK (dilute 2 µL stock in 1 mL medium with serum, final concentration 10 mM).
5. Incubate cells 20 min at 37°C in CO_2 incubator.
6. Spin down cells and wash 1× in 1 mL PBS.
7. Resuspend in 0.5 mL PBS.
8. Analyze by flow cytometry.

3.2.3. Antibody to Active Caspase-3 with or Without Reporter Gene (see **Fig. 5**)

If using EGFP reporter gene start with **step 1.** Otherwise, skip this step.

1. When transfecting your cells using your favorite method, co-transfect pEGFPN-Spectrin as a 1/10 fraction of the total DNA. Allow at least 8 h before analysis, so sufficient EGFP can be synthesized.
2. Harvest ~10^6 cells; if cells are adherent, be sure to collect cells that have detached.
3. Spin down cells and wash in 1 mL PBS with 1% FBS.
4. Resuspend cells in 250 µL Cytofix/Cytoperm solution.
5. Incubate cells 20 min at room temperature.
6. Spin down cells and wash 2× in 1 mL Perm/Wash solution.
7. Remove as much supernatant as possible, and resuspend cells in 50 mL Perm/Wash.
8. Add 2 µL anti-active caspase-3 Mab, which as been diluted to 0.125 mg/µL in Perm/Wash (0.25 µg/sample).
9. Incubate 20 min at room temperature, protected from light.
10. Wash cells 1× in 1 mL Perm/Wash.
11. Remove as much supernatant as possible, and resuspend cells in 50 µL Perm/Wash.
12. Add 2 µL goat F(ab')$_2$ anti-rabbit IgG-PE, which as been diluted to 0.125 mg/mL in Perm/Wash (0.25 µg/sample).
13. Incubate 20 min at room temperature, protected from light.
14. Wash cells 1× in 1 µL Perm/Wash.
15. Resuspend in 300 µL Perm/Wash.
16. Analyze by flow cytometry.

3.3. Plasma Membrane Changes

3.3.1. Annexin-V (see **Fig. 6**)

Annexin-V is a phospholipdid-binding protein with a high affinity for phosphotidylserine (PS). When PS relocates from the cytoplasmic face to the outer leaflet during apoptosis *(9,10),* it becomes available for annenin-V binding.

1. Treat cells with your favorite apoptosis inducer.
2. Harvest ~10^6 cells; if cells are adherent, be sure to collect cells that have detached.
3. Spin down and resuspend in 0.5 mL binding buffer (from annexin V-EGFP kit).

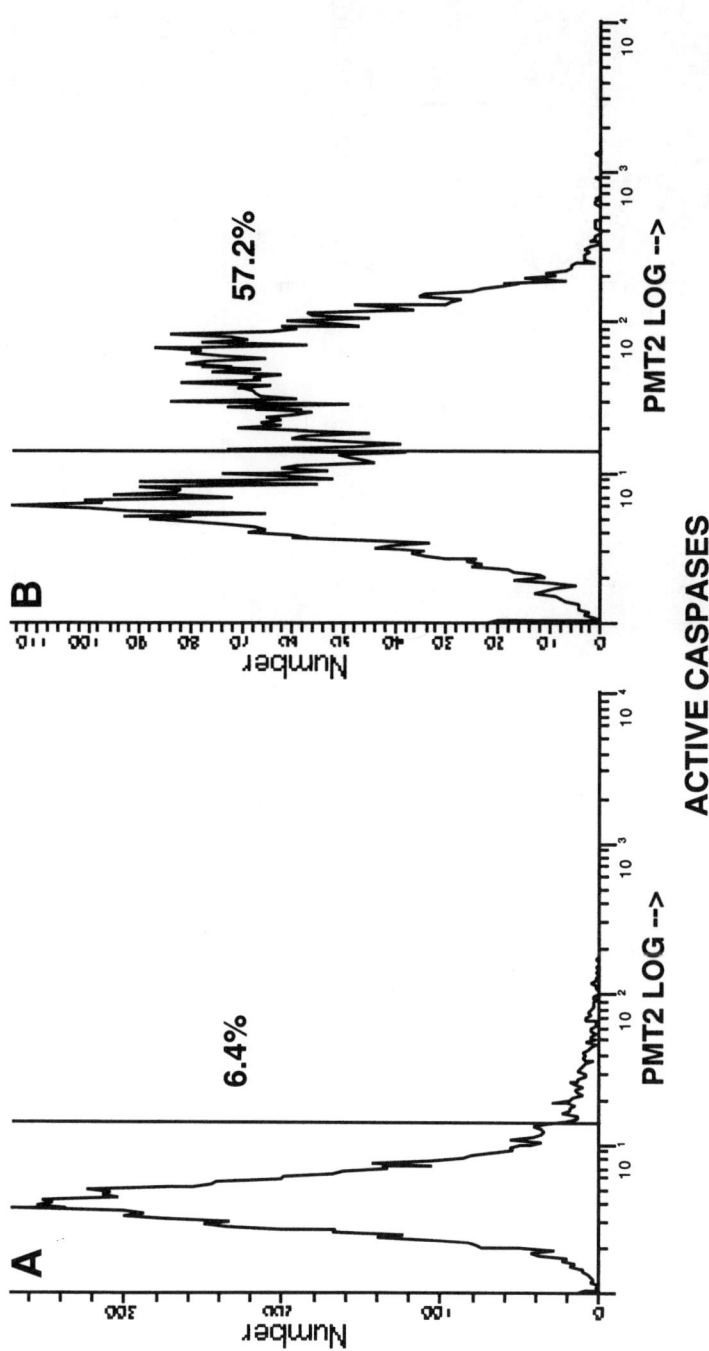

Fig. 4. SW480 cells (**A**) control or (**B**) treated with 20 ng/mL TRAIL for 4 h, then assayed with FITC-VAD-FMK.

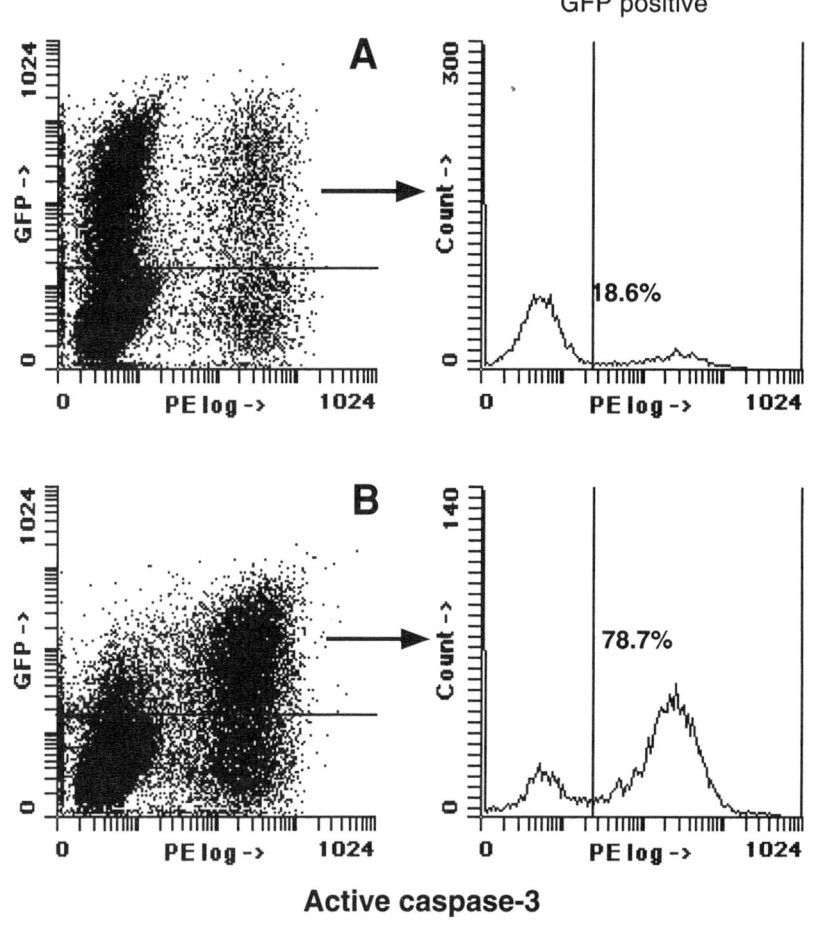

Fig. 5. SW480 cells transfected with EGFP and (**A**) vector alone or (**B**) KILLER(DR5).

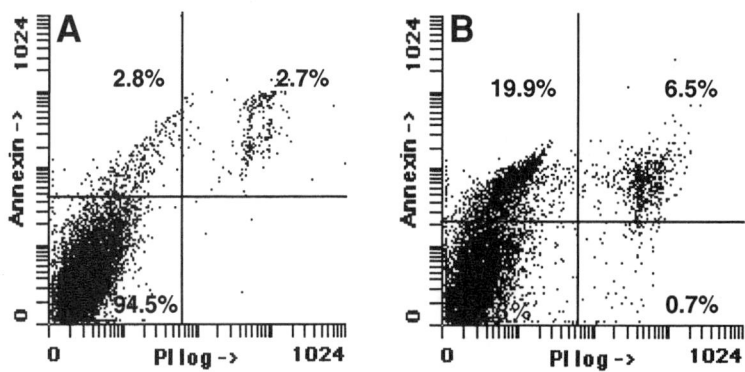

Fig. 6 Jurkat cells (**A**) control or (**B**) treated with 20 ng/mL TRAIL for 5 h, then assayed with annexin V -EGFP.

4. Add 5 µL annexin V-EGFP (from annexin V-EGFP kit) + 5 µL PI solution.
5. Incubate at room temperature for 5 min protected from light.
6. Analyze by flow cytometry within 60 min.

3.4. Mitochondrial Changes

3.4.1. Apo2.7 Assay (see **Fig. 7**)

Apo2.7 is an antibody to the 38-kDa mitochondrial membrane protein 7A6 antigen, which appears to be exposed in cells undergoing apoptosis *(11)*.

1. Treat cells with your favorite apoptosis inducer.
2. Harvest ~10^6 cells; if cells are adherent, be sure to collect cells that have become detached.
3. Spin down cells and wash in 1 mL PBS with 2.5% FBS.
4. Resuspend in 100 mL cold 100-µg/mL digitonin, diluted freshly from stock with PBS.
5. Incubate on ice for 20 min.
6. Wash 1× PBS + 2.5% FBS.
7. Resuspend in 80 mL PBS + 2.5% FBS, add 20 µL Apo2.7-PE.
8. Incubate at room temperature for 15 min.
9. Add 1 mL PBS + 2.5% FBS, spin down, and resuspend in 0.5 µL PBS + 2.5% FBS.
10. Analyze by flow cytometry.

3.4.2. DiOC$_6$—Membrane Potential (see **Fig. 8**)

An early event in apoptosis is a decrease in the mitochondrial transmembrane potential ($\Delta\Psi_m$) *(12)*. This DY$_m$ disruption is a constant feature in apoptosis in many different cell types *(13)*.

1. Treat cells with your favorite apoptosis inducer in a 6-well plate.
2. Harvest ~10^6 cells; if cells are adherent, be sure to collect cells that have detached.
3. Spin down cells, and resuspend in 300 µL 100 nM DiOC$_6$(3) in medium.
4. Incubate at 37°C for 30 min.
5. Analyze by flow cytometry.

References

1. Darzynkiewicz, Z., Juan, G., Li, X., Gorczynca, W., Murakami, T., and Traganos, F. (1997) Cytometry in cell necrobiology: anaysis of apoptosis and accidental cell death. *Cytometry* **27,** 1–20.
2. Arends, M. J., Morris, R. G., and Wyllie, A. H. (1990) Apoptosis: the role of endonuclease. *Am. J. Pathol.* **136,** 593-608.
3. Compton, M. M. (1992) A biochemical hallmark of apoptosis: internucleosomal degradation of the genome. *Cancer Metast. Rev.* **11,** 105–119.
4. Wyllie, A. H. (1992) Apoptosis and the regulation of cell numbers in normal and neoplastc tissues. An overview. *Cancer Metast. Rev.* **11,** 95–103.
5. Li, X. and Darzynkiewicz, Z. (1995) Labelling DNA strand breaks with BrdUTP. Detection of apoptosis and BrdUTP incorporation. *Cell Prolif.* **28,** 572–579.
6. Kalejta, R. F., Shenl, T., and Beavis, A. S. (1997) Use of a membrane-localized green fluorescent protein allows simultaneous identification of transfected cells and cell cycle analysis by flow cytometry. *Cytometry* **29,** 286–291.
7. Thornberry, N. A. and Lazebnik, Y. (1998) Caspases: enemies within. *Science* **291,** 1312–1316.

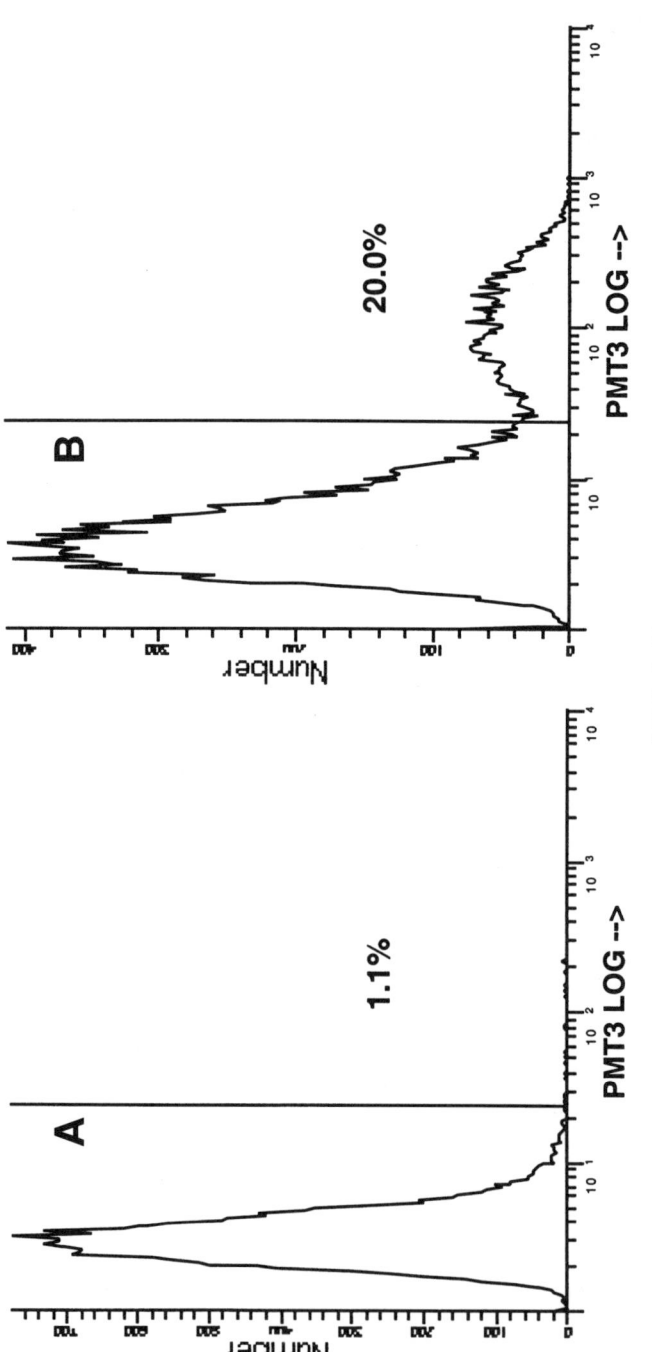

Fig. 7. H460 cells (**A**) control (**B**) irradiated with 100 J/m^2 UV, then stained with Apo 2.7-PE.

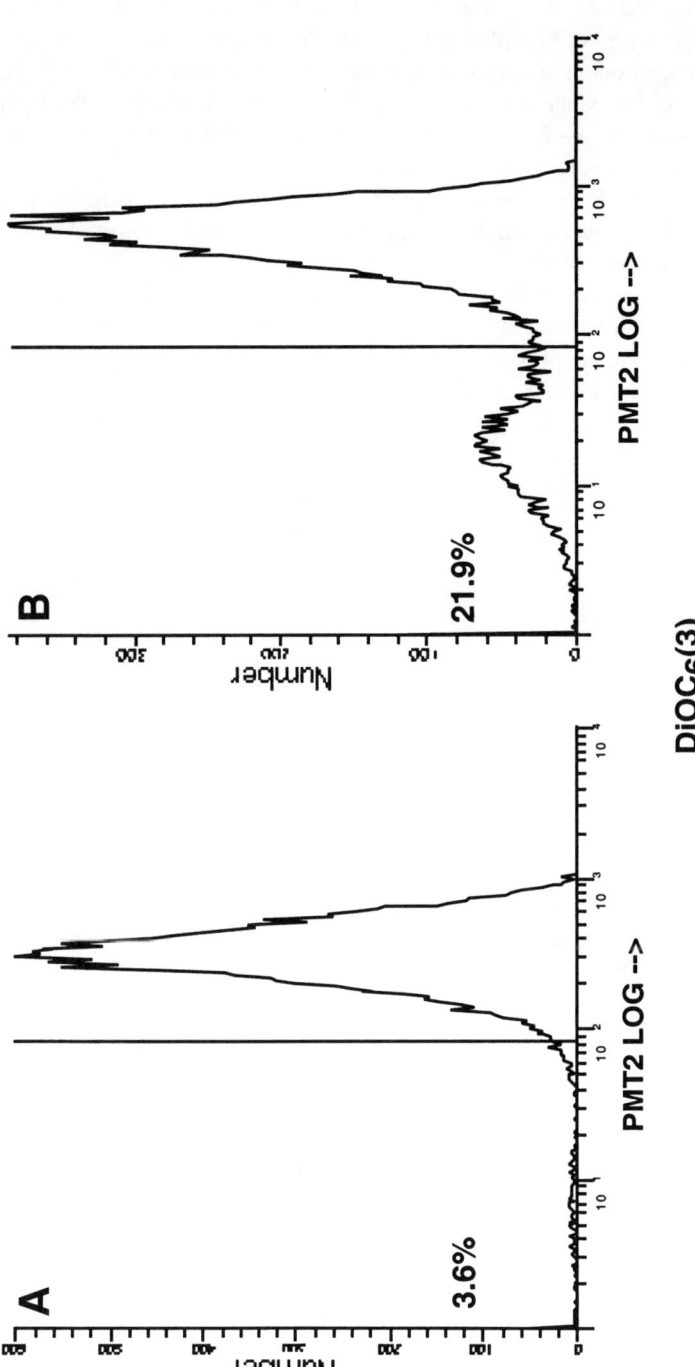

Fig. 8. H460 cells (**A**) control or (**B**) irradiated with 150 J/m^2 UV, then stained with DiOC$_6$(3).

8. Komoriya, A., Packard, B. Z., Brown, M. J., Wu, M.-L., and Henkart, P. A. (2000) Assessment of caspase activities in intact apoptotic thymocytes using cell-permeable fuorogenis caspase substrates. *J. Exp. Med.* **191,** 1819–1828.

9. Fadok, V. A., VoelKer, D. R., Cammpbell, P. A., Cohen, J. J., Bratton, D. L., and Henson, P. M. (1992) Exposure of phosphatidylserine on the surface of apoptotic lymphocytes triggers specific recognition and removal by macrophages. *J. Immunol.* **148,** 2207–2216.

10. Van Engeland, M., Nieland, L., Ramaekers, F., Schutte, B., and Reutelingsperger, C. (1998) Annexin V-affinity assay: a review on an apoptosis detection system based on phosphatidylserine exposure. *Cytometry* **31,** 1–9.

11. Zhang, C., Ao, Z., Seth, A., and Schlossman, S. F. (1996) A mitochondrial membrane protein defined by a novel monoclonal antibody is preferentially detected in apoptotic cells. *J. Immunol.* **157,** 3980–3987.

12. Petit, X., Lecouer, H., Zorn, E., Dauguet, C., Mignotte, B., and Gougeon, M.-L. (1997) Alterations in mitochondrial structure and function are early events of dexamethasone-induced thymocyte apoptosis. *J. Cell Biol.* **130,** 157–167.

13. Marchetti, P., Decaudin, D., Macho, A., et al. (1997) Redox regulation of apoptosis: impact of thiol oxidation status on mitochondrial function. *Eur. J. Immunol.* **27,** 289–296.

23

Functional Analysis of Tumor Suppressor Genes in Mice

Stephen N. Jones and Lawrence A. Donehower

1. Introduction

The utilization of mice as a model for human cancer goes back to the generation of inbred laboratory strains in the early twentieth century. Some inbred mouse strains were genetically predisposed to certain types of cancer, providing an early indication that genetic manipulation of a species could provide novel insights into the processes of tumorigenesis *(1,2)*. The development of transgenic and knockout technologies has extended our ability to manipulate the mouse genome and has greatly facilitated modeling of the genetic and biologic aspects of cancer *(3–6)*.

The popularity of inbred mice as cancer models arises from a number of attributes. Mice have a close evolutionary relationship to humans. For example, the completion of the rough draft of the human genome has indicated that humans contain only 300 genes not found in the mouse *(7)*. Moreover, the tumor types arising in mice and humans and the alterations of growth signaling pathways giving rise to them are highly similar. There are economic considerations as well. Mice are among the smallest mammals and can be housed in large numbers relatively inexpensively. They have comparatively short generation times (3 mo) and life spans (2–3 yr), allowing tumor studies to be completed with in 1–2 yr *(1)*. The embryology, pathobiology, aging, and tumor biology of most inbred strains have been extensively studied *(1,2,8–14)*. With the availability of the mouse genome sequence *(15)*, the plethora of mouse polymorphic genetic markers *(16,17)*, and the vast numbers of available mutant mouse strains *(12,18,19)*, the genetics of the inbred mouse is in a very advanced state. The inbred state itself facilitates genetic analyses, since genetic variation outside the desired mutation is eliminated. Thus, among those animal model systems that actually develop cancer (i.e., vertebrates), the experimental manipulability of the inbred mouse is unsurpassed.

Analysis of tumor suppressor function has benefited greatly from knockout mouse engineering techniques. These tumor suppressor-deficient mice have provided important insights into the biologic role of tumor suppressors in embryonic development and tumorigenesis unobtainable through conventional biochemical and cell culture methodologies *(3,4,6,20)*. Moreover, biologic interactions between different tumor suppressors

From: *Methods in Molecular Biology, Vol. 223: Tumor Suppressor Genes: Regulation, Function, and Medicinal Applications.* Edited by: Wafik S. El-Deiry © Humana Press Inc., Totowa, NJ

or between tumor suppressors and oncogenes can be examined readily in a biologic context by the appropriate crossing of genetically altered mice. Cells derived from mice altered in a single tumor suppressor gene are ideal substrates for analysis of affected signaling pathways through conventional biochemical methods, or through the newly emerging microarray expression profiling and proteonomics methods.

The number of different tumor suppressor-deficient mice that has been generated is quite extensive *(3,4,6,20,21)*. A large number of mice deficient in DNA repair genes are also susceptible to early spontaneous tumors *(22)*. Moreover, genes that had not previously been identified as tumor suppressors in humans (or mice) have been shown to confer increased susceptibility to tumorigenesis when inactivated in the mouse germline *(20)*. However, since it is not the purpose here to detail exhaustively all of the tumor susceptible knockout mice that have been generated, we direct the reader to a number of reviews on the subject *(3,4,6,20)*. In addition, there are databases that provide genotypic and phenotypic information on the various knockout strains *(21,23,24)*. The more widely used tumor suppressor-deficient mice are available from the Jackson Laboratory Induced Mutant Resource (www.jax.org) and commercial mouse vendors such as Taconic (www.taconic.com). The National Cancer Institute-funded Mouse Models of Human Cancer Consortium (MMHCC) has also begun to make available newly generated mutant lines designed to show tumor susceptibility in specific tissues (mmhcc.nci.nih.gov).

This chapter focuses on the mouse knockout technologies and their utilization in the analysis of tumor suppressor function. We describe some of the newer methods for generating mice in which the targeted gene can be more subtly altered in sequence or in expression. We also outline some of the ways to exploit the tumor suppressor-deficient mice for different types of cancer modeling studies. A variety of molecular, genetic, and biologic assays that can be performed on normal cells and tumor cells from the tumor suppressor-deficient mice are examined. In some examples, we describe results obtained on the p53-deficient mouse, in part because of its widespread use, but also because of our extensive familiarity with this particular model. We do not provide detailed protocols for each method, but instead refer the reader to the appropriate references with the necessary experimental detail.

2. Generation of Tumor Suppressor-Deficient Mice

The advent of gene targeting technology using mouse pluripotent embryonic stem (ES) cells has greatly facilitated analysis of gene function in mice over the past 10 years. More recently, methods of gene targeting in ES cells have been further refined, permitting the generation of mutations in the mouse ranging in size from megabase deletions to single nucleotide alterations. As the widespread use of ES cell technology has led to an ever-increasing number of variations on targeting strategies, a review of all gene targeting methodologies and modifications would require multiple chapters. Instead, this section examines some of the more common strategies used by researchers to analyze tumor suppressor gene functions in mice. Researchers wishing to obtain more detailed information for generating transgenic mice either through microinjection of oocytes or blastocysts, or the culturing of embryonic stem cells, should consult one of the excellent manuals on these subjects *(10,25–28)*.

2.1. Embryonic Stem Cells

ES cells are derived from the inner cell mass of a mouse blastocyst-stage embryo harvested at d 3.5 of gestation *(29,30)*. The embryo is composed of two cell types at this stage of development: inner cell mass (ICM) cells and trophectoderm cells. The ICM is the source of ES cells, which will subsequently give rise to all embryonic tissues during development. In order to retain their pluripotency, researchers harvest and grow these ES cells in vitro under conditions that mimic the embryonic environment of the inner cell mass. ES cells are plated and grown on a feeder layer of cells, typically either established lines or embryonic fibroblasts, in order to simulate the cell–cell contacts shared between ES cells and trophectoderm in vivo. Leukemia inhibitory factor (LIF) has been found to be important for maintenance of ES cell pluripotency *(31,32)*. Thus, either the feeder cells are transduced with a LIF transgene, or LIF is supplied exogenously to the culture media. In addition, the feeder cells must also contain genes providing resistance to the various drugs that will be employed in the gene targeting. Regardless of whether the feeder cells are derived from embryonic tissue (mouse embryonic fibroblasts) or are a modified cell line such as STO cells *(33)*, the feeder cells must be mitotically inactivated in order to prevent the cells from dividing faster than the ES cells and subsequently overwhelming the culture. Feeder cells are growth-arrested by exposure either to the DNA cross-linking drug mitomycin C, or to γ-irradiation. As an alternative to using feeder cells, some ES cell lines have been adapted to grow on tissue culture plasticware in the absence of feeders, with recombinant LIF being supplied to the culture medium *(34)*. This avoids the need for mitotically inactivated drug-resistant feeders in order to culture and target the ES cells. However, the majority of researchers using ES cells continue to utilize a feeder cell layer, as the feeders may be supplying additional factors to the ES cells and the general consensus is that use of feeders increases the likelihood of the targeted ES cells contributing to the germline in mice.

The utility of ES cells is due not only to their ability to contribute to all tissues during development, including the germ cell layer *(35–38)*, but also to the fact that large numbers of these cells may be grown with relative ease in culture. The growth of large numbers of ES cells means that very rare events such as homologous recombination between exogenously added DNA and an endogenous gene (approximately 1 in 10^6–10^7 cells) can be identified *(39–43)*. Furthermore, ES cell clones harboring these rare events can be sufficiently expanded to facilitate detailed analysis of the genetic changes in the clone. By maintaining and targeting the ES cells under rigorous culture conditions, the expanded ES cell clones will retain their pluripotency and will contribute to both the somatic and germ cell lineage upon reintroduction into blastocysts *(44,45)*.

2.2. Gene Targeting in ES Cells: Knockouts and Knockins

A common approach to ablate or "knock out" gene functions in mice has been to utilize a gene replacement vector for gene targeting experiments in ES cells. The targeting vector typically is composed of short regions (2–4 Kbp) of mouse genomic DNA flanking the portion of the gene to be targeted. A drug selection marker, typically a gene encoding resistance to neomycin, hygromycin, or puromycin, is included in the construct in order to identify those ES cells that incorporate the transduced vector into the genome *(46–50)*. Depending on the regions of target gene homology present in the

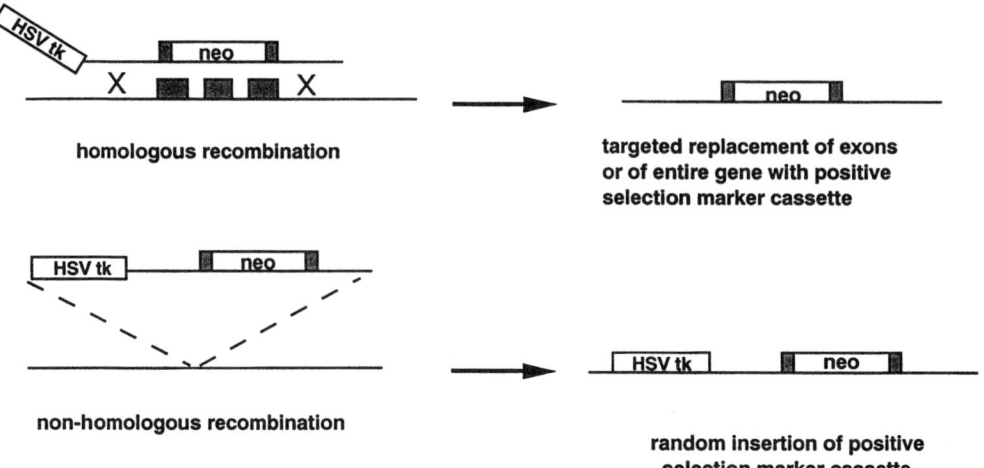

Fig. 1. Generation of a loss-of-function mutation by gene targeting. The gene replacement vector has coding sequences (filled boxes) replaced within the region of shared homology with a positive selection marker (neomycin cassette). A negative selection marker (herpes simplex virus thymidine kinase cassette) is included at one end of the homologous regions of the vector. Recombination between the left and right arms of sequence homology on the vector and the endogenous gene is depicted by X, and results in the replacement of exon sequences with the *neo* cassette and loss of the *tk* gene. These cells will survive G418 drug selection and selection against the presence of the *tk* gene using gancyclovir or FIAU (2′-deoxy-2′-fluoro-B-D-arabinosfuranosyl-5-iodouracil). Nonhomologous recombination of the targeting vector into ES cell DNA results in random insertion of the vector into the genome, with retention of the *tk* gene. Thus, these cells would not survive the positive–negative selection.

flanking arms of the vector, recombination between the vector and the targeted locus can result in deletion of some or all of the target coding sequences, which are replaced by the drug selection marker gene (**Fig. 1a**). The "gene replacement" vector is electroporated or transfected into ES cells, which are then selected in the appropriate drug to recover those clones that integrated the positive selection marker. A gene replacement vector may integrate into the ES cell genome either at the desired target locus via homologous recombination or randomly in the host genome (**Fig. 1**). In general, the frequency of targeted integration is several orders of magnitude less than the frequency of random integration *(48)*. This ratio of targeted integration to random integration will determine the ease with which a properly targeted clone can be identified. Factors that have been found to influence the frequency of homologous recombination include the length of homology shared between the vector and the targeted gene, the degree of polymorphic variation between the vector and the target sequences, and the confluency of the ES cells at the time of transduction *(42,43,46)*. Although the frequency of homologous recombination is low (10^{-6}), the ratio of targeted events to random integrants can be enriched by including a negative selection marker in the vector (**Fig. 1**). Placing this negative selection marker at either end of the homologous sequences permits selection against random integration events that favor inclusion of the negative marker into the host genome *(48)*. Proper homologous recombination between the targeting vector and the endogenous gene should result in loss of the negative selection cassette, usually the herpes simplex virus thymidine kinase (HSV-*tk*), yielding clones that are *tk*− and therefore survive the negative drug selection (gancyclovir).

Following growth of the ES cell clones in the appropriate selection media, colonies are usually picked from the plates and arrayed into 96-well tissue culture plates that were pre-seeded with inactivated feeder cells. These clones are expanded and a portion of each clone is frozen, while the rest of the cells are passed onto plates lacking the feeder cell layer to avoid subsequent DNA contamination from the feeder cells. Genomic DNA is prepared from the ES cell clones for Southern or polymerase chain reaction (PCR)-based analysis. The use of PCR has the advantage that a large number of clones may be screened and that fewer cells are needed per clone to give adequate DNA for screening. Furthermore, the PCR analysis may be done on pools of clones *(51)*. Typically, one primer for the PCR would anneal to drug marker DNA sequences included in the targeting vector and the other primer of the pair would recognize genomic sequences that are outside (or external) to the flanking sequences present in the vector *(52–54)*. However, these assays run the risk of having significant numbers of false positives and negatives, making it difficult to identify properly targeted clones. Often, the relative targeting frequencies of genes in ES cells are sufficient to permit Southern analysis to be performed on genomic DNA isolated from each clone—either in a 96-well format or following expansion of the clones to larger wells. By using a portion of the targeted locus that is not present on the targeting vector, it is possible to screen for the presence of a junction fragment of anticipated size following endonuclease digestion of the genomic DNA *(50)*. This method of analysis is more reliable than PCR-based methods and requires only a simple restriction map of the locus. However, it does require more DNA, and most researchers use a combination of both approaches to analyze their clones.

A variation on the gene targeting knockout strategy is to engineer the gene replacement vector so that a reporter gene encoding an easily assayed product (such as green fluorescence protein or β-galactosidase), but lacking eukaryotic promoter elements, is introduced into the targeted gene locus along with the positive selection marker. Proper integration of this vector by homologous recombination serves not only to ablate the targeted gene function, but also places the reporter gene included in the construct under transcriptional control of the endogenous gene promoter elements (**Fig. 2**). Assaying the pattern of marker gene expression in mice derived from these ES cells containing a reporter gene "knockin" can greatly facilitate promoter analysis of the targeted gene in addition to generating a knockout allele *(55,56)*.

Knockin strategies have proven very useful in placing gene coding sequences under the transciptional control of specific heterologous promoter regions. A cDNA encoding a protein of interest may be included in the knockin construct in lieu of a reporter gene. Recombination of the knockin vector with the target locus would result in expression of the cDNA being regulated by the target locus *(57)*. Although more time-consuming and laborious than the classic pronuclear microinjection approach to generating transgenic mice, the ES cell knockin strategy is superior in that transgene copy number is strictly regulated and expression of the transgene is not altered by positional effects of integration. Furthermore, all upstream and downstream elements that might contribute to regulation of gene expression are present at the knockin locus, whereas only those elements included in the transgene construct will contribute to regulation of gene expression in the transgenic mice (**Fig. 2**).

Using a similar approach, researchers have performed knockin experiments to replace portions of the endogenous targeted gene with a similar region of the gene encoding several base-pair changes or even a single-point mutation. This strategy facilitates the intro-

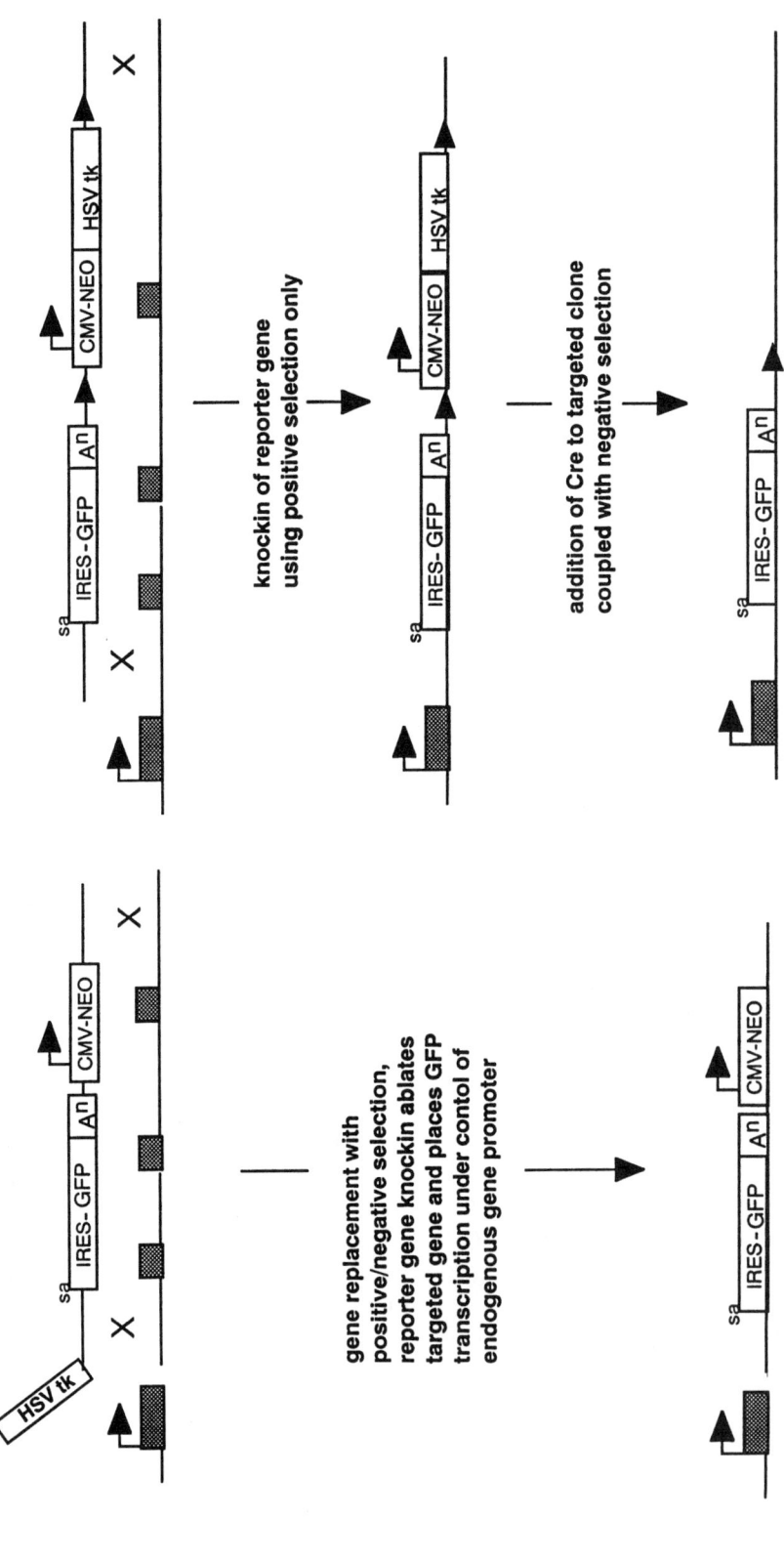

Fig. 2. Gene replacement to generate a null allele, coupled with the knockin of a gene reporter cassette. Positive–negative drug selection will enrich for clones that have undergone gene targeting via homologous recombination. Properly targeted clones will not only be deleted for one functional copy of the gene, but the targeting event will place a green fluorescent protein reporter gene under the transcriptional control of the targeted locus. Transcription from the endogenous promoter is depicted by the arrow. SA = splice acceptor signal, IRES = internal ribosome entry signal, An = polyadenylation signal, CMV-Neo = a neomycin gene driven by a cytomegalovirus immediate-early promoter enhancer sequence. In the right-hand panel, proper homologous recombination between the vector and the targeted gene results in the replacement of the exons with the GFP reporter gene. The CMV-Neo is paired with an HSV-*tk* cassette, and both cassettes are flanked by loxP sites (arrowheads). Following positive selection, the properly targeted clones are transduced with a Cre expression vector. Recombination of the loxP sites by Cre deletes the *tk* gene, and these clones will survive negative selection against tk. This also excises the neo cassette from the allele.

duction of single nucleotide alterations into a target gene without altering the chromosomal location of the gene, thus expression levels should precisely copy that of the endogenous wild-type allele, thereby permitting mouse modeling of point mutations observed in human patients (for example, *see* **ref. *58***). Furthermore, the use of ES cell knockin technology allows structure–function experiments to be performed in vivo in ES cells and in mice *(59)*.

2.3. Cre-Lox Technology in Gene Targeting

One caveat to employing knockin vectors in gene targeting experiments is that the targeting vector must also contain a positive selection marker cassette to permit drug selection of transduced ES cell clones. The positive selection marker typically contains a polyadenylation signal as well as cryptic splice sites encoded in the drug-resistance gene that may negatively impact production of full-length transcripts from the targeted allele. Furthermore, the drug selection marker often utilizes a strong mammalian gene promoter such as CMV or PGK to drive expression of the drug-resistance gene, and the presence of this promoter element may also influence expression of the reporter gene or cDNA knocked in to the target locus. To avoid these problems, many researchers to take advantage of Cre-loxP technology to excise the positive selection cassette from the recombinant locus once the properly targeted ES cell clone is identified. Cre is a 38-kDa protein encoded by the bacteriophage P1 *cre* (cyclization recombination) gene, which induces site-specific recombination of DNA *(60)*. The Cre enzyme recognizes a 34-bp bacteriophage DNA motif known as a loxP site (locus of "X"-over of P1) and mediates reciprocal recombination between two loxP sites. The presence of the non-palindromic core sequence gives the loxP site directionality, and two loxP sites inserted into genomic DNA in the same orientation will, in the presence of Cre, recombine by deleting DNA sequences lying between the two sites. Thus, flanking the positive selection marker cassette in the knockin vector with loxP sites will permit Cre-mediated excision of the marker in the ES cell clone prior to injection of the ES clone into mouse embryos (**Fig. 3**). Once the properly targeted ES cell clone is identified by its ability to survive drug selection, and proper targeting is confirmed by Southern analysis of genomic DNA isolated from the clone, a Cre cassette is electroporated into the ES cell clone in order to excise the positive selection cassette. The Cre cassette usually contains a eukaryotic promoter driving expression of the Cre gene, which encodes for Cre fused to a eukaryotic nuclear localization signal. Transient expression of *cre* in ES cells mediates highly efficient recombination of the loxP sites, thereby deleting the drug selection cassette from the targeted locus. Following confirmation of proper Cre-mediated excision of the marker (either by Southern analysis or PCR experiments using ES clone genomic DNA), the ES cells are used to generate mice. This methodology allows the introduction of subtle mutations into a targeted locus without additional alterations arising from nearby placement of a strong promoter or marker gene. It also allows for subsequent utilization of the excised marker in other gene targeting vectors if a second round of gene targeting in the ES clone is desired *(61)*.

Addition of the Cre cassette to targeted ES cells necessitates a second round of electroporation, which increases the risk that the ES cells might lose their pluripotency. Furthermore, since approximately 1–5% of ES cells are transduced following electroporation, many of the electroporated cells will lack the desired excision of the

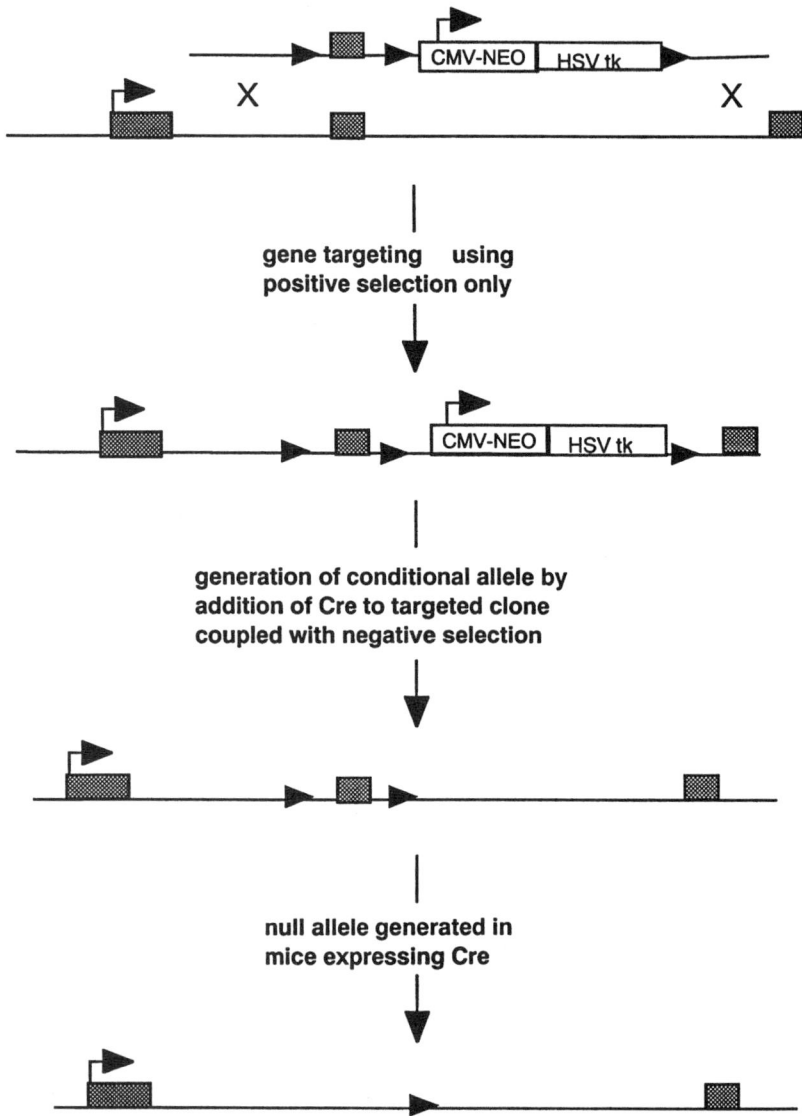

Fig. 3. Creating a conditional allele by gene targeting and use of Cre-lox technology. An endogenous gene is targeted in a manner that leaves the coding sequences intact, but "floxed" (flanked by lox
P). Following positive drug selection, identification, and expansion of properly targeted clones, Cre
is introduced into the ES cells to excise the drug selection markers. Selection against the *tk* gene will
result in loss of either the drug cassettes only, or loss of both the selection cassettes and the nearby
exon. The first possibility creates a conditional allele, whereas the second event would generate a loss
of function allele. The conditional allele has two loxP sites flanking the exon, and will excise the exon
upon subsequent transduction of Cre in the cells of tissues of the mice.

positive selection marker. To circumvent this problem, researchers often include a negative selection marker such as the HSV *tk* gene in addition to the positive selection
cassette flanked by loxP sites. Once the targeted clone is electroporated with the Cre
cassette, the cells are placed on negative drug selection media (gancyclovir) to select
only for those clones that have expressed Cre and excised sequences between the loxP
sites. While this greatly enhances the likelihood of recovery of properly targeted and

Cre-excised clones, the use of this approach means that the HSV *tk* gene can no longer be used as a negative selection cassette in the initial gene targeting. Recently, ES cells bearing a stable, integrated Cre transgene under transcriptional control of the mouse Protamine gene promoter *(62)* have been used in gene targeting experiments to introduce subtle gene mutations into the p53 gene *(59)*. Because Cre did not have to be added exogenously to the ES cells, the negative selection marker could be used at the end of one arm of homology to aid in identification of properly targeted clones. Once ES clones were identified that contained the subtle mutations (as well as the positive selection marker flanked by loxP sites), introduction of the ES cells into mice resulted in excision of the positive selection cassette in the sperm of males, where expression of the protamine transgene was induced. Alternatively, the ES cells bearing the floxed positive marker cassette could be used to generate chimeric mice, which are then mated with transgenic mice that express Cre, in order to excise the marker gene *(63,64)*. However, this approach requires production of an additional generation of mice in order to obtain the properly targeted locus.

Cre-lox technology has also been used in ES cells to generate large-scale deletions of gene clusters *(65,66)*. The targeting of one locus with a knockin vector that contains half of a functional hypoxanthine phospho-ribosyltransferase (*hprt*) minigene and a loxP site is followed by the targeting a second gene lying (in cis) on the same chromosome with a knockin vector that introduces the other half of the *hprt* minigene, plus a second loxP site. Subsequent addition of a Cre vector to these cells results in deletion of sequences lying between the loxP sites and results in the two halves of the *hprt* minigene being brought together. Thus, hprt− ES cell can be selected in HAT media to select for those clones that contain the reconstructed *hprt* minigene. This type of approach has been used recently to generate ES cells deleted for megabase pair length regions of DNA *(66)*. The use of these modified ES cells to generate mice haploinsufficient for large regions of DNA should assist in identifying tumor suppressor genes that lie within the region in question. Haploinsufficient mice can be mutagenized with chemicals such as ethylnitrosourea (ENU), or infected with retroviruses to insertionally inactivate genes *(67,68)*. Tumorigenesis in these mice would likely be a by-product of mutation of genes in the remaining wild-type allele corresponding to the region of DNA excised in the ES cells by Cre.

Other Cre-mediated DNA rearrangements are possible; recombination between two loxP sites placed in inverted orientation will result in an inversion of the intervening DNA sequences. While this methodology has been explored as a means of generating balancer chromosomes to permit mutational scanning of defined regions of the chromosome in mice *(69),* these types of genetic rearrangements have been used less frequently in functional analysis of tumor suppressors to date. In addition, other site-specific recombinases such as the FLP-frt system from yeast have been adapted for ES cells *(70).* While Cre-lox remains the most commonly used system, development of other recombinase systems has expanded our repertoire of methodologies for ES cell targeting.

2.4. Generation of Conditional Alleles

Often, generation of knockout mice results in an early embryonic lethal phenotype. While such lethality may be informative for the role of the targeted gene in development, analysis of the role of the gene in organogenesis, aging, and cancer may be unobtainable. To circumvent this problem, researchers have begun utilizing Cre-lox technology to generate mice bearing conditional alleles, and ablating the gene function in a temporal or tissue-specific manner *(71–76)*. To this end, loxP sites are introduced into the target locus flanking as much as the entire gene or as little as one exon encoding key functions. In addition, a positive selection marker must also be used in the targeting vector, and must also be flanked by loxP sites in order to permit subsequent removal of the drug cassette following identification of a properly "floxed" (flanked by loxP) gene. This is especially critical in generating conditional alleles, as expression of the floxed gene must remain unaffected by the addition of loxP sites. The loxP sites themselves may be safely inserted into introns as well as upstream or downstream of the gene, as the 34 bp loxP site has not been found to alter expression levels of the targeted gene. However, loxP sites do contain several ATG nucleotides, and researchers often prefer to introduce the loxP sites in reverse orientation with respect to transcription of the host gene to ensure that the loxP sites themselves do not induce aberrant initiation of translation.

Addition of Cre to the targeted ES cells will result in some clones that excise the drug cassette, but retain the loxP sites that flank key endogenous sequences **(Fig. 3)**. The ES cells are used to generate mice following identification via Southern analysis of genomic DNA of those ES clones with correct targeting. Subsequent addition of Cre to these mice will induce excision of the intervening sequences to produce a null allele. Cre may be added to the mice either by infection of the mice with recombinant viral vectors encoding Cre *(72,73)*, or by breeding the mice to transgenic mice expressing Cre in a known temporal or spatial manner *(74–76)*. Thus, Cre-mediated gene ablation is controlled by regulating expression of Cre in the mice bearing the floxed allele. The availability of mice with established patterns of Cre transgene expression is crucial, and the numbers of Cre transgenic mouse lines increases each year.

2.5. Other Methods of Introducing Mutations: Hit-and-Run Gene Targeting

Other methods for generating single nucleotide alterations in mice using ES cell technology have been used by researchers studying tumor suppressor gene functions. One such method, called "hit and run," involves the creation of a targeting vector that contains a mutation within the region of homology to the endogenous gene *(77)*. The targeting vector also contains a positive selection cassette and a negative selection cassette in the backbone of the plasmid **(Fig. 4)**. The vector is linearized within the region of homology and transfected into ES cells. Proper targeting of the vector to the endogenous gene results in the insertion of the vector into the target locus and creates a gene duplication event. These cells may be identified by Southern analysis following positive drug selection. Clones bearing the targeted event can then be expanded in the absence of drug selection, permitting the duplication event to resolve via intrachromosomal recombination or sister chromatid exchange. Revertant clones can be isolated by growing the ES cells under negative drug selection, as the negative selection cassette will be lost during resolution of the duplication. A portion of the revertants will retain the mutation event and can be identified by PCR or Southern analysis of the ES clone DNA.

Hit-and-run gene targeting was developed prior to the establishment of Cre-lox technology in ES cells, and by and large has fallen out of favor since that time, as the run step in the approach would sometimes result in large scale rearrangements of the targeted gene. However, a variation of the hit and run approach has been recently used suc-

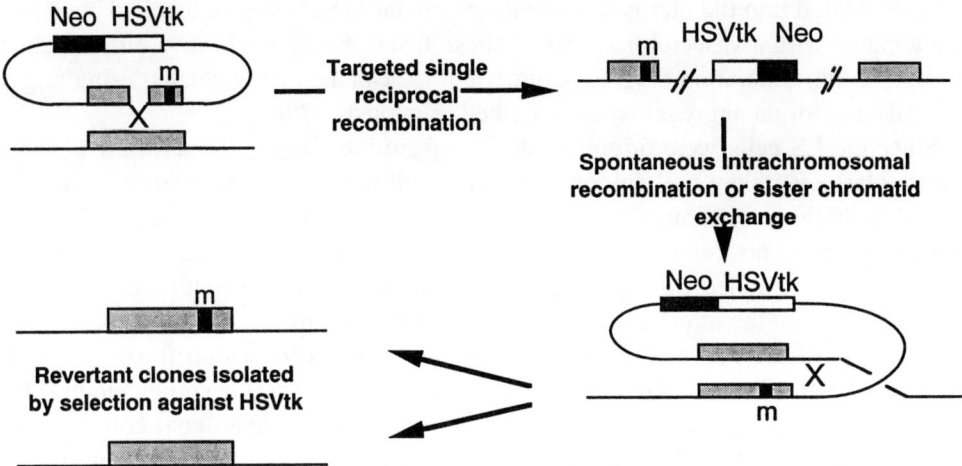

Fig. 4. Generation of subtle mutations in ES cells or mice by "hit-and-run" gene targeting. An insertional recombination event is induced by cleavage of the targeting vector in the region of homology shared between the vector and the endogenous gene. The vector also bears the desired mutation (**m**) in the target homology, as well as a positive selection marker (neomycin cassette) and a negative selection marker (HSV *tk* cassette) in the plasmid backbone. Recombinant clones are selected in G418 and screened by conventional methods for the targeted event (the hit). For the second step (the run), the targeted clones are expanded without selection. The duplication event generated by the insertion of the vector will spontaneously resolve via intrachromosomal recombination or sister chromatid exchange, resulting in the loss of both the positive and negative selection marker. Thus, revertant clones may be isolated by growing the cells in gancyclovir. These clones will either lose or retain the desired mutation, depending on the point of resolution of the intrachromosomal recombination event. Those clones retaining the mutation may then be used to generate mice. Alternatively, the hit step may be performed in ES cells, and properly targeted hit ES cells used to generate mice. Resolution of the gene duplication event caused by the hit may be used to model spontaneous mutation of the allele in individual cells in mice, as the run step may result in the mutant exon being transcribed instead of the wild-type coding sequences.

cessfully to generate a conditional allele in an oncogene in mice *(78)*. Hit-and-run targeting of the K-Ras gene in ES cells was used to create a duplication in an allele in ras, with one gene bearing an oncogenic mutation that was not expressed unless intrachromosomal recombination occurred and resolved the gene duplication. Rather than selecting for this revertant in ES cell culture, mice were generated using the "hit" ES cells, and the "run" portion was allowed to occur spontaneously in the mice, inducing expression of a mutated gene in place of a wild-type copy. Thus, generation of a mutated gene via intrachromosomal recombination was used in this system to mimic independent, spontaneous mutation of the endogenous allele, producing a much more realistic model of oncogene activation and tumorigenesis in mice.

2.6. Generation of Chimeric Mice Using Targeted ES Cells

Two different approaches have been used successfully to introduce targeted mutations in ES cells into the germline of mice. In the first method, blastocyst embryos are harvested from pregnant female mice at d 3.5 postcoitus (p.c.) by flushing the uterus with medium *(79)*. Approximately 15 targeted ES cells are microinjected into each blastocyst using a finely drawn glass needle. Following microinjection, the blastocysts are surgically implanted into the uterus of a pseudo-pregnant female mouse that had been previously mated with a vasectomized male. These foster mothers will give birth 17 d after blastocyst transplantation (**Fig. 5**). Chimeric mice can also be generated using targeted ES cells by morula aggregation as described previously *(80)*.

Since the ES cells were originally derived from the inner cell mass of a 129-strain mouse blastocyst, they will colonize the inner cell mass of the injected blastocyst and contribute to the formation of the resulting mouse. Since the inner cell mass of the recipient blastocyst is not removed, the resulting mouse will be chimeric, with endogenous and exogenously added ES cells contributing to formation of the embryo. The amount of contribution of the injected ES cells to the resulting chimeric mouse can be estimated by the color of the fur: 129-strain mice are homozygous for agouti (brown) coat coloration, whereas the recipient blastocysts are usually harvested from nonagouti (black) C57BL/6 mice. Therefore, the chimeric mice will have a mixed coat color, with the amount of agouti present in the coat of the chimeric mice serving as a rough indicator of the degree of chimerism. Germline contribution of the targeted and injected ES cells can be determined by crossing the chimeric mice with a C57BL/6 mouse and scoring the offspring for the presence of agouti mice, as agouti is a dominant trait. The agouti F1 mice have a 50% chance of inheriting the targeted allele from the chimeric parent, and can be scored for the presence of the targeted allele by Southern or PCR-based analysis of genomic DNA isolated from tail biopsy. Mice that are heterozygous for the targeted allele can be intercrossed to obtain homozygous targeted mice.

3. Characterization of Tumor Suppressor-Deficient Mice

The successful introduction of a tumor suppressor mutation into the germline of mice opens up a huge array of experimental opportunities. While the obvious first approach is to monitor such mice for early spontaneous tumors, a number of other methods to study tumor suppressor function in both normal tissues and tumors present themselves. Some of these methodologies and approaches are outlined below.

3.1. Assessment of Developmental Abnormalities

Once mice heterozygous for a germline tumor suppressor mutation are generated, they are crossed to each other in an attempt to generate offspring nullizygous for the mutation. Mutation of both germline tumor suppressor alleles often confers embryonic lethality *(6)*. The development of embryonic lethality can be late in the 19-d gestation period (e.g., *Rb* at 12.5–14.5 d postcoitus) *(81,82)* or early (e.g., some *Brca1*-null mutants show lethality before 7.5 d) *(83,84)*. In the later-stage embryos, abnormalities may affect only specific organ systems, such as in *Rb*-deficient embryos where neurogenesis and hematopoiesis are affected or in Wilms'-null embryos where early kidney development is abnormal *(85)*. In other cases, development of null embryos may be unaffected, as is usually the case for p53, p19[ARF], and p16 *(86–88)*. However, even a subset of p53-null embryos exhibit exencephaly, a failure of the neural tube to close properly, suggesting that penetrance of an embryonic lethality phenotype may vary *(89,90)*. This is particularly evident where the genetic background or other mutations can influence the onset of lethality *(91,92)*.

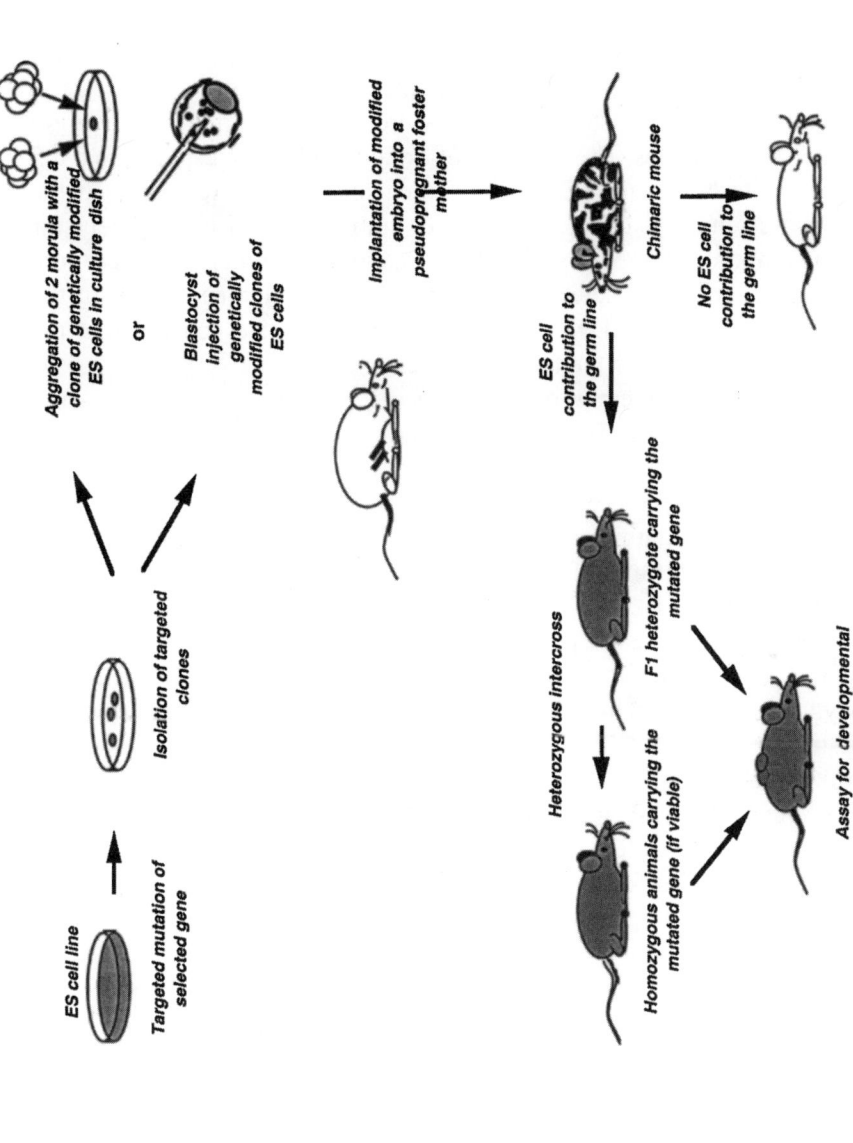

Fig. 5. Creation of genetically modified mice using embryonic stem cells bearing targeted mutations in tumor suppressor genes. Properly targeted ES cell clones containing the desired mutational event are microinjected into blastocyst-stage embryos or aggregated with morula that have had the zona pellucida removed. The resulting embryos are surgically implanted into pseudo-pregnant foster mothers and give rise to chimeric mice. Breeding of the chimeras with wild-type mice results in the production of F1 heterozygotes that carry the mutated allele if the genetically modified ES cells contribute to the germline of the chimeric mice. The heterozygous mice may be intercrossed to produce homozygous null mice to assess the role of the tumor suppressor gene in development. Both the heterozygous and homozygous mice (if viable) may be monitored for tumor development.

The initial test of an embryonic lethal condition is simply to determine the genotypes of the offspring of heterozygote crosses at weaning, assuming there are no obvious morphologic abnormalities among the weanlings. If there is a significant reduction in the numbers of nullizygotes obtained, or no nullizygotes among the offspring, and the neonates are consistently healthy, then an embryonic lethal condition exists. If the neonates do die frequently, this may be indicative of a perinatal lethality effect. In this case, a thorough pathologic analysis of both abnormal and normal control neonates should be undertaken. This should be done in collaboration with a skilled veterinary pathologist or mouse embryologist. Reference books on the anatomy of the various mouse organ systems can be helpful *(8–10)*.

If viable nullizygotes are not obtained, the next step is to analyze progressively earlier-stage embryos. We usually sacrifice pregnant mothers in heterozygous crosses at d 10.5–12.5 postcoitus (p.c.). Days postcoitus are measured from a morning observation of vaginal plugs in mating females. The presence of a vaginal plug corresponds to d 0.5 (at noon) of embryonic development *(10)*. Once the pregnant mothers are sacrificed, the individual embryos are removed and examined after cutting a small tail segment for DNA analyses. The embryos can then be fixed and sectioned in order to detect more subtle defects in organogenesis. If defects are observed in roughly one quarter of the embryos, then PCR genotyping of DNA from the tail biopsies can confirm that the defective embros are indeed null for the targeted gene.

If no defective or nullizygous embryos are observed at midgestation, then earlier embryo analyses must be performed. Typically, we examine embryos at d 8.5 p.c. because the dissections are easily performed, the embryo readily separated from the maternal deciduum, and the yolk sac (which is embryonic in origin) yields enough DNA for PCR or even Southern analysis, facilitating genotyping of each embryo. Furthermore, defects in gastrulation (d 6.5 p.c.) or turning of the embryo (d 8 p.c.) are often reflected by an aberrant morphology at d 8.5 *(93)*. If no null embryos are recovered at d 8.5, one is faced with a series of dissections that must be performed at various days earlier in the gestation of the embryo in order to document the phenotype. Initially, we favor flushing the uteri of pregnant heterozygous females that have been mated with heterozygous males in order to recover blastocyst-stage embryos at d 3.5 p.c. Although genotyping of embryos at this stage requires a PCR-based method and the use of the entire embryo to generate sufficient genomic DNA for use in the PCR reaction, the embryos are easily removed from the females and may be photo-documented or allowed to develop further in vitro before genotyping. Absence of null blastocysts or inability of the blastocysts to adhere to tissue-culture plasticware and expand in culture indicates a very early phenotype, such as we observed recently in *Ini1* tumor suppressor-null embryos *(94)*. The presence of healthy null blastocysts would provide a rationale for harvesting and scoring embryos for nullizygosity between d 4 (implantation) and d 8.5 p.c.

Egg cylinder-stage embryos may be harvested from female partners of timed matings at d 6.5 and d 7.5 and dissected free of the maternal deciduum with some practice. An ectoplacental cone, an extraembryonic tissue that is fetal in origin and will give rise to the placenta, caps these embryos. This structure may be removed and genotyped by PCR to identify any null embryos. Embryos harvested between d 4 and d 6.5 p.c. are more difficult to free from the surrounding maternal decidua without damaging embryonic morphology, and these embryos are generally left in their individual deciduum during harvest. The entire embryo is usually fixed and sectioned for histologic analysis, precluding any definitive genotyping of the embryo. However, if one-quarter of the embryos display an aberrant phenotype, it is customary to assume that these embryos are the nulls. If uncertainty persists, some sections may be taken from the embryos and genotyped on the slide via PCR. While the d 4–6 p.c. timepoint in development is probably the most challenging time to analyze the phenotype of a null embryo, ablation of many genes involved in DNA recombination and repair, tumor suppression, and modifiers of tumor suppressor functions result in lethality during this period of development *(6,91)*. However, information can be gleaned, even from these early lethal embryos, by using a combination of careful structural analysis (histology), DAPI staining or TUNEL assays for aberrant apoptosis, BrdU incorporation to document cell proliferation, and antibody staining to observe the formation of the three tissue lineages *(9,10,95,96)*.

3.2. Assessment of Tumor Susceptibility

Once germline heterozygotes for a mutant tumor suppressor gene are obtained, sufficient numbers should be maintained for tumor monitoring over their life span. While 20–30 mice of a given mutant genotype are usually sufficient for the generation of reliable tumor-free survival curves, this number may be insufficient for accurate determination of tumor spectrum if the tumor types that arise are quite varied. Mutant mice of both sexes along with wild-type littermate controls should be monitored at least up to the age of 18 mo and preferably 24 mo, if not through their complete life span. We have found that some weakly tumor-susceptible mutant lines will rarely develop tumors before 18 mo of age, yet will develop tumors between 18 and 24 mo at a significantly higher frequency than their wild-type counterparts (L.A.D., unpublished data). However, even normal C57BL/6 mice begin to exhibit tumors after the age of 18 mo *(12–14)*, so significant numbers of these controls need to be monitored to obtain a reliable background tumor incidence. If spontaneous tumors do not arise significantly faster in the mutant line than in wild-type controls, this does not necessarily mean that the mutant line lacks increased tumor susceptibility. Alternative strategies to uncover subtle increases in tumor susceptibility will be discussed below.

If the original mice chimeric for the mutation were crossed to C57BL/6 mice, these mice will be a mixture of 129/Sv (the strain origin of most current ES cell lines) and C57BL/6. Some laboratories will intercross the F1 mice from the chimeric outcross, but others may backcross two or more times into C57BL/6 before tumor monitoring. Initially, tumor monitoring can be performed on N1 or N2 mice (offspring of intercrossing between mice backcrossed once or twice into C57BL6). Ultimately, it will be desirable to obtain congenic mice backcrossed 10 or more generations into C57BL/6 as well as other inbred lines. One shortcut to obtaining an inbred line with the mutation is to cross the chimeric mice to 129/Sv inbred mice. All offspring will have a pure 129/Sv background, assuming the original ES cells were of 129/Sv origin.

Placing the tumor suppressor mutation on multiple genetic backgrounds can often be quite useful, as tumor incidence and spectrum can be profoundly dependent on strain background *(97,98)*. Such differences have been exploited to identify strain-specific tumor-modifier loci. Once classic example of this approach involves the *min* mice (deficient in the *Apc* tumor suppressor) developed by Dove and colleagues *(99)*. On a C57BL/6 background the min mice develop large numbers of intestinal polyps rather early *(100)*. However, on an AKR background, the polyps were significantly fewer, suggesting a modifier gene that influenced the effects of the *min* mutation *(100)*. After extensive mapping and positional cloning experiments, the primary modifier gene was shown to be the secretory phospholipase 2 group 2a gene *(101)*.

Unlike humans, who usually develop a variety of carcinomas with age, most inbred lines of mice tend to develop tumors of mesenchymal and hematopoietic origin *(12,14,102)*. Some tumor suppressor-deficient mice will develop unusual or unexpected tumor types, such as early pituitary adenomas in the Rb+/− mice *(81,82)*. Others, such as the p53+/− mice, develop tumor types often seen in wild-type mice of the same strain, only much sooner *(103,104)*. However, p53−/− mice largely develop thymic lymphomas, a tumor type rarely observed in p53+/− mice or p53+/+ controls. As mentioned earlier, introducing a tumor suppressor mutation into a different background strain can alter the tumor spectrum. For example, almost half of the female p53+/− mice of Balb/c background develop mammary adenocarcinomas *(98),* a tumor virtually never observed in C57BL/6 p53+/− females *(98,103,104)*. Thus, tumor spectrum can be affected by which gene is altered, the nature of the mutation in the altered gene, mutant gene dosage, and strain background.

Identification of mice with tumors is usually a straightforward process dependent on regular and careful observations of each mouse in the monitored groups. In our colony, observations of the mice are made daily by animal care technicians, and these are supplemented by weekly observations by the scientist in charge of the experiment. Evidence for a tumor-bearing animal includes overt growths, a swollen abdomen (indicative of internal tumors) or swellings in other areas, cachexia or wasting, heavy breathing, ruffled fur indicative of a failure to groom, sluggishness in the animal's movement when prodded, and hindlimb paralysis (in the p53+/− mice, this often indicates an osteosarcoma occluding the spinal column). Once suspected tumor-bearing animals are identified, they are sacrificed and subjected to a full necropsy. Ideally, the necropsy should be done in collaboration with a qualified veterinary pathologist. However, for those without access to a veterinary pathologist, a number of useful references detail the necropsy and histopathology procedures *(105–107)*. Any evidence of abnormalities should be noted through a gross pathological examination of all tissues and abnormal growths. Photographs of interesting morphologic abnormalities or tumors should be taken. Blood and urine can also be obtained at this time and subjected to automated biochemical analyses common to most veterinary pathology laboratories. Sometimes smaller primary tumors or metastatic lesions can be observed, and these lesions should be carefully noted. When storing frozen tumor tissue, it is critical that as much skin, connective tissue, blood, and other normal tissue as possible be removed prior to freezing. This will reduce the likelihood of normal tissue as possible contamination of the tumor, which could compromise the integrity of subsequent molecular assays.

Tissues and tumors can be stored in one of a variety of fixatives, depending on the types of analyses that will be performed. Neutral buffered 10% formalin is the fixative commonly used in our laboratory, as specimens can be fixed overnight and left indefinitely. In addition, part of any tumor and parts of one or two normal organs (e.g., a kidney or liver lobe) should be flash-frozen in an Eppendorf tube in a dry-ice ethanol bath and stored long term in liquid nitrogen or in a −80°C freezer. Frozen tissues can then be used as substrates for a variety of DNA, RNA, and protein analyses. The formalin-fixed tissues are then trimmed into tissue cassettes before delivery to the histology laboratory for embedding. Tissue trimming is straightforward and can be performed as previously described *(108)*. The histology laboratory can then embed the tissues in paraffin, prepare 4-μm-thick sections on slides and stain them in hematoxylin and eosin. The stained slides of normal tissues and tumors should then be examined by a veterinary pathologist or a clinical pathologist with mouse expertise. References for the pathology and tumor pathology of the laboratory mouse are widely available *(11,13,109,110)*.

Once the mice of each genotype have been monitored for tumors for a designated length of time and all the tumors have been collected and examined by histopathology, tumor spectum graphs and tumor incidence curves can be constructed. Tumor spectrum can be presented as a table listing the tumor types and the number of times each tumor type is observed for each mouse genotype. For a more informative presentation, tumor spectrum can be presented as a pie graph, in which the percentages of each tumor type are indcated as proportionally sized pie slices. Tumor incidences are best displayed as a Kaplan-Meier plot, where the percentage of mice remaining tumor-free is plotted as a function of age in days, weeks, or months. If the tumor suppressor-deficient mice show dramatically accelerated tumorigenesis compared to controls, then statistical analyses of tumor incidence are usually not necessary. However, in cases where multiple mutant genotypes are being compared, some tumor incidence curves may not differ by a large degree. In such cases, the log-rank test (also called the Cox-Mantel test) is typically used to determine whether two curves differ in a statistically significant manner. A number of good biostatistics texts describe these and other related survival analysis methods *(111,112)*. The analyses can be performed by a qualified statistician or through the use of standard computer statistical software. Our laboratory uses Statview, (SAS Institute, Inc.), which will construct Kaplan-Meier plots for each set of survival curves and compare statistical significance for the resulting curves. In the assumption of a statistically significant difference in tumor incidence between two genotypes, it is important that all other variables such as sex, environment, diet, bedding, and genetic background be kept as similar as possible.

3.3. Alternative Ways to Assess Tumor Susceptibility

Some tumor suppressor-deficient mice remain refractory to spontaneous tumors *(6)*. The *Brca1* and *Brca2* heterozygous mice are a good example of this phenomenon, despite the fact that women with germline mutations in one *BRCA1* or *BRCA2* allele are highly susceptible to breast cancer *(113)*. Since early embryonic lethality in the *Brca1*– and *Brca2*-null mice precludes an examination of tumorigenesis in the nullizygous state, other methods have been employed to enhance tumor susceptibility in the mutant mice. One obvious technique for uncovering subtle tumor susceptibilities is to treat the mutant (and control mice) with carcinogens or mitogenic hormones. Mice that are already susceptible to early spontaneous tumors, such as the p53+/− mice, often develop tumors even sooner in response to treatment with mutagens such as ionizing radiation or chemical carcinogens *(114)*. In another example, mice null for the xeroderma pigmentosum group C *(Xpc)* gene did not develop overt spontaneous tumors *(115)*, despite the fact that humans deficient in this gene have a greatly enhanced susceptibility to skin cancer following sun exposure. By exposing the skin of *Xpc*-null mice to daily ultraviolet radiation, greatly enhanced skin carcinoma incidence was obtained *(115,116)*.

In the case of *Brca1* and *Brca2* heterozygous mice, treatment with the estrogenic compound diethylstibestrol induced morphologic abnormalites in the mammary glands, but did not appear to enhance tumorigenesis *(117)*. A second approach that did show enhancement of mammary tumorigenesis was through crossing of p53-deficient mice to *Brca1*-deficient mice. There was an increase in the number of mammary tumors in the *Brca1+/−*, p53−/− mice compared to the p53−/− mice *(118)*. Moreover, when *Brca1+/−*, p53+/− mice were treated with ionizing radiation, a higher incidence of mammary tumors was also noted compared to irradiated p53+/− mice *(118)*.

Probably the most efficient method of generating the desired mammary tumors in the absence of *Brca1* was described by Deng, Hennighausen, and colleagues, who used the Cre-LoxP system described earlier *(119)* to engineer a Cre-mediated excision of exon 11 of the Brca1 allele in the mouse mammary epithelial cells *(74)*. Specific mammary tumors did occur in these mice that were deficient in both *Brca1* alleles, but after a relatively long latency and in only 22% of animals. Crossing of these mice to make them p53+/− greatly accelerated mammary tumor formation and increased the frequency of animals developing mammary cancers to 75% *(74)*. Another elegant approach was demonstrated by Shibata et al. *(73)*, who engineered loxP sites into the germline *Apc* tumor suppressor gene flanking exon 14. Mice null for the *Apc*-LoxP insertion were normal, but when the colons of these mutant mice were exposed to a Cre-expressing adenovirus, the mice developed adenomas within 4 wk. The adenomas showed deletion of both *Apc* exon 14 alleles, consistent with Cre-mediated deletion mutagenesis of the *Apc* gene. Thus, these types of tissue-specific allele inactivations have shown themselves to be a very useful way to circumvent the constraints imposed by nullizygous embryonic lethality.

3.4. Assessment of Tumor Suppressor Cooperativity In Vivo

Cancer is the result of multiple genetic lesions. Even a mouse null for a tumor suppressor gene such as p53 will usually require several months for a single clonal tumor to appear. The rate-limiting step for the appearance of these tumors may be the development of a requisite number of further oncogenic lesions. If one or more of these lesions can be engineered into the germline, then the tumor incidence should be further accelerated in conjunction with the original tumor suppressor mutation. Acceleration of tumorigenesis by intercrossing two tumor susceptible mouse strains is a common way to measure gene cooperativity in vivo. If cooperativity (i.e., accelerated tumorigenesis) is observed in the double-mutant mice compared to their singly mutant parents, then this suggests that these genes both contribute to the cancer process in a nonoverlapping way, presumably by affecting different signaling pathways. In addition to effects on tumorigenesis, such crosses have provided novel insights into the role of tumor suppressors in the developmental process. In this section, we provide some examples of interesting approaches to studying cooperativity utilizing the p53-deficient mice. The *Brca1*-deficient mice crossed to p53-deficient mice cited earlier *(74,118)* illustrate how the p53-deficient mice have been widely used as an tumor accelerator in a wide variety of other models.

Two signaling pathways that are almost invariably affected in human tumors are the p53 and Rb pathways *(120,121)*. These prototypic tumor suppressors are also the critical proteins inactivated by many DNA tumor viruses in the transformation process. We and the Jacks laboratory examined the cooperativity of Rb and p53 at the organismal level by crossing Rb- and p53-deficient mice *(122,123)*. Not surprisingly, the combined deficiency of Rb and p53 accelerated tumorigenesis. Because Rb nullizygosity causes embryonic lethality, we compared tumorigenesis in p53−/− Rb+/− mice to p53−/−

Rb+/+ mice or tumorigenesis in p53+/− Rb+/− mice to p53+/− Rb+/+ mice. In both cases, deficiency in both alleles accelerated tumorigenesis. These results were consistent with a model in which Rb loss abrogates normal growth control and p53 loss abrogates a normal apoptotic response to the aberrantly dividing cell. Surprisingly, the tumor spectrum in the double-mutant mice consisted largely of neuroendocrine tumors not seen in either of the parental Rb- or p53-deficient mice. This indicated that the susceptibility of a particular tissue to oncogenic transformation is critically dependent on the nature and combination of genetic lesions. This principle was recently illustrated by DePinho and colleagues *(124)*, who showed that late-generation telomerase-deficient mice often developed carcinomas in a p53+/− background, as opposed to the usual lymphomas and sarcomas seen in p53+/− mice. DePinho hypothesized that high turnover rates in epithelial lineages would lead to more rapid telomere attrition in this cell type, followed by chromosomal instability and further oncogenic lesions, particularly in the absence of the apoptotic response supplied by p53 *(102)*.

Another rationale for intercrossing two mutant mouse lines is to facilitate mechanistic studies of tumorigenesis. We had found that the diversity of tumor types in the p53-deficient mice inhibited our ability to obtain sufficient numbers of tumors derived from a specific tissue type. Moreover, corresponding control tumors that were p53+/+ were difficult to obtain. To circumvent these limitations, p53-deficient mice were crossed to *Wnt-1* transgenic mice *(125)*. The *Wnt-1* transgenic mice contained a *Wnt-1* oncogene with a mammary gland-specific promoter, and these mice developed mammary adenocarcinomas in a relatively short amount of time (3–12 mo) *(126)*. When intercrossed to p53-deficient mice, we could compare mammary tumorigenesis in *Wnt-1* transgenic females with a p53−/−, p53+/−, or p53+/+ background. Importantly, all of the double-mutant female mice developed mammary adenocarcinomas, and we had sufficient numbers of control p53+/+ tumors to compare to p53−/− tumors. Mammary tumors arise subcutaneously and can be monitored for tumor growth from a very early stage using caliper measurements. In these studies, the *Wnt-1*-initiated mammary tumors in the absence of p53 arose sooner, grew faster, exhibited more malignant histopathology, showed higher cell division rates, and showed more genomic instability than their counterparts with normal p53 alleles *(125,127)*. The tumors were also subjected to a number of molecular and genetic assays and these assays have facilitated our understanding of p53 loss in promoting tumorigenesis in a well controlled tumor model system *(125,127,128)*.

In addition to the effects on tumor spectrum and incidence that can be observed through intercrossing of tumor-susceptible mutant mice, interesting effects on development can also be obtained through appropriate crosses. For example, intercrossing of p53-deficient mice with *mdm2*-deficient mice by our laboratories and the Lozano laboratory led to a surprising developmental phenotype *(91,129)*. Mdm2 is an oncoprotein that binds p53 and mediates its degradation. Mice null for *mdm2* exhibit embryonic lethality at roughly 6.5 d embryogenesis, possibly due to an excess accumulation of p53 which is no longer degraded appropriately and may induce apoptosis in the embryo. However, when double-null p53 and *mdm2* mice were generated, these mice developed normally and were no different morphologically than their p53−/− or wild-type littermates *(130–132)*. This result suggests that a key role of mdm2 in development is to regulate p53 levels. Interestingly, this rescue effect of p53 deficiency has been partially recapitulated in crosses with *Brca1-* and *Brca2*-deficient mice. Double-null p53 and *Brca1-* or *Brca2* mice survive 1–2 d longer during embryogenesis than do the *Brca1-*

or *Brca2*-null embryos in the presence of wild-type p53 *(130–132)*. This suggests that the lack of p53-induced apoptotic response in the repair-deficient *Brca1–* and *Brca2*-null embryos may allow a longer survival time.

3.5. Biological Assays of Tumors

Aside from the histopathologic assays to identify tumor type, a number of assays can be utilized to measure various biologic activities within the tumor itself. Some solid tumor types, such as mammary tumors, arise on the surface of the animal, and are identifiable at a relatively early stage by palpation or careful observation. The volume of these tumors can be determined once or twice a week by measuring the diameter of the tumor in two dimensions with a set of calipers. The volume of the tumor can then be estimated by the equation $TV = (A^2 \times B)/2$, where A is the shorter diameter (in mm), B is the longer diameter (in mm), and TV indicates tumor volume (in mm^3) *(133)*. Measurement of tumor volumes over a period of 4–5 wk can then provide an index of relative tumor growth rates.

In many cases, mouse tumors arise internally and are not measurable with calipers. However, cell proliferation rates in tumor sections can be measured by a number of techniques. One straightforward method involves mitotic figure counting in hematoxylin-and-eosin-stained slides, which is an indicator of the percentage of mitotic cells within a tumor cell population. The aligned and condensed chromosomes of a mitotic cell are easily distinguishable from cells in other phases of the cell cycle, and these cells are counted in each of 10 random microscopic fields (usually at a magnification of 400×) from a hemotoxylin-and-eosin-stained tumor section. The mean number of mitotic figures per microscopic field provides an index of cell proliferation in comparing tumors *(134)*.

Another means of measuring tumor cell proliferation is through the bromodeoxyuridine (BrdU) incorporation assay for measurement of the fraction of tumor cells in S phase. Commercially available kits can facilitate these assays. The tumor-bearing animal is injected intraperitoneally with BrdU solution 3 h prior to sacrifice. The tumor is harvested, fixed in methacarn fixative, paraffin-embedded, and unstained slides prepared and processed as previously described *(135)*. The slides are exposed to an anti-BrdU antibody followed by incubation with a peroxidase-coupled antibody specific for the primary anti-BrdU antibody. Cells that undergo peroxidase staining following treatment with a chromogenic peroxidase substrate are in S phase. Again, stained cells in each of 10 microscopic fields are counted and averaged as a relative indicator of cells undergoing DNA replication *(135)*.

Another method for measuring cell proliferation that we have utilized is flow cytometry for DNA content on fixed tumor sections, based on a modification of Hedley's method *(127,136)*. Briefly, five 50-µm-thick paraffinized sections of a tumor are deparaffinized and nuclear suspensions of the fixed tumor cells subjected to flow cytometry following incubation of the nuclei in propidium iodide. Percentages of cells in G0/G1, S phase, and G2/M can then be determined by standard flow cytometer analyses. However, it should be noted that the tumor processing steps involved in this particular methodology are extensive, and not all flow cytometry cores may be equipped to perform these types of analyses. Moreover, tumor aneuploidy may complicate interpretation of the cell cycle parameters in the flow cytometry profile.

The growth of a tumor is influenced by the rate of cell death as well as cell division, and we have found that apoptosis rates in tumors can sometimes be quite high. There are many commercially available apoptosis assays, and a number of these can be applied to tumor sections. A common assay for apoptotic cells in a tumor section utilizes the TUNEL assay *(127,137)*. The TUNEL assay specifically detects and allows staining free 3′ DNA ends that are a hallmark of apoptotic cells *(137)*. Again, counting the average number of stained apoptotic cells per microscopic field in a tumor is a way to quantitate relative rates of apoptosis. Another assay for apoptosis is the DNA fragmentation assay, in which total tumor DNA is run on an agarose gel and low-molecular-weight DNA quantitated as an indicator of relative numbers of apoptotic cells *(127)*. There are also a number of assays that detect specific protein products of apoptosis which can be used as indicators *(137)*.

One critical measurement of the malignant potential of a tumor is its ability to metastasize. Generally, we have found that most tumors that arise in p53-deficient and other tumor suppressor-deficient mice do not metastasize to a high degree. However, recently Lozano and colleagues *(138)* have reported a p53-mutant mouse with a codon 172 mutation that did produce tumors with high metastatic potential. Thus, in any new model generated, aside from the primary tumor, any evidence for metastatic lesions should be carefully examined during the gross necropsy in the lungs, brain, liver, kidneys, spleen, and bone marrow for nodules of discoloration or other small abnormal growths. These areas should then be examined histopathologically by a qualified pathologist. In addition, the normal organs from the tumor-bearing animal should be examined by histopathology for any evidence of micrometastatic lesions not visible to the naked eye during necropsy.

3.6. Genetic Assays of Tumors

If mice heterozygous for a mutant tumor suppressor allele are susceptible to early tumors, an immediate question is whether the remaining wild-type allele is mutated or lost. According to the "two-hit" model for tumor suppressors, first formulated by Knudson *(139),* loss of function of both tumor suppressor alleles is a prerequisite for tumor formation. Virtually all of the tumors from Rb+/− mice show a deleted Rb allele (in addition to engineered mutated allele), conforming to the two hit model *(82,140)*. In contrast, about half of the tumors in p53+/− mice exhibit a structurally and functionally intact wild-type p53 allele, suggesting that mere reduction in dosage of p53 may be sufficient to promote tumorigenesis *(141)*. When assaying for loss of the wild-type tumor suppressor allele, initial examination is usually performed by Southern blot analysis with allele-specific probes. In many cases, the wild-type allele will be deleted. In those cases where it is not, further sequencing of the wild-type allele may be required to assay for more subtle mutations. If the allele is structurally intact, it may be useful to look at RNA and protein expression of the tumor suppressor in the tumor. If expression is normal or near normal, tumor suppressor functional assays could be performed. In p53+/− tumors with intact p53, we showed that p53 was functional by irradiating the tumor-bearing animals and demonstrating that tumor cells upregulated prototypical p53 transcriptional targets *(141)*.

Identification of genetic lesions that cooperate with the targeted tumor suppressor allele in the generation of a tumor is often a high priority. There are a number of methods for identifying such alleles. The most direct way is to perform a candidate gene search based on historical associations or interconnected signaling pathways. For example, it was known in that in breast tumors from women with germline BRCA1 mutations, p53 inactivation occurred at a very high frequency *(142)*. Moreover, when Deng and colleagues analyzed the mammary tumors in their mammary gland-targeted Brca1-

knockout mice, they found that p53 was silent or rearranged in two-thirds of their tumors *(74)*. Thus, genes likely to be altered in combination with the targeted tumor suppressor are usually the first to be assayed. This can be done by a number of methods, starting with Southern mapping of candidate genes, or sequencing of the relevant candidate cDNAs or exon segments known to be "hot spots" for mutation. RNA expression analyses may also reveal whether a gene is transcriptionally silenced.

Genome screening approaches have also been used to identify potential cooperating lesions. Mapping of mouse strain-specific polymorphisms has provided a valuable tool to perform genome-wide scans for loss of heterozygosity (LOH) in tumors *(143,144)*. This approach is facilitated by an extensive array of available PCR primers that flank specific short microsatellite repeats interspersed throughout the genome. Because the sizes of these short repeats vary from strain to strain, PCR with primers adjacent to these repeats will generate fragments of a size specific for each strain. If the tumors being analyzed are from a mouse heterozygous at many polymorphic loci (e.g., F1 offspring of two different inbred strains), then alleles that are heterozygous in normal tissues, but show LOH in the tumors, indicate a deletion of this region during the tumorigenesis process. If tumors of the same type consistently show LOH in the same chromosomal region, then this would be presumptive evidence of a region containing a tumor suppressor gene frequently mutated in this tumor type. Several laboratories have productively used PCR-based mapping methods to identify possible tumor suppressor regions or tumor modifiers in some of their mouse cancer models *(145–147)*.

Another approach, comparative genomic hybridization (CGH), has been developed by Gray, Pinkel, and collaborators to map regions of DNA copy number gain (e.g., amplifications) or loss (e.g., chromosomal deletions) in tumors *(148)*. Widely used in analyzing human tumors, it has also been applied to mouse tumor models by us and others *(125,141,149)*. This technique involves the co-hybridization of normal and tumor DNA from the same animal to mouse cell metaphase chromosome spreads. Because one of the hybridizing DNAs is labeled with flourescein (which fluoresces green) and the other DNA is labeled with rhodamine (which fluoresces red), laser scanning of the hybridized spreads produces a series of red:green fluorescence ratios along the length of each chromosome. Deviation from the normal 1:1 fluorescence ratios is usually indicative of a DNA copy number gain or loss. We have used this assay not only to identify frequent regions of copy number changes in p53-deficient tumors, but also found that the frequency of these chromosomal gains and losses increased as wild-type p53 dosage decreased, confirming the role of p53 in influencing genomic stability *(125,141)*.

Recently, the CGH methods have been increased in resolution through the implementation of CGH array hybridization *(150,151)*. The methods resemble those of CGH, but rather than metaphase chromosomes attached to slides, hundreds of mouse genomic DNA clones from bacterial artificial chromosomes (BACs) or cDNA clones are arrayed on slides. If enough clones can be screened, a higher resolution can be attained, allowing detection of smaller regions of amplification or deletion. Currently, Spectral Genomics (Houston, TX) markets BAC arrays and the tools to analyze the hybridization results.

Other methods of chromosome analyses include standard cytogenetics of cells cultured from tumors. However, since mouse chromosomes are acrocentric and of similar size, this particular approach requires a skilled cytogeneticist. A recent cytogenetic method of great promise for characterizing chromosomes in mouse tumors is the spectral karyotyping method *(152)*. Using combinations of different chromosome-specific fluorescent probes and imaging techniques, Ried and colleagues have been able to paint each mouse chromosome with a specific color. A number of laboratories have used this procedure to show various types of chromosomal aberrations, including reciprocal and nonreciprocal translocations and dicentric chromosomes *(152–155)*.

3.7. Other Molecular Assays of Tumors

Analysis of RNA and protein expression patterns can provide important insights into the tumorigenic process. The cDNA microarray, oligonucleotide array, and serial analysis of gene expression (SAGE) technologies have been very useful in classifying tumors and identifying target genes of tumor suppressors such as p53 *(156–159)*. Their potential for analyzing mouse tumor models is obvious. Not only can changes in gene expression be identified as a normal cell progresses to a cancer cell, but differences between gene expression patterns in tumors arising in the presence or absence of a particular tumor suppressor may further shed light on the role of that tumor suppressor in regulating oncogenesis. As an example of the latter approach we have compared gene expression patterns in *Wnt-1* transgenic mammary tumors that were either p53+/+ or p53−/− *(128)*. Using candidate gene approaches, differential display PCR, and commercially available cDNA microarrays, we identified seven genes that showed significant differences in expression between p53−/− and p53+/+ tumors. However, because of the heterogeneity between tumors even of the same genotype, it was necessary to confirm our initial screens by Northern blot and Western blot analysis on at least 6–8 tumors of each p53 genotype. The differentially expressed genes fell into two categories, either growth-regulatory genes or differentiation markers. As expected, p53−/− tumors showed lower levels of negative growth regulators and differentiation markers, consistent with their higher proliferation rate and undifferentiated histopathology *(128)*.

Performance of cDNA microarray and SAGE technologies on mouse tumor models will be greatly facilitated by the current availability of slides and kits from a number of companies. However, because of the inherent danger of false positive of false negative results in these assays, a number of guidelines should be observed when doing these assays with tumors. First, it is important to use tumor specimens as free as possible from surrounding normal tissue. Second, where possible, care should be taken that preneoplastic tissues or tumors of the same histopathologic type are compared to reduce gene expression differences resulting from pathologic status rather than genotypic status. Third, as many tumors as possible in each category should be examined, to reduce the effects of tumor heterogeneity on the analyses. Fourth, differential expression levels of a particular RNA should be substantial, preferably 2.5-fold or greater. Fifth, evidence of differential expression results identified through cDNA microarray or SAGE screening should be confirmed and quantitated by Northern blot analysis and even Western blot analysis and immunohistochemistry if possible.

3.8. In Vitro Studies of Cells Derived from Mutant Mice

One of the more useful aspects of the mice with targeted germline mutations is the ability to derive cells from virtually any tissue of the mutant mouse to study the effects of the deficiency under standard cell culture conditions. In those situations where the null animal is viable (e.g., p53−/−, p19ARF−/−, or p16−/− mice), the derivation of cultured cells is straightforward and has been performed for a wide array of cell types in the case of the p53−/− mice *(114,160)*. In fact, some cell types that do not grow well in culture, such as neonatal astrocytes and hepatocytes, grow significantly better if they are lacking p53 *(161,162)*. Moreover, p53−/− cells often immortalize and sometimes undergo oncogenic transformation *(161,162)*.

Since many tumor suppressor-deficient null mice exhibit embryonic lethality, deriving cultured cells from such mice can sometimes be problematical. Less differentiated cell types (e.g., embryonic fibroblasts) can often be obtained from null mice that undergo late embryonic lethality, such as the Rb-null mice *(82)*. However, null mice displaying early embryonic lethality, such as the *Brca1*- and *Brca2*-null mice, may present greater obstacles in obtaining viable cells in culture. One way to circumvent this particular problem is to derive embryonic stem cells from null 3.5-d blastocysts after heterozygote crosses. In some cases, however, even this approach is impossible if the nullizygous state induces a cell lethal effect. We have observed this phenotype in *Chk1*−/− cells *(163)*. Even *Chk1*-null ES cells undergo rapid apoptosis.

One of the most widely studied cell types from tumor suppressor-deficient mice is the embryo fibroblast. These are obtained from the mincing of mid-gestation mouse embryos (11–14 d postcoitus) and plating the fragments on plastic tissue culture dishes under standard medium conditions. Fibroblasts grow out and can be grown for about 10–25 passages before they decline in growth rate. Details of the embryo fibroblast derivation and culturing methods have been published elsewhere *(160,164)*. Once these cells are obtained, basic biologic characterizations include standard growth curves, cell cycle analyses by flow cytometry, transformation susceptibility, response to DNA damage, genomic stability assays, and senescence assays. These null cells and their wild-type counterparts afford an excellent opportunity to compare the effects of a relevant tumor suppressor on growth signaling pathways through standard biochemical assays.

4. Future Prospects

In this review, we have attempted to outline some of the many experimental opportunities that the tumor suppressor-deficient mice have afforded the investigator. Not all methods for generating and analyzing mouse cancer models could be included. For further and more detailed information on the methods and concepts described here and other methods not described, an array of useful reviews, databases, and books can be consulted *(3,4,6,10,11,20–28,75,76,79,105–110,165,166)*.

A number of new developments should expand the usefulness of these mouse models in the near future. These include the development of more sophisticated gene targeting technologies to induce subtle manipulations of tumor suppressor gene function and expression, both temporally and spatially. The concurrent development of more extensive random mutagenesis protocols for the mouse should lead to identification of novel cancer-related genes and phenotypes, which will complement the current models generated by targeted approaches *(67,68)*. The completion of the mouse genome sequence will extend our understanding of mouse gene structure and may result in elucidation of new cancer related genes whose function can be explored experimentally.

The explosive development of gene expression microarray and proteonomics technologies should provide opportunities to identify new genes, proteins, and signaling pathways. Ultimately, such knowledge will lead to a much broader understanding of the cancer cell, so that cancer can be attacked in the patient in a much more rational and targeted manner.

References

1. Silver, L. M. (1995) *Mouse Genetics.* Oxford University Press, New York.
2. Snell, G. D., ed. (1941) *Biology of the Laboratory Mouse.* Blakistan, New York.
3. Clarke, A. R. (2000) Manipulating the germline: its impact on the study of carcinogenesis. *Carcinogenesis* **21**, 435–441.
4. Bradley, A., ed. (1996) Mutant Mice in Cancer Research. Semin. Cancer Biol. **7**, 227–306.
5. Christofori, G. and Hanahan, D. (1994). Molecular dissection of multi-stage tumorigenesis in transgenic mice. *Semin. Cancer Biol.* **5**, 3–12.
6. Ghebranious, N. and Donehower, L. A. (1998) Mouse models in tumor suppression. *Oncogene* **17**, 3385–3400.
7. Venter, J. C. (2001) The sequence of the human genome. *Science* **291**, 1304–1351.
8. Rugh, R. (1990) *The Mouse: Its Reproduction and Development.* Oxford Science Publications, Oxford, UK.
9. Kaufman, M. H. and Bard, J. B. L. (1999) *The Anatomical Basis of Mouse Development.* Academic Press, New York.
10. Hogan, B., Beddington, R. Constantini, F., and Lacy, E. (1994) *Manipulating the Mouse Embryo: A Laboratory Manual,* 2nd ed. Cold Spring Harbor Laboratory Press, Plainview, NY.
11. Bannasch, P. and Gossner, W., eds. (1994) *Pathology of Neoplasia and Preneoplasia in Rodents.* Schattauer, Stuttgart, Germany.
12. Altman, P. L. and Katz, D., eds. (1979) Inbred and Genetically Defined Strains of Laboratory Animals: Part 1, Mouse and Rat. FASEB, Bethesda, MD.
13. Mohr, U., Dungworth, D. L., Capen, C. C., Carlton, W. W., Sundberg, J. P., and Ward, J. M., eds. (1996) *Pathobiology of the Aging Mouse.* ILSI Press, Washington, D.C.
14. Bult, C. J., Krupke, D. M., Sundberg, J. P., and Eppig, J. T. (2000) Mouse Tumor Biology Database (MTB): enhancements and current status. *Nucleic Acids Res.* **28**, 112–114.
15. Pennisi, E. (2001) What's next for the genome centers? *Science* **291**, 1204–1207.
16. Dietrich, W. F., Copeland, N. G., Gilbert, D. J., Miller, J. C., Jenkins, N.A., and Lander, E. S. (1995) Mapping the mouse genome: current status and future prospects. *Proc. Natl. Acad. Sci. USA* **92**, 10849–10853.
17. Blake, J. A., Eppig, J. T., Richardson, J. E., Davisson, M. T., and the Mouse Genome Database Group. (2000) The Mouse Genome Database (MGD): expanding genetic and genomic resources forthe laboratory mouse. *Nucleic Acids Res.* **28**, 108–111.
18. Beck, J. A., Lloyd, S., Hafezparast, M., Lennon-Pierce, M., Eppig, J. T., Festing, M. F., and Fisher, E. M. (2000) Genealogies of mouse inbred strains. *Nat. Genet.* **24**, 23–25.
19. Festing, M. F. (1998) Inbred strains of mice and rats. The Jackson Laboratory Mouse Genome Informatics Web site. www. informatics.jax.org/external/festing.
20. Palapattu, G. S., Bao, S., Kumar, T. R., and Matzuk, M. M. (1998) Transgenic mouse models for tumor suppressor genes. *Cancer Detect. Prev.* **22**, 75–86.
21. Jacobson, D. and Anagnostopoulos, A. V. and The Jackson Laboratory. (2001) TBASE (The Transgenic/Targeted Mutation Database). http://tbase.jax.org.
22. Friedberg, E. C. and Meira, L. B. (2000) Database of mouse strains carrying targeted mutations in genes affecting cellular responses to DNA damage. Version 4. Mutat. Res. **459**, 243–274.

23. Siamak Tabibzadeh, S. and Elzie, J. L. (2001) Database of Gene Knockouts. http://www.bioscience.org/knockout/knochome.htm.

24. BioMedNet Mouse Knockout Database (2001) http://biomednet.com/db/mkmd.

25. Ramirez-Solis, R., Davis, A., and Bradley, A. (1993) Gene targeting in mouse and embryonic stem cells, in *Methods in Enzymology: Guide to Techniques in Mouse Development* (Wasserman, P.M., and DePamphilis, M.L., eds.). Academic Press, San Diego, CA. pp. 855–877.

26. Tymms, M. J. and Kola, I. (2001) *Gene Knockout Protocols. Methods in Molecular Biology*, *Vol. 158.* Humana Press, Totowa, NJ.

27. Kmiec, E. B. (2000) Gene Targeting Protocols. Methods in Molecular Biology, Vol. 137. Humana Press, Totowa, NJ.

28. Joyner, A. L. (2000) *Gene Targeting: A Practical Approach,* 2nd ed. Oxford University Press, Oxford, UK.

29. Evans, M. J. and Kaufman, M. H. (1981) Establishment in culture of pluripotential cells from mouse embryos. *Nature* **292**,154–156.

30. Martin, G. R. (1981) Isolation of a pluripotent cell line from early mouse embryos cultured in medium conditioned by teratocarcinoma stem cells. *Proc. Natl. Acad. Sci. USA* **78**, 7634.

31. Williams R. L., Hilton, D. J., Pease, S., et al. (1988) Myeloid leukaemia inhibitory factor maintains the developmental potential of embryonic stem cells. *Nature,* **336,** 684–687.

32. Smith, A. G., Heath, J. K., Donaldson, D. D., et al. (1988) Inhibition of pluripotential embryonic stem cell differentiation by purified polypeptides. *Nature*, **336**, 688–690.

33. Smith, T. A. and Hooper, M. L. (1983) Medium conditioned by feeder cells inhibits the differentiation of embryonal carcinoma cultures. *Exp. Cell Res*. **145**, 458-462.

34. Pease, S., Braghetta, P., Gearing, D., Grail, D., and Williams, R. L. (1990) Isolation of embryonic stem (ES) cells in media supplemented with recombinant leukemia inhibitory factor (LIF). *Dev. Biol.* **141**, 344–352.

35. Bradley, A., Evans, M., Kaufman, M. H., and Robertson, E. J. (1984) Formation of germline chimaeras from embryo-derived teratocarcinoma cell lines. *Nature* **309**, 255–256.

36. Gossler, A., Doetschman, T., Korn, R., Serfling, E., and Kemler, R. (1986) Transgenesis by means of blastocyst-derived embryonic stem cell lines. *Proc. Natl. Acad. Sci. USA* **83**, 9065–9069.

37. Robertson, E., Bradley, A., Kuehn, M., and Evans, M. (1986) *Nature* **323**, 445.

38. Kuehn, M., Bradley, A., and Robertson, E. (1987) A potential animal model for Lesch-Nyhan syndrome through introduction of HPRT mutations into mice. *Nature* **326**, 295–298.

39. Lin, F. L., Sperle, K., and Sternberg, N. (1985) Recombination in mouse L cells between DNA introduced into cells and homologous chromosomal sequences. *Proc. Natl. Acad. Sci. USA* **82**, 1391–1395.

40. Smithies, O., Gregg, R. G., Boggs, S. S., Koralewski, M. A., and Kucherlapati, R. S. (1985) Insertion of DNA sequences into the human chromosomal beta-globin locus by homologous recombination. *Nature* **317**, 230–234.

41. Doetschman, T., Gregg, R. G., Maeda, N., et al. (1987) Targetted correction of a mutant HPRT gene in mouse embryonic stem cells. *Nature* **330**, 576–578.

42. Thomas, K. R. and Capecchi, M. R. (1987) Site-directed mutagenesis by gene targeting in mouse embryo-derived stem cells. *Cell* **51**, 503–512.

43. Hasty, P., Rivera-Perez, J., and Bradley, A. (1991) Site-directed mutagenesis by gene targeting in mouse embryo-derived stem cells. *Mol. Cell. Biol.* **11**, 5586–5591.

44. Robertson, E. J. (1987) Embryo-derived stem cell lines, in *Teratocarcinomas and Embryonic Stem Cells: A Practical Approach* (Robertson, E. J., ed.). IRL Press, Oxford, UK, pp. 71–112.

45. Bradley, A. (1987) Production and analysis of chimaeric mice, in *Teratocarcinomas and Embryonic Stem Cells: A Practical Approach* (Robertson, E. J., ed.). IRL Press, Oxford, UK, pp. 113–152.

46. Hasty, P., Rivera-Perez, J., Chang, C., and Bradley, A. (1991). Target frequency and integration pattern for insertion and replacement vectors in embryonic stem cells. *Mol. Cell. Biol.* **11**, 4509–4517.

47. te Riele, H., Maandag, E. R., and Berns, A. (1992) Highly efficient gene targeting in embryonic stem cells through homologous recombination with isogenic DNA constructs. *Proc. Natl. Acad. Sci. USA* **89**, 5128–5132.

48. Mansour, S. L., Thomas, K. R., and Capecchi, M. R. (1988) Disruption of the proto-oncogene int-2 in mouse embryo-derived stem cells: a general strategy for targeting mutations to non-selectable genes. *Nature* **336**, 348–352.

49. Thompson, S., Clarke, A. R., Pow, A., Hooper, M. L., and Melton, D. W. (1989) Germ line transmission and expression of a corrected HPRT gene produced by gene targeting in embryonic stem cells. *Cell* **56**, 313–321.

50. Ramirez-Solis, R., Rivera, J., Wallace, J. D., WIms, M., Zheng, H., and Bradley, A. (1992) Genomic DNA microextraction: a method to screen numerous samples. *Anal. Biochem.* **201**, 331–335.

51. Joyner, A. L., Skarnes, W. C., and Rossant, J. (1989) Production of a mutation in mouse En-2 gene by homologous recombination in embryonic stem cells. *Nature* **338**,153–156.

52. McMahon, A. P. and Bradley A. (1990) The Wnt-1 (int-1) proto-oncogene is required for development of a large region of the mouse brain. *Cell* **62**, 1073–1085.

53. Soriano, P. Montgomery, C., Geske, R., and Bradley, A. (1991) Targeted disruption of the c-src proto-oncogene leads to osteopetrosis in mice. *Cell* **64**, 693–702.

54. Thompson, S., Clarke, A. R., Pow, A., Hooper, M. L., and Melton, D. W. (1989) Germ line transmission and expression of a corrected HPRT gene produced by gene targeting in embryonic stem cells. *Cell* **56**, 313–321.

55. Vaulont, S., Daines, S., and Evans, M. (1995) Disruption of the adenosine deaminase (ADA) gene using a dicistronic promoterless construct: production of an ADA-deficient homozygote ES cell line. *Transgenic Res.* **4**, 247–255.

56. Tajbakhsh, S., Bober, E., Babinet, C., Pournin, S., Arnold, H., and Buckingham, M. (1996) Gene targeting the myf-5 locus with nlacZ reveals expression of this myogenic factor in mature skeletal muscle fibres as well as early embryonic muscle. *Dev. Dyn.* **206,** 291–300.

57. Wang, Y., Schnegelsberg, P. N., Dausman, J., and Jaenisch, R. (1996) Functional redundancy of the muscle-specific transcription factors Myf5 and myogenin. *Nature* **379**, 823–825.

58. Wang, Y., Spatz, M. K., Kannan, K., et al. (1999) A mouse model for achondroplasia produced by targeting fibroblast growth factor receptor 3. *Proc. Natl. Acad. Sci. USA* **96**, 4455–4460.

59. Jimenez, G. S., Nister, M., Stommel, J. M., et al. (2000) A transactivation-deficient mouse model provides insights into Trp53 regulation and function. *Nat. Genet.* **26**, 37–43.

60. Sauer, B. (1998) Inducible gene targeting in mice using the Cre/lox system. *Methods: A Companion to Methods in Enzymology* **14**, 381–392.

61. Abuin, A. and Bradley, A. (1996) Recycling selectable markers in mouse embryonic stem cells. *Mol.Cell. Biol.* **16**, 1851–1856.

62. O'Gorman, S., Dagenais, N. A., Qian, M., and Marchuk, Y. (1997) Protamine-cre recombinase transgenes efficiently recombine target sequences in the male germline of mice, but not in embryonic stem cells. *Proc. Natl. Acad. Sci USA* **94**, 14602–14607.

63. Schwenk, F., Baron, U., and Rajewsky, K. (1995) A cre-transgenic mouse strain for the ubiquitous deletion of loxP-flanked gene segments including deletion in germ cells. *Nucleic Acids Res.* **23**, 5080–5081.

64. Lakso, M., Pichel, J. G., Gorman, J. R., et al. (1996) Efficient in vivo manipulation of mouse genomic sequences at the zygote stage. *Proc. Natl. Acad. Sci. USA* **93**, 5860–5865.

65. Ramirez-Solis, R., Liu, P., and Bradley, A. (1995) Chromosome engineering in mice. *Nature* **378**, 720–724.

66. Zheng, B., Sage, M., Sheppeard, E. A., Jurecic, V., and Bradley, A. (2000) Engineering mouse chromosomes with Cre-loxP: range, efficiency, and somatic applications. *Mol. Cell. Biol.* **2**, 648–655.

67. Justice, M. J., Zheng, B., Woychik, R. P., and Bradley, A. (1997) Using targeted large deletions and high-efficiency N-ethyl-N-nitrosourea mutagenesis for functional analyses of mammalian genome. *Methods* **13**, 423–436.

68. Justice, M. J. (2000) Capitalizing on large-scale mouse mutagenesis screens. *Nat. Rev. Genet.* **1,**109–115.

69. Zheng, B., Sage, M., Cai, W. W., et al. (1999) Engineering a mouse balancer chromosome. *Nat. Genet.* **22**, 375–378.

70. Dymecki, S. M. (1996) Flp recombinase promotes site-specific DNA recombination in embryonic stem cells and transgenic mice. *Proc. Natl. Acad. Sci. USA* **93**, 6191–6196.

71. Gu, H., Marth, J. D., Orban, P. C., Mossmann, H. and Rajewsky, K. (1994) Deletion of a DNA polymerase beta gene segment in T cells using cell type-specific gene targeting. *Science* **265**, 103–106.

72. Wang, Y., Krushel, L.A., and Edelman, G.M. (1996) Targeted DNA recombination in vivo using an adenovirus carrying the cre recombinase gene. *Proc. Natl. Acad. Sci. USA* **93**, 3932–3936.

73. Shibata, H., Toyama, K., Shioya, H., et al. (1997) Rapid colorectal adenoma formation initiated by conditional targeting of the Apc gene. *Science* **278**, 120–123.

74. Xu, X., Wagner, K. U., Larson, D., et al. (1999) Conditional mutation of Brca1 in mammary epithelial cells results in blunted ductal morphogenesis and tumour formation. *Nat. Genet.* **22**, 37–43.

75. Meuwissen, R., Jonkers, J., and Berns, A. (2001) Mouse models for sporadic cancer. *Exp. Cell Res.* **264**, 100–110.

76. Resor, L., Bowen, T. J., and Wynshaw-Boris, A. (2001) Unraveling human cancer in the mouse: recent refinements to modeling and analysis. *Hum. Mol. Genet.* **10**, 669–675.

77. Hasty, P., Ramirez-Solis, R., Krumlauf, R., and Bradley, A. (1991) Introduction of a subtle mutation into the Hox-2.6 locus in embryonic stem cells. *Nature* **350**, 243–246.

78. Johnson, L., Mercer, K., Greenbaum, D., et al. (2001). Somatic activation of the K-ras oncogene causes early onset lung cancer in mice. *Nature* **410**, 1111–1116.

79. Ledermann, B. (2000) Embryonic stem cells and gene targeting. *Exp. Physiol.* **85,** 603–613.

80. Wood, S. A., Allen, N. D., Rossant, J., Auerbach, A., and Nagy, A. (1993) Non-injection methods for the production of embryonic stem cell-embryo chimaeras. *Nature* **365**, 87–89.

81. Lee, E. Y., Chang, C. Y., Hu, N., et al. (1992) Mice deficient for Rb are nonviable and show defects in neurogenesis and haematopoiesis. *Nature* **359**, 288–294.

82. Jacks, T., Fazeli, A., Schmitt, E. M., Bronson, R. T., Goodell, M. A., and Weinberg, R. A. (1992) Effects of an Rb mutation in the mouse. *Nature* **359**, 295–300.

83. Liu, C.Y., Flesken-Nikitin, A., Li, S., Zeng, Y., and Lee, W.H. (1996) Inactivation of the mouse Brca1 gene leads to failure in the morphogenesis of the egg cylinder in early postimplantation development. *Genes Dev.* **10,**1835–1843.

84. Hakem, R., de la Pompa, J. L., Sirard, C., et al. (1996) The tumor suppressor gene Brca1 is required for embryonic cellular proliferation in the mouse. *Cell* **85**, 1009–1023.

85. Kreidberg, J. A., Sariola, H., Loring, J. M., et al. (1993) WT-1 is required for early kidney development. *Cell* **74**, 679–691.

86. Donehower, L. A., Harvey, M., Slagle, B. L., et al. (1992) Mice deficient for p53 are developmentally normal but susceptible to spontaneous tumours. *Nature* **356**, 215–221.

87. Kamijo, T., Zindy, F., Roussel, M. F., et al. (1997) Tumor suppression at the mouse INK4a locus mediated by the alternative reading frame product p19ARF. *Cell* **91**, 649–659.

88. Serrano, M., Lee, H., Chin, L., Cordon-Cardo, C., Beach, D., and DePinho, R. A. (1996) Role of the INK4a locus in tumor suppression and cell mortality. *Cell* **85**, 27–37.

89. Sah, V. P., Attardi, L. D., Mulligan, G. J., Williams, B. O., Bronson, R. T., and Jacks. T. (1995) A subset of p53-deficient embryos exhibit exencephaly. *Nat. Genet.* **10**, 175–180.

90. Armstrong, J. F., Kaufman, M. H., Harrison, D. J., and Clarke, A. R.(1995) High-frequency developmental abnormalities in p53-deficient mice. *Curr. Biol.* **5**, 931–936.

91. Jones, S. N., Roe, A. E., Donehower, L. A., and Bradley, A. (1995) Rescue of embryonic lethality in Mdm2-deficient mice by absence of p53. *Nature* **378**, 206–208.

92. Hakem, R., de la Pompa, J. L., Elia, A., Potter, J., and Mak, T. W. (1997) Partial rescue of Brca1 (5-6) early embryonic lethality by p53 or p21 null mutation. *Nat. Genet.* **16**, 298–302.

93. Yamaguchi,T., Bradley, A., McMahon, A. E., and Jones, S. (1999) A Wnt5a pathway underlies outgrowth of multiple structures in the vertebrate embryo. *Development* **126**, 1211–1223.

94. Guidi, C., Turner, T., Smith, T., et al. (2001) Disruption of Ini1 leads to peri-implantation lethality and tumorigenesis in mice. *Mol. Cell. Biol.* **21**, 3598–3603.

95. Kaufman, M. H. (2000) Methods for handling mouse embryos for anatomy studies: gross and microscopic anatomy of the mouse, in *Pathology of Genetically Engineered Mice* (Ward, J. M., ed.). Iowa State University Press, Ames, IA, pp. 63–88.

96. Maronpot, R. R., Flagler, N. D., Thai-Vu, T., Foley, J. F., and Goldsworthy, T. L. (2000) Measurement of cell replication and apoptosis in mice, in *Pathology of Genetically Engineered Mice* (Ward, J. M., ed.). Iowa State University Press, Ames, IA, pp. 45–62.

97. Harvey, M., McArthur, M. J., Montgomery, C. A., Bradley, A., Donehower, and L. A. (1993) Genetic background alters the spectrum of tumors that develop in p53-deficient mice. *FASEB J.* **7**,938–943.

98. Kuperwasser, C., Hurlbut, G. D., Kittrell, F. S., et al. (2000) Development of spontaneous mammary tumors in BALB/c p53 heterozygous mice. A model for Li-Fraumeni syndrome. *Am. J. Pathol.* **157**, 2151–2159.

99. Su, L. K., Kinzler, K. W., Vogelstein, B., et al. (1992) Multiple intestinal neoplasia caused by a mutation in the murine homolog of the APC gene. *Science* **256**, 668–670.

100. Dietrich, W. F., Lander, E. S., Smith, J. S., et al. (1993) Genetic identification of Mom-1, a major modifier locus affecting Min-induced intestinal neoplasia in the mouse. *Cell* **75**, 631–639.

101. Cormier, R. T., Hong, K. H., Halberg, R. B., et al. (1997) Secretory phospholipase Pla2g2a confers resistance to intestinal tumorigenesis. *Nat. Genet.* **17**, 88–91.

102. DePinho, R. A. (2000) The age of cancer. *Nature* **408**, 248–254.

103. Venkatachalam, S., Tyner, S., and Donehower, L. A. (2001) The p53-deficient mouse as a cancer model, in *Tumor Models in Cancer Research* (Teicher, B.A., ed.). Humana Press, Totowa, NJ 247–261.

104. Donehower, L. A., Harvey, M., Vogel, H., et al. (1995). Effects of genetic background on tumorigenesis in p53-deficient mice. *Mol. Carcinog.* **14**, 16–22.

105. Sundberg, J. P. and Boggess, D., eds. (1999) *Systematic Approach to Evaluation of Mouse Mutations*. CRC Press, Boca Raton, FL.

106. Suckow, M. A., Danneman, P., and Brayton, C. (2001) *The Laboratory Mouse*. CRC Press, Boca Raton, FL.

107. Brayton, C. A., Justice, M., and Montgomery, C. A. (2001) Evaluating mutant mice: anatomic pathology. *Vet. Pathol.* **38**, 1–19.

108. Relyea, M. J., Miller, J., Boggess, D., and Sundberg, J. P. (1999) Necropsy methods for laboratory mice: biological characterization of a new mutation, in *Systematic Approach to Evaluation of Mouse Mutations*. (Sundberg, J. P., and Boggess, D., eds.) CRC Press, Boca Raton, FL, pp. 57–89.

109. Ward, J. M. (2000) Pathology phenotyping of genetically engineered mice, in *Pathology of Genetically Engineered Mice* (Ward, J.M. ed.). Iowa State University Press, Ames, Iowa, pp. 155–160.

110. Ward, J. M., Anver, M. R., Mahler, J. F., and Devor-Henneman, D. E. (2000) Pathology of mice commonly used in genetic engineering (C57BL/6; 129; B6,129; and FVB/N), in *Pathology of Genetically Engineered Mice* (Ward, J. M., ed.). Iowa State University Press, Ames, IA, pp. 161–182.

111. Lang, T. A. and Secic, M. (1997) How to Report Statistics in Medicine: Annotated Guidelines for Authors, Editors, and Reviewers. American College of Physicians, Philadelphia.

112. Glantz, S. A. (1997) *Primer of Biostatistics,* 4th ed. McGraw-Hill, New York.

113. Arver, B., Du, Q., Chen, J., Luo, L., and Lindblom, A. (2000) Hereditary breast cancer: a review. *Semin. Cancer Biol.* **10**, 271–288.

114. Donehower, L. A. (1996) The p53-deficient mouse: a model for basic and applied cancer studies. *Semin. Cancer Biol.* **7**, 269–278.

115. Sands, A. T., Abuin, A., Sanchez, A., Conti, C. J., and Bradley, A. (1995) High susceptibility to ultraviolet-induced carcinogenesis in mice lacking XPC. *Nature* **377**, 162–165.

116. Friedberg, E. C., Bond, J. P., Burns, D. K., et al. (2000) Defective nucleotide excision repair in xpc mutant mice and its association with cancer predisposition. *Mutat. Res.* **459**, 99–108.

117. Bennett, L. M., McAllister, K. A., Malphurs, J., et al. (2000). Mice heterozygous for a Brca1 or Brca2 mutation display distinct mammary gland and ovarian phenotypes in response to diethylstilbestrol. *Cancer Res.* **60**, 3461–3469.

118. Cressman, V. L., Backlund, D. C., Hicks, E. M., Gowen, L. C., Godfrey. V., and Koller, B. (1999) Mammary tumor formation in p53- and BRCA1-deficient mice. *Cell Growth Differ.* **10**, 1–10.

119. Le, Y. and Sauer, B. (2000) Conditional gene knockout using cre recombinase. *Meth. Mol. Biol.* **136**, 477–485.

120. Sherr, C. J. (2000) The Pezcoller lecture: cancer cell cycles revisited. *Cancer Res.* **60**, 3689–3695.

121. Stewart, S. A., Weinberg, R. A. (2000) Telomerase and human tumorigenesis. *Semin. Cancer Biol.* **10**, 399–406.

122. Williams, B. O., Remington, L., Albert, D. M., Mukai, S., Bronson, R. T., and Jacks, T. (1994) Cooperative tumorigenic effects of germline mutations in Rb and p53. *Nat. Genet.* **7**, 480–484.

123. Harvey, M., Vogel, H., Lee, E. Y., Bradley, A., and Donehower, L. A. (1995) Mice deficient in both p53 and Rb develop tumors primarily of endocrine origin. *Cancer Res.* **55**, 1146–1151.

124. Artandi, S. E., Chang, S., Lee, S. L., et al. (2000) Telomere dysfunction promotes non-reciprocal translocations and epithelial cancers in mice. *Nature* **406**, 641–645.

125. Donehower, L. A., Godley, L. A., Aldaz, C. M., et al. (1995). Deficiency of p53 accelerates mammary tumorigenesis in *Wnt-1* transgenic mice and promotes chromosomal instability. *Genes Dev.* **9**, 882–895.

126. Tsukamoto, A. S., Grosschedl, R., Guzman, R. C., Parslow, T., Varmus, and H. E. (1988) Expression of the int-1 gene in transgenic mice is associated with mammary gland hyperplasia and adenocarcinomas in male and female mice. *Cell* **55**, 619–625.

127. Jones, J. M., Attardi, L., Godley, L. A., et al. (1997) Absence of p53 in a mouse mammary tumor model promotes tumor cell proliferation without affecting apoptosis. *Cell Growth Differ.* **8**, 829–838.

128. Cui, X. S. and Donehower, L. A. (2000) Differential gene expression in mouse mammary adenocarcinomas in the presence and absence of wild type p53. *Oncogene* **19**, 5988–5996.

129. Montes de Oca Luna, R., Wagner, D. S., and Lozano, G. (1995) Rescue of early embryonic lethality in mdm2-deficient mice by deletion of p53. *Nature* **378**, 203–206.

130. Hakem, R., de la Pompa, J. L., Elia, A., Potter, J., and Mak, T. W. (1997) Partial rescue of Brca1 (5-6) early embryonic lethality by p53 or p21 null mutation. *Nat. Genet.* **16**, 298–302.

131. Ludwig, T., Chapman, D. L., Papaioannou, V. E., and Efstratiadis, A. (1997) Targeted mutations of breast cancer susceptibility gene homologs in mice: lethal phenotypes of Brca1,

Brca2, Brca1/Brca2, Brca1/p53, and Brca2/p53 nullizygous embryos. *Genes Dev.* **11**, 1226–1241.

132. Brugarolas, J. and Jacks, T. (1997) Double indemnity: p53, BRCA and cancer. p53 mutation partially rescues developmental arrest in Brca1 and Brca2 null mice, suggesting a role for familial breast cancer genes in DNA damage repair. *Nat. Med.* **3**, 721–722.

133. Miller, B. E., Miller, F. R., Wiburn, D., and Heppner, G. H. (1988) Dominance of a tumor subpopulation in mixed heterogeneous mouse mammary tumors. *Cancer Res.* **48**, 5747–5753.

134. Jones, J. M., Cui, X. S., Medina, D., and Donehower, L. A. (1999) Heterozygosity of p21WAF1/CIP1 enhances tumor cell proliferation and cyclin D1-associated kinase activity in a murine mammary cancer model. *Cell Growth Differ.* **10**, 213–222.

135. Papadopoulos, I., Rudolph, P., Wirth, B., and Weichert-Jacobsen, K. (1996) p53 expression, proliferation marker Ki-65, DNA content and serum PSA; possible biopotential markers. *Urology* **48**, 261–268.

136. Hedley, D. W. (1989) Flow cytometry using paraffin-embedded tissue: Five years on. *Cytometry* **10**, 229–241.

137. Stadelmann, C. and Lassmann, H. (2000) Detection of apoptosis in tissue sections. *Cell Tissue Res.* **301**, 19–31.

138. Liu, G., McDonnel, T. J., Montes de Oca Luna, R., et al. (2000) High metastatic potential in mice inheriting a targeted p53 missense mutation. *Proc. Natl. Acad. Sci. USA* **197**, 4174–4179.

139. Knudson, A. G. (1971) Mutation and cancer: statistical study of retinoblastoma. *Proc. Natl. Acad. Sci. USA* **68**, 820–823.

140. Hu, N., Gutsmann, A., Herbert, D. C., Bradley, A., Lee, W. H., and Lee, E. Y. (1994) Heterozygous Rb-1 delta 20/+ mice are predisposed to tumors of the pituitary gland with a nearly complete penetrance. *Oncogene* **9**, 1021–1027.

141. Venkatachalam, S., Shi, Y. P., Jones, S. N., et al. (1998) Retention of wild-type p53 in tumors from p53 heterozygous mice: reduction of p53 dosage can promote cancer formation. *EMBO J.* **17**, 4657–4667.

142. Schuyer, M. and Berns, E. M. (1999) Is TP53 dysfunction required for BRCA1-associated carcinogenesis? *Mol. Cell. Endocrinol.* **10**, 155, 143–152.

143. Dietrich, W. F., Copeland, N. G., Gilbert, D. J., Miller, J. C., Jenkins, N. A., and Lander, E. S. (1995) Mapping the mouse genome: current status and future prospects. *Proc. Natl. Acad. Sci. USA* **92**, 10849–10853.

144. Devereux, T. R. and Kaplan, N. L. (1998) Use of quantitative trait loci to map murine lung tumor susceptibility genes. *Exp. Lung Res.* **24**, 407–417.

145. Dietrich, W. F., Lander, E. S., Smith, J. S., et al. (1993) Genetic identification of Mom-1, a major modifier locus affecting Min-induced intestinal neoplasia in the mouse. *Cell* **75**, 631–639.

146. Dietrich, W. F., Radany, E. H., Smith, J. S., Bishop, J. M., Hanahan, D., and Lander, E. S. (1994) Genome-wide search for loss of heterozygosity in transgenic mouse tumors reveals candidate tumor suppressor genes on chromosomes 9 and 16. *Proc. Natl. Acad. Sci. USA* **91**, 9451–9455.

147. Nagase, H., Bryson, S., Cordell, H., Kemp, C. J., Fee, F., and Balmain, A. (1995) Distinct genetic loci control development of benign and malignant skin tumours in mice. *Nat. Genet.* **10**, 424–429.

148. Kallioniemi, A., Kallioniemi, O. P., Sudar, D., et al. (1992) Comparative genomic hybridization for molecular cytogenetic analysis of solid tumors. *Science* **258**, 818–821.

149. Shi, Y. P., Naik, P., Dietrich, W. F., Gray, J. W., Hanahan, D., and Pinkel, D. (1997) DNA copy number changes associated with characteristic LOH in islet cell carcinomas of transgenic mice. *Genes Chromosomes Cancer* **19**, 104–111.

150. Pinkel, D., Segraves, R., Sudar, D., et al. (1998) High resolution analysis of DNA copy number variation using comparative genomic hybridization to microarrays. *Nat. Genet.* **20**, 207–211.

151. Pollack, J. R., Perou, C. M., Alizadeh, A. A., et al. (1999) Genome-wide nalysis of DNA copy-number changes using cDNA microarrays. *Nat. Genet.* **23**, 41–46.
152. Liyanage, M., Coleman, A., du Manoir, S., et al. (1996) Multicolour spectral karyotyping of mouse chromosomes. *Nat. Genet.* **14**, 312–315.
153. McCormack, S. J., Weaver, Z., Deming, S., et al. (1998) Myc/p53 interactions in transgenic mouse mammary development, tumorigenesis and chromosomal instability. *Oncogene* **16**, 2755–2766.
154. Zimonjic, D. B., Pollock, J. L., Westervelt, P., Popescu, N. C., and Ley, T. J. (2000) Acquired, nonrandom chromosomal abnormalities associated with the development of acute promyelocytic leukemia in transgenic mice. *Proc. Natl. Acad. Sci. USA* **97**, 13306–13311.
155. Artandi, S. E., Chang, S., Lee, S. L., et al. (2000) Telomere dysfunction promotes nonreciprocal translocations and epithelial cancers in mice. *Nature* **406**, 641–645.
156. Polyak, K., Xia, Y., Zweier, J. L., Kinzler, K. W., and Vogelstein, B. (1997) A model for p53-induced apoptosis. *Nature* **389**, 300–305.
157. Zhao, R., Gish, K., Murphy, M., et al. (2000) Analysis of p53-regulated gene expression patterns using oligonucleotide arrays. *Genes Dev.* **14**, 981–993.
158. Yu, J., Zhang, L., Hwang, P. M., Rago, C., Kinzler, K. W., and Vogelstein, B. (1999) Identification and classification of p53-regulated genes. *Proc. Natl. Acad. Sci. USA* **96**, 14517–14522.
159. Golub, T. R., Slonim, D. K., Tamayo, P., et al. (1999) Molecular classification of cancer: class discovery and class prediction by gene expression monitoring. *Science* **286**, 531–537.
160. Harvey, M., Sands, A. T., Weiss, R. S., et al. (1993) In vitro growth characteristics of embryo fibroblasts isolated from p53-deficient mice. *Oncogene* **8**, 2457–2467.
161. Yahanda, A. M., Bruner, J. M., Donehower, L. A., and Morrison, R. S. (1995) Astrocytes derived from p53-deficient mice provide a multistep in vitro model for development of malignant gliomas. *Mol. Cell. Biol.* **15**, 4249–4259.
162. Bellamy, C. O., Clarke, A. R., Wyllie, A. H., and Harrison, D. J. (1997) p53 Deficiency in liver reduces local control of survival and proliferation, but does not affect apoptosis after DNA damage. *FASEB J.* **11**, 591–599.
163. Liu, Q., Guntuku, S., Cui, X. S., et al. (2000) Chk1 is an essential kinase that is regulated by Atr and required for the G(2)/M DNA damage checkpoint. *Genes Dev.* **14**, 1448–1459.
164. Freshney, I. R. (1983) *Culture of Animal Cells: A Manual of Basic Technique.* Allan R. Liss, New York.
165. Mak, T. W. (1998) *The Gene Knockout Factsbook.* Academic Press, San Diego, CA.
166. Ward, J. M., ed. (2000) *Pathology of Genetically Engineered Mice*, Iowa State University Press, Ames, IA.

24

The Tissue Microenvironment as an Epigenetic Tumor Modifier

Meredith Unger and Valerie M. Weaver

1. Introduction

The transition from normal to cancerous tissue frequently takes years to occur. Epithelial tumors in particular can exist as premalignant lesions for prolonged periods of time, and even then only a low proportion of benign tumors undergo malignant transformation *(1,2)*. Once tumors have formed, their behavior is often erratic, so that determining a treatment strategy based on current clinical criteria can be challenging. For example, some tumors may grow and invade aggressively, while tumors of similar grade can experience extended periods of dormancy *(3–5)*. Tumor metastasis and drug responsiveness are also just as difficult to predict *(6–8)*. This behavioral variability despite similarity in tumor phenotype demonstrates that cancer is a complex disease that is likely regulated by multiple pathways.

Studies in humans and animals have shown that malignant transformation occurs through a series of increasingly abnormal morphologic stages *(9)*. These tissue morphologies have been histologically and behaviorally classified as benign, dysplastic, and premalignant, and they culminate with the transformation of the premalignant lesion into an invasive and often metastatic tumor *(10)*. Extensive molecular and genetic data has shown that the progressively altered precancerous tissues harbor an increasingly perturbed and unstable genome. This has led to the formulation of the "multihit gene model" of tumor progression. The multihit gene model maintains that tumors arise through a combination of hereditary genetic alterations and the accumulation of incremental and sequential somatic changes in the target genome *(11–13)*. Central to this paradigm is the concept that cumulative mutations, amplifications, and/or deletions occur in critical genes. These mutations release the targeted cells from their normal growth and survival constraints, and permit them to invade into the surrounding tissue and survive, grow, and metastasize *(14)*. Tumor suppressor proteins, by virtue of their critical role in cell-cycle checkpoint control and as regulators of the apoptosis/stress response, play an important role in this process. Accordingly, much effort has been extended to delineate how tumor suppressors influence the pathogenesis of cancer.

From: *Methods in Molecular Biology, Vol. 223: Tumor Suppressor Genes: Regulation, Function, and Medicinal Applications.* Edited by: Wafik S. El-Deiry © Humana Press Inc., Totowa, NJ

Familial cancers have provided fertile ground for studying the interrelationship between tumor suppressor genes and cancer initiation and progression. Data show that mutations in tumor suppressors genes such as *p53*, retinoblastoma gene (*RB1*), *BRCA1*, *BRCA2*, and adenomatous polyposis coli gene (*APC*) exert a vital role in promoting and driving tumorigenesis *(15)*. The best example of the integral role that tumor suppressors play in cancer progression is given by the model of colon cancer, in which high incidence and penetrance of the disease are directly related to truncations in or loss of the APC gene *(16)*. Indeed, colon cancer has served as a model template to understand and classify the functional role of tumor suppressors in cancer progression. This model has led to the formulation of tumor suppressor categories defined as growth monitor or "gatekeeper suppressors," genomic stabilizer or "caretaker suppressors," and effectors or "landscaper suppressors" *(17,18)*. The tumor suppressor classifications have been further subdivided to include "initiation gatekeepers," "progression gatekeepers," and "metastasis gatekeepers," and examples of tumor suppressors that fall into one or more category have been described *(15)*. Despite the greater than 90% penetrance rate documented for *APC* and colon cancer, germline mutations in other tumor suppressors fail to exhibit such a robust relationship. For example, in the case of breast cancer, the penetrance rates for *BRCA1* and *BRCA2* range anywhere from 35 to 80% depending on the population studied (Peterson, 1994,) *(19–21)*. Variable and incomplete penetrance has also been documented for hereditary nonpolyposis suppressor cancer (*HNPCC*), in which the *ATM* 1853N variant was found to modulate effects on the mismatch repair tumor suppressor genes *MLH1* and *MSH2* *(22)*. Similarly, germline mutations in specific regions of the *RB1* gene are associated with low-penetrant retinoblastoma phenotypes *(23,24)*. These findings indicate that additional factors must be operating to modulate the effects of tumor suppressors on cancer progression.

The lack of concordance between germline mutations in tumor suppressor genes and cancer penetrance has led to appreciation of the importance of tumor modifiers, tissue microenvironmental factors, and epigenetic pathways as critical regulators of malignancy. Such an enlightened view of the complexity of cancer has sparked an interest in understanding how epigenetic and tissue microenvironmental factors function to maintain tissue-specific differentiation, and how their dysfunction could contribute to cancer. For instance, there has been renewed attention paid to the role played by hyper- and hypomethylation of DNA CpG genome islands in cancer progression (for review, *see* **ref. 25**). Additionally, the tissue microenvironment is an important regulator of differentiation and tissue homeostasis and its role in tumor progression has recently become recognized (for reviews, *see* **refs. 26–29**).

The tumor tissue microenvironment broadly encompasses the vasculature, cells of the immune system, and the extracellular stroma *(30–33)*. The extracellular stroma consists of cellular (fibroblasts, endothelial cells, chrondrocytes, immune cells, and adipocytes) and noncellular components (extracellular matrix [ECM], and soluble factors such as growth factors). Mounting evidence has accumulated to show that the stroma is dynamically regulated during tumor progression *(34),* and can act to promote malignant transformation *(35)* or to drive expression of the malignant phenotype of the epithelial cells within a tissue *(28,36,38)*. Studies have also demonstrated that tempering the stromal response reduces the frequency of malignant transformation in vivo *(39)*. Rescuing abnormal stromal–epithelial interactions by altering integrins can revert the malignant phenotype of epithelial tumors ex vivo and repress their malignancy in vivo *(40,41)*.

This shows that the stromal ECM and its associated adhesion-linked pathways can function as tumor suppressors and may constitute an important and unexplored epigenetic tumor-modifier mechanism. Although the molecular basis for such a concept remains obscure, considering the critical role of tumor suppressors in cancer progression, it seems reasonable to predict that the ECM and its receptors collaborate functionally with tumor suppressors to inhibit malignant transformation.

In this chapter, we summarize experimental evidence showing that the ECM and adhesion molecules dramatically influence tumor progression and expression of the malignant phenotype. We then cite and discuss data that support the concept of a dynamic and reciprocal link between cell adhesion and tumor suppressor pathways. Cancer is a disease of altered organ homeostasis, therefore a firm understanding of tumor progression and the role of tumor suppressors must be appreciated within the context of the tissue microenvironment and the three dimensional (3-D) organization of the diseased organ. Accordingly, we have attempted to formulate a model of tumor progression that encompasses the interrelationship between tumor suppressors and cell adhesion within the context of the 3-D architecture of the tissue. We hope that this review will encourage research to examine the functional association between tumor suppressors and cell adhesion and to explore the underlying mechanisms by which such a relationship could operate. Given the incredible progress that has been made to date on the Genome Projects, and the availability of 3-D culture systems and sophisticated molecular, cell biology, and genomic tools, this goal is readily achievable. Indeed, the field clearly stands prepared to advance our understanding of how complex interactions at the tissue level interface with gene expression to regulate tissue homeostasis and to drive diseases such as cancer.

2. Stromal–Epithelial Interactions and Tissue Homeostasis:

Stromal–epithelial interactions direct tissue organization and specify cell fate during embryonic and fetal development *(42–48)*. The stroma also maintains epithelial organ homeostasis and tissue-specific differentiation in the adult organism *(49–51)*. Numerous studies including those using keratinocytes, MECs, endothelial cells, and salivary epithelial cells have demonstrated that the ECM, which is the insoluble proteinaceous component of the stroma, can directly influence cell proliferation, survival, migration, and differentiation *(29,52–56)*.

The ECM is comprised of a variety of secreted proteins including collagen, elastins, proteoglycans, glycosaminoglycans, and glycoproteins that form multidomain structures by either multimerizing or by interacting with each other. Functionally the ECM provides mechanical support and biochemical cues to the cells within the tissue, and serves as a reservoir for soluble mitogens and morphogens. There are two broad classes of ECM in tissues. The first is the interstitial matrix (which also functions as a provisional matrix in wounded tissues), which surrounds the mesenchymal cells (including fibroblasts, adipocytes and chondrocytes) and is composed of fibrillar collagens, fibronectin, and tenascin with minor amounts of vitronectin and fibrinogen. The second is the basement membrane (BM), on which epithelial and endothelial cells abut, and this ECM consists of laminins 1 and 5 as well as collagen IV, nidogen/entactin and proteoglycans, and some fibronectin. There is a considerable distribution in the molecular characteristics of these two types of ECMs between tissues, such that isoforms vary, expression

Table 1
Vertebrate αβ Integrin Heterodimers

β Subunit	α Partner
β1	α1, α2, α3, α4, α5, α6, α7, α8, α9, α10, α11, αv
β2	$α_D$, $α_L$, $α_M$, $α_X$
β3	αv, αIIb
β4	α6
β5	αv
β6	αv
β7	α4, $α_E$
β8	αv

levels change, and composition and protein organization are altered from organ to organ, and studies have emphasized that such differences can and do profoundly influence cell behavior *(53,57)*.

Cell–ECM interactions are mediated by adhesion receptors that include: integrins, dystroglycan, cell surface proteoglycan receptors such as syndecans, CD44, and thrombomodulin, and a novel subfamily of receptor tyrosine kinases called discoidin-domain receptors *(58–60)*. Of these receptors, integrins have emerged as the most prevalent and the best characterized. Integrins are a large family of transmembrane heterodimeric proteins consisting of an alpha (α) and a beta (β) subunit. At present no less then 24 integrin heterodimer combinations have been described, and these heterodimer combinations recognize components of the ECM including laminins, collagens, and fibonectin, as well as counterreceptors on the surfaces of neighboring cells (*see* **Table 1**; for review, *see* **refs. 58**, **61**, and **62**). The importance of the ECM and its receptors in development and tissue homeostasis is underscored by their close evolutionary conservation among vertebrates and invertebrates *(63–65)*. Studies using transgenic mice also attest to the critical role of ECM proteins and their receptors in dictating cell fate during development, and for maintaining tissue homeostasis in the adult organism *(58,66,67)*.

Integrins mediate their cellular effects via the formation of multimolecular complexes that link integrins to the cytoskeletal network and to intracellular signaling pathways. Integrins form complex cytoskeletal interactions with intracellular cytoskeletal proteins including α-actinin (an actin-bundling protein) and talin and filamin (actin-binding proteins), and these associations play a role in integrin-mediated cell spreading, migration, and matrix assembly by facilitating connections with the actin cytoskeleton *(68)*. Integrins also interact with numerous cytoplasmic proteins including focal adhesion kinase (FAK), cytohesin-1, integrin-linked kinase (ILK), β3-endonexin, integrin cytoplasmic domain-associated protein-1, and receptor for activated protein kinase c (Rack1), to regulate outside-in and inside-out signaling *(69)*. Thus, while integrins lack intrinsic catalytic activity, following their ligation-induced activation and ECM-mediated clustering they recruit multiple proteins, many of which themselves possess catalytic activity (for reviews, *see* **refs. 70** and *71*). For example, following activation of α5β1 integrin by fibronectin FAK, paxillin, tensin, and p130 become recruited and rapidly tyrosine-phosphorylated, which leads to the association of new proteins such as PI-3 kinase and Src by creating binding sites for SH2- and SH3-containing proteins. This newly formed complex then recruits additional proteins and modifies other molecules and eventually induces alterations in the organization of the actin cytoskeleton to drive focal adhesion formation and results in changes in gene expression that support cell proliferation, survival, or differentiation *(72)*. Although such integrin-linked signaling is likely responsible for eliciting most of the behavioral effects of integrins, just how these signals mediate complex cellular decisions to regulate growth, survival, and differentiation still remains largely unresolved.

Studies have shown that many integrins can bind several ECM ligands, and often one ECM ligand is recognized by several integrin heterodimers. This underscores an important question, which is how integrins modulate signaling specificity in a cell. At least part of the answer may reside in the uniqueness of the different alpha subunit cytoplasmic domains of the integrin heterodimers. Data suggest that subgroups of α-cytoplasmic domains differentially regulate integrin-mediated biological responses. Studies have indicated that each cytoplasmic α-tail modifies integrin function by either directly eliciting unique signaling events or by altering the functions of the β-integrin subunit. For example, Sastry and co-workers reported that the cytoplasmic domains of the α5 and α6A differentially regulated cell cycle withdrawal and differentiation in myocytes. Wary and colleagues showed that the transmembrane domain of the α5 and α1 integrins could physically and functionally interact with tyrosine kinases such as Fyn and recruit the adaptor protein Shc, to regulate MAP kinase signaling and cell cycle progression via interactions with caveolin-1 (reviewed in **ref. 73**). Integrin function can also be modified by cell surface proteins that interact with regions of the extracellular domain such as those of the transmembrane-4 superfamily (TM4SF, or tetraspans), or by transmembrane-associated proteins such as the integrin-associated protein (IAP or CD47) or CD147 (basigin), or by glycophosphatidylinositol (GPI)-linked proteins such as CD87 (or urokinase plasminogen activator receptor (UPAR) and such interactions may profoundly influence integrin behavior and effects (for reviews, *see* **refs. 74–76**). The mechanical properties of the ECM (which reflect ECM composition and organization) can also profoundly alter integrin signaling to modify cell behavior. Thus the mechanical stiffness of the ECM can regulate integrin expression and cell spreading, which in turn can alter cell shape and cell function via effects on integrin activation, the organization of molecular signaling platforms, or other as yet unidentified mechanisms (Weaver et al., unpublished observations; [77]). Thus, Meyer and colleagues showed that mechanical stresses applied to endothelial cell surface integrins can alter the cyclic AMP cascade and change gene expression by modulating integrin signaling in a G-protein-dependent manner. Similarly, studies conducted in either nonadherent fibroblasts or rounded and attached endothelial cells showed that cells could not transit the G1/S boundary, nor could they increase cyclin D1 protein or downregulate the cell cycle inhibitor p27^{kip1}, or phosphorylate pRb in response to growth factors such as PDGF or EGF and FGF (for reviews, *see* **refs. 78** and **79**). Integrin function therefore can be profoundly influenced by the mechanical composition of the ECM microenvironment and an array of interacting proteins.

Much of what is presently known about integrin signaling and its effects on cell growth, differentiation, and survival has been derived from experiments using simplified, defined systems such as immortalized 3T3 fibroblasts, transformed CHO cells, and platelets in combination with purified ECM components or bioactive ECM peptides such as RGD. These systems have been and continue to be instrumental in defining integrin biochemistry and for identifying important molecular components of integrin signaling *(80)*. Yet in-vivo cells reside within complex multicellular organs and are integrated into 3-D tissue structures. So how do integrins regulate the behavior of cells that are incorporated into multicellular tissues, and does tissue structure itself influence how integrins function? Can we extrapolate current models of integrin signaling to complex 3-D systems, and if so, are there additional levels of regulatory complexity? Data derived from primary cultures, epithelial cell lines and 3-D tissuelike cultures have illustrated that integrin function is additionally regulated by crosstalk with growth factor receptors, cell–cell junctions, and by different integrin heterodimers, and that such interactions play a vital role in determining cell fate *(81)*. For example, Russell and colleagues reported that normal wound healing in primary and immortalized nonmalignant keratinocytes is a coordinated process that requires integrated and dynamic crosstalk between α3β1-integrin and α6β4-integrin via RhoGTPases and the actin cytoskeleton (Russell et al., submitted). Our studies in nonmalignant and tumorigenic MECs have demonstrated that MEC morphogenesis and differentiation depend on a temporal interplay between β1- and β4-integrins, epidermal growth factor receptors (EGFRs), and E cadherin adherens junctions ([*40,82,83*]; Weaver, unpublished observations). To summarize then, integrin-ECM interactions likely influence cell behavior via complex interactions between cytoskeletal and cytosolic molecules, which elicit mechanical and biochemical signals but which can be modified by numerous interacting proteins, and altered by the actions of different integrin heterodimers, growth factor receptors, cell–cell adhesion molecules, and the mechanical characteristics of the tissue microenvironment. Thus, how a cell responds to cues from the ECM in vivo is most likely a highly integrated process that can be modified at multiple levels.

3. The Tissue Stroma as a Tumor Promoter

Cancer is a disease of altered tissue homeostasis. Studies examining breast, prostate, and colon cancers have consistently revealed that the stromal tissue surrounding these tumors exhibit dramatic phenotypic changes. These changes are referred to as the desmoplastic response *(27,84,85)*. Fibroblast proliferation, migration and transdifferentiation, inflammatory cell infiltration, changes in ECM composition, integrity, and organization, and increases in blood vessel density "angiogenesis" are all characteristic features of a desmoplastic stroma. The importance of the stromal response to tumor formation has been explored in breast cancer (reviewed in **refs. *34*** and ***86***). During breast cancer progression it was reported that the stroma exhibits desmoplastic changes even in early benign fibrocystic lesions, and that these alterations become progressively pronounced in association with increasing perturbations in the ductal epithelium (for review, *see* **ref. *34***). Experiments have also shown that a reactive stroma can drive the transformation of mammary gland fibroblasts to tumor-associated myofibroblasts *(87)*. This suggests that the stroma actively contributes to the tumor phenotype and may additionally influence tumor behavior.

The desmoplastic tumor stroma is similar in character to the stroma present in an injured tissue. However, a tumor is a wound that never heals *(88, 89)*. The essential difference between a tumor and a tissue that undergoes normal healing is that cancer cells do not perceive extracellular cues as an integrated tissue unit and therefore do not respond appropriately. Whereas the cells in a nonmalignant tissue efficiently titrate the secretion of cytokines, growth factors, and metalloproteinases and control the turnover of their ECM and integrin proteins, such feedback control mechanisms are either absent or no longer functioning adequately in malignant cells.

Wound repair is a complex process that is mediated by coordinated interactions among epithelial cells, fibroblasts, endothelial cells, and inflammatory cells. The process of wound repair is orchestrated by a continuously evolving network of secreted growth factors and cytokines that act within a dynamically evolving ECM microenvironment to re-form the differentiated tissue (for review, *see* **refs. 90** and **91**). Following injury, the ECM of the wounded tissue undergoes dramatic remodeling, replacing the initial fibrin(ogen) clot with an early granulation tissue that is rich in hyaluronic acid, tenascin, and SPARC proteins. This provisional matrix is subsequently replaced by late granulation tissue that abounds in type I and II collagens and proteoglycans. The transitional matrix eventually resolves to form the mature interstitial matrix that encompasses the mesencymal cells and the basement membrane (BM) that surrounds the basal domain of the regenerated epithelium.

The process of wound repair is driven by the concerted action of growth factors (GF) and cytokines such as PDGF, basic FGF, IGFs, TNF-α, Scatter factor, VEGF, TGFβ, KGF, and EGF. These growth factors and cytokines are secreted or released into the injured microenvironment by resident epithelial and stromal cells and by newly recruited inflammatory cells. The combined action of these soluble factors stimulates endothelial and fibroblast cell proliferation and migration and immune cell invasion into the site of wound repair. These "activated" cells then jointly synthesize and secrete ECM proteins, metalloproteinases, and motility factors needed to remodel the ECM, to repopulate the wound by epithelial cells, and to ensure contraction of the injured matrix so that proper healing can take place. The microenvironment of a wound therefore is one that promotes angiogenesis, cell proliferation, and migration. Because transformed cells cannot temper their response to extracellular cues such as these, the microenvironment of a wound is compatible with tumor promotion. In fact, more then a decade ago, Bissell and colleagues demonstrated that the microenvironment of a wounded tissue could dramatically enhance tumor formation in vivo *(92–94)*.

Over the last few years compelling evidence has accumulated to show that the "activated" tissue stroma can directly promote malignant transformation and drive the oncogenic behavior of "genetically primed" cells (for reviews, *see* **refs. 27, 28, 31, 34, 38,** and **95**). It was reported that co-injection of tumor cells with reconstituted BM (rBM) or collagen I ECM protein enhanced tumor formation in nude mice, and permitted the recapitulation of primary tumor histology in vivo *(96)*. Studies also showed that tumor formation in nude mice is significantly enhanced by direct co-injection of transformed mammary cells with stromal fibroblasts *(26),* and that tumor-associated fibroblasts can significantly reduce the latency of breast tumor progression in vivo *(97)*. A direct tumor-promoting effect of the stroma was provided by experiments conducted by Olumi and colleagues (1999). These studies demonstrated that prostate carcinoma-associated fibroblasts could induce malignant transformation of SV40 immortalized human prostate epithelial cells in nude mice, provided the two cell types were co-cultured in collagen I plugs under the kidney capsule *(98)*. Similarly, immortalized murine mammary epithelial cells (MECs) harboring genetic abnormalities exhibited malignant behavior when injected into previously irradiated and "stromally active" mammary fat pads, whereas the same cells recapitulated normal branching morphogenesis when injected into nonirradiated glands *(36)*.

The mechanism by which the stroma can modify tumor formation remains elusive. Most investigators have emphasized the contribution of stromally derived soluble factors to tumor progression, and have focused their studies on delineating the contribution of these components to epithelial cell proliferation and/or survival ([98]; for reviews, *see* **refs. *32*, *37*,** and ***99***). However, the integrity and composition of the ECM surrounding the primary tumor cells is also significantly altered during malignancy *(27)*. Indeed, changes in cell–ECM interactions and ECM turnover appear to be critical for tumor cell growth, survival, and migration (reviewed in **refs. *100–103***) [Hoffmann, 2000, #4838]). Experiments conducted using the HPV16 transgenic mouse skin cancer model elegantly illustrate the essential contribution of stromally induced ECM degradation to cancer progression. Although born normal, the HPV16 mouse rapidly develops hyperplastic skin lesions that progress from angiogenic dysplasias to invasive squamous cell carcinomas within a short time span. The stroma surrounding the developing malignant lesions in the HPV16 mouse exhibits classical desmoplastic changes, including elevated expression of metalloproteinases such as MMP9. Coussens and co-workers showed that HPV16 mice crossed with MMP9-null mice had significantly reduced incidences of malignant transformation, which were restored to normal levels following bone marrow reconstitution with MMP9-expressing immune cells ([*39*]; reviewed in **ref. *31***). The critical role of ECM integrity during tumor progression, however, was definitively established in studies using transgenic mice engineered to express stromelysin-1 in their mammary glands (another metalloproteinase that has been shown to degrade ECM protein). In these experiments, overexpression of stromelysin-1 in the mammary glands of mice induced a marked desmoplastic response and premature branching morphogenesis followed by development of premalignant and malignant lesions with multiple genetic alterations *(35)*. These two studies irrefutably show that a reactive stroma not only promotes malignant transformation and tumor formation, but can also induce genomic instability. Thus an understanding of the contribution of the ECM and its receptors to tumor progression is warranted. Delineating the underlying mechanisms by which the ECM and its receptors regulate expression of the malignant phenotype in turn depends on the use of tissue engineered tractable model systems that can be easily manipulated ex vivo and in vivo *(104–105)*. Two such systems will be discussed in the following section.

4. Stromal–Epithelial Interactions and Repression of the Malignant Phenotype

Despite the incredible advances being made in diagnostics and pharmacogenomics, the characterization and identification of a malignant lesion still depends largely on a subjective macroscopic and microscopic evaluation. The major criteria used to diagnose and stage cancer is histologic evidence of a loss of normal tissue architecture with cell invasion, and nuclear changes that include anaplasia, large and multiple nucleoli, and chromatin asymmetry *(106)*. Because tissue architecture is mediated by the coordinated action of cell–cell and cell–ECM interactions *(107–109)* that in turn influence nuclear architecture *(110)*, it is not surprising that invasive tumors have an altered ECM and show changes in their cell adhesion receptors. Numerous studies have shown that tumors of diverse tissue types, including those of the breast, kidney, liver, pancreas, lung, and prostate, consistently exhibit changes in the composition, organization, and integrity of their ECM *(111–117)*. Indeed, the very definition of a tumor maintains that BM integrity must be compromised to permit invasion of the transformed cells into the surrounding tissue. Integrin expression and organization are also perturbed in tumors of the breast, kidney, thyroid, lung, liver, and colon, among others *(118–123),* and disruptions in tumor cell–ECM responsiveness have been directly demonstrated using isolated tumor cells and a rBM assay *(52,124,125).*

We have been studying the role of cell–ECM interactions and integrins in malignant transformation of the breast using a mammary epithelial cell (MEC) tumor progression model called HMT-3522, and 3-D ECM assays. The HMT-3522 tumor progression series was established from MECs isolated from mammary reduction mammoplasty tissue derived from a patient with fibrocystic breast disease *(126).* Continued passage and selection of these cells in defined media led eventually to the malignant transformation of the cells following their injection into nude mice *(127).* Subsequent studies have shown that this tumor progression continuum, which consists of early, nonmalignant S-1 EGF-dependent passages 50–400, premalignant EGF independent S-2 passages 160–225, and the tumorigenic EGF-independent T4-2 passages 5–50, sustain numerous chromosomal amplifications and deletions as they progress toward malignancy, yet remain essentially diploid *(128).* These cells also obtain a predominant p53 mutation by passage 50 *(129),* and the S-2 and T4-2 cells sustain amplifications in c-myc, ErB2, and EGF receptor (EGFR) and exhibit altered growth regulation *(130).* Using the rBM assay we found that cell–ECM interactions become compromised in these cells, prior to malignant transformation (reviewed in **refs.** *54* and *105*); Weaver et al., in preparation). In response to rBM the nonmalignant S-1 cells in this series first proliferate and then undergo morphogenesis to form growth-arrested, polarized 3-D acini-like structures (analogous to primary MECs and terminal ductal lobular units found in the breast in vivo *(124).* In contrast, their T4-2 tumorigenic counterparts fail to undergo morphogenesis, and instead form continuously proliferating, disorganized, and unpolarized invasive colonies in response to these same ECM cues. The premalignant S-2 cells exhibit a heterogenous response to rBM, such that after 10 d within rBM, some colonies growth-arrest and polarize while a minor population grows continuously to eventually form large cystlike structures, reminiscent of premalignant benign lesions in vivo *(40)* Weaver et al., in preparation). Studies are now in progress to determine if this minor population of S-2 cells gave rise to the T4-2 cells.

Upon examination of the integrin profiles of the cells within this tumor progression series we determined that the T4-2 cells had elevated expression and enhanced signaling through β1 integrin and EGFR. Studies showed that whereas inhibition of β1 integrin or EGFR signaling induced apoptosis in the S-1 cells, the T4-2 cells remained viable. Instead of dying, the T4-2 cells underwent a phenotypic and behavioral reversion characterized by reorganization of the cytoskeleton, reassembly of adherens and tight junctions, and re-formation of polarized, growth-arrested acini structures. Moreover, the reverted T4-2 cells (T4-R) were no longer anchorage-independent for their growth and survival, and exhibited reduced tumor formation in vivo *(40,82).* Phenotypic reversion was found to be reversibly dependent on the tissue ECM microenvironment and adhesion/growth factor-linked signaling. Furthermore, phenotypic reversion occurred despite the presence of numerous genomic abnormalities, as revealed by comparative genomic hybridization (CGH) and genomic array analysis (unpublished observations). Interestingly, inhibiting α6β4 integrins in the S-1 cells perturbed morphogenesis, such that the S-1 cells formed continuously proliferating and nonpolarized colonies in response to rBM

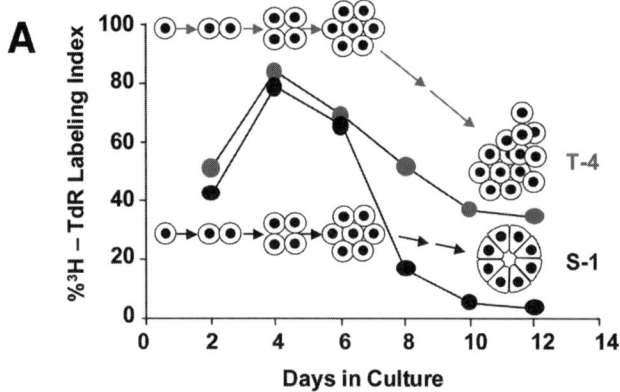

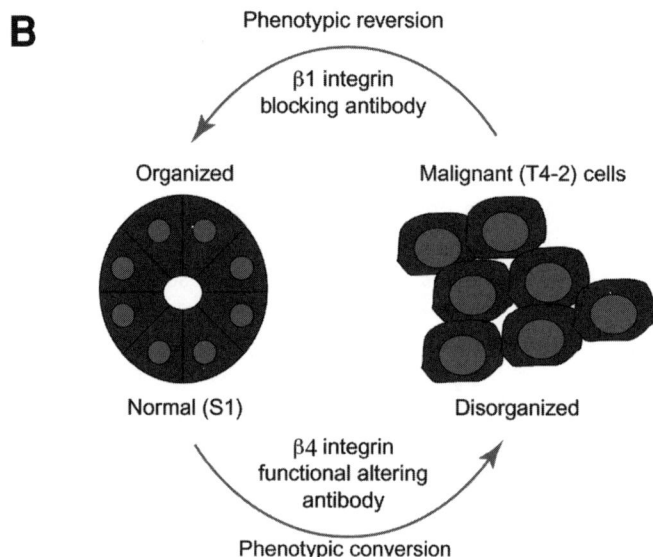

Fig. 1. Different growth and morphologic responses of early-passage nonmalignant mammary epithelial HMT-3522 S-1 cells and their tumorigenic progeny T4-2 in response to culture for 12 d within a reconstituted laminin-rich basement membrane (rBM) matrix and alterations in integrin signaling. Although the S-1 cells growth-arrest within 8 d of culture within a 3-D rBM and form polarized acini-like structures, T4-2 cells fail to growth-arrest and instead form continuously proliferating, unpolarized, and invasive colonies of cells (**A**). S-1 cells treated with a function-blocking antibody to α6 integrin, however, fail to growth-arrest and instead form continuously proliferating and unpolarized structures, while T4-2 cells treated with a function-blocking antibody to β1 integrin undergo a phenotypic reversion, and form growth-arrested and polarized, differentiated, acini-like structures (**B**). Thus cell behavior and tissue morphogenesis are dictated by basement membrane-induced integrin signaling.

(*see* **Fig. 1**). These studies demonstrated that cell adhesion could direct tissue architecture and regulate cell behavior regardless of genotype.

Similarly, Javaherian and colleagues (1998) reported that the malignant behavior of genomically abnormal Ras transformed keratinocytes could be repressed by driving the reformation of normal tissue architecture *(131)*. In these experiments, the investigators used submerged organ cultures of keratinocytes consisting of a feeder layer of collagen I containing dermal fibroblasts overlaid with a single layer of proliferating epithelial

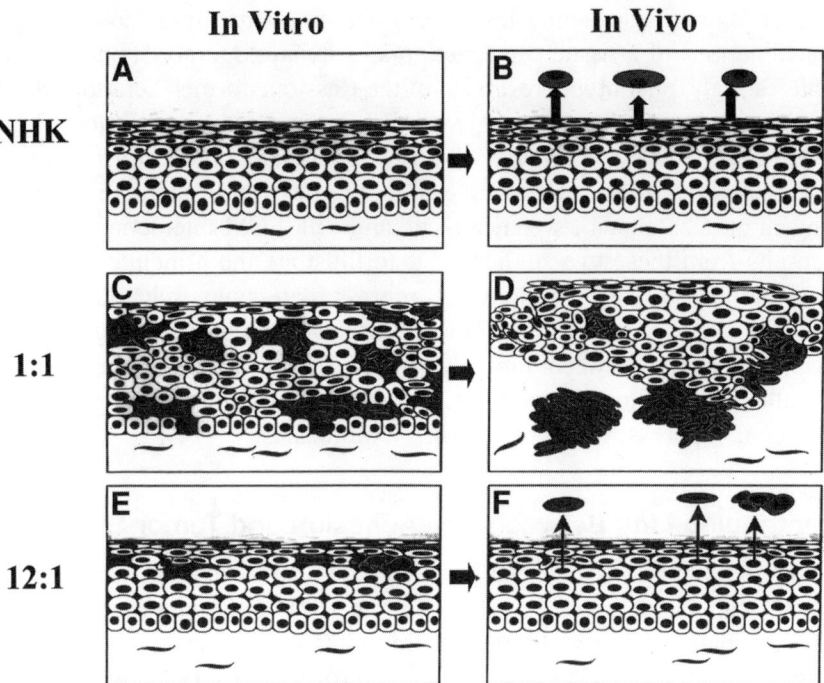

Fig. 2. Normal human keratinocytes suppress the growth of adjacent intraepithelial tumor cells in a 3-D engineered tissue model of human stratified epithelium. Normal human keratinocytes (NHK) seeded as a single layer onto an interstial collagen I matrix with fibroblasts proliferate to form a stratified epithelium that differentiates (**A,** clear cells), and undergoes "normal" exfoliation upon grafting of the engineered tissue onto the back of a nude mouse. (**B,** see darkened shed cells). In the absence of significant numbers of normal keratinocytes (1:1), premalignant Ras-transformed keratinocytes proliferate continuously and fail to differentiate (**C,** darkened cells); instead they form invasive tumors in nude mice (**D,** darkened cells). If the Ras-transformed cells are co-cultured with a high number of normal keratinocytes (12:1); under conditions where numerous cell–cell interactions between normal and transformed keratinocytes can be established, then the malignant phenotype of the Ras keratinocytes is repressed and the cells do differentiate (**E,** darkened cells surrounded by clear cells) and become exfoliated in vivo (**F,** darkened shed cells).

cells. Under these conditions nonmalignant keratinocytes proliferate until they form a multilayered continuum of cells, whereupon they growth-arrest and terminally differentiate (*see* **Fig. 2a**). Upon grafting of this engineered skin tissue onto the back of a nude mouse, the differentiated keratinocytes eventually become exfoliated (*see* **Fig. 2b**). In their studies, the researchers showed that in the absence of sufficient numbers of nonmalignant keratinocytes (e.g., 1:1), Ras-transformed keratinocytes neither growth-arrest nor form a normal skinlike architecture (*see* **Fig. 2c**). Instead, Ras-transformed keratinocytes proliferate uncontrollably down into the basal layer and form invasive tumors in nude mice (*see* **Fig. 2d**). If, however, the Ras transformed cells are co-cultured with a high ratio of nonmalignant keratinocytes (e.g., 12:1), under conditions where numerous cell–cell interactions between nonmalignant and transformed cells can be established, then the transformed Ras phenotype can be repressed. Under these conditions the transformed keratinocytes reform normal skin architecture, undergo growth-arrest, and differentiate, as demonstrated by expression of the differentiated skin protein fillagrin. Moreover, when these engineered mixed tissues are grafted onto the backs of nude mice, the tumor cells are subsequently lost by exfoliation, elegantly demonstrating how the coordinated actions of a tissue can function as a dynamic tumor suppressor (*see* **Figs. 2e, f**). Interestingly, phenotypic reversion of the Ras-transformed keratinocytes does not occur if these cultures are pretreated with the protein kinase activator 12-O-tetrade-canoylphorbol-13-acetate *(132)*. Because PKC stimulates integrin activation and promotes cell spreading, these studies suggest that repression of the malignant phenotype is contingent on a balance between cell–cell and cell–ECM interactions.

The results from these two studies serve to illustrate the principle that tissue architecture, as induced by cell adhesion, can repress malignancy regardless of an altered genotype. Although the mechanism for these effects remains obscure, given evidence of reciprocal interactions between tumor suppressors and integrins, it seems reasonable to predict that tissue architecture represses malignancy through a functional collaboration between tumor suppressors and cell adhesion molecules.

5. The Intriguing Link Between Cell Adhesion and Tumor Suppressors

Cell adhesion, cell growth, and the cellular stress/survival response are dynamically linked in normal cells (reviewed in **refs.** *133* and *134*). Malignant transformation is frequently accompanied by a disruption of this interdependency, as exemplified by the acquisition of anchorage independence for cell growth and survival in tumors. Just how integrins regulate normal cell growth and survival and the mechanism by which tumors are able to circumvent this process has yet to be delineated. What has become evident from recent studies is that there exists an intriguing functional link between tumor suppressors and cell adhesion in the regulation of cell growth and survival (*see* **Table 2** and **Fig. 3**). Studies have found that cell adhesion is essential for cell cycle progression (80), and that tumor suppressors function by regulating cell cycle transit *(135)*. In the last few years, links between these pathways have become increasingly apparent. Data now show that integrin-linked signaling can regulate G1 cell cycle transition by influencing the transcription and translation of cyclin D1, a critical cell cycle protein that modulates the activity of the tumor suppressor pRb by regulating its phosphorylation status (reviewed in **ref.** *80*). Experiments have revealed that the $\beta1$ integrin-dependent survival of cultured prostate cells and regressing prostate epithelial cells is dependent on a functional pRb to E2F pathway *(136)*. More recently it was reported that integrin-directed pRb phosphorylation is necessary for initiating E2F transcription and cell cycle transition *(137)*. The relevance of this interaction was emphasized by studies that examined the role of cyclin D1 expression on oncogene-induced mammary tumor formation. It was reported that the Ras and Neu oncogenes, which likely cooperate with adhesion-linked pathways to circumvent G1 cell cycle arrest to thwart anchorage-dependent survival, were unable to induce mammary tumor formation in the absence of a common target, cyclin D1 *(138)*.

TGFβ induces growth arrest in normal cells and can repress the malignant behavior of early benign lesions. TGFβ mediates its cellular effects by activating cell surface receptors and inducing signaling via Smads *(139–141)*. Colorectal cancers frequently show loss of Smad4/DPCC, and studies have shown that targeted ablation of the Smad4 gene induces malignant transformation of colorectal cancer cells *(142,143)*. Therefore, signaling via Smads likely mediates the tumor suppressor and growth inhibitory actions of TGFβ. Consistent with the idea that cell adhesion pathways intersect functionally with tumor suppressor proteins to control cell fate, studies have shown that TGFβ and

Table 2
Examples of Links Between Tumor Suppressors and Cell Adhesion

Tumor suppressor gene	Mutation-associated tumors	Associated cancer syndrome	References
I. Examples of tumor suppressors modulated by cell adhesion			
BRCA1	Breast and ovarian	Familial breast and ovarian cancer	O'Connell, 2000 Marquis, 1995
DPC4	Pancreatic, colon, hamartomas	Juvenile polyposis	Colonge, 1999 Song, 1998
NF1	Neurofibromas, sarcomas, and gliomas	Von Recklinghausen, neurofibromatosis	Koivunen, 2000
p53	Sarcomas, breast and brain tumors	Li-Fraumeni	Tlsty, 2000 Nigro, 1997
Rb1	Retinoblastoma, osteosarcoma	Familial retinoblastoma	Yu, 2001 Spancake, 1999
II. Examples of tumor suppressors modulated by cell–cell interactions			
APC	Colon	Familial adenomatous polyposis	Tetsu, 1999 Barth, 1997 Munemistu, 1995
E-cadherin	Breast, colon, skin, and lung	Familial gastric cancer	Guildford, 1998 Kemler, 1993
PTEN	Glioblastoma, prostate and breast tumors	Cowden syndrome	Persad, 2001
HL	Hemangliomas, renal, pheochromocytoma	Von-Hippel-Lindau	Kondo, 2001 Kamada, 2000
Wt1	Nephroblastoma	Wilms tumor	Tetsu, 1999 Barth, 1997 Munemistu, 1995
III. Examples of tumor suppressors that modulate cell–ECM interactions			
INK4a	Melanoma, pancreatic	Familial melanoma	Plath, 2000 Fahraeus, 1999
p53	Sarcomas, breast, and brain tumors	Li-Fraumeni	Alexandrova, 2000 Vitale, 1999
Rb1	Retinoblastoma, osteosarcoma	Familial retinoblastoma	Day, 1997 Day, 1999

Smads can influence integrin and ECM expression and/or function. For example, Dalton and colleagues (1999) showed that in the absence of adhesion, fibroblasts lose cell-surface $\alpha5\beta1$ integrin expression and die (apoptosis), but that treatment with TGFβ enhances $\alpha5\beta1$ integrin expression and permits cell survival *(144)*. Similarly, experiments with osteoblasts demonstrated that $\alpha v\beta5$ integrin expression, which is essential for osteoblast differentiation and survival, are transcriptionally regulated by Smads 3 and 4 *(145)*. These and other findings now suggest that in the absence of specific Smads, as occurs frequently in tumors, TGFβ may not be able to modulate integrin and ECM expression to adequately regulate cell function. Such perturbed transcriptional regulation could explain the paradoxical observation that TGFβ suppresses the growth of normal and premalignant cells, but enhances tumor cell proliferation and invasion.

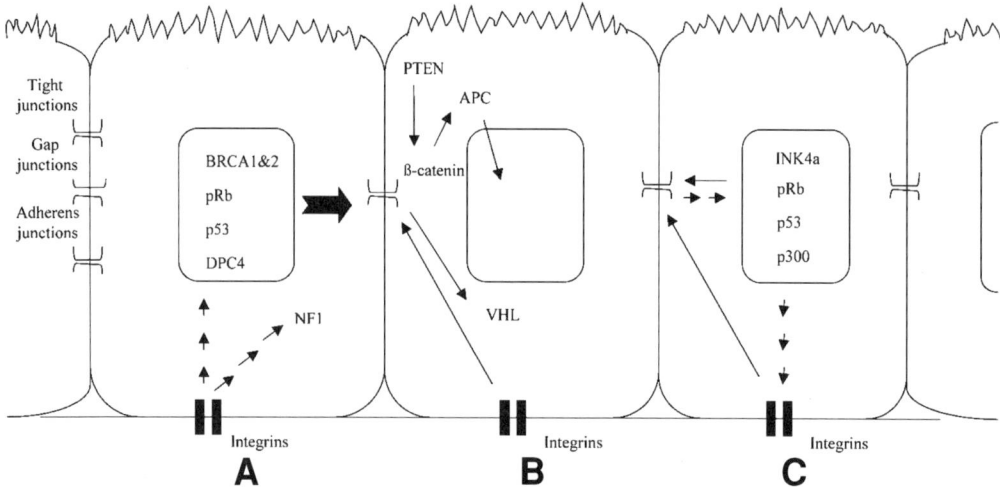

Fig. 3. Cell–extracellular matrix interactions and integrin-linked signaling modulate the functions of classic tumor suppressor proteins (**A**). Cell–cell adherens and tight junction plaque molecules also directly and indirectly alter the function of tumor suppressors and are themselves regulated by integrin-linked signaling (**B**). Recent evidence shows that tumor suppressors can directly and indirectly regulate cell–ECM interactions and adherens and tight junctions and alter cell adhesion and extracellular matrix expression (**C**). Thus complex interactions among the extracellular matrix microenvironment, cell adhesion molecules, and tumor suppressors influence cell growth, survival, and differentiation in tissues in vivo to maintain tissue-specific differentiation and tissue homeostasis.

Cell–matrix interactions influence the cells stress response by regulating the activity of several signaling molecules including members of the JNK and NFkappa β pathways *(146,147)*. For example, ligation of αvβ3 integrin stimulates nuclear translocation of NFkappa β p65 to permit the survival of rat aorta-derived endothelial cells *(148),* and loss of cell–matrix interactions and integrin signaling induces Jun-N-terminal kinase or stress-activated kinase activation *(149)*. Similarly, p53 is also regulated by cell adhesion. The p53 gene is frequently altered in a wide variety of human tumors, and targeted ablation of p53 gene expression leads to tumor formation in mice. Reduced p53 protein expression and/or activity, either by dysregulated proteolytic degradation or by functional inactivation, can also have a major effect on cell function and tumor incidence (for reviews, *see* **refs. *150*** and ***151***). Experiments conducted with primary human keratinocytes demonstrated that in the absence of cell adhesion, p53 protein was rapidly lost and p53 functional activity was decreased *(152)*. Because p53 plays a key role in genome surveillance, the ramifications of this observation are clear. That is, alterations in cell–ECM interactions and/or integrin signaling could contribute to tumor progression by influencing genomic integrity through modification of the cell's response to exogenous damaging agents *(153)*.

A relationship between cell adhesion and the tumor suppressors BRCA1 and BRCA2 has also emerged. Mutations in *BRCA1* and *2* have been implicated in breast and/or ovarian cancers *(154)*. Studies have shown that BRCA1 and BRCA2 are coordinately regulated during mammary gland development. Thus BRCA 1 and BRCA2 are dramatically induced during pregnancy but repressed during lactation, when the mammary gland is maximally differentiated *(155,156)*. Because the mammary epithelium undergoes rapid proliferation and branching morphogenesis in concert with ECM remodeling during pregnancy, and mammary differentiation depends on cell–ECM interactions, this indicates that BRCA1 and BRCA2 expression is coordinately regulated in association with adhesion *(29)*. In support of this concept, recent studies by O'Connell and co-workers (2000) showed that repression of BRCA1 expression following differentiation was strictly dependent on MEC-BM interactions *(154)*. Indeed, induction of BRCA1 depends on a functional pRb-E2F pathway *(157)*, and BRCA1 protein expression is downregulated by TGFβ-induced pRb dephosphorylation *(158)*. Furthermore, both of these pathways are regulated by integrins (see above). The relevance of this relationship becomes clearer when one considers that ligation of BM integrins is necessary for MEC survival (reviewed in **ref.** *159*), and that loss of BM integrin interactions enhances stress response signaling and expression of stress response genes (see above; unpublished observations). It follows that induction of BRCA1 in the absence of BM integrin interactions may constitute part of an essential stress response mechanism that protects the cells when they are exposed to challenging microenvironments or when they are no longer part of a stable differentiated tissue structure *(160)*. This possibility is strongly supported by evidence demonstrating that BRCA1 is associated with the DNA repair/stress response apparatus *(161)*.

Additional data now indicate that tumor suppressors may influence cell cycle progression and stress control by directly modifying cell–ECM response pathways and integrin function. For example, studies showed that von Hippel-Lindau (VHL), a gene linked to renal cell carcinoma, that affects cell proliferation and angiogenesis via enhancing proteolysis of HIF-1 alpha *(162–164)*, can influence actin and vinculin organization and promote the assembly of focal adhesions to inhibit cell motility *(165,166)*. The cell cycle kinase inhibitor p16^{INK4a} can suppress the anchorage-independent growth of human pancreatic carcinoma and melanoma cells by upregulating the fibronectin receptor α5β1 integrin *(167,168)*. Fahraeus and Lane (1999) determined that p16 peptide inhibitors could specifically repress αvβ3 integrin-dependent spreading on vitronectin, indicating that p16 may induce some of its effects by directly influencing integrin function through inside-out activation *(69,168)*.

Inactivation of p16 INK4a is deemed necessary to circumvent the agonesence block that occurs during MEC immortalization ex vivo *(169)*. This raises the question as to whether alterations in the crosstalk between adhesion-linked signaling and tumor suppressors might be essential for cell immortalization and tumor progression. In support of this idea, a dialogue between cell adhesion and tumor suppressors in cell cycle control and tumor inhibition was illustrated by studies that examined effects of pRb expression on ECM-induced cell differentiation. In these studies, investigators used primary "normal" human MECs and showed that loss of functional pRb, by overexpression of the viral protein E7, inhibited G1 cell cycle exit and prevented the cells from completely differentiating in response to cues from a rBM *(170)*.

6. Cooperative Interactions Between Cell Adhesion and Metastasis Suppressors

Tumor metastasis is actively controlled by a group of genes called metastasis suppressors. The metastatic spread of tumors is likely mediated by soluble factors secreted into the stroma that activate tumor integrins and degrade the ECM, thereby promoting cell migration and extravasation *(103)*. Data now suggest that the effectiveness of some suppressors to inhibit tumor metastasis may be mediated by their ability to modify integrin function and to restore tumor–ECM responsiveness. In an intriguing set of experiments, Howlett and co-workers (1994) showed that overexpression of the nm-23 H1 metastasis suppressor protein, in the highly aggressive and metastatic breast cancer cell line MD MDA 435, not only repressed the metastatic phenotype of these cells in vivo, but also permitted the cells to respond appropriately to ECM cues and to undergo morphogenesis. In these experiments the parental MDA 435 cells formed large, continuously growing and invasive colonies 10 d after embedment within rBM. In contrast, MDA 435 cells transfected with the nm23 H1 gene were able to partially differentiate to form acini-like structures in response to the same rBM cues. These cells were able to reassemble cell–cell interactions, acquire markers of apical and basal polarity, assemble an endogenous BM composed of laminin-1 and collagen IV, and achieve near-growth-arrest *(125)*. Although these studies failed to investigate the role of integrins in this phenotypic reversion, recent work reported that over expression of nm23-H1 was sufficient to inhibit the anchorage-independent growth and adhesion behavior of prostate carcinoma cells, suggesting integrin signaling must be linked to this process *(171)*.

The newly described metastasis suppressor CD82/KAI-1 illustrates a more direct connection between metastasis suppressor proteins and integrins. CD82/KAI-1 is a member of the tetraspanin superfamily of transmembrane proteins that were originally identified and characterized as integrin-associated proteins ([*172*]; for a review of tetraspanins, *see* **ref.** *76*). Ectopic expression of CD82/KAI-1 suppresses tumor cell migration and invasion; processes that employ integrins *(103,173)*. Recent data suggest that the ability of CD82/KAI-1 to suppress tumor metastasis may be owing to its physical interaction with the α3β1 integrin and members of the PKC family *(174)*. CD82/KAI-1 can also associate with the epidermal growth factor receptor (EGFR), and data indicate that this interaction can desensitize EGFR signaling *(175)*. Because activation of EGFR can drive cell migration and invasion through cooperative interactions with PKC and integrins *(176),* it is possible that CD82/KAI-1 inhibits tumor metastasis by interfering with the function of this pathway.

7. Cooperative Interactions Between the Nucleus, Cell–Cell, and Cell–ECM Dictate Tissue Homeostasis

How might tumor suppressors collaborate with cell adhesion pathways to influence tumor progression? Perhaps a more instructive question to ask is how do tumor suppressors cooperate with cell adhesion to maintain the differentiated state of the tissue, and what are the mechanisms by which this tissue-specific differentiation is achieved?

More than two decades ago, the concept that form and function are linked was proposed (reviewed in **refs.** *29* and *177*). Since then numerous studies have demonstrated that tissue-specific gene expression depends on the coordinated interplay between cell–matrix and cell–cell interactions, and that this cooperation is linked to the cytostructural tension within the cell and the spatial organization of the tissue. Reviews and research articles have repeatedly emphasized the critical role played by both the physicomechanical and biochemical properties of the ECM microenvironment in the control of cell differentiation, and established that many of these effects depend on biomechanical and biochemical signaling pathways. For example, experiments using malleable rBMs have demonstrated that expression of the salivary gland protein cystatin strictly depends on the formation of a 3-D acini-like salivary gland structure *(52)*. Similarly, keratinocyte differentiation and behavior, including expression of the differentiated skin protein fillagrin and epithelial exfoliation, are only recapitulated faithfully using 3-D organ co-culture models and in-vivo skin grafts *(131)*.

Experiments using cultured primary and immortalized MECs and designer microenvironments have been instrumental in delineating the underlying mechanisms whereby ECM-induced cell shape changes and adhesion-directed 3-D tissue organization regulate tissue-specific gene expression. These studies have shown that cell shape, BM-induced integrin signaling, and cytoskeletal reorganization, as well as formation of a 3-D polarized mammosphere, dictate hierarchies of differentiated gene expression in MECs. Data show that although cell rounding and the presence of cortical actin are sufficient for lactoferrin gene expression, β-casein gene expression also requires the presence of lactogenic hormones and signaling via the BM integrins α6, β1, and β4 ([*83*], reviewed in **ref. 29**). WAP gene expression additionally requires the formation of a 3-D polarized acini, indicating that cell–cell interactions and spatial organization of the tissue (e.g., tissue polarity/tight junction formation) also regulate tissue-specific gene expression (reviewed in **ref. 29**).

Data obtained using MECs and rECM assays have revealed that tissue-specific gene expression is regulated by dynamic and reciprocal communication between the nucleus and cell–ECM adhesion plaques, and that epigenetic alterations in chromatin organization are likely involved. Myers and co-workers (1998) conducted promoter analysis of the beta casein gene and identified an ECM response element termed BCE-1. Their studies showed that BCE-1 promoter activity strictly depends on BM-induced alterations in histone acetylation and implied that changes in chromatin organization were required *(178)*. Whether cell–ECM interactions and integrin signaling does in fact induce changes in chromatin organization, and if these effects require reorganization of nuclear architecture, have yet to be determined. However, studies by Lelievre and colleagues (1998) illustrated that rBM-directed MEC morphogenesis and acquisition of tissue polarity are associated with a definable reorganization of chromatin structure (histone acetylation) and changes in nuclear architecture. Using immunofluorescence confocal microscopy, these researchers documented the dramatic reorganization of the nuclear mitotic apparatus protein (NuMA), a nuclear structural protein, and tumor suppressor pRb occurred as MECs undergo BM-directed differentiation (*see* **Fig. 4**). Moreover, directly perturbing nuclear organization using C-terminus-specific NuMA antibodies not only disrupted nuclear architecture and histone acetylation, but also activated membrane-associated metalloproteinases, degraded the endogenous BM, and induced alterations in cell–ECM and cell–cell interactions *(110)* (unpublished observations). These studies add credibility to the concept that tissue-specific gene expression is linked to ECM–integrin-induced changes in nuclear organization. However, they also demonstrate for the first time the existence of reciprocal signaling from the nucleus to sites of cell–ECM interaction.

Integrins can modify cell–cell junctions to alter cell fate by directly altering the activity of cell adhesion molecules or by indirectly influencing gene expression. Data show that crosstalk between integrins and cell–cell junctions can either promote cell differentiation *(40)*, or enhance malignant transformation and tumor invasion, depending on

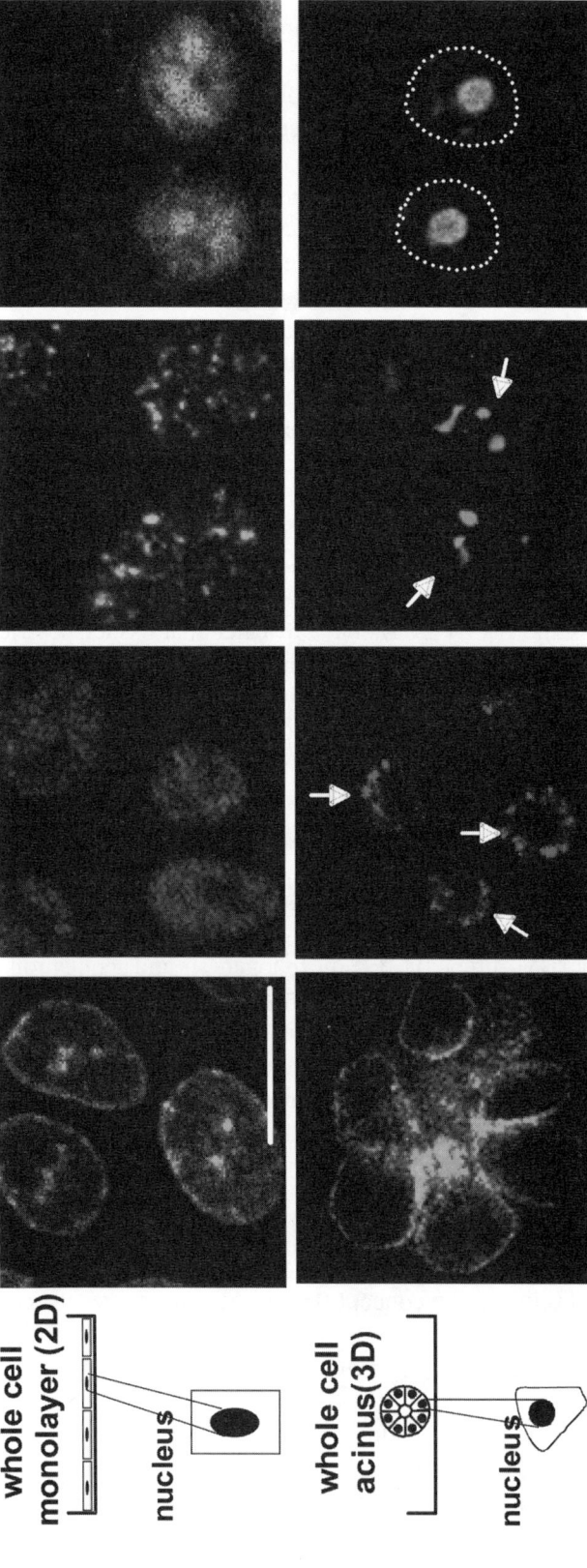

Fig. 4. Redistribution of nucleur proteins in human mammary epithelial cells after reconstituted basement membrane (rBM)-induced acinar morphogenesis and differentiation. Confocal fluorescence images (0.2-μm optical sections) of lamin B, nuclear mitotic apparatus protein (NuMA), splicing factor Srm160 (SRm160), and the retinoblastoma protein (pRb) in cells grown as 2-D monolayers (top) and within 3-D rBM for 12 d (bottom). NuMA was diffusely distributed in the nuclei of cells grown as monolayers (NuMA top) but reorganized into large nuclear foci in cells induced to undergo morphogenesis (acini formation; NuMA bottom). Srm160 was distributed as multiple nuclear speckles in cells cultured as monolayer (SRm160 top), whereas it was concentrated into fewer and larger speckles in the acini (SRm160 bottom). pRb was diffusely distributed in the nucleus of growth-arrested in monolayer (Rb top) but coalesced into a central, single nuclear focus in the rBM-induced growth-arrested and differentiated mammary acini (Rb bottom). Arrows indicate nuclei found within the plane of the section. (Bar = 10 μm).

the nature of the signal and the differentiation state of the cell *(179)*. Inhibiting β1 integrin signaling in malignant MECs, for example, induces reformation of cell–cell adherens *(40)* and tight junctions (unpublished observations), and activation of α3β1 integrin by laminin 5 enhances connexin 43-mediated gap junction communication in keratinocytes *(180)*. On the other hand, αv integrin activation induces disassembly of E cadherin adherens junctions and promotes tumor invasion in breast cancer cells *(181)*.

Some cell–cell adhesion plaque proteins can shuttle between cell adhesion complexes and the nucleus where they can regulate gene expression, as is the case with β-catenin. β-Catenin was first identified as a cell adhesion protein that forms complexes with α-catenin, cadherins, and the actin cytoskeleton so that adherens junctions that mediate stable cell–cell contacts can be assembled *(182,183)*. It is now well established that β-catenin also acts as an architectural transcription factor, which translocates to the nucleus and forms transcriptional complexes with proteins such as Lef-1 to regulate the expression of genes such as cyclin D1, c-myc, and E cadherin *(184–187)*. A recent article by Somasiri and colleagues illustrated how integrins can modulate cell behavior, by influencing the integrity of cell–cell junctional complexes and modifying the activity of adhesion plaque proteins such as β-catenin. In these studies, constitutively active integrin-linked kinase (a serine threonine kinase whose activity is activated by β1 and β3 integrin) induced the disassembly of adherens junctions, increased the transcriptional activity of β-catenin/LEF-1, repressed E cadherin expression, and promoted an epithelial-to-mesenchymal transition in MECs *(179,188)*. The catenin protein p120 also shuttles between junctional complexes and the nucleus, where it interacts with the transcription factor Kaiso *(189),* and the tight junction protein ZO-1 interacts with a Y-box transcription factor (ZO-1-associated nucleic acid-binding protein) in the nucleus to regulate ErbB-2 gene expression *(190)*. Moreover, a member of the trithorax family of nuclear chromoproteins ASH-1 was recently found to shuttle between the nucleus and tight junctions *(191–193)*. Thus numerous molecular mechanisms exist whereby integrins could regulate gene expression by modulating the integrity of cell–cell adhesions.

Whether alterations in cell–ECM interactions do in fact influence tumor progression by regulating the transcriptional behavior of cell–cell adhesion plaque proteins awaits direct experimental proof. However, pathologists have documented that changes in tissue polarity and the stromal microenvironment do occur early and in tandem during malignant transformation of the breast (reviewed in **ref. 54**). Our studies using the HMT-3522 MEC tumor model indicate that tight junction integrity and adherens junction organization are lost early during tumor progression, and occur in association with alterations in integrins and ECM protein secretion (Weaver et al., in preparation). Phenotypic reversion of the tumor cells in this series (through inhibition of β1 integrin) is associated with reformation of adherens and tight junctions and normalization of integrin and ECM expression *(40,82,194)*. Similarly, expression of the gap junction protein connexin 43 reversed the transformed phenotype of human glioblastoma cells in association with nuclear accumulation of connexin 43 protein *(195)*. In addition, the overexpression of the nervous system adhesion molecule protein zero triggered reversion of HeLa cells to an epithelioid phenotype, and induced assembly of epithelial junctions via upregulation of desmoplaskin and N-cadherin *(196)*. Clearly, then, understanding how integrins regulate bidirectional nuclear shuttling of adhesion molecules and chromo proteins should provide insight into the mechanism by which the tissue microenvironment maintains tissue homeostasis or drives transformation.

8. Toward a Unifying Theme: Do Cell Adhesion/Tumor Suppressor "CATS" Nodes Exist?

Our proposal that a signaling nexus between the nucleus, cell–ECM, and sites of cell–cell adhesion exists suggests that alterations in key molecules that integrate these circuits would compromise tissue homeostasis. Mutations or loss in the expression of pathway integrators may constitute a necessary prelude to cell transformation. Indeed, the ability to coordinately sense and integrate extracellular cues is an essential feature of normal tissues, and loss of this function defines the malignant phenotype *(28)*. The inability to integrate extracellular cues could ensure that stimuli received from the "reactive stromal matrix" would be of sufficient magnitude to induce aberrations in cell behavior and could promote tumor formation. Whether tumor suppressors function alone or collaboratively to inhibit tumor formation by acting as such cell adhesion/tumor suppressor "CATS" nodes, and if they function to integrate nuclear to cell–ECM and cell–cell signaling to maintain tissue homeostasis is an open question (*see* **Fig. 5**). However, given accumulating evidence of functional crosstalk between cell adhesion molecules and tumor suppressors, this is a realistic possibility. The fact that mutations in the tumor suppressor APC exhibit such high penetrance for colorectal cancer argues that APC may constitute one such "CATS" node protein. APC controls β-catenin function by regulating its degradation via the ubiquitin-proteosome system *(197–199)*. Loss of functional APC leads to stabilization and accumulation of β-catenin and enhanced wnt signaling, which leads to increased expression of oncogenes such as cyclin D1 and c-myc *(187,200)*. Deregulation of the APC–β-catenin–wnt pathway, either through loss or mutations of APC or mutations in β-catenin, plays an important role in colorectal cancer *(201,202)*.

The importance of strict regulation of the β-catenin cell adhesion pathway to tumor progression is underscored by studies that show that the function of β-catenin is intimately interconnected with the fate of several tumor suppressors besides APC, including E cadherin, PTEN, BRCA1, and p53 *(138,203,204)*. Work by Persad and collageues, for instance, showed that the tumor suppressor PTEN inhibits the nuclear accumulation of the β-catenin transcriptional activation complex independently of APC [204], while APC-mediated degradation of β-catenin and the tumor suppressor p53 are functionally linked via the Siah-1 protein *(205)*.

Although the APC gene is expressed in most tissues including the lung, liver, kidney, and mammary gland, individuals with germline mutations in APC rarely develop tumors other than colorectal and thyroid carcinomas *(16)*. This indicates that proteins involved in the epigenetic regulation of gene expression, such as methylases and histone acetylases, might influence the tissue-specific function of cell adhesion plaque/architectural transcription factors such as β-catenin *(206,207)*. In this regard, p300 was recently shown to interact with β-catenin and to facilitate its transcriptional activity *(208)*. p300 was first identified as an E1A-interacting protein and later shown to function as a transcriptional coactivator and to possess histone acetyl transferase activity (reviewed in **refs.** *209* and *210*). Because p300 can bind to several known tumor suppressors including p53, Smads 2, 3, and 4, BRCA1 and pRb, and the cell adhesion plaque protein β-catenin, this indicates that p300 retains the potential to profoundly affect cell behavior and may function as a cooperative tumor suppressor *(208,211–215)*. Indeed, mice heterozygous for p300 exhibit multiple developmental abnormalities (reviewed in **ref.** *214*).

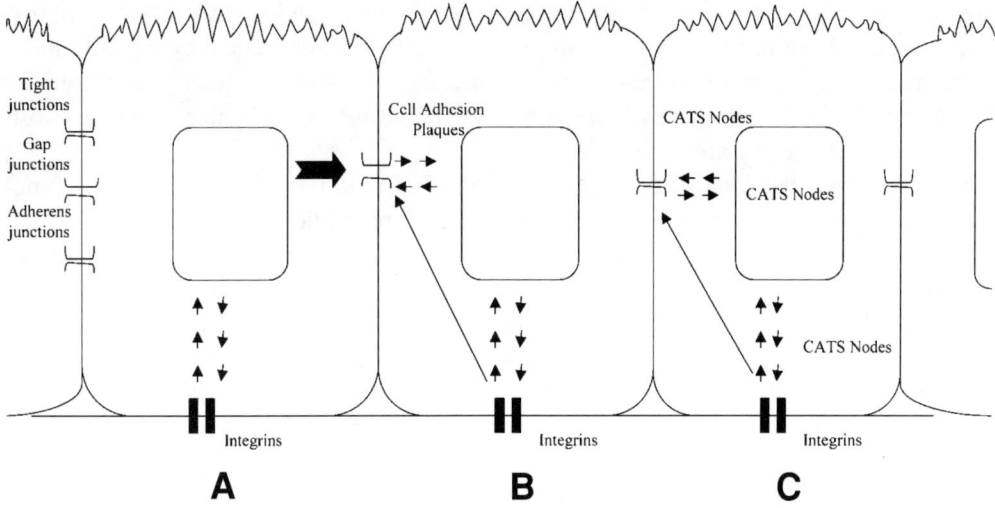

Fig. 5. A dynamic and reciprocal signaling nexus between the nucleus and sites of cell–ECM (**A**) and cell–cell adhesion (**B**) regulates tissue homeostasis in multicellular organs in vivo, and the integrity of this circuit is maintained and coordinated by cell adhesion/tumor suppressor molecules or "CATS" nodes (**C**). Mutations in or loss of critical CATS nodes molecules therefore disrupt the fidelity of this nexus and compromise the cells' ability to coordinately sense and integrate extracellular cues, leading to the loss of tissue homeostasis and alterations in tissue architecture, and enhanced susceptibility to (micro)environmental tumor promoters.

Data now show that p300 and integrin-mediated cell spreading are functionally linked and that p300 regulates ECM expression. For example, Eid and co-workers (2000) demonstrated that interactions between the protooncogene SYT and p300 promoted α5β1 integrin-mediated cell adhesion and spreading on fibronectin *(216)*, while over-expression of p300 repressed laminin A3 and C2 gene expression (components of laminin 5) in MCF10A human MECs *(217)*. Therefore, p300 embodies the essential qualities of a potential cell adhesion/tumor suppressor "CATS" node protein. Because studies have shown that the co-repressor protein CtBp can regulate the malignant phenotype by altering cell adhesion, this suggests that other chromatin modifier/coactivator proteins might also function as novel "CATS" nodes *(218)*.

Chromo modulating proteins alter chromatin structure and regulate the expression of multi gene families and loci. Trithorax (trx) and polycomb (Pc) genes constitute one group of chromo modulating proteins that regulate the expression of homeobox (hox) genes during development *(191)*. Spatially restricted expression of homeobox (hox) genes in turn determines position specific identities in the embryo *(219)*. Trx and Pc genes also stimulate enhancer-dependent gene expression and establish gene silencing after development has ceased, and hox genes regulate tissue behavior in the adult organism. Accumulating evidence attests to the existence of a dynamic and reciprocal link between integrins and hox genes, and data now indicate that this relationship may have significance for tissue homeostasis and tumor progression (M. Unger, in preparation). During angiogenesis, for example, hox D3 regulates expression of αvβ3 integrin and urokinase plasminogen activator to drive ECM adhesion, invasion, and migration of endothelial cells, and chronic overexpression of hox D3 leads to endotheliomas *(220)*. However, these effects are normally countered by the actions of hox B3, which promotes capillary morphogenesis by altering expression of the ephrin A1 ligand *(221)*. Similarly, hox A1 and B4 are induced in the mammary gland following pregnancy when branching morphogenesis and ECM remodeling are present, but are repressed by cues from the BM when MECs differentiate *(222)*. In breast tumors where BM interactions are lost, however, hox A1 levels are high *(223)*.

Corroborating the idea that hox genes might facilitate tissue homeostasis by functioning as cell adhesion/tumor suppressor proteins are studies that have examined the role of cell–ECM–hox gene regulation in intestinal cells. For example, cdx-1 increases integrin signaling via Ras, Rho, and PI3 kinase and induces anchorage-independent growth and survival in intestinal epithelial cells *(224)*. However, cdx-1's effects are countered by laminin-1 (a BM component)-induced expression of cdx2, which induces differentiation by modifying expression of cell–cell and cell–ECM molecules including β4 integrin, E cadherin, and ICAM-1 *(225,226)*. Whether such a dynamic and reciprocal regulatory loop exists for other hox genes or chomatin modifying/co-activator proteins has yet to be determined. Regardless, since trx, Pc, and hox genes appear to mediate their effects on tissue homeostasis via coordination between repressor and inducer effects, it is possible that an imbalance in these circuits could lead to malignancy and alterations in cell adhesion. Thus chromo repressor proteins might constitute a new class of tumor suppressor protein. In conclusion, exploring potential functional links between cell adhesion and chromo regulatory proteins may lead to the identification of a novel class of tumor suppressor or "CATS" node proteins, whose function is to integrate extracellular cues so that gene expression and cell behavior can be efficiently regulated within the context of the multicellular tissue.

Acknowledgments

We wish to thank Associate Professor J. Garlick, Department of Oral Biology and Pathology, School of Dental Medicine, Westchester Hall, SUNY at Stony Brook, Stony Brook, New York 11794-8702, for contributing material for **Fig. 2,** and Assistant Professor S. Lelièvre, Basic Medical Sciences Department, Purdue University, West Lafayette, IN 4737-1246, for contributing material for **Fig. 4.** This work was supported by funds from the National Cancer Institute (CA 78731) and the DOD Breast Cancer Research Program (DAMD1701-1-0368) and the American Cancer Society (IRG-78-002-23) to V.M.W. and the Susan G. Komen Association (000284) to M.U.

References

1. Hwang, E. and Esserman, L. (1999) Management of ductal carcinoma in situ. *Surg. Clin. N. Am.* **79**(5), 1007–1030.
2. Bergers, G. and Coussens, L. (2000) Extrinsic regulators of epithelial tumor progression: metalloproteinases. *Curr. Opin. Genet. Dev.* **10**, 120–127.
3. Demicheli, R., Retsky, M. W., Swartzendruber, D. E, and Bonadonna, G. (1997) Proposal for a new model of breast cancer metastatic development. *Ann. Oncol.* **8**(11), 1075–1080.
4. Riethmuller, G. and Klein, C. (2001) Early cancer cell dissemination and late metastatic relapse: clinical reflections and biological approaches tot he dormancy problem in patients. *Semin. Cancer Biol.* **11**(4), 307–311.
5. Tsao, H., Cosimi, A., and Sober, A. (1997) Ultra-late recurrence (15 years or longer) of cutaneous melanoma. *Cancer* **79**(12), 2361–2370.

6. Chambers, A., Naumov, G. N., Varghese, H. J., Nadkami, K. V., MacDonald, I. C., and Groom, A. C. (2001) Critical steps in hematogenous metastasis: an overview. *Surg. Oncol. Clin. N. Am.* **10**(2), 243–255.

7. Hart, I. (1999) Perspective: tumour spread—the problems of latency. *J. Pathol.* **187**(1), 91–94.

8. Kostler, W., Brodowicz, T., Hejna, M., Wiltschke, C., and Zielinski, C. C. (2000) Detection of minimal residual disease in patients with cancer: a review of techniques, clincal implications, and emerging therapeutic consequences. *Cancer Detect. Prev.* **24**(4), 376–403.

9. Kinzler, K. W. and Vogelstein, B. (1996) Lessons from hereditary colorectal cancer. *Cell* **87**, 159–170.

10. Hanahan, D. and Weinberg, R. (2000) The hallmark of cancer. *Cell* **100**, 57–70.

11. Compagni, A. and Christofori, G. (2000) Recent advances in research on multistage tumorigenesis. *Br. J. Cancer* **83**(1), 1–5.

12. Weinberg, E. (1989) Oncogenes, antioncogenes, and the molecular bases of multistep carcinogenesis. *Cancer Res.* **49**, 3713–3721.

13. Farber, E. (1984) The multistep nature of cancer development. *Cancer Res.* **44**, 4217–4223.

14. Ponder, B. (2001) Cancer genetics. *Nature* **411**(6835), 336–341.

15. Macleod, K. (2000) Tumor suppressor genes. *Curr. Opin. Genet. Dev.* **10**, 81–93.

16. Fearnhead, N., Britton, M., and Bodmer, W. (2001) The ABC of APC. *Hum. Mol. Genet.* **10**(7), 721–733.

17. Kinzler, K. W. and Vogelstein, B. (1997) Cancer-susceptibility genes. Gatekeepers and caretakers [news; comment]. *Nature* **386**(6627), 761–763.

18. Kinzler, K. W. and Vogelstein, B. (1998) Landscaping the cancer terrain. *Science* **280**, 1036–1037.

19. Easton, D. F., Bishop, D. T., Ford, D., and Crockford, G. P. (1993) Genetic linkage analysis in familial breast and ovarian cancer: results from 214 families. The Breast Cancer Linkage Consortium. *Am. J. Hum. Genet.* **52**(4), 678–701.

20. Struewing, J. P., Tarone, R. E., Brody, L. C., Li, F. P., and Boice, J. D., Jr. (1996) *BRCA1* mutations in young women with breast cancer [letter]. *Lancet* **347**(9013), 1493.

21. Fodor, F. H., Weston, A., and Bleiweiss, I. J. (1998) Frequency and carrier risk associated with common *BRCA1* and *BRCA2* mutations in Ashkenazi Jewish breast cancer patients. *Am. J. Hum. Genet.* **63**(1), 45–51.

22. Maillet, P., Chappuis, P. O., Vaudan, G., et al. (2000) A polymorphism in the ATM gene modulates the penetrance of hereditary non-polyposis colorectal cancer. *Int. J. Cancer* **88**(6), 928–931.

23. Alonso, J., Garcia-Miguel, P., Abelairas, J., et al. (2001) Spectrum of germline RB1 gene mutations in Spanish retinoblastoma patients: phenotypic and molecular epidemological implications. *Hum. Mutat.* **17**(5), 412–422.

24. Ahmad, N., et al. (1999) A possible hot spot in exon 21 of the retinoblastoma gene predisposing to a low penetrant retinoblastoma phenotype? *Ophthal. Genet.* **20**(4), 225–231.

25. Baylin, S., Esteller, M., Rountree, M. R., Bachman, K..E, Schuebel, K., and Herman, J. G. (2001) Aberrant patterns of DNA methylation, chromatin formation and gene expression in cancer. *Hum. Mol. Genet.* **10**(7), 687–692.

26. Elenbaas, B., Spiro, L., Koerner, F., et al. (2001) Human breast cancer cells generated by oncogenic transformation of primary mammary epithelial cells. *Genes Dev.* **15**(1), 50–65.

27. Schor, S. and Schor, A. (2001) Phenotypic and genetic alterations in mammary stroma: implications for tumour progression. *Breast Cancer Res.* **3**, 373–379.

28. Radisky, D., Hagios, C. and Bissell, M. (2001) Tumors are unique organs defined by abnormal signaling and context. *Semin. Cancer Biol.* **11**, 87–95.

29. Bissell, M., Weaver, V. M. Lelievre, S. A., Wang, F., Peterson, O. W., and Schmeichel, K. L. (1999) Tissue structure, nuclear organization, and gene expression in normal and malignant breast. *Cancer Res.* **59**(7 suppl.), 1757–1763.

30. Barcellos-Hoff, M. (2001) It takes a tissue to make a tumor: epigenetics,cancer and the microenvironment. *J. Mammary Gland Biol. Neoplasia* **6**(2), 213–221.
31. Coussens, L. and Werb, Z. (2001) Inflammatory cells and cancer: think different! *J. Exp. Med.* **193**(6), F23–F26.
32. Liotta, L. and Kohn, E. (2001) The microenvironment of the tumour-host interface. *Nature* **411**, 375–379.
33. Park, C., Bissell, M. J., and Barcellos-Hoff, M. H. (2000) The influence of the microenvironment on the malignant phenotype. *Mol. Med. Today* **6**, 324–329.
34. Ronnov-Jessen, L., Peterson, O., and Bissell, M. (1996) Cellular changes involved in the conversion of normal to malignant breast: importance of the stromal reaction. *Physiol. Rev.* **76**(1), 69–125.
35. Sternlicht, M., Lochter, A., Sympson, C. J., et al. (1996) The stromal proteinase MMP3/stromelysin-1 promotes mammary carcinogenesis. *Cell* **98**(2), 137–146.
36. Barcellos-Hoff, M. and Ravani, S. (2000) Irradiated mammary gland stroma promotes the expression of tumorigenesis potential by unirradiated epithelial cells. *Cancer Res.* **60**(5), 1254–1260.
37. Tlsty, T. and Hein, P. (2001) Know thy neighbor: stromal cells can contribute oncogenic signals. *Curr. Opin. Genet. Dev.* **11**, 54–59.
38. Tlsty, T. (2001) Stromal cells can contribute oncogenic signals. *Cancer Biol.* **11**, 97–104.
39. Coussens, L., Tinkle, C. L., Hanahan, D., and Werb, Z. (2000) MMP-9 supplied by bone marrow-derived cells contributes to skin carcinogenesis. *Cell* **103**(3), 481–490.
40. Weaver, V., Petersen, O. W., Wang, F., et al. (1997) Reversion of the malignant phenotype of human breast cells in three-dimensional culture and in vivo by integrin blocking antibodies. *J. Cell Biol.* **137**, 231–245.
41. Zutter, M., Santoro, S. A., Staatz, W. D., and Tsung, Y. L. (1995) Re-expression of the alpha 2 beta 1 integrin abrogates the malignant phenotype of breast carcinoma cells. *Proc. Natl. Acad. Sci. USA* **92**(16), 7411–7415.
42. Clegg, D., Mullick, L. H., Wingerd, K. L., et al. (2000) Adhesive events in retinal development and function: the role of integrin receptors. *Results Probl. Cell Differ.* **31**, 141–156.
43. Davies, J. and Davey, M. (1999) Collecting duct morphogenesis. *Pediatr. Nephrol.* **13**(6), 535–541.
44. Anderson, G., Harman, B. C., Hare, K. J., and Jenkinson, E. J. (2000) Microenvironmental regulation of T cell development in the thymus. *Semin Immunol.* **12**(5), 457–464.
45. Pires Neto, M., Braga-de-Souza, S., and Lent, R. (1999) Extracellular matrix molecules play diverse roles in the growth and guide central nervous system axons. *Braz. J. Med. Biol. Res.* **32**(5), 633–638.
46. Zagris, N. (2001) Extracellular matrix in development of the early embryo. *Micron* **32**(4), 427–438.
47. De Arcangelis, A., and George-Labouesse, E. (2000) Integrin and ECM functions: roles in vertebrate development. *Trends Genet.* **16**(9), 389–395.
48. Tarone, G., Hirsch, E., and Brancaccio, M. (2000) Integrin function and regulation in development. *Int. J. Dev. Biol.* **44**(6 spec. no.), 725–731.
49. Sakakura, T., Nishizuka, Y., and Dawe, C. (1976) Mesenchyme-dependent morphogenesis and epithelium-specific cytodifferentiation in mouse mammary gland. *Science* **19**, 1439–1441.
50. Hayward, S., Cunha, G., and Dahiya, R. (1996) Normal development and carcinogenesis of the prostate, a unifying hypothesis. *Basis Cancer Management* **784**, 50–63.
51. Haigios, C., Lochter, A., and Bissell, M. J. (1998) Tissue architecture: the ultimate regulator of epithelial function? *Phil. Trans. R. Soc. Lond., B: Biol. Sci.* **353**, 857–870.

52. Hoffman, M., Kibbey, M. C., Letterio, J. J., and Kleinman, H. K. (1996) Role of laminin-1 and TGF-beta 3 in acinar differentiation of a human submandibular gland cell line (HSG). *J. Cell Sci.* **109**(8), 2013–2021.

53. Boudreau, N. and Bissell, M. (1996) Regulation of gene expression by the extracellular matrix, in *Extracellular Matrix; Molecular Components and Interactions* (Comper, E., ed.). Harwood, Amsterdam, pp. 246–261.

54. Weaver, V., Fischer, A. H., Petersen, O. W., and Bissell, M. J. (1996) The importance of the microenvironment in breast cancer progression: recapitulation of mammary tumorigenesis using a unique human mammary epithelial cell model and a three-dimensional culture assay. *Biochem. Cell Biol.* **74**, 833–851.

55. Howe, A., Aplin, A. E., Alahari, S. K., and Juliano, R. L. (1998) Integrin signaling and cell growth control. *Curr. Opin. Cell Biol.* **10**, 220–231.

56. Martins-Green, M. and Bissell, M. (1995) (Cell-ECM interactions in development. *Semin. Dev. Biol.* **6**, 149–159.

57. Schwarzbauer, J. and Sechler, J. (1999) Fibronectin fibrillogenesis: a paradigm for extracellular matrix assembly. *Curr. Opin. Cell Biol.* **11**, 622–627.

58. Hynes, R. and Hodivala-Dilke, K. (1999) Insights and questions arising from studies of a mouse model of Glanzmann thrombasthenia. *Thromb. Haemost.* **82**(2), 481–485.

59. Henry, M. and Campbell, K. (1999) Dystroglycan inside and out. *Curr. Opin. Cell Biol.* **11**(5), 602–607.

60. Rapraeger, A. (2000) Syndecan-regulated receptor signaling. *J. Cell Biol.* **149**(5), 1143–1155.

61. Boudreau, N. and Bissell, M. (1998) Extracellular matrix signaling: integration of form and function in normal and malignant cells. *Curr. Opin. Cell Biol.* **10**, 640–646.

62. De Arcangelis, A. and Georges-Labouesse, E. (2000) Integrins and ECM functions: roles in vertebrate development. *TIG* **16**(9), 389–395.

63. Hynes, R. and Zhao, Q. (2000) The evolution of cell adhesion. *J. Cell Biol.* **150**(2), F89–F95.

64. Ashkenas, J., Damsky, C. H., Bissell, M. J., and Werb, Z. (1994) Integrins, signaling, and the remodeling of the extracellular matrix, in *Integrins*. Academic Press, New York, 79–109.

65. Brown, N. (2000) Cell-cell adhesion via the ECM: integrins genetics in fly and worm. *Matrix Biol.* **19**(3), 191–201.

66. Aszodi, A., Bateman, J. F., Gustafsson, E., Boot-Handford, R., and Fassler, R. (2000) Mammalian skeletogenesis and extracellular matrix: what can we learn from knockout mice? *Cell Struct. Funct.* **25**(2), 73–84.

67. Gustafsson, E. and Fassler, R. (2000) Insights into the extracellular matrix functions from the mutant mouse models. *Exp. Cell Res.* **261**, 52–68.

68. Critchley, D. (2000). Focal adhesions-the cytoskeletal connection. *Curr. Opin. Cell Biol.* **12**(1), 133–139.

69. Faull, R. and Ginsberg, M. (1996) Inside-out signaling through integrins. *J. Am. Soc. Nephrol.* **7**(8), 1091–1097.

70. Kumar, C. (1998) Signaling by integrin receptors. *Oncogene* **17**(11), 1365–1373.

71. Danen, E., Lafrenie, R. M., Miyamoto, S., and Yamada, K. M. (1998) Integrin signaling: cytoskeletal complexes, MAP kinase activation, and regulation of gene expression. *Cell Adhes. Commun.* **6**(2–3), 217–224.

72. LaFlamme, S. and Auer, K. (1996) Integrin signaling. *Semin. Cancer Biol.* 7, 111–118.

73. Giancotti, F. and Ruoslahti, E. (1999) Integrin signaling. *Science* **285**(5430), 1028–1032.

74. Brown, E. and Frazier, W. (2001) Integrin-associated protein (CD47) and its ligands. *Trends Cell Biol.* **11**(3), 130–135.

75. Brown, D. and London, E. (1998) Functions of lipid rafts in biological membranes. *Annu. Rev. Cell Dev. Biol.* **14**, 111–136.

76. Maecker, H., Todd, S., and Levy, S. (1997) The tetraspanin superfamily: molecular facilitators. *FASEB J.* **11**(6), 428–442.

77. Burack, W. and Shaw, A. (2000) Signal transduction: hanging on a scaffold. *Curr. Opin. Cell Biol.* **12**, 211–216.

78. Ingber, D. (1998) In search of cellular control: signal transduction in context. *J. Cell. Biochem. Suppl.* **30–31**, 232–237.

79. Huang, S. and Ingber, D. (2000) Shape-dependent control of cell growth. differentiation, and apoptosis: switching between attractors in cell regulatory networks. *Exp. Cell Res.* **261**(1), 91–103.

80. Assoian, R. and Schwartz, M. (2001) Coordinate signaling by integrins and receptor tyrosine kinases in the regulation of G1 phase cell-cycle progression. *Curr. Opin. Genet. Dev.* **11**(1), 48–53.

81. Aplin, A., Howe, A., and Juliano, R. (1999) Cell adhesion molecules, signal transduction and cell growth. *Curr. Opin. Cell Biol.* **11**, 737–744.

82. Wang, F., Weaver, V. M., Petersen, O. W., et al. (1998) Reciprocal interactions between beta1-integrin and epidermal growth factor receptor in three-dimensional basement membrane breast cultures: a different perspective in epithelial biology. *Proc. Natl. Acad. Sci. USA* **95**(25), 14821–14826.

83. Muschler, J., Lochter, A., Roskelley, C. D., Yurchenco, P., and Bissell, M. J. (1999) Division of labor among the alpha6beta4 integrin, beta1 integrins, and an E3 laminin receptor to signal morphogenesis and beta-casein expression in mammary epithelial cells. *Mol. Biol. Cell* **10**(9), 2817–2828.

84. Seljelid, R., Jozefowski, S., and Sveinbjornsson, B. (1999) Tumor stroma. *Anticancer Res.* **19**(6A), 4809–4822.

85. Lochter, A. and Bissell, M. (1995) Involvement of extracellular matrix constituents in breast cancer. *Semin. Cancer Biol.* **6**(3), 165–173.

86. Walker, R. (2001) The complexities of breast cancer desmoplasia. *Breast Cancer Res.* **3**, 143–145.

87. Ronnov-Jessen, L., Petersen, O. W., Koteliansky, V. E., and Bissell, M. J. (1995) The origin of the myofibroblasts in breast cancer. Recapitulation of tumor environment in culture unravels diversity and implicates converted fibroblasts and recruited smooth muscle cells. *J. Clin. Invest.* **95**(2), 859–873.

88. Dolberg, D., Hollingsworth, R., Hertle, M., and Bissell, M. J. (1985) Wounding and its role in RSV-mediated tumor formation. *Science* **230**(4726), 676–678.

89. Sieweke, M., Stoker, A., and Bissell, M. (1989) Evaluation of the cocarcinogenic effect of wounding in Rous sarcoma virus tumorigenesis. *Cancer Res.* **49**(22), 6419–6424.

90. Gailit, J. and Clark, R. (1994) Wound repair in the context of extracellular matrix. *Curr. Opin. Cell Biol.* **6**(5), 717–725.

91. Eckes, B., Zigrino, P., Kessler, D., et al. (2000) Fibroblast-matrix interactions in wound healing fibrosis. *Matrix Biol.* **19**(4), 325–332.

92. Sieweke, M. and Bissell, M. (1994) The tumor-promoting effect of wounding: a possible role for TGF-beta in stromal alterations. *Crit. Rev. Oncogene.* **5**(2–3), 297–311.

93. Sieweke, M., Thompson, N. L., Sporn, M. B., and Bissell, M. J. (1990) Mediation of wound-related Rous sarcoma virus tumorigenesis by TGF-beta. *Science* **248**(4963), 1656–1660.

94. Martins-Green, M. and Bissell, M. (1990) Localization of 9E3/CEF-4 in avain tissues: expression is absent in Rous Sarcoma virus-induced tumors but is stimulated by injury. *J. Cell Biol.* **110**(3), 581–595.

95. Cunha, G. (1994) Role of mesenchymal-epithelial interactions in normal and abnormal development of the mammary gland and prostate. *Cancer* **74**(3 suppl.), 1030–1044.

96. Yang, J., Guzman, R., and Nandi, S. (2000) Histomorphologically intact primary human breast lesions and cancers can be propagated in nude mice. *Cancer Lett.* **159**, 205–210.

97. Schor, S. L. (1995) Fibroblast subpopulations as accelerators of tumor progression: the role of migration stimulating factor. *EXS* **74**, 273–296.
98. Olumi, A., Grossfeld, G. D., Hayward, S. W., Carroll, P. R., Tisty, T. D., and Cunha, G. R. (1999) Carcinoma-associated fibroblasts direct tumor progression of initiated human prostatic epithelium. *Cancer Res.* **59**(19), 5002–5011.
99. Elenbaas, B. and Weinberg, R. (2001) Heterotypic signaling between epithelial tumor cells and fibroblasts in carcinoma formation. *Exp. Cell Res.* **264**(1), 169–184.
100. Kleinman, H., Kablinski, J., Lee, S., and Engbring, J. (2001) Role of basement membrane in tumor growth and metastasis. *Surg. Oncol. Clin. N. Am.* **10**(2), 329–338, ix.
101. Bemis, L. and Schedin, P. (2000) Reproductive state of rat mammary gland stroma modulates human breast cancer cell migration and invasion. *Cancer Res.* **60**(13), 3414–3418.
102. Ruoslahti, E. (1999) Fibronectin and its integrin receptors in cancer. *Adv. Cancer Res.* **76**, 1–20.
103. Ivaska, J. and Heino, J. (2000) Adhesion receptors and cell invasion: mechanisms of integrin-guided degradation of extracellular matrix. *Cell. Mol. Life* **57**(1), 16–24.
104. Weaver, V. and Bissell, M. (1999) Functional culture models to study mechanisms governing apoptosis in normal and malignant mammary epithelial cells. *J. Mammary Gland Biol. Neoplasia* **4**(2), 193–201.
105. Schmeichel, K., Weaver, V., and Bissell, M. (1998) Structural cues from the tissue microenvironment are essential determinants of the human mammary epithelial cell phenotype. *J.Mammary Gland Biol. Neoplasia* **3**(2), 201–213.
106. Robbins, S. L., Angell, M., and Kumar V. (1981) *Basic Pathology*, 3rd ed. Saunders, Philadelphia.
107. Gumbiner, B. (1996) Cell adhesion: the molecular basis of tissue architecture and morphogenesis. *Cell* **84**(3), 345–357.
108. Hagios, C., Lochter, A., and Bissell, M. (1998) Tissue architecture: the ultimate regulator of epithelial function? *Phil. Trans. R. Soc. Lond.* **353**, 857–870.
109. Yeaman, C., Grindstaff, K., and Nelson, W. (1999) New perspectives on mechanisms involved in generating epithelial cell polarization. *Physiol. Rev.* **79**(1), 73–98.
110. Lelievre, S., Weaver, V. M., Nickerson, J. A., et al. (1998) Tissue phenotype depends on reciprocal interactions between the extracellular matrix and the structural organization of the nucleus. *Proc. Natl. Acad. Sci. USA* **95**(25), 14711–14716.
111. Lohi, J., Leivo, I., Oivula, J., Lehto, V. P., and Virtanen, I. (1998) Extracellular matrix in renal cell carcinomas. *Histol. Histopathol.* **13**(3), 785–796.
112. Jones, P. (2001) Extracellular matrix and tenascin-C in pathogenesis of breast cancer. *Lancet* **357**(9273), 1992–1994.
113. De Iorio, P., Midulla, C., Pisani, T., et al. (2001) Implication of laminin and collagen type IV expression in the progression of breast carcinoma. *Anticancer Res.* **21**(2B), 1395–1399.
114. Zvibel, I., et al.(2000) The role of the liver environment in the regulation of colon cancer metastasis. *Isr. Med. Assoc. J.* **2**(1), 48–51.
115. Ellenrieder, V., Adler, G., and Gress, T. (1999) Invasion and metastatsis in pancreatic cancer. *Ann. Oncol.* **10**(suppl.), 46–50.
116. Rowley, D. (1998–1999) What might a stromal response mean to prostate cancer progression? *Cancer Metastasis Rev.* **17**(4), 411–419.
117. Catusse, C., Polette, M., Coraux, C., Burlet, H., and Birembaut, P. (2000) Modified basement membrane composition during bronchopulmonary tumor progression. *J. Histochem. Cytochem.* **48**(5), 663–669.
118. Korhonen, M., Latinen, L., Yianne, J., Gould, V. E., and Virtanen, I. (1992) Integrins in developing, normal and malignant human kidney. *Kidney Int.* **41**(3), 641–644.

119. Zutter, M., Sun, H., and Santoro, S. (1998) Altered integrin expression and the malignant phenotype: the contribution of multiple integrin receptors. *J. Mammary Gland Biol. Neoplasia* **3**(2), 191–200.

120. Shaw, L. (1999) Integrin function in breast carcinoma progression. *J. Mammary Gland Biol. Neoplasia* **4**(4), 367–376.

121. Agrez, M. and Bates, R. (1994) Colorectal cancer and the integrin family of cell adhesion receptors: current status and future. *Eur. J. Cancer* **30A**(14), 2166–2170.

122. Kemperman, H., Driessens, M. H., La Riviere, G., Meijne, A. M., and Roos, E. (1995) Adhesion mechanisms in liver metastasis formation. *Cancer Surv.* **24**, 67–79.

123. Weinacker, A., Ferrando, R., Elliott, M., Hogg, J., Balmes, J., and Sheppard, D. (1995) Distribution of integrins alpha v beta 6 and alpha 9 beta 1 and their known ligands, fibronectin, tenascin, in human airways. *Am. J. Respir. Cell Mol. Biol.* **12**(5), 547–556.

124. Petersen, O. W., Ronnov-Jessen, L., Howlett, A. R., and Bissell, M. J. (1992) Interaction with basement membrane serves to rapidly distinguish grwoth and differentiation pattern of normal and malignant human breast epithelial cells. *Proc. Natl. Acad. Sci. USA* **89**, 9064–9068.

125. Howlett, A., Petersen, O. W., Steeg, P. S., and Bissell, M. J. (1994) A novel function for the nm23-H1 gene: Overexpression in human breast carcinoma cells leads to the formation of basement membrane and growth arrest. *J. Natl. Cancer Inst.* **86**(24).

126. Briand, P., Peterson, O., and Van Deurs, B. (1987) A new diploid nontumorigenic human breast epithelial cell line isolated and propagated in chemically defined medium. *In Vitro Cell Dev Biol.* **123**(3), 181–188.

127. Briand, P., Nielsen, K. V., Madsen, M. W., and Petersen, O. W. (1996) Trisomy 7p and malignant transformation of human breast epithelial cells following epidermal growth factor withdrawal. *Cancer Res.* **56**(9), 2039–2044.

128. Nielsen, K., Niebuhr, E., Ejlertsen, B., et al. (1997) Molecular cytogenetic analysis of a nontumorigenic human breast epithelial cell line that eventually turns tumorigenic: validation of an analytical approach combining karyotyping, comparative genomic hybridization, chromosome painting, and single-locus fluorescence in sity hybridization. *Genes Chromosomes Cancer* **20**(1), 30–37.

129. Villadsen, R., Nielsen, K. V., Bolund, L., and Briand, P. (2000) Complete loss of wild-type TP53 in a nontransformed human epithelial cell line preceded by a phase during which a heterozygous TP53 mutant effectively outgrows the homozygous wild-type cells. *Cancer Genet. Cytogenet.* **116**(1), 28–34.

130. Madsen, M., Lykkesfeldt, A. E., Laursen, I., Nielsen, K. V., and Briand, P. (1992) Altered gene expression of c-myc, epidermal growth factor receptor, transforming growth factor-alpha, and c-erb-B2 in an immortalized human breast epithelial cell line, HMT-3522, is associated with decreased growth requirements. *Cancer Res.* **52**(5), 1210–1217.

131. Javaherian, A., Vaccariello, M., Fusenig, N. E., and Garlick, J. A. (1998) Normal keratinocytes suppress early stages of neoplastic progression in stratified epithelium. *Cancer Res.* **58**, 2200–2208.

132. Karen, J., Wang, Y., Javaherian, A., Vaccariello, M., Fusenig, N. E., and Garlick, J. A. (1999) 12-O-Tetradecanoylphorbol-13-acetate induces clonal expansion of potentially malignant keratinocytes in a tissue model of early neoplastic progression. *Cancer Res.* **59**, 474–481.

133. Schwartz, M. (1997) Integrins, oncogenes, and anchorage independence. *J. Cell Biol.* **139**(3), 575–578.

134. Frisch, S. and Ruoslahti, E. (1997) Integrins and anoikis. *Curr. Opin. Cell Biol.* **9**, 701–706.

135. O'Connor, P. (1997) Mammalian G1 and G2 checkpoints. *Cancer Surv.* **29**, 151–182.

136. Day, M., Foster, R. G., Day, K. C., et al. (1997) Cell anchorage regulated apoptosis through the retinoblastoma tumor suppressor/E2F pathway. *J. Biol. Chem.* **272**(13), 8125–8128.

137. Tsuboi, N., Yoshida, H., Kawamura, T., Furukawa, Y., Hosoya, T., and Yamada, H. (2000) Three-dimensional matrix suppresses E2F-controlled gene expression in glomerular mesangial cells. *Kidney Int.* **57**, 1581–1589.

138. Yu, Q., Geng, Y., and Sicinski, P. (2001) Specific protection against breast cancers by cyclin D1 ablation. *Nature* **411**, 1017–1021.

139. Ghosh, A., Yuan, W., Mori, Y., Chen, S., and Varga, J. (2001) Antagonistic regulation of type 1 collagen gene expression by interferon-gamma and transforming growth factor-beta. *J. Biol. Chem. 276*(14), 11041–11048.

140. Pasche, B. (2001) Role of transforming growth factor beta in cancer. *J. Cell. Physiol.* **186**, 153–168.

141. Hata, A. (2001) TGF-beta signaling and cancer. *Exp. Cell Res.* **264**, 111–116.

142. Zhou, S., Buckhaults, P., Zawel, L., et al. (1998) Targeted deletion of SMAD4 shoes it is required for transforming growth factor beta and activin signaling in colorectal cancer cells. *Proc. Natl. Acad. Sci. USA* **95**, 2412–2416.

143. Riggins, G., Kinzler, K. W., Vogelstein, B., and Thiagalingam, S. (1997) Frequency of Smad gene mutations in human cancers. *Cancer Res.* **57**(13), 2578–2580.

144. Dalton, S., Scharf, E., Davey, G., and Assoian, R. K. (1999) Transforming growth factor-beta overrides the adhesion requirement for surface expression of alpha(5)beta(1) integrin in normal rat kidney fibroblasts. A necessary effect for induction of anchorage independent growth. *J. Biol. Chem.* **274**(42), 30139–30145.

145. Lai, C., Feng, X., Nishimura, R., et al. (2000) Transforming growth factor-beta up-regulates the beta5 integrin subunit expression via sp1 and smad signaling. *J. Biol. Chem.* **275**(46), 36400–36406.

146. Weitzman, J. (2000) Quick guide.Jnk. *Curr. Biol.* **10**(8), R290.

147. Almeida, E., Ilic, D., Han, Q., et al. (2000) Matrix survival signaling: from fibronectin via focal adhesion kinase to c-Jun NH(2)-terminal kinase. *J. Cell Biol.* **149**(3), 741–754.

148. Scatena, M., Almeida, M., Chaisson, M. L., Fausto, N., Nicosia, R. F., and Giachelli, C. M. (1998) NF-kappaB mediates alpha five beta three integrin-induced endothelial cell survival. *J. Cell Biol.* **141**(4), 1083–1093.

149. Frisch, S., Vuori, K., Kelaita, D., and Sicks, S. (1996) A role for Jun-N-terminal kinase in anoikis; suppression by bcl-2 and crmA. *J. Cell Biol.* **135**(5), 1377–1382.

150. Mc Donald, E. R. and El-Deiry, W. (2001) Checkpoint genes in cancer. *Ann. Med.* **33**(2), 113–122.

151. Ryan, K., Phillips, A., and Vousden, K. (2001) Regulation and function of the p53 tumor suppressor protein. *Curr. Opin. Cell Biol.* **13**, 332–337.

152. Nigro, J., Aldape, K. D., Hess, S. M., and Tisty, T. D. (1997) Cellular adhesion regulates p53 protein levels in primary human keratinocytes. *Cancer Res.* **57**, 3635–3639.

153. Tlsty, T. (1998) Cell-adhesion-dependent influences on genomic instability and carcinogenesis. *Curr. Opin. Biol.* **10**(5), 647–653.

154. O'Connell, F. and Martin, F. (2000) Laminin-rich extracellular matrix association with mammary epithelial cells suppresses Brca1 expression. *Cell Death Different.* **7**(4), 360–367.

155. Marquis, S.T., Rajan, J. V., Wynshaw-Boris, A., et al. (1995) The developmental pattern of Brca1 expression implies a role in differentiation of the breast and other tissues. *Nature Genet.* **11**(1), 17–26.

156. Rajan, J.V., Wang, M., Marquis, S. T., and Chodash, L. A. (1996) Brca2 is coordinately regulated with Brca1 during proliferation and differentiation in mammary epithelial cells. *Proc. Natl. Acad. Sci. USA* **93**(23), 13078–13083.

157. Wang, A., Schneider-Broussard, R., Kumar, A. P., MacLeod, M. C., and Johnson, D. G. (2000) Regulation of BRCA1 expression by the Rb-E2F pathway. *J. Biol. Chem.* **275**, 4532–4536.

158. Satterwhite, D., Matsunami, N., and White, R. (2000) TGF-beta1 inhibits BRCA1 expression through a pathway that requires pRb. *Biochem. Biophysic. Res. Commun.* **276**, 686–692.
159. Chrenek, M., Wong, P., and Weaver, V. (2001) Integrins and cell adhesions as modulators of mammary cell survival and transformation. *Breast Cancer Res.* **3**(4), 224–229.
160. Boudreau, N., Werb, Z., and Bissell, M. (1996) Suppression of apoptosis by basement membrane requires three-dimensional tissue organization and withdrawl from the cell cycle. *Proc. Natl. Acad. Sci. USA* **93**(8), 3509–3913.
161. Welcsh, P. L., Owens, K. N., and King, M. C. (2000) Insights into the functions of BRCA1 and BRCA2. *Trends Genet.* **16**(2), 69–74.
162. Okamoto, A., Demetrick, D. J., Spillare, E. A., et al. (1994) Mutations and altered expression of p16INK4 in human cancer. *Proc. Natl. Acad. Sci. USA* **91**, 11045–11049.
163. Serrano, M. (1997) The tumor suppressor protein p16INK4A. *Exp. Cell Res.* **237**, 7–13.
164. Stolle, C., Glenn, G., Zbar, B., et al. (1998) Improved detection of germline mutations in the von Hippel-Lindau disease tumor suppressor gene. *Hum. Mutat.* **12**(6), 417–423.
165. Kondo, K. and Kaelin, W. (2001) The von Hippel-Lindau tumor suppressor gene. *Exp. Cell Res.* **264** (117–125).
166. Kamada, M., Suzuki, K., Kato, Y., Okuda, H., and Shuin, T. (2001) von Hippel-Lindau protein promotes the assembly of actin and vinculin and inhibits cell motility. *Cancer Res.* **61**, 4184–4189.
167. Plath, T., Detjen, K., Welzel, M., et al. (2000) A novel function for the tumor suppressor p16INK4a: Induction of anoikis via upregulation of the alpha-five beta-one Fibronectin Receptor. *J. Cell Biol.* **150**(6), 1467–1477.
168. Fahraeus, R. and Lane, D. (1999) The p16INK4a tumour suppressor protein inhibits alpha-v beta-3 integrin-mediated cell spreading on vitronectin by blocking PKC-dependent localization of alpha-v beta-3 to focal contacts. *EMBO J.* **18**(8), 2106–2118.
169. Romanov, S., Kozakiewicz, B. K., Holst, C. R., Stampfer, M. R., Haupt, L. M., and Tisty, T. D. (2001) Normal human mammary epithelial cells spontaneously escape senescence acquire genomic changes. *Nature* **409**(6820), 633–637.
170. Spancake, K. M., Anderson, C. B., Weaver, V. M., Matsunami, M., Bissell, M. J. and White, R. L. (1999) E7-transduced human breast epithelial cells show partial differentiation in three-dimensional culture. *Cancer Res.* **59**, 6042–6045.
171. Lim, S., Lee, H., and Lee, H. (1998) Inhibition of colonization and cell-matrix adhesion after nm23-H1 transfection of human prostate carcinoma cells. *Cancer Lett.* **133**(2), 143–149.
172. Kreidberg, J. (2000) Functions of alpha3beta1 integrin. *Curr. Opin. Cell Biol.* **12**(5), 548–553.
173. Hofmann, U., Westphal, J. R., Van Muijen, G. N., and Ruiter, D. J. (2000) Matrix metalloproteinases in human melanoma. *J. Invest. Dermatol.* **115**(3), 337–344.
174. Zhang, X., Bontrager, A., and Hemler, M. (2001) Transmembrane-4 Superfamily Proteins Associate with Activated Protein Kinase C (PKC) and Link PKC to Specific beta1 Integrins. *J. Biol. Chem.* **276**(27), 25005–25013.
175. Odintsova, E., Sgiura, T., Berditchevski, F. (2000) Attenuation of EGF receptor signaling by a metastasis suppressor, the tetraspanin CD82/KAI-1. *Curr. Biol.* **10**(16), 1009–1012.
176. Ruoslahti, E. (1994) Cell adhesion and tumor metastasis. *Princess Takamatsu Symp.* **24**, 99–105.
177. Huang, S. and Ingber, D. (1999) The structural and mechanical complexity of cell-growth control. *Nat. Cell Biol.* **1**, E131–E138.
178. Myers, C., Schmidhauser, C., Mellentin-Michelotti, J., et al. (1998) Characterization of BCE-1, a transcriptional enhancer regulated by prolactin and extracellular matrix and modulated by the state of histone acetylation. *Mol. Cell. Biol.* **18**(4), 2184–2195.
179. Somasiri, A., Howarth, A., Goswami, D., Dedhar, S., and Roskelley, C. D. (2001) Overexpression of the integrin-linked kinase mesenchymally transforms mammary epithelial cells. *J. Cell Sci.* **114**(pt. 6), 1125–1136.

180. Lampe, P., Nguyen, B. P., Gil, S., et al. (1998) Cellular interaction of integrin alpha3beta1 with laminin 5 promotes gap junctional communication. *J. Cell. Biol.* **143**(6), 1735–1747.

181. von Schlippe, M., Marshall, J. F., Perrry, P., Stone, M., Zhu, A. J., and Hart, I. R. (2000) Functional interaction between E-cadherin and alphav-containing integrin carcinoma cells. *J. Cell Sci.* **113**(pt. 3), 425–437.

182. Guilford, P., Hopkins, J. B., Grady, W. M., et al. (1998) E-cadherin germline mutations in familial gastric cancer. *Nature* **392**, 402–405.

183. Kemler, R. (1993) From cadherins to catenins: cytoplasmic protein interactions and regulation of cell adhesion. *Trends Genet.* **9**(9), 317–321.

184. Peifer, M. (1993) The product of *Drosophila* segment polarity gene armadillo is part of a multi-protein complex resembling the vertebrate adherens junction. *J. Cell Sci.* **105**(pt. 4), 993–1000.

185. Wodarz, A. and Nusse, R. (1998) Mechanisms of Wnt signaling in development. *Ann. Rev. Cell Dev. Biol.* **14**, 59–88.

186. He, T., Sparks, A. B., et al. (1998) Identification of cMyc as a target of the APC pathway. *Science* **281**, 1509–1512.

187. Tetsu, O. and McCormick, F. (1999) Beta-catenin regulates expression of cyclin D1 in colon carcinoma cells. *Nature* **398**, 422–426.

188. Novak, A., Hsu, S. C., Leung-Hagesteijn, C., et al. (1998) Cell adhesion and the integrin-linked kinase regulate the LEF-1 and beta-catenin signaling pathways. *Proc. Natl. Acad. Sci. USA* **95**(8), 4374–4379.

189. Daniel, J. and Reynolds, A. (1999) The catenin p120(ctn) interacts with Kaiso, a novel BTB/POZ domain zinc-finger transcription factor. *Mol. Cell Biol.* **19**(5), 3614–3623.

190. Balda, M. and Matter, K. (2000) The tight junction protein ZO-1 and an interacting transcription factor regulate ErbB-2 expression. *EMBO J.* **19**(9), 2024–2033.

191. Caldas, C. and Aparicio, S. (1999) Cell memory and cancer-the story of trithorax and Polycomb group genes. *Cancer Metastasis Rev.* **18**, 313–329.

192. Nakamura, T., Blechman, J., Tada, S., et al. (2000) huASH1 protein, a putative transcription factor encoded by a human homologue of the *Drosophila* ash 1 gene, localizes to both nuclei and cell-cell tight junctions. *Proc. Natl. Acad. Sci. USA* **97**(13), 7284–7289.

193. Brock, H. and van Lohuizen, M. (2001) The Polycomb group-no longer an exclusive club? *Curr. Opin. Genet. Dev.* **11**, 175–181.

194. Chen, H., Schmeichel, K. L., Mian, I. S., Lelievre, S., Petersen, O. W., and Bissell, M. J. (2000) AZU-1: a candidate breast tumor suppressor and biomarker for tumor progression. *Mol. Biol. Cell* **11**(4), 1357–1367.

195. Huang, R., Fan, Y., Hossain, M. Z., Peng, A., Zeng, Z. L., and Boynton, A. L. (1998) Reversion of the neoplastic phenotype of human glioblastoma cells by connexin 43 (cx43). *Cancer Res.* **58**, 5089–5096.

196. Doyle, J., Stempak, J. G., Cowin, P., Colman, D. R., and D'Urso, D. (1995) Protein zero, a nervous system adhesion molecule, triggers epithelial reversion in host carcinoma cells. *J. Cell Biol.* **131**(2), 465–482.

197. Aberle, H., Bauer, A., Stappert, J., Kispert, A., and Kemler, R. (1997) Beta-catenin is a target for the ubiquitin-proteosome pathway. *EMBO J.* 1997. **16**(13), 3797–3804.

198. Orford, K., Crokett, C., Jensen, J. P., Weissman, A. M., and Byers, S. W. (1997) Serine phosphorylation-regulated ubiquitination and degradation of beta-catenin. *J. Biol. Chem.* **272**(40), 24735–24738.

199. Salomon, D., Sacco, P. A., Roy, S. G., et al. (1997) Regulation of beta-catenin levels and localization by overexpression of plakoglobin and inhibition of the ubiquitin-proteosome system. *J. Cell Biol.* **139**(5), 1325–1335.

200. Munemitsu, S., Albert, I., Souza, B., Rubinfeld, B., and Polakis, P. (1995) Regulation of intracellular beta-catenin levels by the adenomatous polyposis coli (APC) tumor-suppressor protein. *Proc. Natl. Acad. Sci. USA* **192**, 3046–3050.

201. Morin, P., Sparks, A. B., Korinek, V., et al. (1997) Activation of beta-catenin-Tcf signaling in colon cancer by mutation in beta-catenin or APC. *Science* **275**, 1787–1790.

202. Barth, A., Nathke, I., and Nelson, W. (1997) Cadherins, catenins and APC protein: interplay between cytoskeletal complexes and signaling pathways. *Curr. Opin. Cell Biol.* **9**(5), 683–690.

203. Day, M., Zhao, X., Vallorosi, C. J., et al. (1999) E-cadherin mediates aggregation-dependent survival of prostate and mammary epithelial cells through the retinoblastoma cell cycle control pathway. *J. Biol. Chem.* **274**(14), 9656–9664.

204. Persad, S., Troussard, A. A., McPhee, T. R., Mulholland, D. J., and Dedhar, S. (2001) Tumor suppressor PTEN inhibits nuclear accumulation of beta-catenin and T cell/lymphoid enhancer factor 1-mediated transcriptional activation. *J. Cell Biol.* **153**(6), 1161–1173.

205. Liu, J., Stevens, J., Rote, C. A., et al. (2001) Siah-1 mediates a novel beta-catenin degradation pathway linking p53 to the adenomatous polyposis coli protein. *Mol. Cell* **7**, 927–936.

206. Wolfee, A. and Matzke, M. (1999) Epigenetics: regulation through repression. *Science* **286**, 5439.

207. Cheung, W., Briggs, S., and Allis, C. (2000) Acetylation and chromosomal functions. *Curr. Opin. Cell Biol.* **12**(3), 326–333.

208. Sun, Y., Kolligs, F. T., Hottiger, M. O., Mosavin, R., Fearon, E. R., and Nabel, G. J. (2000) Regulation of beta-catenin transformation by the p300 transcriptional coactivator. *Proc. Natl. Acad. Sci. USA* **97**(23), 12613–12618.

209. Marmorstein, R. (2001) Structure and function of histone acetyltransferases. *Cell. Mol. Life Sci.* **58**(5–6), 693–703.

210. Lee, J., Lee, Y. C., Na, S. Y., Jung, D. J., and Lee, S. K. (2001) Transcriptional coregulators of the nuclear receptor superfamily: coactivators and corepressors. *Cell. Mol. Life Sci.* **58**(2), 289–297.

211. Pao, G., Janknecht, R., Ruffner, H., Hunter, T., and Verma, I. M. (2000) CBP/p300 interact with and function as transcriptional coactivators of BRCA1. *Proc. Natl Acad. Sci. USA* **97**(3), 1020–1025.

212. Miyagishi, M., Fujii, R., Hatta, M., et al. (2000) Regulation of Lef-mediated transcription and p53-dependent pathway by associating beta-catenin with CBP/p300. *J. Biol.Chem.* **275**(45), 35170–35175.

213. Ali, S. and DeCaprio, J. (2001) Cellular transformation by SV40 large T antigen: interaction with host proteins. *Semin. Cancer Biol.* **11**(1), 15–23.

214. Goodman, R. and Smolik, S. (2000) CBP/p300 in cell growth/transformation, and development. *Genes Dev.* **14**(13), 1553–1577.

215. de Caestecker, M., Yahata, T., Wang, D., et al. (2000) The Smad4 activation domain (SAD) is a proline-rich, p300-dependent transcriptional activation domain. *J. Biol. Chem.* **275**(3), 2115–2122.

216. Eid, J., Kung, A. L., Scully, R., and Livingston, D. M. (2000) p300 interacts with the nuclear proto-oncoprotein SYT as part of the active control of cell adhesion. *Cell* **102**(6), 839–848.

217. Miller, K., Chung, J., Lo, D., Jones, J. C., Thimmapaya, B., and Weitzman, S. A. (2000) Inhibition of laminin-5 production in breast epithelial cells by overexpression of p300. *J. Biol. Chem.* **275**(11), 8176–8182.

218. Grooteclaes, M. and Frisch, S. (2000) Evidence for a function of CtBP in epithelial gene regulation and anoikis. *Oncogene* **19**(33), 3823–3828.

219. Cillo, C., Cantile, M., Faiella, A., and Boncinelli, E. (1999) Homeobox genes and cancer. *Exp. Cell Res.* **248**(1), 1–9.

220. Boudreau, N., Andrews, C., Srebronx, A., Ravanpay, A., and Cheresh, D. A. (1997) Hox D3 induces an angiogenic phenotype. *J. Cell Biol.* **139**, 257–264.

221. Myers, C., Charboneau, A., and Boudreau, N. (2000) Homeobox B3 promotes capillary morphogenesis and angiogenesis. *J. Cell Biol.* **148**(2), 343–351.

222. Srebrow, A., Friedman, Y., Ravanpay, A., Daniel, C. W., and Bissell, M. J. (1998) Expression of HOXA-1 and HOXB-7 is regulated by ECM dependent signals in mammary epithelial cells. *J. Cell Biochem.* **69**(4), 377–391.

223. Chariot, A. and Castronova, V. (1996) Detection of HOXA1 expression in human breast cancer. *Biochem. Biophys. Res. Commun.* **222**(2), 292–297.

224. Soubeyran, P., Haglund, K., Garcia, S., Barth, B. U., Iovanna, J., and Dikic, I. (2001) Homeobox gene Cdx1 regulates Ras, Rho, and PI3 kinase pathways leading to transformation and tumorigenesis of intestinal epithelial cells. *Oncogene* **20**(31) 4180–4187.

225. Soubeyran, P., Mallo, G. V., Moucadel, V., Dagom, J. C., and Iovanna, J. L. (2000) Overexpression of Cdx1 and Cdx2 homeogenes enhances expression of HLA-I in HT-29 cells. *Mol. Cell. Biol. Res. Commun.* **3**(5), 271–276.

226. Lorentz, O., Duluc, I., Arcangelis, A. D., Simon-Assmann, P., Kedinger, M., and Freund, J. M. (1997) Key role of the Cdx2 homeobox gene in extracellular matrix-mediated in cell differentiation. *J. Cell Biol.* **139**(6), 1553–1565.

25

Analyzing the Function of Tumor Suppressor Genes Using a *Drosophila* Model

Raymond A. Pagliarini, Ana T. Quiñones, and Tian Xu

1. Introduction

With 1600 eyes, a pair of antennae, 6 legs, and an open circulatory system, the fruit fly *Drosophila melanogaster* may seem an unlikely model for the host of pathologies resulting from human cancers. However, the results of a century of research in *Drosophila* only accents the fundamental similarities between many biologic processes in both flies and humans. And as genetic analysis in yeast lent crucial insights into the conserved mechanisms of cell division and cell cycle control *(1,2)*, genetic studies in a relatively simple multicellular organism such as *Drosophila* can help us understand how mutations in tumor suppressor genes and oncogenes affect organs and tissues, and also help us to find new genes functioning in the processes related to cancer biology. The goal of this chapter is to review how one can use *Drosophila* as a model to study the functions of tumor suppressor or oncogene homologs, and to identify novel genes involved in tumorigenic processes. We discuss why *Drosophila* is a relevant model for cancer development in mammals, and why studies in *Drosophila* offer advantages over a number of other model systems. We review the history of studying cancer in *Drosophila*, and explain the powerful genetic techniques that allow for refined in-vivo studies of cancer-causing genes.

1.1. Conservation of Genes and Pathways Between Flies and Humans

Nearly a century of research in *Drosophila* has led to the identification of numerous genes and pathways in various developmental and cellular processes. Since many of these genes are common to higher eukaryotes, research in the fruit fly often provides initial insight into studies of their mammalian counterparts. Indeed, preliminary analysis of the *Drosophila* and human genome projects revealed that humans and flies share more than 40% of their genes. In addition, a comprehensive survey of human disease genes has found that 62% of human disease genes appear to be conserved in *Drosophila* *(3)*. This includes genes implicated in cancer, cardiovascular, neurologic, and endocrine disease as well as those involved in renal, innate immunity, metabolic, and hematologic diseases.

From: *Methods in Molecular Biology, Vol. 223: Tumor Suppressor Genes: Regulation, Function, and Medicinal Applications.* Edited by: Wafik S. El-Deiry © Humana Press Inc., Totowa, NJ

Similarity does not stop with gene identities. Several processes important to the development of cancer in humans are well conserved from flies to humans. For example, studies of *Drosophila* embryonic patterning have elucidated signaling pathways such as the Wingless and Hedgehog pathways, which have turned out to be important to the understanding of human colon and skin cancers, respectively *(4–6)*. In humans, mutations in the orthologue of the *ptc* gene, PTCH, cause nevoid basal cell carcinoma syndrome (NBCCS). NBCCS patients develop multiple basal cell carcinomas as well as medulloblastomas and jaw cysts *(7)*. The *ptc* gene was first identified from a screen for genes essential during embryonic development *(8)*. The interdependent secretions of the morphogens wingless and hedgehog control this process. Interestingly, the mammalian homologs of a number of genes in both the wingless and hedgehog pathways have also been shown to play critical roles in the development of tumors. In addition to this example, the cell cycle of some *Drosophila* tissues contains checkpoints similar to those in humans, and is controlled by a set of cyclins and their cyclin-dependent kinase partners, as well as by key cell cycle regulators such as Rb and E2F *(9–11)*. The study of cell growth in *Drosophila* has demonstrated the fundamental similarities of the insulin signaling pathway and PTEN function throughout evolution, as well as indicated how the tuberous sclerosis complex (TSC) tumor suppressor genes may be exerting their function through this same pathway *(12–14)*. DNA repair in *Drosophila* is controlled by proteins homologous to DNA repair proteins in humans, such as Rad51 family proteins, and recQ helicases *(15)*. *Drosophila* contains a number of proteins that govern programmed cell death (PCD) such as caspases and IAPs, and studies of PCD in *Drosophila* have not only shown similarity to the situation in humans, but have also indicated novel proteins that couple PCD to MAPK signaling and processes such as determination of overall tissue size during development *(16–21)*. The Notch pathway, which is mainly defined by work in flies, also contains components that contribute to cancer and other diseases *(22)*.

1.2. Conservation of Biological Processes

Conservation extends beyond basic cellular processes, such as transcriptional regulation and protein signaling, to higher-order processes such as development, behavior, and physiologic response. For example, *Drosophila* epithelial tissues such as imaginal discs are remarkably similar to epithelial layers in humans, with distinct apical and basal polarity, cell junctions, and basement membrane *(23,24)*. These tissues grow and divide similarly to mammalian tissues, with distinct G1, S, G2, and M phases. Evolutionary conservation is evident not only in the similarities in cell type and behavior, but also in the behavior of whole organs. Entire signaling pathways are conserved that regulate the genetic control of limb *(25)*, eye *(26)*, and heart development *(27)*. These studies suggest a fundamental unity in development among organisms, which makes *Drosophila* an important model for studying many aspects of mammalian biology.

1.3. What Makes Drosophila a Powerful Genetic Organism?

Despite the amazing degree of conservation throughout metazoan evolution, what gives *Drosophila* an advantage over other model organisms that have physiology even more closely related to that of humans, such as mice? The true advantage of *Drosophila* is the ease with which genetic experiments can be conducted. A number of attributes make *Drosophila* a superior organism for genetics, some of which are discussed below.

First, the fruit fly has a relatively simple genome that has been almost completely sequenced *(28,29)*. The *Drosophila* genome consists of one sex chromosome and three autosomes, which are cytologically and genetically characterized in exhaustive detail, allowing for fine mapping of genetic aberrations simply by either viewing the large polytene chromosomes in the larval salivary gland, or by mapping mutations with chromosomes containing multiple visible marker mutations, respectively *(30)*. The availability of the *Drosophila* genome sequence greatly facilitates many aspects of genetic analysis, such as positional cloning. Not only is the genome sequence a wonderful tool for geneticists, it also reveals many important characteristics regarding the molecular nature of the fruit fly's biology. For example, many *Drosophila* genes are present in a single copy, as opposed to a higher degree of genetic redundancy in what is known from the sequenced mammalian genomes. This comparatively simple genome facilitates the ease of genetic analysis. The results of the genome project can be viewed at the Berkeley *Drosophila* Genome Project web page (http://www.fruitfly.org/annot/index.html).

Second, *Drosophila* have a short generation time, large progeny size, and are relatively cheap and easy to maintain as inbred stocks under laboratory conditions. This allows one to do quite detailed genetic analysis. For example, imagine the ease of mapping a human disease gene if the family was completely inbred, except for the mutation in the gene of interest, and the progeny consisted of hundreds to thousands of individuals that could be easily checked for the relevant disease phenotype. The ability to keep interesting mutations, even lethal mutations, as an inbred stock is greatly facilitated by the presence of specialized chromosomes called balancer chromosomes *(31)*. The canonical balancer chromosome contains multiple chromosomal inversions and dominant visible marker mutations. The chromosomal inversions prevent meiotic recombination, allowing one to maintain a mutated chromosome in its original state, and also prevent the viability of animals homozygous for the balancer chromosome. The dominant marker mutation allows one easily to identify animals heterozygous for the mutation of interest. This results in a stock of viable flies heterozygous for an interesting mutation, which can then be mated to produce progeny homozygous for the mutation of interest. As such, the isolation of a broad spectrum of visible and lethal phenotypes, even ones that are manifested in the second or third generations of mutagenized individuals, is quite easy. A number of balancers exist in each chromosome, and these balancers can differ in their visible markers as well as in the regions of chromosome in which meiotic recombination is prevented.

Third, transgenics in *Drosophila* are highly refined, and facilitate stable ectopic gene expression, mutagenesis, and positional cloning of mutations. In *Drosophila*, transposon-based methods, such as P-element insertion, make it possible to alter, reintroduce, and manipulate DNA fragments throughout the genome. Many engineered P-elements typically contain a dominant marker as well as a bacterial origin of replication, which facilitates cloning of the flanking genomic DNA *(32)*. These vectors are defective in that they do not produce transposase, so that once they are inserted into the genome, they do not normally remove themselves from their initial insertion site. P-elements are introduced into the genome by injecting DNA into the developing germline of embryos in the presence of a transposase source, and then collecting transformant progeny, as visualized by the dominant marker *(33)*. P-elements can be remobilized by crossing a P-element strain to a transposase-producing line. This allows one to do a number of experiments, which are discussed in **Subheading 3.2.**

Fourth, powerful genetic tools can be combined with the detailed knowledge of developmental biology in *Drosophila*. Over a century of research has described the development of both normal and mutant fruit flies. The wealth of available literature allows one to understand how to visualize even very subtle perturbations in the fly's development or behavior. This facilitates one's ability to pinpoint the biologic processes and underlying molecular mechanisms perturbed, and subsequently allows for an in-vivo system in which to study interactions among different genes involved in the same biological process.

1.4. History of Identifying Tumor Suppressor Genes in Drosophila

Melanotic tumorlike granules were first discovered in mutant *Drosophila* larvae in the early 1900s *(34)*. However, these abnormal growths lacked some of the basic features of tumors: *in situ* overproliferation, altered cell morphology, inability to differentiate, invasiveness, and transplantability. As a result, it was speculated that invertebrate cells were unable to become truly neoplastic.

In 1967, Gateff, Schneiderman, and others *(35,36)* conclusively demonstrated the existence of neoplastic tumors in invertebrates with the isolation of a spontaneous, recessive lethal mutation. The new mutation was the fourth allele of the well-known *Drosophila* gene, *lethal (2) giant larvae (lgl)*, originally discovered by Bridges in 1930. This was the first evidence that inactivation of a tumor suppressor gene could directly lead to neoplastic transformation in vivo. Subsequent screens for larval or pupal homozygous lethal mutants resulted in the identification of multiple genes whose inactivation leads to overproliferation of various tissues *(37,38)*. Approximately half of these genes have been cloned and characterized at the molecular level, and in most cases at least one human and/or mouse homolog has been discovered. Interestingly, many of these genes had not been previously implicated in the control of cell proliferation.

With technical advancements, such as the ability to efficiently produce genetically mosaic animals (discussed in **Subheading 2.2.**), it was for the first time feasible to look for mutations in genes that could cause the production of tumorlike overgrowths in otherwise normal adult flies, modeling the situation in cancer patients. In 1994, such a gene was discovered. Called the *large tumor suppressor* gene (*lats*), mutations caused dramatic tumors that could reach one-fifth of the size of the entire animal, with characteristics remarkably similar to human tumors *(39)*. Mammalian homologs of *lats* were subsequently identified with a fly *lats* probe, and it was shown that the human homolog of *lats*, LATS1, when introduced into the fly, could fully rescue the defect caused by mutations of the endogenous fly *lats* gene, suggesting a functional conservation of the *lats* gene throughout evolution *(40)*. Knockout of the mouse *lats* homolog (LATS1) resulted in the production of ovarian tumors and soft-tissue sarcomas in homozygous mutant mice, showing that the biologic consequences of loss of lats function were also conserved *(41)*. Further experiments showed that the lats protein binds to cdc2 and the cdc2/lats complex is inactive for histone H1 kinase activity. These experiments underscore the ability of *Drosophila* to identify genes important to our understanding of cancer biology, genes that otherwise may be overlooked.

However, the ability of *Drosophila* to enhance our understanding of cancer biology is not limited to finding novel tumor suppressor genes. *Drosophila* has also been used successfully to elucidate the function of human tumor suppressor genes, most commonly by studying their fly homologs. Examples include the study of the fly homologs of molecules such as β-catenin and patched, which play crucial roles in regulating patterning events during development *(4,42)*. Other, more recent experiments have elucidated the function of genes such as *PTEN*, *TSC1*, and *TSC2* in regulating cell size and proliferation through conserved pathways *(43–48)*. These conclusions would have been difficult to visualize and prove rigorously with other model systems.

With that, the study of tumor suppressor genes in *Drosophila* today takes two forms; (a) using forward genetics to mutate genes and look for phenotypes that resemble cancer phenotypes in mammals; and (b) using reverse genetics to study the fly homologs of mammalian tumor suppressor genes. We now discuss how these two approaches are accomplished.

2. Identifying Novel Tumor Suppressor Genes in *Drosophila*

In *Drosophila*, traditional mutagenesis screens have been used successfully to isolate mutations in many genes affecting growth and proliferation. Many of the isolated genes have turned out to be novel or previously unimplicated in processes related to cancer biology, suggesting that screens in *Drosophila* can shed light on previously unappreciated aspects of cancer biology. In a genetic screen, the goal is to mutagenize as many genes as possible and then look for mutations that can cause the desired phenotype. This provides a presumably unbiased way of finding genes that can cause a given phenotype. After screening thousands of mutagenized flies, one can hope to isolate a small number of mutations affecting the process in question, depending on the stringency of the screening criteria.

2.1. Traditional (F2) Screen

In a traditional mutagenesis screen, male flies are treated with a mutagen in order to induce mutations in the rapidly dividing germline cells. Typically used mutagens include EMS, ENU, X-rays, and P-elements. EMS and ENU are alkylating agents that efficiently induce point mutations. These mutations tend to affect single nucleotides in coding regions, thereby creating nonsense or missense point mutations that can provide structure–function information about the gene product. X-rays are capable of inducing chromosome breaks, which are then sometime repaired incorrectly, producing translocations, deletions, transpositions, and inversions, as well as point mutations. These rearrangements often can be detected on genomic Southern blots *(49)*. P-element mutagenesis occurs when P-elements are inserted in the genome. *De novo* P-element mutagenesis is more labor-intensive than the other two methods, due to its tendency to insert into "hot spots" either in noncoding regions or in the 5′ upstream sequence of genes *(50,51)*. Nevertheless, existing lethal P-element lines can be obtained at stock centers (see below) and used to screen for interesting phenotypes.

Mutagenized males are crossed to females carrying marked balancer chromosomes (this is the P0, or parental generation). Individual progeny from these males (the F1, or first filial generation) will presumably carry mutations in different genes when the mutagenized males are used in crosses for no more than 4 d. Individual F1 heterozygous animals resulting from this cross are segregated and crossed to the marked balancer strain again, allowing the individual mutant lines to be maintained as stocks (the F2 generation). Heterozygotes within a stock can then be crossed to each other to generate animals that are homozygous for the newly induced mutations. Phenotypic analysis of the homozygous animals can then be used to gain insight into gene function. Since mutations in tumor suppressor genes often cause lethality before adulthood, imaginal discs and other proliferating tissues in larval stages are often examined.

2.2. Screens Using Genetically Mosaic Animals (Mosaic Screens)

The general form of fly genetic screens has proven to be tremendously useful. However, there are a number of limitations in the traditional screens. Technical innovations have led to a few specific variations of genetic screens that are particularly suited to bypass this problem. They are therefore well suited to the study of cancer. The mosaic screen allows the fly to mimic cancer development in humans more closely. Cancer in humans can be viewed as a clonal phenomenon in which a small population of cells in an otherwise healthy individual loses growth control by mutation in one or more cancer-related genes. Likewise, in a genetically mosaic animal, cells of a distinct genotype are generated within an otherwise healthy organism.

In a mosaic screen, typically one chromosome arm is targeted for mutation, and mutant phenotypes are observed using the FLP/FRT system. The chromosome arm to be mutated contains a centromeric insertion of an FRT sequence. Males are mutagenized as in a traditional screen, and then crossed to females that have an FRT inserted at the same location, as well as a source of FLP recombinase. The F1 generation then contains flies homozygous for the FRT, heterozygous for a mutation, and expressing a source of FLP in tissues of interest. When FRTs are present in trans on sister chromosome arms, and a source of FLP recombinase is present, recombination can occur between homologous chromosome arms during mitosis. This results in the creation of one homozygous mutant and one homozygous wild-type daughter cell. Subsequent cell division perpetuates these genotypes, producing a clone of homozygous mutant cells, and another clone of wild-type tissue (referred to as the "twin spot"). Mutant clones can then be observed in adult flies (or in developing tissues), and flies with interesting phenotypes can be crossed to balancers in order to maintain the mutation as a stock *(52–54)*.

The mosaic screen offers several advantages over traditional screens. Most prominently, many mutations can cause lethality before any proliferative or otherwise cancer-related phenotypes are observed, especially for those phenotypes observed in adults. In mosaic screens such phenotypes can be observed, and the causative mutations can be recovered. Second, fly embryos develop away from their mothers; thus, many gene products are maternally deposited in the eggs, which often mask the mutant phenotypes, even when the embryos are zygotically mutant. In mosaic screens, phenotypes are typically being observed in later stages, that is, in the adult, after the period where maternal contributions no longer affect the organism. Third, the traditional screens require the establishment of individual lines before examining potential phenotypes, which is quite labor-intensive, even for flies. In mosaic screens, phenotypes are observed in the F1 generation, and only those with interesting phenotypes are kept to make stocks. This saves a tremendous amount of time and effort.

3. Inactivating *Drosophila* Tumor Suppressor Gene Homologs

3.1. Finding Existing Mutants

To study the functions of tumor suppressor gene homologs in *Drosophila*, one must first generate mutations in the gene of interest. As there are many curated mutant strains of *Drosophila*, the first order of business is to see if mutations are already present in this gene. Homology searches can be used to compare *Drosophila* gene sequences against the human protein of interest (http://www.fruitfly.org/blast). If the homologous *Drosophila* gene has been mutated and characterized, the location of useful stocks can be found at flybase (http://flybase.bio.indiana.edu). Fly stocks can then be obtained from individual laboratories or public stock centers. Otherwise, the chromosomal region surrounding the gene can be searched on GadFly, the *Drosophila* genome annotation database (http://www.fruitfly.org/cgi-bin/annot/query) for nearby P-elements. P-elements have many uses, some of which are now discussed.

3.2. P-Element Mutants

Many homozygous lethal mutant lines have been generated that contain a single P-element insert; the P-element presumably disrupts the function of a single gene . If lethal P-elements exist in or near the gene of interest, they can be found at GadFly. If the P-element is in the gene, new mutant alleles can be created by crossing these flies to a source of transposase (Δ2-3 containing strains) and selecting for flies that have lost the P-element (visualized by the marker in P-elements, such as w^+). If the P-element excises precisely from the genome, the lethality should be reverted, but if the P-element excises imprecisely, it will sometimes take fragments of surrounding genomic DNA, creating small deletions in the gene of interest *(55)*. Deletions are also created by irradiating the flies with X-rays, then selecting for loss of the P-element marker *(56)*. Such strains have deletions in nearby genomic DNA in addition to the marker gene (this is particularly useful in cases where the P-element has obtained a mutation that prevents its excision by transposase). Another P-element-based method that can be used is male recombination *(57)*. In addition to its use as a tool for precisely mapping the location of mutations between P-element insertions, male recombination can also be used to create deletions. The advantage to this method is that strains can be selected that contain deletions to only one side of the P-element, or rather, only to the side of the P-element that contains the gene of interest. Therefore, deletions will have one of their endpoints precisely mapped to genomic DNA sequence. Deletions are created by first making a strain that has a visible marker mutation to either side of the P-element. Normally there is no meiotic recombination in *Drosophila* males. However, by generating males with P-element and transposase, transposase can express in the germline cells and produce breaks at the P-element ends and, thus, induce recombinations between homologous arms. Such recombination events often associate with deletions near the P-element ends. By selecting proper genetic markers on the recombinant chromosomes, one can recover these recombinant chromosomes that most likely contain deletions in the desired direction. The extent of created deletions can be determined by polymerase chain reaction (PCR) or

Southern-based methods, and if the gene of interest has a developmentally important function, deletions in the gene would presumably cause lethality or a visible phenotype.

If there are no P-elements in the gene of interest, P-elements inserted near the gene can be useful. By crossing these lines to a line expressing transposase, and selecting for flies that have retained the P-element, one selects for flies in which the P-element has "hopped" to a new location *(58)*. It has been noted that when this is done, P-elements typically reinsert close to their original location (referred to as "local hopping"). Lines in which the P-element has hopped can be screened by Southern or PCR-based methods to find insertions in the gene of interest.

3.3. Deficiency Screens

If there are no fly stocks with existing mutations or P-element insertions in the gene of interest, there are still more methods to create mutations. There are a large number of *Drosophila* stocks that contain chromosomal deletions (deficiencies) at different places in the genome. These deficiencies are viable as heterozygous stocks, but lethal as homozygotes. If deficiencies are found that delete the gene of interest (by finding deficiency strains that delete the cytological location of the gene of interest), one should look for the smallest available deficiency that uncovers the gene of interest. Wild-type flies can then be randomly mutagenized [with X-rays or chemical mutagens such as ethyl methane sulphonate (EMS)] and crossed to flies containing the deficiency. If the new mutation crossed to the deficiency is lethal, the newly mutagenized stock likely contains a mutation in a gene uncovered by the deficiency. The mutations in such lines can then be mapped and sequenced in order to determine if they lie in the gene of interest.

3.4. Transgenes with Dominant Negative Mutations

If there are data on the molecular function of the human tumor suppressor gene, then it is also possible to create a dominant negative transgene and introduce it into the fly. The dominant negative construct can be cloned into a P-element vector and transformed into flies. This construct can then be expressed, which will presumably allow it to interfere with the function of the wild-type gene product. Methods to express such transgenes are discussed in **Subheading 4.**

3.5. RNAi

Despite the usefulness of these techniques, actually targeting mutations to a gene of interest is still laborious. However, two methods have been developed for *Drosophila* that may solve this problem. RNAi, which has proven so successful for functionally inactivating genes in *Caenorhabditis elegans (59)*, also works quite well on *Drosophila* embryos and cell lines *(60–62)*. While injecting double-stranded RNA into flies during later developmental stages does not appear to work well, there are methods to express double-stranded RNA in adults using transgenics *(63)*. This promises the advantage of allowing heritable transmission of an inhibitory double-stranded RNA that can be induced at a variety of developmental timepoints.

3.6. Drosophila *Knockouts*

In addition to RNAi, the method of homologous recombination has recently reported for *Drosophila (64,65)*. In this procedure, three transgenic constructs are involved. The

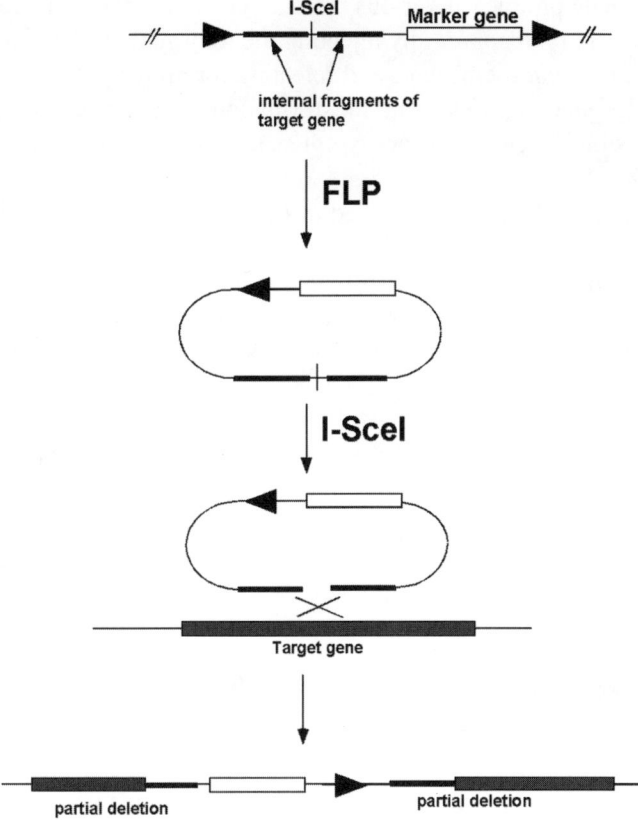

Fig. 1. Gene knockout in *Drosophila*. A construct is inserted into the genome that carries sequences homologous to internal regions of the target gene, a restriction site of the yeast-specific enzyme I-SceI, and a marker gene, all flanked by FRT sites. In the presence of FLP recombinase, the construct is excised from the genome. The construct is linearized by expression of the I-SceI enzyme. Homologous recombination occurs in an "ends-in" fashion, destroying the functional copy of the gene. Recombinant flies are selected by the visible marker gene.

first expresses a yeast site-specific recombinase (FLP) under the control of an inducible promoter. The second contains the knockout or knockin construct flanked by the FLP recombination target (FRT) sites. This construct also contains a restriction site for a yeast-specific endonuclease, I-SceI. The final construct expresses the I-SceI enzyme under an inducible promoter. These three constructs are crossed into the same fly, and the transgenes are induced. This allows excision of the construct from the genome by FLP and linearization of the construct by I-SceI. This linearized DNA can then recombine with homologous sequence in the genome (**Fig. 1**). This procedure may prove successful with many genes and increase the ease of reverse genetics in *Drosophila*.

4. Ectopic Expression of Tumor Suppressor Genes

Inactivating tumor suppressor genes is a valuable way to understand their normal functions. However, to gain more insight into the mechanisms of how a given gene functions, it is often necessary also to find out what the consequences are of overexpressing either wild-type gene product or various mutant versions of the gene. Stable transgenic lines can be created in *Drosophila* by injecting the transgene in a P-element vector into the germline of *Drosophila* embryos, and selecting for progeny that have integrated the P-element into the genome. P-element transformation has allowed *Drosophila* to be well suited for expression of genes in a variety of tissues, in temporally and spatially specific manner.

As many genes have been characterized in *Drosophila*, there exist a large number of genes with characterized promoters. Expression of cDNAs (most transgenes in *Drosophila* are made from cDNA rather than genomic DNA) from these promoters can be easily accomplished with P-element gene transfer. Promoters from genes such as actin, tubulin, and ubiquitin allow for ubiquitous expression of a cDNA. Many other characterized promoters exist that express in specific body parts, specific cell types (i.e., neurons or proliferating cells), or at specific times in development *(66–68)*. This allows one to bypass the lethality that would seem likely from ubiquitous expression of a tumor suppressor gene or oncogene. It also allows for the study of tissue-specific effects. Another useful promoter commonly used for overexpression is the *hsp70* promoter *(69)*. This allows one to induce expression of the gene product at any desired time during development by "heat shocking" the fly.

4.1. The Gal4/UAS System

Although transgenes can be made by fusing cDNAs directly to a promoter, this becomes cumbersome when one wants express the gene in many different scenarios. In addition, stocks may not be easily obtained or maintained if the transgene expresses a highly toxic product. One of the most widely used solutions to this problem is the Gal4/UAS system *(70)*. A large number of enhancer traps have been created that express the yeast Gal4 transcriptional activator in known tissues and/or at known developmental times (a searchable index of many Gal4 enhancer traps can be found at FlyView [http://pbio07.uni-muenster.de/]). In addition, there also exist many transgenic lines that express Gal4 under the control of characterized promoters (**Fig. 2**). Overexpression of Gal4 protein, for the most part, does not interfere with normal developmental processes. The cDNA being studied is cloned downstream of the sequence recognized by the Gal4 protein, called the upstream activating sequence (UAS). Expression of this UAS-controlled transgene is off unless the fly is crossed to a line expressing Gal4. The UAS line can be crossed to many different Gal4 lines that already exist, allowing for conditional expression of a gene in a variety of locations and at a variety of developmental timepoints.

4.2. Overexpression Using EP Elements

It has been noted that P-elements commonly insert into the 5′ region of genes, and one specialized P-element vector takes advantage of this *(71)*. Called EP elements, they contain a UAS promoter, which, when inserted 5′ to a gene in the proper orientation, allow expression of the gene in the presence of Gal4. Many fly stocks have been created with single EP insertions, analogous to the situation with lethal P elements. One can check GadFly for EP insertions in the gene of interest, as much time otherwise spent making a transgenic UAS line can be saved. EP elements can also be used to screen for novel overexpression phenotypes.

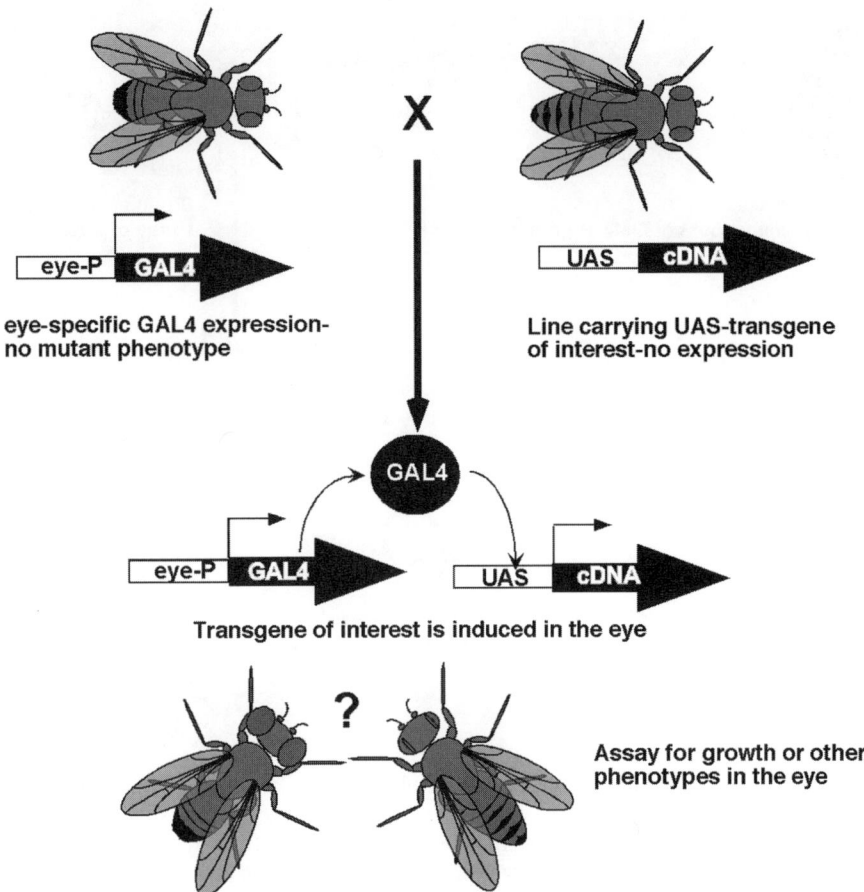

Fig. 2. The Gal4/UAS system. In this example, an eye-specific promoter is fused to a gene encoding the yeast-specific Gal4 protein. Expression of this protein causes no visible phenotype. This fly is crossed to a fly bearing a transgene containing the gene of interest cloned downstream of an upstream activating sequence (UAS). In the absence of Gal4, gene transcription does not occur. Progeny that contain both constructs will allow Gal4-mediated gene activation. These flies can be assayed for interesting phenotypes in the eye.

4.3. "FLP-Out"-Mediated Transgene Expression

Another method that refines analysis of Gal4/UAS-mediated gene expression is the "FLP-out" technique *(72)*. A transgene is made in which a ubiquitous promoter such as actin is cloned upstream of one FRT, and the cDNA of interest is cloned after a second one. These two FRTs are separated by the coding region of a visible marker gene, such as one affecting cuticle or bristle phenotype. Transcription of the cDNA of interest is stopped because transcription from the actin or marker gene promoter does not proceed past the polyadenylation site of the marker gene. When a fly containing this transgene is crossed to a transgene expressing the FLP recombinase, the marker gene is excised through recombination at the FRT sites, and the cDNA can then be expressed under the strong actin promoter (**Fig. 3**). This allows for the region of tissue expressing the cDNA of interest to be precisely marked by loss of the marker gene. This is important when addressing issues of gene autonomy, when one wants to know if a given genotype cor-

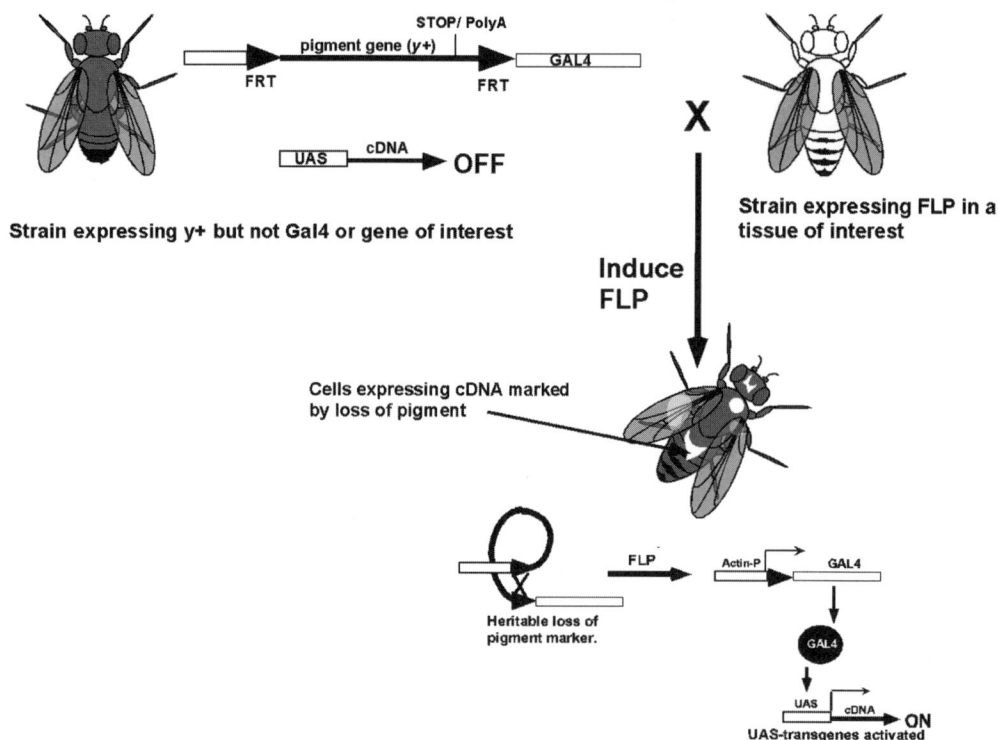

Fig. 3. The "FLP-out" system. This system allows one to make marked clones overexpressing a gene of interest. An intervening marker gene (here, a pigment gene) prevents ubiquitous expression of Gal4 through its sites of transcription termination. This marker is flanked by FRT sites. In the presence of FLP, the intervening sequence is excised and lost, and Gal4 can be heritably expressed through subsequent cell divisions. This results in clones of tissue that express Gal4, and subsequently, the gene of interest. These clones are visualized through loss of the pigment gene, and can be analyzed for interesting phenotypes.

responds directly with phenotype. For example, if the product of the cDNA can diffuse or relay a signal through a diffusable protein, then mutant phenotypes could be observed outside of the region actually expressing the cDNA.

5. Phenotypic Characterization

Once one obtains or creates mutations and transgenic lines, how does one go about determining the resulting mutant phenotypes? This is a very large topic, even for an organism as small as *Drosophila*, and as such, only characterization techniques that study cancer-related phenotypes are discussed here.

5.1. Analysis of Homozygous Mutant Animals

In the most general sense, there are two ways in which inactivating or loss-of-function mutations are studied. The first is to observe the development of the homozygous animal. For example, many *Drosophila* tumor suppressor genes were found by studying the growth of various homozygous mutants relative to their heterozygous siblings. Animals homozygous for many of these mutations live to the larval or pupal stages, and exhibit visible overgrowth of tissues. When mutations are made in target genes, they can be studied in a similar manner; namely, observing alterations in the development of homozygous animals relative to heterozygous siblings.

5.2. Analysis of Mosaic Animals

However, what if the mutation exhibits an early embryonic defect? If this is the case, the animal may die before any cancer-related phenotypes are observed. This is bypassed by creating genetically mosaic animals. The most commonly used method to create genetically mosaic animals is through the use of mitotic recombination with FLP recombinase and FRT chromosomes. This allows for an animal that is, for the most part, heterozygous for a mutation, and therefore viable. A subset of tissues, however, is induced to undergo recombination during cell division, creating homozygous tissues. This method is useful for understanding many facets of developmental biology, such as issues of cell autonomy. It is, however, particularly relevant to the study of cancer because it models the two-hit or LOH hypothesis of tumor formation by tumor suppressor genes. This method allows for the visualization of cancer-related phenotypes such as hyperplasia and even tumor outgrowth, which would otherwise have been missed due to lethality of homozygous animals.

5.2.1. Using the FLP/FRT System for Mosaic Analysis

Mitotic recombination in *Drosophila* can be induced with X-rays, creating DNA double-strand breaks while tissues are undergoing mitosis. However, a more efficient system for creating mosaics again uses the FLP/FRT system *(52,53,73)*. Stocks exist that contain FRT elements near the centromeres of all but the small fourth chromosome. The mutation to be studied is recombined distally to a chromosome carrying the appropriate FRT, and then crossed to flies carrying an FRT on the same chromosome, as well as a source of FLP recombinase. Resulting progeny generate mutant and twin spot clones in an otherwise heterozygous individual. The twin spot creates an internal control, such that characteristics such as growth of the mutant clone relative to the twin spot clone can be observed in the same animal (**Fig. 4a**).

5.2.2. Marking Mutant Clones in Genetically Mosaic Animals

When mosaic animals are made, they need cell markers (as in the FLP-out system) that allow for visualization of mutant tissue. One way to do this is to have a visible marker distal to one of the FRT chromosomes (i.e., on the chromosome trans to the one bearing the mutation) for eye color, cuticle color, or bristle shape (**Fig. 4b**). For the study of internal tissues, a set of chromosomes has been created with epitope-tagged markers. This allows for antibody staining of internal tissues. Heterozygous and twin spot tissues stain for the epitope, and these genotypes can actually be distinguished based on the relative intensity of the stain *(52)*. Homozygous mutant tissue, in this case, does not stain. Another method exists to mark the mutant tissue exclusively. This system, called the MARCM system (for Mosaic Analysis with a Repressible Cell Marker), utilizes the Gal4/UAS system *(74,75)*. Instead of an epitope tag, the chromosome in trans to the chromosome bearing the mutation contains a transgene ubiquitously expressing Gal80, a repressor of UAS-mediated gene transcription. In the genetic background are a Gal4 transgene (Actin-Gal4, for

A

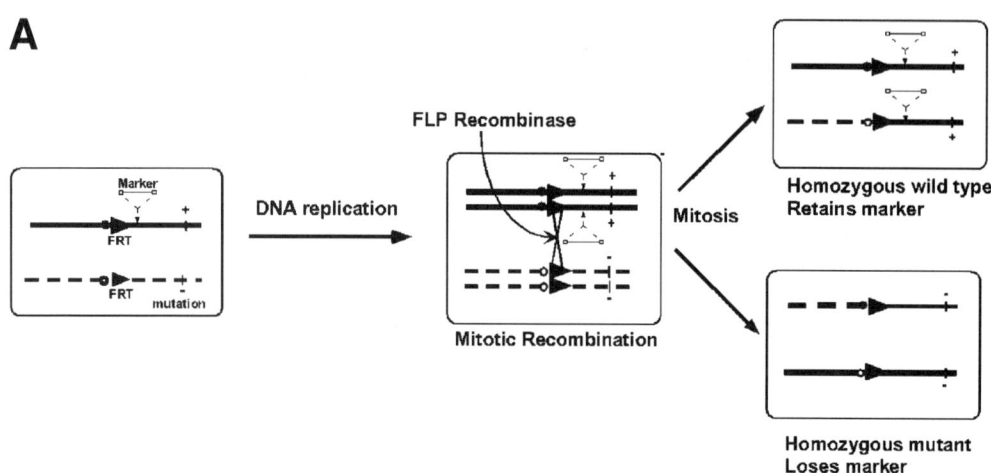

B   **Marker = w+ eye pigment**

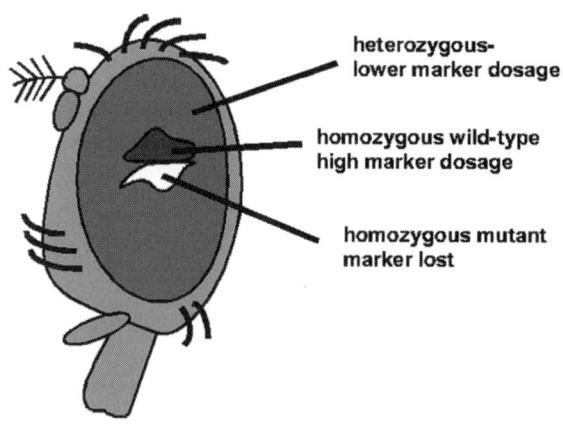

Fig. 4. Generating mosaic animals using the FLP/FRT system. Using FRT-bearing chromosomes, tissues with a homozygous loss-of-function mutation can be generated in an otherwise heterozytgous organism. This provides a relevant model for tumor suppressor gene function. (**A**) Generation of homozygous clones. Here, a pair of homologous chromosomes contains centromerically placed FRT sites. Distal to one is a recessive mutation in the gene of interest. Distal to the other is a marker (endogenous mutation or transgene) and a wild-type copy of the gene of interest. When FLP recombinase is expressed, recombination can occur during mitosis, generating ahomozygous mutant and homozygous wild-type cells. These cells produce clones of mutant and "twin spot" tissue. (**B**) Visibly marked tissues. In this case, the marker is an eye pigment transgene. Heterozygous tissue is light red. Homozygous mutant tissue is visualized by loss of the pigment (white tissue), and twin spot tissue is darker red due to double dosage of the marker. In the case of epitope-tagged chromosomes, the situation is similar after antibody staining of tissues. (**C**) The MARCM system. In this case, the marker is a ubiquitously expressed (tubulin-promoter) Gal80 transgene. The Gal80 protein prevents Gal4-mediated gene expression; in this case, expression of GFP is kept off. When clones are generated, Gal80 protein is retained I heterozygous and twin spot tissue. In homozygous mutant tissue, however, Gal80 is lost, and Gal4-mediated GFP expression is activated. This results in animals with GFP-marked mutant tissue and unmarked heterozygous and twin spot tissue.

C

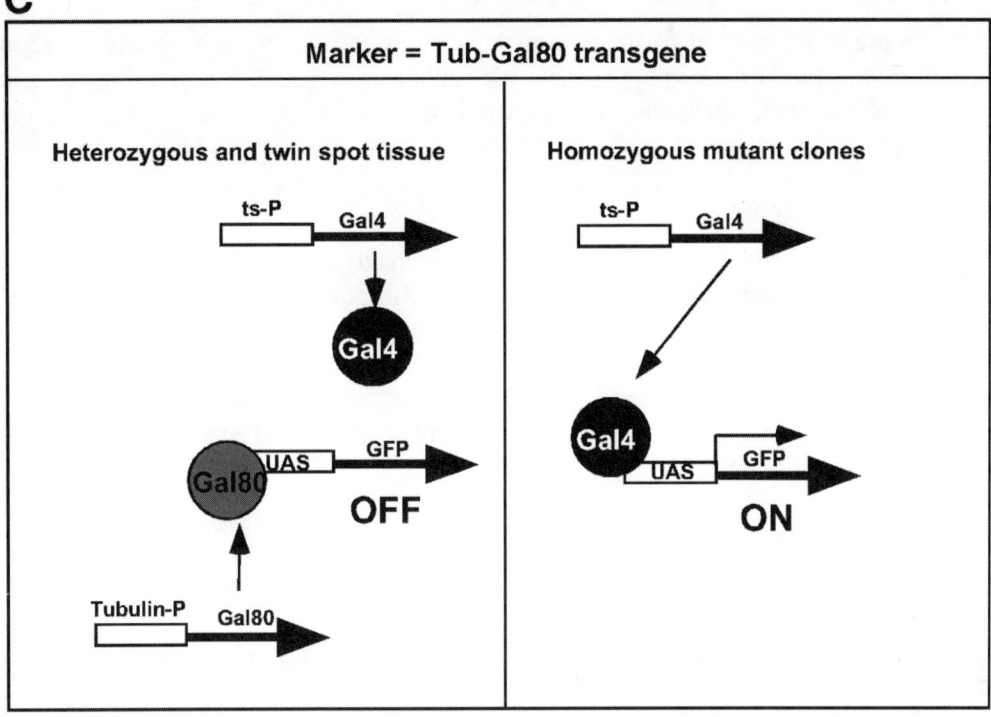

Marker = Tub-Gal80 transgene

Fig. 4. *(continued)*

example) and a UAS transgene driving expression of a marker (i.e., UAS-GFP). The marker is kept off by presence of Gal80 protein. When recombination occurs, the tissue becomes homozygous for the mutations and loses the Gal80 transgene. This allows for the marker gene to transcribe product. The MARCM system is useful for following the growth and morphology of mutant tissues in living animals (**Fig. 4c**).

In addition, the proliferation defects due to mutations or transgenes can be monitored by inducing mitotic recombination or FLP-out, respectively, dissecting discs at known time points, and then counting the number of cells produced in various mutants relative to wild-type controls. In this case, mutant cells are marked with a UAS-GFP construct, and counted using confocal microscopy. Cell doubling times are then calculated as

$$\frac{\log 2}{\log N} \quad (h)$$

where N = the number of cells in a clone and h = the time in hours from FLP induction *(76)*.

Since most cells of imaginal discs do not divide synchronously, specific cell cycle blocks are not as easily observed as in early embryos. However, fluorescence-activated cell sorting (FACS) has been adapted for imaginal discs. Mutant cells are labeled with a UAS-GFP construct, and discs are dissected from the animal. The disc cells are dissociated with trypsin and sorted by the machine. This allows for the detection of cell populations blocked at specific stages of the cell cycle based on their DNA content *(76)*.

5.2.3. Generating Clones with a Mutant Growth Advantage

There are more methods of mosaic analysis that either give a growth advantage to mutant tissue or ablate heterozygous and twin spot tissue. This is important because in some cases, mutant tissue has a severe growth disadvantage, and mutant clones may die before they can be observed due to growth competition against wild-type tissue.

5.2.3.1. MINUTE TECHNIQUE

One technique used to give mutant clones a growth advantage is the *Minute* technique *(77)*. *Minutes* are dominant mutations with a growth disadvantage. *Minute* mutations have been isolated on all chromosome arms, and have been placed distal to FRT chromosomes. If such chromosomes are in trans to an FRT chromosome bearing a mutation in the gene of interest, and clones are made, all homozygous *Minute* twins spots will die, and heterozygous tissue will have a severe growth disadvantage, allowing the homozygous mutant tissue to proliferate.

5.2.3.2. FLP-DFS METHOD

Another method, called the FLP-DFS (dominant female sterile) technique, allows for the selection of homozygous mutant tissue in the female germline *(78)*. A mutation in the gene of interest is placed distal to an FRT. The homologous chromosome arm also contains an FRT, and distal to this FRT is a dominant mutation, ovoD1, that results in female sterility. In the presence of FLP, clones homozygous for the mutated gene of interest will be produced. Heterozygous and homozygous twin spot tissue will not survive due to the presence of ovoD1, allowing homozygous mutant tissue to be analyzed. The FLP-DFS technique is useful in understanding whether the gene of interest plays a role in female fertility and oogenesis. In addition, this technique is used to create embryos that lack maternally deposited gene products, which allows one to study whether a gene plays a role in early development.

5.2.3.3. CELL ABLATION TECHNIQUE USING HID

The final technique discussed here is conceptually similar to the FLP-DFS technique, but utilizes the adult eye instead of the female germline *(79)*. This is particularly useful because the eye is both dispensable for viability and fertility, and has a highly repetitive structure in which even subtle perturbations can be observed. One chromosome contains an FRT, distal to which are a recessive cell-lethal mutation and a transgene expressing the cell death gene *head involution defective* (*hid*) under the control of an eye specific promoter. Flies heterozygous for this chromosome have no eyes, and homozygous flies die. The mutation of interest is placed distal to an FRT on the sister chromosome. In the presence of FLP, mutant clones are made, and these will be the only tissues that survive, since heterozygous and twin spot tissues have both retained the *hid* transgene.

5.3. Considerations with Transgenes

When studying transgenic constructs, as shown above, methods such as the Gal4/UAS and FLP-out systems allow one to ectopically express genes at various times and locations during development. These procedures, like mosaic analysis, allow one to bypass lethal phenotypes that ubiquitous overexpression of a tumor suppressor gene would likely produce. Researchers have not only used these techniques to express *Drosophila* gene products. Human genes have also been cloned into UAS constructs and expressed in *Drosophila*, and this can be used both to generate observable phenotypes as well as to demonstrate functional conservation of a gene. One important consideration when expressing transgenes is to carefully select where and when the transgene should be expressed in order to obtain meaningful results. For example, if transgene expression is induced in a location where the endogenous gene is never expressed, one may create artificial phenotypes that are irrelevant to the normal biologic function of the gene. Another important consideration is that the genomic insertion site of a transgene can affect its expression. This problem is resolved by creating multiple transgenic lines for the same construct, and assaying for the phenotypes common to all transgenic lines.

5.4. Expected Phenotypes with Tumor Suppressor Genes

When studying tumor suppressors and other genes related to cancer in flies, what tissues are commonly inspected for phenotypes? This section compiles some of the cancer-related phenotypes observed in *Drosophila*, the tissues most commonly inspected, and the techniques used to further characterize phenotypes. As physiologic roles for tumor suppressor genes may lie outside direct control over cell proliferation, other possible phenotypes are explored. One should keep in mind that the phenotype could be unexpected or not obviously related to cancer. Nevertheless, one should have an open mind about such phenotypes; the strength of the fly is that one can find the biological functions of a gene, however unexpected they may be.

5.4.1. Cell Proliferation

5.4.1.1. EMBRYONIC PHENOTYPES

If the gene of interest is presumed to play an essential role in the cell division machinery, one would expect that homozygous mutations in this gene would be embryonic lethal. However, early defects due to mutations in cell cycle genes are clouded by the fact that many gene products are maternally deposited into embryos, often bypassing any zygotic deficits *(80)*. If necessary, this problem can be resolved using the FLP-DFS technique. There are distinct advantages to studying mutant phenotypes in the embryo. First, there are thirteen synchronous rounds of mitosis during early embryonic development. This allows for easy visualization of perturbations in the cell cycle. In addition, these 13 rounds of mitosis occur without cytokinesis. This gives one the ability to observe organismal phenotypes using microinjection experiments with various proteins, RNA, and toxins such as colchicine and cyclohexamide *(81,82)*. Mutant phenotypes can be observed by immunostaining embryos for phospho-histon3 (PH3), which labels mitotic nuclei, as well as for incorporation of the nucleotide analog Brdu, which labels nuclei in S phase. There are also GFP fusion proteins that can label the mitotic apparatus for in-vivo experiments (UAS-*tau*:GFP, UAS-*h2*:GFP). Stained or labeled nuclei are subsequently observed by confocal microscopy *(82,83)*.

5.4.1.2. IMAGINAL DISC PHENOTYPES

Despite the advantages of analyzing embryonic cells, they have cell cycles somewhat divergent from mammalian cells. This is not the case with *Drosophila* imaginal disc cells. Imaginal discs are sacs of epithelial monolayers initially set aside during embryonic development. These cells proliferate during larval stages and differentiate during pupal stages to produce adult structures. While other larval cells become increasingly polypoid throughout development, imaginal discs proliferate with regular G1, S, G2, M, and cytokinetic phases, modeling the cell cycle of mammalian tissues. The epithelial cells of the disc have a morphology characteristic of mammalian epithelial cells, with distinct apical and basal polarity, as well as apically located adherens junctions, and septate junctions, the invertebrate equivalent of tight junctions. Imaginal discs are also surrounded by a basement membrane made of collagen IV, laminin, and other components similar to vertebrate basement membrane components *(23)*.

As in embryos, imaginal discs can be stained with anti-PH3, or labeled for Brdu incorporation to look for cells in mitosis or S phase, respectively. This is particularly useful in the larval eye disc, where there is a wave of cells that go through one synchronized round of cell division before differentiating. Differentiating cells posterior to this wave do not stain appreciably for PH3 or Brdu incorporation, and defects in cell proliferation control can be easily observed here *(84,85)*. Mutations in many *Drosophila* tumor suppressor genes are homozygous lethals that show either hyperplastic or neoplastic overgrowth of the imaginal disc *(35)*. These categories are now explored in further detail.

5.4.1.3. HYPERPLASTIC PROLIFERATION MUTANTS

Most of the *Drosophila* tumor suppressor genes so far isolated are classified as hyperplastic. In these mutants, disc tissue retains most of its epithelial structure and is able to differentiate adult structures when transplanted into a larval host *(36,37,86)*. Lethal mutations at the *fat* locus cause hyperplastic overgrowth of all the imaginal discs, with abnormal junctional complexes connecting epithelial cells. The *fat* gene product is a transmembrane protein with protein homology to cadherins *(87–89)*. Mutations in *expanded*, an SH3-containing protein similar to proteins in the Protein 4.1 family, result in enlarged wings, and fattened, deformed legs. Expanded protein localizes to the apical side of epithelial cells, suggesting an association with adherens junctions *(90–92)*. Mutations in *lethal (3) discs overgrown (dco)*, a casein kinase homolog, result in overproliferation of imaginal discs during an extended larval period. These mutants have defects in communication through gap junctions *(93)*.

5.4.1.4. NEOPLASTIC PROLIFERATION MUTANTS

In neoplastic overgrowth mutants, imaginal discs lose their cellular and tissue morphology as well as their ability to differentiate after transplantation into a larval host. Transplanted neoplastic tissues also proliferate uncontrollably, and invade host tissues, eventually killing the host *(36,86,94)*.

The prototypical *Drosophila* tumor suppressor gene *lgl* causes neoplastic overgrowth of imaginal discs and brain tissues. These animals can grow to almost twice the size of normal larvae during an extended larval period. The overgrowing tissues, which are normally singe-layered epithelia, become disorganized masses *(35,36)*. Interestingly, *lgl* mutant larvae have increased levels of type IV collagenase, a hallmark of invasive tumors in mammals *(95)*. The novel Lgl protein product localizes to the plasma membrane, and associates with nonmuscle myosin II heavy chain *(96,97)*. Recessive mutations in the *lethal(1) discs large (dlg)* gene also cause neoplastic overgrowth of larval imaginal discs, and the subsequent death of the animal in the early pupal stage. In the imaginal discs of larvae carrying a *dlg* loss-of-function mutation, apical-basal polarity is almost completely lost and septate junctions are missing. Dlg protein is a prototypical PDZ-containing protein in the MAGUK family, and its human homolog can associate with the APC/β-catenin complex *(98–100)*. *Dlg* appears to interact with *lgl* and *scribble*, a PDZ-domain protein that also localizes to the septate junction, in a pathway that governs both proliferation as well as epithelial polarity and organization *(101,102)*. This highlights a pathway as yet unidentified in mammalian systems that can control both proliferation and epithelial structure.

Mutations in the *lats* gene appear to be one notable exception to the strict distinction between hyperplastic and neoplastic phenotypes. Tumors formed by *lats* mutant cells grow beyond a simple epithelial monolayer and lose their ability to differentiate normal structures, but do not invade surrounding tissues *(39,73)*.

5.4.1.5. BRAIN TUMORS

Several genes have been found to be necessary to control proliferation in the developing central nervous system of the larva. In wild-type larvae, the central nervous system is composed of two brain hemispheres and a ventral ganglion. A single layer of glial cells together with an extracellular matrix, the neurilemma, surrounds the brain hemispheres. The outer cortical region of the brain hemispheres contains several cell types: neuroblasts, ganglion mother cells, neurons, neurosecretory cells, and glial cells. The mass of axons makes up the neuropile in the center of the hemispheres. The lateral part of each hemisphere is occupied by the inner and outer proliferation centers. These layers of neuroblasts give rise to the optic lobes during metamorphorsis *(103)*.

Mutations in several tumor suppressor genes cause dramatic enlargement of the two brain hemispheres, usually with no noticeable effect on the ventral ganglion. In three cases that have been investigated in detail, *lethal (2)giant larvae (lgl); brain tumor (brat)*; and *lethal (3) malignant brain tumor [l(3)mbt]*, the mutant brain hemispheres contain large numbers of undifferentiated cells, interpreted as small undifferentiated neuroblasts (*brat*) and ganglion mother cells [*lgl* and *l(3)mbt*]. The *lgl* brain hemispheres contain all of the cell types found in wild-type animals, but they are intermingled rather than separated into well-defined layers and proliferation centers *(104,105)*. In *l(3)mbt*, the brain is disorganized and the excess growth seems to obliterate at least part of the neuropile and causes fragmentation of the remaining neuropile *(106)*. In all three mutants, pieces of the mutant larval brain continue to grow by cell proliferation when transplanted into the abdomens of wild-type adults. Additionally, tissues from *lgl* and *brat* mutant brains invade host tissues such as the ovaries, fat body, gut, and thoracic muscles *(94)*. *brat* encodes an RNA-binding protein involved in translation repression, and *l(3)mbt* is a zinc-finger protein, suggesting that both of these proteins play more general roles in proliferation control than *lgl (107–110)*.

5.4.1.6. OVARIAN/FEMALE STERILE PHENOTYPES

Another tissue shown to have overgrowth phenotypes is the ovary. This is important because the germarium, where oocytes are formed, is a well-defined stem cell niche. Stem cells in the ovary regularly divide to produce one stem cell and one germline cell, which goes on to divide and differentiate, forming female gametes. In the normal ovary each egg chamber or cyst contains one oocyte and 15 nurse cells, which together represent the progeny from four successive divisions of a single cystoblast *(111,112)*. Ovarian tumor mutants are characterized by the production of large numbers of poorly differentiated germ cells in the adult ovary, resulting in female sterility.

The *bag of marbles (bam)* gene encodes a novel protein that, when mutated, prevents the differentiation of stem cells, resulting in ovarian tumors *(113)*. Mutations in another gene, *benign(2) gonial cell neoplasm (bgcn)*, have a similar phenotype and can synergize genetically with mutations in *bam (114–116)*. These genes function in a pathway that governs the maintenance of the ovarian stem cell niche. The product of another gene, *ovarian tumor (otu)*, results in a similar overgrowth phenotype. The *otu* protein product localizes to the cytoplasm and associates with the cytoskeleton *(117)*.

5.4.1.7. DROSOPHILA LEUKEMIA

Aside from solid tissue phenotypes, mutants could result in overproliferation of *Drosophila* blood cells. Mutations at a number of loci lead to a characteristic larval phenotype including hyperplastic hematopoietic organs, excess blood cells in circulation, and melanizing aggregates formed by mutant blood cells. Melanotic "tumors" are believed to be due to an abnormal immune response from the overproliferated blood cells, as *Drosophila* blood cells normally sequester foreign matter into melanotic bodies. It should be noted that melanotic "tumors" are not tumors, but rather a result of an abnormal immune response. It must be determined whether these melanotic masses are a result of overproliferated blood cells behaving abnormally, or alternatively, deregulated control of the immune response *(118,119)*.

Mutations in the *hopscotch (hop)* gene cause a distinct underproliferation of larval blood cells. Interestingly, a dominant mutation at this locus, called Tumorous-lethal (hop^{Tum-l}), causes overproliferation of the larval blood cells and the formation of melanotic bodies. The *hopscotch* gene encodes a Jak kinase, which is instrumental in regulation of the cell cycle in flies as well as mammals *(120–122)*.

Animals mutant for the air8 locus show a generalized slowing of larval growth, but also exhibit gross hypertrophy, hyperplasia and melanization of the larval lymph glands. Blood cells are overproduced and their differentiation is altered such that lamellocytes, not normally produced until the late larval or pupal stage, are found both in the lymph glands and circulation. These lamellocytes aggregate and form melanotic tumors. The air8 locus encodes a *Drosophila* homolog of the S6 ribosomal protein *(123)*.

5.4.2. Cell Growth

As alluded to above, it is important to distinguish exactly how a gene affects tissue size. Cell growth and cell proliferation are two separate phenomena, and should be addressed independently *(124)*. A few recessive mutations have been shown to affect the overall size of an organism by altering cell size, notably the *Drosophila* homologues of the *myc* oncogene (*dMyc*) *(125)* and components of the insulin signaling pathway. Three components of this pathway turn out to be homologs of the known human tumor suppressor genes, PTEN, TSC1, and TSC2 *(43–48,126)*. Most mutations affecting cell growth, however, would be expected to result in zygotic lethality. So, as in mutants affecting cell proliferation, mosaic analysis in imaginal discs is well suited to study mutants expected to alter cell growth. In adults, clones in the eye can be inspected for larger than normal compound eye segments, called ommatidia, under a dissecting scope, or using scanning electron microscopy. Larger ommatidia have been shown to result from increased cell size. This can be confirmed by histologic section of the adult eye *(127)*. Clones in the wing can be monitored for the density and size of wing bristles, one of which corresponds to one cell (i.e., lower bristle density may mean larger cells). Again, clones of such genes may not affect overall tissue size in the background of wild-type tissue, but cell ablation techniques such as the *hid* technique can produce entirely mutant tissues that are larger than normal, as seen with mutants in the *Drosophila* homologs of PTEN and TSC1. Mutant cell size can be quantitated by expressing GFP in these cells, then using FACS to sort for cell size.

5.4.3. Cell Death

During embryonic development, a subset of cells is fated to die. These dying cells occur at predictable locations and at predictable times. As a result, some of the *Drosophila* genes involved in the regulation of apoptosis were originally found by staining homozygous mutant embryos with the vital dye acridine orange, which selectively labels dying cells, and screening for mutants with altered patterns of apoptosis *(16,19–21)*. Therefore, if the gene of interest plays a role in apoptosis, it could also alter this pattern of acridine orange staining. Although acridine orange can stain tissues in living animals, these animals can also be fixed and stained by TUNEL labeling, which labels DNA fragmentation *(128)*. Ectopic cell death can be analyzed in vivo by combining the mutant with a transgene that expresses the baculovirus caspase inhibitor p35 *(129)*. If the gene plays a regulatory role in apoptosis, its phenotype will likely be suppressed by expression of p35. Another test involves exposing the animal to a source of X-rays or UV light, then assaying for the ability of a mutation or transgene to prevent radiation-induced apoptosis *(130)*.

Clonal analysis of cell death can be efficiently accomplished in the developing adult eye, which, like the embryo, also contains a number of cells normally fated to die *(84)*. If animals have mutant clones that result in ectopic cell death, it is possible that clones in the adult may not be recovered. In this case, only twin spot tissue will be observed. A regulatory role in apoptosis can be determined by expression of p35 in the mutant background, or imaginal discs can be dissected earlier in development and stained with acridine orange or TUNEL labeling.

5.4.4. Cell Differentiation

As stated before, some of the genes involved in epithelial and colon cancers were first discovered in *Drosophila* as genes important for embryonic patterning. Likewise, the Ras/Egfr signaling pathway was fleshed out in both *Drosophila* and *C. elegans* by studying eye and vulval cell differentiation, respectively *(131,132)*. Therefore, it could be likely that many genes interesting to the cancer researcher are normally involved in early development and differentiation. Many of the genes involved in the patched and wingless pathways have severe embryonic phenotypes *(133,134)*. The cuticles of embryos are divided into multiple segments, and defects in this regular segmentation pattern can be visualized by making cuticle preparations of such embryos *(135)*. Again, maternal contributions can mask such phenotypes, and this can be bypassed by using the FLP-DFS technique. Another phenotype, known to be due to EGFR signaling, is that of dorsal appendage placement. The dorsal appendages are two stalks that sprout from the dorsal anterior side of the embryo, used to provide air to the developing organism. Alteration of EGFR signaling can distort the placement of these appendages *(136–138)*.

Clonal and FLP-out analysis with genes involved in differentiation can have similar effects on body patterning and differentiation. Phenotypes can include doubling or loss of body segments in the adult legs and wings *(72,139)*. Again, because the eye is dispensable for viability and fertility, it is particularly suited for developmental studies. The ommatidia are formed from eight photoreceptor cells arranged in a regular, repeating pattern. These cells are surrounded by other accessory cells such as pigment cells. Major alterations in this pattern can be observed under the dissecting scope as a "rough" or disorganized eye phenotype. More specific characterizations of pattern defects can be identified using histologic sections, or by immunostaining with antibodies against the elav protein (a marker of neural differentiation) *(127,140)*. For example, signaling through Ras can affect the differentiation of the R7 photoreceptor cell.

5.4.5. Cell Adhesion

Many of the cell adhesion molecules important to mammalian epithelial cell–cell and cell–substratum adhesion are structurally and functionally conserved in *Drosophila (141)*.

Loss of cadherin-dependent cell–cell adhesion causes many embryonic structures to fall apart, including the cuticle, malphigian tubules (the *Drosophila* kidney), tracheal branches, and dorsal vessel (the *Drosophila* heart) *(142,143)*. The latter three structures are epithelial tubes, and defects in these organs indicate that loss of cadherin-dependent cell–cell adhesion has strong effects on tissues experiencing mechanical stress during organogenesis. Defects in internal epithelial structures can be visualized in embryos using enhancer traps and antibodies that stain these structures.

Integrin-dependent cell–substratum adhesion is equally important in *Drosophila* development *(144–146)*. Loss of integrin function in homozygous mutants results in a number of defects in morphogenesis. One notable phenotype is the inability of muscles to attach to their proper sites in the larvae. These phenotypes are observed in a variety of integrin and integrin ligand mutations. Again, these phenotypes are visualized by using appropriate enhancer traps and antibodies.

Drosophila imaginal discs are surrounded by a basement membrane similar to that in contact with mammalian epithelial cells. During pupal development, the basement membrane can degrade and reform, allowing the tissue movements that accompany differentiation of adult structures from the imaginal discs *(147)*. Mosaic analysis of integrin and other mutants underscore the importance of this dynamic cell–substratum contact. Mutant clones in adult wings can result in the formation of wing blisters, an overtly visible phenotype due to the inability of the dorsal and ventral surfaces of the wing to adhere *(148)*.

5.4.6. Cell Migration and Invasion

A number of cell types in *Drosophila* have stereotyped movements during development *(149,150)*. If the gene of interest is suspected to play a role in altering migratory or invasive phenotypes, it may alter some of the processes described below. Although there are other types of characterized cell migrations in *Drosophila*, the ones below have been studied and utilized the most.

In the early syncytial embryo, primordial germ cells (PGCs) are some of the first cells to form. Fated to become the adult germline, PGCs are first seen as approximately 10 cells in the posterior of the embryo. These cells divide, and eventually acquire the capability to migrate actively through an epithelial layer, later associating with mesoderm *(151)*. At this point, PGCs have also been shown to migrate in vitro along different types of extracellular matrix components *(152)*. PGCs eventually split into two laterally displaced groups of cells and reside at abdominal segment 5 in stage 14 embryos. PGCs can be specifically labeled in the embryos with antibodies to the vasa protein, an RNA helicase required for PGC formation *(153)*.

In *Drosophila* ovaries, a group of 6–10 cells called border cells has been intensively studied as a model for cell migration. These cells dissociate from a follicular epithelium surrounding each developing egg chamber, and migrate through a cluster of germ cells before associating anterior to the oocyte. A number of mutations can affect border cell migration, highlighting the effects of hormones, E-cadherin, small GTPases, the actin cytoskeleton, and focal adhesion kinase, among others, on this process *(154–160)*. Border cells can be visualized by staining egg chamber nuclei with DAPI or by staining for F-actin with rhodamine-phalloidin. Also, a number of enhancer traps are available that specifically label border cells.

Aside from affecting developmentally programmed cell migrations, some *Drosophila* mutations and transgenes can create invasive phenotypes. For example, overexpression of the *Drosophila* PAK homolog has been shown to disturb the normal projection of axons from the retina into the optic lobe. However, when higher levels of PAK are expressed, not only axons, but also cell bodies, can migrate into the optic lobe *(161)*. These phenotypes can be observed in histologic sections of the adult optic lobe.

In addition, it is possible to culture *Drosophila* tissues by transplantation into adult or larval hosts. Wild-type imaginal discs do not grow beyond their expected size when transplanted. However, a number of mutant tissues can not only grow, but also invade host tissues, eventually killing the host *(94)*. Interestingly, one of these mutants, *lethal giant larvae*, is known to express higher than normal levels of a type IV collagenase, as do many malignant human tumors *(95)*. Phenotypes can be visualized by transplanting mutant tissue tagged with a reporter such as lacZ, then staining either whole-mount or sectioned adults after tumors have had a chance to grow and spread.

5.4.7. DNA Repair

Mutations in double-strand break repair can have a number of phenotypes in *Drosophila*, including patterning defects and female sterility. Mutants may also show no overt phenotype unless exposed to agents that induce double-strand breaks *(15,162,163)*.

Defects in mitotic repair can be observed by exposing mutants to X-rays or methylmethanesulfonate (MMS). Viability will be lowered if mutants cannot perform double-strand break repair functions. These defects can also be assayed by the somatic mutation and recombination test (SMART). In this test, the fly is heterozygous for a recessive mutation that causes defect in eye pigmentation. In the presence of a mutagen, double-strand breaks and mitotic recombination will occur, resulting in the presence of eye cells homozygous for the eye pigmentation defect. This is easily observed as patches of white ommatidia in an otherwise red eye. With increasing mutagen dosage, more white patches are observed. This dose–response pattern can be altered in the background of a mutation or transgene that affects mitotic DNA repair *(164)*.

Meiotic defects can be easily observed in a number of ways (for an example, *see* **ref. 165**). In one assay, females are heterozygous for a number of visible marker mutations along one chromosome. The progeny of these females will normally include a number of recombinants due to crossing over with the wild-type chromosome during meiosis. The amount of recombinant progeny is proportional to the distance of these mutations away from each other along the chromosome arm. If the mutation or transgene being studied affects this process, it may lower the proportion of recombinant progeny. Another test assays for the presence of X-chromosome nondisjunction in a mutant background. In this case, mutants also have marked X or Y chromosomes. This allows for one to detect the presence of XO males and XXY females (the sex of *Drosophila* is determined by the ratio of X chromosomes to autosomes rather than the presence or absence of a Y chromosome). Percent nondisjunction is calculated by the equation

$$\%_{nondisjunction} = \frac{2N_{nondisjunction}}{N_{total}} \times 100$$

where $N_{nondisjunction}$ = the number of progeny with X-chromosome nondisjunction and N_{total} = the total number of progeny.

6. Building Pathways Using Genetic Analysis

One of the greatest advantages to using *Drosophila* is that genetic interactions can be easily studied in vivo, and these genetic interactions can be used to order genes into pathways. This is done by either crossing the original line to lines that contain mutations in candidate genes, or by screening newly induced mutations with the original line, and then selecting those that can alter the initial phenotype. Such genes are called modifiers, and these can be further grouped into enhancers (which make the original mutant phenotype worse) and suppressors (which make the mutant phenotype less severe).

6.1. Testing Candidate Modifiers

If one has either a loss-of-function mutation or a transgene that displays an interesting phenotype, i.e., overproliferation, one can cross mutations in candidate genes, i.e., genes involved in the cell cycle, to see if they have a modifying effect on the initial phenotype. Flies heterozygous for most genes do not display mutant phenotypes on their own. However, in a mutant background where the signal of a given pathway is already reduced, removal of one copy of a gene in the same pathway may change the mutant phenotype, indicating that these genes may interact.

6.2. Modifier Screens

Crossing in candidate genes is an easy way to correlate the function of the gene of interest with genes in known pathways. This approach, however, is biased toward genes that have already been characterized. One may miss important interacting genes. An alternative is to generate new mutations randomly, then screen for those that modify the original phenotype. This technique, called a modifier screen, is extremely useful in that it is an unbiased approach to find new genes that interact with the gene of interest. In a dominant modifier screen, newly induced mutations (X-ray, chemical, or P-element mutations) are crossed to a tester strain with an observable phenotype (i.e., a transgene in the gene of interest that affects tissue size). Practically all of these mutations will be loss of function, and have no observable heterozygous phenotype. However, if this heterozygous mutations alters (i.e., enhances or suppresses) the tester phenotype, this is a strong genetic interaction that implies the gene uncovered by the mutation is in the same pathway as the gene of interest (for an example, *see* **ref.** *166*).

6.3. Epistasis Analysis

After mutations in interacting genes are identified, they can be further analyzed and actually ordered in a linear pathway (at least in the case where gene products function in a cascade, as in signal transduction) by means of epistasis tests *(167,168)*. In an epistasis test, two mutations with opposite phenotypes (i.e., one overproliferation and one underproliferation) are crossed to each other, creating a double-mutant animal. If loss-of-function mutations are used, it is essential for the results of this test that the mutations are null alleles. If the genes are not in the same pathway, then one would expect an intermediate phenotype. However, if these genes function together in a linear pathway, then only the phenotype of one mutation will be observed. In this case, the gene defined by this mutation is said to be epistatic to the other. If the mutations being analyzed are both null alleles, then the epistatic mutation is likely to be downstream of the other. The same applies for analyzing two gain-of-function alleles or transgenes with opposite phenotypes.

Analyzing two null mutations with similar phenotypes will not give meaningful results unless gain-of-function alleles in one of the genes are available. In this case, transgenic constructs are essential to order genes in a pathway. A fly strain is made with one of the null mutants and a transgene for the other. If the loss-of-function mutation is epistatic, then it is downstream of the other. If the transgenic construct is epistatic, then it may be downstream, or it may function in a parallel pathway. This can be resolved by making a fly with the other null mutant and a transgene for the first gene. In this case, if the loss-of-function gene is epistatic, then it is the downstream gene. If the transgenic construct is epistatic, then the genes probably function in a parallel pathway.

As mutants with similar phenotypes are found, they can all be ordered into a pathway by performing epistasis analysis with null mutants and transgenes. However, if genes do not function in linear or strictly parallel pathways, epistasis tests may not yield easily interpretable results. One specialized case is that of nonallelic noncomplementation. In this case, two mutations with similar phenotypes can result in a mutant phenotype if both mutations are heterozygous in the same animal. This is normally observed only if mutations are in the same gene (called noncomplementation). However, if this occurs with mutations in two separate genes, a very rare occurrence, it implies that the two gene products may function as binding partners. If gene products function in a multiprotein complex or in multiple signaling pathways, epistasis results may become increasingly confusing. This sort of situation is beyond the range of what genetic analysis can do on its own. However, when combined with biochemical data, genetic analysis in such situations still provides important in-vivo data that can confirm in-vitro results.

7. Limitations of the *Drosophila* System

Despite the knowledge gained by genetic experiments in *Drosophila*, there are, of course, a number of limitations when interpreting results. The astounding similarity of molecular mechanisms conserved throughout evolution should not blur the fundamental differences among organisms. Following are a few examples.

Although many genes are conserved throughout evolution, there are some genes important to cancer in humans that have no convincing homologs in *Drosophila (3)*. In addition, physiologic processes can also differ greatly among species. For example, questions such as why people with mutations in Rb, a gene apparently crucial to organismal cell cycle regulation, are primarily susceptible to retinoblastomas are very difficult for the *Drosophila* system to address *(169)*.

In addition, while *Drosophila* does have a vascular system for delivery of oxygen, blood cells float in an open circulatory system. The tracheal system has been used as a model for vascular development, and indicates genes that regulate this process, as well as the role of hypoxia in the induction of vascularization *(170–172)*. However, the tracheal network is an epithelial one, and appears to differ from the endothelial vasculature in mammals. *Drosophila* does not have convincing homologs of a number of genes involved in endothelial development and function *(141)*. This, to some degree, prevents one from studying genes involved in tumor vascularization and angiogenesis.

Because of the availability of cell lines derived from normal and cancerous tissues, it has been much easier to do biochemical analyses or cell cycle characterization in mammalian cell culture systems. Creation of *Drosophila* cell lines has not been as easy a task, and available cell lines behave somewhat differently from mammalian ones *(173)*. This limits the ability to relate tissue culture experiments in flies with experiments using mammalian cells.

8. Conclusion

With the completion of the human genome sequence, it now falls on researchers to analyze the functions of discovered genes both in normal development as well as in pathologic processes, and to order these genes into complex signaling networks. The use of genetically tractable model organisms provides us with a forum to address these issues. Equipped with the depth and breadth of genetic analytical tools described above, the *Drosophila* model has proved, and will continue to prove, in combination with the powerful techniques used in mammalian systems, an important player in unraveling the in-vivo functions of genes and signaling pathways.

References

1. Russell, P. (1998) Checkpoints on the road to mitosis. *Trends Biochem. Sci.* **23**, 399–402.
2. Wassmann, K. and Benezra, R. (2001) Mitotic checkpoints: from yeast to cancer. *Curr. Opin. Genet. Dev.* **11**, 83–90.
3. Rubin, G. M., Yandell, M. D., Wortman, J. R., et al. (2000) Comparative genomics of the eukaryotes. *Science* **287**, 2204–2215.
4. Hahn, H., Wojnowski, L., Miller, G., and Zimmer, A. (1999) The patched signaling pathway in tumorigenesis and development: lessons from animal models. *J. Mol. Med.* **77**, 459–468.

5. Booth, D. R. (1999) The hedgehog signalling pathway and its role in basal cell carcinoma. *Cancer Metastasis Rev.* **18,** 261–284.

6. Siegfried, E. and Perrimon, N. (1994) *Drosophila* wingless: a paradigm for the function and mechanism of Wnt signaling. *Bioessays* **16,** 395–404.

7. Hahn, H., Wicking, C., Zaphiropoulous, P. G., et al. (1996) Mutations of the human homolog of *Drosophila* patched in the nevoid basal cell carcinoma syndrome. *Cell* **85,** 841–851.

8. Nusslein-Volhard, C., Wieschaus, E., and Kluding, H. (1984) Mutations affecting the pattern of the larval cuticle in *Drosophila melanogaster. Roux's Arch. Dev. Biol.* **193,** 267–282.

9. Vidwans, S. J. and Su, T. T. (2001) Cycling through development in *Drosophila* and other metazoa. *Nat. Cell Biol.* **3,** E35–E39.

10. Edgar, B. A. and Lehner, C. F. (1996) Developmental control of cell cycle regulators: a fly's perspective. *Science* **274,** 1646–1652.

11. Orr-Weaver, T. L. (1994) Developmental modification of the *Drosophila* cell cycle. *Trends Genet.* **10,** 321–327.

12. Oldham, S., Bohni, R., Stocker, H., Brogiolo, W., and Hafen, E. (2000) Genetic control of size in *Drosophila. Phil. Trans. R. Soc. Lond. B, Biol. Sci.* **355,** 945–952.

13. Stocker, H. and Hafen, E. (2000) Genetic control of cell size. *Curr. Opin. Genet. Dev.* **10,** 529–535.

14. Potter, C. J. and Xu, T. (2001) Mechanisms of size control. *Curr. Opin. Genet. Dev.* **11,** 279–286.

15. Sekelsky, J. J., Brodsky, M. H., and Burtis, K. C. (2000) DNA repair in *Drosophila:* insights from the *Drosophila* genome sequence. *J. Cell Biol.* **150,** F31–F36.

16. Abrams, J. M. (1999) An emerging blueprint for apoptosis in *Drosophila. Trends Cell Biol.* **9,** 435–440.

17. Steller, H., Abrams, J. M., Grether, M. E., and White, K. (1994) Programmed cell death in *Drosophila. Phil. Trans. R. Soc. Lond. B, Biol. Sci.* **345,** 247–250.

18. Vernooy, S. Y., Copeland, J., Ghaboosi, N., Griffin, E. E., Yoo, S. J., and Hay, B. A. (2000) Cell death regulation in *Drosophila*: conservation of mechanism and unique insights. *J. Cell Biol.* **150,** F69–F76.

19. Meier, P., Finch, A., and Evan, G. (2000) Apoptosis in development. *Nature* **407,** 796–801.

20. Lee, C. Y. and Baehrecke, E. H. (2000) Genetic regulation of programmed cell death in *Drosophila. Cell Res.* **10,** 193–204.

21. Bangs, P. and White, K. (2000) Regulation and execution of apoptosis during *Drosophila* development. *Dev. Dyn.* **218,** 68–79.

22. Joutel, A. and Tournier-Lasserve, E. (1998) Notch signalling pathway and human diseases. *Semin. Cell Dev. Biol.* **9,** 619–625.

23. Cohen, S. M. (1993) Imaginal disc development, in *The Development of Drosophila Melanogaster* (Bate, M., and Martinez Arias, A., eds.). Cold Spring Harbor Laboratory Press, Cold Spring Harbor, NY, vol. 2, pp. 747–842.

24. Bryant, P. J. and Schmidt, O. (1990) The genetic control of cell proliferation in *Drosophila* imaginal discs. *J. Cell Sci. Suppl.* **13,** 169–189.

25. Capdevila, J. and Johnson, R. L. (2000) Hedgehog signaling in vertebrate and invertebrate limb patterning. *Cell. Mol. Life Sci.* **57,** 1682–1694.

26. Gehring, W. J. (1996) The master control gene for morphogenesis and evolution of the eye. *Genes Cells* **1,** 11–15.

27. Chen, J. N. and Fishman, M. C. (2000) Genetics of heart development. *Trends Genet.* **16,** 383–388.

28. Myers, E. W., Sutton, G. G., Delcher, A. L., et al. (2000) A whole-genome assembly of *Drosophila. Science* **287,** 2196–2204.

29. Adams, M. D., Celniker, S. E., Holt, R. A., et al. (2000) The genome sequence of *Drosophila melanogaster. Science* **287,** 2185–2195.

30. Ashburner, M. (1989) Chromosomes, in *Drosophila—A Laboratory Manual* (Ashburner, M., ed.). Cold Spring Harbor Laboratory Press, Cold Spring Harbor, NY, vol. 2, pp. 21–72.

31. Ashburner, M. (1989) Balancers and other special chromosomes, in *Drosophila—A Laboratory Manual* (Ashburner, M., ed.). Cold Spring Harbor Laboratory Press, Cold Spring Harbor, NY, vol. 2, pp. 529–548.

32. Huang, A. M., Rehm, E. J., and Rubin, G. M. (2000) Recovery of DNA sequences flanking P-element insertions: inverse PCR and plasmid rescue, in *Drosophila Protocols* (Sullivan, W., Ashburner, M., and Hawley, R. S., eds.). Cold Spring Harbor Laboratory Press, Cold Spring Harbor, NY, pp. 429–438.

33. Spradling, A. C. (1986) P-element mediated transformation, in *Drosophila: A Practical Approach* (Roberts, D., ed.). IRL Press, Oxford, UK, pp. 175–198.

34. Stark, M. B. (1918) An hereditary tumor in *Drosophila. J. Cancer Res.* **3,** 279–301.

35. Gateff, E. (1978) Malignant neoplasms of genetic origin in *Drosophila melanogaster. Science* **200,** 1448–1459.

36. Gateff, E. and Schneiderman, H. A. (1969) Neoplasms in mutant and cultured wild-type tissues of *Drosophila. Natl. Cancer Inst. Monogr.* **31,** 365–397.

37. Watson, K. L., Justice, R. W., and Bryant, P. J. (1994) *Drosophila* in cancer research: the first fifty tumor suppressor genes. *J. Cell Sci. Suppl.* **18,** 19–33.

38. Gateff, E. and Mechler, B. M. (1989) Tumor-suppressor genes of *Drosophila melanogaster. Crit. Rev. Oncog.* **1,** 221–245.

39. Xu, T., Wang, W., Zhang, S., Stewart, R. A., and Yu, W. (1995) Identifying tumor suppressors in genetic mosaics: the *Drosophila lats* gene encodes a putative protein kinase. *Development* **121,** 1053–1063.

40. Tao, W., Zhang, S., Turenchalk, G. S., et al. (1999) Human homologue of the *Drosophila melanogaster* lats tumour suppressor modulates CDC2 activity. *Nat. Genet.* **21,** 177–181.

41. St John, M. A., Tao, W., Fei, X., et al. (1999) Mice deficient of Lats1 develop soft-tissue sarcomas, ovarian tumours and pituitary dysfunction. *Nat. Genet.* **21,** 182–186.

42. Waltzer, L. and Bienz, M. (1999) The control of beta-catenin and TCF during embryonic development and cancer. *Cancer Metastasis Rev.* **18,** 231–246.

43. Huang, H., Potter, C. J., Tao, W., et al. (1999) PTEN affects cell size, cell proliferation and apoptosis during *Drosophila* eye development. *Development* **126,** 5365–5372.

44. Tapon, N., Ito, N., Dickson, B. J., Treisman, J. E., and Hariharan, I. K. (2001) The *Drosophila* tuberous sclerosis complex gene homologs restrict cell growth and cell proliferation. *Cell* **105,** 345–355.

45. Potter, C. J., Huang, H., and Xu, T. (2001) *Drosophila* Tsc1 functions with Tsc2 to antagonize insulin signaling in regulating cell growth, cell proliferation, and organ size. *Cell* **105,** 357–368.

46. Gao, X., Neufeld, T. P., and Pan, D. (2000) *Drosophila* PTEN regulates cell growth and proliferation through PI3K-dependent and -independent pathways. *Dev. Biol.* **221,** 404–418.

47. Gao, X. and Pan, D. (2001) TSC1 and TSC2 tumor suppressors antagonize insulin signaling in cell growth. *Genes Dev.* **15,** 1383–1392.

48. Goberdhan, D. C., Paricio, N., Goodman, E. C., Mlodzik, M., and Wilson, C. (1999) *Drosophila* tumor suppressor PTEN controls cell size and number by antagonizing the Chico/PI3-kinase signaling pathway. *Genes Dev.* **13,** 3244–3258.

49. Ashburner, M. (1989) Mutation and mutagenesis, in *Drosophila—A Laboratory Handbook* (Ashburner, M., ed.). Cold Spring Harbor Laboratory Press, Cold Spring Harbor, NY, vol. 2, pp. 299–418.

50. Cooley, L., Kelley, R., and Spradling, A. (1988) Insertional mutagenesis of the *Drosophila* genome with single P elements. *Science* **239,** 1121–1128.

51. Spradling, A. C., Stern, D. M., Kiss, I., Roote, J., Laverty, T., and Rubin, G. M. (1995) Gene disruptions using P transposable elements: an integral component of the *Drosophila* genome project. *Proc. Natl. Acad. Sci. USA* **92,** 10824–10830.

52. Xu, T. and Rubin, G. M. (1993) Analysis of genetic mosaics in developing and adult *Drosophila* tissues. *Development* **117,** 1223–1237.

53. Xu, T. and Harrison, S. D. (1994) Mosaic analysis using FLP recombinase. *Meth. Cell Biol.* **44,** 655–681.

54. Golic, K. G. (1991) Site-specific recombination between homologous chromosomes in *Drosophila*. *Science* **252,** 958–961.

55. Geyer, P. K., Richardson, K. L., Corces, V. G., and Green, M. M. (1988) Genetic instability in *Drosophila melanogaster*: P-element mutagenesis by gene conversion. *Proc. Natl. Acad. Sci. USA* **85,** 6455–6459.

56. Margulies, L. and Griffith, C. S. (1991) The synergistic effect of X-rays and deficiencies in DNA repair in P-M hybrid dysgenesis in *Drosophila melanogaster*. *Genet. Res.* **58,** 15–26.

57. Chen, B., Chu, T., Harms, E., Gergen, J. P., and Strickland, S. (1998) Mapping of *Drosophila* mutations using site-specific male recombination. *Genetics* **149,** 157–163.

58. Golic, K. G. (1994) Local transposition of P elements in *Drosophila melanogaster* and recombination between duplicated elements using a site-specific recombinase. *Genetics* **137,** 551–563.

59. Fire, A., Xu, S., Montgomery, M. K., Kostas, S. A., Driver, S. E., and Mello, C. C. (1998) Potent and specific genetic interference by double-stranded RNA in *Caenorhabditis elegans* [see comments]. *Nature* **391,** 806–811.

60. Kennerdell, J. R. and Carthew, R. W. (1998) Use of dsRNA-mediated genetic interference to demonstrate that frizzled and frizzled 2 act in the wingless pathway. *Cell* **95,** 1017–1026.

61. Misquitta, L. and Paterson, B. M. (1999) Targeted disruption of gene function in *Drosophila* by RNA interference (RNA-i): a role for nautilus in embryonic somatic muscle formation. *Proc. Natl. Acad. Sci. USA* **96,** 1451–1456.

62. Caplen, N. J., Fleenor, J., Fire, A., and Morgan, R. A. (2000) dsRNA-mediated gene silencing in cultured *Drosophila* cells: a tissue culture model for the analysis of RNA interference. *Gene* **252,** 95–105.

63. Kennerdell, J. R. and Carthew, R. W. (2000) Heritable gene silencing in *Drosophila* using double-stranded RNA. *Nat. Biotechnol.* **18,** 896–898.

64. Rong, Y. S. and Golic, K. G. (2000) Gene targeting by homologous recombination in *Drosophila*. *Science* **288,** 2013–2018.

65. Rong, Y. S. and Golic, K. G. (2001) A targeted gene knockout in *Drosophila*. *Genetics* **157,** 1307–1312.

66. Hama, C., Ali, Z., and Kornberg, T. B. (1990) Region-specific recombination and expression are directed by portions of the *Drosophila* engrailed promoter. *Genes Dev.* **4,** 1079–1093.

67. Hauck, B., Gehring, W. J., and Walldorf, U. (1999) Functional analysis of an eye specific enhancer of the eyeless gene in *Drosophila*. *Proc. Natl. Acad. Sci. USA* **96,** 564–569.

68. Kim, Y. J. and Baker, B. S. (1993) The *Drosophila* gene *rbp9* encodes a protein that is a member of a conserved group of putative RNA binding proteins that are nervous system-specific in both flies and humans. *J. Neurosci.* **13,** 1045–1056.

69. Bonner, J. J., Parks, C., Parker-Thornburg, J., Mortin, M. A., and Pelham, H. R. (1984) The use of promoter fusions in *Drosophila* genetics: isolation of mutations affecting the heat shock response. *Cell* **37,** 979–991.

70. Brand, A. H. and Perrimon, N. (1993) Targeted gene expression as a means of altering cell fates and generating dominant phenotypes. *Development* **118,** 401–415.

71. Rorth, P., Szabo, K., Bailey, A., et al. (1998) Systematic gain-of-function genetics in *Drosophila*. *Development* **125**, 1049–1057.
72. Struhl, G. and Basler, K. (1993) Organizing activity of wingless protein in *Drosophila*. *Cell* **72**, 527–540.
73. St John, M. A. and Xu, T. (1997) Understanding human cancer in a fly? *Am. J. Hum. Genet.* **61**, 1006–1010.
74. Lee, T. and Luo, L. (2001) Mosaic analysis with a repressible cell marker (MARCM) for *Drosophila* neural development. *Trends Neurosci.* **24**, 251–254.
75. Lee, T. and Luo, L. (1999) Mosaic analysis with a repressible cell marker for studies of gene function in neuronal morphogenesis. *Neuron* **22**, 451–461.
76. Neufeld, T. P., de la Cruz, A. F., Johnston, L. A., and Edgar, B. A. (1998) Coordination of growth and cell division in the *Drosophila* wing. *Cell* **93**, 1183–1193.
77. Morata, G. and Ripoll, P. (1975) Minutes: mutants of *Drosophila* autonomously affecting cell division rate. *Dev. Biol.* **42**, 211–221.
78. Chou, T. B. and Perrimon, N. (1996) The autosomal FLP-DFS technique for generating germline mosaics in *Drosophila melanogaster*. *Genetics* **144**, 1673–1679.
79. Stowers, R. S. and Schwarz, T. L. (1999) A genetic method for generating *Drosophila* eyes composed exclusively of mitotic clones of a single genotype. *Genetics* **152**, 1631–1639.
80. Foe, V. E., Odell, G. M., and Edgar, B. A. (1993) Mitosis and morphogenesis in the *Drosophila* embryo: point and counterpoint, in *The Development of Drosophila Melanogaster* (Bate, M., and Martinez Arias, A., eds.). Cold Spring Harbor Laboratory Press, Cold Spring Harbor, NY, vol. 1, pp. 149–300.
81. Su, T. T., Campbell, S. D., and O'Farrell, P. H. (1999) *Drosophila* grapes/CHK1 mutants are defective in cyclin proteolysis and coordination of mitotic events. *Curr. Biol.* **9**, 919–922.
82. Grosshans, J. and Wieschaus, E. (2000) A genetic link between morphogenesis and cell division during formation of the ventral furrow in *Drosophila*. *Cell* **101**, 523–531.
83. Hazelrigg, T. (2000) GFP and other reporters, in *Drosophila Protocols* (Sullivan, W., Ashburner, M., and Hawley, R. S., eds.). Cold Spring Harbor Laboratory Press, Cold Spring Harbor, NY, pp. 313–344.
84. Wolff, T. (2000) Histological techniques for the *Drosophila* eye. Part I: larva and pupa, in *Drosophila Protocols* (Sullivan, W., Ashburner, M., and Hawley, R. S., eds.). Cold Spring Harbor Laboratory Press, Cold Spring Harbor, NY, pp. 201–228.
85. Blair, S. S. (2000) Imaginal discs, in *Drosophila Protocols* (Sullivan, W., Ashburner, M., and Hawley, R. S., eds.). Cold Spring Harbor Laboratory Press, Cold Spring Harbor, NY, pp. 159–174.
86. Woods, D. F., Wu, J. W., and Bryant, P. J. (1997) Localization of proteins to the apico-lateral junctions of *Drosophila* epithelia. *Dev. Genet.* **20**, 111–118.
87. Garoia, F., Guerra, D., Pezzoli, M. C., Lopez-Varea, A., Cavicchi, S., and Garcia-Bellido, A. (2000) Cell behaviour of *Drosophila* fat cadherin mutations in wing development. *Mech. Dev.* **94**, 95–109.
88. Mahoney, P. A., Weber, U., Onofrechuk, P., Biessmann, H., Bryant, P. J., and Goodman, C. S. (1991) The fat tumor suppressor gene in *Drosophila* encodes a novel member of the cadherin gene superfamily. *Cell* **67**, 853–868.
89. Bryant, P. J., Huettner, B., Held, L. I., Jr., Ryerse, J., and Szidonya, J. (1988) Mutations at the fat locus interfere with cell proliferation control and epithelial morphogenesis in *Drosophila*. *Dev. Biol.* **129**, 541–554.
90. Boedigheimer, M. J., Nguyen, K. P., and Bryant, P. J. (1997) Expanded functions in the apical cell domain to regulate the growth rate of imaginal discs. *Dev. Genet.* **20**, 103–110.
91. Boedigheimer, M. and Laughon, A. (1993) Expanded: a gene involved in the control of cell proliferation in imaginal discs. *Development* **118**, 1291–1301.

92. Blaumueller, C. M. and Mlodzik, M. (2000) The *Drosophila* tumor suppressor expanded regulates growth, apoptosis, and patterning during development. *Mech. Dev.* **92,** 251–262.
93. Jursnich, V. A., Fraser, S. E., Held, L. I., Jr., Ryerse, J., and Bryant, P. J. (1990) Defective gap-junctional communication associated with imaginal disc overgrowth and degeneration caused by mutations of the *dco* gene in *Drosophila. Dev. Biol.* **140,** 413–429.
94. Woodhouse, E., Hersperger, E., and Shearn, A. (1998) Growth, metastasis, and invasiveness of *Drosophila* tumors caused by mutations in specific tumor suppressor genes. *Dev. Genes Evol.* **207,** 542–550.
95. Woodhouse, E., Hersperger, E., Stetler-Stevenson, W. G., Liotta, L. A., and Shearn, A. (1994) Increased type IV collagenase in lgl-induced invasive tumors of *Drosophila. Cell Growth Differ.* **5,** 151–159.
96. Strand, D., Jakobs, R., Merdes, G., et al. (1994) The *Drosophila* lethal(2)giant larvae tumor suppressor protein forms homo-oligomers and is associated with nonmuscle myosin II heavy chain. *J. Cell Biol.* **127,** 1361–1373.
97. Mechler, B. M., McGinnis, W., and Gehring, W. J. (1985) Molecular cloning of lethal(2)giant larvae, a recessive oncogene of *Drosophila melanogaster. EMBO J.* **4,** 1551–1557.
98. Woods, D. F. and Bryant, P. J. (1989) Molecular cloning of the lethal(1)discs large-1 oncogene of *Drosophila. Dev. Biol.* **134,** 222–235.
99. Makino, K., Kuwahara, H., Masuko, N., et al. (1997) Cloning and characterization of NE-dlg: a novel human homolog of the *Drosophila* discs large (dlg) tumor suppressor protein interacts with the APC protein. *Oncogene* **14,** 2425–2433.
100. Hanada, N., Makino, K., Koga, H., et al. (2000) NE-dlg, a mammalian homolog of *Drosophila* dlg tumor suppressor, induces growth suppression and impairment of cell adhesion: possible involvement of down-regulation of beta-catenin by NE-dlg expression. *Int. J. Cancer* **86,** 480–488.
101. Bilder, D., Li, M., and Perrimon, N. (2000) Cooperative regulation of cell polarity and growth by *Drosophila* tumor suppressors. *Science* **289,** 113–116.
102. Bilder, D. and Perrimon, N. (2000) Localization of apical epithelial determinants by the basolateral PDZ protein Scribble. *Nature* **403,** 676–680.
103. Truman, J. W., Taylor, B. J., and Awad, T. A. (1993) Formation of the adult nervous system, in *The Development of Drosophila Melanogaster* (Bate, M., and Martinez Arias, A., eds.). Cold Spring Harbor Laboratory Press, Cold Spring Harbor, NY, vol. 2, pp. 1245–1276.
104. Peng, C. Y., Manning, L., Albertson, R., and Doe, C. Q. (2000) The tumour-suppressor genes *lgl* and *dlg* regulate basal protein targeting in *Drosophila* neuroblasts. *Nature* **408,** 596–600.
105. Ohshiro, T., Yagami, T., Zhang, C., and Matsuzaki, F. (2000) Role of cortical tumour-suppressor proteins in asymmetric division of *Drosophila* neuroblast. *Nature* **408,** 593–596.
106. Gateff, E., Loffler, T., and Wismar, J. (1993) A temperature-sensitive brain tumor suppressor mutation of *Drosophila melanogaster*: developmental studies and molecular localization of the gene. *Mech. Dev.* **41,** 15–31.
107. Wismar, J., Loffler, T., Habtemichael, N., et al. (1995) The *Drosophila melanogaster* tumor suppressor gene *lethal(3)malignant brain tumor* encodes a proline-rich protein with a novel zinc finger. *Mech. Dev.* **53,** 141–154.
108. Koga, H., Matsui, S., Hirota, T., Takebayashi, S., Okumura, K., and Saya, H. (1999) A human homolog of *Drosophila* lethal(3)malignant brain tumor (l(3)mbt) protein associates with condensed mitotic chromosomes. *Oncogene* **18,** 3799–3809.
109. Arama, E., Dickman, D., Kimchie, Z., Shearn, A., and Lev, Z. (2000) Mutations in the beta-propeller domain of the *Drosophila* brain tumor (brat) protein induce neoplasm in the larval brain. *Oncogene* **19,** 3706–3716.

110. Sonoda, J. and Wharton, R. P. (2001) *Drosophila* brain tumor is a translational repressor. *Genes Dev.* **15,** 762–773.

111. Spradling, A. C. (1993) Developmental genetics of oogenesis, in *The Development of Drosophila Melanogaster* (Bate, M., and Martinez Arias, A., Eds.). Cold Spring Harbor Laboratory Press, Cold Spring Harbor, NY, vol. 1, pp. 1–70

112. McKearin, D. and Christerson, L. (1994) Molecular genetics of the early stages of germ cell differentiation during *Drosophila* oogenesis. *Ciba Found. Symp.* **182,** 210–219.

113. McKearin, D. and Ohlstein, B. (1995) A role for the *Drosophila* bag-of-marbles protein in the differentiation of cystoblasts from germline stem cells. *Development* **121,** 2937–2947.

114. Gonczy, P., Matunis, E., and DiNardo, S. (1997) bag-of-marbles and benign gonial cell neoplasm act in the germline to restrict proliferation during *Drosophila* spermatogenesis. *Development* **124,** 4361–4371.

115. Lavoie, C. A., Ohlstein, B., and McKearin, D. M. (1999) Localization and function of Bam protein require the benign gonial cell neoplasm gene product. *Dev. Biol.* **212,** 405–413.

116. Ohlstein, B., Lavoie, C. A., Vef, O., Gateff, E., and McKearin, D. M. (2000) The *Drosophila* cystoblast differentiation factor, benign gonial cell neoplasm, is related to DExH-box proteins and interacts genetically with bag-of-marbles. *Genetics* **155,** 1809–1819.

117. King, R. C. and Storto, P. D. (1988) The role of the otu gene in *Drosophila* oogenesis. *Bioessays* **8,** 18–24.

118. Ghelelovitch, S. (1969) Melanotic tumors in *Drosophila melanogaster. Natl. Cancer Inst. Monogr.* **31,** 263–275.

119. Mathey-Prevot, B. and Perrimon, N. (1998) Mammalian and *Drosophila* blood: JAK of all trades? *Cell* **92,** 697–700.

120. Binari, R. and Perrimon, N. (1994) Stripe-specific regulation of pair-rule genes by hopscotch, a putative Jak family tyrosine kinase in *Drosophila. Genes Dev.* **8,** 300–312.

121. Hanratty, W. P. and Dearolf, C. R. (1993) The *Drosophila* tumorous-lethal hematopoietic oncogene is a dominant mutation in the hopscotch locus. *Mol. Gen. Genet.* **238,** 33–37.

122. Harrison, D. A., Binari, R., Nahreini, T. S., Gilman, M., and Perrimon, N. (1995) Activation of a *Drosophila* Janus kinase (JAK) causes hematopoietic neoplasia and developmental defects. *EMBO J.* **14,** 2857–2865.

123. Watson, K. L., Konrad, K. D., Woods, D. F., and Bryant, P. J. (1992) *Drosophila* homolog of the human S6 ribosomal protein is required for tumor suppression in the hematopoietic system. *Proc. Natl. Acad. Sci. USA* **89,** 11302–11306.

124. Su, T. T. and O'Farrell, P. H. (1998) Size control: cell proliferation does not equal growth. *Curr. Biol.* **8,** R687–R689.

125. Johnston, L. A., Prober, D. A., Edgar, B. A., Eisenman, R. N., and Gallant, P. (1999) *Drosophila* myc regulates cellular growth during development. *Cell* **98,** 779–790.

126. Ito, N. and Rubin, G. M. (1999) gigas, a *Drosophila* homolog of tuberous sclerosis gene product-2, regulates the cell cycle. *Cell* **96,** 529–539.

127. Wolff, T. (2000) Histological techniques for the *Drosophila* eye. Part II: adult, in *Drosophila Protocols* (Sullivan, W., Ashburner, M., and Hawley, R. S., eds.). Cold Spring Harbor Laboratory Press, Cold Spring Harbor, NY, pp. 229–244.

128. Sweeney, S. T., Hidalgo, A., deBelle, J. S., and Keshishian, H. (2000) Functional cell ablation, in *Drosophila Protocols* (Sullivan, W., Ashburner, M., and Hawley, R. S., eds.). Cold Spring Harbor Laboratory Press, Cold Spring Harbor, NY, pp. 449–478.

129. Hay, B. A., Wolff, T., and Rubin, G. M. (1994) Expression of baculovirus P35 prevents cell death in *Drosophila. Development* **120,** 2121–2129.

130. Brachmann, C. B., Jassim, O. W., Wachsmuth, B. D., and Cagan, R. L. (2000) The *Drosophila* bcl-2 family member dBorg-1 functions in the apoptotic response to UV-irradiation. *Curr. Biol.* **10,** 547–550.

131. Han, M. (1992) Ras proteins in developmental pattern formation in *Caenorhabditis elegans* and *Drosophila*. *Semin. Cancer Biol.* **3**, 219–228.

132. Wassarman, D. A., Therrien, M., and Rubin, G. M. (1995) The Ras signaling pathway in *Drosophila*. *Curr. Opin. Genet. Dev.* **5**, 44–50.

133. Miller, J. R. and Moon, R. T. (1996) Signal transduction through beta-catenin and specification of cell fate during embryogenesis. *Genes Dev.* **10**, 2527–2539.

134. Currie, P. D. (1998) Hedgehog's escape from Pandora's box. *J. Mol. Med.* **76**, 421–433.

135. Stern, D. L. and Sucena, E. (2000) Preparation of larval and adult cuticles for light microscopy, in *Drosophila Protocols* (Sullivan, W., Ashburner, M., and Hawley, R. S., eds.). Cold Spring Harbor Laboratory Press, Cold Spring Harbor, NY, pp. 601–616.

136. Casci, T. and Freeman, M. (1999) Control of EGF receptor signalling: lessons from fruitflies. *Cancer Metastasis Rev.* **18**, 181–201.

137. Nilson, L. A. and Schupbach, T. (1999) EGF receptor signaling in *Drosophila* oogenesis. *Curr. Top. Dev. Biol.* **44**, 203–243.

138. Wasserman, J. D. and Freeman, M. (1998) An autoregulatory cascade of EGF receptor signaling patterns the *Drosophila* egg. *Cell* **95**, 355–364.

139. Basler, K. and Struhl, G. (1994) Compartment boundaries and the control of *Drosophila* limb pattern by hedgehog protein. *Nature* **368**, 208–214.

140. Robinow, S. and White, K. (1988) The locus *elav* of *Drosophila melanogaster* is expressed in neurons at all developmental stages. *Dev. Biol.* **126**, 294–303.

141. Hynes, R. O. and Zhao, Q. (2000) The evolution of cell adhesion. *J. Cell Biol.* **150**, F89–F96.

142. Tepass, U. (1999) Genetic analysis of cadherin function in animal morphogenesis. *Curr. Opin. Cell Biol.* **11**, 540–548.

143. Takeichi, M., Nakagawa, S., Aono, S., Usui, T., and Uemura, T. (2000) Patterning of cell assemblies regulated by adhesion receptors of the cadherin superfamily. *Phil. Trans. R. Soc. Lond. B, Biol. Sci.* **355**, 885–890.

144. Brown, N. H. (2000) Cell-cell adhesion via the ECM: integrin genetics in fly and worm. *Matrix Biol.* **19**, 191–201.

145. Brown, N. H., Gregory, S. L., and Martin-Bermudo, M. D. (2000) Integrins as mediators of morphogenesis in *Drosophila*. *Dev. Biol.* **223**, 1–16.

146. Brower, D. L., Brabant, M. C., and Bunch, T. A. (1995) Role of the PS integrins in *Drosophila* development. *Immunol. Cell Biol.* **73**, 558–564.

147. Murray, M. A., Fessler, L. I., and Palka, J. (1995) Changing distributions of extracellular matrix components during early wing morphogenesis in *Drosophila*. *Dev. Biol.* **168**, 150–165.

148. Walsh, E. P. and Brown, N. H. (1998) A screen to identify *Drosophila* genes required for integrin-mediated adhesion. *Genetics* **150**, 791–805.

149. Forbes, A. and Lehmann, R. (1999) Cell migration in *Drosophila*. *Curr. Opin. Genet. Dev.* **9**, 473–478.

150. Montell, D. J. (1999) The genetics of cell migration in *Drosophila melanogaster* and *Caenorhabditis elegans* development. *Development* **126**, 3035–3046.

151. Gomperts, M., Wylie, C., and Heasman, J. (1994) Primordial germ cell migration. *Ciba Found. Symp.* **182**, 121–134.

152. Jaglarz, M. K. and Howard, K. R. (1995) The active migration of *Drosophila* primordial germ cells. *Development* **121**, 3495–3503.

153. Saffman, E. E. and Lasko, P. (1999) Germline development in vertebrates and invertebrates. *Cell Mol. Life Sci.* **55**, 1141–1163.

154. Bai, J., Uehara, Y., and Montell, D. J. (2000) Regulation of invasive cell behavior by taiman, a *Drosophila* protein related to AIB1, a steroid receptor coactivator amplified in breast cancer. *Cell* **103**, 1047–1058.

155. Chen, J., Godt, D., Gunsalus, K., Kiss, I., Goldberg, M., and Laski, F. A. (2001) Cofilin/ ADF is required for cell motility during *Drosophila* ovary development and oogenesis. *Nat. Cell Biol.* **3,** 204–209.

156. Fox, G. L., Rebay, I., and Hynes, R. O. (1999) Expression of DFak56, a *Drosophila* homolog of vertebrate focal adhesion kinase, supports a role in cell migration in vivo. *Proc. Natl. Acad. Sci. USA* **96,** 14978–14983.

157. Lee, T., Feig, L., and Montell, D. J. (1996) Two distinct roles for Ras in a developmentally regulated cell migration. *Development* **122,** 409–418.

158. Lee, T. and Montell, D. J. (1997) Multiple Ras signals pattern the *Drosophila* ovarian follicle cells. *Dev. Biol.* **185,** 25–33.

159. Murphy, A. M. and Montell, D. J. (1996) Cell type-specific roles for Cdc42, Rac, and RhoL in *Drosophila* oogenesis. *J. Cell Biol.* **133,** 617–630.

160. Niewiadomska, P., Godt, D., and Tepass, U. (1999) DE-Cadherin is required for intercellular motility during *Drosophila* oogenesis. *J. Cell Biol.* **144,** 533–547.

161. Hing, H., Xiao, J., Harden, N., Lim, L., and Zipursky, S. L. (1999) Pak functions downstream of Dock to regulate photoreceptor axon guidance in *Drosophila*. *Cell* **97,** 853–863.

162. Henderson, D. S. (1999) DNA repair defects and other (mus)takes in *Drosophila melanogaster*. *Methods* **18,** 377–400.

163. Morris, J. and Lehmann, R. (1999) *Drosophila* oogenesis: versatile spn doctors. *Curr. Biol.* **9,** R55–R58.

164. Vogel, E. W. and Nivard, M. J. (1993) Performance of 181 chemicals in a *Drosophila* assay predominantly monitoring interchromosomal mitotic recombination. *Mutagenesis* **8,** 57–81.

165. Ghabrial, A., Ray, R. P., and Schupbach, T. (1998) *okra* and *spindle-B* encode components of the RAD52 DNA repair pathway and affect meiosis and patterning in *Drosophila* oogenesis. *Genes Dev.* **12,** 2711–2723.

166. Karim, F. D., Chang, H. C., Therrien, M., Wassarman, D. A., Laverty, T., and Rubin, G. M. (1996) A screen for genes that function downstream of Ras1 during *Drosophila* eye development. *Genetics* **143,** 315–329.

167. Avery, L. and Wasserman, S. (1992) Ordering gene function: the interpretation of epistasis in regulatory hierarchies. *Trends Genet.* **8,** 312–316.

168. Guarente, L. (1993) Synthetic enhancement in gene interaction: a genetic tool come of age. *Trends Genet.* **9,** 362–366.

169. Giordano, A. and Kaiser, H. E. (1996) The retinoblastoma gene: its role in cell cycle and cancer. *In Vivo* **10,** 223–227.

170. Metzger, R. J. and Krasnow, M. A. (1999) Genetic control of branching morphogenesis. *Science* **284,** 1635–1639.

171. Zelzer, E. and Shilo, B. Z. (2000) Cell fate choices in *Drosophila* tracheal morphogenesis. *Bioessays* **22,** 219–226.

172. Jarecki, J., Johnson, E., and Krasnow, M. A. (1999) Oxygen regulation of airway branching in *Drosophila* is mediated by branchless FGF. *Cell* **99,** 211–220.

173. Echalier, G. (1997) *Drosophila* continuous cell lines, in *Drosophila Cells in Culture* (Echalier, G., ed.). Morgan Kaufmann, San Francisco, CA, pp. 131–187.

26

Assembling a Tumor Progression Model

Mark Redston

1. Introduction

Tumor progression can very broadly be defined as the progression from a less advanced to a more advanced neoplasm, due to the acquisition of genetic or other cell biologic alterations (**Fig. 1**). Implicit in this definition is the hypothesis that these acquired cellular alterations endow neoplastic cells with a growth advantage that is subsequently selected for in the evolution of the more advanced lesion *(1)*. As such, these acquired alterations are the answers to many critical questions regarding tumor biology. As an example, consider the transformation that occurs when a benign neoplasm progresses to malignancy. Delineating the events that occur during this process could also yield new clinical tools for diagnosis and treatment, while understanding the forces that drive the acquisition of these alterations could yield insights into putative preventive targets.

A host of observational, descriptive, and epidemiologic studies have provided abundant evidence for tumor progression in human neoplasia *(2)*. In recent years, studies linking the spectrum of neoplastic lesions observed with an increasing array of genetic alterations have provided conclusive proof of the importance of this process in the development of a malignancy. This chapter focuses primarily on models of tumor progression and approaches to dissecting and studying this phenomenon. Finally, several essential methods for accruing and analyzing tissue samples from preinvasive neoplastic lesions are described.

1.1. Pathology of Neoplastic Progression

1.1.1. Overview

Neoplastic progression is an extremely complex process with subtle as well as major differences between organ sites and subtypes of cancer. Despite this complexity, several generalizations can be made. The most important of these is that the classic depiction of stepwise morphologic progression accompanied by accumulation of genetic alterations is a phenomenon observed primarily in adult epithelial neoplasms *(2)*. While tumor progression in other systems is briefly discussed, most of this chapter focuses on and uses examples from epithelial neoplasms.

From: *Methods in Molecular Biology, Vol. 223: Tumor Suppressor Genes: Regulation, Function, and Medicinal Applications.* Edited by: Wafik S. El-Deiry © Humana Press Inc., Totowa, NJ

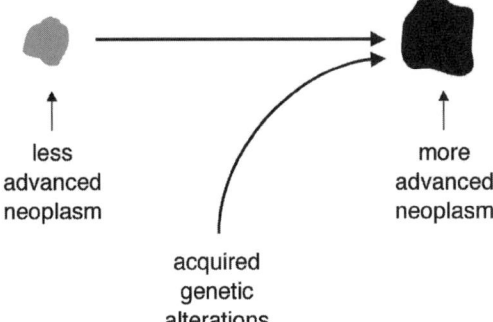

Fig. 1. Simple model of neoplastic progression. Less advanced neoplasms acquire genetic or epigenetic alterations that are selected for during tumor growth, resulting in clonal evolution.

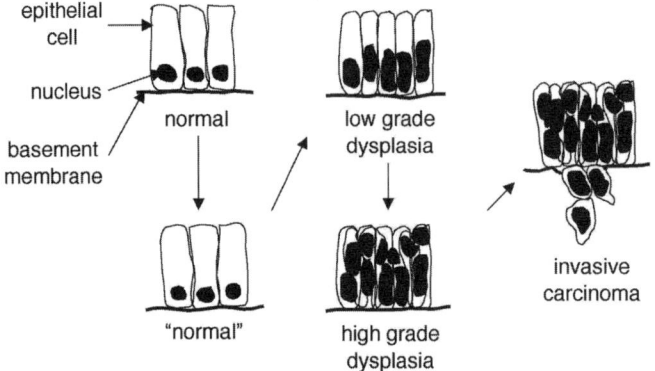

Fig. 2. Morphologic stages of intraepithelial neoplasia and early invasive carcinoma. Normal cells can undergo a number of cellular alterations regarded as preneoplastic, but not necessarily associated with morphologic abnormalities ("normal"). Dysplasia is associated with a number of characteristic cytomorphologic changes that may ultimately lead to invasion.

1.1.2. Simple (Linear) Model of Epithelial Neoplasia

For most of the past century, pathologists have recognized a host of morphologic abnormalities that could make up a continuum of lesions typical of neoplastic progression. Convincing evidence of direct neoplastic evolution from low- to high-grade neoplasms first came from careful histopathologic and epidemiologic studies of cervical neoplasia *(3)*. Based on this model, preinvasive epitheial neoplasia is now broadly termed intrapithelial neoplasia, meaning that it is a neoplastic proliferation of cells that is confined within, and does not penetrate through, the underlying basement membrane.

Intraepithelial neoplasms have a spectrum of cytomorphologic abnormalities, with increasing nuclear to cytoplasmic ratio, increasing nuclear hyperchromasia and irregularity, and increasing lack of maturation and cellular polarity (**Fig. 2**). Termed dysplasia, these changes (a) are the hallmark of intraepithelial neoplasms, (b) resemble the cytomorphologic abnormalities found in invasive cancers, and (c) can be graded according to their severity *(2)*. Often described as mild, moderate, and severe in older literature, most classification systems now utilize simply low- and high-grade dysplasia. The most severe forms of intraepithelial dysplasia are also referred to as carcinoma-*in-situ* in some organs and some older classification schemes. In some tissues, intraepithelial neoplasms often form distinctive masses or tumors, such as the adenoma (or "polyp") in the gastrointestinal tract, while in others, most notably the bladder, lung and cervix, intraepithelial neoplasms are distinctly "flat," and often difficult to recognize by eye. This is a very important point for later discussion of tissue accrual (*see* **Subheading 3.5.**). Finally, the architectural organization of some other tissues, including the breast, prostate, and pancreas, among others, is such that intraepithelial neoplasms may be papillary or nodular, but do not lend themselves to visual examination.

In simple or linear models of epithelial neoplastic progression, increasing severity of dysplasia is eventually followed by invasion through the basement membrane, the hallmark of a malignant tumor. Following invasion, a number of different morphologic abnormalities are similarly recognized, and these too are often thought to represent stages of neoplastic progression. For instance, cancers may be well, moderately, or poorly differentiated, which reflects the degree to which they resemble their normal tissue counterparts. They may have limited or extensive local spread. They may invade blood vessels and perineural spaces, and finally, they may disseminate to regional lymph nodes, as well as distant lymph nodes and other organs.

1.1.3. Complex Model of Epithelial Neoplasia

Although the continuum of morphologic lesions raises the possibility of a stepwise progression through all stages of neoplasia, studies of human tumors suggest a much more complex picture. For instance, it is clear that only a subset of low-grade dysplasias appears to be at significantly increased risk for progression to high-grade dysplasia, and many high-grade dysplasias appear to arise *de novo (3)*. Furthermore, most poorly differentiated cancers do not seem to progress from well-differentiated cancers, and even well-differentiated cancers without significant local invasion can have associated metastases *(2)*. These observations also bring into question the frequently used terms "early" and "late" neoplasms, which may be more accurately described as low-grade or high-grade neoplasms, without inference to their timing within a neoplastic progression pathway (see also discussions of cross-sectional vs longitudinal studies of neoplastic progression, **Subheading 1.3.**).

1.1.4. Nonepithelial Neoplasia

Compared to epithelial neoplasms, progression of nonepithelial tumors is not nearly as well defined or understood. Although several neoplasms have the characteristic trait of progression from a relatively indolent malignancy to a much more aggressive and/or drug-resistant cancer, progression from clear-cut benign precursors is not well recognized. This may in part reflect the fact that nonepithelial neoplasms are often driven by translocational activation of major oncogenes *(2)*. For these reasons, except for a few specific progression events, nonepithelial neoplasms have not lent themselves nearly as well to investigations of tumor progression.

1.2. Defining Tumor Progression

There may be a broad spectrum of progression events within any single tumor model, but most investigative studies address a single key transition point. The nature of these neoplastic events during tumor progression is such that they often require accrual from very different patient populations, necessitating careful design and specimen acquisition.

1.2.1. Neoplastic Initiation

Tumor initiation is of critical importance in human neoplasia, but it is perhaps the most difficult event to study, owing to the fact that clinically inapparent neoplasms are not often recognized, let alone sampled. Furthermore, significantly less advanced neoplasms are often no longer apparent in patients with fully established cancers. Although approaches to accrue incipient neoplasms and preneoplastic abnormalities are emerging, typically they require specific tissue sampling in order to harvest appropriate research samples. Preneoplastic tissues are further addressed separately (see **Subheadings 1.5.** and **3.5.**).

1.2.2. Intraepithelial Neoplasia

Development of a tumor progression model wherein cancer arises from intraepithelial neoplasia suggests that eradication of intraepithelial neoplasia should eradicate cancer. In tissues where screening for intraepithelial neoplasia is practical at a population level, such as the cervix and colon, ablation and/or removal of intraepithelial neoplasms results in decreased cancer mortality. A direct spin-off of this clinical application of the tumor progression model has been the collection of a large number of intraepithelial neoplasms that can be utilized in studies of tumor progression. Therefore investigations of the genetic events underlying intraepithelial neoplasias have been widely performed in several human tissues. Importantly, uncovering these alterations could lead to the development of novel targets for precancer intervention.

1.2.3. Invasion (Malignant Transformation)

Like intraepithelial neoplasia, the process of malignant transformation is of great interest in tumor progression models. The cellular alterations accompanying malignant transformation are prime targets for diagnostic and therapeutic interventions. Furthermore, many cancer resection specimens lend themselves well to dissection of the cancer and the associated adjacent cancer precursor for analysis of longitudinal tumor progression (*see* **Subheadings 1.3.** and **3.5.**).

1.2.4. Postinvasive Progression

Following the advent of invasion, but prior to dissemination, a number of fascinating tumor progression events are still known to occur. Cancers can continue to evolve by developing foci that are less well differentiated, they can evolve several distinct histologic subtypes, they induce a variety of host stromal and inflammatory reactions, they can become locally invasive, and they can develop resistance to chemotherapy. All of these processes could be considered components of tumor progression.

1.2.5. Metastasis

Metastasis is usually at the root of patient mortality, and is thus the most studied of all the tumor progression events. A number of different stages are recognized, including lymphatic and blood vessel invasion in the primary cancer, regional lymph node metas-

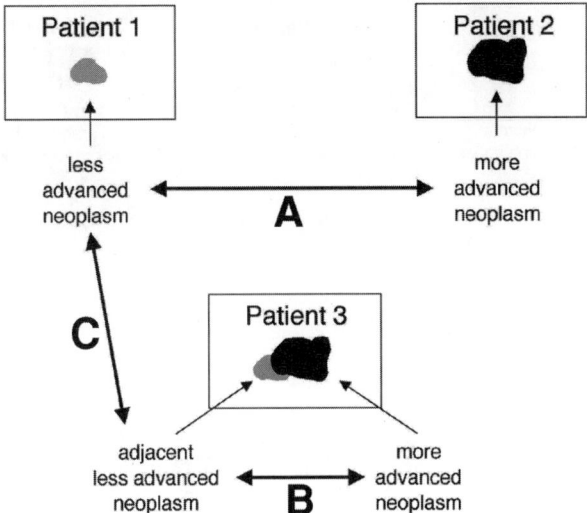

Fig. 3. Approaches to the investigation of neoplastic progression. Samples may be compared (**A**) between patient populations with different stages of disease (cross-sectional studies), (**B**) between different stages of disease within one patient (longitudinal studies), or (**C**) by a combination of these approaches.

tases, and distant organ dissemination. Understanding the metastatic process and developing effective therapies would have a profound effect in oncology. Longitudinal studies of metastasis are complicated by the fact that primary and metastatic tumors are often diagnosed and biopsied at widely dispersed time points, and tissue samples of metastatic tumors are typically very limited in quantity.

1.3. Cross-Sectional Versus Longitudinal Models

1.3.1. Overview

In simple terms, studies of neoplastic progression aim to delineate the alterations associated with progression from a less advanced to a more advanced neoplasm (as depicted in **Fig. 1**). However, the realities of working with patient samples often dictate more indirect approaches. Typically, the less advanced neoplasms of one series of patients are compared to more advanced neoplasms of another series of patients (**Fig. 3A**). I refer to these studies, which tend to be the most commonly undertaken in human cancer research, as cross-sectional. Although these studies often yield clues as to the alterations associated with a specific stage of progression, there is no direct evidence to causally link the alterations to neoplastic progression (see also the examples below under **Subheadings 1.3.2.** and **1.3.3.**).

In comparison, some patients are identified in whom two separate stages of neoplastic evolution are identified in a single tissue. Importantly, while these could be separate neoplasms, when they occur immediately adjacent to one another, the more advanced neoplasm is usually the result of direct clonal evolution. In this scenario, comparison of the alterations in the less advanced and more advanced neoplasms yields direct information of the biologic events selected during clonal evolution (**Fig. 3B**). I refer to these studies as longitudinal.

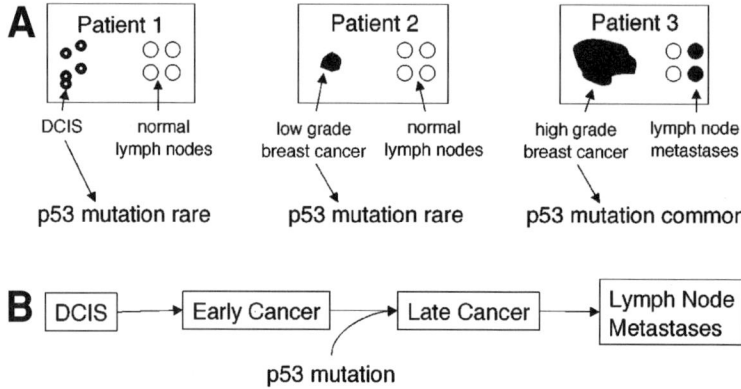

Fig. 4. Cross-sectional model of p53 in breast cancer progression. (**A**) Mutations are rare in patients with less advanced lesions, but common in patients with advanced disease. (**B**) These findings suggest a model where p53 mutations are a late event, associated with the development of advanced cancers.

Finally, the less advanced neoplasms that occur in isolation can be compared to less advanced neoplasms that are associated with more advanced neoplasms (**Fig. 3C**). These studies can yield clues as to the genetic alterations associated with increased risks of neoplastic progression in two morphologically similar tumors. These types of studies have components of both cross-sectional and longitudinal models. As should be readily apparent, the conclusions for all three types of studies are vastly different, highlighting experimental design as one of the most important considerations for developing a tumor progression model.

1.3.2. The Role of p53 Inactivation in Breast Neoplastic Progression

p53 is a major human tumor suppressor gene that is involved in a cell fate checkpoint (division vs apoptosis) in response to DNA damage *(4)*. Early studies of p53 in breast cancers revealed that inactivating mutations were rare in low-grade axillary node-negative tumors and ductal carcinomas *in situ* (DCIS), but were relatively common in high-grade metastatic breast cancers (which also tended to be larger, and thus pesumably more "advanced"; **Fig. 4A; ref. 5**). These cross-sectional studies of breast cancer progression suggested that p53 mutations were a "late" occurrence in breast neoplasia, often associated with progression from a low-grade nonmetastatic cancer to a high-grade metastatic cancer (**Fig. 4B**).

Subsequent longitudinal studies of p53 inactivation in breast cancer revealed very different findings (**Fig. 5A**). Studies of invasive cancers and associated adjacent DCIS revealed that p53 mutations were always present in the adjacent DCIS associated with p53 mutant breast cancers *(6)*. Thus rather than being a "late" genetic alteration, p53 mutations typically occurred prior to the development of invasion in these cases. Furthermore, p53 mutant DCIS was typically high-grade, a finding also confirmed in DCIS not associated with invasive carcinoma *(7)*. Therefore the longitudinal studies revealed a radically different tumor progression model (**Fig. 5B**), with p53 mutations occurring in association with the development of high-grade DCIS, and being associated with a neoplastic pathway more likely to progress into a high-grade and metastatic breast cancer.

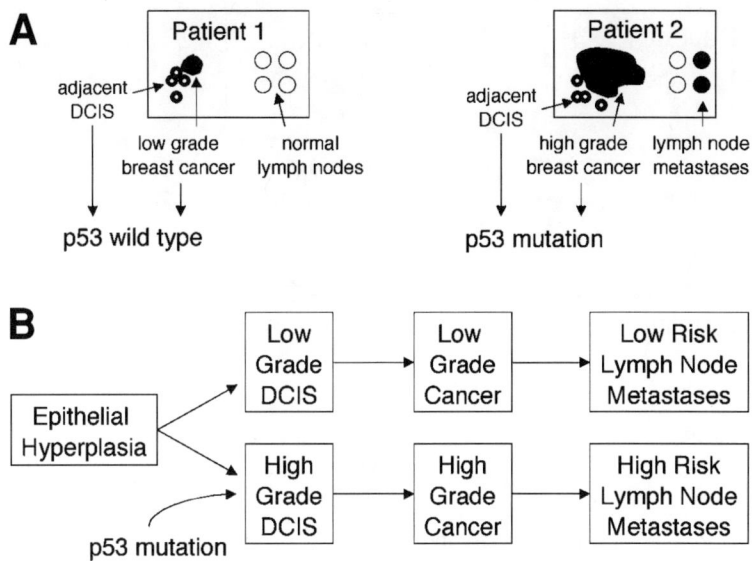

Fig. 5. Longitudinal model of p53 in breast cancer progression. (**A**) Mutations are rare in patients with less advanced lesions, but when they are found in patients with advanced disease, they are present in less advanced neoplasms from the same patients. (**B**) These findings suggest a model in which p53 mutations usually occur prior to invasion, and are associated with progression to high-grade DCIS and high-grade cancer.

1.3.3. The Role of DNA Mismatch Repair Deficiency in Colorectal Neoplastic Progression

Defective DNA mismatch repair in colorectal cancers results in a distinctive type of genomic instability called microsatellite instability *(8)*. Early cross-sectional studies found that microsatellite instability was relatively rare in isolated colorectal adenomas, but was present in about 15% of colorectal cancers (**Fig. 6A; ref. 9**). These findings led to two possible models, in which mismatch repair inactivation occurred (a) during the progression from adenoma to carcinoma, or (b) during progression from an unknown nonadenomatous precursor (**Fig. 6B**).

Subsequent longitudinal studies of a series of colorectal cancers and their adjacent neoplastic precursors revealed that colorectal cancers with microsatellite instability were usually accompanied by adjacent adenomas with microsatellite instability (**Fig. 7A; ref. 10**). Furthermore, it was found that the adjacent precursor was occasionally a mismatch repair-deficient hyperplastic polyp, shattering the dogma that hyperplastic polyps were not precursors to cancer *(11)*. Thus a different understanding of mismatch repair-deficient colorectal neoplasia progression emerged (**Fig. 7B**), with microsatellite instability developing prior to invasion, in the evolution of an adenomatous or hyperplastic polyp precursor. In addition, based on the relative absence of microsatellite instability in isolated adenomas from cross-sectional studies, it seems likely that mismatch repair-deficient colorectal precursors evolve rapidly to invasive carcinomas *(8)*. Further evidence that mismatch repair-deficient neoplasms evolve along a unique pathway is found in studies describing the natural history of these tumors. Colorectal cancers with microsatellite instability are less likely to metastasize to regional lymph nodes *(12)*, have

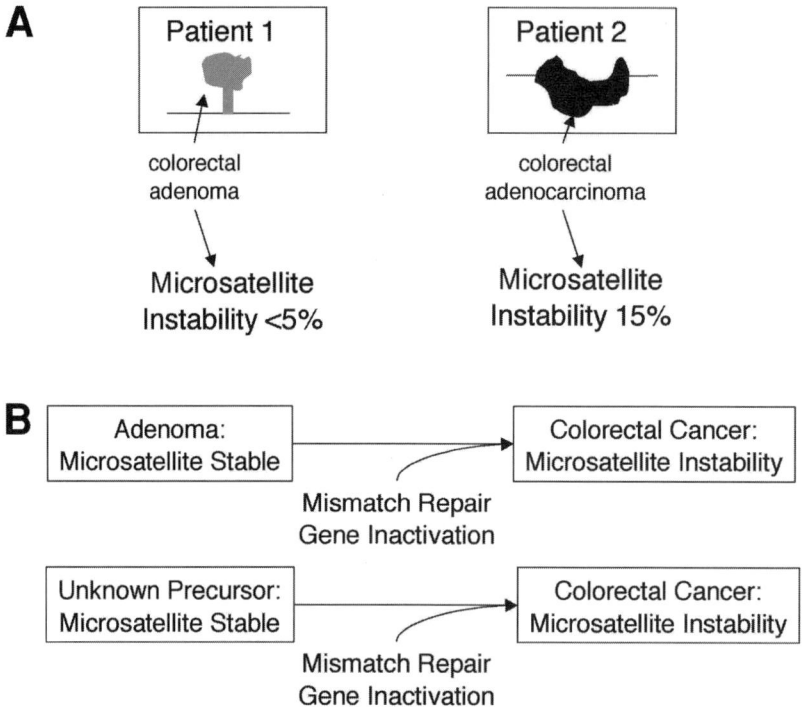

Fig. 6. Cross-sectional model of mismatch repair gene inactivation in colorectal neoplasia. (**A**) Microsatellite instability, the hallmark of mismatch repair gene inactivation, is rare in adenomas, but relatively frequent in carcinomas. (**B**) These findings suggest a model in which mismatch repair gene inactivation is a late event, associated with the progression of an adenoma (or other neoplastic precursor) to cancer.

a better overall survival even when controlled for stage *(12)*, and when they do metastasize, are less likely to involve the liver (unpublished data).

1.3.4. Pathways of Neoplastic Progression

These examples highlight the importance of longitudinal studies in the development of a tumor progression model. Longitudinal investigations are particularly suited to recent models of tumorigenic "pathways," wherein early acquired alterations determine later alterations and morphology, rather than occurring randomly across a spectrum of lesions *(13)*. This type of pathway progression is seen in both of the previous examples, where p53 mutations and mismatch repair gene inactivation have a major influence in determining the subsequent molecular and morphologic progression of the neoplasm.

1.3.5. Experimental Models of Tumor Progression

Although this chapter focuses almost exclusively on studies that utilize human tissue samples to study neoplastic progression, it should also be noted that several experimental models of tumor progression also exist. This includes both models to examine experimental manipulations of human neoplastic cells, and models of tumor progression in genetically altered and/or carcinogen-treated mice. Most commonly studied are models to look at metastasis, primarily the effects of manipulation of gene expression in

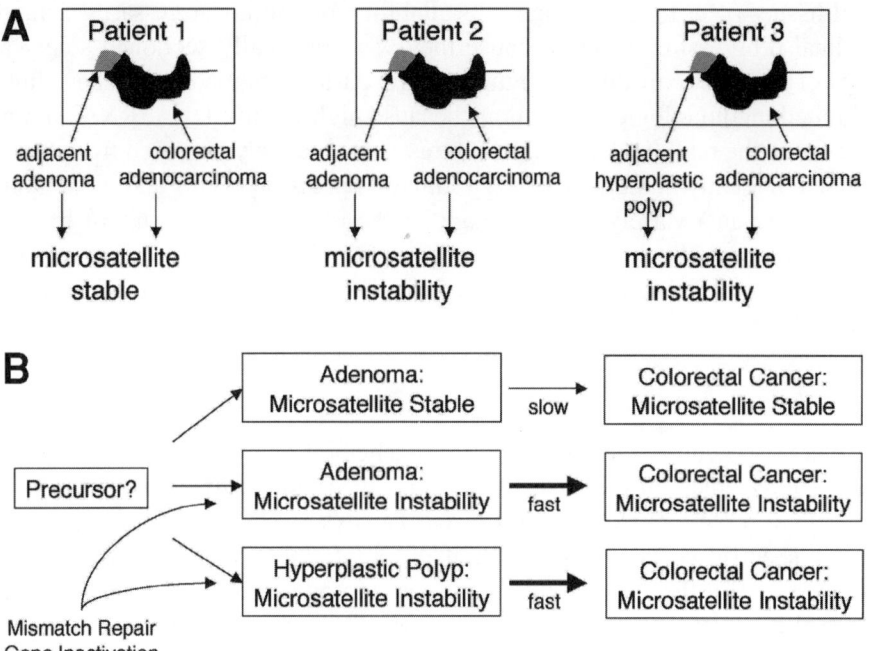

Fig. 7. Longitudinal model of mismatch repair gene inactivation in colorectal neoplasia. (**A**) Microsatellite instability is always present in the precursor adenoma when present adjacent to a carcinoma with microsatellite instability. Furthermore, occasional mismatch repair-deficient cancers are seen to arise from hyperplastic polyps with microsatellite instability. (**B**) These findings suggest a model in which mismatch repair gene inactivation occurs prior to the development of invasion, and is associated with the development of a mismatch repair-deficient adenoma or hyperplastic polyp.

human cancer cells on the metastatic phenotype in mice. Tumor–stromal interactions have also been studied in cell culture experiments, as well as the effects of manipulation of stromal cells in mice. Finally, several investigators have attempted to establish cell culture and other models of noninvasive neoplastic cells by a variety of techniques. These experimental systems are beyond the scope of this chapter.

1.4. Approaches

1.4.1. Overview

Studies aimed at assembling a tumor prgoression model require a variety of tissue samples that encompass the progression event, and that ideally allow longitudinal studies of neoplastic progression. In general, these tissues can be obtained from archival paraffin-embedded tissues taken at the time of tissue biopsy or excision, or they can be obtained by prospective accrual of fresh tissue specimens. The reader is directed to the National Institutes of Health for information regarding human subjects research.[1]

---

[1]Office of Human Subjects Research, http://ohsr.od.nih.gov; Guidelines for the Conduct of Research Involving Human Subjects at the National Institutes of Health, http://ohsr.od.nih.gov/guidelines.php3; Bioethics Resources on the Web, http://www.nih.gov/sigs/bioethics/index.html.

Archival tissues offer the advantage of availability, but do not necessarily demonstrate longitudinal progression events because they were originally sectioned solely for the purposes of diagnostic evaluation. Furthermore, paraffin-embedded tissues allow relatively limited methodologic applications because high-quality DNA, RNA, and protein samples cannot be readily obtained. In contrast, prospectively procured fresh tissues can be obtained to ideally demonstrate longitudinal progression events and yield high-quality tissue for a wider variety of technologic methods. However, accrual of fresh tissues requires significant effort and planning, regardless of whether paraffin or fresh tissues are ultimately used for analyses. Some methodologic principles are outlined under **Subheading 3.5.**

1.4.2. In-Vitro Approaches

There are two general methodologic approaches for analyzing tumor progression in human tissue samples. In the first, tissue samples are microdissected and subjected to nucleic acid or protein purification for in-vitro analyses. This microdissection approach allows for specific comparisons between different stages of neoplastic progression. There are several different methods for tissue microdissection. Protocols for paraffin and fresh tissue dissection are included below. These methods allow relatively rapid dissection of reasonably pure cell populations suitable for a wide range of assays. Methods are also available for dissection of ultrapure cell populations, most notably laser capture microdissection (LCM; **ref. *14***). Although it is an extremely powerful method for analyzing paraffin-embedded and fresh tissues *(15,16)*, LCM requires specialized equipment, operator training, and is often run as a core facility with established local protocols to suit variables such as slide adherence and lab humidity. Additional reading and protocols are available from the National Institutes from Health,[2] and methods will not be further discussed. Finally, there are a number of additional protocols for tissue dissection and analysis that are relevant to specific tissues (such as aberrant crypt foci and crypt epithelial preparations of the colon). These are beyond the scope of this chapter.

1.4.3. In-Situ Approaches

An alternative to tissue dissection and in-vitro analyses is to perform investigative assays on whole tissue sections. This offers the remarkable advantage of maintaining architectural integrity, and has the potential for longitudinal analysis of mutlistage tumor progression in a single assay on a single slide. This approach is currently limited by methodologies, and is applicable only for protein expression by immunohistochemistry, mRNA expression by *in situ* hybridization, genomic copy number and translocations by fluorescent *in situ* hybridization, and *in situ* polymerase chain reaction (PCR). Several of these methods are described elsewhere in this series. As in the tissue dissection approaches to analyze tumor progression, careful tissue accrual and experimental planning (for instance, cross-sectional vs longitudinal studies) are required when outlining an investigative question using these approaches.

---

[2]Laser Capture Microdissection, follow links from http://cgap-mf.nih.gov/

1.5. Models of Preneoplasia and Incipient Neoplasia

The discussion so far has focused on morphologically recognizable neoplasms. In recent years, investigations have increasingly recognized abnormalities in tissues that do not meet diagnostic criteria for dysplasia (and thus neoplasia) or that even appear morphologically normal. As an example, patients with chronic inflammatory bowel disease are at increased risk for cancer, and often develop multiple cancers. Careful examination of apparently normal epithelium adjacent to and away from the cancer often reveals areas of abnormal proliferation. Molecular investigations of these foci sometimes demonstrate clonal mutations of *KRAS* and *p53 (17,18)*. These lesions are interpreted as representing proliferative expansions of growth-advantaged mutant clones, and may even be considered to be very low-grade neoplasms that do not meet classic histomorphologic criteria. In addition, a variety of alterations in gene expression are identified in non-neoplastic tissue associated with neoplasia. Several studies have also found evidence of accumulating DNA damage in non-neoplastic tissues. Finally, some neoplasms develop out of fields of metaplasia, which can also harbor gene mutations and other clonal abnormalities *(19)*. Further studies of these types of changes require abundant tissue samples of grossly normal tissue, either non-neoplastic tissue adjacent to cancers or intraepithelial neoplasms, or non-neoplastic tissue from patients with variable risks for cancer. Ultimately, these studies will be helpful in understanding the earliest events in neoplasia, and will be instrumental in developing biomarkers for cancer risk. See also tissue banking for discussion of accrual of specimens (*see* **Subheading 3.5.**).

2. Materials

2.1. Slide Microdissection of Paraffin-Embedded Tissue

1. Paraffin tissue block.
2. Standard microscopy slides.
3. Permanent histology marker pens.
4. Scalpel blades, size 11.
5. Sharps container.
6. Sterile 1.5-mL microcentrifuge tubes.

2.2. DNA Extraction from Paraffin-Embedded Tissue—QIAamp Method

1. QIAamp DNA Mini Kit.
2. Sterile 1.5-mL microcentrifuge tubes.
3. Sterile 1.5-mL screw-cap tubes with rubber gaskets.
4. Xylene.
5. Absolute ethanol.
6. 95% Ethanol.
7. Proteinase K, 20 mg/mL.

Xylene is a hazardous chemical that must be stored in a vented storage cabinet. All xylene liquid waste must be disposed of in a halogenated solvent waste can.

2.3. DNA Extraction from Paraffin-Embedded Tissue—Rapid Method

1. Sterile 0.5- or 1.5-mL microfuge tubes.
2. Lysis buffer: 10 mM Tris-HCl, pH 8, 100 mM KCl, 2.5 mM MgCl$_2$, 0.45% Tween-20.

3. Proteinase K, 20 mg/mL.

2.4. Microscope-Directed Slide Microdissection and DNA Extraction

1. Paraffin tissue block.
2. Standard microscopy slides.
3. Permanent histology marker pens.
4. Xylene.
5. Absolute ethanol.
6. 95% Ethanol.
7. Hematoxylin staining solution.
8. 18-Gage needles.
9. Sharps container.
10. Sterile 0.5-mL microcentrifuge tubes.
11. Digestion buffer: 50 mM Tris-HCl, 1 mM EDTA, 0.5% Tween-20, 200 μg/mL proteinase K.

Xylene is a hazardous chemical that must be stored in a vented storage cabinet. All xylene liquid waste must be disposed of in a halogenated solvent waste can.

2.5. Tissue Banking to Study Tumor Progression

1. Sterile 2.0-mL cryostorage screw-cap tubes with rubber gaskets.
2. Cryomolds (Tissue-Tek, Sakura Finetek).
3. Frozen tissue embedding compound (Tissue-Tek O.C.T. Compound, Sakura Finetek).
4. Histology processing cassettes.
5. Cryostorage facilities.
6. Tissue processing and staining facilities.

2.6. Cryostat Microdissection of Frozen Tissue

1. Cryostat sectioning facility.
2. 50-mL chilled polypropylene tubes.
3. Scalpel blades.

3. Methods
3.1. Slide Microdissection of Paraffin-Embedded Tissue

This protocol is detailed in **Fig. 8**.

1. Cut one 4-μm section for hematoxylin and eosin (H&E) staining and ten 10-μm sections for tissue dissection (**Fig. 8A**; *see* **Note 1**).
2. Review H&E-stained section and circle selected tissue samples for microdissection (**Fig. 8B**; *see* **Note 2**). This slide will be used as the template for guiding microdissection.
3. Overlay H&E-stained slide onto one unstained noncover-slipped slide, being careful to align the tissue section properly (**Fig. 8C**; *see* **Note 3**).
4. Flip the aligned slides over, and transcribe the circles onto the back of the unstained 10-μm slide (**Fig. 8D**; *see* **Note 4**). This will serve as the guide for tissue scraping. Repeat for all 10 slides to be used in microdissection.
5. Scrape the marked area into a 1.5-mL sterile centrifuge tube using a scalpel blade (**Fig. 8E**; *see* **Note 5**). All of the tissue from a single lesion should be combined from all scraped slides into a single tube.

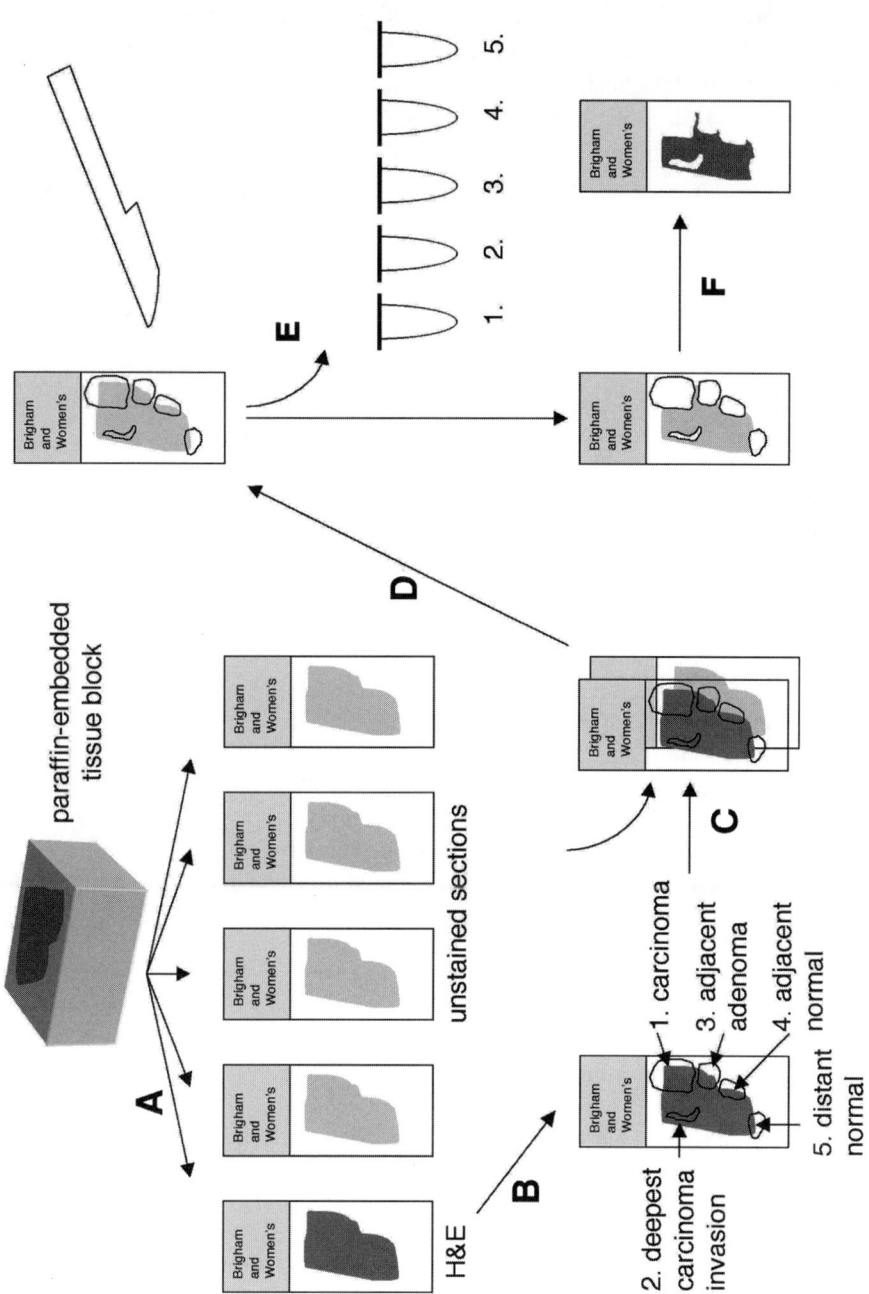

Fig. 8. Slide microdissection of paraffin-embedded, formalin-fixed tissue. See text for details.

6. The postdissection slide can be stained with H&E to confirm that appropriate tissue has been dissected (**Fig. 8F**).

3.2. DNA Extraction from Paraffin-Embedded Tissue—QIAamp Method

This method provides abundant pure DNA suitable for long-term storage and a variety of applications including automated sequencing. Accurate quantitation by fluorimetry is also possible. Unfortunately, variables outside of the control and knowledge of the investigator (such as time to fixation, quality of fixation, fixative pH, quality of processing, etc.) will determine the DNA quality. Usually it is possible to routinely amplify 100-bp products from almost all material, and up to 300 bp for well-fixed and well-processed tissue.

1. Add 1.0 mL of xylene to paraffin dissected tissue in a 1.5-mL centrifuge tube and mix on a rotator for 15 min. Vortex to eliminate clumps of tissue if necessary. Centrifuge at 20,000 g for 2 min. Remove supernatant. Repeat × 1.
2. Add 1.0 mL of 100% ethanol and mix on a rotator for 10 min. Vortex to eliminate clumps of tissue if necessary. Centrifuge at 20,000 g for 2 min. Remove supernatant. Repeat × 1.
3. Incubate tubes with caps open at 55°C for 10 min to evaporate off any residual ethanol.
4. Add 180 μL of Buffer ATL™ and mix tube to resuspend tissue.
5. Add 20 μL of (20-mg/mL) proteinase K. Mix well.
6. Incubate at 55°C overnight. Mix tube as frequently as possible.
7. In the morning, evaluate tubes to determine if additional proteinase K is required. No thin threadlike strands should be visible in tube. If required, add an additional 20 μL of (20-mg/mL) proteinase K. Keep track of the total volume in tubes. Mix well.
8. Incubate at 55°C for an additional 2–4 h. If digestion begins early in the day, **step 7** can be done at the end of the first day. Tubes may be evaluated in the morning to determine if second addition of 20 μL of proteinase K is required.
9. Add an equal volume of Buffer AL™. Vortex to mix. Incubate at 70°C for 10 min.
10. Add an equal volume of 95% ethanol. A precipitate should be seen at this point. Vortex to mix.
11. Place spin column in collection tube provided. Load column carefully with mixture from **step 10**. If volume in **step 10** is greater than 600 μL, apply only half the volume to the column.
12. Place column and collection tube in centrifuge and spin at 6000 g for 1 min. If volume in **step 10** is greater than 600 μL and only half the mixture is applied to the column, spin for only 15 s. Remove eluate, apply remaining mixture to column, and spin for 1 min.
13. Add 500 μL of Buffer AW1™ to column. Place column and collection tube in centrifuge and spin at 6000 g or 1 min. Remove eluate.
14. Add 500 μL of Buffer AW2™ to column. Place column on collection tube provided and place in centrifuge and spin at 20,000 g for 3 min.
15. Discard collection tube and eluate and place column in a new collection tube.
16. Add 100 μL of Buffer AE™ (prewarmed to 70°C) to column. Buffer AE™ must be diluted 5-fold with sterile filtered dH$_2$O. Place column and new collection tube in 70°C water bath and incubate for 5 min. Centrifuge at 6000 g for 1 min.
17. Apply the same 100 μL of Buffer AE™ (now containing DNA) to the column a second time. Measure volume and adjust volume to 100 μL with Buffer AE™ if necessary. Incubate at 70°C for 5 min. Centrifuge at 6000 g for 1 min. Transfer DNA to a prelabeled 500-μL screw-cap tube with a rubber gasket.

3.3. DNA Extraction from Paraffin-Embedded Tissue—Rapid Method

This method provides rapid, PCR-grade DNA suitable for many applications. It is particularly suited to multiple dissections in which each is tested for only a few assays

(20). Because of impurities, some difficult products may not amplify well. In addition, long-term storage is not recommended.

1. Add 50–100 μL of lysis buffer to paraffin-dissected tissue in microcentrifuge tube and mix well (*see* **Note 6**).
2. Incubate at 95°C for 10 min.
3. Add 15–35 μL of proteinase K. Mix well.
4. Incubate at 55°C overnight. Use directly in PCR (*see* **Note 7**).

3.4. Direct Microscope Slide Microdissection and DNA Extraction

For microscopic lesions, the tissue dissection can be performed under direct microscopic visualization. This is necessary for smaller samples, for which template matching of circled areas on a slide cannot be relied upon (such as ductal carcinoma *in situ* and some other intraepithelial neoplasms; **ref. 6**). The slide is lightly counterstained with hematoxylin, providing light nuclear staining. This allows identification of tissue architecture under low-magnification microscopy.

1. Cut one 4-μm section for H&E staining and ten 5-μm sections for tissue dissection.
2. Review H&E-stained section and circle selected tissue samples for microdissection. This slide will be used as an aid to identify lesional tissue.
3. Wash slides in xylene to dissolve paraffin.
4. Wash slides in in increasing dilutions of ethanol to rehydrate.
5. Stain slides in hematoxyline for 30 s. Wash in water.
6. Dissect lesional tissue sample under direct microscopic visualization utilizing two 18-gage needles and transfer to micofuge tube (*see* **Note 8**).
7. Add 40 μL/mm^2 digestion buffer.
8. Incubate at 55°C for 24 h. Use directly in PCR (*see* **Note 7**).

3.5. Tissue Banking to Study Tumor Progression

Recent growth in translational research has led to an increase in tissue banking efforts at many major academic medical centers. Putting together an excellent tumor bank is clearly a multidisciplinary effort, involving surgeons, pathologists, oncologists, geneticists, epidemiologists, basic scientists, and ethicists *(21–23)*. There are a large number of issues that cannot be considered here, including patient consent, staffing, storage, access, and distribution, to name a few *(21–23)*. Although there have been many institutional efforts to build tumor tissue banks, care must be taken to sample tissue appropriately in order to accrue the tissue resources necessary for investigations of tumor progression. Every human tissue and cancer type has unique issues related to tissue banking, but several very broad principles are illustrated in the following protocol, which have general applicability. This example is taken from a protocol for banking colorectal cancers in the setting of inflammatory bowel disease. The protocol is diagrammed in **Fig. 9**. The guiding principle of tissue banking to study tumor progression is that the low-grade intraepithelial neoplasm or incipient/preinvasive neoplastic precursor lesion may not be grossly visible, and there is no practical way to identify and diagnose it accurately at the time of tissue banking. Thus tissue samples are taken liberally of surrounding "normal" tissues. In order to piece the progression model back together afterwards, the exact locations of all tissue samples must be carefully recorded on a "map," and many sections sent for formalin fixation, paraffin-embedding, and

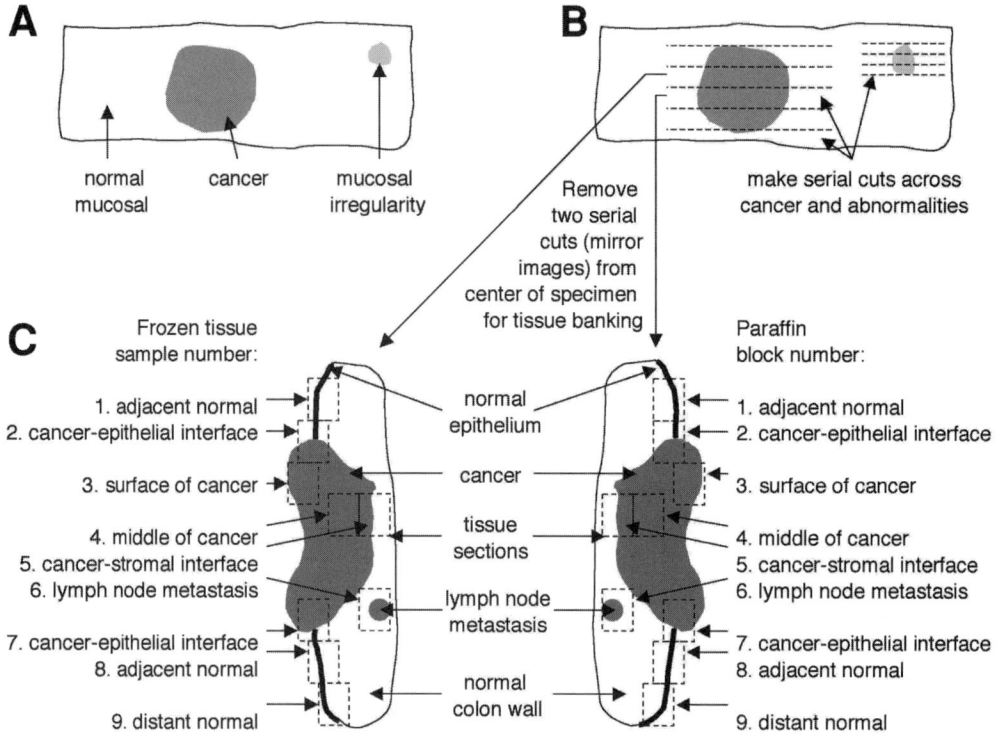

Fig. 9. Colon tumor banking to obtain tissue for neoplastic progression studies. (**A**) The colon is opened, and all lesions carefully examined. (**B**) Serial cuts are made across the lesions. (**C**) Opposing slices of tissue are submitted for processing and routine sectioning and H&E staining, with matching mirror images of the tissue submitted into the frozen tissue bank. Abundant samples of adjacent tissue, which may contain preneoplastic lesions, are harvested with careful correlation to a "map" of the specimen. Hatched squares indicate tissue sampling sites.

H&E slides made to diagnose all tissue abnormalities present in the frozen samples. Finally, although frozen tissue is taken, many assays can be performed on paraffin sections or from paraffin tissue microdissection, which offers significant time saving compared to working with the frozen tissue.

1. Open the specimen and carefully examine the entire mucosal surface, recording the locations, measurements, and descriptions of all lesions on the Banking Worksheet Map (*see* **Note 9**).
2. Breadloaf all lesions, and take measurements of thickness, depth of invasion, and distance to closest resection margin. Take representative section of deepest/thickest tumor for permanent (diagnostic) section. Bank opposing (mirror-image) tissue sections in cryovials and or as frozen tissue chucks (*see* **Note 10**). Record exact location of each vial of tissue on the Banking Worksheet Map, noting which permanent section has the mirror image (*see* **Note 11**).
3. Take representative sections that span the lesion and the adjacent tissue. These sections often harbor lower-grade intraepithelial neoplasia (such as adenoma or dysplasia).
4. Take a liberal sampling of tissue surrounding all neoplastic lesions. These areas may harbor low-grade intraepithelial neoplasia or incipient/preinvasive neoplasia. Note the locations of all samples on the Banking Worksheet Map so that any clonal lesions can be pieced back together (like a puzzle). If desired, different colors of ink can be used to identify tissue orientation.

3.6. Cryostat Microdissection of Frozen Tissue

This method requires frozen section slides to be cut using a cryostat *(18,24)*. These are available in clinical pathology deparments, and are also sometimes found in histology research cores at some institutions.

1. Remove frozen tissue samples from bank, and freeze onto a cryostat chuck using optimum cutting temperature (OCT).
2. Cut one 5-μm section for H&E stain (*see* **Note 12**). Review and circle areas for further dissection.
3. Allign circled slide with frozen tissue on chuck, and cut away unwanted tissue (*see* **Note 13**).
4. Cut another 5-μm section for H&E stain. Review to assess cellularity of selected sample.
5. Repeat **steps 4** and **5** as required until only the desired sample remains on the chuck.
6. Cut 10–50 sections into a prechilled 50-mL tube. Cut one 5-μm section for H&E every 10 sections to recheck tissue sample. Redissect unwanted tissue as necessary.
7. Nucleic acids can be extracted and cell lysates made for protein analysis by a wide variety of methods.

4. Notes

1. Sections can be cut by the investigator, but this requires access to a microtome (predominantly in pathology departments, with restricted access for clinical use only), and moderate training to produce adequate-quality sections. We have favored the use of a histology core facility, either within a clinical pathology department or a research facility. Most cores require that slides are provided by the investigator. While 10-μm slides decrease the number of slides that need to be scraped, standard 4–5-μm slides can also be used. Thicker (20–50-μm) slides can be substituted, although these are often more difficult to cut (resulting in tissue "wasteage" for wrinkled sections), and less accurate for tissue dissection. Additional 4-μm sections should be cut after every ten 10-μm slides so that an H&E can be reviewed to confirm the tissue sample to be dissected. The nature of the sample in question can change quite significantly 100 μm into the block.
2. For the study of human neoplastic progression, microscopic review is best performed by a pathologist or pathology trainee with expertise and interest in the system under examination. Areas are selected based on the research question, purity of a morphologic lesion, and estimated neoplastic cellularity (proportion of lesional cells in the dissected sample compared to total cells, including inflammatory cells, stromal cells, and nonlesional epithelial cells). The required neoplastic cellularity depends on the assays to be undertaken on the sample. In general, standard analyses of loss of heterozygosity require samples of at least 70% neoplastic cellularity, mutation screening by sequencing requires 40% neoplastic cellularity, and mutation screening by SSCP requires 10–20% neoplastic cellularity. The "average" neoplastic cellularity that can be readily dissected using this approach varies with tumor types. For instance, most colon cancers readily yield samples with at least 70% neoplastic cellularity, while most pancreatic cancers yield samples less than 50% neoplastic cellularity (due to stromal reaction).
3. Be aware that the proper alignment of tissue sections requires attention and practice. The tissue section on some slides may be inverted (and occasionally flipped right over) relative to others.
4. If more than five 10-μm sections are to be used in the dissection, a second H&E obtained after the unstained sections should be used for tissue circling and microdissection template.
5. Some operators find it easiest to scrape the tissue from several slides onto a clean piece of filter paper, then deposit it in the tube.

6. The amount of lysis buffer and proteinase K are adjusted to suit the mount of tissue in the dissection. Lower-end values are used for small (<20-mm^2) dissection from only 1–2 slides, and scaled up as needed.

7. The final DNA concentration is variable, but usually quite low. Template volume must be adjusted accordingly in PCR reactions (up to 10 or even 20 μL may be required for microscopic dissections). Because of the impurities and low concentration, accurate DNA quantitation is usually not possible.

8. The tip of the needle can be inserted directly into the buffer in the microfuge tube in order to release the tissue into the buffer.

9. Correlation of morphology and topographic location of all lesions is vital in mapping clonal evolution of precursors. All descriptive findings and all sections taken should be carefully indicated on a "map" of the specimen on the Banking Worksheet. A standard form can be constructed outlining the essential anatomy of the tissue to be banked, and lesions, dimensions, block, and vial numbers are simply added.

10. There are two different tissue storage methods in general use. The fastest method is to deposit tissue in a cryostorage vial. However, if tissue morphology needs to be assessed and dissected prior to sample analysis, this method becomes difficult to work with, as the tissue must be removed from the tubes and reembedded for frozen sections. Alternatively the tissue can be frozen in a cryomold in embedding media oriented for ease of later frozen section analysis. This approach is more time-consuming up front, and more cumbersome compared to cryovials for storage, but greatly facilitates subsequent processing.

11. A major difficulty in using tissue banks to study tumor progression is knowing which (of many) frozen samples has neoplastic or preneoplastic tissue of interest. Cutting a frozen section off all samples for an H&E slide to review the morphology is often not practical. One approach is to take a permanent section that mirrors the frozen sample. In this way, a review of all slides from the permanent sections can be used to direct choice of frozen samples to identify neoplastic tissues. As a compromise, a mixture of histologically controlled and randomly banked tissue samples can be obtained, since permanent sections also have an associated processing cost.

12. Cryostat sectioning requires training and experience. Cryostat microdissection is best performed with two individuals because of the need for tissue management, sectioning, staining, and microscopic review. It is often best to hire the services of an experienced histology technologist.

13. Unlike the slide microdissection, in this method the unwanted tissue is removed, leaving only lesional tissue, which can then be harvested. Depending on the tissue sample, sometimes it is easier to cut the tissue in halves or thirds at the start and refreeze the unwanted portions. Unwanted tissue can also be collected in a 50-mL tube and stored, or it can be discarded.

References

1. Nowell, P. C. (1976) The clonal evolution of tumor cell populations. *Science* **194**, 23–28.
2. Cotran, R. S., Kumar, V., Collins, T., and Robbins, S. L., Eds. (1999) *Robbins Pathologic Basis of Disease*. Saunders, Philadelphia.
3. Ponten, J. and Guo, Z. (1998) Precancer of the human cervix. *Cancer Surv.* **32**, 201–229.
4. Wahl, G. M. and Carr, A. M. (2001) The evolution of diverse biological responses to DNA damage: insights from yeast and p53. *Nat. Cell. Biol.* **3**, E277–E286.
5. Davidoff, A. M., Kerns, B. J., Pence, J. C., Marks, J. R., and Iglehart, J. D. (1991) p53 alterations in all stages of breast cancer. *J. Surg. Oncol.* **48**, 260–267.

6. Done, S. J., Arneson, N., Ozcelik, H., Redston, M., and Andrulis, I. L. (1998) p53 mutations in mammary ductal carcinoma in situ but not in epithelial hyperplasias. *Cancer Res.* **58**, 785–789.

7. Done, S. J., Eskandarian, S., Bull, S., Redston, M., and Andrulis, I. L. (2001) p53 missense mutations in microdissected high-grade ductal carcinom in situ of the breast. *J. Natl. Cancer Inst.* **93**, 700–704.

8. Peltomaki, P. (2001) Deficient DNA mismatch repair: a common etiologic factor for colon cancer. *Hum. Mol. Genet.* **10**, 735–740.

9. Young, J., Leggett, B., Gustafson, C., et al. (1993) Genomic instability occurs in colorectal carcinomas but not in adenomas. *Hum. Mutat.* **2**, 351–354.

10. Iino, H., Simms, L., Young, J., et al. (2000) DNA microsatellite instability and mismatch repair protein loss in adenomas presenting in hereditary non-polyposis colorectal cancer. *Gut* **47**, 37–42.

11. Jass, J. R., Young, J., and Leggett, B. A. (2000) Hyperplastic polyps and DNA microsatellite unstable cancers of the colorectum. *Histopathology* **37**, 295–301.

12. Gryfe, R., Kim, H., Hsieh, E. T. K., et al. (2000) Tumor microsatellite instability and patient survival in a population-based series of young colorectal cancer patients. *N. Engl. J. Med.* **342**, 69–77.

13. Kern, S. E. (2001) Progressive genetic abnormalities in human neoplasia, in *The Molecular Basis of Cancer* (Mendelsohn, J., Howley, P. M., Israel, M. A., and Liotta, L. A., eds.). Saunders, Philadelphia, pp. 41–69.

14. Emmert-Buck, M. R., Bonner, R. F., Smith, P. D., et al. (1996) Laser capture microdissection. *Science* **274**, 998–1001.

15. Simone, N. L., Paweletz, C. P., Charbonneau, L., Petricoin, E. F. 3rd, and Liotta, L. A. (2000) Laser capture microdissection: beyond functional genomics to proteomics. *Mol. Diagn.* **5**, 301–307.

16. Craven, R. A. and Banks, R. E. (2001) Laser capture microdissection and proteomics: possibilities and limitation. *Proteomics* **1**, 1200–1204.

17. Kern, S. E., Redston, M., Seymour, A. B., et al. (1994) Molecular genetic profiles of colitis-associated neoplasms. *Gastroenterology* **107**, 420–428.

18. Redston, M. S., Papadopoulos, N., Caldas, C., Kinzler, K. W., and Kern, S. E. (1995) Common occurrence of APC and K-Ras mutations in the spectrum of lesions in colitis-associated neoplasia. *Gastroenterology* **108**, 383–392.

19. Barrett, M. T., Sanchez, C. A., Prevo, L. J., et al. (1999) Evolution of neoplastic cell lineages in Barrett oesophagus. *Nat. Genet.* **22**, 106–109.

20. Mirabelli-Primadahl, L., Gryfe, R., Kim, H., et al. Beta-catenin mutations are common in colorectal carcinomas with microsatellite instability but occur in endometrial carcinomas irrespective of mutator pathways. *Cancer Res.* **59**, 3346–3351.

21. Haimowitz, M. D. (1997) Practical issues in tissue banking. *Am. J. Clin. Pathol.* **107**, S75–S81.

22. Burke, H. B. and Henson, D. E. (1998) Specimen banks for prognostic factor research. *Arch. Pathol. Lab. Med.* **122**, 871–874.

23. Grizzle, W. E., Aamodt, R., Clausen, K., LiVolsi, V., Pretlow, T. G., and Qualman, S. (1998) Providing human tissues for research: how to establish a program. *Arch. Pathol. Lab. Med.* **122**, 1065–1076.

24. Fearon, E. R., Hamilton, S. R., and Vogelstein, B. (1987) Clonal analysis of human colorectal tumors. *Science* **238**, 193–197.

27

Conformation-Sensitive Gel Electrophoresis for Detecting BRCA1 Mutations

Smita Lakhotia and Kumaravel Somasundaram

1. Introduction

Detection of mutations is of central importance in the study of genetic and malignant diseases. Mutation detection helps us in understanding the protein structure, function, and expression. More than that, it is also important for presymptomatic/antenatal diagnosis, confirmation of the genetic cause of the disease and the mode of inheritance of a disease in a particular family, the prediction of clinical phenotype, and the potential for diagnostic analysis in the case of families with incomplete pedigrees or with new mutations. Therefore, the importance of direct mutation analysis cannot be understated.

Breast cancer is one of the most common malignancies affecting women worldwide. Genetic predisposition for familial early onset of breast cancer accounts for approximately 5–10% of all breast cancers. Mutations in two autosomal dominant genes, *BRCA1* and *BRCA2*, have been linked to familial breast cancer *(1–4)*. Mutations in these two genes were predicted to account for 85–90% of hereditary breast and ovarian cancer cases. Genetic screening might contribute to cancer control by identifying a group of at-risk patients who could benefit from augmented screening or prophylactic measures that accompany genetic susceptibility testing for cancer. Several techniques are available for detecting mutations in *BRCA1*. This chapter describes a relatively new technique of conformation sensitive gel electrophoresis (CSGE).

1.1. Overview

A rapidly increasing number of methods is available for detection of mutations. Choosing a mutation detection method requires careful consideration of many factors. The tests must be easy to use and cost-effective. Accuracy in the 99.9% range and day-to-day reliability are the basic requirements for the acceptance of new technologies, but attributes such as cost and convenience are also important.

The general strategy adopted for detecting mutations is to amplify the gene in small fragments, analyze the fragments and then sequence only those fragments positive for mutation by a rapid technique to locate and identify the nature of mutation. The very commonly used rapid techniques are single-stranded conformation polymorphism (SSCP) *(5)*, enzymatic or chemical cleavage of mismatched base pair *(6–12)*, and differential unfolding of homoduplexes and heteroduplexes by denaturing gradient gel

From: *Methods in Molecular Biology, Vol. 223: Tumor Suppressor Genes: Regulation, Function, and Medicinal Applications.* Edited by: Wafik S. El-Deiry © Humana Press Inc., Totowa, NJ

electrophoresis (DGGE) *(13,14)*. Among these techniques, single-stranded conformation polymorphism analysis of polymerase chain reaction (PCR) products (PCR-SSCP) is very commonly used for detecting mutations. However, the use of radioactivity and loss of mutation detection sensitivity for fragments larger than 200 bp with SSCP technique warrants the need for newer techniques. Conformation-sensitive gel Electrophoresis (CSGE), described in this chapter for detecting *BRCA1* mutations, is very simple, very effective for larger-size fragments, and does not involve radioactivity. This method basically identifies heteroduplex bands, which arise due to annealing of complementary strands, one each from mutant and wild-type alleles.

There are at least two principles on which the CSGE method works. First, single-base mismatches can produce conformational changes in the double-stranded DNA, leading to the differential migration of heteroduplex and homoduplex. Second, mildly denaturing solvents in an appropriate buffer can intensify the conformational changes produced by single-base mismatches, resulting in the increased differential migration of heteroduplexes and homoduplexes. The CSGE method involves heteroduplex analysis of PCR products in a novel, mildly denaturing polyacrylamide gel matrix using a different cross-linker, bis-acrolyl-piperazine, instead of the conventional bis-acrylamide *(15,16)*. Essentially, the protocol involves amplification of the entire coding region in small fragments and analyzing them by CSGE. The presence of additional slow or differentially migrating bands in comparison to normal sample indicates the presence of heteroduplex bands, which are suggestive of presence of mutation (**Fig. 1**). The samples that show heteroduplex bands have to be sequenced to locate and identify the nature of mutations.

There are several advantages to using the CSGE method. The CSGE method works fine with fragments of size up to 450 bp without loss of sensitivity. Optimization of gel running conditions is very simple as against SSCP. Another advantage is that no radioactivity is needed as compared to SSCP. It is also possible to put more than one sample (of varying sizes) in a single lane *(17)*, which is usually difficult with the SSCP method due to the presence of multiple bands per sample. It is also possible to carry out several steps of CSGE by automatic processes *(17)*.

One critical question about any technique is whether it will detect all the mutations in various sequence contexts. The CSGE conditions described by Ganguly et al. *(15)* for PCR products ranging from 200 to 800 bp were able to detect 60 of 63 single-base mismatches. Recently, Korkko et al. *(18)* developed newer CSGE conditions for PCR products of size up to 450 bp, which was able to detect 73 of 73 mutations studied.

2. Materials

1. 40% Acrylamide/BAP solution: Dissolve 39.6 g of acrylamide (Gibco BRL, cat. no. 15512-049), and 0.4 g of 1,4-bis (acrolyl) piperazine (Fluka, cat. no. 14470) in 60 mL of double-distilled water. Make up the volume to 100 mL with double-distilled water. Store in a dark glass bottle or a regular bottle wrapped with aluminum foil, at 4°C.
2. Ethylene glycol (Sigma, cat. no. E9129). Store at room temperature.
3. Deionized formamide (Sigma, cat. no. T9037). Store in aliquots for 2–3 uses at –20°C.
4. 20× TTE. Disslove 43.04 g Tris-HCl (Gibco BRL, cat. no. 15504-038), 14.24 g taurine (Sigma, cat. no. T9931), and 0.28 g EDTA (Gibco BRL, cat. no. 15576-028) in 130 mL of double-distilled water and adjust the pH to 9.0. Make up the volume to 200 mL with double-distilled water. At the time of use, dilute to 0.5× (44.4 mM Tris-HCl, 14.25 mM taurine, 0.1 mM EDTA, pH 9.0) with double-distilled water. *Do not autoclave the TTE buffer.*

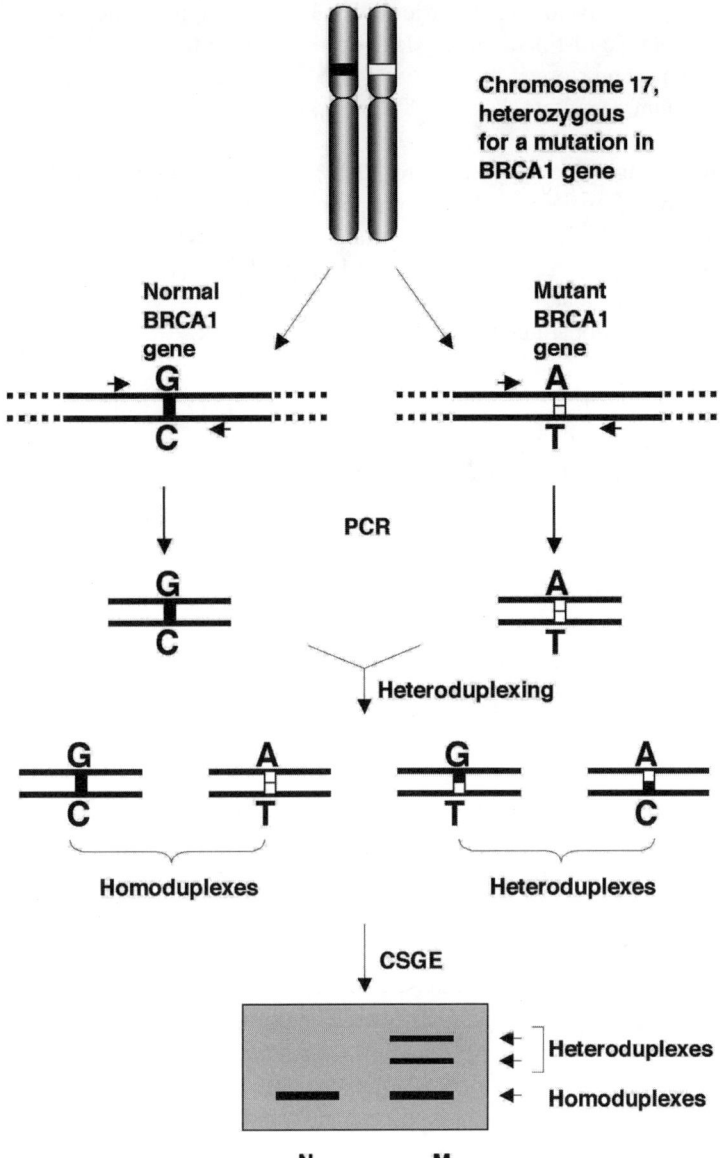

Fig. 1. Detection of mutations by using conformation-sensitive gel electroporesis (CSGE). Genomic DNA was isolated from blood sample of a patient (P) and a mornal person (N). With respect to BRCA1 screening, the patient refers to a person who has developed breast and/or ovarian cancer and belongs to a breast and/ovarian cancer family. Different regions of the coding region of BRAC1 gene are synthesized by polymerase chain reaction. After amplification, heterduplexing was carried out (see text for details) and electrophoreses in a CSGE gel. Please note that if the patient carries BRCA1 mutation, it is presumed that he or she is heterozygous for the mutation. One band appears in the normal sample, while additional bands appear in the case of the patient. The additional slow or differentially migrating bands represent heterodulplx DNA species.

5. ACD solution: Dissolve 0.48 g of citric acid, 1.32 g of sodium citrate, and 1.47 g of glucose in 50 mL of double-distilled water. Make up the volume to 100 mL with double-distilled water. Autoclave and store at 4°C.
6. 10% Ammonium persulfate. Keep at 4°C for up to 2 wk.
7. TEMED. Keep at 4°C.
8. 10× DNA loading buffer: 30% glycerol, 0.25% bromophenol blue, 0.25% xylene cyanol. Dissolve 30 mL of glycerol, 0.25 g bromophenol blue, and 0.25 g xylene cyanol in 50 mL of double-distilled water. Make up the volume to 100 mL with double-distilled water.
9. Ethidium bromide (Sigma, cat. no. E7637): 10 mg/mL in water.
10. dNTPS (Gibco BRL, cat. no. 10297-018).
11. Electrophoretic gel apparatus and high-voltage power pack.

3. Methods

3.1. Blood Collection and DNA Isolation

Collect 5 mL of blood from the patient using a sterile syringe and a needle and transfer the blood to a sterile tube containing 0.5 mL of sterile ACD solution. Mix gently by inverting the tube 3–4 times. The blood can be stored at –70°C indefinitely or at –20°C for several weeks before DNA isolation (*see* **Note 1**). We isolate the genomic DNA from 200 μL of blood by using the QIAamp DNA Mini Kit (Qiagen, cat no. 51304). The yield is typically about 5–10 μgs.

3.2. Amplification of BRCA1 Coding Region

The sequences of primers for genomic DNA amplification of BRCA1 are described in **Table 1**. Since the exon 11 is very large, 13 PCR reactions are carried out to generate successive overlapping fragments to cover the entire exon 11. The noncoding exons 1 and 4 are usually not included for the study. A total of 34 PCR reactions are carried out to cover the entire BRCA1 coding region. The sequence for primers for each exon is chosen such that they are located at least 50 bp away from exon–intron boundaries in order to detect the splice junction mutations. The composition of the 10× buffer is 100 mM Tris-HCl (pH 9.0), 15 mM MgCl$_2$, 500 mM KCl, 0.1% gelatin (*see* **Note 2**). PCR is performed in 25 μL solutions containing 100 ng of genomic DNA, 1× PCR buffer, 100 μM dNTPs, 10 pmol of each primer, and 0.5 U of *Taq* DNA polymerase. The reactions are carried out in a thermal cycler (Perkin-Elmer) as follows: 94°C for 5 min, 35 cycles (94°C for 1 min, 58°C for 1 min, and 72°C for 30 s), and 72°C for 10 min.

3.3. Heteroduplexing

The heteroduplex formation is enhanced by heating PCR products in the PCR machine as follows. Samples are heated at 98°C for 5 min and then followed by incubation at 68°C for 1 h. This step can be carried out during the last cycle of PCR if the samples are to be used for CSGE immediately. Alternatively, heteroduplexing reaction can be done just before loading the samples onto the gel.

3.4. CSGE

3.4.1. Pouring the Gel

1. Clean the glass plates thoroughly with a detergent and rinse with double-distilled water.

Table 1.
Sequence of PCR Primers (Forward and Reverse Primers) for Amplification of Various Exons of BRCA1 Coding Region[a]

Exon[b]	Size[c]	Forward primer	Reverse primer
2	258	5'-GAAGTTGTCATTTATAAACCTTT	5'-TGTCTTTTCTTCCCTAGTATGT
3	339	5'-TCCTGACACAGCAGAAGACATTTA	5'-TTGGATTTCGTTCTCACTTA
5	404	5'-CTCTTAAGGGCAGTTGTGAG	5'-GCATTAGAGAAAGGCAGTAAGTTTC
6	326	5'-CTTATTTTAGTGTCCTTAAAGG	5'-GGTGTGAGACCAGTGGGAGTAA
7	353	5'-CACAACAAAGAGCATACATAGGG	5'-GGCAGGAGGACTGCTTCTAGCCTG
8	268	5'-TGTTAGCTGACTGATGATGGT	5'-ATCCAGCAATTATTATTAAATAC
9	333	5'-CCACAGTAGATGCTCAGTAAATA	5'-ACACCAAATCCCAAGTCGTGTG
10	241	5'-TGGTCAGCTTTCTGTAATCG	5'-GTATCTACCCACTCTCTTCTTCAG
11.1	468	5'-CCACCTCCAAGGTGTATGAA	5'-TTCTTTTCTCTCACACAGGGGATC
11.2	385	5'-ACAGCCTGGCTTAGCAAGGAG	5'-ATGAGGATCACTGGCCAGTAAGTC
11.3	406	5'-AGAAACTGCCATGCTCAGAGAATC	5'-TGTGGCTCAGTAACAAATGCTCC
11.4	483	5'-TGTATTGGACGTTCTAAATGAGGT	5'-TCTATTGGGTTAGGATTTTTCTCA
11.5	488	5'-CAAACGGAGCAGAATGGTCA	5'-AGCTCTGGGAAAGTATGCTGTC
11.6	480	5'-TCCACAATTCAAAAGCACCTAAAA	5'-TTTGCAAAACCCTTTCTCCACTTA
11.7	437	5'-TTGTCAATCCTAGCCTTCCAAGAG	5'-GCGCTTGAAACCTTGAATGTAT
11.8	464	5'-GCACTCTAGGGAAGGCAAAAACAG	5'-TTTGGCATTATCAACTGGCTTATC
11.9	426	5'-CCACTCTGGGTCCTTAAAGAAACAAAGTCC	5'-TTACGGCTAATTGTGCTCACTG
11.10	439	5'-TCTGCTAGAGAAACTTTG	5'-AGATGCATGACTACTTCCCATAGG
11.11	325	5'-TCCTGGAAGTAATTGTAAGCATCC	5'-GGCCCCTCTTCGGTAACC
11.12	410	5'-AGGTTTGTTCTGAGACACCTGA	5'-AGATGCCTTTGCCAATATTACCTG
11.13	430	5'-TGCTACCGAGTGTCTGTCTAAGAA	5'-GCTCCCCAAAAGCATAAACA
12	265	5'-GTCCTGCCAATGAGAAGAAA	5'-TGTCAGCAAACCTAAGAATGT
13	320	5'-AATGGAAAGCTTCTCAAAGTA	5'-ATGTTGGAGCTAGGTCCTTAC
14	312	5'-CTAACCTGAATTATCACTATCA	5'-GTGTATAAATGCCTGTATGCA
15	338	5'-TGGCTGCCCAGGAAGTATG	5'-AACCAGAATATCTTTATGTAGGA
16	450	5'-AATTCTTAACAGAGACCAGAAC	5'-AAAACTCTTTCCAGAATGTTG
17	263	5'-GTGTAGAACGTGCAGGATTG	5'-TCGCCTCATGTGGTTTTA
18	352	5'-GGCTCTTTAGCTTCTTAGGAC	5'-GAGACCATTTTCCCAGCATC

Table 1. (continued)
Sequence of PCR Primers (Forward and Reverse Primers) for Amplification of Various Exons of BRCA1 Coding Region[a]

Exon[b]	Size[c]	Forward primer	Reverse primer
19	249	5'-CTGTCATTCTTCCTGTGCTC	5'-CATTGTTAAGGAAAGTGGTGC
20	401	5'-ATATGACGTGTCTGCTCCAC	5'-GGGAATCCAAATTACACAGC
21	298	5'-AAGCTCTTCCTTTTTGAAAGTC	5'-GTAGAGAAATAGAATAGCCTCT
22	297	5'-TCCCATTGAGAGGTCTTGCT	5'-GAGAAGACTTCTGAGGCTAC
23	255	5'-CAGAGCAAGACCCTGTCTC	5'-ACTGTGCTACTCAAGCACCA
24	280	5'-ATGAATTGACACTAATCTCTGC	5'-GTAGCCAGGACAGTAGAAGGA

[a]Ganguly et al., 1997 (*20*).
[b]Exons 1 and 4 are left out because they are noncoding exons; exon 11 is covered by 13 overlapping PCR (11.1–11.13) fragments.
[c]Refers to the size of the PCR fragment generated.

2. Clean the inner surface of the plates with 70% alcohol and wipe it dry.
3. Apply petroleum jelly to the spacers (1.0-mm) and assemble the glass plates.
4. Put the clamps all around and pour the gel mix (see below for the gel composition) slowly from a side, avoiding any air bubbles.
5. Place the comb between the two glass plates after pouring the gel.
6. Let the gel polymerize for 1.5 h.
7. The size of the plate, volume of gel mix needed vary with different electrophoresis system. We normally use the large size format (31.0 × 38.5 cm) Gibco BRL (S2) gel electrophoresis system.

The gel composition (for 130 mL) is as follows:

Reagent	Final concentration	Volume
40% Acrylamide/BAP (99:1)	10%	32.50 mL
20× TTE buffer	0.5×	3.25 mL
dH$_2$O		60.45 mL
Ethylene glycol	10%	13.00 mL
Formamide	15%	19.50 mL
10% Ammonium persulfate		1.30 mL
TEMED		74.10 μL

3.4.2. Running the Gel

1. After gel is polymerized, remove lower spacer and comb, attach plates to the electrophoresis apparatus.
2. Pour 0.5× TTE buffer into upper and lower chambers. Check for leaks.
3. Clean wells and pre-run for 1 h at 750 V.
4. Clean wells again (*see* **Note 3**).
5. Load the gel with appropriate amount (typically 6–10 μL) of PCR product mixed with 2–4 μL of loading dye. An appropriate standard marker in the size range of 500–1000 bp also needs to be loaded. We normally use pBR322 plasmid DNA digested with *Hinf* 1.
6. Run the gel for about 16 h at 400 V for 300–500 base fragments (for 38.5-cm glass plate).

3.4.3. Reading the Gel

1. Remove plates from gel apparatus and separate the small plate from the large plate.
2. Stain the gel with in ethidium bromide solution for 5–7 min in a large staining tray. Minimize background by not overstaining.
3. Destain the gel for 10 min in water.
4. Visualize bands with a hand-held UV torch in dark room. Due to the small amount of DNA and reflection from plate, bands may be hard to see by naked eye. Cut the relevant portion of the gel with a scalpel, transfer onto the UV transilluminator by wetting with water. One can also transfer the entire or large portion of the gel to the top of the UV transilluminator for detection of bands if the UV torch is not available.
5. Immediate photography is very important not only for the purpose of record and also to identify the heteroduplexes in cases where the heteroduplexes move very close to the homoduplexes. We usually use a Polaroid MP4 camera with film type 667.

3.5. Interpretation

The band(s) generated by PCR fragments of patient DNA are compared to that of a normal control individual. Any additional slow or differentially migrating bands in the patient sample in comparison to that of normal control sample indicates the presence of

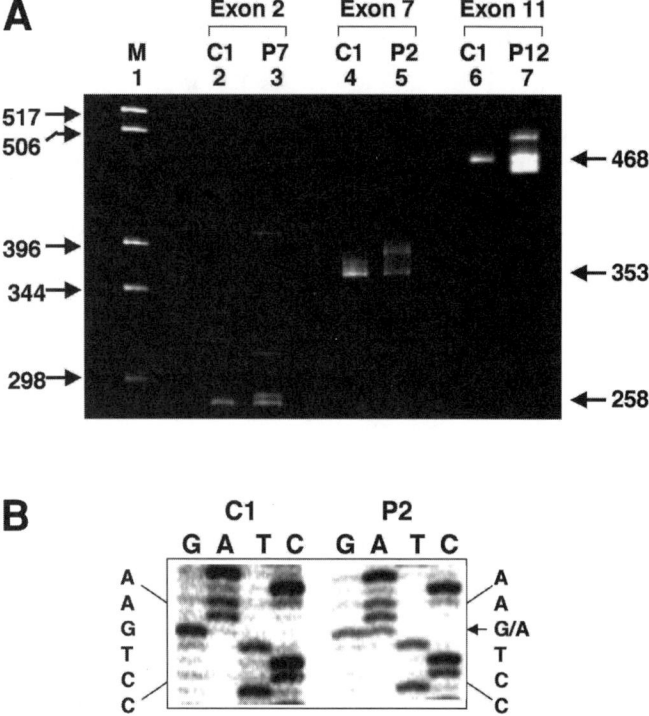

Fig. 2. **(a)** An ethidium bromide-stained CSGE gel showingt the heteroduplex bands. PCR products of exon 2, exon 7, and exon 11 of both normal person (C1) and relevant patients were analyzed by CSGE. C1 refers to a normal person, P7, P2, and P12 refer to individual patients. **(b)** DNA sequence of exon 7 PCR products from normal person (C1) and patient (P2) who is showing herteroduplex bands (*see* **Fig. 2a**). Patient P2 has a missense mutation (G–A). **a, b**, modified from Kumar et al. with permission *(19)*.

Table 2
Composition of PCR Optimization Buffers[a]

No.	Composition
Buffer #1	100 mM Tris-HCl (ph 8.3), 15 mM MgCl$_2$, 250 mM KCl
Buffer #2	100 mM Tris-HCl (ph 8.3), 15 mM MgCl$_2$, 750 mM KCl
Buffer #3	100 mM Tris-HCl (ph 8.3), 35 mM MgCl$_2$, 250 mM KCl
Buffer #4	100 mM Tris-HCl (ph 8.3), 35 mM MgCl$_2$, 750 mM KCl
Buffer #5	100 mM Tris-HCl (ph 8.8), 15 mM MgCl$_2$, 250 mM KCl
Buffer #6	100 mM Tris-HCl (ph 8.8), 15 mM MgCl$_2$, 750 mM KCl
Buffer #7	100 mM Tris-HCl (ph 8.8), 35 mM MgCl$_2$, 250 mM KCl
Buffer #8	100 mM Tris-HCl (ph 8.8), 35 mM MgCl$_2$, 750 mM KCl
Buffer #9	100 mM Tris-HCl (ph 9.2), 15 mM MgCl$_2$, 250 mM KCl
Buffer #10	100 mM Tris-HCl (ph 9.2), 15 mM MgCl$_2$, 750 mM KCl
Buffer #11	100 mM Tris-HCl (ph 9.2), 35 mM MgCl$_2$, 250 mM KCl
Buffer #12	100 mM Tris-HCl (ph 9.2), 35 mM MgCl$_2$, 750 mM KCl

[a]Schoettlin et al., 1993 *(21)*.

heteroduplex DNA (*see* **Notes 4** and **5**). **Figure 2a** shows the results of CSGE analysis of PCR products of BRCA1 exons 2, 7, and 11.1 from patients belonging to different breast cancer families *(19)*. All the three patients shown in this figure have heteroduplex bands. PCR products from patient samples showing heterduplexes are sequenced to locate the site and nature of the mutation. **Figure 2b** shows the results of sequencing of one of the patient sample (P2) shown in **Figure 2a**. This patient (P2) has a missense mutation (G to A) in exon 7, which results in the conversion of glutamic acid to lysine at codon 116 *(19)*.

3.6. Conclusion

The association of BRCA1 mutations with hereditary breast and/or ovarian cancer is very well established. Genetic testing could potentially offer different management options to BRCA1 mutation carrying high-risk individuals. That brings us to the requirement of a simple, reliable testing methodology for detecting BRCA1 mutations. Conformation-sensitive gel electrophoresis described here is a simple method with no requirement of any special instrument. The procedure is easy to carry out with little standardization. We routinely check the samples by CSGE for BRCA1 mutations. Only those PCR products that show heteroduplex pattern are subjected to DNA sequencing to identify the mutation. For a simple and effective mode of screening hereditary breast cancer samples for BRCA1 mutations, we suggest an initial scanning with CSGE method and subsequent sequencing of only those products showing heteroduplexes.

4. Notes

1. Do not store the blood sample for long at –20°C. It is better to keep the blood in 200 μL aliquots in a –70°C freezer. Blood transportation should be done on ice or dry ice.
2. Different DNA samples vary in purity, GC content, chemical modification, secondary structure, and other parameters that can inhibit PCR. In cases where the PCR for a particular sample does not work, a set of different buffers (**Table 2**) with different pH and salt composition can be tried. This provides an easy way of determining which PCR buffers are most effective for a specific PCR template and primer set.
3. It is very important for the sharpness of the DNA bands that the wells are thoroughly cleaned before the pre-run and again after the pre-run prior to loading the samples.
4. It is a good idea to have a DNA sample carrying mutation to be included in the gel at the time of standardizing the CSGE technique as well as with each subsequent gel as an internal control.
5. If a sample shows heteroduplex band(s), the same region should be reamplified from genomic DNA and the heteroduplex band pattern verified by CSGE.

References

1. Hall, J. M., Friedman, L., Guenther, C., et al. (1990) Linkage of early-onset familial breast cancer to chromosome 17q21. *Science* **250**, 1684–1689.
2. Miki, Y., Swensen, J., Shattuck-Eidens, D., et al. (1994) A strong candidate for breast and ovarian susceptibility gene BRCA1. *Science* **266**, 66–71.
3. Wooster, R., Neuhausen, S. L., Mangoin, J., et al. (1994) Localization of a breast cancer susceptibility gene, BRCA2, to chromosome 13q12-13. *Science* **265**, 2088–2090.
4. Wooster, R., Bignell, G., Lancaster, J., et al. (1995) Identification of breast cancer susceptibility gene BRCA2. *Nature* **378**, 789–792.

5. Orita, M., Iwahana, H., Kanazawa, H., Hayashi, K., and Sekiya,T. (1989) Detection of polymorphisms of human DNA by gel electrophoresis as single-strand conformation polymorphisms. *Proc. Natl. Acad. Sci. USA* **86**, 2766–2770.

6. Cotton, R. G. H. (1993) Current methods of mutation detection. *Mutat. Res.* **285**, 125–144.

7. Novack, D. F., Casna, N. J., Fischer, S. G., and Ford, J. P. (1986) Detection of single base-pair mismatches in DNA by chemical modification followed by electrophoresis in 15% polyacrylamide gel. *Proc. Natl. Acad. Sci. USA* **83**, 586–590.

8. Myers, R. M., Larin, Z., and Maniatis, T. (1985) Detection of single base substitutions by ribonuclease cleavage at mismatches in RNA:DNA duplexes. *Science.* **230**, 1242–1246.

9. Cotton, R. G. H., Rodrigues, N. R., and Campbell, R. D. (1988) Reactivity of cytosine and thymine in single-base-pair mismatches with hydroxylamine and osmium tetroxide and its application to the study of mutations. *Proc. Natl. Acad. Sci. USA* **85**, 4397–4401.

10. Ganguly, A., Rooney, J. E., Hosomi, S., Zeiger, A., and Prockop, D. J. (1989) Detection and location of single-base mutations in large DNA fragments by immunomicroscopy. *Genomics* **4**, 530–538.

11. Ganguly, A. and Prockop, D. J. (1990) Detection of single-base mutations by reaction of DNA heteroduplexes with a water-soluble carbodiimide followed by primer extension: application to products from the polymerase chain reaction. *Nucleic Acids Res.* **18**, 3933–3939.

12. Youil, R., Kemper, B. W, and Cotton, R. G. (1995) Screening for mutations by enzyme mismatch cleavage with T4 endonuclease VII. *Proc. Natl. Acad. Sci. USA* **92**, 87–91.

13. Myers, R. M., Maniatis, T., and Lerman, L. S. (1987) Detection and localization of single base changes by denaturing gradient gel electrophoresis *Meth. Enzymol.* **155**, 501–527.

14. Sheffield, V. C., Cox, D. R., Lerman, L. S., and Myers, R. M. (1989) Attachment of a 40-base-pair G + C-rich sequence (GC-clamp) to genomic DNA fragments by the polymerase chain reaction results in improved detection of single-base changes. *Proc. Natl. Acad. Sci. USA* **86**, 232–236.

15. Ganguly, A., Rock, M. J., and Prockop, D. J. (1993) Conformation-sensitive gel electrophoresis for rapid detection of single-base differences in double-stranded PCR products and DNA fragments: evidence for solvent-induced bends in DNA heteroduplexes. *Proc. Natl. Acad. Sci. USA* **90**, 10325–10329.

16. Williams, C. J., Rock, M., Considene, E., McCarron, S., Gow, P., and Ladda, R. (1995) Three new point mutations in type II procollagen (COL2A1) and identification of a fourth family with the COL2A1 Arg519 to Cys base substitution using conformation sensitive gel electrophoresis. *Hum. Mol. Genet.* **4**, 309–312.

17. Ganguly, T., Dhulipala, R., Godmilow, L., and Ganguly, A. (1998) High throughput fluorescence-based conformation-sensitive gel electrophoresis (F-CSGE) identifies six unique BRCA2 mutations and an overall low incidence of BRCA2 mutations in high-risk BRCA1-negative breast cancer families. *Hum. Genet.* **102**, 549–556.

18. Korkoo, J., Pihlajamaa, T., Prockop, D. J., and Ala Kokko, L. (1996) Comparision of conformation sensitive gel electrophoresis with denaturing gel electrophoresis for detection of single base mutations in PCR products. *Sixth International Conference on Matrix Biology*, III-3.

19. Kumar, B. V., Lakhotia, S., Ankathil, R., Madhavan, J., Jayaprakash, P. G., Nair, K., and Somasundaram, K. (2002) Germline BRCA1 mutation analysis in Indian breast/ovarian cancer families. *Cancer Biology and Therapy*, in press.

20. Ganguly, A., Leahy, K., Marshall, A., Dhulipala, R., Godmilow, L., and Ganguly, T. (1997) Genetic testing for breast cancer susceptibility: frequency of BRCA1 and BRCA2 mutations. *Genet. Testing* **1**, 85–90.

21. Schoettlin, W., Nielson, K. B., and Mathur, E. (1993) *Strategies* **6**, 43–44.

II

MEDICINAL APPLICATIONS OF TUMOR SUPPRESSORS

28

Conversion Technology and Cancer Predispositions

Nickolas Papadopoulos

1. Introduction

The field of diagnostics has gathered the first crop of the genetic insurgency in the identification of the genetic basis of common and uncommon hereditary diseases. Discoveries of genes that predispose to cancer have led to the development of genetic tests. In turn, genetic tests can provide information valuable for the management of individuals who belong to families with history of hereditary disease and lead to important changes in clinical practice of cancer predispositions. The promise of cancer gene testing is to decrease morbidity and mortality through screening and hopefully prevention. Besides the promise, however, there are problems associated with genetic testing. Genetic testing is associated with psychologic problems and ethical concerns, which are not going to be discussed here. What is often overlooked are the technical problems associated with genetic testing, which lead to tests with sensitivities lower than what is required in the clinical environment. Genetic testing here refers strictly to testing of germline mutations in familial predisposition to cancer. This chapter focuses on the utilization of haploid templates for increasing the sensitivity of genetic testing.

2. Sensitivity of Genetic Testing

Because of the aforementioned promise and the impact that the result of a genetic test will have on the life of a person, the expectations have been set high. Good diagnostic tests should have sensitivities and specificities approaching 100% *(1)*. However, for many tests, especially for cancer predispositions, this is not the case. In some cases the genes are not known and it is difficult to set up a genetic test that will be specific and sensitive. In this cases the test may at best rely on haplotype associations and family history.

Even when the mutated gene is known, the gene test may fail to identify a mutation. Among common cancer predispositions, hereditary breast and ovarian cancer (genes: *BRCA1*, *BRCA2*) , hereditary nonpolyposis colerectal cancer (HNPCC) (major genes: *hMSH2*, *hMLH1*), familial adenomatus polyposis (FAP) (gene: *APC*), hereditary melanoma (gene: *p16*), neurofibromatosis type 2 (gene: *NF2*) now have commercially available gene tests, but none of them has 100% sensitivity *(2–5)*.

From: *Methods in Molecular Biology, Vol. 223: Tumor Suppressor Genes: Regulation, Function, and Medicinal Applications.* Edited by: Wafik S. El-Deiry © Humana Press Inc., Totowa, NJ

When a mutation predisposing to cancer is identified in an individual, this is considered a positive result. In this case, any other member of the family with the same mutation either is at high risk or already affected. The members of the same family who test negative for the same mutation are true negatives, and their risk for getting the disease is the same as that of the general population. This information is then used by the clinicians in order to better manage the individuals. If the sensitivity of the test is low, genetic testing that fails to identify mutations is inconclusive. Absence of mutation in members of a family with predisposition to cancer does not necessarily mean that the individual is at low risk when the mutation in the family has not been identified. Interpretation of such genetic testing results is not as straightforward either. In a recent study *(4)*, in 31.6% of the cases the physician misinterpreted the test results. Studies have shown that people deal far better with either positive or negative results than with receiving inconclusive or no results at all *(6,7)*. Because of the above, genetic testing should not be offered without appropriate genetic counseling.

3. Technical Challenges

Technical challenges associated with genetic testing in cancer predisposition had been overlooked for a while. Now, it is clear that if we would like to have tests that are sensitive, we have to overcome such challenges. There are three major factors that can contribute to false negatives in genetic tests.

3.1. The Number of Genes That Predispose to the Particular Cancer

It is clear that multiple genes can contribute to the predisposition of a single disease. Two notable examples are hereditary breast cancer and HNPCC. In the case of FAP there is only one gene, *APC*, linked to the disease. It is obvious that in diseases for which not all of the genes have been identified, it is not possible to have a test that can identify all of the mutations that predispose to the disease. Absence of mutations in a family cannot exclude the possibility that other loci are involved. Linkage in individual cases, however, can increase the confidence that the most likely gene has been checked for mutations. Even when all genes involved in the predisposition to a certain disease are known, analysis is complicated and the chance of missing mutations is increased.

3.2. Heterogeneity of Mutations

The distribution and the nature of the mutations along the length of a gene can be diverse. Some of the mutations will be easy to identify and some more difficult. Based on the ease of identification, mutations were classified as compliant and refractory *(8)*. Unlike diseases such as sickle cell anemia or Huntington's disease, for which the number of different mutations is small and usually take place on the same region of the gene, the mutation spectrum of *hMSH2* is very different *(9)*. *hMSH2* mutations that predispose to HNPCC include point mutations, splice site changes, deletions of variable lengths, genomic rearrangements, nonsense mutations, and mutations that affect the expression of the gene. This is not unique to *hMSH2*; *BRCA1* and *BRCA2* are just two of the genes with diverse mutation spectra. Depending on their nature, mutations can be identified by some techniques but missed by others. For example, when a point mutation and a small deletion can be found by sequencing of polymerase chain reaction

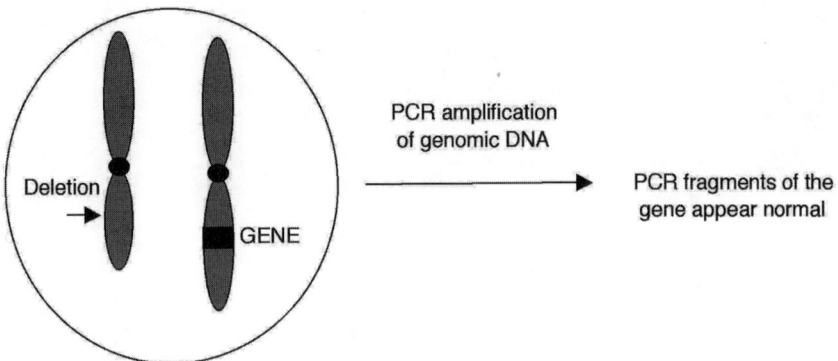

Fig. 1. Certain types of mutations, such as the deletion shown here, can be missed by sequencing. The amplified products are generated from the gene copy present in the diploid cell.

(PCR)-amplified genomic exons of a gene, a large deletion will be missed (**Fig. 1**). There are a number of reports indicating that mutations have been missed by one technique but identified by another. Mutations of the *hMSH2* gene that were missed by genomic sequencing were identified by Southern blotting *(10)*. Recently it has been reported that a deletion on *BRCA1* has been missed by a commercial genomic sequencing platform. Upon employment of color bar coding on combed DNA, the mutation was identified *(11)*. There is a range of technologies that are utilized in mutation analysis. Some of them are direct, such as sequencing that reveals the exact nature of the mutation, and some are indirect, such as SSCP, which requires a subsequent direct method for the identification of the exact change. The degree of easiness, cost effectiveness, and sensitivity varies among these techniques (for review of techniques used in genetic testing, *see* **ref. 12**). However, it is difficult to design a single assay that can detect all of these type of mutations. So, a series of different technologies is employed for a single study *(13)*. Although in some cases such combination may be successful, it adds cost and time, and complicates clinical testing.

3.3. Humans Are Diploid

The previous section focused on the sensitivity of the techniques in the identification of diverse type of mutations in diploid samples. There is another view. Techniques would be more sensitive if the template were haploid. In cancer predispositions mentioned above, one allele is inherited as wild-type and the other as mutant. This means that mutations are identified in the background of the normal allele and thus they can be masked. For example, a deletion of multiple exons could be missed by sequencing, and a mutation that affects expression can be missed almost by any technique that uses diploid templates. There are some indirect ways to get around the diploidy problem, such as semiquantitative assays that try to discriminate between two copies or analysis software that can eliminate some of the background by subtracting wild-type signal to enhance signals from variant sequences. Probably the best way is to use direct techniques for the isolation/separation of alleles, so that mutation analysis can be done on an allele in the absence of the other allele(s) of the same chromosome. This should unmask mutations (*see* **Fig. 2**).

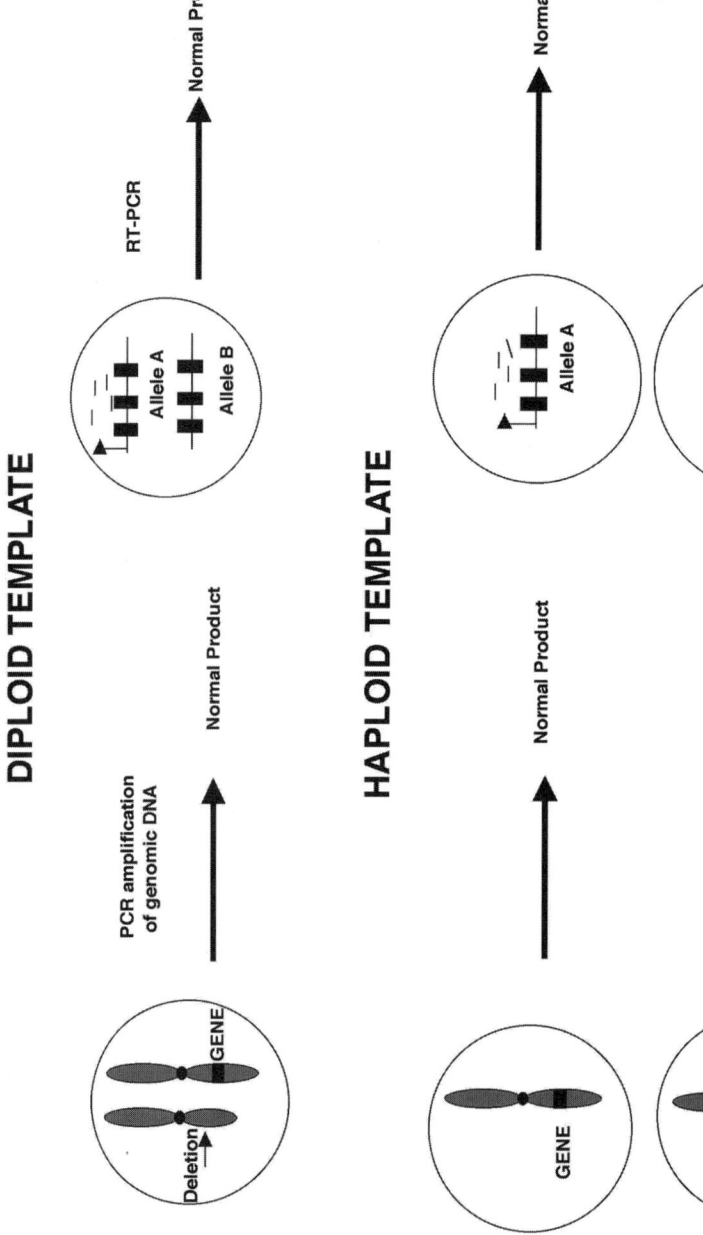

Fig. 2. Two examples of mutations that are masked in diploid templates. Left is a deletion: PCR of genomic DNA will reveal the deletion only in haploid templates. Right is a mutation affecting expression of a gene: RT-PCR will reveal the lack of expression only in haploid templates.

4. The Value of Haploid Template in Genetic Testing

Fusions of human cells to rodent cells for the creation of somatic cell hybrids has been used in the past for a variety of reasons including genetic diagnosis *(14–17)*. It is well known that human chromosomes are unstable and cannot be retained for many generations in somatic cell hybrids unless there is selective pressure. The generation of somatic cell hybrids with stably retained human chromosomes for the identification of mutations in genes that predispose to cancer was initially described as monoallelic mutation analysis (MAMA). Fusions were performed between human blood cells and rodent cell lines acting as recipients for specific human chromosomes. Each rodent cell line was defective for a specific gene. Therefore, the hybrids were not viable under selective conditions unless they contained the human chromosome that could complement the defect. This meant that, for example, for the isolation of human chromosome 2, a rodent cell line deficient in a gene whose human ortholog is present in human chromosome 2 was required.

Monoallelic mutation analysis was initially developed to unmask mutations in FAP and HNPCC patients. The proof of principle was demonstrated by the identification of a masked mutation in hMSH2. The mutation was a deletion of exon 1 of hMSH2 and resulted in the absence of expression from one allele *(16)*. Subsequently, MAMA utilized to increase the sensitivity of FAP testing. Nine probands from FAP families in which no mutations were identified with the protein truncation test, the mainstay of genetic testing in FAP, were used for the study. MAMA revealed mutations that affected the expression of the APC gene in six of the nine individuals. In the cases that linkage was available, lack of expression of APC was from the linked allele, indicating that a mutation segregating with disease affected the expression of the gene *(18)*. In one of the cases, reduced expression of APC was observed produced from one allele *(18)*. Subsequent studies suggested that this quantitative change in expression is likely the alteration that predisposes the family to FAP *(19)*. In similar types of experiments a mutation in the CSA gene was confirmed for homozygosity *(17)*, while germline deletion mutations were identified in the NF1 gene *(20)*. In all of the examples, the haploid template augmented the sensitivity of the technique used for the identification of the mutations.

5. Conversion Technology

The applicability of MAMA to other genes and diseases depends on the availability or development of a panel of cell lines that could retain specific human chromosomes. Alternatively, somatic cell hybrids with transient retention of specific human alleles have been utilized for the generation of haploid templates *(20)*. Either way, it is obvious that none of the alternatives is a robust way of generating hybrids for every human chromosome. For these reasons a robust system involving a new universal recipient cell line, which, when fused to human cells, generates hybrids that can contain any human autosome after a single fusion was developed. This approach of converting diploid templates to haploid templates is called Conversion Technology *(21)*.

This cell line (E2) is used to produce hybrids from fusions with human blood cells (**Fig. 3**). This E2 line is HPRT- and therefore HAT-sensitive. It also contains a neo cassette that enables Geneticin resistance. Human lymphocytes are HAT-resistant and Geneticin-sensitive. The human HPRT gene is on chromosome X and therefore human

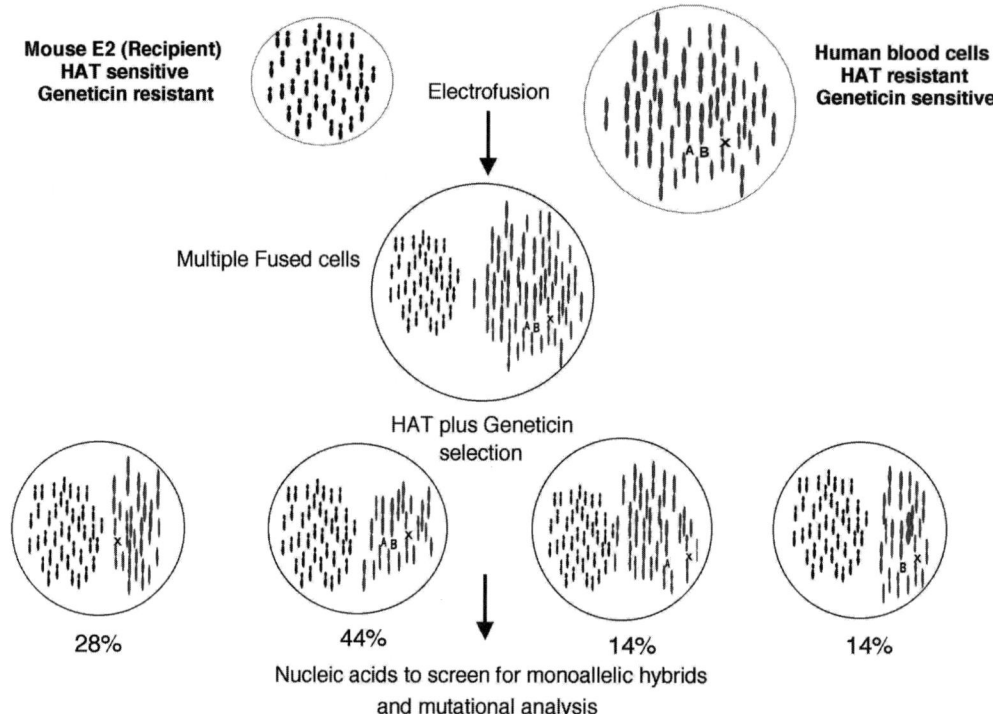

Fig. 3. Conversion technology: Mouse and human cells are fused. Initially the fused cells contain both sets of chromosomes. After selection, hybrids retain a random set of human autosomes and the human X chromosome. For any given chromosome (allele A and B), 14% of the hybrids retain allele A, 14% allele B, 48% both alleles, and 28% neither A nor B.

chromosome X is required for the survival of hybrids in selection media containing HAT and Geneticin. The somatic cell hybrids contain not only the human X chromosome but a subset of human chromosomes. Each hybrid contains approximately 12–15 human chromosomes stably for at least 90 generations without selection for their retention. FISH analysis has shown that the human chromosomes are intact and do not present any rearrangements or incorporate into mouse chromosomes *(22,23)*. Seven to nine of these human chromosomes are monoallelic. Analysis of hundreds of hybrids *(19,23)* has not shown any bias against the retention of any human chromosome. On average, from one fusion, for a given chromosome, 14% of the hybrids contain allele A, 14% allele B, 44% both alleles, and 28% do not contain any of the alleles. Analysis of 90 hybrids from four individuals showed that 24 hybrids are enough for isolating at least one monoalleleic hybrid for each of the human chromosomes *(23)*. The somatic cell hybrids are genotyped with human-specific DNA markers for the identification of the human alleles present in the hybrids. Polymorphic markers that can distinguish the two alleles in the diploid cells indicate which hybrids are monoallelic for which of the two alleles, bi-allelic, or null for the chromosome in question.

The hybrids were also tested for expression of specific genes. The genes that predispose to HNPCC, FAP, and BRCA were tested by RT-PCR and antibodies specific for human proteins. Expression of the genes in hybrids derived from blood cells of nonaffected individuals was obvious in every single hybrid containing the appropriate allele.

This implied that the expression of these genes was conserved in the hybrids and made possible the analysis of their expression from single alleles.

Conversion is a robust way for creating somatic cell hybrids that contain subsets of human chromosomes and enables the isolation of all human autosomes from a single fusion event.

6. Conversion Increases the Sensitivity of Genetic Testing in HNPCC

The utility of conversion was tested in a study for the identification of mutations in HNPCC families. A cohort of 48 families meeting Amsterdam criteria and exhibiting microsatellite instability (MSI) was used for the identification of mutations (**ref. *24*** and references within). Mutation analysis was initially performed by cDNA sequencing from diploid templates. Mutations were identified in 34 kindreds, with 31 kindreds having mutations in the hMSH2 and hMLH1 genes. From the 31 mutations, 10 mutations could not have been found by genomic sequencing. These 10 mutations were 9 deletions of single or multiple exons and one insertion between exons 6 and 7. However, mutations were not identified in the rest of the kindreds. Material for conversion was available from 22 kindreds. From these, 12 mutations had been identified previously, while 10 kindreds did not have any mutation. Lymphoblastoid cells from these kindreds were fused to the E2 cell line and somatic cell hybrids containing chromosome 2 and 3 alleles were identified by genotyping with human chromosome 2-and 3-specific markers. Hybrids monoallelic for these chromosomes were expanded and nucleic acids were prepared. RT-PCR was performed either for hMSH2 or hMLH1. Upon analysis of the RT-PCR products it was obvious that 6 cases did not have detectable transcript from one of the alleles. It is important to note that every hybrid with the same allele examined exhibited the same pattern of expression for the genes tested, without exception. In three of these cases the mutation was deletion of multiple exons, and in three of the cases the mutations affected expression of the gene. The last four mutations were identified by sequencing of the RT-PCR products. In this study, Conversion Technology resulted in the identification of mutations in all of the samples raising the sensitivity of mutation detection in the cohort of 22 samples to 100% *(21)*.

So, it appears that a combination of Conversion Technology , RT-RCR, and sequencing of RT-PCR products is a very sensitive test for HNPCC. RT-PCR in monoallelic hybrids should immediately identify mutations that result in truncation or absence of transcripts. This will include deletions, mutations that affect expression, splice-site mutations, and chromosomal rearrangements. Subsequent sequencing of the same products should reveal other type of mutations. In addition, this test will be more economical because the amplification and sequencing of cDNA requires fewer reactions as opposed to the amplification of all individual exons from genomic DNA.

7. Applications

The application of conversion should not be restricted only to the identification of mutations that predispose to HNPCC and FAP. Theoretically, conversion is universal to all genetic diseases. Its best application will be in diseases in which the mode of mutation includes high percentages of noncomplaint mutations such as deletions. The value of conversion is in the generation of haploid templates. This should increase the sensitivity of any subsequent test. Among other genes, good candidates for conversion are *BRCA1* and *BRCA2*, *PKD1* and *PKD2*.

Besides the identification of mutations in predispositions, converted samples can be useful in fine mapping of genetic events. For example, it is easy to envision that the exact location of a translocation can be easily identified if a derivative chromosome is isolated from the normal chromosomes and the other derivative. Mapping in this case does not require the need of polymorphic markers. In the same way, converted samples can help in the definition of a deletion breakpoint in cancer cells.

Recently, the value of conversion technology has been realized in human molecular genetics in the experimental determination of haplotypes that are useful in association studies *(24)*. Furthermore, haploid templates from somatic cell hybrids were the preferred templates for amplification of DNA fragments used in hybridizations to DNA chips for resequencing of the genomic regions *(25)*.

Technical innovations such as conversion can one day lead to the development of diagnostic tests with the desired sensitivity.

References

1. Eng, C. and Vijg, J. (1997) Genetic testing: the problems and the promise. *Nat. Biotechnol.* **15**, 422.
2. Serova, O. M., Mazoyer, S., Puget, N., et al. (1997) Mutations in BRCA1 and BRCA2 in breast cancer families: are there more breast cancer-susceptibility genes? *Am. J. Hum. Genet.* **60**, 486.
3. Hubbard, R. and Lewontin, R. C. (1996) Pitfalls of genetic testing. *N. Engl. J. Med.* **334**, 1192–1194.
4. Giardello, F. M., Bresinger, J. D., Petersen, G. M., et al. (1997) The use and interpretation of commercial APC gene testing for familial adenomatous polyposis. *N. Engl. J. Med.* **336**, 823–827.
5. Cho, M. K., Sankar, P., Wolpe, P. R., and Godmilow, L. (1999) Commercial APC gene testing for familial adenomatous polyposis. *Am. J. Med. Genet.* **83**, 157.
6. Dubok de Wit, A. C., Tibben, A., Duivenvoorden, H. J., et al. (1998) Predicting adaptation to presymptomatic DNA testing for late onset disorders: who will experience distress? Rotterdam Leiden Genetics Workgroup. *J. Med. Genet.* **35**, 745–754.
7. de Wert, G. (1998) Ethics of predictive DNA-testing for hereditary breast and ovarian cancer. *Patient Educ. Couns.* **35**, 43–52.
8. Yan, H., Kinzler, K. W., and Vogelstein, B. (2000) Genetic testing—present and future. *Science* **289**, 1890–1892.
9. Papadopoulos, N. and Lindblom, A. (1997) molecular basis of HNPCC: mutations of MMR genes. *Hum. Mut.* **10**, 89–99.
10. Wijnen, J., van der Klift, H., Vasen, H., et al. (1998) MSH2 genomic deletions are a frequent cause of HNPCC. *Nat. Genet.* **20**, 326–328.
11. Gad, S., Scheuner, M. T., Pages-Berhouet, S., et al. (2001) Identification of the BRCA1 gene using colour bar code on combed DNA in an American breast/ovarian cancer family previously studied by direct sequencing. *J. Med. Genet.* **38**, 388.
12. van Ommen, G. J. B., Bakker, E., and den Dunnen, J. T. (1999) The Human Genome Project and the future of diagnostics, treatment, and prevention. *Lancet* **354**, si5–si10.
13. Fidalgo, P., Almeida, M. R., West, S., et al. (2000) Detection of mutations in mismatch repair genes in Portuguese families with hereditary non-polyposis colorectal cancer (HNPCC) by multi-method approach. *Eur. J. Hum. Genet.* **8**, 49–53.

14. Harris, H. (1985) Suppression of malignancy in hybrid cells: the mechanism. *J. Cell Sci.* **79**, 83–94.
15. Schultz, R. A., Saxon, P. J., Glover, T. W., and Friedberg, E. C. (1987) Microcell-mediated transfer of a single human chromosome complements xeroderma pigmentosum group A fibroblasts. *Proc. Natl. Acad. Sci. USA* **84**, 4176–4179.
16. Papadopoulos, N., Leach, F. S., Kinzler, K. W., and Vogelstein, B. (1995) Monoalleleic mutation analysis (MAMA) for identifying germline mutations. *Nat. Genet.* **11**, 99–102.
17. McDaniel, L. D., Legerski, R., Lehmann, A. R., Friedberg, E. C., and Schultz, R. A. (1997) Confirmation of homozygosity foe a single mucleotide substitution mutation in Cockayne syndrome patient using monoallelic mutation analysis in somatic cell hybrids. *Hum. Mutat.* **10**, 317–321.
18. Laken, S. J., Papadopoulos, N., Petersen, G., et al. (1999) Analysis of masked mutations in familial adenomatous polyposis. *Proc. Natl. Acad. Sci. USA* **96**, 2322–2326.
19. Yan, H., Dobbie, Z., Gruber, S. B., et al. (2001) Small changes in expression affect predisposition to tumorigenesis. *Nat. Genet.* **30**, 25–26.
20. Kayes, L. M., Burke, W., Riccardi, V. W., et al. (1994) Deletions spanning the neurofibromatosis 1 gene: identification and phenotype of five patients. *Am. J. Hum. Genet.* **54**, 424–436.
21. Yan, H., Papadopoulos, N., Marra, G., et al. (2000) Conversion of diploidy to haploidy. *Nature* **403**, 723–724.
22. Langer, S., Jentsch, I., Gangnus, R., Yan, H., Lengauer, C., and Speicher, M. R. (2001) Facilitating haplotype analysis by fully automated analysis of all chromosomes in human-mouse hybrid cell lines. *Cytogenet. Cell Genet.* **93**, 11–15.
23. Douglas, J. A., Boehnke, M., Gillanders, E., Trent, J. M., and Gruber, S. B. (2001) Experimentally-derived haplotypes substantially increase the efficiency of linkage disequilibrium studies. *Nat. Genet.* **28**, 361–364.
24. Liu, B., Parsons, R., Papadopoulos, N., et al. (1996) Analysis of mismatch repair genes in hereditary non-Polyposis colorectal cancer patients. *Nat. Med.* **2**, 169–174.
25. Patil, N., Berno, A. J., Hinds, D. A., et al. (2001) Blocks of limited haplotype diversity revealed by high-resolution scanning of human chromosome 21. *Science* **294**, 1719–1723.

29

Discovering Novel Anticancer Drugs

Practical Aspects and Recent Advances

Cristina T. Lewis and Patrick M. O'Connor

1. Introduction

There has never been a more exciting time than the present for discovering anti-cancer drugs. The past decade has witnessed the nearly complete sequencing of the human genome, major advances in the multidisciplinary application of new genomic technologies, and the integration of combinatorial chemistry into the drug discovery process. This review is intended for cancer researchers who are not familiar with the details of the drug discovery process in the pharmaceutical industry. As such, it offers an introduction to the procedure, providing a description of each step of a representative drug discovery process that integrates combinatorial chemistry with structure-based drug design into the design of development compounds for subsequent evaluation in clinical trials. The impact of new genomics technologies and the sequencing of the human genome on drug discovery is described. A strong emphasis is placed on the pragmatic considerations that guide a multidisciplinary, iterative small-molecule drug discovery program, as well as the recent technological advances that permit greater efficiency and productivity. Because of the focus on structure-based drug design, the iterative processes described are somewhat biased toward the inhibition of enzymes and proteins, rather than receptors, by small molecules. However, most of the general practices apply equally well to drug discovery executed in the absence of protein structures. Problems and applications specific to the discovery of broad classes of anti-cancer drugs are described, although, again, general principles may apply to other therapeutic areas.

A drug discovery program is typically divided into multiple stages, defined by the activities associated with that stage, and thus also by the resources applied during each period. **Figure 1** illustrates the typical sub-divisions of a drug discovery program, and the time typically associated with each stage. The duration of a program can obviously vary, depending on the difficulties encountered, but a reasonable duration for a successful program is targeted for less than 5 years from target validation to nomination of a clinical candidate.

From: *Methods in Molecular Biology, Vol. 223: Tumor Suppressor Genes: Regulation, Function, and Medicinal Applications*. Edited by: Wafik S. El-Deiry © Humana Press Inc., Totowa, NJ

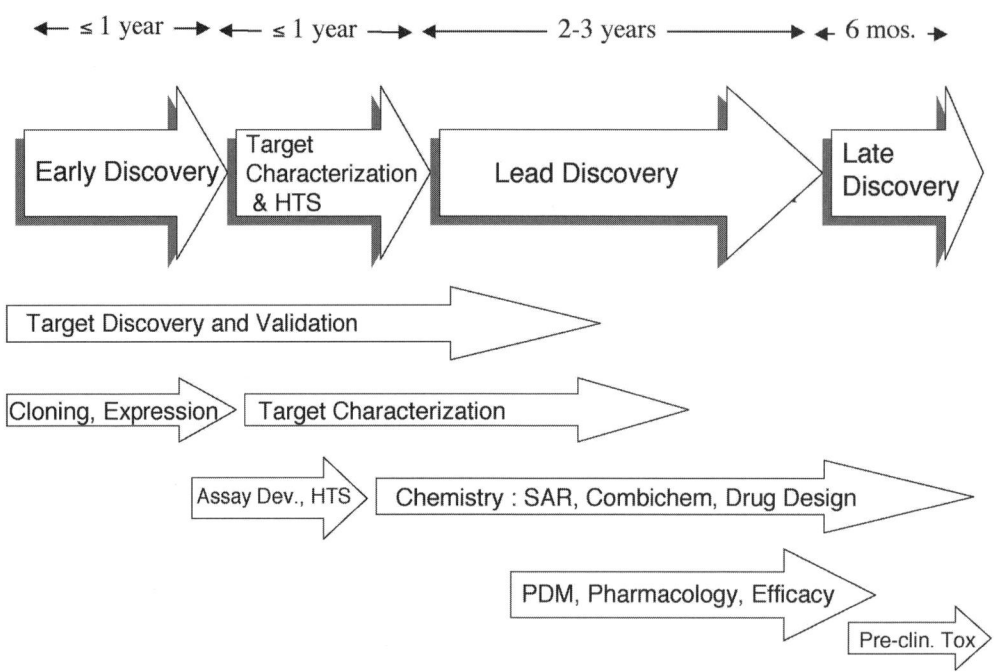

Fig. 1. Approximate timelines for the stages of drug discovery (see text for details). HTS = high-throughput screening; PDM = pharmacokinetics and drug metabolism; SAR = structure-activity realtionships.

2. Target Selection

2.1. Mining for New Target Genes: The Impact of Genomics

The identification and subsequent validation of new targets for cancer chemotherapy remains one of the greatest creative challenges facing cancer researchers today. With the completion of the draft sequence of the entire human genome, the concept of the human cell as a vessel of thousands of mysterious enzymes, unknown proteins, and undiscovered pathways will transition into a deeper understanding of biologic processes and thus more pivotal drug targets. The availability of the human genome sequence has resulted in a paradigm shift in the pharmaceutical industry, from the time-honored approach of characterizing a protein with known pathophysiologic relevance, to the challenge of mining tractable pharmaceutic targets from a myriad of gene sequences that encode proteins with unknown biological actions. The list of all theoretically possible human drug targets (all human genes) is nearly complete; one of the most important challenges for cancer researchers today is to harvest this wealth of information to pinpoint those targets with strong relevance to the disease, and to pursue those targets for which pharmaceutical intervention will lead to a cure for the disease. Some of the considerations and methods being applied in this new paradigm are described in the following sections.

The simultaneous publication of the draft sequence of the human genome by the public consortium Human Genome Project *(1)* and Celera Genomics *(2)* was an extraordinary landmark in biology and medicine. The published draft covers 96% of the genome, with complete sequencing to 99.9% accuracy of the entire genome expected by 2003. The estimated number of human genes currently stands at approximately 30,000–40,000, considerably lower than earlier estimates of over 100,000 genes. There is a surprisingly small difference between the number of protein coding genes in the human genome and the genomes of other organisms, such as the mustard weed *Arabidopsis thaliana* (26,000), the nematode worm *Caenorhabditis elegans* (~19,500), or even the laboratory fruit fly *Drosophila melanogaster* (~13,600). This difference is even more remarkable considering that the human genome size is 30 times larger than the worm and fly genomes *(1)*. Nevertheless, this estimated number of human genes is in agreement with independent estimates based on extrapolation from the completed sequencing of human chromosomes 21 and 22 *(3,4)*, and on expressed sequence tags (ESTs; DNA sequences read from the ends of cDNA molecules) *(5)*. Less than 2% of the human genome represents protein-coding DNA, and exons of the worm, fly, and human are all of similar size. In contrast, human intron size is significantly more variable in humans and there are more "long introns," resulting in an average intron length up to 10-fold longer in the human than in the worm and fly *(1)*.

2.2. Computational Genomics

With the certainty that all human genes have been mapped, an initial task of discovering relevant targets for cancer therapy might be simply to identify genes that are homologous to proteins known to play a role in oncogenesis, or to search for human counterparts to signal transduction proteins identified from smaller organisms. However, the sheer volume of sequences within the human non-protein-coding regions complicates the initial task of simply identifying genes. Computer-based algorithms provide a first step in predicting genes from genomic sequences, using either *ab initio* prediction of genes based on splice sites or coding regions that are expected to be associated with genes (GenScan and Grail), or comparisons of genomic sequences to homologous protein sequences (Blast, Procrustes, Genewise). While these methods provide a valuable first look and can satisfy initial questions, some *ab initio* methods yield false positives, or overpredictions, of gene numbers *(6)*. On the other hand, the *in silico* search for novel targets might be disappointingly unproductive, particularly for intensively studied areas of cancer research. For example, an initial search of the draft human genome sequence for genes homologous to known tumor suppressors identified no novel genes *(7)*. A similar search for the cyclins and cyclin-dependent kinases (CDKs) that regulate the cell cycle also revealed only three novel cyclins and no new CDKs. Examination of the completely sequenced genomes of several organisms for homologs to the yeast genes known to regulate the spindle checkpoint machinery of cell division (Mad1-3, Bub1, Bub3) also yielded no novel genes *(8)*. Whether these results are due to the inherent limitations of computational screening, or to extensive prior research on conserved protein families, remains to be seen. Disease susceptibility or protection genes can also be identified computationally through single nucleotide polymorphisms (SNPs). These single-base-pair differences in the DNA sequence from two individuals are being assigned along with sequences; thus far 1.42 million SNPs have been mapped on the human genome *(9)*. This information provides the foundation for a grand-scale human genetics experiment, as the identity and frequency of SNPs from patients and controls will permit correlation with drug action and susceptibility or resistance to different disease states. Furthermore, the design of novel drugs that either exploit SNPs as a means of selectivity or are "designed around" their presence to allow greater usage will launch a new era in drug discovery. However, in order for SNPs to have a relevant impact on drug discovery efforts, much deeper databases will need to be created. Indeed, one is likely to need ~100 individualized DNA samples to have the accuracy of predicting a SNP in a gene of interest with >95% confidence. Diverse ethnic backgrounds are required in the matrix and should be representative of the intended treatment population of interest. The SNP needs to have a relevant impact on the conformation of the protein to alter the design team's opportunities, and one must have tests (genetic screening & response biomarkers) in treated patients to correlate drug action.

2.3. High-Throughput Gene Sequencing and DNA Microarrays

Fortunately, the challenge of sorting through the vast complexity of the human genome sequence, of which only 1.1–1.4% represents protein-coding DNA, is facilitated by the parallel high-throughput sequencing of many diverse cDNA libraries. Each expressed sequence tag derived from the cDNA libraries represents only a portion of a complete gene, but computational approaches are used to assemble the ESTs into contiguous sequences (contigs) to define large libraries of genes. To date, millions of ESTs from different species have been identified and deposited into public databases such as dbEST (over 3.5 million human ESTs accessible at http://www.ncbi.nlm.gov/dbEST/). Other consortiums with long-standing EST projects continue to provide new data, and a list of databases providing gene and protein sequences is available *(10)*. Although the EST data are by definition incomplete, are more subject to error than genomic sequence data, and are biased by the expression levels of the mRNA in the tissue of origin, they nevertheless have proved valuable to the interpretation of genomic data, and to the drug discovery process. Perhaps the most powerful aspect of the EST databases is their utility in defining and comparing genes from a variety of cDNA libraries. Thus, genes from different tissues and organs, from normal or diseased tissues, and from different organisms are all accessible for iterative comparisons, mining for homologs, or tentative assignment of gene functions. Certainly the ability to mine gene sequences from other organisms is of undisputed value to cancer research, given, for example, the tremendous contribution of yeast genetics to our current understanding of the cell division cycle, or the identification in nematode of the bcl2 protein and subsequent elaboration of its role in regulating apoptosis.

Tissue-specific cDNA libraries are the core of another new technology that can provide insight into gene function: the DNA chip or microarray. This relatively new technology seeks to exploit the inherent value of quantifying expression levels of gene messages, applied in a high-throughput format. DNA oligonucleotides or polymerase chain reaction (PCR) products from a defined cell or tissue type are deposited on a glass plate at a high sample density of approximately 1000 discrete DNA samples/cm^2. Experimental RNA samples are labeled (using fluorescent or radioactive labels) during first-strand cDNA synthesis, and the cDNA probes are hybridized to the arrays. Fluorescence or phosphorimager intensity measurement at each coordinate provides a quantitative readout of all genes represented in the array. A mRNA expression range of 2–3 orders of magnitude can be detected with as little as 50 ng up to 100 μg of total RNA sample, depending on the detection method *(11)*. Comparison of gene expression in normal vs diseased tissue, or tissue subjected to some external stimulus, offers perhaps the greatest advantage for cancer research. For example, a small DNA microarray was used to identify all genes that were induced by phorbol ester treatment compared to untreated cells. In addition to two genes known to be upregulated by phorbol ester treatment, one previously unidentified gene was discovered to be moderately induced *(12)*. The initiative within the National Cancer Institute's Cancer Genome Anatomy Projects intends to catalog all cancers by gene expression level and pattern (http://www.ncbi.nlm.nih.gov/ncicgap/). Although the questions that can be addressed by microarray-based assays are virtually unlimited, these experiments are enormously data-intensive, and will require judicious experimental design, as well as universal quality standards to enable comparison and, ideally, integration of data from different laboratories.

High-technology approaches such as those described above are complemented by lower-throughput techniques such as real-time quantitative PCR *(13,14)* and positional cloning, in which an unknown gene is identified based on knowledge of its chromosomal location and association with a disease state. Positional cloning is likely to be most useful as a complementary early approach, for illuminating the critical components of complex pathways, rather than for direct target identification.

The power available from combining these different genomic approaches was recently exemplified by Monni et al. *(15)* This study combined molecular, genomic, and microarray technologies to perform a comprehensive analysis of the genes expressed within a 4-Mb genomic contig from the chromosomal region 17q23, known to be amplified in breast cancer and associated with poor prognosis of patients. Boundaries of the amplified region were mapped using fluorescence *in situ* hybridization to breast tissue microarrays. The expression levels of all of the genes and ESTs identified within this region were then analyzed in six different breast cancer cell lines using cDNA microarrays. Several overexpressed genes were confirmed from previous studies, as well as one unidentified EST. Similarly, DNA microarrays were used to identify the specific target genes regulated by the EGR1 transcription factor, which is overexpressed in prostate cancer *(16)*.

2.4. Proteomics

Target mining strategies that focus on DNA and mRNA offer the advantage of being highly amenable to high-throughput data collection; however, the complementary strategy of focusing on the protein product of the gene is of growing importance to assigning function to sequenced genes. The new field of proteomics, or the characterization of the expressed proteins of a genome, has begun to integrate with functional genomics to provide a powerful tool for filling in the gaps that are not covered by ESTs and DNA microarrays. In some cases, the concentration of a given mRNA and its cognate protein are not correlated. For example, signaling events triggered by protein release, intracellular trafficking, protein–protein interactions, multiprotein complex formations, degradation, or posttranslational modification are expected to be invisible at the level of mRNA analysis. Frequently, the relatively simple task of identifying the partners of protein–protein interactions, as exemplified by the yeast two-hybrid approach, can shed light on the function of an unknown protein, or can reveal components of a signaling pathway. The major technique that has accelerated proteomics is that of two-dimensional gel electrophoresis in tandem with mass spectrometry. Separation of proteins first by charge using isoelectric focusing, followed by denaturing SDS-PAGE to separate by size, yields a complex but highly reproducible map of spots. Fluorescent dyes are now available with greater sensitivity than that of the traditional Coomassie blue dye, yet that avoid the protein modifications sustained by silver staining. Commercial image analysis software packages are available that assist in the mapping of over 2000 resolved spots per gel. Two-dimensional electrophoresis is an established tool for analyzing proteins of complex mixtures *(17)*, but the advances in analytical mass spectrometry have transformed the technique to one suitable to the scale of proteomics. Typical detection methods are matrix-assisted laser desorption ionization–time-of-flight (MALDI-TOF) mass spectrometry, with a mass accuracy of up to 10 ppm. Electrospray MS can provide partial sequence information for smaller peptides, and can also generate information about posttranslational modifications. The improved sensitivity of these instruments now permits detection of only several hundred femtomoles of protein, thus allowing a lower limit for detection of proteins with approximately 1000 copies per cell for samples of 10^8–10^9 cells *(18)*. Although an accurate mass of a protein is useful, the identity of the protein can be confirmed by proteolytic digestion and analysis of peptide fragments, combined with EST database searching. The automated database searching and computational methods provide tremendous power to this approach in their ability to suggest the identity of a protein based on mass and predicted mass of derived peptides. Moreover, sample purity is not an issue, as masses of several proteins can frequently be distinguished in a single analysis. Frequently, two-dimensional gels are blotted, subjected to tryptic digestion *in situ*, and the masses of the resulting peptides are measured and compared against the computer-generated predicted tryptic fragments of a protein database. These techniques are increasingly subjected to automation with robot-guided spot excision and multiple-well sample analysis for MS. In a familiar theme, public databases of two-dimensional SDS-PAGE maps are available, cataloging over 772 proteins in 31 reference gels from different species (http://www.expasy.ch/ch2d/).

Advances in microdissection have also proved of particular utility to oncologists struggling to generate samples for proteomic analysis. The technique of laser capture microdissection (LCM) offers a simplified method of isolating specific cell populations (e.g., cancerous vs normal) from tissue specimens or ctyologic smears, by permitting laser-assisted capture of microscopic fields of cells onto film, for subsequent analysis of DNA, mRNA, or protein *(19)*.

These proteomics approaches typically complement genomic surveys of comparative samples, such as control vs drug-treated cells, or cancerous vs normal tissue. For example, Myers et al. used two-dimensional gel electrophoresis to profile the effect of drug treatment on the proteome of the 60-cell line panel of the National Cancer Institute *(20)*. The list of examples of proteomic projects applied to cancerous tissues continues to grow *(19)*.

2.5. Structural Genomics

In the natural evolution of the investigation of a gene, mRNA, and protein primary structure, the characterization of protein secondary and tertiary structure has advanced into the field of structural genomics. Given the inherent time and labor involved in determining a protein's crystal structure, application of this technology to the fast-paced world of genomics may seem incongruous. However, as in many other disciplines faced with repetitive tasks, the initiation of broad arrays of crystallization trials is now accelerated by robotics. In addition, improved software for molecular replacement is under continual development, and more powerful X-ray beams are online at multiple synchrotron centers. To date, nearly a dozen academic centers, consortiums, or biotechnology companies have formed to focus on structural genomics *(21)*. Although there are examples of using a protein's structure to determine function *(22)*, it is likely that the greatest contribution of the structural genomics effort will be the cataloging of secondary structures and folds with the ultimate goal of refining structure prediction algorithms, as well as the simplification of homology modeling for defined classes of proteins. The number of solved protein, peptide, or virus structures in the protein databank (PDB; *[23]*; www.rcsb.org/pdb/index.html) is currently at 13,263, and the content of the PDB appears to be growing exponentially. Interestingly, the number of new protein folds reported also grew rapidly and then declined after 1997. Have we detected most of the existing protein folds, or does this plateau reflect solution of only the readily accessible protein structures? Only continued structural genomics efforts will answer this question.

2.6. Functional Genomics

A purist might argue that the enormous amount of data generated by the different high-throughput and medium-throughput genomic approaches are no substitute for a cleverly designed biologic experiment that focuses on a particular defect or pathway in an effort to pinpoint a target for pharmaceutical intervention. Certainly this is a valid argument, and has given rise to some of the more elegant and efficient strategies of mining targets, in which a particular function or phenotype is the readout of a screen. The concept of "synthetic lethal screening" in yeast is one such example. This experimental approach asks, "What proteins exist that, when mutated or inactivated, do not themselves cause lethality, but when combined with another primary mutation, result in cell death?" In this case, the primary mutation may be in one of a number of conserved genes that are frequently mutated in human cancer, such as a gene involved in DNA repair or growth factor signaling *(24)*. Yeast strains bearing a selection of different primary defects have been constructed, and can be used to screen for the synthetic lethal mutations. This strategy points the way toward the discovery of cancer drugs that are more selective against cancer cells, rather than targeting all proliferating cells. Similar strategies define a specific phenotypic assay or readout in model cellular or organism systems, and then screen for interference with this pathway using antisense RNA *(25)*, ribozymes, mutagenized DNA *(26)*, expressed random peptides, genetic suppressor elements *(27)* or even chemical libraries *(28)*. Such strategies are limited only by the creativity of the scientist in posing the original question and defining the appropriate phenotypic assay.

2.7. Target Tractability

When the genomics and functional screens have revealed an interesting protein or functional pathway for intervention, the next critical question that cannot be ignored addresses the pharmaceutic tractability of a target. How feasible will it be to inhibit the function of this target using a small-molecule inhibitor? Drug discovery may proceed with or without structure-based drug design, depending on the target. In fact, cell membrane receptors are believed to constitute approximately 45% of drug discovery targets *(29,30)*, and the difficulty inherent in solving the structure of such receptors typically prevents rational drug design. Enzymes, in contrast, are readily amenable to structure-based drug design, and these targets comprise 28% of the total. For drug discovery efforts in which structure-based drug design plays a central role in advancing leads, the ease with which a protein can be overexpressed, produced in up to hundred-milligram quantities, and crystallized for structural determination is of prime significance. A key goal of the medicinal chemist is to avoid proteins with the "aircraft carrier syndrome"— a large, flat active site with few nooks and crevices that provides little or no clues for achieving binding affinity or specificity. Of course, it is frequently difficult to predict such features from a previously undiscovered protein; chemists must frequently rely on past precedent, protein families, homology modeling, and so on. For those targets that are not readily amenable to structure-based drug design, high-throughput screening combined with judicious combinatorial chemistry may suffice to provide adequate potency and selectivity.

Ligand- or substrate-binding enzyme active sites are typically of a defined, smaller shape and are less accessible to competing interactions with solvent, in contrast to protein–protein interaction sites, which can comprise large, very hydrated clefts on the surface of proteins. A recent analysis of numerous different protein–protein binding sites that had been subjected to alanine-scanning mutagenesis and for which structural information and binding affinity were known, revealed that the major contributors to binding energy consisted of "hot spots" of a small subset of residues clustered together at the center of the heterodimer interface. More than 70% of these hot spots were protected from bulk solvent by a surrounding "O-ring" of energetically inconsequential binding contacts *(31)*. It is the challenge of re-creating this "O-ring" with small molecules that can also simultaneously make the high-energy, primary binding interactions that has proved quite daunting. Thus, targets requiring inhibition of protein–protein binding are typically regarded as considerably higher-risk targets. Nevertheless, a balanced portfolio should include a variety of strategies, and high-risk, high-yield targets can be aggressively pursued, particularly if early discovery strategies reveal a binding site suitable for small-molecule binding, and inhibitors demonstrating even moderate potency. Of course, many other considerations enter into the overall prioritization of new targets, ranging from scientific feasibility (e.g., adaptability of cell-based assays, availability of animal models) to marketing assessments.

3. Target Validation

Frequently, discovery of a target overlaps with experiments designed to demonstrate that functional inhibition of a target results in the desired therapeutic effect. All of the aforementioned genomics techniques can be used to provide additional evidence that a target is linked with cancer. Because the target validation efforts begin during target discovery and can continue through "proof of concept" in humans, each validation experiment merely serves to provide additional supportive evidence for pursuit of a target; the final or deliverable "proof" is a drug that can meet regulatory requirements for approval, which in the United States involves (in most cases) demonstration of a survival benefit over standard-of-care treatments. A "surrogate effect" for survival includes objective responses (reduction in tumor burden) and prolonged time to disease progression. Clearly, however, the more supportive evidence that is collected from testing in-house compounds or competitors' compounds, the more likely a pharmaceutical company will be to invest the millions of dollars necessary for a full-scale drug discovery and development effort. Thus, the target validation experiments help to prioritize different projects within a pipeline, and minimal validation milestones are frequently required before a project can advance to a more resource-intensive stage. Knowledge that a target is overexpressed in cancerous tissue compared to normal tissue, or that overexpression is correlated with disease progression or a negative clinical prognosis, can provide a compelling first step in target validation. Similarly, demonstration that overexpression of a protein can re-create the malignant phenotype is a valuable standard, particularly in conjunction with experiments that attempt to mimic reversible inhibition of the target.

Typical of these types of target validation strategies include antisense oligonucleotides or ribozymes that attempt to deplete the cell of the target mRNA. Success with these approaches typically requires more groundwork in development of optimal ribozymes or oligonucleotides than in executing the experiment demonstrating the phenotype associated with loss of function. In each case, the sequence chosen for hybridization must be of appropriate affinity, (binding neither too tightly nor too weakly in the case of ribozymes) and of appropriate composition so as to avoid complementary regions with higher-order mRNA structures. However, both of these techniques are relatively rapid, and are adaptable to different cell types *(32)*.

Direct interference with the target protein is the strategy exemplified by antibody neutralization using microinjection or intracellular antibody fragments ("intrabodies") *(33)*, or by inhibitory peptides conjugated to a protein fragment that permits intracellular delivery (e.g., penetratin). Single-chain antibodies (scFv's) can be introduced by transfection with recombinant cDNA encoding only the variable region of the immunoglobulin heavy and light chains *(34)*. Such intrabodies were used, for example, to profile the cell types that were susceptible to apoptosis during expression of an adenovirus encoding an anti-erbB-2 scFv *(35)*. Conventionally generated antibodies are of course most suitable for interfering with signaling pathways by binding to extracellular receptors, as exemplified by the anti-VEGF antibodies. However, microinjection techniques permit neutralization of intracellular targets as well. Dominant-negative enzymes effect a similar strategy, as overexpression can serve as a surrogate for chemical inhibition. Signaling pathways can sometimes be clarified by focusing downstream from the target, for example by creating nonphosphorylatable substrates of protein kinases. Similarly, knowledge of a naturally occurring inhibitory or counteracting protein can prove especially useful. For example, overexpression of the highly selective cyclin-dependent kinase 4 (CDK4) inhibitor $P16^{INK4a}$ causes cell cycle arrest and inhibits tumor growth (e.g., **ref. 36**). Although these experiments served as partial validation of this kinase as a target at the time, subsequent experiments demonstrated that although $p16^{INK4a}$ bound specifically to CDK4, overexpression of this inhibitory protein displaced other nonselective protein inhibitors (for example, p27Kip1) from CDK4, resulting in nonselective simultaneous inhibition of other cyclin-dependent kinases *(37)*. In retrospect, those experiments illustrate a possible pitfall of validation experiments—overinterpretation of the results caused by an incomplete understanding of a complex pathway or process.

All of the aforementioned strategies are advantageous in that they can be quickly executed shortly after target discovery. In contrast, gene knockout experiments using transgenic animals require up to 12–18 months before conclusive results can be interpreted. In the rapid flow of a discovery pipeline, such a time frame may be prohibitive. Furthermore, ablation of a gene may be lethal at the embryonic or early developmental stage, or the phenotypic observations of the animal may be inconclusive. Some of these drawbacks can be overcome, however, by using the relatively new technologies that permit regulation of gene expression in an animal in a tissue-specific manner. The tetracycline transcriptional regulation systems permit either repression or induction of gene expression by tetracycline or a tetracycline analog, following integration of the appropriate constructs into the mouse genome *(38)*. Site-specific recombinases such as the FLP/FLT and CRE-loxP systems also permit insertion of different genes into the genome, as well as deletion or inversion of specific sequences. These highly versatile systems permit reversible manipulation of gene expression, tissue-specific gene knockouts, and options for temporal control *(39,40)*.

Target discovery and validation by pharmaceutical companies ideally takes full advantage of all of the technologies described. In the genomics era, entire divisions or departments are focused on target discovery and validation. Many pharmaceutical efforts also involve outside collaborations with the relatively new biotechnology companies or consortiums that have arisen specifically to address these new technologies and capabilities.

4. Target Elaboration: The Early Stages of Discovery

The iterative progression of a typical anticancer drug discovery endeavor is illustrated in **Fig. 2**. The paradigm described here operates on the following basic principles: (a) Each drug discovery project proceeds with a fully integrated combination of combinatorial chemistry and structure-based drug design. (b) Several processes function through iterative cycles of improvement (**Fig. 2 A–C**), with maximal flexibility to solve specific problems, or accelerate certain steps as needed. (c) Advancement through the stages of early discovery, high-throughput screening, lead discovery, and late discovery can take from 3 to 6 yr or longer, depending on the tractability of the target, or the difficulties encountered along the way. In addition, each project that successfully nominates a development candidate must consider the inclusion of a "backup" effort—a continuation of a project that attempts to improve upon any obvious shortcomings of the first development compound. (d) Virtually all scientific data originates from within a multidisciplinary project team, with representatives from each discipline contributing to the project throughout its lifetime. The drug discovery effort is briefly described within these contexts, with particular focus on the evolving technologies within each discipline that drive progress.

4.1. Target Cloning and Overexpression

The task of cloning and overexpressing a new target constitutes the first step in a drug discovery program. The structure-based component of drug design can place an extra burden on the molecular biologist or biochemist, who must not only generate protein sufficient for high-throughput screening, but must also provide very pure protein that reproducibly generates high-quality crystals with sufficient diffractive resolution for solution of its three-dimensional structure. A structure-based effort can consume hundreds of milligrams of protein from early crystallization trials to routine solution of protein–ligand complex structures. Although the demand for protein typically subsides once suitable crystallization conditions have been optimized, the requirement for pro-

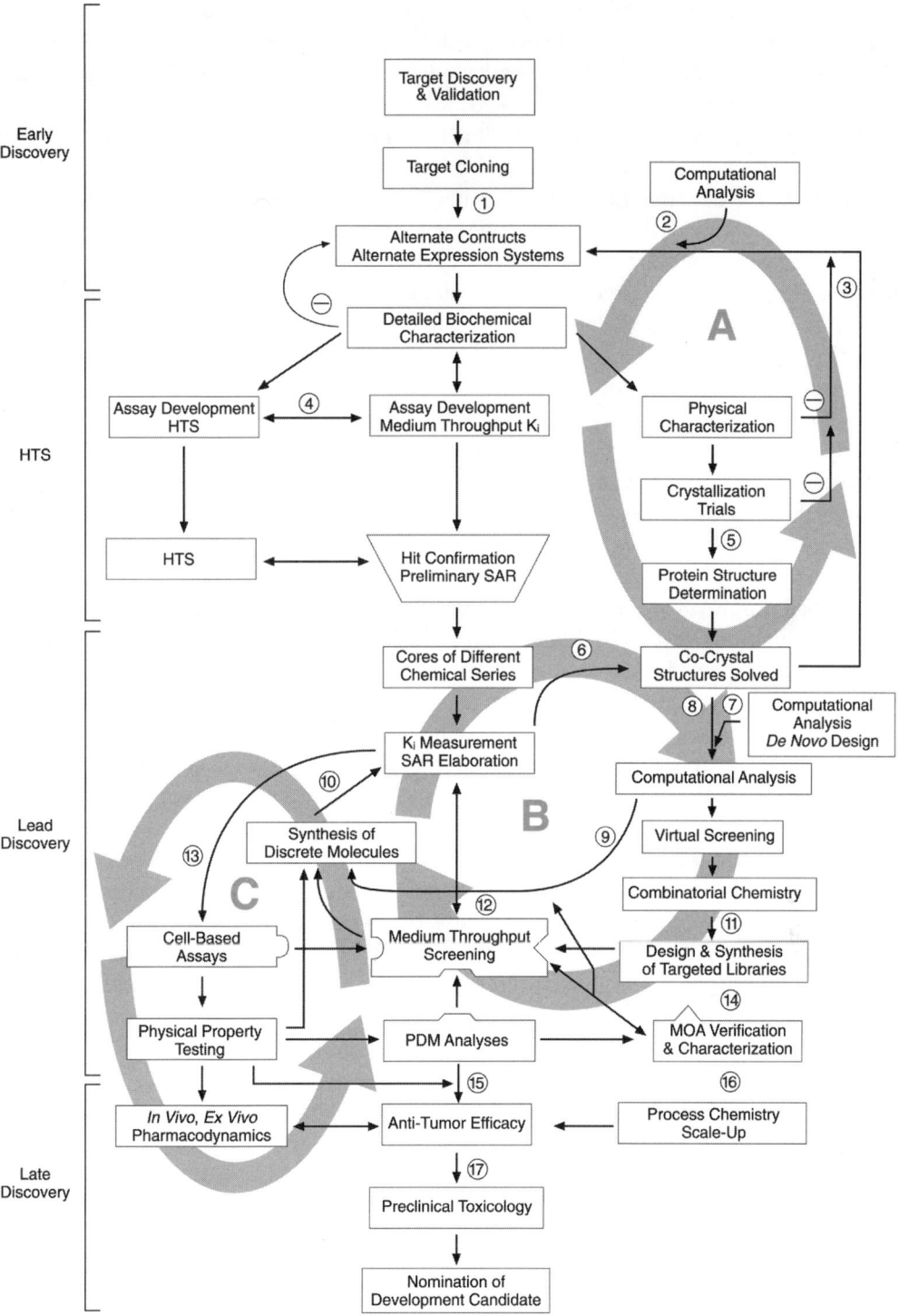

Fig. 2. The iterative cycles of a drug discovery effort that combines high-throughput screening with structure-based drug design. The general stages of a drug discovery effort are shown on the left. Circled numbers represent specific processes as described in the text. Circled negative signs represent failed or negative outcomes that require additional research at a previous step. Different iterative cycles are illustrated by gray arrowheads and letters A–C. Circled numbers appear in braces ({}) when referred to in the text.

tein can continue for several years, or as long as structure-based design remains a component of the discovery or the backup effort. Maximizing protein expression level thus becomes the first challenge, immediately following or concurrent with selection of the host expression system.

Protein expression in *Escherichia coli* is the first choice for simplicity, speed of production, and cost efficiency. Plasmid-based expression systems are particularly amenable to the new parallel cloning strategies that have been developed for high- or medium-throughput optimization of expression conditions, although these ultimately permit expression in mammalian and insect cells as well *(41,42)*. Given the historical success of overexpressing recombinant proteins in *E. coli*, there is an expansive foundation of empirical knowledge to draw upon. Techniques such as lowering bacterial growth temperature, optimizing induction conditions, co-expressing chaperone proteins, expressing domains as fusion proteins, expressing proteins in protease-deficient strains of *E. coli*, supplementing growth media with cofactors required for protein stability or activity, and opting for periplasmic secretion are all broadly applied as needed *(43)*. Recovery of the recombinant protein as a soluble, properly folded, and nonaggregated entity is the ideal outcome, whereas purification from bacterial inclusion bodies is more labor-intensive and requires refolding of the solubilized, denatured protein. In addition, the optimization of the refolding protocol, somewhat specific for each protein, must be determined empirically, although commercially available refolding screens now exist (Foldit, Hampton Research, Laguna Niguel, CA). Certainly, bacterially expressed recombinant proteins have been used with great success in drug discovery efforts, and the speed of the experiments makes this expression system an obvious first choice.

Some proteins, however, are simply not suitable for bacterial expression systems, perhaps because they require eukaryotic processing machinery for accurate folding or processing. In these cases, insect cell or mammalian expression systems will typically solve the problem. A variety of baculovirus vectors are available for protein expression in cell lines derived from the army worm *Spodoptera frugiperda*, and, unlike mammalian cells, cell culturing can occur at 27°C. without requirements for CO_2. Protein yields can be quite high (from 1 to 500 mg protein/L cells), but overall yields are typically lower than for bacterial expression systems. Moreover, most posttranslational modifications, including phosphorylation, glycosylation, acylation, amidation, and assembly, are supported by the insect cells *(44)*. This system can be particularly advantageous for cases in which proteins or cofactors required for posttranslational assembly or modification are not completely understood, and simple mixing or incubation of lysates can yield the desired protein product. Although the production and maintenance of baculovirus stocks can be time-consuming, the chance of obtaining properly folded, soluble, protein is much higher than for bacterial expression systems. Thus, large-scale insect cell expression systems are in place at most pharmaceutical companies, and are frequently the method of choice for production of proteins for oncology programs.

More important than robust protein production, however, is the need for flexibility. A target as discovered may not be suitable for structure-based drug design—it may be a membrane-bound receptor, or a multidomain protein, or the protein may simply be refractory to crystallization. There is unfortunately no mechanism for *a priori* prediction of optimal crystallization conditions, and as discussed below, it is now accepted that the best way to ensure structure determination is to analyze multiple protein constructs or mutants. Truncation mutants may be designed in order to focus only on the relevant catalytic domain, or deletion, insertion, or site-directed mutagenesis may be needed to obtain reproducible crystals. The required construct for an oncology drug discovery effort is of course the human isoform, but expression of the gene from alternative species may aid in crystallization, or may be required for later assessment of species-specific differences in the target, such as during selection of mouse models, or analyses of toxicologic effects. Insertion of a "tag" on the N- or C-terminus of the protein during cloning provides a tremendous advantage during purification, particularly if the expression level is low. Numerous tags have been used successfully *(43)*, but the $(His)_6$ tag, used with a chelated metal affinity column, is probably the most common.

4.2. Target Purification and Characterization

The choice of the protein tag and the purification strategy must be considered in the context of structural biology. The effect of even a small peptide tag on protein crystallization remains a topic of debate, although clearly crystal structures of tagged proteins have been successfully determined. Tag-based affinity chromatography constitutes an excellent first purification step, and most tags have engineered proteolytic cleavage sites that permit removal from the purified protein. Protein purity as assessed by SDS-PAGE is a weak criterion; heterogeneity with respect to folding species or posttranslational modifications can be far more important to crystallographic analyses. Purification protocols that include ion-exchange, hydrophobic interaction, and gel filtration chromatography are particularly powerful for evaluating or purifying different subsets of the target protein. Protein supplied for crystallographic trials should be of the highest concentration possible without creating aggregates, in order to approach the saturation limit. Here the burden falls on the biochemist to execute rigorous assessment of purity using, for example, denaturing and native electrophoresis, isoelectric focusing, analytical gel filtration, dynamic light scattering, limited proteolysis, mass spectrometry, metal content analysis, and critical tracking of total and specific enzymatic activity. Consideration of the unique function of each protein can provide valuable insight into possible roadblocks to crystallization. For example, the protein surface may have exposed hydrophobic domains if it typically exists in a complex with other cellular proteins. Sequence alignment within protein families may indicate the presence of a nonconserved and unstructured loop, which might be excised or subjected to mutagenesis. Function in a signaling pathway suggests the possibility of microheterogeneity caused by posttranslational modifications. As illustrated in **Fig. 2A** {1}–{3}, the design, purification, and characterization of alternative protein constructs is necessarily an iterative process in collaboration with molecular biologists, crystallographers, and biochemists. Computational modeling of similar proteins or protein families can assist in defining minimal soluble domains, eliminating flexible domains or linkers that may hinder protein crystallization, or even evaluating potential crystal contacts from protein surface residues.

4.3. Assay Development and High-Throughput Screening

The precise definition of structure–activity relationships (SAR) is at the heart of every drug discovery effort. Therefore, rigorous measurements of ligand binding affinity are needed to enable detection of subtle changes in affinity, as well as to permit measurement of a broad range of ligand affinity. Early discovery efforts attempt to identify different classes of relatively weak inhibitors (K_d's typically from 0.5 to 50 μM), using the less quantitative methods of high-throughput screening (HTS). As the discovery effort progresses, much more potent inhibitors (K_d's as low as 10^{-12} M) must typically be characterized using a more quantitative assay with low or medium throughput. In addition, affinity binding measurements are always interpreted (either retrospectively or prospectively) within the context of computational docking programs and quantification of all binding interactions as detected by crystallographic structures of protein and ligand. The computational chemist can rely on affinity data only to define, or even help refine, quantitative SAR (QSAR) for a system if the assay data reflect thermodynamic and kinetic binding parameters as closely as possible. Furthermore, every drug discovery program must assess selectivity of ligands for a specific target over other related targets within a protein family. These measurements should be directly comparable across targets and also provide some realistic consideration of in-vivo conditions. Finally, it is extremely valuable to clinicians, and indeed all researchers, to be able to compare directly the affinity of drugs from different pharmaceutical programs. It is for these reasons that measurements of dissociation constants (K_d's) or inhibition constants for enzymes (K_i's) are preferable to IC_{50}'s, which assess half-maximal inhibitor concentrations only as defined under specific assay conditions. Detailed kinetic characterization of a novel enzyme is required during assay development to ensure routine measurement of initial rates and steady-state kinetics *(45)*. Choice of substrates, experimental conditions, and assay formats are optimized to provide for the efficient turnaround of rigorously measured K_i's. Diverse assay platforms are employed and may rely on the more traditional enzymatic assays (e.g., continuous or discontinuous radioactive, fluorometric, spectrophotometric, or chromatographic techniques); but the overall goal is to optimize quality of the quantitative data. These assays need only keep pace with the synthesis of discrete molecules or small focused libraries from a team of chemists; thus, formats beyond 96-well templates are rarely necessary.

In contrast, the goal of a high-throughput screening assay is to test hundreds of thousands of molecules for nominal binding affinity in the short period of time of only several weeks to months. Only relative assessments of affinity are necessary, because the best "hits" from a chemical library are subsequently retested using a purified, discrete chemical entity and a precise measurement of K_i. Design and development of a high-throughput assay format must be rigorous, however, to permit assaying under initial rate conditions, to provide for high precision (low coefficient of variation), to eliminate false positive and false negative results, and in general to permit maximum agreement between high-throughput screens and low-throughput quantitative assays. Even with a validated quantitative assay in place, development and validation of an HTS assay, with a typically quite different format or detection system, can add several months to a project in early discovery.

The major challenge of high-throughput screening assays is preventing signal interference from samples of organic compounds, which may be colored, or fluorescent, or contain minor impurities that interfere with functional assays. By taking advantage of the recent technological advances in HTS assay formats and robotics, these challenges can be addressed once a framework of optimal experimental conditions or substrates have been defined (**Fig. 2**, {4}).

The last five years have witnessed an explosion in the advancement of screening technologies. Formats of 384-well assays are routine, and increased miniaturization to 1536-

well formats is demonstrating increased acceptance *(46)*. The radiometric scintillation proximity assay (SPA; Amersham Pharmacia Biotech) and Flashplate (NEN Life Sciences) technologies were among the first to be adapted for high-throughput screening. These assays take advantage of the requirement for radioactive β-particles to be proximal to scintillant for efficient transfer of energy and emission of light from the scintillant. Receptors or enzyme substrates are immobilized onto scintillant-impregnated beads or 96- or 384-well scintillant-coated plates using passive adsorption, or tight-binding linkers. Binding of the radioligand to the receptor, or incorporation of radioactivity onto the immobilized substrate, results in an increase in signal due to proximity of the radioactive moiety to the scintillant. Free, unbound radioligands in solution are too far away to generate a significant signal. The proximity assays are thus amenable to automation because they can avoid additional separation steps. Disadvantages of the radiometric formats include the need for additional operator safety, disposal of radioactive waste, quenching by colored compounds, and relatively long time periods required to count plates *(47)*. However, new imaging plate readers use high-sensitivity charge-coupled device (CCD) cameras that can quantify light emission from even high-density plates much more rapidly *(48)*, and this will likely extend the use of radiometric assays for high-throughput- and ultrahigh-throughput screening.

Fluorescence-based assays have traditionally been viewed as a convenient alternative to radioactive assays; the sensitivity is at least as good as radiometric assays, and there are no safety or disposal issues. Historically, these assays were limited by interference from chemical library samples either by quenching of fluorescence, or generation of an interfering signal from intrinsic fluorescence. The development of new fluorophores with unusually long excitation and emission wavelengths, and long lifetime emissions, has revolutionized this industry. Chelates of lanthanides such as europium (Eu^{3+}), terbium (Tb^{3+}), samarium (Sm^{3+}), and dysprosium (Dy^{3+}) display unusual spectral properties: upon fluorescent excitation of the chelate and transfer of energy, the lanthanides emit high-intensity fluorescence with a long (hundreds of milliseconds) lifetime. The emission peak is sharp and is several hundred nanometers removed from the excitation peak (large Stokes shift) *(49)*. As a result, interference from library compounds, which typically display a short lifetime and a lower emission wavelength, can be avoided by observing only a narrow emission wavelength above 600 nm, and by delaying the observation time to 400–800 ms after excitation (time-resolved fluorescence, or TRF). Fluorescence resonance energy transfer (FRET) permits detection of binding or catalytic events by monitoring proximity of donor–acceptor fluorophore pairs. Various permutations on this technology exist (e.g., DELFIA and LANCE assays from Wallac/Perkin-Elmer), and the methods continue to evolve *(50)*. As with scintillation proximity assays, unbound ligands are too far away for efficient energy transfer, and these assays thus offer the potential for avoiding separation steps.

Similar exploitation of novel fluorophores with high quantum yields and long excitation and emission wavelengths has been applied to fluorescence polarization (FP) assays. Excitation of a fluorophore with plane-polarized light results in emission of a polarized signal, and the extent of polarization depends on the speed of rotation of the fluor during its excited state lifetime. The rotational speed of a small reporting ligand is substantially slowed upon binding to a macromolecule, thus causing a measurable change in the fluorescence polarization signal upon binding. Because FP assays directly measure relative fluorescence (parallel vs perpendicular emission), they do not require high quantum yields, and they are less sensitive to interfering background signals from fluorescent organic library compounds *(51)*. Nevertheless, fluorescence quenching by inhibiting compounds can decrease the signal-to-noise ratio, and so the new fluorophores with high intensity and longer emission wavelengths can avoid these complications.

The latest technology applied to HTS is fluorescence correlation spectroscopy, or FCS. These methods use confocal optics and femtoliter observation volumes to monitor fluctuations in fluorescence due to single molecules. New methods to measure molecular brightness, such as fluorescence intensity distribution analysis, or FIDA *(52)*, can be applied to simple binding assays, FRET-based interactions, or catalytic events *(53)*. The small volumes render FCS particularly amenable to miniaturization and ultra-high-throughput screening, but these newer technologies are not yet in routine use throughout the pharmaceutical industry.

Most of the HTS assays developed today incorporate some aspect of time-resolved fluorescence or FP. Needless to say, new instrumentation has evolved with the technology, and 96-well and 384-well plate readers that measure fluorescence, time-resolved fluorescence, fluorescence polarization, and chemiluminescence are routinely sold by several different vendors. A variety of different robotic workstations incorporate these multifunctional detection systems into platforms that also house pipetting stations, plate-washing devices, and carousels for compound and assay plates. With these tools, screens of more than 300,000 compounds can be executed against a target in a matter of weeks to a few months. The cost of reagents can be quite high, no matter which assay format is chosen, ranging from a few cents to several dollars per well.

How many compounds should be screened for inhibition of a new target? The simplest approach is to take no prisoners and to screen every available compound from corporate collections or commercial sources. Corporate compound collections number in the hundreds of thousands to millions, including discrete molecules and combinatorial chemistry libraries. However, as discussed below, recent history has shown that broad screening of hundreds of thousands to millions of molecules has not yet resulted in a universal increase in the speed of drug discovery and development. Each screening process is entirely target-dependent, and the ideal screening strategy would be optimized for each target, and for the goals of the project. For example, screening against a target from a protein family (e.g., kinases, proteases, or G-protein-coupled receptors) can build on existing pharmacophores or on small, focused libraries known to have at least moderate activity against certain family members. Consideration of the patent status of library products is also relevant. The goal of HTS is not to identify every possible inhibitor of a target, but rather to identify several diverse chemical classes of inhibitors that are suitable for further elaboration. Rescreening targets with poor initial HTS hits against a much larger compound file to exhaust opportunities for success on the target in question is not unreasonable. Alternatively, the use of ultra-high-throughput screening approaches to decrease screening time shows increasing promise.

The confirmation of HTS hits proceeds by reassaying the most potent inhibitors at multiple concentrations. The secondary screening typically includes assay modifications that will eliminate potential causes of interference, or false positives, and may also begin to define SAR. If discrete, purified representatives of a library are available, K_i measurement using a medium-throughput assay provides additional confirmation, as well as defining quantitative structure–activity relationships. Additional kinetic experiments are required for initiating characterization of the mechanism of inhibition—competitive, noncompetitive, slow-binding, tight-binding, and so on.

4.4. Crystallization and Structure Determination

The most common method of protein–ligand structural determination in the pharmaceutical industry is X-ray crystallography, and that is the primary focus reviewed here. High-resolution X-ray crystallography data typically display precision (root-mean-square deviation) approximately 1.5–2 times better than that of NMR solution structures, as seen by a comparison of NMR and X-ray structures of the same protein *(54)*. Crystal structures can be routinely obtained with a resolution of better than 2 Å, and a 1.8-Å crystal structure is typically measured with a precision of 0.2–0.3 Å *(55)*. For iterative drug design, the medicinal chemist wishes to observe even subtle changes in aromatic ring positions or amino acid side chain placements; X-ray crystallography readily offers the necessary resolution in a format conducive to rapid repetition. NMR spectroscopy has seen increasing utility in screening applications, however, in which NMR signals from either the protein or ligand are monitored. By measuring the 2 D ^{1}H-^{15}N spectra of the target protein, binding events from libraries of compounds (10^3–10^4) can be assessed and subsequently deconvoluted. This "SAR by NMR" approach *(56)* requires prior solution structure of the protein (currently limited to proteins < 30 kDa), and can require hundreds of milligrams of ^{15}N-labeled protein. Monitoring of ligand signals eliminates these concerns, and these strategies are gaining increased use in the pharmaceutical industry *(57)*.

Crystallization trials and protein structure solution can proceed in parallel to high-throughput screening (**Fig. 2**, {5}). As discussed previously, protein preparation and construct design is necessarily an iterative process with collaboration between biochemists, molecular biologists, computational chemists, and crystallographers (**Fig. 2A**). The time span for cycling through several protein constructs to identify those amenable to rapid crystallization is being accelerated by parallel cloning and expression systems *(43)*, but application of these methods to structure-based drug design in the pharmaceutical industry is lagging behind the structural genomics effort. However, the painstaking process of setting up dozens of trays of crystallization drops has improved substantially, with the availability of many commercially available screening solutions and conditions (e.g., Hampton Research, Laguna Niguel, CA) as well as robots that automate this somewhat tedious task. Currently hundreds of crystallization trials can be initiated over a period of several days, with the help of automation.

Perhaps the greatest advances in crystallography have come in the improvements in hardware, software, and computational speed. Crystallographers are increasingly relying on synchrotrons as the source of high-intensity X-rays, rather than the weaker sources typical of internal systems. There are now 35 synchrotron sources worldwide, and pharmaceutical companies typically "rent" time on these beamlines for the most rapid and sensitive collection of crystal data. Moreover, collection of the crystallographic diffraction data now utilizes image plates or charge-coupled devices as detection systems, which are much faster and more sensitive than the older multiwire area detectors or diffractometers *(58)*.

Once high-quality diffraction data are collected, the crystallographer is faced with the challenge of translating the data into a map of the electron density of the entire protein.

The detection devices simply measure the X-rays that are diffracted by the protein's electrons; the Fourier synthesis that must be applied to reconstruct the electron density map requires knowledge of frequency, amplitude, and phase of the wave equation. While the frequency and amplitude are readily calculated from the X-ray source and the intensity of the reflection, the phases must be determined experimentally using heavy atom or isomorphous replacement, anomalous dispersion, or molecular replacement. Soaking a crystal with salts of electron-rich "heavy" atoms may provide estimates of phase information by assessing the slight perturbations in the diffraction pattern caused by the heavy atoms, provided that these atoms do not disturb the protein conformation or the crystal packing. This process can be quite time-consuming, as several heavy-atom derivatives must be prepared and analyzed. Recently, a database of heavy-atom derivatizations was constructed and published, with the goal of improving the success rate of experimentation *(59)*. However, the technique of *multiwavelength anomalous dispersion* (MAD) has provided the greatest boost to phase assignments from heavy-atom-derivatized crystals. MAD phasing relies on the fact that heavy atoms can absorb X-rays of a specific wavelength, causing unequal reflections or anomalous scattering. Synchrotron sources offer X-ray sources that can be tuned to the absorption wavelength of the atom of interest, and this capability, combined with advancements in software that abstracts phasing information from the MAD data, have made this an increasingly popular technique *(60)*. Moreover, soaking of crystals with ions of heavy atoms such as Hg or Pt or Au, which may perturb protein structure, may not be necessary for MAD experiments. A common technique is to measure anomalous scattering of an intrinsic metal or of selenomethionine, which has been metabolically substituted for methionine during protein production. Finally, the ability to now routinely cryofreeze crystals protects against damage and permits more data collection from a single crystal. The most straightforward solution of protein structure can be gained by using the solved structures of a homologous protein to provide initial phase estimates. This *molecular replacement method* has been commonly applied to kinases and serine proteases, or any pair of proteins with > 30% sequence identity. New software that automates the procedure of fitting the known structure into different positions in the unsolved crystal has helped to automate this process *(58)*.

Solution of the phase problem is followed by model building, crystallographic refinement, and a critical evaluation of the structure. The overall process can proceed much more rapidly than in the past; solution of a novel structure may take from a few days to several months. While structure solution of the apo-enzyme may be initially satisfying, the essential goal is the solution of multiple crystal structures of protein–ligand complexes. Ideally, concentrated solutions of ligands may be soaked into existing crystals to generate complex structures repetitively and efficiently. This may not always be possible, however, particularly if ligand binding proceeds with a protein conformational change. If crystal soaks do not yield the desired data, co-crystallization of ligand–protein complexes is necessary, possibly requiring modified crystallization conditions. In some cases, the iterative process of construct design, production, characterization, and crystallization trials may need to begin all over again if problems arise in crystallization of protein–ligand complexes (**Fig. 2A**). Once crystallization conditions have been identified, however, iterative solution and refinement of new protein–ligand crystal structures may require only hours to several days.

5. Lead Discovery: Iterative Combinatorial Chemistry and Structure-Based Drug Design

With conditions in hand for routine co-crystal structure solution, and several chemically distinct inhibitors identified and confirmed from high-throughput screening, the process of iterative structure-based drug design and combinatorial chemistry can begin in earnest (**Fig. 2B**, {6}–{13}). At this stage, a project is elevated to a more advanced stage of drug discovery, as medicinal chemists join the project. Also at this time, the product profile for the intended drug should already be fully and specifically defined. Questions as to how the drug will be used (single agent vs combination, acute vs chronic administration, preferred route of administration, acceptable therapeutic window, expected mechanism of action and expected mechanism-based toxicities, etc.) are defined as carefully as possible to direct the drug discovery effort.

After K_i's are measured for the first-generation inhibitors identified from HTS, the kinetic mechanism of the inhibitors is clarified. Do the molecules display inhibition that is competitive with the expected substrate? Is inhibition due to reversible binding? Is metal chelation involved? Is compound solubility limiting for assays or crystallization? If solubility permits, even relatively weak inhibitors (>10–50 μ*M*) can be subjected to co-crystallization trials to determine the structure of the protein–ligand complex (**Fig. 2**, {6}). As the binding modes of these first inhibitors are revealed, the contributions of structure-based drug design can begin to be realized.

5.1. Structure-Based Drug Design

What advantages does the known structure of a target offer? Without a picture of the three-dimensional shape of the active site, medicinal chemists are "flying blind," required to deduce the shape of the pocket from the binding properties of a collection of inhibitors. At a minimum, knowledge of protein structure minimizes the number of compounds that must be synthesized and assayed to explore the structure–activity relationships. Indeed, an ideal case describes the use of structure based drug design to improve ligand binding from micromolar- to nanomolar-range affinity with only three molecules *(61)*. Available hydrogen bond donors or acceptors, charged residues, or aromatic residues of the protein can be exploited by designing a complementary moiety in the ligand. Ligand design based on protein structure is exploited whenever possible throughout the pharmaceutical industry, and it has made substantial contributions to successful drug discovery *(62)*.

The optimal practice of structure-based drug design requires high-resolution structural information and multiple protein–ligand structures generated in an iterative process (**Fig. 2**). Whereas an electron density map calculated with X-ray diffraction data at 3.0 Å resolution is missing detail and will provide only a poor-quality image of side chains and modeled main chains, a 2.0 Å map is quite detailed, with side chains clearly defined and main chain carbonyls delineated *(58)*. For diffraction data collected with resolution better than 2 Å, now routinely accessible, many associated water molecules can be defined as well. These high-resolution structures are necessary to detect bridging water molecules that could contribute favorably to ligand binding energy, or the displacement of water molecules from the binding site, or subtle structural differences in each bound ligand. But just as the structure of an apo-enzyme is not as informative as a structure of a protein–ligand complex, a single co-crystal structure does not provide a complete picture of the three-dimensional space accessible for ligand design. Ligands can bind in different orientations, and several examples have demonstrated multiple binding modes for a given ligand or family of ligands *(63)*. Proteins are flexible, and active sites may become more or less compact with different bound ligands. Flexible loops may change shape and become ordered or disordered in the presence of different ligands, and in some cases substantial conformational changes may occur upon binding of certain ligands, in an illustration of "induced fit." More important, an ultimate goal of an integrated drug discovery effort is to correlate predicted binding modes and pre-dicted binding energies with experimental measurements of binding affinity, and exper-imentally determined structures. A single picture of protein–ligand binding interactions cannot provide information about some of the entropic contributions to binding energy derived from protein or ligand flexibility, or solvation of protein and ligand *(64)*. As described below, current computational methods are also limited in their ability to accu-rately predict binding energies derived from many of these entropic contributions *(65)*. Thus the enhancement of ligand binding affinity remains a somewhat empirical process, and this is efficiently accomplished by the iterative cycle of modeling designed ligands, synthesizing and measuring binding affinity, confirming the ligand binding mode, and using this new information to design new ligands (**Fig. 2B**; *(61)*. It is widely accepted in the pharmaceutical industry that structure-based drug design also lends itself partic-ularly well to improving upon "me-too" compounds. Chemical analogs of known drugs or patented compounds can be designed within the protein binding pocket so as to retain or improve upon binding potency while simultaneously introducing novel (patentable) chemical entities or overcoming limitations of earlier compounds such as pharmacoki-netic, metabolism, or plasma protein-binding issues.

It is the combination of structure-based drug design and high-throughput screen-ing/combinatorial chemistry that provides the most powerful approach to drug discov-ery programs. Historically, the first steps of synthetic chemistry might involve modification of the natural ligand or substrate, or synthetic alteration of a (typically highly complex) inhibitory natural product. HTS and hit follow-up can now yield a col-lection of structurally diverse chemical inhibitors with potency in the nanomolar to micromolar range. The structure of these protein–ligand complexes will likely reveal unfilled binding pockets and/or poorly fulfilled binding interactions. The initial weakly binding inhibitors can serve as binding templates or scaffolds, on which optimal bind-ing substituents can be added using combinatorial chemistry. Several successful exam-ples of this combined approach have already been published *(63)*, and more are likely to follow as new drugs become public knowledge. This strategy is of particular impor-tance as the genome is defined and target families are identified. As our structural knowledge of protein families grows, and larger sets of template molecules or pharma-cophores are defined, the use of combinatorial chemistry and structure-based drug design will accelerate at an even greater pace.

5.2. Combinatorial Chemistry

The advent of combinatorial chemistry is considered one of the most important recent advances in medicinal chemistry, and its use has unequivocally increased the efficiency with which medicinal chemists can generate new small molecules. The origins of the technology are found in the automated solid-phase synthesis of peptides and oligonu-cleotides. The concept of expanding the combinations of synthetically available peptide sequences by applying a "split-and-mix" synthetic strategy *(66)* was applied to drug dis-covery in the early 1990s *(67)*. Expansion into small-molecule libraries followed shortly thereafter *(68)*, exploiting the general principle of applying a common synthetic method to ensembles of reactant molecules in a simultaneous manner.

The solid-phase synthesis method, in which molecules are covalently attached to a support or bead during the synthetic sequence, offers the advantage of improved yields and improved purity. Because reaction products remain attached to the beads, excess reactants can be used to drive a reaction to completion, and can then be readily washed away from the beads. A variety of companies offer automated instruments for solid-phase organic synthesis, typically in the form of blocks of multiple pins as solid sup-ports that are compatible with 96-well reaction vessels. Solvent and reagent delivery are controlled by pressurized valves, and temperature and atmospheric conditions can be controlled within the reaction manifold *(69)*. Solid-phase chemistry can yield 1000–10,000 spatially separated compounds in a synthetic run, with yields up to 50 μmol per pin. Larger or more complex libraries can be accessed through a split-and-mix approach, in which beads are separately processed through distinct reactions, recom-bined and mixed, and subjected to several rounds of this synthesis and pooling process to generate a large number of possible combinations. However, the complications involved in screening pools of several impure compounds and "deconvoluting" hits by laborious resynthesis and rescreening are a huge disadvantage compared to the one-well, one-compound approach. A diverse repertoire of organic reactions has been adapted to solid-phase synthesis approaches *(70)*, including multicomponent condensations such as the Ugi reaction that combine three or more reactants in a single step *(71)*.

There are some limitations to the solid-phase approach, however. First, each library component must include a point of attachment to the bead, which means that every reac-tion product in the library must incorporate this moiety. New examples of "traceless linkers" have been devised *(72)*, but this does place more constraints on the synthetic reaction and the reactants. In addition, reactions must be carefully optimized for high yields, particularly for multistep syntheses. Reaction yields are typically lower for the inherently more difficult nonpeptidic syntheses than for peptide-based solid-phase syn-thesis, and thus the purification advantage offered by the ability to wash the beads becomes less important in multistep reactions. New spectroscopic techniques for mon-itoring reaction intermediates on beads have been devised to address this problem, including magic angle-spinning spectroscopy and single-bead Fourier transform infrared spectroscopy *(69)*. Finally, the compatibility of solvents, reagents, and reaction temperature with various components of the solid support (including the cleavable linker) must be considered as well, all of which can lengthen the synthetic process.

The wealth of accumulated knowledge of synthetic organic reactions in solution, and the desire for greater flexibility in library synthesis, has driven the recent trend toward solution-phase combinatorial chemistry. By avoiding both the necessity of attaching the starting material to the solid support and the final cleavage step, a wider possibility of synthetic reactions can be explored. Reactions that proceed in a single step with high yield are obviously ideally suited for these libraries. However, with increased focus on automated parallel solution phase synthesis, new techniques have evolved to expand the repertoire of suitable reactions. The historical bias against solution-phase combinatorial chemistry was the purification required for each reaction, many of which may not have proceeded with the same yield. Advanced phase-separation methods have been explored to preferentially partition either reactant or products into different liquid phases, particularly per-fluorinated compound extraction into fluorinated solvents. By attaching scavenger reagents to solid supports or polymers, unreacted starting materials can be covalently bound and thus removed (73). Perhaps the most powerful tool has been the advancement of automated systems for compound purification or characterization using liquid chromatography-mass spectrometry (74). Because reaction mixtures are typically not complex (reactant, product, side product, reagent), simplifications in chromatography can dramatically shorten analysis time. Several vendors now sell HPLC-MS instruments that trigger fraction collection based on UV signal or appropriate mass, and multichannel HPLCs are coupled with mass spectrometers capable of very rapid data collection rates. These types of systems will permit, for example, analysis of the contents of a 96-well microtiter plate in 1 h using a 5-min HPLC gradient (75).

The latest trends in combinatorial chemistry are moving away from generating very large libraries. So far, it is clear throughout the industry that simply producing extraordinarily large numbers of compounds has not significantly shortened drug development timelines, although this was clearly the goal (76). Certainly the diversity of molecules examined for appropriate biological activity has increased in a much more efficient manner, and perhaps this will lead to *better*, if not more rapid, drug candidates for development. Only detailed analyses of failed vs successful drug candidates will begin to address this question. Greater emphasis is now placed on the content and purity of library components. As described below, (**Fig. 2** {12} libraries of high purity can be applied in plate format to multiple enzyme assays, and even straightforward cell-based assays or toxicity screens. This rapid application of multiple assays for even a small library can briskly accelerate selection of development candidates. Screening efforts that provide answers more sophisticated than yes/no, such as estimation of even approximate structure–activity relationships, require libraries of high purity, with only nonreactive contaminants. Rather than focusing simply on synthetic feasibility to design combinatorial libraries, chemists today incorporate multiple criteria into library design. This may include libraries of molecules with desirable druglike properties, or target-based libraries that incorporate a recognition element specific for certain receptors, such as hydroxamic acids for metalloproteases (77), secondary or tertiary amines for G-protein-coupled receptors (78), or hydrogen bond donor–acceptor patterns for kinases.

5.3. De Novo *Drug Design and Virtual Screening*

The challenge of designing "intelligent" target-based compound libraries has united the computational chemist and the synthetic organic chemist in a uniquely powerful collaboration. Before combinatorial compound libraries were widely available, computational chemists were limited to drug design *de novo*—that is, the attempt to identify small molecule inhibitors "from scratch," given the known structure of a protein-binding site and, in some cases, given large "virtual" libraries of proprietary compound collections or commercially available compounds such as the Available Chemicals Directory (ACD; MDL Incorporated, San Leandro, CA), Chembridge Corporation collection (San Diego, CA), or BioSPECS library (Rijswijk, The Netherlands). These computational approaches use commercial and/or proprietary software to perform multiple docking experiments of a collection of ligands into an empty binding site, and rank each docked molecule in terms of binding energy (**Fig. 2** {7}). This is not a trivial exercise, however, as multiple parameters of ligand binding energy must be considered, including electrostatics, hydrophobic interactions, hydrogen bonding, van der Waals forces, geometries, water molecules that may provide bridging hydrogen bonds, ligand and protein desolvation, and so on. The problem of protein flexibility, multiple ligand-binding modes, and the difficulty of accurately calculating the entropic contributions to binding energy from ligand and protein solvation make it very difficult for scoring programs to yield practical results. Different docking and scoring programs emphasize different components of binding energies, but no one docking program is yet capable of the kind of accurate predictions needed for discovering multiple classes of even moderately potent inhibitors from collections of thousands of molecules in a timely manner *(64)*. Throughout the pharmaceutical industry, *de novo* drug design has proved difficult in practice, and it has yet to yield the number of structurally diverse, potent inhibitors that are discovered through HTS.

The marriage of combinatorial chemistry and structure-based drug design has provided the computational chemists with a more defined starting point, in the form of a moderately potent inhibitor (typically an HTS screening hit) with a known binding mode within the target binding site (**Fig. 2**, {6}–{8}. As described previously, multiple co-crystal structures of distinct chemical classes are an ideal starting point, but even the structure of the target and knowledge of inhibitor potency can provide the foundation of a virtual screening effort. The synthetic possibilities theoretically feasible with combinatorial chemistry efforts are almost always much greater than is practical for automated parallel synthesis. Computational chemists can begin to evaluate all of the theoretical synthetic possibilities by performing automated docking procedures of thousands of molecules (**Fig. 2**, {11}). Search techniques, including evolutionary programming, genetic algorithms, and Monte Carlo methods, are used to sample the three-dimensional configuration space available to a ligand within an active site, and molecules are ranked based on binding-energy scoring functions *(79,80)*. Filters can be applied to eliminate molecules that yield poor scoring functions, perhaps due to spatial limitations within the active site, electrostatic repulsion, or weak hydrogen bonding potentials. The docking and scoring exercise should provide only an approximation of appropriate compounds (or more accurately, elimination of inappropriate compounds), but should also permit maximal molecular diversity within the available binding space. Docking is not limited simply to the target receptor; in many cases valuable information about compound selectivity may be surmized by performing automated docking and scoring iterations to multiple proteins within a family. For these approximations, crystal structures of each protein family member may not be necessary, if homology models have been prepared.

In addition to docking, more straightforward filters can be applied, with the goal of making the ultimate library compounds more druglike. Limitations may be placed on molecular weight, physical properties such as the partitioning coefficient between water and octanol (log P), or avoidance of certain chemical moieties known to be unstable, induce toxicity, and so on. Lipinski's "rule of five"—a rubric applied to define molecules as druglike or not based on such features as molecular weight, number of heteroatoms, and number of hydrogen bond donors (see below; *[81]*)—is frequently used to limit synthesis to the best molecules. In this manner, large libraries, originally restricted only by synthetic feasibility, become "targeted libraries" designed for a specific protein and optimized to yield druglike molecules. These smaller libraries, perhaps less than 2000 compounds, can be subjected to medium-throughput screening (**Fig. 2**, {12}) optionally independent of robotics systems. Screening is followed by synthesis of the most promising compounds as discrete molecules (**Fig. 2**, {10}) and assaying for potency of target inhibition.

Iterative structure determination of protein–ligand co-crystal structures and measurements of binding affinity serve as a reality check for the virtual screening and computational docking exercises. As described previously, many docking programs must simplify the entropic contributions to binding energy for computational efficiency, such as by treating the protein as a relatively rigid structure. Highly refined, high-resolution crystal structures and accurately measured binding constants provide snapshots that may reflect more realistically the protein or ligand conformational changes that occur, or the entropic contributions to binding. As a drug discovery program progresses, these growing data sets may help guide the evolution of more sophisticated docking and scoring programs, or at least the application of certain docking programs to a specific receptor.

5.4. Cell-Based Assays and Verification of Mechanism of Action

Repetition of the iterative cycle of structure-based design, synthesis, and assessment of biologic activity yields compounds with improved potency for enzyme inhibition or receptor binding. Once a critical threshold of inhibitor potency is reached, compounds are subjected to cellular assays to assess cytotoxicity, antiproliferative activity, or other, more specific target-based assays that monitor effects at the cellular level (**Fig. 2**, {13}). Nonselective assays are commonly applied, which simply assess cell viability in microtiter plate formats by relying on the ability of viable cells to reduce colored dyes, utilize glucose, or maintain ATP levels. These assays are particularly well suited for medium-throughput screening assays, as illustrated in **Fig. 2** {12}. **Figure 3** illustrates an example of a decision tree or flowchart that might guide the evaluation of hundreds of molecules, from synthesis to ultimate tests of antitumor efficacy in vivo. A combinatorial library of sufficient quality will permit simultaneous screening for inhibition at the protein and cellular levels. Thus, only the subset of molecules displaying inhibition in both assays might be subjected to resynthesis for further evaluation.

More specific cell-based assays will have typically evolved from the target validation process, and should provide a direct readout of the outcome of target inhibition in a particular cell line or group of cell lines. These assays serve to quantify the cellular activity of optimal enzyme inhibitors while verifying a specific mechanism of action predicted for the target. Use of a potent and selective small-molecule inhibitor for target validation provides a compelling addition to the results generated in earlier target validation studies using antisense, dominant negative, or other molecular means to disrupt the function of the target in a cell. Examples of more specific cell-based assays include Western blotting of lysates from treated cells to assess modification of the target substrate, demonstrating modulation of signaling pathways by assaying for downstream effects, or surveying specific cellular phenotypes induced by target modulation. Analysis of specific effects on the cell growth and division cycle are assessed by more selective assays for DNA synthesis, mitosis, or flow cytometry. Similarly, numerous assays are available for assessing apoptosis, including staining of DNA, the terminal

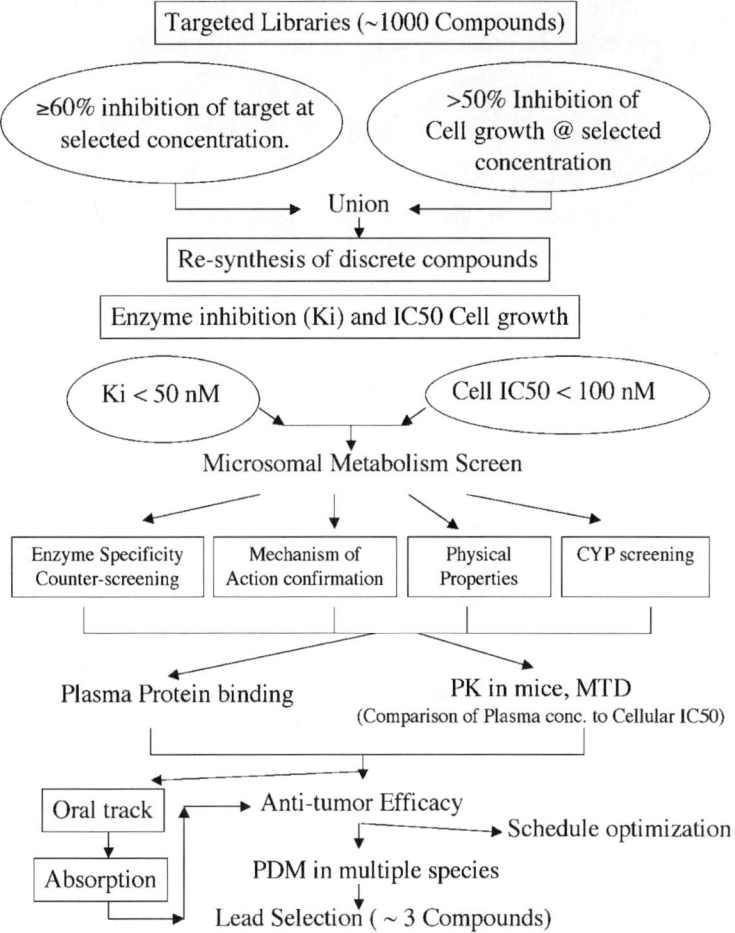

Fig. 3. Example of a decision tree used to guide selection of the optimal compound. One challenge of the drug discovery process is the channeling of resources and the design of experiments to guide a scientifically rigorous yet efficient selection of the single best compound from a collection of thousands. This flowchart illustrates a representative example of how experiments might be chosen to guide this selection. Targeted libraries may be subjected to enzyme or receptor-based inhibition assays and simultaneous cellular assays in a medium-throughput format; only compounds passing criteria from both tests are subjected to resynthesis. The choice of subsequent experiments is arranged according to the limitations of a particular chemical series (obvious problems screened earlier), or according to resource limitations and throughput capabilities. Compounds slated for oral delivery must undergo additional in-vitro and in-vivo absorption testing prior to efficacy testing using the oral dosing schedule.

deoxynucleotidyl transferase (TdT)-mediated dUTP nick end labeling (TUNEL) assay, flow cytometry, PARP cleavage, and caspase activation, to name a few. The COMPARE analysis by the National Cancer Institute and genomics efforts to characterize genetic defects in different cell lines can aid description of the mechanism(s) of action of drugs *(82)*, and can provide genetic background correlations to compound response profiles *(83)*. Given the complexity of cancer signaling pathways, and our incomplete understanding of these pathways, demonstrating that a certain chemical inhibitor functions through a defined mechanism is critical. In addition, proof of concept in humans is an ultimate goal of a drug discovery process, and more detailed cellular assays may be translated to biomarker studies that are applicable for clinical use.

5.5. Optimization of Pharmacokinetics, Drug Metabolism, and Physical Properties

The drug discovery process up to the point of testing cellular activity and verifying mechanism of action (**Fig. 2**, {14}) is relatively straightforward. Hypotheses for improving binding to the target are tested, structure–activity relationships are fully defined, and optimal compounds are readily selected. The challenge of identifying the most promising drug from a collection of potent inhibitors is much more formidable. What constitutes a good drug? The question may perhaps be more easily considered in the context of unfavorable clinical outcomes: what prevents efficacy? The possibilities are many: a drug intended for intravenous formulation may be poorly soluble; a drug intended for oral delivery may be poorly absorbed; a compound may be rapidly metabolized requiring large and/or frequent dosing; or a drug may exhibit mechanism-based or unrelated toxicity that does not provide an adequate therapeutic window; the absorption, distribution, metabolism, and excretion (ADME) of the compound may not follow expected kinetics, resulting in unpredictable accumulation of the drug; metabolic processing may occur with too much variability between subjects; metabolic processing may result in drug–drug interactions; metabolic processing may generate a toxic metabolite; rapid clearance (efflux pumps), poor distribution (very high plasma protein binding), and so on. The safety and bioavailability standards for anticancer drugs are somewhat different than those for drugs for diseases that are not life-threatening. A cytotoxic agent might be administered only acutely to patients with advanced disease. Even if such a drug displayed a toxicity that could be monitored and controlled, the drug would be accepted with less reservations than a chronically administered drug with the same toxicity. Cancer patients are frequently hospitalized, and intravenous administration of anticancer chemotherapeutics is more common than in other therapeutic areas, so oral bioavailability may not be as great a concern for development of certain anticancer drugs. However, there is a trend toward more oral therapy options for patients. This is driven both by the requirement for chronic dosing as well as the ease of delivery for the patient. This is a complex issue, and the field is still evolving through hurdles that include the lack of reimbursement through Medicare for novel oral anticancer agents. Of course, the next generations of anticancer agents are focusing on novel mechanisms of action, so each project must set goals for the desirable pharmacokinetic characteristics within the framework of their predefined product profile.

The assessment of pharmacokinetic properties, ADME, and physicochemical properties of molecules are undertaken in part to eliminate molecules that would fail in the clinic due to these types of shortcomings. This process is a difficult challenge for several reasons. First, the criteria for a chemical to function as an acceptable drug may be obvious, but the mechanisms for identifying molecules that satisfy all of the criteria are somewhat empirical. The chemical structural features that provide a good pharmacokinetic profile and low toxicity are generally not predictable, and few rules exist to assign parameters for these criteria to each chemical moiety. Second, as a result of the requirement for empirical testing, compounds must be put through a battery of tests and experiments that are traditionally low-throughput. Moreover, with the advent of combinatorial chemistry, the number of molecules to be evaluated within a project has increased substantially, placing an even greater burden on the development pharmacologist. In addition, a multivariate optimization is required, and optimal properties do not always overlap. This dilemma has frequently been described with a Venn diagram, wherein large areas may describe compounds fulfilling criteria of target potency, cellular potency, and defined mechanism of action, but smaller, distinct circles describe compounds with optimal solubility, good absorption, and low toxicity. Needless to say, the overlap or union of all defined criteria sets can become infinitesimally small. This is compounded by the frequent observation that changing the chemical structure of a drug to improve one property may negatively affect several other properties in an unpredictable way. For example, modifying a drug to improve metabolic stability at one site of the molecule can alter metabolic stability at another site. Similarly, compounds with improved metabolic stability may suffer poor enzymatic or cellular activity, resulting in a backward step in the drug discovery process!

Historically, drug discovery project teams were separated from drug development teams, resulting in a finite "hand-off" of a lead compound for which little pharmacokinetic or toxicology data had been gathered. This approach provided very little feedback to the synthetic organic chemist for improving upon metabolic stability or physical properties, and also left the development scientist with the awkward task of trying to "fix" suboptimal lead compounds. As a result of the high attrition and slow development timelines with this approach, as well as the greater numbers of compounds that must be evaluated, pharmaceutical companies are now making a strong effort to integrate development scientists and discovery scientists early in the discovery project, prior to selection of the best lead compound (**Fig. 2C**) *(84,85)*. This evolving paradigm that incorporates members of a pharmaceutical development division (such as development pharmacologists, toxicologists, process chemists, and analytical or formulations chemists), into drug discovery teams provides for a much more efficient course of action. Medicinal chemists and even early discovery project members gain greater exposure to the challenges of drug development, and with feedback from data analysis, the team is better able ultimately to select the best molecule for drug development.

One strategy that pharmaceutical companies are exploiting to assist in compound evaluation is the predictive analysis of metabolism, toxicity, carcinogenicity, and physicochemical and "druglike" properties *in silico*. Various computational methods have been developed that calculate log P values based on molecular size and shape *(86)*; the log P values in turn are used to estimate membrane permeability. Lipinski's "rule of five" describes four rules for identifying compounds displaying poor absorption or permeability: a compound that has more than five hydrogen-bond donors; a molecular weight greater than 500, a log P greater than 5, or more than 10 hydrogen-bond acceptors is more likely to be poorly absorbed *(81)*. Other computational methods attempt to predict physicochemical parameters such as solubility, metabolism, or permeability using either knowledge-based systems or statistical definitions of chemical descriptors derived from the molecular structure *(87)*. As the collection of ADME data grows, future efforts will be focused on developing training data sets to supply additional predictive value.

There is no substitute for experimental data, however, and in-vitro assay models that can serve as surrogates for more costly and time-consuming in-vivo experiments are employed whenever possible. For example, cultures of Caco-2 cells (derived from human colon cancer cells) are used as an *in vitro* test for membrane permeability and absorption *(88)*. When cultured on porous membranes, Caco-2 cells differentiate into monolayers with many of the functional characteristics of human intestinal cells. Measurement of compound passage or transport through the monolayer has proved to be a good predictor of absorption in the human intestine *(89)*. Madin-Darby canine kidney (MDCK) cells have also been used as a tool for membrane permeability screening *(90)*, but these time-consuming assays are not high-throughput by any means. A more straightforward in-vitro alternative involves measuring the partitioning of molecules on immobilized artificial membranes (IAMs) using HPLC-based methods *(91)*. Physicochemical properties that influence drug absorption, such as log P, solubility, and pK_a, are measured experimentally with pure compounds, but only a few medium-throughput or predictive tools are in use *(92)*. Concerns about a molecule's ability to be absorbed are linked with the potential problem of active efflux out of cells, for example by the P-glycoprotein transporter *(93)*. This potential liability can be addressed by comparing model cell lines with different expression of the P-glycoprotein, but again, these are not high-throughput assays. A drug that is tightly bound by serum proteins may suffer from poor distribution or unfavorable pharmacokinetics. Traditional equilibrium dialyses experiments are typically performed to address this potential shortcoming, although capillary electrophoresis and "frontal analysis," in which free drug concentration is estimated from the shape of the frontal peak in the electropherogram, have been recently advanced for this purpose *(87)*.

As illustrated in **Fig. 2C**, the inclusion of in-vitro pharmacokinetics, dynamics, and metabolism (PDM) experiments, as well as physicochemical experiments, within the iterative process of medium-throughput testing is an ideal situation that may permit the development scientists to keep pace with the discovery scientists. Mechanistic screens may be devised that permit yes/no answers from small targeted libraries (**Fig. 2 {14}**), even though these may be more time-consuming or difficult to format than enzyme assays. Although the PDM and physical property testing must be performed on purified compounds, rather than impure libraries, these types of in-vitro assays are also evolving for more rapid throughput. For example, compound metabolism can be assessed by using liquid chromatography plus mass spectrometry (LC/MS) to monitor disappearance of the parent drug during incubation with human liver microsomes or hepatocytes. LC/MS technology is highly sensitive and is adaptable to rapid throughput. It is thus becoming an increasingly popular tool in the discovery setting for monitoring the rate of drug metabolism, as well as for detecting and identifying metabolites. Just as high-throughput screening for enzyme inhibition relies on percent inhibition for a given drug concentration rather than a full dose–response, a more rapid-throughput metabolism screen can be executed by measuring the fraction of parent compound remaining at a given time, rather than a complete timecourse. These types of assays are quite useful for parallel comparisons of series of compounds; more detailed assays can be performed on a smaller number of lead compounds. Pulsed ultrafiltration-electrospray mass spectrometry has been recently introduced as a novel high-throughput technique for metabolite screening *(94)*. Hepatic microsomes containing the cytochrome P450-metabolizing enzymes are trapped in an ultrafiltration device, through which drugs and enzyme cofactors are injected. Metabolites, including glutathione adducts, are detected by online mass spectrometry.

The extent of compound breakdown during first-pass metabolism by the intestine or liver can determine the oral bioavailability of a drug. It is the cytochrome P450 (CYP) enzymes that perform most of the oxidative metabolic reactions, and the P450 3A isoenzyme family is believed to metabolize 50–60% of all currently administered drugs. High-throughput fluorometric and radiometric assays for the five major human metabolizing enzymes (CYP1A2, 2C9, 2C19, 2D6, and 3A4) have recently been developed to aid in compound screening *(95)*. In addition, the availability of crystal structures of several bacterial P450 enzymes will continue to expand knowledge of the structural basis of isoenzyme specificity. These studies, in turn, will guide database compilation and programs used to predict sites of metabolism for given compounds *(96)*.

6. Late-Stage Discovery: Selecting the Development Compound

As the iterative cycles of drug discovery proceed, the best compounds advance through the battery of tests as described in **Figs. 2** and **3**, and a relatively small collection of optimal compounds are available for more detailed tests of efficacy. In today's pharmaceutical world, the primary distinction of a late-stage lead discovery (or, more accurately, development candidate selection) program is the decreased number of new molecules being subjected to more complex experiments that focus on animal models. At this stage of the project, all relatively straightforward in-vitro analyses are completed, and the remaining studies must rely on in-vivo experimentation to assess the pharmacokinetic parameters of the drug and the efficacy. As illustrated in **Fig. 2C**, there is still ample opportunity for feedback to medicinal chemists from the outcome of the in-vivo experiments, as the discovery and development scientists remain fully integrated.

6.1. In-Vivo *Pharmacokinetics*

The best assessment of the bioavailability of a potential drug can only be obtained by administering low doses of the compound to rodents and higher animals, and carefully monitoring serum plasma samples over time to assess distribution and metabolism of the molecule. The goals of such studies are to determine the pharmacokinetic profile associated with optimal efficacy in the preclinical species of interest, and the likelihood that such will be achievable in humans at a reasonably delivered dose by assessment of the compound in multiple species (rats, mice, dogs, monkeys). Compounds are dosed using the route and schedule that will be selected for first tests of efficacy. Intravenous and oral administration are always included to permit calculation of clearance, volume of distribution, elimination half-life, and oral bioavailability. Two approaches have been applied in recent years in attempts to improve throughput and/or reduce the animal requirements and expense of these studies. The first method, defined as cassette or "N-in-1" dosing, makes use of administering more than two chemicals to an animal at once *(97)*. Serum plasma is then analyzed by LC/MS, which will permit detection of multiple analytes per sample provided none of the parent molecules or metabolites have the same mass ion. The detection method thus introduces the first caveat, which is that groups of chemicals must be carefully chosen prior to administration. Sets of structurally distinct molecules would be preferable, but late-stage projects are typically focused on only one, or at most two, distinct chemical series for compound selection. Complications can also arise if one or more chemicals inhibit CYP enzymes, resulting in drug–drug interactions and an inaccurate readout of bioavailability for every other chemical in the mixture. This can be partially addressed by including a reference compound, but ultimately verification by single dosing is ideal. Although the amount of each chemical required for cassette dosing is less than for single-molecule dosing, the possibility of adverse effects in the animal is greater. Nevertheless, the analysis of fewer samples and the resultant savings in compound, time, and animals have made cassette dosing a widely tested concept *(98)*. An alternative time-saving approach chosen to eliminate problems of compound interference involves pooling of different plasma samples at each time point for testing by LC/MS, after administration of individual chemicals in single doses. This results in the analysis of fewer samples, but since sample pooling also requires additional time, the overall time-saving benefit is not clear. A similar method attempts to obtain time-averaged concentrations for estimation of the area under the concentration–time curve (AUC) *(99)*.

The administration of low doses of a chemical in both rodent and nonrodent species to study the pharmacokinetic profile provides another important level of compound optimization. For advanced, highly optimized compounds, a full data package of in-vitro and in-vivo ADME data in multiple species is gathered. However, it has been emphasized that there is a very poor correlation between the absolute bioavailability of drugs in rodents, dogs, and primates compared to the absolute bioavailability reported in humans *(85)*, due primarily to physiologic and metabolic differences between species. Analysis of pharmacokinetics in animal models is of primary importance for correlation with efficacy and toxicologic studies in the same animals, rather than for drawing conclusions about human bioavailability. Prior to efficacy studies, some preliminary estimation of the model animal's ability to tolerate higher doses of the compound is obtained, typically initiated with a dose escalation guided by ADME data. These preliminary studies provide early estimates of the maximal tolerated dose (MTD), and may provide insights into obvious toxicities. Given that poor pharmacokinetic properties are the single greatest cause for withdrawal of drugs from development *(85)*, it becomes very important to fully explore a compound's PDM properties in the context of its product profile.

6.2. Compound Scale-up: Process Chemistry

As more in-vitro studies are performed, the requirement for purified, solid material increases from just a few milligrams up to a hundred milligrams. As in-vivo pharmacokinetic and MTD assessments begin, the necessary amount of solid material can exceed 1 g, depending on the dosing regimen. Medicinal chemists collaborate with process chemists to provide this material in a form that is pure with respect to contaminants and stereoisomers (**Fig. 2**, {16}). As final compound selection draws near, analytical chemists also form part of this collaboration, as different salt forms are tested, compound crystallization is assessed, and stability is examined. Process chemists begin to seek any limitations to large-scale synthesis, such as long linear synthetic schemes, low-yield reactions, limited availability of starting materials, or high cost of materials. Complicated syntheses can make for a challenging development program, and a synthesis route of greater than 15 steps is typically considered problematic. Optimization of the synthesis route becomes an important procedure, as chemists evaluate each step for methods to improve yield and/or efficiency. Interestingly, parallel synthesis approaches are also applied here, with the goal of rapidly assessing multiple synthesis conditions of the same molecule, rather than applying the identical reaction conditions to different molecules. A wide range of reaction variables may be assessed to improve reaction yields, such as reaction temperature and pressure, rate of reagent addition, solvents, choice of bases, mixing, or searches for a new catalyst *(100)*.

6.2.1. Assessment of Efficacy: In-Vivo *Tumor Models*

The exact experiments used to assess efficacy of an anticancer agent will of course be defined by the mechanistic approach and the predefined product profile of the project. However, whether a target-based strategy focuses on antiangiogenesis, signal transduction pathways, cell cycle effects, or combination therapies, there will almost certainly be a requirement for preclinical testing of a molecule's ability to inhibit the growth of tumors growing in a living animal. Murine tumor models provide one option (e.g., the murine tumor panel used some time ago by the National Cancer Institute), but these models have been found to have poor predictive value and to yield an undesirable number of false positives *(101)*. There can be substantial differences between mice and humans in tumor doubling time, drug dosing, and even genetic status of the cancer cell line. Human solid tumors grown as xenografts in immunocompromised mice are currently the most common approach to examining efficacy in the whole animal. The mutant nu/nu ("nude") strain of mice are genetically immune deficient due to lack of a thymus. Although these animals require special care such as a completely sterilized environment (food, bedding, handling), their permanent lack of immunity has made them the most popular host of human tumor xenografts. Whereas the nude mice have a functional B-lymphocyte system, the severe combined immunodeficient (SCID) mice lack both B- and T-cell function, and this murine strain has found use as hosts for human lymphoma cell lines, which grow poorly in nude mice *(102)*.

Human tumors are implanted subcutaneously as fragments in the flank of the immunocompromised mice. Tumor growth can be readily visualized and carefully monitored by caliper measurements that permit automated data storage. Once tumors have reached a minimum size, for example 5 mm diameter or approximately 100 mm^3 volume, chemotherapy is initiated in large, randomized groups, according to the specified regimen. Inhibition of tumor growth is monitored as (a) tumor weight change of treated animals as a percent of control animals ($T/C \times 100$); (b) percent tumor growth delay, as a measure of median time a treated tumor reaches a specified size compared to control tumor. The growth delay is expressed as a ratio of the rate of growth of the control tumor ($[(T - C)/T)] \times 100$) to account for different growth rates of tumors; and/or (c) an estimate of the number of cells killed, or net log cell kill *(102)*. All animals are monitored for weight loss and for dose-related deaths throughout the duration of the study as a means of assessing gross toxicity.

The initial development of the xenograft can be painstaking, requiring serial passaging of tumors in vivo, with growth to the appropriate size before use. Depending on the rate of tumor growth (and the efficacy of the drug), each individual efficacy study can require 3–4 wk. This constraint, as well as the cost of the study and the requirement for additional drug solid, severely limit the throughput of this assay. A more rapid estimate of antitumor efficacy has been developed at the National Cancer Institute, where tumor cells are grown for several days within small, hollow glass fibers implanted subcutaneously or intraperitoneally in immunocompromised mice. Efficacy is measured by a colorimetric cell viability assay upon removal of the fiber *(102,103)*. This assay is more cost-efficient but less complete, and can provide preliminary screening data for a larger number of compounds. However, the number of compounds fulfilling all of the necessary criteria for in-vivo efficacy assessment is typically low (**Fig. 3**), and it is usually more efficient overall to proceed with a complete antitumor efficacy test using the human tumor xenograft model. The hollow-fiber model is also not applicable for some mechanistic approaches, such as antiangiogenesis or antimetastatic agents.

A wide variety of tumor models derived from different cancer cell lines have been used successfully in xenograft models *(102)*. The particular tumor model can be chosen based on genetic status, doubling time or growth rate, take rate, hormone requirement, or other practical considerations. However, it is generally accepted that the human tumor xenograft models are not necessarily predictive of responsive cancer types in patients, nor in the overall clinical response *(104)*. Thus, the best assessment of efficacy is sought by testing compounds in multiple diverse tumor models. The goal is to measure how broad the compound efficacy is in different models, and within each model how deep that effectiveness extends at well-tolerated doses. It is not unreasonable to assess efficacy in 5 independent tumor models and expect to achieve activity in only 3 models. Activity of a compound is usually encouraging if associated with >50% growth inhibition of the tumor at or below the maximum tolerated dose on a particular schedule of administration.

6.2.2. Pharmacodynamics and Biomarkers

Many of the current drugs advancing through discovery and development target novel mechanisms, exploiting the latest knowledge in angiogenesis, signal transduction, or cell-cycle mechanism pathways. As these drugs progress into animal models, it becomes essential to apply the cell-based experiments that were used to characterize mechanism of action to the in-vivo tumor studies, and ultimately to the "proof of concept" in the human. Demonstration that inhibition of tumor growth is correlated with the predicted effect on the targeted pathway may be achieved simply by modifying the cellular mechanistic assay for analysis of tumor tissue. Alternatively, many of the molecular techniques described previously, including proteomics, quantitative PCR, Western blotting, or gene microarray technology may be applied. At this stage of the project, significant effort may be expended in identifying simple assays or surrogate markers that permit analysis of human samples obtained by noninvasive techniques, especially given the difficulty of tumor sampling during human clinical trials. Imaging techniques such as positron emission tomography or magnetic resonance imaging permit monitoring of uptake of labeled drugs, tumor metabolism, or vascular changes *(105)*. Research addressing the challenge of identifying suitable biomarkers will probably undergo the greatest growth in coming years, given the pressure to define readily determined clinical endpoints for the new compounds that display a novel mechanism of action, as well as the ability to use new genomics and proteomics technologies to help identify surrogate affected genes *(106)*.

6.3. Pre-clinical Toxicology

Approximately 10–20% of the drugs that fail during preclinical development fail because of unacceptable toxicity *(85)*. As described earlier, it is the attempt to decrease the high attrition rate of drugs in development, and improve the efficiency of development timelines, that has resulted in a blurring of the boundaries between drug discovery and drug development. One strategic approach applied during lead discovery uses computational methods in an attempt to identify toxic chemical moieties *in silico*. Commercially available programs such as DEREK, TOPKAT, and MULTICASE use either rule-based systems built on large training sets of data or QSAR methods to predict toxicity based on chemical structure. However, each program has specific limitations, and none of the current programs provides sufficient accuracy for use as a true filter for preselecting compounds. Better assessments can be obtained when results from multiple programs are compared, but further validation is required *(107)*. Again, in-vivo experimentation is unavoidable. Typically, a few superior compounds proceed with preliminary toxicology experiments in rodents, using simple dose escalation followed by clinical observation and necropsy. These experiments are not intended to define a complete toxicology package, nor certainly to determine any mechanism of toxicity, but merely to provide the first hints of potential problems. Such studies may also permit a final selection of molecules based on maximal tolerated dose or therapeutic window.

Once initial toxicities are identified in a potential development compound, the methods used to characterize the observed toxicity are also now beginning to exploit the genomics frontiers. Tissue-specific DNA microarray chips or proteomics approaches can be used to assess the genes that are induced or repressed following drug treatment *(106)*. Careful experimental design is required, with low doses of drug and appropriate timecourses, in order to decipher direct drug effects. As use of these technologies increases, it is likely that different toxicity signatures will become more obvious. The human SNPs associated with different adverse drug effects are also being aggressively pursued as a means to identify patients who might be particularly sensitive to certain therapies *(108)*.

7. Summary: The Multidisciplinary Challenge of Drug Discovery

The final selection of a single compound for advancement into drug development, starting from a first search for a valid oncology target, is clearly a multidisciplinary and labor-intensive effort. Teams of up to 30 participants may be required in part-time or full-time effort, depending on the difficulty and chronology of the project. Success of a program relies on the expertise and best scientific judgment of a team of scientists with quite diverse training. Each scientific representative is challenged with developing and utilizing the latest technologies in his or her field to accelerate the process, to increase the efficiency of drug discovery without sacrificing the necessary rigor and attention to detail. From the first steps of target cloning, biochemical characterization, highthroughput screening, and protein crystallization, teams of biologists work together to advance a project beyond early discovery. With successful completion of each hurdle, larger teams of medicinal and computational chemists become involved in the beginning stages of lead discovery. As the quality of the compounds improve, the project gains momentum as inhibitors demonstrate improved potency and selectivity. This chemistry-intensive stage gives rise to more qualitative biologic testing. Teams of pharmacologists become central, as more complex cell-based assays and animal models are used to select the most promising compounds.

Of course, the final nomination of a compound for development does not guarantee a drug; numerous additional hurdles, requiring scientific approaches that are frequently quite distinct from discovery efforts, are required for advancement of a novel chemical entity into final approval and launch of a new drug. This path from nominated development compound through preclinical and clinical development will be the topic of a forthcoming review.

References

1. Consortium, I. H. G. S. (2001) Initial sequencing and analysis of the human genome. *Nat. Genet.* **409**, 860–921.
2. Venter, J. C., Adams, M. D., Myers, E. W., et al. (2001) The sequence of the human genome. *Science* **291**, 1304–1351.
3. Hattori, M., Fujiyama, A., Taylor, T. D., et al. (2000) The DNA sequence of human chromosome 21. *Nature* **405**, 311–319.
4. Dunham, I., Hunt, A. R., Collins, J. E., et al. (1999) The DNA sequence of human chromosome 22. *Nature* **402**, 489–495.
5. Ewing, B. and Green, P. (2000) Analysis of expressed sequence tags indicates 35,000 human genes. *Nat. Genet.* **25**, 232–234.
6. Guigo, R., Agarwal, P., Abril, J. F., Burset, M., and Fickett, J. W. (2000) An assessment of gene prediction accuracy in large DNA sequences. *Genome Res.* **10**, 1631–1642.
7. Futreal, P. A., Kasprzyk, A., Birney, E., Mullikin, J. C., Wooster, R., and Stratton, M. R. (2001) Cancer and genomics. *Nature* **409**, 850–852.
8. Murray, A. W. and Marks, D. (2001) Can sequencing shed light on cell cycling? *Nature* **409**, 844–846.
9. Group, T. I. S. M. W. (2001) A map of human genome sequence variation containing 1.42 million single nucleotide polymorphisms. *Nature* **409**, 928–933.
10. Burks, C. (1999) Molecular biology database list. *Nucleic Acids Res.* **27**, 1–9.
11. Duggan, D. J., Bittner, M., Chen, Y., Meltzer, P., and Trent, J. M. (1999) Expression profiling using cDNA microarrays. *Nat. Genet: The Chipping Forecast,* **21**, 10–14.
12. Schena, M., Shalon, D., Heller, R., Chai, A., Brown, P. O., and Davis, R. W. (1996) Parallel human genome analysis: microarray-based expression monitoring of 1000 genes. *Proc. Natl. Acad. Sci. USA* **93**, 10614–10619.
13. Gibson, U. E. M., Heid, C. A., and Williams, P. M. (1996) A novel method for real time quantitative RT-PCR. *Genome Res.* **6**, 995–1001.
14. Heid, C. A., Stevens, J., Livak, K. J., and Williams, P. M. (1996) Real time quantitative PCR. *Genome Res.* **6**, 986–994.
15. Monni, O., Bärlund, M., Mousses, S., et al. (2001) Comprehensive copy number and gene expression profiling of the 17q23 amplicon in human breast cancer. *Proc. Natl. Acad. Sci. USA* **98**, 5711–5716.
16. Svaren, J., Ehrig, T., Abdulkadir, S. A., Ehrengruber, M. U., Watson, M. A., and Milbrandt, J. (2000) EGR1 Target genes in prostate carcinoma cells identified by Microarray Analysis. *J. Biol. Chem.* **275**, 38524–38531.
17. O'Farrell, P. H. (1975) High resolution two-dimensional electrophoresis of proteins. *J. Biol. Chem.* **250**, 4007–4021.
18. Blackstock, W. P. and Weir, M. P. (1999) Proteomics: quantitative and physical mapping of cellular proteins. *Trends Biotechnol.* **17**, 121–127.
19. Jain, K. K. (2000) Applications of proteomics in oncology. *Pharmacogenomics* **1**, 385–393.
20. Myers, T. G., Anderson, N. L., Waltham, M., et al. (1997) A protein expression database for the molecular pharmacology of cancer. *Electrophoresis* **18**, 647–653.
21. Service, R. F. (2000) Structural genomics offers high-speed look at proteins. *Science* **287**, 1954–1956.

22. Zarembinski, T. I., Hung, L.-W., Mueller-Dieckmann, H.-J., et al. (1998) Structure-based assignment of the biochemical function of a hypothetical protein: a test case of structural genomics. *Proc. Natl. Acad. Sci. USA* **95**, 15189–15193.

23. Berman, H. M., Westbrook, J., Feng, Z., et al. (2000) The Protein Data Bank. *Nucleic Acids Res.* **28**, 235–242.

24. Hartwell, L. H., Szankasi, P., Roberts, C. J., Murray, A. W., and Friend, S. H. (1997) Integrating genetic approaches into the discovery of anticancer drugs. *Science* **278**, 1064–1068.

25. Taylor, M. F., Wiederholt, K., and Sverdrup, F. (1999) Antisense oligonucleotides: a systematic high-throughput approach to target validation and gene function determination. *Drug Discov. Today* **4**, 562–567.

26. Nolan, P. M. (2000) Generation of mouse mutants as a tool for functional genomics. *Pharmacogenomics* **1**, 243–255.

27. Ossovskaya, V. S., Mazo, I. A., Chernov, M. V., et al. (1996) Use of genetic suppressor elements to dissect distinct biological effects of separate p53 domains. *Proc. Natl. Acad. Sci. USA* **93**, 10309–10314.

28. Remy, I. and Michnick, S. W. (1999) Clonal selection and in vivo quantitation of protein interactions with protein-fragment complementation assays. *Proc. Natl. Acad. Sci. USA* **96**, 5394–5399.

29. Drews, J. (2000) Drug discovery: a historical perspective. *Science* **287**, 1960–1963.

30. Landro, J. A., Taylor, I. C. A., Stirtan, W. G., et al. (2000) HTS in the new millennium: the role of pharmacology and flexibility. *J. Pharm. Toxicol. Meth.* **44**, 273–289.

31. Bogan, A. A. and Thorn, K. S. (1998) Anatomy of hot spots in protein surfaces. *J. Mol. Biol.* **280**, 1–9.

32. Jones, D. A. and Fitzpatrick, F. A. (1999) Genomics and the discovery of new drug targets. *Curr. Opin. Chem. Biol.* **3**, 71–76.

33. Rondon, U. and Marasco, W. A. (1997) Intracellular antibodies (intrabodies) for gene therapy of infectious disease. *Annu. Rev. Microbiol.* **51**, 257–283.

34. Cochet, O., Gruel, N., Fridman, W.-H., and Teillaud, J.-L. (1999) Ras and p53 intracellular targeting with recombinant single-chain Fv (scFv) fragments: a novel approach for cancer therapy? *Cancer Detect. Prev.* **23**, 506–510.

35. Wright, M., Grim, J., Deshane, J., et al. (1997) An intracellular anti-erbB-2 single-chain antibody is specifically cytotoxic to human breast carcinoma cells overexpressing erbB-2. *Gene Ther.* **4**, 317–322.

36. Jin, X., Nguyen, D., Zhang, W. W., Kyritsis, A. P., and Roth, J. A. (1995) Cell cycle arrest and inhibition of tumor cell proliferation by the p16INK4 gene mediated by an adenovirus vector. *Cancer Res.* **55**, 3250–3253.

37. Sherr, C. J. and Roberts, J. M. (1999) CDK inhibitors: positive and negative regulators of G1-phase progression. *Genes Dev.* **13**, 1501–1512.

38. Rosenberg, M. P. (1997) Gene knockout and transgenic technologies in risk assessment: the next generation. *Mol. Carcinogen.* **20**, 262–274.

39. Fukushige, S. and Sauer, B. (1992) Genomic targeting with a positive-selection lox integration vector allows highly reproducible gene expression in mammalian cells. *Proc. Natl. Acad. Sci. USA* **89**, 7905–7909.

40. St. Onge, L., Furth, P. A., and Gruss, P. (1996) Temporal control of the Cre recombinase in transgenic mice by a tetracycline responsive promoter. *Nucleic Acids Res.* **24**, 3875–3877.

41. Liu, Q., Li, M. Z., Leibham, D., Cortez, D., and Elledge, S. J. (1998) The univector plasmid-fusion system, a method for rapid construction of recombinant DNA without restriction enzymes. *Curr. Biol.* **8**, 1300–1309.

42. Novy, R., Yaekger, K., Monsma, S., and Scott, M. (1999) pTriEx-1 multisystem vector for protein expression in E.coli, mammalian, and insect cells. *inNOVAtions* **10**, 1–5.

43. Stevens, R. C. (2000) Design of high-throughput methods of protein production for structural biology. *Structure (Lond.)* **8**, R177–R185.

44. King, L. A. and Possee, R. D. (1992) *The Baculovirus Expression System*, Chapman & Hall, London.

45. Allison, R. D. and Purich, D. L. (1983) Practical considerations in the design of initial velocity enzyme rate assays, in *Contemporary Enzyme Kinetics and Mechanism* (Purich, D. L., ed.). Academic Press, New York, pp. 33–52.

46. Hertzberg, R. P. and Pope, A. J. (2000) High-throughput screening: new technology for the 21st century. *Curr. Opin. Chem. Biol.* **4**, 445–451.

47. Hill, D. C. (1998) Trends in development of high-throughput screening technologies for rapid discovery of novel drugs. *Curr. Opin. Drug Discov. Dev.* **1**, 92–97.

48. Ramm, P. (1999) Imaging systems in assay screening. *Drug Discov. Today* **4**, 401–410.

49. Alpha, B., Lehn, J.-M., and Mathis, G. (1987) Energy transfer luminscence of europium(III) and terbium(III) cryptates of macrobicyclic polypyridine ligands. *Angew. Chem. Int. Ed. Engl.* **26**, 266–267.

50. Hemmilä, I. and Webb, S. (1997) Time-resolved fluorometry: an overview of the labels and core technologies for drug screening applications. *Drug Discov. Today* **2**, 373–381.

51. Sportsman, J. R., Lee, S. K., Dilley, H., and Bukar, R. (1997) Fluorescence polarization, in *High-Throughput Screening: The Discovery of Bioactive Substances* (Devlin, J. P., ed.). Marcel Dekker, New York, pp 389–399.

52. Kask, P., Palo, K., Ullmann, D., and Gall, K. (1999) Fluorescence-intensity distribution analysis and its application in biomolecular detection technology. *Proc. Natl. Acad. Sci. USA* **96**, 13756–13761.

53. Moore, K. J., Turconi, S., Ashman, S., et al. (1999) Single molecule detection technologies in high throughput screening: fluorescence correlation spectroscopy. *J. Biomol. Screening* **4**, 335–353.

54. Gronenborn, A. M. and Clore, G. M. (1995) Structures of protein complexes by multidimensional heteronouclear magnetic resonance spectroscopy. *Critical Rev. Biochem. Mol. Biol.* **30**, 351–385.

55. Brünger, A. T. (1997) X-ray crystallography and NMR reveal complementary views of structure and dynamics. *Nat. Struct. Biol.* **4**, 862–865.

56. Shuker, S. B., Hajduk, P. J., Meadows, R. P., and Fesik, S. W. (1996) Discovering high-affinity ligands for proteins: SAR by NMR. *Science* **274**, 1531–1534.

57. Moore, J. M. (1999) NMR screening in drug discovery. *Curr. Opin. Biotechnol.* **10**, 54–58.

58. Shoichet, B. K. and Bussiere, D. E. (2000) The role of macromolecular crystallography and structure for drug discovery: advances and caveats. *Curr. Opin. Drug Discov. Dev.* **3**, 408–422.

59. Islam, S. A., Carvin, D., Sternberg, M. J. E., and Blundell, T. L. (1998) HAD, a data bank of heavy-atom binding sites in protein crystals: a resource for use in multiple isomorphous replacement and anomalous scattering. *Acta Crystallogr. D*, **54**, 1199–1206.

60. Walsh, M. A., Evans, G., Sanishvili, R., Dementieva, I., and Joachimiak, A. (1999) MAD data collection—current trends. *Acta Crystallogr. D*, **55**, 1726–1732.

61. Appelt, K., Bacquet, R. J., Bartlett, C. A., et al. (1991) Design of enzyme inhibitors using iterative protein crystallographic analysis. *J. Med. Chem.* **34**, 1925–1934.

62. Gubernator, K. and Bohm, H.-J., (1998). *Structure-Based Ligand Design*. Wiley-VCH, Weinheim, Germany.

63. Antel, J. (1999) Integration of combinatorial chemistry and structure-based drug design. *Curr. Opin. Drug Discov. Dev.* **2**, 224–233.

64. Salemme, F. R., Spurlino, J., and Bone, R. (1997) Serendipity meets precision: the integration of structure-based drug design and combinatorial chemistry for efficient drug discovery. *Structure (Lond.)*, **5**, 319–324.

65. Carlson, H. A. and McCammon, J. A. (2000) Accommodating protein flexibility in computational drug design. *Mol. Pharmacol.* **57**, 213–218.
66. Furka, A., Sebestyen, F., Asgedom, M., and Dibo, G., (1988). Cornucopia of peptides by synthesis., in *145th Int. Congr. Biochem.* Prague, **5**, 47.
67. Houghten, R. A., Pinilla, C., Blondelle, S. E., Appel, J. R., and Cuervo, J. H. (1991) Generation and use of synthetic peptide combinatorial libraries for basic research and drug discovery. *Nature* **354**, 84–86.
68. Bunin, B. A. and Ellman, J. A. (1992) A general and expedient method for the synthesis of 1,4-benzodiazepine derivatives. *J. Am. Chem. Soc.* **114**, 10997–10998.
69. Thompson, L. A. and Ellman, J. A. (1996) Synthesis and applications of small molecule libraries. *Chem. Rev.* **96**, 555–600.
70. Felder, E. R. and Poppinger, D. (1997) Combinatorial compound libraries for enhanced drug discovery approaches. *Adv. Drug Res.* **30**, 111–199.
71. Armstrong, R. W., Brown, S. D., Keating, T. A., and Tempest, P. A. (1997) Combinatorial synthesis exploiting multiple-component condensations, microchip encoding, and resin capture. *Comb. Chem.* 153–190.
72. Reitz, A. B. (1999) Recent advances in traceless linkers. *Curr. Opin. Drug Discov. Dev.* **2**, 358–364.
73. Myers, P. L. (1997) Will combinatorial chemistry deliver real medicines? *Curr. Opin. Biotechnol.* **8**, 701–707.
74. Zeng, L., Bruton, L., Yung, K., Shushan, B., and Kassel, D. B. (1998) Automated analytical/preparative high-performance liquid chromatography-mass spectrometry system for the rapid characterization and purification of compound libraries. *J. Chromatogr. A,* **794**, 3–13.
75. Kyranos, J. N., Cai, H., Wei, D., and Goetzinger, W. K. (2001) High-throughput high-performance liquid chromatography/mass spectrometry for modern drug discovery. *Curr. Opin. Biotechnol.* **12**, 105–111.
76. Ausman, D. J. (2001) Screening's age of insecurity. *Modern Drug Discov.* **4**, 33–39.
77. Rockwell, A., Melden, M., Copeland, R. A., Hardman, K., Decicco, C. P., and DeGrado, W. F. (1996) Complementarity of combinatorial chemistry and structure-based ligand design. Application to the discovery of novel inhibitors of matrix metalloproteinases. *J. Am. Chem. Soc.* **118**, 10337–10338.
78. Dolle, R. E. (1998) Comprehensive survey of chemical libraries yielding enzyme inhibitors, receptor agonists and antagonists, and other biologically active agents: 1992 through 1997. *Mol. Diversity* **3**, 199–233.
79. Blaney, J. M. and Dixon, J. S. (1993) A good ligand is hard to find: automated docking methods. *Perspect. Drug Discov. Des.* **1**, 301–319.
80. Verkhivker, G. M., Rejto, P. A., Gehlhaar, D. K., and Freer, S. T. (1996) Exploring the energy landscapes of molecular recognition by a genetic algorithm: analysis of the requirements for robust docking of HIV-1 protease and FKBP-12 complexes. *Proteins* **25**, 342–353.
81. Lipinski, C. A., Lombardo, F., Dominy, B. W., and Feeney, P. J. (1997) Experimental and computational approaches to estimate solubility and permeability in drug discovery and development settings. *Adv. Drug Deliv. Rev.* **23**, 3–25.
82. Weinstein, J. N., Myers, T. G., O'Connor, P. M., et al. (1997) An information-intensive approach to the molecular pharmacology of cancer. *Science* **275**, 343–349.
83. Scherf, U., Ross, D. T., Waltham, M., et al. (2000) A gene expression database for the molecular pharmacology of cancer. *Nat. Genet.* **24**, 236–244.
84. Venkatesh, S. and Lipper, R. A. (2000) Role of the development scientist in compound lead selection and optimization. *J. Pharm. Sci.* **89**, 145–154.
85. Sinko, P. J. (1999) Drug selection in early drug development: screening for acceptable pharmacokinetic properties using combined in vitro and computational approaches. *Curr. Opin. Drug Discov. Dev.* **2**, 42–48.

86. Buchwald, P. and Bodor, N. (1998) Octanol-water partition: searching for predictive models. *Curr. Med. Chem.* **5**, 353–380.

87. Caldwell, G. W. (2000) Compound optimization in early- and late-phase drug discovery: acceptable pharmacokinetic properties utilizing combined physicochemical, in vitro and in vivo screens. [Erratum to document cited in CA132:259990]. *Curr. Opin. Drug Discov. Dev.* **3**, 250.

88. Delie, F. and Rubas, W. (1997) A human colonic cell line sharing similarities with enterocytes as a model to examine oral absorption: Advantages and limitations of the Caco-2 model. *Crit. Rev. Ther. Drug Carrier Syst.* **14**, 221–286.

89. Artusson, P., Palm, K., and Luthman, K. (1996) Caco-2 monolayers in experimental and theoretical predictions of drug transport. *Adv. Drug Deliv. Rev.* **22**, 67–84.

90. Irvine, J. D., Takahashi, L., Lockhart, K., et al. (1999) MDCK (Madin-Darby canine kidney) cells: a tool for membrane permeability screening. *J. Pharm. Sci.* **88**, 28–33.

91. Yang, C. Y., Cai, S. J., Liu, H., and Pidgeon, C. (1997) Immobilized artificial membranes-screens for drug-membrane interactions. *Adv. Drug Deliv. Rev.* **23**, 229–256.

92. Smith, D. A. and Van de Waterbeemd, H. (1999) Pharmacokinetics and metabolism in early drug discovery. *Curr. Opin. Chem. Biol.* **3**, 373–378.

93. Spahn-Langguth, H., Baktir, G., Radschuweit, A., et al. (1998) P-glycoprotein transporters and the gastrointestinal tract: evaluation of the potential in vivo relevance of in vitro data employing talinolol as model compound. *Int. J. Clin. Pharmacol. Ther.* **36**, 16–24.

94. Nikolic, D., Fan, P. W., Bolton, J. L., and van Breemen, R. B. (1999) Screening for xenobiotic electrophilic metabolites using pulsed ultrafiltration-mass spectrometry. *Comb. Chem. High Throughput Screening* **2**, 165–175.

95. Crespi, C. L., Miller, V. P., and Penman, B. W. (1998) High throughput screening for inhibition of cytochrome P450 metabolism. *Med. Chem. Res.* **8**, 457–471.

96. Erhardt, P. W. (1999) Drug Metabolism: Databases and High-Throughput Testing During Drug Design and Development. IUPAC and Blackwell Publishers, Toledo, OH.

97. Berman, J., Halm, K., Adkison, K. K., and Shaffer, J. (1997) Simultaneous pharmacokinetic screening of a mixture of compounds in the dog using API LC/MS/MS analysis for increased throughput. *J. Med. Chem.* **40**, 827–829.

98. Frick, L. W., Adkison, K. K., Wells-Knecht, K. J., Woollard, P., and Higton, D. M. (1998) Cassette dosing: rapid in vivo assessment of pharmacokinetics. *Pharm. Sci. Tech. Today* **1**, 12–18.

99. Bajpai, M. and Adkison, K. K. (2000) High-throughput screening for lead optimization: a rational approach. *Curr. Opin. Drug Discov. Dev.* **3**, 63–71.

100. Cannarsa, M. J., Uno, T., and Larsen, C. (2000) High-throughput screening applied to drug synthesis process development. *Curr. Opin. Drug Discov. Dev.* **3**, 743–749.

101. Winograd, B., Boven, E., Lobbezoo, M. W., and Pinedo, H. M. (1987) Human tumor xenografts in the nude mouse and their value as test models in anticancer drug development (review). *In Vivo,* **1**, 1–14.

102. Plowman, J., Dykes, D. J., Hollingshead, M., Simpson-Herren, L., and Alley, M. C. (1997) Human tumor xenograft models in NCI drug development, in *Anticancer Drug Development Guide: Preclinical Screening, Clinical Trials, and Approval* (Teicher, B., ed.). Humana Press, Totowa, NJ, pp. 101–125.

103. Bibby, M. C. (1999) Making the most of rodent tumor systems in cancer drug discovery. *Br. J. Cancer* **79**, 1633–1640.

104. Corbett, T., Valeriote, F., LoRusso, P., et al. (1995) Tumor models and the discovery and secondary evaluation of solid tumor active agents. *Int. J. Pharmacogn.* **33**, 102–122.

105. Phelps, M. E. (2000) Positron emission tomography provides molecular imaging of biological processes. *Proc. Natl. Acad. Sci. USA* **97**, 9226–9233.

106. Debouck, C. and Goodfellow, P. N. (1999) DNA microarrays in drug discovery and development. *Nat. Genet.* **21**, 48–50.
107. Durham, S. K. and Pearl, G. M. (2001) Computational methods to predict drug safety liabilities. *Curr. Opin. Drug Discov. Dev.* **4**, 110–115.
108. Debouck, C. and Metcalf, B. (2000) The impact of genomics on drug discovery. *Annu. Rev. Pharmacol. Toxicol.* **40**, 193–208.

30

Targets in Apoptosis Signaling

Promise of Selective Anticancer Therapy

Ramadevi Nimmanapalli and Kapil Bhalla

1. Introduction

The human body is composed of approximately 10^{14} cells, each of which is capable of committing suicide by apoptosis. Normally, the processes of cell division and cell death are tightly coupled, so that no net increase in cell numbers occurs. However, alterations in the expression or function of the genes that control cell division and cell death can upset this delicate balance, contributing to expansion of neoplastic cells causing cancers. While the existing, conventional anticancer drugs either cause cell cycle perturbation or DNA damage, they do not interact directly with the intracellular machinery for apoptosis. Tumor selectivity of such agents is due largely to the increased sensitivity to apoptosis of tumor cells following DNA damage or cell cycle perturbation. The study of the molecular basis of cancer has generated the promise of identifying new targets for more selective and molecularly focused anticancer therapies. This chapter describes several novel therapeutic agents or strategies that target critical regulators or effectors of apoptosis. These agents or strategies have the potential to exert selective cytotoxicity against cancer cells and are currently under investigation.

Caspases are the central engines of apoptosis. These are proteases that exist as inactive zymogens and are activated by proteolytic cleavage of their proforms in response to a variety of death stimuli (explained below). This processing occurs at conserved aspartic acid residues, thus generating the enzymatically active caspases *(1)*. Caspase activation is organized as a cascade, with upstream initiator and downstream effector caspases *(2)*. Upstream initiator caspases contain large prodomains that interact with specific proteins involved in triggering the cascade. The downstream caspases, which function as the ultimate effectors of apoptosis, possess small prodomains and are activated predominantly by proteolytic cleavage by upstream caspases. The irreversible cleavage of specific protein death substrates by the downstream effector caspases accounts directly or indirectly for the biochemical and morphological changes that are recognized as apoptosis.

At present at least three pathways of caspase activation leading to apoptosis have been elucidated: (a) receptor-initiated apoptosis pathway, in which the tumor necrosis factor (TNF) family of cytokine receptors activates initiator caspases such as caspase-8; (b) the

From: *Methods in Molecular Biology, Vol. 223: Tumor Suppressor Genes: Regulation, Function, and Medicinal Applications.* Edited by: Wafik S. El-Deiry © Humana Press Inc., Totowa, NJ

Table 1
Effectors of Apoptosis Pathways as Targets for Cancer Therapy

Receptor-mediated pathway	Mitochondrial pathway targets or agents	Common pathway
Targets and agents	Targets and agents	N-terminus SMAC peptide
At the death receptors	Anti Bcl-2 protein	Anti-IAP family of proteins
Apo-2L/TRAIL	Antisense oligos to Bcl-2	Antisense oligos to survivin
OPG-RANKL	BH3 peptides	
Anti-DR5 antibody	HA14-1	
At death-inducing		
signaling complex		
Anti-FLIP		
NFκB inhibitors: PSC-341		
Demethylating agents		
CDDO		

mitochondria-initiated apoptosis pathway, in which cytochrome (cyt) c is released from mitochondria into cytosol that results in activation of caspase-9; (c) a third pathway of caspase activation, which involves a serine protease, granzyme-B, that directly cleaves and activates several caspases including procaspase 3 (*see* **Table 1**).

1.1. Pathways to Apoptosis

1.1.1. Death Receptor-Initiated Apoptotic Signaling

Several TNF family receptors are known to transduce signals that result in apoptosis. These include TNFR1 (CD120a), CD95 (APO-1/Fas/DR2), DR3 (Apo-3/TRAMP/ LARD), DR4 (TRAIL-R1), and DR5 (TRAIL-R2). These receptors, also called death receptors, are characterized by the presence of a *death domain* (DD) within their cytoplasmic region, and have been shown to trigger apoptosis upon binding to their cognate ligands or specific agonist antibody *(3)*. The activating ligands for these death receptors are structurally related molecules that also belong to the TNF gene superfamily, such as Fas/CD95 ligand (FasL), TNF, and Apo2L/TRAIL.

Ligation of death receptors results in receptor trimerization, and formation of *death inducing signaling complex* (DISC) *(4)*. This is composed of Fas, FADD (*Fas-associating protein with death domain*), and procaspase-8, an apical signaling complex that mediates receptor-induced apoptosis *(4)*. FADD binds directly to FasR, DR4, or DR5, and indirectly to TNF-R1 via TRADD. FADD is essential for cell death signaling from all three receptors. FADD (MORT-1) *(5,6)* interacts through its C-terminal DD to cross-link CD95, DR4, or DR5 receptors, and recruits procaspase-8 (FLICE/MACH α1/ MACH 5) and procaspase-10, or TNF receptor-associated death domain protein (TRADD), through its N-terminal *death effector domain* (DED) to the DISC (**Fig. 1**) *(7,8)*. Oligomerization of caspase-8 within the DISC results in a high local concentration of the zymogen. The induced proximity under these crowded conditions generates low levels of intrinsic proteolytic activity of caspase-8, enough to allow the various proenzyme molecules to mutually cleave each other *(2)*. Processing of caspase-8 removes the DED-containing prodomain,

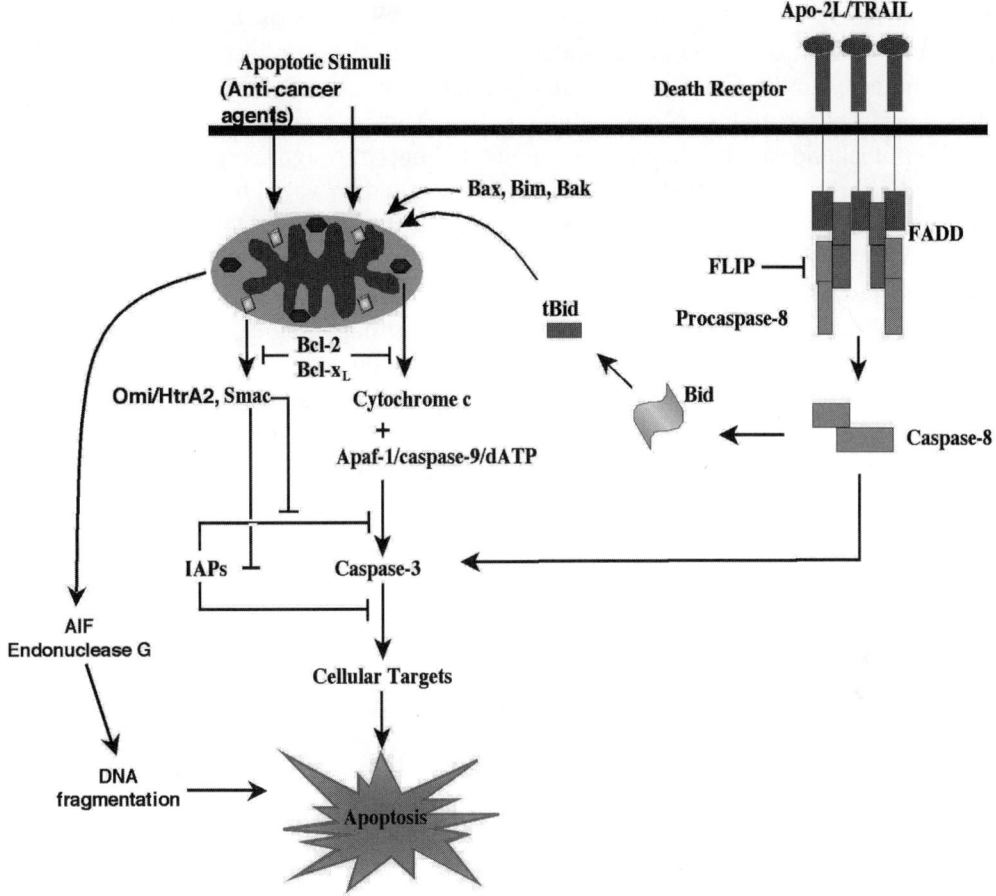

Fig. 1. Schematic representation of apoptotic pathways.

thus releasing the activated protease into the cytosol, where it can cleave and activate other downstream procaspases. In certain types of cells (type I), enough caspase 8 is activated by the activation of the death receptor to cause apoptosis *(9)*. In these cells, caspase inhibitor, but not the overexpression of Bcl-2, can block apoptosis. In other cell types, such as hepatocytes (type II), a high level of caspase-8 activation is not achieved and the apoptotic signal is amplified by the mitochondrial pathway. The active caspase-8 cleaves the cytosolic p22Bid into a BH3-only domain-containing, proapoptotic, truncated p15 tBid fragment, which translocates to the mitochondria and triggers the release of cyt c into the cytosol *(10,11)*. Recently, cleavage by caspase-8 has been shown to cause exposure of a glycine residue on p15Bid that is myristoylated, thereby targeting p15 Bid to the mitochondria *(12)*. Thus, N-myristoylation acts as an activating switch that enhances tBid-induced release of cyt c and apoptosis. The ability to cleave Bid may not be limited to caspase-8. Other caspases, such as caspase-3, as well another proteases, such as granzyme B and lysosomal proteases, have been shown to activate Bid *(10,13)*. This indicates that Bid probably serves to amplify the caspase cascade rather than to initiate it (**Fig. 1**).

The Fas ligation has also been shown to initiate an alternative pathway to apoptosis, involving activation of c-Jun N-terminal kinase/stress-activated protein kinase (JNK/SAPK) *(14,15)*. After ligation, Fas recruits an adaptor protein called Daxx, which interacts with the apoptosis signal-regulating kinase 1 (ASK1), activating the transcription factors AP-1 and ATF-2. Once activated, ASK-1 launches a phosphorylation cascade that culminates in the activation of c-Jun N terminal kinase (JNK) *(16)*. Activated JNK phosphorylates substrates such as c-Jun and p53. This, in turn, has been shown to trigger the mitochondrial or death receptor-initiated apoptotic signaling *(16)*.

RANK (for receptor activator of NFκB) is a newly discovered member of the TNFR family that bears high similarity in its extracellular domain to CD40. The intracellular domain of RANK does not show homology to any of the known TNFR family members. Although RANK is ubiquitously expressed in human tissue, RANK ligand, a type 2 trans-membrane protein, is expressed primarily on primary T-cells, and activated T- and B-cells in lymphoid tissues. RANKL is also expressed on osteoblast cell surface and has been shown to play a major role in osteoclast differentiation in cooperation with macrophage colony-stimulating factor (M-CSF) *(17)*. RANKL binds to RANK and initiates a cascade of signaling events that leads to osteoclast differentiation. This signaling activates c-Jun N-terminal kinase (JNK) and through the adapter molecules of the TRAF family, NFκB, PI3-kinase/AKT, and c-Src. *(18)* (**Fig. 2**). Inhibitors of RANKL-RANK interaction may trigger cell death by abrogating the NFκB-mediated survival signaling in leukemic cells, where the RANKL-RANK pathway is active. Inhibition of RANK-RANKL interaction may also block osteoclast differentiation and activity, thereby reducing its bone-destructive effects.

1.1.2. Mitochondria-Initiated Apoptotic Signaling

The mitochondrion is not only a cell's powerhouse but also its arsenal. Mitochondria sequester a potent cocktail of proapoptotic proteins. These proteins promote apoptosis either by activating caspases, by cyt c, or nucleases, such as endonuclease G or apoptosis-inducing factor (AIF), or by neutralizing cytosolic inhibitors of apoptosis (SMAC or Omi/Htr 2A) *(19–23)*. Upon induction of apoptosis, unlike other proapoptotic proteins, mitochondrial AIF translocates to the nucleus to induce DNA fragmentation and apoptosis *(21)*. Cyt c is a well-known component of the mitochondria electron transfer chain. Cyt c is released from the mitochondria during apoptosis *(19)*. After release into the cytosol, cyt c binds to Apaf-1 (apoptosis-activating factor), a cytosolic adapter protein that contains a caspase recruitment domain (CARD), a nucleotide-binding domain, and multiple WD-40 repeats *(24)*. Binding of cyt c to Apaf-1 increases its affinity for dATP or ATP by approximately 10-fold. It also triggers the oligomerization of Apaf-1 into a multimeric Apaf-1/cyt c complex, also called the apoptosome *(25)*. This exposes the CARD domain of Apaf-1, which recruits several molecules of procaspase-9, inducing their auto-activation *(25)*. Only the caspase-9 bound to the apoptosome is able to efficiently cleave and activate downstream executioner caspase, caspase-3 *(26)*.

Smac/DIABLO is a 25-kDa mitochondrial protein that is released from mitochondria into the cytosol during apoptosis *(22)*. Smac contains mitochondria-targeting sequence at its N-terminus. This sequence is removed upon import into the mitochondria, generating the mature Smac protein. In the mitochondrial and the common effector pathways of apoptosis, the processing and proteolytic activity of caspase-9 followed by caspases-3 and -7 are inhibited by the IAP (inhibitor of apoptosis) family of proteins *(27–29)*. IAP family members include XIAP, cIAP1, cIAP2, and survivin *(27–29)*. All IAPs contain at least one BIR domain, although some contain three *(30)*. Another region, the RING

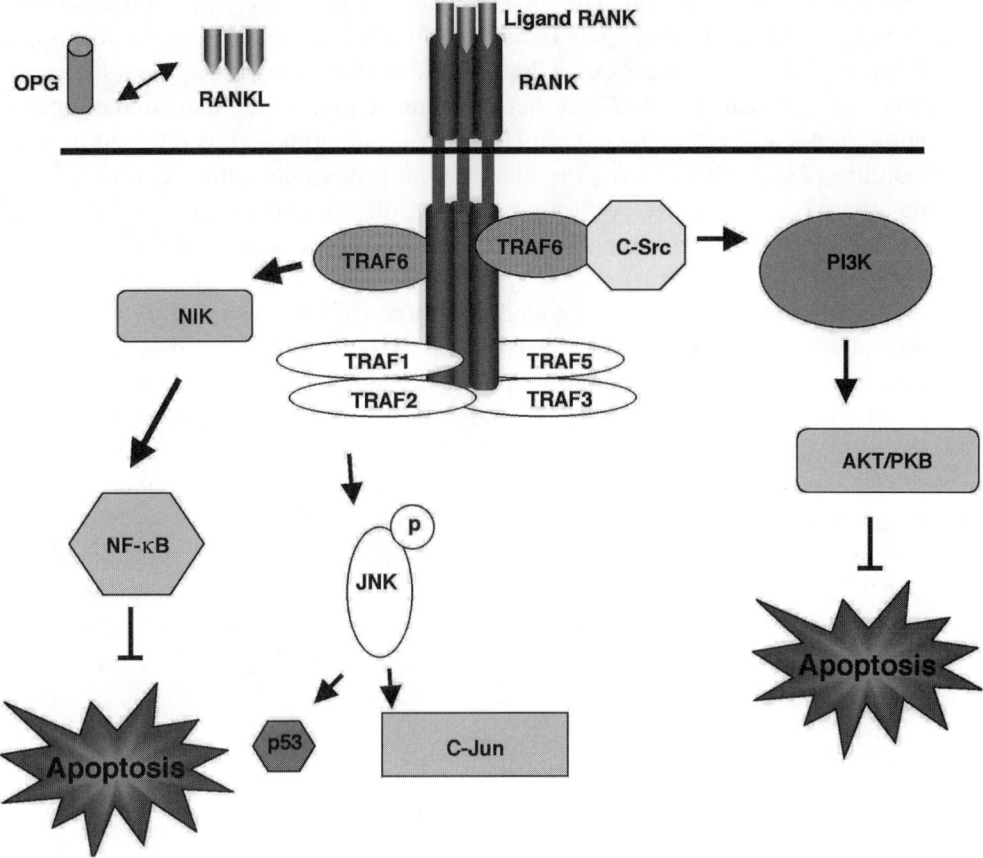

Fig. 2. Schematic representation of RANK and RANKL signaling pathway.

domain, has ubiquitin ligase activity and promotes the self-degradation of IAPs through the proteasomes in response to some apoptotic stimuli *(31)*. Furthermore, during Fas death receptor-mediated apoptosis, XIAP is cleaved by activated caspase-3 into the amino-terminal BIR1 and 2 and BIR3-RING finger fragments *(32)*. The latter specifically inhibits caspase-9 *(32,33)*. Recently, the crystal structures of caspase-3 or -7 complexed with XIAP has been elucidated *(34–36)*. This has revealed that the N-terminal linker peptide of the BIR2 domain of XIAP makes the most contact with the catalytic substrate-binding cleft of these caspases, and sterically hinders its substrate binding, while the BIR2 domain is required only to align and stabilize this inhibitory interaction between the linker peptide and the catalytic cleft *(34,35)*. Overexpression of XIAP inhibits anticancer drug-induced caspase activity and apoptosis *(37)*. In contrast, downregulation of XIAP sensitizes cancer cells to apoptosis induced by anticancer drugs *(38)*. The antiapoptotic activity of NFκB has also been shown to be mediated by the induction of IAPs *(39,40)*. The first four amino acids of the mature Smac, Ala-Val-Pro-Ile (AVPI), bind to the BIR (baculovirus repeat) domain of IAP (inhibitor of apoptosis protein) proteins *(41)*. The four amino acids (AVPI) of Smac that bind to the BIR3 domain of XIAP are similar to the XIAP-binding sequence of active caspase-9 *(41)*. In addition to the BIR 3 domain of XIAP, Smac has also been shown to form a stable complex with the BIR2 domain of XIAP *(34)*. The linker sequence immediately preceding the BIR2 domain is involved in XIAP-mediated binding and inhibition of caspase-3 and caspase-7 *(34)*. Smac binds to the linker-BIR2 domain and presumably disrupts its inhibition of the active caspase-3 and caspase-7 by steric hindrance *(34)*. Upon induction of apoptosis, another mitochondrial protein, Omi/HtrA2 (a serine protease), is released from the mitochondria *(23,42)*. Like Smac, the mature Omi protein contains a conserved IAP-binding motif (AVPS) at its N-terminus *(23)*. Omi complexes with XIAP at high stochiometry and promotes apoptosis by neutralizing the caspase inhibitory effect of XIAP *(23,42)*.

The Bcl-2 family of proteins is divided into three subfamilies *(43,44)*. Members of the subfamily, which includes Bcl-2, Bcl-x_L, and Mcl-1, inhibit apoptosis *(43,44)*. The Bax subfamily members that promote apoptosis, including Bax and Bak, share with Bcl-2 three of the four Bcl-2 homology domains, BH1 to BH3 *(43,45)*. A C-terminus hydrophobic tail is responsible for the localization of these proteins to the outer membranes of the mitochondria and endoplasmic reticulum. The third BH3-only subfamily of proteins, which includes Bid, Bim, and Bad, also promotes apoptosis by binding and inactivating pro-survival Bcl-2 family members *(46)*. However, recent studies have shown that BH3-only proteins require Bax and Bak to trigger mitochondrial apoptotic signaling *(47,48)*. For example, the exposed BH3 domain of tBid oligomerizes with Bak and Bax, causing their mitochondrial membrane insertion and cyt c release *(49,50)*.

2. Selective Antitumor Agents or Strategies

2.1. At Death Receptor

2.1.1. Recombinant Apo-2L/TRAIL

Although Apo-2L/TRAIL is a member of the TNF family, it has some striking differences from TNF and FasL. Unlike Fas, whose expression is limited to certain tissues Apo-2L/TRAIL receptors are widely expressed *(51,52)*. Therefore, most tissues and cell types may be Apo-2L/TRAIL targets. In a study by Ashkenazi et al., soluble Apo-2L/TRAIL was shown to induce apoptosis of a variety of leukemia and solid tumor cell types *(53)*. In contrast, normal tissues including prostate, colon, fibroblasts, and smooth muscle cells were unaffected by exposure to Apo-2L/TRAIL, while TNF or FasL induced fulminant massive liver damage *(54)*. When introduced systemically, Apo-2L/TRAIL exhibited no detectable cytotoxicity in mice *(55)* and monkeys *(53)*. Apo-2L/TRAIL has also been shown to induce apoptosis of cancer cells regardless of their p53 status, which is mutant or dysfunctional in about 50% of all human tumors. Since p53 has been shown to be a trans-activator of Apo-2L/TRAIL receptor (DR5) expression, the functional status of p53 may have a role in Apo-2L/TRAIL sensitivity in vivo *(56)*. Collectively, these findings suggest that homotrimeric Apo-2L/TRAIL may potentially be tumor-selective. However, human clinical studies of homotrimeric Apo-2L/TRAIL have not yet been launched and are eagerly awaited. The overexpression of inhibitory molecules such as cFLIP *(57)* or decoy receptors TRAIL-R3/TRID/DcR1 and TRAIL-R4/DcR2 *(53)* may confer resistance to Apo-2L/TRAIL. Preclinical studies suggest that augmenting Apo-2L/TRAIL induced DISC activity through diverse mechanisms (as detailed below) may enhance the Apo-2L/TRAIL-induced apoptosis. This may help to overcome in-vivo resistance of tumors to Apo2L/TRAIL.

2.1.2. Anti-DR5 Receptor

A recent report described a novel DR5 monoclonal antibody (TRA-8) that induces apoptosis in Apo-2L/TRAIL-sensitive tumor cells both in vitro and in vivo *(58)*. DR5 expression is seen high in tumor cells compared to normal tissues. TRA-8 did not trigger apoptosis of normal hepatocytes *(58)*. Therefore, a selective, specific targeting of DR5 with a specific agonistic antibody might also be an effective strategy for cancer therapy.

2.2. At DISC

2.2.1. c-FLIP Antagonists

An endogenous intracellular protein c-FLIP (also known as FLAME-1, CASH, CLARP, MRIT, I-FLICE, and Usurpin), which has N-terminus FADD homology and C-terminus caspase homology domains without caspase activity, has a dominant-negative effect against caspase-8 and caspase-10. Therefore, c-FLIP can inhibit the Apo-2L-induced DISC activity. This has been recently tested in preclinical studies in which downregulation of c-FLIP levels through inhibition of PI3/AKT kinases were shown to enhance the antitumor effect of Fas in endothelial cells *(59)*. PPARγ is a member of the steroid/retinoid superfamily of ligand-activated transcription factors. PPARγ agonists have been reported to suppress the growth of human colon and breast cancer cell lines in vitro and in vivo in the mouse *(60,61)*. PPARγ modulators have been shown to selectively reduce levels of c-FLIP through proteasomal degradation and enhance Apo-2L/TRAIL induced apoptosis *(62)*. This suggests that the neutralization of FLIP might be useful in the treatment of cancers that have acquired resistance to Fas or Apo-2L/TRAIL due to reduced assembly of DISC. This has been tested as a strategy against cholangiocarcinoma *(63)*. Several reports have indicated that treatment with chemotherapeutic agents or γ-irradiation increases DISC formation due to FasL or Apo-2L/TRAIL *(64,65)*. This may occur through either upregulation of FasL or Apo-2L/TRAIL or their receptors *(64,65)*. Consequently, pretreatment with a chemotherapeutic agent has been shown to increase Apo-2L/TRAIL-induced apoptosis of cancer and leukemia cells. This raises the possibility that downmodulation of c-FLIP may further potentiate antitumor interactions between chemotherapy and/or death ligands.

2.3. CDDO

Similar to downregulation of c-FLIP, other manipulations or agents that have the potential to activate the receptor-mediated pathway by activating caspase-8 may successfully overcome blocks to apoptosis due to reduced assembly of DISC. Activity of such compounds would be independent of death receptor expression or ligation. One such agent is 2-cyano-3,12-dixolean-1,9-dien-28-oic acid (CDDO), which is a potent analog of a naturally occurring triterpenoid. CDDO induces apoptosis by activating caspase-8 *(66)*. The mechanism by which CDDO causes activation of caspase-8 is unclear. Preclinical studies have shown promising differentiation activity in cellular models of human myeloid leukemia cells and in osteosarcoma and neuroblastoma models *(67)*. The mechanisms underlying the differentiating effects of CDDO are also poorly understood. CDDO has also been shown to inhibit the proliferation of several of human tumor cell types. Recently, CDDO was shown to induce apoptosis of human myeloid leukemia cells through a caspase-8-dependent mechanism *(66)*. Overexpression of the caspase-8 inhibitor CrmA blocked CDDO-induced cyt c release and apoptosis of human myeloid leukemia cells. Overexpression of the antiapoptotic Bcl-x_L protein blocked CDDO-induced cyt c release, but only partly inhibited caspase-3 activation and apoptosis. These findings suggest that CDDO activates caspase-8, which leads directly to the activation of caspase-3 by a cyt c-independent mechanism, as well as induces cyt c release by caspase-8-dependent cleavage of Bid *(66)*. This suggests that CDDO may be useful as a therapeutic agent either alone or in combination with Fas L or Apo-2L/TRAIL, in which Fas L- or Apo-2L/TRAIL-induced apoptosis is blocked because of lack of receptor expression or ligation.

2.4. Demethylating Agents

In addition to genetic alterations, epigenetic modifications such as DNA methylation have been shown to downregulate the molecular modulators of apoptosis, such as caspase-8 and Apaf-1, in cancer cells *(68,69)*. Loss of caspase-8 expression, by DNA hypermethylation of the CpG islands in the promoter region of caspase-8, has been shown to cause resistance against Apo-2L/TRAIL-induced apoptosis of tumor cells *(70–72)*. This was shown in neuroblastoma cells with N-Myc amplification and in peripheral neuroectodermal tumor cell types *(70–72)*. Treatment with the demethylating agent 5'-aza-2'-deoxycytdine reversed hypermethylation of caspase-8 DNA, resulting in a time- and concentration-dependent reexpression of caspase-8 mRNA and protein. This resensitized the cells to Apo-2L/TRAIL-induced apoptosis. This suggests that cotreatment with a DNA-demethylating agent may be a useful strategy to augment epigenetically repressed effectors of apoptosis pathways, such as caspase-8 or Apaf-1, to enhance the antitumor effects of death ligands and/or chemotherapeutic agents.

2.5. OPG or RANKL

The soluble receptor osteoprotegrin (OPG) is capable of sequestering RANKL and inhibits its ability to bind to cell surface RANK. Systemic administration of OPG, or an inhibitor that specifically blocks the interaction of RANK/RANKL, could inhibit RANK mediated antiapoptotic signaling through NFκB. In metastatic tumor models, in which the tumor causes increased osteoclastogenesis and bone destruction, systemic administration of OPG was shown to reduce tumor-mediated bone destruction and bone pain *(73)*. Bone marrow from patients with multiple myeloma or adult T-cell leukemia (ATL) cells from patients with hypercalcemia show high expression of RANKL *(74)*. This has been shown to induce in-vitro differentiation of human hematopoietic precursor cells into osteoclasts in the presence of M-CSF by inhibiting RANKL/RANK interaction and resulting osteoclast differentiation *(75)*. OPG has the potential to inhibit osteoclastic bone destruction in tumor metastasis, thus improving bone mass, as was shown recently in a hypercalcemic nude mouse carrying tumors associated with humoral hypercalcemia of malignancy *(76)*. Therefore, inhibitors of RANKL-RANK interaction may have two salutary effects. They may trigger cell death of tumor cells by abrogating the NFκB-mediated survival signaling, as well as reduce bone destruction and improve bone mass in metastatic cancers to the bone or in multiple myeloma, where the RANKL-RANK pathway is active.

2.6. Bcl-2 Family of Proteins as Targets for Anticancer Drug Design

Several studies have demonstrated that overexpression of Bcl-2 or Bcl-x_L confers resistance against apoptosis induced by chemotherapeutic drugs or γ-irradiation *(77–80)*. Other studies have shown that a temporary inhibition of Bcl-2 is not toxic to normal cells *(81)*. Also, inhibition of Bcl-2 in cancer cells does not induce a compensatory upregulation of the other antiapoptotic Bcl-2 family of proteins but instead sensitizes cells to apoptotic stimuli. Collectively, these observations indicate that downregulation of Bcl-2, directly or indirectly, would lower the threshold for apoptosis due to anticancer agents. This would be a useful therapeutic manipulation, particularly against tumor cell types in which the threshold has been set higher due to increased intracellular levels or antiapoptotic activity of Bcl-2 or Bcl-x_L.

2.6.1. Antisense to Bcl-2

One strategy to inhibit Bcl-2 expression and hence its antiapoptotic activity is by antisense deoxyoligonucleotide, G-3139 (Genta). Liposomal antisense Bcl-2 oligonucleotide (Bcl-2-ASODN) has been shown to cause apoptosis in many types of cancer cells, including lung *(82)*, malignant glioma *(83)*, bladder *(84)*, renal cell carcinoma *(85)*, prostate cancer *(86)*, and chronic lymphocytic leukemia cells *(87)*. Bcl-2-ASODN also sensitized tumor cells to γ-irradiation and chemotherapeutic drugs such as Ara-C, cyclophosphamide or doxorubicin *(88–90)*. Bcl-2-ASODN was also shown to be effective in chemosensitizing several types of mouse xenograft models of human prostate cancer *(91,92)*, transitional cell cancer *(93)*, and melanoma to different anticancer agents *(94)*. In one clinical Phase I/II study, 21 patients with Bcl-2-positive relapsed non-Hodgkin's lymphoma (NHL) received a 14-d subcutaneous infusion of G-3139 (Genasense). Eight cohorts of patients received doses between 4.6 and 195.8 mg/m^2/d. No significant systemic toxicity was seen at doses up to 110.4 mg/m^2/d *(95)*. At this dose, the half-life of Bcl-2-ASODN was 7.5 h and the maximally tolerated dose was 147.2 mg/m^2/d. Dose-limiting toxicities were thrombocytopenia and hypotension. Plasma levels of >4 μg/mL were observed associated with these dose-limiting toxicities. Clinical responses included one complete remission and two minimal responses *(96)*. More recently, a combination of a chemotherapeutic agent with Bcl-2-ASODN has been used for the treatment of melanoma or prostate cancer with better clinical outcome. In the melanoma clinical trial, 14 patients with advanced melanoma were treated with Bcl-2-ASODN at escalating doses of 0.6–6.5 mg/kg in combination with standard decarbozine therapy. Toxicities were tolerable, and 6 patients showed antitumor responses (1 complete, 2 partial, and 3 minor). Compared to the baseline, a maximal reduction of 40% was observed in Bcl-2 protein levels concomitantly with increased rate of apoptosis in the melanoma cells *(97,98)*. With the goal of simultaneously targeting both Bcl-2 and Bcl-x_L with a single compound, a bispecific antisense ODN has been designed. A report by Uwe et al. described a bispecific antisense oligonucleotide inhibiting both Bcl-2 and Bcl-x_L expression *(99)*. The authors designed a 2′-O-methoxy-ethoxy-modified phosphorothioate antisense oligonucleotide targeting a region of high homology shared between the Bcl-2 and Bcl-x_L mRNAs. Three bispecific ODNs were used against small-cell and non-small-cell lung carcinoma cell types. One of these three compounds, 4625, efficiently downregulated Bcl-2 and Bcl-x_L mRNA and protein levels and induced apoptosis of both types of human lung cancer cells *(99)*. In vivo, the bispecific ODN 4625 inhibited growth and induced tumor cell apoptosis of breast and colorectal carcinoma xenografts.

2.6.2. BH3 Peptide

Another strategy to create Bcl-2 inhibitors has focussed on developing small molecules that mimic the action of the endogenous Bcl-2-binding death agonists. Compounds that mimic the BH3-only class of death agonists, such as BAD, inhibit the survival proteins Bcl-2 and Bcl- x_L, but do not appear to have independent proapoptotic activity. Holinger et al. *(100)* created a cell-permeable version of BH3 Bak by fusing it to the internalization domain of the Antennapedia (Ant) protein as a transporter. Fusion Ant-BH3 peptide inhibited Bcl-2 and induced cell death and apoptosis in HeLa cells *(100)*. Recently, Wang et al. *(101)* designed a cell permeable Bcl-2-binding peptide by chemically attaching the N-terminus of synthetic BH3 domain of BAD with the decanoic fatty acid CPM ($CH_3[CH_2]_8CO$), which can shuttle peptides across cell membranes without untoward toxicity. One such peptide, termed CPM-1285, has been shown to enter human myeloid leukemia HL-60 cells and induced the activation of caspase-3 and triggered apoptosis *(101)*. CPM-1285 was also shown to slow human myeloid leukemia growth in severe combined immunodeficient (SCID) mice *(101)*. When administered intraperitoneally into SCID mice, this peptide suppressed the growth of tumor xenografts without untoward toxicity *(101)*. Degterev et al. identified two classes of novel small-molecule cell-permeable inhibitors of the Bcl-x_L-BH3 domain (BH3I). Their studies demonstrated that BH3Is induce apoptosis by preventing BH3 domain-mediated interaction between proapoptotic and antiapoptotic members of the Bcl-2 family *(102)*.

Two natural products have been suggested to antagonize the antiapoptotic function of Bcl-2 or Bcl-x_L. Nakashima et al. reported that Tetrocarcin A inhibited mitochondrial functions of Bcl-2 and suppressed its antiapoptotic activity *(103)*. In another report, Antimycin-A was shown to mimic activity of BH3 peptides and selectively induce apoptosis in cell lines overexpressing Bcl-x_L *(104)*.

In contrast to the random screening of natural products, another approach to drug discovery is computer-aided design of specific ligands, based on the considerations of the high-resolution three-dimensional structure of the target receptor. Wang et al. modeled the binding of Bak's BH3 domain to BH1, BH2, BH3 pockets of Bcl-2 *(105)*. Then, using a computer simulation of the binding, they screened a virtual library of 193,833 compounds for their potential ability to bind Bcl-2. From these, 28 compounds were selected for physical testing. Of these 28, one synthetic compound, HA14-1, a small molecule and nonpeptidic ligand of Bcl-2 surface pocket, effectively induced apoptosis of cells overexpressing Bcl-2 protein. This was associated with a decrease in mitochondrial membrane potential and activation of caspase-9 followed by caspase-3. The discovery of this cell-permeable molecule provides a chemical probe to study Bcl-2 regulated apoptotic pathways in vitro and in vivo. Several groups have now embarked upon the drug discovery mission to identify and study the activity of related anti-Bcl-2 small-molecule inhibitors.

2.7. Antisense to Survivin

Survivin is a 16-kDa protein that contains a single BIR domain, followed by a C-terminal, long α-helical region that is required for its binding to the microtubules in the mitotic spindle *(29,30)*. Survivin is expressed in cancer but not in normal terminally differentiated adult tissue *(106)*. Survivin is phosphorylated on Thr34 by CDK1-cyclin B1, and survivin levels and activity increase during G_2/M phase in a cell cycle-dependent manner *(29,107,108)*. Overexpression of survivin inhibits caspase activity and apoptosis induced by Fas, Bax, and anticancer drugs *(109)*. Abrogation of the expression/function of survivin causes increased caspase-3 activity at the G_2/M phase of the cell cycle and apoptosis. Inhibition of survivin also results in the dysregulation of mitotic progression, with supernumerary centrosomes, aberrant mitotic spindles and polyploidy *(110–112)*. Survivin expression has also been correlated with poor clinical outcome in acute myelogenous leukemia (AML) and other cancers *(113)*. It has been reported that the antisense to survivin causes spontaneous apoptosis in lung cancer cells, malignant melanomas, and other cancer cell types *(111)*. Therefore, it is possible that the antisense oligonucleotides directed against survivin may induce selective apoptosis of cancer cells.

2.8. Neutralization of IAPs and Potentiation of Caspase-9 and -3 Activities by Smac and Omi/Htr2A

Smac relieves inhibition of caspase-9 by XIAP, by disrupting the interaction between the BIR3 domain of XIAP and the N-terminus small subunit of caspase-9 *(114,115)*. The latter residues share homology with the N-terminal tetrapeptide in Smac, which can also bind to BIR3. Thus, binding to BIR3 by two conserved peptides (one from Smac and one from caspase-9) is mutually exclusive and has an opposing effect on caspase-9 activity and apoptosis *(114,115)*. Smac also promotes the proteolytic activation of procaspase-3 and the enzymatic activity of mature caspase-3 *(116)*. A seven-residue peptide derived from the amino terminus of Smac promotes procaspase-3 activation *(116)*. Recently, consistent with this, during dexamethasone-induced apoptosis of multiple myeloma cells, the cytosolic Smac was shown to neutralize the inhibitory effect of XIAP on caspase-9 *(117)*. Indeed, recent studies in human acute leukemia cells show that the ectopic overexpression of the full-length Smac or treatment with N-terminus Smac tetrapeptide enhances Apo-2L/TRAIL-induced effector caspase activity and apoptosis *(117)*. A recent report by Deng et al. demonstrated that Bax-dependent release of Smac, not cyt c from mitochondria, mediates the mitochondrial involvement in Apo-2L/TRAIL-induced apoptosis *(118)*. These findings suggest that the cytosolic Smac or its N-terminus peptide may bypass Bcl-2 or Bcl-x_L inhibition and promote caspase-3 activation by Apo-2L/TRAIL.

During apoptosis, Omi, a recently identified mitochondrial protein, is also released into the cytosol along with cyt c and Smac. Like Smac, Omi (through its N-terminal AVPS motif) binds and neutralizes XIAP, thereby potentiating the caspase-dependent pathway of apoptosis. More importantly, Omi has also been shown to induce cell death via its serine protease activity, independent of caspases, Apaf-1, or IAPs *(23)*. Hence, treatment with Omi peptide may be effective against tumors in which Apaf-1 is not expressed because of promoter-based hypermethylation and gene silencing.

2.8.1. NFκB Inhibition

NFκB protects cells from apoptosis by inducing the expression of survival factors, including the IAP family (c-IAP1, c-IAP2, XIAP) *(39,40,119)* and the Bcl-2 homologs, Bfl-1/A1 and Bcl-x_L *(120,121)*. Proteasome inhibitors arrest or retard cancer progression by interfering with ordered, temporal degradation of cell-cycle regulatory proteins as well as the degradation of the inhibitory protein IκB, which is required for activation of NFκB *(122)*. Inhibition of proteasome-mediated IκB degradation may limit metastasis via the attenuation of NFκB-dependent cell adhesion molecule expression, making the dividing cancer cells more sensitive to apoptosis *(123–125)*. By blocking the activation of IκB kinase complex, inhibition of NFκB has been shown to sensitize tumor cells to chemotherapeutic agents and Apo-2L/TRAIL-induced apoptosis *(126)*. Several naturally occurring and synthetic proteasome inhibitors have been discovered, including the boronic acid peptides such as PS-341; the natural product lactacystin, a streptomyces metabolite; MG0132; and a synthetic peptide aldehyde. PS-341 is the compound being most widely studied in vitro and in vivo. This dipeptidyl boronic acid is a specific and selective inhibitor of 26S proteasome *(127)*. In-vitro studies have demonstrated marked synergy between PS-341 and CPT-11 (topoisomerase I inhibitor irinotecan) *(128)*. In addition to inhibiting NFκB, PS-341 was also shown to overcome the protective effect of Bcl-2 and induces apoptosis of Jurkat cells *(125)*. PS-341 is currently in clinical trials for a variety of solid tumors, leukemias, and lymphomas. In a Phase I trial in patients with hematologic malignancies, PS-341 was administered as a twice-weekly bolus injection for 4 consecutive weeks followed by a 2-wk rest period *(129)*. The maximum pharmacodynamic effect of PS-341 on 26S proteasome activity in the peripheral blood was observed at 1 h postdose *(130)*. The compound has been well tolerated at the levels of 26S inhibition of >60% that are considered essential for its antitumor efficacy. Toxicities encountered to date including dose-limiting neuropathy as well as fatigue, gastrointestinal symptoms, and thrombocytopenia. Results of clinical trials demonstrating the efficacy of PS-341 against a variety of tumors are eagerly awaited.

3. Conclusions

It is now evident that there are several strategies that can be devised to target various effector molecules in the death receptor, mitochondria, and common pathways of apoptosis. These targeted therapies are being tested in the preclinical and clinical Phase I and II studies against specific tumor types. The challenge is to identify those targets that mediate death signals in the cancer but not in host cells. These targets may be different for different cancer cell types. In addition, strategies to target effectors of apoptosis pathway may also be combined with, and complement, the conventional chemotherapeutic agents or ′-irradiation in enhancing apoptosis of cancer cells. c-DNA and tissue microarray techniques, coupled with proteomic analyses, are now allowing us to characterize the molecular phenotype of the various cancer cell types, thereby helping to identify targets in the apoptosis pathways. The emerging concept of specific therapeutic targets in specific cancer types needs to be validated as an effective therapeutic approach in the new era of postgenomic cancer research and therapy. Undeniably, the initial experience with this emerging anticancer strategy appears to be quite promising.

4. References

1. Salvesen, G. S. and Dixit, V. M. (1997) Caspases: intracellular signaling by proteolysis. *Cell* **91**(4), 443–446.
2. Salvesen, G. S. and Dixit, V. M. (1999) Caspase activation: the induced-proximity model. *Proc. Natl. Acad. Sci. USA* **96**(20), 10964–10967.

3. Ashkenazi, A. and Dixit, V. (1998) Death receptors: signaling and modulation. *Science* **281**, 1305–1308.

4. Kischkel, F. C., Hellbardt, S., Behrmann, I., et al. (1995) Cytotoxicity-dependent APO-1 (Fas/CD95)-associated proteins form a death-inducing signaling complex (DISC) with the receptor. *EMBO J.* **14**(22), 5579–5588.

5. Boldin, M. P., Varfolomeev, E. E., Pancer, Z., Mett, I. L., Camonis, J. H., and Wallach, D. (1995) A novel protein that interacts with the death domain of Fas/APO1 contains a sequence motif related to the death domain. *J. Biol. Chem.* **270**(14), 7795–7798.

6. Chinnaiyan, A. M., O'Rourke, K., Yu, G. L., et al. (1996) Signal transduction by DR3, a death domain-containing receptor related to TNFR-1 and CD95. *Science* **274**(5289), 990–992.

7. Boldin, M. P., Goncharov, T. M., Goltsev, Y. V., and Wallach, D. (1996) Involvement of MACH, a novel MORT1/FADD-interacting protease, in Fas/APO-1- and TNF receptor-induced cell death. *Cell* **85**(6), 803–815.

8. Muzio, M., Chinnaiyan, A. M., Kischkel, F. C., et al. (1996) FLICE, a novel FADD-homologous ICE/CED-3-like protease, is recruited to the CD95 (Fas/APO-1) death-inducing signaling complex. *Cell* **85**(6), 817–827.

9. Scaffidi, C., Fulda, S., Srinivasan, A., et al. (1998) Two CD95 (APO-1/Fas) signaling pathways. *EMBO J.* **17**(6), 1675–1687.

10. Lou, X., Budihardjo, I., Zou, H., Slaughter, C., and Wang, X. (1998) Bid, a Bcl2 interacting protein, mediates cytochrome c release from mitochondria in response to activation of cell surface death receptors. *Cell* **94**, 481–490.

11. Gross, A., Yin, X., Wang, K., et al. (1999) Caspase cleaved BID targets mitochondria and is required for cytochrome c release, while Bcl-xL prevents this release but not tumor necrosis factor-R1/Fas death. *J. Biol. Chem.* **274**, 1156–1163.

12. Zha, J., Weiler, S., Oh, K., Wei, M., and Korsmeyer, S. (2000) Posttranslational N-myristoylation of Bid as a molecular switch for targeting mitochondria and apoptosis. *Science* **290**, 1761–1765.

13. Walczak, H., Miller, R. E., Ariail, K., et al. (1999) Tumoricidal activity of tumor necrosis factor-related apoptosis-inducing ligand in vivo. *Nat. Med.* **5**, 157–163.

14. Yang, X., Khosravi-Far, R., Chang, H. Y., and Baltimore, D. (1997) Daxx, a novel Fas-binding protein that activates JNK and apoptosis. *Cell* **89**, 1067–1076.

15. Ko, Y.-G., Kim, E.-K., Kim, T., et al. (2001) Glutamine-dependent antiapoptotic interaction of human glutaminyl-trna synthetase with apoptosis signal-regulating kinase 1 *J. Biol. Chem.* **276**, 6030–6036.

16. Chang, H. Y., Yang, X., and Baltimore, D. (1999) Dissecting Fas signaling with an altered-specificity death-domain mutant: requirement of FADD binding for apoptosis but not Jun N-terminal kinase activation. *Proc. Natl. Acad. Sci. USA* **96**, 1252–1256.

17. Teitelbaum, S.L. (2000) Osteoclasts, integrins, and osteoporosis. *J. Bone Miner. Metab.* **18**(6), 344–349.

18. Darnay, B. G. and Aggarwal, B. B. (1999) Signal transduction by tumour necrosis factor and tumour necrosis factor related ligands and their receptors. *Ann. Rheum. Dis.* **58** (Suppl. 1), I2–I13.

19. Liu, X., Kim, C. N., Yang, J., Jemmerson, R., and Wang, X. (1996) Induction of apoptotic program in cell-free extracts: Requirement for dATP and cytochrome c. *Cell* **86**, 147–157.

20. Li, L. Y., Luo, X., and Wang, X. (2001) Endonuclease G is an apoptotic DNase when released from mitochondria. *Nature* **412** (6842), 95–99.

21. Susin, S. A., Lorenzo, H. K., Zamzami, N., et al. (1999) Molecular characterization of mitochondrial apoptosis-inducing factor. *Nature* **397**(6718), 441–446.

22. Du, C., Fang, M., Li, Y., Li, L., and Wang, X. (2000) Smac, a mitochondrial protein that promotes cytochrome c-dependent caspase activation by eliminating IAP inhibition. *Cell* **102**, 33–42.

23. Hegde, R., Srinivasula, S. M., Zhang, Z., et al. (2002) Identification of Omi/HtrA2 as a mitochondrial apoptotic serine protease that disrupts inhibitor of apoptosis protein-caspase interaction. *J. Biol. Chem.* **277**(1), 432–438.

24. Zou, H., Henzel, W. J., Liu, X., Lutschg, A., and Wang, X. (1997) Apaf-1, a human protein homologous to *C. elegans* CED-4, participates in cytochrome c-dependent activation of caspase-3. *Cell* **90**(3), 405–413.

25. Zou, H., Li, Y., Liu, X., and Wang, X. (1999) An APAF-1 cytochrome c multimeric complex is a functional apoptosome that activates procaspase-9. *J. Biol. Chem.* **274**, 11549–11556.

26. Rodriguez, J. and Lazebnik, Y. (1999) Caspase-9 and APAF-1 form an active holoenzyme. *Genes Dev.* **13**, 3179–3184.

27. Deveraux, Q., Roy, N., Stennicke, H. R., et al. (1998) IAPs block apoptotic events induced by caspase-8 and cytochrome c by direct inhibition of distinct caspases. *EMBO J.* **17**, 2215–2223.

28. Deveraux, Q. L., Takahashi, R., Salvesen, G. S., and Reed, J. C. (1997) X-linked IAP is a direct inhibitor of cell-death proteases *Nature* **338**, 300–304.

29. Li, F., Ambrosini, G., Chu, E., et al. (1998) Control of apoptosis and mitotic spindle checkpoint by survivin. *Nature* **396**, 580–584.

30. Reed, F. and Bischoff, J. (2000) BIRinging chromosomes through cell division—and survivin' the experience. *Cell* **102**, 545–548.

31. Yang, Y., Fang, S., Jensen, J., Weissman, A., and Ashwell, J. (2000) Ubiquitin protein ligase activity of IAPs and their degradation in proteasomes in response to apoptotic stimuli. *Science* **288**, 874–877.

32. Deveraux, Q., Leo, E., Stennicke, H., Welsh, K., Salvesen, G., and Reed, J. (1999) Cleavage of human inhibitor of apoptosis protein XIAP results in fragments with distinct specificities for caspases. *EMBO J.* **18**, 5242–5251.

33. Bratton, S., Walker, G., Srinivasula, S., et al. (2001) Recruitment, activation and retention of caspase -9 and -3 by Apaf-1 apoptosome and associated XIAP complexes. *EMBO J.* **20**, 998–1009.

34. Chai, J., Shiozaki, E., Srinivasula, S., et al. (2001) Structural basis of caspase -7 inhibition by XIAP. *Cell* **104**, 769–780.

35. Riedl, S., Renatus, M., Schwarzenbacher, R., et al. (2001) Structural basis for the inhibition of caspase-3 by XIAP. *Cell* **104**, 791–800.

36. Huang, Y., Park, Y., Rich, R., Segal, D., Myszka, D., and Wu, H. (2001) Structural basis of caspase inhibition by XIAP: differential roles of the Linker versus the BIR domain. *Cell* **104**, 781–790.

37. Datta, R., Oki, E., Endo, K., Biedermann, V., Ren, J., and Kufe, D. (2000) XIAP regulates DNA damage-induced apoptosis downstream of caspase-9 cleavage. *J. Biol. Chem.* **275**, 31733–31738.

38. Sasaki, H., Sheng, Y., Kotsuji, F., and Tsang, B. (2000) Down-regulation of X-linked inhibitor of apoptosis protein induces apoptosis in chemoresistant human ovarian cancer cells. *Cancer Res.* **60**, 5659–5666.

39. Wang, C. Y., Mayo, M. W., Korneluk, R. G., Goeddel, D. V., and Baldwin, A. S., Jr. (1998) NF-kappaB antiapoptosis: induction of TRAF1 and TRAF2 and c-IAP1 and cIAP2 to suppress caspase-8 activation. *Science* **281**, 1680–1683.

40. Stehlik, C., Martin, R. D., Kumabashiri, M., Binder, B. R., and Lipp, J. (1998) NFκB regulated X-chromosome-linked IAP gene expression protects endothelial cells fron TNFα induced apoptosis. *J. Exp. Med.* **188**, 211–216.

41. Srinivasula, S. M., Hegde, R., Saleh, A., et al. (2001) A conserved XIAP-interaction motif in caspase-9 and Smac/DIABLO regulates caspase activity and apoptosis. *Nature* **410**, 112–116.

42. Verhagen, A. M., Silke, J., Ekert, P. G., et al. (2002) HtrA2 promotes cell death through its serine protease activity and its ability to antagonize inhibitor of apoptosis proteins. *J. Biol. Chem.* **277**(1), 445–454.

43. Gross, A., McDonnell, J., and Korsmeyer, S. (1999) Bcl-2 family members and the mitochondria in apoptosis. *Genes Dev.* **13**, 1899–1911.

44. Adams, J. M. and Cory, S. (1998) The Bcl-2 protein family, arbiters of cell survival. *Science* **281**, 1322–1326.

45. Hengartner, M. (2000) The biochemistry of apoptosis. *Nature* **407**, 770–775.

46. Huang, D. and Strasser, A. (2000) BH3-only proteins—essential initiators of apoptotic cell death. *Cell* **103**, 839–842.

47. Zong, W. X., Lindsten, T., Ross, A., MacGregor, G., and Thompson, C. (2001) BH3-only proteins that bind pro-survival Bcl-2 family members fail to induce apoptosis in the absence of Bax and Bak. *Genes Dev.* **15**, 1481–1486.

48. Wei, M., Zong, W. X., Cheng, E., et al. (2001) Proapoptotic BAX and BAK, A requisite gateway to mitochondrial dysfunction and death. *Science* **292**, 727–730.

49. Eskes, R., Desagher, S., Antonsson, B., and Martinou, J. C. (2000) Bid induces the oligomerization and insertion of Bax into the outer mitochondrial membrane. *Mol. Cell. Biol.* **20**, 929–935.

50. Wei, M., Lindsten, T., Mootha, V., et al. (2000) tBID, a membrane-targeted death ligand, oligomerizes BAK to release cytochrome c. *Genes Dev.* **14**, 2060–2071.

51. Wiley, S. R., Schooley, K., Smolak, P. J., et al. (1995) Identification and characterization of a new member of the TNF family that induces apoptosis. *J. Immun.* **3**, 673–682.

52. Pitti, R. M., Marsters, S. A., Ruppert, S., Donahue, C. F., Moore, A., and Ashkenazi, A. (1996) Induction of apoptosis by Apo-2L ligand, a new member of the tumor necrosis factor cytokine family. *J. Biol. Chem.* **271**, 12687–12690.

53. Ashkenazi, A., Pai, R. C., Fong, S., et al. (1999) Safety and antitumor activity of recombinant soluble Apo2 ligand. *J. Clin. Invest.* **104**, 155–162.

54. Ogasawara, J., Watanabe-Fukunaga, R., Adachi, M., et al. (1993) Lethal effect of the anti-Fas antibody in mice. *Nature* **364**, 806–809.

55. Walczak, H., Miller, R. E., Ariail, K., et al. (1999) Tumoricidal activity of tumor necrosis factor-related apoptosis-inducing ligand in vivo. *Nat. Med.* **5**, 157–163.

56. Takimoto, R. and El-Diery, W. S. (2000) Wild-type p53 transactivates the KILLER/DR5 gene through an intronic sequence-specific DNA-binding site. *Oncogene* **19**(14), 1735–1743.

57. Irmler, M., Thome, M., Hahne, M., et al. (1997) Inhibition of death receptor signals by cellular FLIP. *Nature* **388**, 190–195.

58. Ichikawa, K., Liu, W., Zhao, L., et al. (2001) Tumoricidal activity of a novel anti-human DR5 monoclonal antibody without hepatocyte cytotoxicity. *Nat. Med.* **7**(8), 954–960.

59. Suhara, T., Mano T., Oliveira, B., and Walsh, K. (2001) Phosphatidylinositol 3-kinase/Akt signaling controls endothelial cell sensitivity to Fas-mediated apoptosis via regulation of FLICE-inhibitory protein (FLIP) *Circ. Res.* **89**, 13–19.

60. Sarraf, P., Mueller, E., Jones, D., et al. (1998) Differentiation and reversal of malignant changes in colon cancer through PPARγ. *Nat. Med.* **4**, 1046–1052.

61. Elstner, E., Mueller, C., Koshizuka, K., et al. (1998) Ligands for peroxisome proliferator-activated receptor´ and retinoid acid receptor inhibit growth and induce apoptosis of human breast cancer cells in vitro and in BNX mice. *Proc. Natl. Acad. Sci. USA* **95**, 8806–8811.

62. Kim, Y., Suh, N., Sporn, M., and Reed, J. (2002) Pharmacological reduction in FLIP protein sensitizes tumor cells to TRAIL-induced apoptosis. *Cancer Res.*, in press.

63. Que, F. G., Phan, V. A., Phan, V. H., et al. (1999) Cholangiocarcinomas express Fas ligand and disable the Fas receptor. *Hepatology* **30**, 1398–1404.

64. Nagane, M., Pan, G., Weddle, J., Dixit, V., Cavenee, W., and Huang, H-S. (2000) Increased death receptor 5 expression by chemotherapeutic agents in human gliomas causes synergistic cytotoxicity with tumor necrosis factor-related apoptosis-inducing ligand in vitro and in vivo. *Cancer Res.* **60**, 847–853.

65. Wen, J., Ramadevi, N., Nguyen, D., Perkins, C., Worthington, E., and Bhalla, K. (2000) Antileukemic drugs increase death receptor 5 levels and enhance Apo-2L-induced apoptosis of human acute leukemia cells. *Blood* **96**, 3900–3906.

66. Ito, Y., Pramod, P., Sporn, M. B., Datta, R., Kharbanda, S., and Kufe, D. (2001) The novel triterpenoid CDDO induces apoptosis and differentiation of human osteosarcoma cells by a caspase-8 dependent mechanism. *Mol. Pharmocol.* **595**, 1094–1099.

67. Suh, N., Wang, Y., Honda, T., et al. (1999) A novel synthetic oleanane triterpenoid, 2-cyano-3,12-dioxoolean-1,9-dien-28-oic acid, with potent differentiating, antiproliferative, and anti-inflammatory activity. *Cancer Res.* **59**(2), 336–341.

68. Tycko, B. (2000) Epigenetic gene silencing in cancer. *J. Clin. Invest.* **105,** 401–407.

69. Herman, J. G. and Baylin, S. B. (2000) Promoter-region hypermethylation and gene silencing in human cancer. *Curr. Top. Microbiol. Immunol.* **249**, 35–54.

70. Hopkins-Donaldson, S., Bodmer, J. L., Bourloud, K., Brognara, C., Tschopp, J., and Gross, N. (2000) Loss of caspase-8 expression in highly malignant human neuroblastoma cells correlates with resistance to tumor necrosis factor-related apoptosis inducing ligand-induced apoptosis. *Cancer Res.* **60**, 4315–4319.

71. Teitz, T., Wei, T., Valentine, M., et al. (2000) Caspase-8 is deleted or silenced preferentially in childhood neuroblastomas with amplification of MYCN. *Nat. Med.* **6**, 529–535.

72. Grotzer, M., Eggert, A., Zuzak, T., et al. (2000) Resistance to TRAIL-induced apoptosis in primitive neuroectodermal brain tumor cells correlates with a loss of caspase-8 expression. *Oncogene* **19**, 4604–4610.

73. Honore, P., Luger, N. M., Sabino, M. A., et al. (2000) Osteoprotegerin blocks bone cancer-induced skeletal destruction, skeletal pain and pain-related neurochemical reorganization of the spinal cord. *Nat. Med.* **6**(7), 521–528.

74. Sezer, O., Heider, U., Jakob, C., Eucker, J., and Possinger, K. (2002) Human bone marrow myeloma cells express RANKL. *J. Clin. Oncol.* **20**, 353–354.

75. Nosaka, K., Miyamoto, T., Sakai, T., Mitsuya, H., Suda, T., and Matsuoka, M. (2002) Mechanism of hypercalcemia in adult T-cell leukemia, overexpression of receptor activator of nuclear factor kappaB ligand on adult T-cell leukemia cells. *Blood* **99**(2), 634–640.

76. Akatsu, T., Murakami, T., Ono, K., et al. (1998) Osteoclastogenesis inhibitory factor exhibits hypocalcemic effects in normal mice and in hypercalcemic nude mice carrying tumors associated with humoral hypercalcemia of malignancy. *Bone* **23**(6), 495–498.

77. Bullock, G., Ray, S., Reed, J. C., et al. (1996) Intracellular metabolism of Ara-C and resulting DNA fragmentation and apoptosis of human AML HL-60 cells possessing disparate levels of Bcl-2 protein. *Leukemia* **10**, 1731–1740.

78. Ibrado, A. M., Huang, Y., Fang, G., Liu, L., and Bhalla, K. (1996) Overexpression of Bcl-2 or Bcl-x$_L$ inhibits Ara-C-induced CPP32/Yama protease activity and apoptosis of human acute myelogenous leukemia HL-60 cells. *Cancer Res.* **56**, 4743–4748.

79. Kim, C. N., Wang, X., Huang, Y., et al. (1997) Overexpression of Bcl-xL inhibits Ara-C-induced mitochondrial loss of cytochrome c and other perturbations that activate the molecular cascade of apoptosis. *Cancer Res.* **57**, 3115–3120.

80. Miyashita, T. and Reed, J. C. (1993) BCL-2 oncoprotein blocks chemotherapy-induced apoptosis in human leukemia cell line. *Blood* **81**, 151–157.

81. Nakayama, K., Nakayama, K., Negishi, I., Kuida, K., Sawa, H., and Loh, D. Y. (1994) Targeted disruption of Bcl-2 alpha beta in mice: occurrence of gray hair, polycystic kidney disease, and lymphocytopenia. *Proc. Natl. Acad. Sci. USA* **26**, 3700–3704.

82. Koty, P. P., Zhang, H., and Levitt, M. L. (1999) Antisense bcl-2 treatment increases programmed cell death in non-small cell lung cancer cell lines. *Lung Cancer* **23**(2), 115–127.

83. Julien, T., Frankel, B., Longo, S., et al. (2000) Antisense-mediated inhibition of the bcl-2 gene induces apoptosis in human malignant glioma. *Surg. Neurol.* **53**(4), 360–369.

84. Duggan, B. J., Maxwell, P., Kelly, J. D., et al. (2001) The effect of antisense Bcl-2 oligonu-cleotides on Bcl-2 protein expression and apoptosis in human bladder transitional cell car-cinoma. *J. Urol.* **166**(3), 1098–1105.

85. Pepper, C., Thomas, A., Hoy, T., Cotter, F., and Bentley, P. (1999) Antisense-mediated sup-pression of Bcl-2 highlights its pivotal role in failed apoptosis in B-cell chronic lympho-cytic leukaemia. *Br. J. Haematol.* **107**(3), 611–615.

86. Uchida, T., Gao, J. P., Wang, C., et al. (2001) Antitumor effect of bcl-2 antisense phospho-rothioate oligodeoxynucleotides on human renal-cell carcinoma cells in vitro and in mice. *Mol. Urol.* **5**(2), 71–78.

87. Gleave, M. E., Miayake, H., Goldie, J., Nelson, C., and Tolcher, A. (1999) Targeting bcl-2 gene to delay androgen-independent progression and enhance chemosensitivity in prostate cancer using antisense bcl-2 oligodeoxynucleotides. *Urology* **54**(6A), 36–46.

88. Campos, L., Sabido, O., Rouault, J. P., and Guyotat, D. (1994) Effects of bcl-2 antisense oligodeoxynucleotides on in vitro proliferation and survival of normal marrow progenitors and leukemic cells. *Blood* **84**, 595–600.

89. Jansen, B., Schlagbauer-Wadl, H., Brown, B., et al. (1998) Bcl-2 antisense therapy chemosensitizes human melanoma in SCID mice. *Nat. Med.* **4**, 232–234.

90. Konopleva, M., Tari, A., Estrov, Z., et al. (2000) Liposomal Bcl-2 antisense oligonu-cleotides enhance proliferation, sensitize acute myeloid leukemia to cytosine-arabinoside, and induce apoptosis independent of other antiapoptotic proteins. *Blood* **95**, 3929–3938.

91. Miyake, H., Monia, B. P., and Gleave, M. E. (2000) Inhibition of progression to androgen-independence by combined adjuvant treatment with antisense BCL-XL and antisense Bcl-2 oligonucleotides plus taxol after castration in the Shionogi tumor model. *Int. J. Cancer* **86**(6), 855–862.

92. Leung, S., Miyake, H., Zellweger, T., Tolcher, A., and Gleave, M.E. (2001) Synergistic chemosensitization and inhibition of progression to androgen independence by antisense Bcl-2 oligodeoxynucleotide and paclitaxel in the LNCaP prostate tumor model. *Int. J. Can-cer* **91**(6), 846–850.

93. Bilim, V., Kasahara,T., Noboru, H., Takahashi, K., and Tomita, Y. (2000) Caspase involved synergistic cytotoxicity of bcl-2 antisense oligonucleotides and adriamycin on transitional cell cancer cells. *Cancer Lett.* **155**(2), 191–198.

94. Jansen, B., Schlagbauer-Wadl, H., Brown, B. D., et al. (1998) bcl-2 antisense therapy chemosensitizes human melanoma in SCID mice. *Nat. Med.* **4**(2), 232–234.

95. Webb, A., Cunningham, D., Cotter, F., et al. (1997) BCL-2 antisense therapy in patients with non-Hodgkin lymphoma. *Lancet* **349**(9059), 1137–1141.

96. Waters, J. S., Webb, A., Cunningham, D., et al. (2000) Phase I clinical and pharmacoki-netic study of bcl-2 antisense oligonucleotide therapy in patients with non-Hodgkin's lym-phoma. *J. Clin. Oncol.* **18**(9), 1812–1823.

97. Jansen, B., Wacheck, V., Heere-Ress, E., et al. (2000) Chemosensitisation of malignant melanoma by BCL2 antisense therapy. *Lancet* **356**(9243), 1728–1733.

98. Miayake, H., Tolcher, A., and Gleave, M. E. (2000) Chemosensitization and delayed andro-gen-independent recurrence of prostate cancer with the use of antisense Bcl-2 oligodeoxynucleotides. *J. Natl. Cancer Inst.* **92**(1), 34–41.

99. Zangemeister-Wittke, U., Leech, S., Olie, R., et al. (2000) A novel bispecific antisense oligonucleotide inhibiting both bcl-2 and bcl-x_L expression efficiently induces apoptosis in tumor cells. *Clin. Cancer Res.* **6**, 2547–2555.

100. Holinger, E. P., Chittenden, T., and Lutz, R. J. (1999) Bak BH3 peptides antagonize Bcl-xL function and induce apoptosis through cytochrome c-independent activation of cas-pases. *J. Biol. Chem.* **274**(19), 13298–13304.

101. Wang, J. L., Zhang, Z. J., Choksi, S., et al. (2000) Cell permeable Bcl-2 binding peptides: a chemical approach to apoptosis induction in tumor cells. *Cancer Res.* **60**(6), 1498–1502.

102. Degterev, A., Lugovskoy, A., Cardone, M., et al. (2001) Identification of small-molecule inhibitors of interaction between the BH3 domain and Bcl-x$_L$. *Nat. Cell Biol.* **3**, 173–182.

103. Nakashima, T., Miura, M., and Hara, M. (2000) Tetrocarcin A inhibits mitochondrial functions of Bcl-2 and suppresses its anti-apoptotic activity. *Cancer Res.* **60**(5), 1229–1235.

104. Tzung, S. P., Kim, K. M., Basanez, G., et al. (2001) Antimycin A mimics a cell-death-inducing Bcl-2 homology domain 3. *Nat. Cell Biol.* **3**(2), 43–46.

105. Wang, J. L., Liu, D., Zhang, Z. J., et al. (2000) Structure-based discovery of an organic compound that binds Bcl-2 protein and induces apoptosis of tumor cells. *Proc. Natl. Acad. Sci. USA* **97**(13), 7124–7129.

106. Deveraux, Q. and Reed, J. C. (1999) IAP family proteins—suppressors of apoptosis. *Genes Dev.* **13**, 239–252.

107. Li, F. and Altieri, D. (1999) Transcriptional analysis of human survivin gene expression. *Biochem. J.* **344**, 305–311.

108. O'Connor, D., Grossman, D., Plescia, J., et al. (2000) Regulation of apoptosis at cell division by p34^{cdc2} phosphorylation of survivin. *Proc. Natl. Acad. Sci. USA* **97**, 13103–13107.

109. Tamm, I., Wang, Y., Sausville, E., et al. (1998) IAP-Family protein survivin inhibits caspase activity and apoptosis induced by Fas (CD95), Bax, caspases, and anticancer drugs. *Cancer Res.* **58**, 5315–5320.

110. Chen, J., Wu, W., Tahir, S. K., et al. (2000) Down-regulation of survivin by antisense oligonucleotides increases apoptosis, inhibits cytokinesis and anchorage-independent growth. *Blood* **2**, 235–241.

111. Olie, R., Simoes-Wust, A. P., Baumann, B., et al. (2000) A novel antisense oligonucleotide targeting survivin expression induces apoptosis and sensitizes lung cancer cells to chemotherapy. *Cancer Res.* **60**, 2805–2809.

112. Li, F., Ackermann, E., Bennett, C. F., et al. (1999) Pleiotropic cell-division defects and apoptosis induced by interference with survivin function. *Nat. Cell Biol.* **1**, 461–466.

113. Adida, C., Recher, C., Raffoux, E., et al. (2000) Expression and prognostic significance of survivin in de novo acute myeloid leukemia. *Br. J. Haematol.* **111**, 196–203.

114. Liu, Z., Sun, C., Olejniczak, E., et al. (2000) Structural basis for binding of Smac/DIABLO to the XIAP BIR3 domain. *Nature* **408**, 1004–1012.

115. Srinivasula, S., Hegde, R., Saleh, A., et al. (2001) A conserved XIAP-interaction motif in caspase-9 and Smac/DIABLO regulates caspase activity and apoptosis. *Nature* **410**, 112–116.

116. Chai, J., Du, C., Wu, J., Kyin, S., Wang, X., and Shi, Y. (2000) Structural and biochemical basis of apoptotic activation by Smac/DIABLO. *Nature* **406**, 855–862.

117. Guo, F., Nimmanapalli, R., Paranawithana, S., et al. (2002) Ectopic overexpression of second mitochondrial-derived activator of caspases (Smac/DIABLO) or co-treatment with N-terminus of Smac peptide potentiates epothilone derivative (BMS 247550) and Apo-2L/TRAIL-induced Apoptosis. *Blood* **99**, in press.

118. Deng, Y., Lin, Y., and Wu, X. (2002) TRAIL-induced apoptosis requires Bax-dependent mitochondrial release of Smac/DIABLO. *Genes Dev.* **16**, 33–45.

119. Wang, C. Y., Mayo, M. W., Korneluk, R. G., Goeddel, D. V., and Baldwin, A. S., Jr. (1998) NFκB antiapoptosis: induction of TRAF1 and TRAF2 and c-IAP1 and c-IAP2 to suppress caspase-8 activation. *Science* **281**, 1680–1683.

120. Wang, C. Y., Guttridge, D. C., Mayo, M. W., and Baldwin, A. S., Jr. (1999) NFκB induces expression of the Bcl-2 homologue A1/Bfl-1 to preferentially suppress chemotherapy-induced apoptosis. *Mol. Cell. Biol.* **19**, 5923–5929.

121. Chen, C., Edelstein, L. C., and Gelinas, C. (2000) The Rel/ NFκB family directly activates expression of the apoptosis inhibitor Bcl-x$_L$. *Mol. Cell. Biol.* **20**, 2687–2695

122. Palombella, V. J., Rando, O. J., Goldberg, A. L., and Maniatis, T. (1994) The ubiquitin-proteasome pathway is required for processing the NF-kappa B1 precursor protein and the activation of NF-kappa B. *Cell* **78**(5), 773–785.

123. Beg, A. A. and Baltimore, D. (1996) An essential role for NF-kappaB in preventing TNF-alpha-induced cell death. *Science* **274**(5288), 782–784.

124. Wang, C. Y., Mayo, M. W., Korneluk, R. G., Goeddel, D. V., and Baldwin, A. S., Jr. (1998) NF-kappaB antiapoptosis: induction of TRAF1 and TRAF2 and c-IAP1 and c-IAP2 to suppress caspase-8 activation *Science* **281**(5383), 1680–1683.

125. An, B., Goldfarb, R. H., Siman, R., and Dou, Q. P. (1998) Novel dipeptidyl proteasome inhibitors overcome Bcl-2 protective function and selectively accumulate the cyclin-dependent kinase inhibitor p27 and induce apoptosis in transformed, but not normal, human fibroblasts. *Cell Death Differ.* **5**(12), 1062–1075.

126. Ravi, R., Bedi, G., Engstrom, L., et al. (2001) Regulation of death receptor expression and TRAIL/Apo-2L-induced apoptosis by NFᶜB. *Nat. Cell Biol.* **3**, 409–416.

127. Adams, J., Palombella, V. J., Sausville, E. A., et al. (1999) Proteasome inhibitors: a novel class of potent and effective antitumor agents. *Cancer Res.* **59**(11), 2615–2622.

128. Lin, Z. P., Boller, Y. C., Amer, S. M., et al. (1998) Prevention of brefeldin A-induced resistance to teniposide by the proteasome inhibitor MG-132: involvement of NF-kappaB activation in drug resistance. *Cancer Res.* **58**(14), 3059–3065.

129. Stinchcombe, T. E., Mitchell, B. S., Depcik, V., et al. (2000) PS-341 is active in multiple myeloma: preliminary report of a Phase I trial of the proteasome inhibitor PS-341 in patients with hematologic malignancies (Abstract 2219). *Blood* **96** (Suppl. 1), 516.

130. Nix, D., Pien, C., Newman, R., et al. (2001) Clinical development of a proteasome inhibitor, PS-341, for the treatment of cancer (Abstract 339). *Proc. ASCO* **20**, 86.

31

Tumor Deprivation of Oxygen and Tumor Suppressor Gene Function

Zhong Yun and Amato J. Giaccia

1. Introduction

Tumor hypoxia originates from the inability of neovasculature to provide an adequate blood supply to accommodate the metabolic demands of the tissue. Hypoxia is operationally defined as a spectrum of reduced levels of oxygen that result in the impairment of tissue function. The upper end of this reduced oxygen spectrum plays important physiologic functions in maintaining blood pressure and stimulating erythropoiesis. The lower end of this spectrum approaches a complete lack of oxygen (defined as anoxia) and is uniquely found in solid tumors. In between the two ends of the spectrum in solid tumors lie a large number of cells at intermediate oxygen tension that have been hypothesized to present the greatest therapeutic problem for radiotherapy and chemotherapy *(1)*.

Tumor hypoxia has been measured directly in a variety of human cancers, including head and neck carcinomas, cervical carcinomas, and soft tissue sarcomas, using pO_2 microelectrodes. Brizel et al. and Nordsmark et al. showed that in head and neck carcinomas, hypoxia correlated with a lower probability of disease-free survival *(2,3)*, and in soft tissue sarcomas, hypoxia was associated with increased incidence of distant metastases *(4)*. Hockel et al. also found that hypoxia in cervical carcinomas resulted in increased local and distant failures *(5)*. Interestingly, hypoxia predicted for distant failure not only in patients treated with radiotherapy but also in those treated with surgery alone. These studies suggest that hypoxia alters fundamental and physiologically important pathways in a wide variety of tumors that result in more aggressive tumor behavior in addition to resistance against conventional forms of treatment such as radiotherapy and chemotherapy.

Analysis of gene and protein changes induced by a low-oxygen environment has provided some insight into the homeostatic pathways that are dysregulated under these pathophysiologic conditions. In fact, the existence of a hypoxic microenvironment is a unique feature of solid tumors and represents both advantages as well as problems in the treatment of cancer. This chapter describes the current methods of detecting tumor hypoxia and how tumor suppressor gene function regulates the response of transformed cells to a low-oxygen environment.

From: *Methods in Molecular Biology, Vol. 223: Tumor Suppressor Genes: Regulation, Function, and Medicinal Applications.* Edited by: Wafik S. El-Deiry © Humana Press Inc., Totowa, NJ

Table 1.
Hypoxia Measurements

Direct	O$_2$ electrode	• Direct in vivo measurement • Invasive • No histological correlation
Indirect	Nitromidazole • EF5 • Pimonidazole	• Histologically applicable • Can be used with other markers • Relies on bioreduction
Surrogate marker	• Glut-1 • CA9 • OPN-1 • HIF-1α	• Noninvasive • Clinically applicable • No clear causal effect

2. Methods of Hypoxia Detection In Vitro and In Vivo

Several techniques have been used to either directly measure or indirectly estimate tumor hypoxia (**Table 1**). The use of oxygen microelectrodes to measure tumor oxygenation allows instant assessment of tumor perfusion. Although microlectrodes provide an instantaneous sampling of tumor pO$_2$ readings, they provide little information on whether the sampling comes from a viable or a necrotic region of the tumor. Certain investigators have biopsied the regions surrounding the Eppendorf track and are able to determine whether low pO$_2$ readings originate from necrotic regions, but the problem of sampling error and heterogeneity have not been addressed *(6)*. An additional question that still remains to be resolved is what is the absolute measure of oxygen values in human tumors. Are they truly as low as in rodent tumors *(7)*? Experimental measurements of oxygenation in transplanted tumors are typically 10–20% of the pO$_2$ values that are found in human tumors. Therefore, are transplanted tumors a good system to study the effects of hypoxia on malignant progression? Although this question has stirred debate, a universal finding has been that tumors have lower median pO$_2$ values than their surrounding tissues.

Although most data on tumor oxygenation have been through microelectrode measurements, other techniques have been highly useful in identifying hypoxic tumor regions through the use of metabolic binding of nitroimidazole compounds such as pimonidazole and EF5 *(8)*. These compounds identify hypoxic cells through their bioreduction and formation of adducts with cellular macromolecules *(9)*. The formation of adducts increases as oxygen decreases and is readily detected by specific monoclonal antibodies. These approaches can be used to detect hypoxia in cell culture experiments as well as in tumors *(10)*, and provide an additional independent manner to calibrate the level of pO$_2$ in the cellular microenvironment. Clearly, the use of nitroimidazole markers has given us insight into the spatial organization of hypoxic regions in solid tumors, and the ability to co-localize changes in gene and protein expression with hypoxia *(11–13)*.

Recently, these traditional approaches have been challenged with newer approaches that measure changes in gene or protein expression induced by changes in tissue oxygenation. The most heavily studied hypoxia-induced gene products in human tissue have been hypoxia-inducible factor 1 (HIF-1), glucose transporter 1 (Glut-1), and carbonic anhydrase 9 (CA9). HIF-1 is a heterodimeric basic helix–loop–helix transcription factor that possesses an oxygen-sensitive HIF-1α subunit and a constitutively expressed HIF-1β subunit. Analysis of HIF-1α staining in clinical samples lends support for a relationship between HIF-1α protein accumulation and malignant progression *(14)*. Interestingly, HIF-1α protein accumulation is associated with *p53* mutation and increased proliferation *(15)*. Two transcriptional targets of HIF-1, *Glut-1* and *CA9*, have also shown promise as endogenous markers of tumor hypoxia. They both have been shown to co-localize with known hypoxia markers *in situ*, in both transplanted and spontaneous tumors *(13,16,17)*. In particular, CA9 has been reported to have prognostic value in squamous cell carcinoma of the head and neck and non-small-cell lung cancer *(18,19)*. To date, there is little clinical evidence on the relationship between CA9 staining and tumor suppressor gene inactivation. However, in a small subset of patients with Barrett metaplasia, Glut-1 is expressed in tumors that possess *p53* mutations and not in normal tissue controls *(20)*.

A noninvasive and more universally available test for tumor hypoxia such as a serum marker that can be readily measured in patients' blood would have greater clinical usefulness. Such a marker can be derived from microarray information on gene and protein regulation by hypoxia or by hypoxia and reoxygenation. One means of gaining insight into potential secreted markers is to analyze changes in mRNA expression in tumor cells that possess mutations in genes that have previously been shown to modulate hypoxia-induced gene expression, such as the *VHL* tumor suppressor gene. VHL is a dominantly inherited genetic condition associated with the development of hemangioblastoma, renal cell carcinoma, and pheochromocytoma *(21,22)*. The VHL protein is part of a multiprotein complex that regulates the oxygen-dependent ubiquitination and proteolysis of HIF-1α *(23–25)*. Loss of VHL function substantially decreases HIF-1α degradation under aerobic conditions, and increases expression of downstream genes such as endothelin 1, differentiated embryo chondrocyte 1 (DEC1), transglutaminase 2, and low-density lipoprotein receptor-related protein 1 *(26)*.

Using linear discriminant analysis (LDA) *(27)* on NCI-60 cancer cell line microarray expression dataset to identify genes with high VHL expression from those with low expression, osteopontin (OPN) was found to be the highest-weighted gene in the profile and its expression inversely correlated with VHL expression (Le et al., *Clin. Cancer Res.*, in press). This analysis suggested that OPN, a secreted calcium binding glycophosphoprotein that had been associated with cell transformation *(28)* and tumor progression *(29,30)*, might be a useful marker for tumor hypoxia.

With the advances in DNA microarray technology, global changes in hypoxia-regulated gene expression in tumors can be readily analyzed, and more surrogate hypoxia markers will be identified. Eventually, the array-base technology may serve as a diagnostic test to detect the state of hypoxia in patients with cancers.

3. Role of p53 and pRb Under Hypoxia in Apoptosis and Cell Cycle Arrest Under Hypoxia

The critical role of *p53* in tumor suppression is underscored by the findings that the *p53* gene is mutated in over 50% of human cancers *(31)* and that mice nullizigous for the *p53* gene develop tumors early in their lifetime *(32)*. Two main functions of p53 that have been proposed to be responsible for its role as a tumor suppressor are the induction of cell-cycle arrest in response to DNA damage and the induction of apoptosis. Previous studies by Graeber et al. showed that hypoxia was a physiologic inducer of the p53 tumor suppressor *(33)* and through the induction of apoptosis, hypoxia can act as a selective pressure for the elimination of cells with wild-type (wt) *p53* and the clonal expansion of cells with mutant or otherwise inactive p53 protein *(34)*. These observations provide a possible explanation for the more aggressive nature of hypoxic tumors compared to well-oxygenated ones and for the frequent occurrence of *p53* mutations in advanced stages of tumor development. Thus, hypoxia joins genotoxic stresses such as ultraviolet light (UV) and ionizing radiation as an inducer of p53-dependent apoptosis.

Activation of p53 following genotoxic damage is achieved by accumulation and posttranscriptional modifications of the p53 protein such as phosphorylation and acetylation (*see* reviews, **refs. *35*** and ***36***). Accumulation of the p53 protein following genotoxic stress involves posttranscriptional mechanisms such as enhanced translation of p53 mRNA and decreased proteolytic degradation of the protein *(37–39)*. Following DNA damage, p53 binds to DNA in a sequence-dependent manner, and recruits the transcriptional co-activator p300/CBP *(40,41)* as well as basal transcription factors such as TFIID *(42–44)*. Through these interactions, p53 induces the transcription of downstream effector genes whose products interact with and inhibit the functions of the proteins involved in cell-cycle regulation or apoptosis (for review, *see* **ref. *37***).

The product of the *MDM-2* oncogene, itself a transcriptional target of p53, binds to the N-terminus of p53, inhibits transactivation properties of p53, and promotes the proteolytic degradation of p53 *(45,46)*. Thus, the interaction between p53 and MDM-2 creates a negative regulatory feedback loop that is responsible for the rapid turnover of the p53 protein in unstressed cells. Hypoxia decreases the expression of the endogenous cellular regulator of p53 protein, the MDM-2 protein, resulting in an inhibition of p53 export from the nucleus to the cytoplasm for degradation *(47)*. Inhibition of p53 protein degradation is a posttranscriptional event, since there is no change in MDM-2 mRNA under hypoxic conditions. Thus, the decrease in MDM-2 protein levels seems to be the major mechanism of p53 protein regulation under hypoxic stress, since ectopic expression of MDM-2 to the level of unstressed controls decreases p53 protein accumulation *(47)*. The regulation of p53 protein stability can be mediated by the ARF protein (p14 in human and p19 in mice) in addition to the autoregulatory loop with the *MDM-2* protein *(48,49)*. Interestingly, hypoxia induces a rapid decrease of both ARF and MDM-2 proteins, suggesting that MDM-2 downregulation by hypoxia is through a yet-unidentified ARF-independent mechanism (**Fig. 1**).

One proposed mechanism in the literature for the accumulation of p53 by hypoxia is through the binding of HIF-1α to p53 *(50)*. Though it is possible that p53 and HIF-1α may interact in vivo, this mechanism alone is not sufficient to explain p53 accumulation under hypoxia. HIF-1α protein accumulates at O_2 levels of about 1%, and therefore one would expect p53 accumulation also to occur at these O_2 levels. However, p53 accumulation occurs at O_2 levels of <0.2%. Agani et al. *(51)* have shown that at 1% O_2 there is HIF-1α accumulation and HIF-1α-dependent transcription without p53 accumulation. Furthermore, p53 accumulation can indeed occur in *HIF-1β*$^{-/-}$ and *HIF-1α*$^{-/-}$ cells *(52)*, contradicting the results reported by An et al. A yeast two-hybrid system approach has also failed to detect any interaction between p53 and HIF-1α (Z. Yun and A. J. Giaccia, unpublished observations). Nevertheless, the marked reduction of MDM-2 protein provides a direct explanation for the increased accumulation of p53 under hypoxic conditions *(47)*.

Posttranslational modification of p53 occurs in response to cellular stress. It has been demonstrated that DNA damage induces phosphorylation of p53 on serine15 (Ser15) *(53–56)*. Such modification results in decreased binding between MDM-2 and p53, which

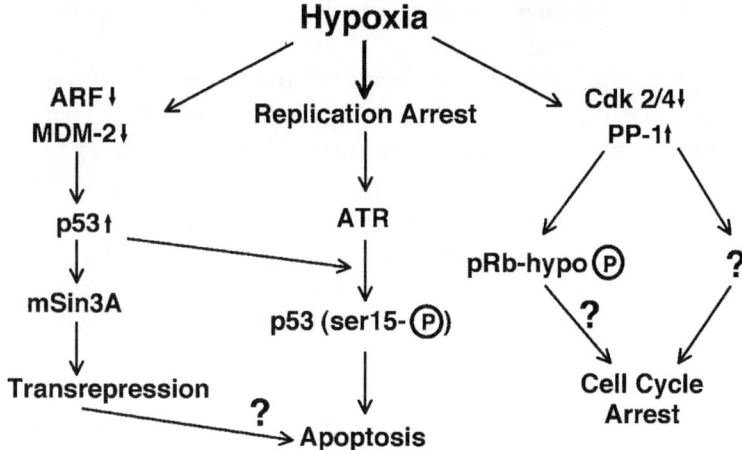

Fig. 1. The function of p53 is regulated under hypoxic conditions by different mechanisms. In contrast to genotoxic stresses, hypoxia does not induce p53-dependent transactivation and cell cycle arrest. Hypoxia induces p53-dependent apoptosis by two perhaps independent mechanisms. First, hypoxia induces p53 accumulation by the downregulation of ARF and MDM-2. p53 interacts with corepressor mSin3 and mediates transrepression, which may lead to apoptosis. Second, p53 becomes phosphorylated at Ser15 by ATR, but not ATM, as a consequence of hypoxia-induced replication arrest. Hypophosphorylation of pRb is observed in hypoxia-induced cell cycle arrest, perhaps due to the decrease in Cdk2 and/or Cdk4 kinase activity as well as enhanced PP-1 phosphatase activity. However, it is likely that pRb plays a minor role, at best, in hypoxia-induced cell cycle arrest.

disrupts this feedback loop and increases the levels of p53 following DNA damage *(55)*. The phosphorylation of p53 on Ser15 is mediated by the ATM family of kinases, such as ATM and ATR. Although sharing common substrates (BRCA1, p53, Chk1, and Chk2) and some overlapping functions, ATM and ATR differ in stress responses. ATM responds primarily to DNA damage throughout the cell cycle, whereas ATR is more involved in stresses that induce the replication arrest. Hammond et al. have recently shown that ATR, but not ATM, is involved in hypoxia-induced Ser15 phosphorylation and stabilization of p53, which correlates with replication arrest in S phase *(57)*. The p53 protein is also extensively phosphorylated at other sites both in vitro and in vivo in response to genotoxic damage *(35,36,58)*. Although some of these posttranslational modifications have been proposed to increase the sequence-specific DNA binding activity of p53 and its transactivation properties in vitro, the exact physiologic significance of these modifications induced by genotoxic damage or hypoxia in vivo remain to be determined.

Since the majority of p53 mutations inhibit its sequence specific DNA-binding activity and p53-signaled apoptosis, it has been proposed that p53 transactivation must be essential for tumor suppression. Although it has been reported that p53-dependent transactivation is necessary for induction of apoptosis in some experimental systems *(59,60)*, under certain circumstances it appears that macromolecular synthesis is completely dispensable for the induction of p53-dependent apoptosis *(61,62)*. Ectopic expression of either wt p53 or p53 mutants that lack transactivation capability is able to induce apoptosis in certain cell lines *(63)*. Furthermore, deletion of the polyproline-rich domain of p53 located between its transactivation and DNA-binding domains abrogates its ability to induce apoptosis without affecting its DNA binding, induction of bax, and cell cycle inhibition *(64)*. Taken together, these data indicate that p53 transactivation is not an absolute requirement to signal apoptosis.

Mounting evidence suggests that p53-dependent transrepression may present a second mechanism that contributes to the induction of apoptosis *(65–67)*. Characterization of *p53* alleles from human tumors has lead to the identification of mutant *p53* genes that still possess transactivation capability, but do not induce apoptosis when ectopically overexpressed *(63)*. Additional evidence to support a role of p53 transrepression in mediating apoptosis stems from the use of viral or oncogenic inhibitors of p53-signaled apoptosis. The bcl-2, WT-1, and E1B-19K proteins all inhibit p53-signaled apoptosis and inhibit p53 transrepression activity, but do not alter p53 transactivation capability *(64,67,68)*. Murphy et al. have demonstrated that wt p53 protein interacts with histone deacetylases (HDACs) through the binding of the co-repressor molecule mSin3, and that inhibition of this interaction by treatment of cells with a histone deacetylase inhibitor reduced p53-signaled apoptosis *(69)*. However, it is difficult to determine the contribution of p53 transrepression to apoptosis when p53 transactivation is also occurring in the same cells.

Interestingly, hypoxia does not induce p53-dependent cell-cycle arrest despite that p53 protein accumulation as well as p53-dependent apoptosis occurs. In contrast to DNA damage, hypoxia does not induce the expression of endogenous p53 target genes or proteins, although ionizing radiation-induced p53 transactivation is not inhibited by hypoxia *(70)*. Furthermore, hypoxia induces an interaction of p53 with co-repressor mSin3A, but not with the co-activator p300 *(70)*. These results indicate that, compared to genotoxic stresses, hypoxia is unique in that it induces only p53 transrepression activity and not p53 transactivation *(70)*.

The emerging picture thus far is that genotoxic stress induces protein accumulation and posttranslational modifications of p53 that give rise to p53 interaction with co-activator and/or co-repressor proteins. Thus, one would expect to find differences in the type of motifs found in p53-induced genes and p53-repressed genes. One hypothesis is that the majority of p53-repressed genes under hypoxia lack TATA boxes and contain downstream promoter elements that bind TBP-associated factors (TAFs). Studies have indicated that overexpression of TAFII40 and TAFII60 inhibits p53 transcriptional activation of GAL-4 reporter genes containing p53 response elements *(43)*. Therefore, one mechanism of p53-mediated repression could be through p53–TAF interactions that result in the recruitment of co-repressor molecules but not co-activator molecules. Ultimately, to prove the hypothesis that p53-mediated gene repression is important for apoptosis, one must find candidate repressed genes whose ectopic overexpression inhibits apoptosis in a p53-dependent manner.

Cell cycle arrest induced by hypoxia is reversible, though the extent of reversibility is reduced overtime under stringent hypoxia *(71)*. In fact, most cells are able to resume DNA synthesis within 10 min to 3 h after reoxygenation *(72)*. However, little information is known about the genetic determinants that govern the reentry of hypoxic cells into the cell cycle, except that recovery of DNA synthesis is blocked by cycloheximide addition in Ehrlich ascites cells *(73,74)* and is associated with the appearance of hyperphosphorylated pRb in T-47D breast cancer cells *(75)*. The biologic functions mediated by pRb may also be important in other types of hypoxia responses in vivo. When subjected to cerebral artery occlusion, *E2F1*$^{+/-}$ mice experience decreased brain infarct as compared to the wild-type controls *(76)*.

Research on more moderate hypoxia (~1% O_2 as opposed to <0.1%; normal atmospheric oxygen is ~21%), under which nucleotide biosynthesis is unlikely to be severely impaired, has supported the hypothesis that hypoxia leads to a cyclin/cdk-dependent, checkpoint-type arrest. Krtolica et al. *(77)* have observed decreased Cdk4 and Cdk2 activity in CV-1P monkey kidney cells treated with ~1% O_2, and this kinase inhibition was associated with an increase in p27^{Kip1} and a decrease in Cdk4, cyclin D, and cyclin E protein levels. They proposed that this decreased kinase activity, coupled with increased PP-1 phosphatase activity, leads to dephosphorylation of pRb and inhibition of cell cycle progression. Similar results were obtained in several other cell types, suggesting a possible common pathway *(78,79)*. Nevertheless, the role of pRb in hypoxia-induced cell cycle arrest remains unresolved. Using *Rb*-deficient cells, Green et al. have found that pRb is not essential to cell cycle arrest in response to hypoxia *(80)*, which appears to be in conflict with previously published studies. However, those studies have been mostly correlative, demonstrating only that cell cycle arrest under hypoxia is associated with the appearance of hypophosphorylated pRb *(81,82)*. Amellem et al. *(81)* find that T-47D breast cancer cells, unlike two *Rb*-null cancer cell lines, could not be stimulated to enter S phase upon deoxynucleotide addition under hypoxia, and they attribute this phenomenon to a secondary arrest pathway dependent on functional *Rb*. However, no matched cell lines that differ only in Rb status were compared in this study *(81)*. It is likely that additional Cdk2 targets, other than pRb, may also be involved in mediating the cell cycle arrest in response to hypoxia (**Fig. 1**).

4. Role of VHL, Hydroxylases, and PTEN in Regulation of HIF-1 Activities

Adaptation of cells to an anaerobic environment is achieved by the transcriptional induction of genes that are involved in glycolysis *(83)*, hematopoiesis *(84,85)*, angiogenesis *(86)*, tissue remodeling *(87)*, and so on. The early insight into the transcriptional regulation of gene expression by hypoxia came from studies on erythropoietin (EPO) gene regulation. Several groups have shown that the hypoxia inducibility of the *EPO* gene is due, in large part, to a hypoxia-responsive element (HRE) or 5′-ACGTG-3′-localized in its 3′-untranslated region (UTR) *(88,89)*. The transcription factor that bound this HRE was designated hypoxia-inducible factor 1 (HIF-1). It is composed of two subunits, an oxygen-sensitive HIF-1α subunit and a constitutively expressed HIF-1β subunit *(90)*. Both HIF-1α and -β are members of the basic helix–loop–helix *Per/Sim/Arnt* (HLH-PAS) family of transcription factors. In contrast to the constitutively expressed HIF-1β subunit, HIF-1α is an oxygen-labile protein that becomes stabilized in response to hypoxia, iron chelators, and divalent cations. As all three inducers of HIF-1α have some direct or indirect relationship with heme, it has been postulated that the cellular oxygen sensor is a heme-associated protein *(84,91)*. To date, there has been no definitive identification of such a protein. Under hypoxic conditions, *HIF-1α* mRNA levels do not change in most cell lines, but HIF-1α protein levels increase *(92,93)*. Although multiple signaling pathways involving MAP kinase *(94)*, PKC *(95)* and PI(3)K *(96,97)* have been shown to affect the accumulation of HIF-1α, the details on how these pathways modulate HIF-1α are still unclear. However, the most critical point concerning the labile nature of HIF-1α under aerobic conditions is that it can be transferred to other proteins *(98,99)*. In fact, fusion of different HIF-1α domains to the yeast Gal4 transcription activator has identified two separable hypoxia-responsive domains. One domain is localized between residues 531 and 575 and is important in modulating HIF-1α protein stability, and thus is referred to as the oxygen-dependent degradation (ODD) domain *(98,100)*.

The second domain is localized between amino acid (aa) residues 786 and 826 (referred to as C-TAD) and is involved in modulating transcriptional activation of HIF-1α under hypoxic conditions. The finding that transfer of aa531–575 to a heterologous protein can confer oxygen sensitivity suggests that the oxygen sensor itself does not have to be associated directly with HIF-1. Interestingly, two additional members of the HIF-1α family, designated HIF-2α (also called EPAS1 or MOP2) *(101–104)* and HIF-3α, have been identified *(105)*. HIF-2α is highly similar to HIF-1α in both structure and function, but exhibits more restricted tissue-specific expression. In contrast to HIF-2α, HIF-3α also exhibits conservation with HIF-1α and HIF-2α in the HLH and PAS domains, but does not possess a similar hypoxia-inducible transactivation domain *(105)*.

Hypoxia regulates HIF-1α protein stability by inhibiting its ubiquitin-mediated degradation *(106,107)*. This concept is highly supported by studies on cell lines derived from tumors that have lost the von Hippel-Lindau (VHL) tumor suppressor gene. These *VHL*-deficient cells display aerobic HIF-1α protein expression *(24,25)*. Tumors such as renal cell carcinomas (RCC) that possess mutations in *VHL* also exhibit high aerobic expression of HIF-1-regulated genes, whereas reintroduction of wt *VHL* substantially reduces the aerobic level of HIF-1α protein to those found in untransformed or transformed cells that express wt *VHL*. In addition, HIF-1α protein stability is separable from its heterodimerization with HIF-1β, as cells that are HIF-1β-deficient still exhibit HIF-1α stabilization *(108)*.

The functional relationship between VHL protein and the ubiquitin system comes from the revelation that VHL protein forms a complex with elongins B and C, Cul2, and the RING-H2 protein Rbx1 *(109–113)*. Significant homology exists between elongin B and ubiquitin, while elongin C, Cul2, and Rbx1 are closely related to the components of the SCF (Skp1.Cul1/F-box) family of the E3 ubiquitin–ligase complexes that are involved directly in the ubiquitination of proteins to be degraded by the 26S proteasomes *(114–116)*. The VHL protein folds into two functional domains, α and β. The α domain of VHL binds to elongin C through aa157–171, and their interaction is stabilized in the presence of elongin B *(113,117,118)*. Frequent natural mutations within aa157–171, underline the importance of the interaction between VHL and elongin C. In contrast, the β domain of VHL is involved in direct interaction with HIF-1α *(23,25,119,120)*. The β domain is also subjected to many natural mutations, with tyrosine 98 as the second most frequently mutated residue in VHL. This and other missense mutations in the β domain often result in diminished interaction with HIF-1α *(25)*.

Deletional analysis indicates that the interaction of VHL with HIF-1α requires aa530–650 of HIF-1α, which overlaps with the previously identified ODD *(25)*. Located in this domain is an important residue, proline 564, that becomes hydroxylated by a family of iron (II)-dependent prolyl hydroxylases that use molecular oxygen as a substrate, and may sense the availability of oxygen directly *(121–123)*. Hydroxylation of proline 564 is necessary and sufficient for the interaction of HIF-1α with VHL (**Fig. 2**).

Although it is not involved in stabilization of HIF-1α, the C-TAD is involved in modulating transcriptional activation of HIF-1α under hypoxic conditions. Under hypoxia conditions, C-TAD is able to interact with transcriptional co-activators such as p300/CBP *(124–126)*. This interaction has been shown to require another oxygen-dependent hydroxylation event, the hydroxylation of the asparagine residue in the conserved domain, YDCEVNV/AP, within C-TAD *(127)*. However, the enzymes that are responsible for the hydroxylation of asparagines remain to be identified. Nevertheless,

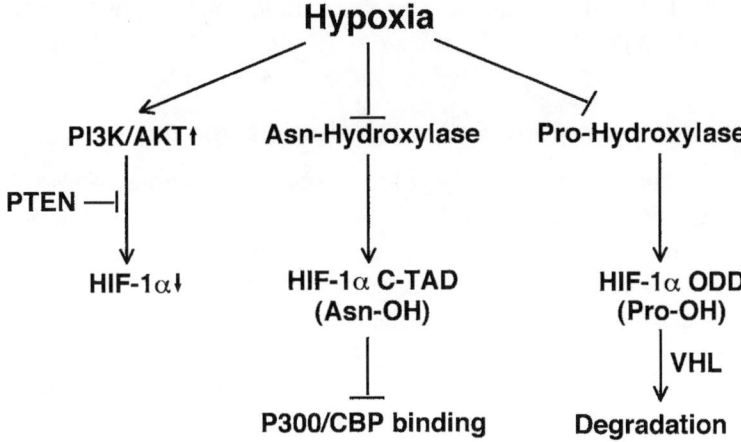

Fig. 2. Hypoxia increases the stability and transcription activity of HIF-1α by at least three mechanisms. Hypoxia activates the PI3K/AKT pathway, which, in turn, activates HIF-1α by a yet-unknown mechanism. Prolyl hydroxylases are found to modify a proline residue(s) located in HIF-1α ODD, which causes the hydroxylated HIF-1α to undergo ubiquitin-dependent degradation through the direct interaction with VHL. The transactivation activity of HIF-1α is modulated through the hydroxylation of an asparagine residue in its C-TAD by an unidentified hydroxylase. Hydroxylation of this asparagine residue renders HIF-1α unable to bind p300/CBP. However, it is not known whether there is any relationship between the PI3K/AKT pathway and this regulation of hydroxylases.

these observations demonstrate that HIF-1 activity and eventually the cellular response to hypoxia are regulated at multiple discrete levels (**Fig. 2**). It is certainly of great interest to identify any natural mutations in the hydroxylases that modify HIF-1α at either proline or the asparagines residues.

HIF-1 transcription activity and the expression of angiogenic factors are subjected to regulation by the phosphatidyl-3-inositol kinase (PI3K)/AKT pathway *(96,97,128, 129)*. Therefore, the tumor suppressor gene *PTEN* could also serve to check the hypoxia-induced stimulation of the PI3K-HIF pathway. The *PTEN* tumor suppressor gene is mutated or inactivated in a large percentage of glioblastomas *(130)*, which possess median pO_2 levels well below those found in other solid tumors. Using *PTEN* -deficient glioblastoma cell lines, Zundel et al. have demonstrated that PTEN inhibits HIF-1 activation and VEGF induction by hypoxia to the same extent as wortmannin, a potent PI3K inhibitor *(97)*. Interestingly, PTEN inhibited the hypoxia-stimulated accumulation of HIF-1α, although not through the direct phosphorylation or interaction with the protein. Similar observations have been made in other tumor cells such as human breast cancer cells *(131,132)* and human prostate cancer cells *(129,133)*. It is interesting to note that the PI3K/AKT pathway preferentially regulates HIF-1α without apparent effect on HIF-2α *(131)* or HIF-1β *(129)*. These observations suggest that tumor cells deficient in PTEN tend to have elevated HIF-1 activity and thus increased production of proangiogenic factors (**Fig. 2**). However, a recent study *(134)* indicates that the effect of the PI3K/AKT pathway on hypoxia-induced activation of HIF-1α may be cell-type-specific. In some cell types (HepG2 and HEK293T), there is no correlation between the activation of the PI3K/AKT pathway and HIF-1α activities under hypoxia. Taken together, these seemingly inconsistent findings suggest the multiple and perhaps redundant regulatory mechanisms modulate HIF-1α activities under hypoxia. It is therefore important to delineate specific pathways leading to the modulation of HIF-1α activities under hypoxia in any given cell type.

5. Role of TSP-1 in Angiogenic Switch Under Hypoxia

Angiogenesis, the recruitment of new blood vessels to regions of low blood supply, is essential for the progression of solid tumors to malignancy *(135–137)*. Increasing evidence supports the hypothesis that tumor angiogenesis is controlled by an "angiogenic switch," a physiologic mechanism involving a dynamic balance of angiogenic factors (i.e., inhibitors and inducers) *(136)*. Numerous pro- and antiangiogenic factors have been identified, including VEGF, cytokines, and inflammatory agents (e.g., TNF-α, IL-8), fragments of circulatory system proteins (e.g., endostatin), and extracellular matrix components (e.g., thrombospondins) *(136,138–142)*. Presumably, this diversity of angiogenic factors reflects a strict requirement for controlling angiogenesis under normal physiologic conditions and in response to oncogenic events. Given that the switch to the proangiogenic phenotype requires a positive net increase in activity of angiogenic inducers, it can be hypothesized that tumor hypoxia promotes angiogenesis by modulating the expression of both angiogenic inducers and inhibitors (**Fig. 3**). This section focuses on the influence of the hypoxic tumor microenvironment on the expression of the antiangiogenic factor, thrombospondin-1 (TSP-1).

TSP-1 is a widely expressed glycoprotein that possesses multiple functional domains and exists as a homotrimer in the extracellular milieu (*see* review by Adams *[143]*). The functions of TSP-1 range from the regulation of cell adhesion and motility to that of angiogenesis. Studies have shown that TSP-1 inhibits neovascularization in vivo, as well as endothelial cell growth and vessel-like formation in vitro (see review by de Fraipont et al. *[144]*). The same property is also shared by TSP-2 *(142)*. The antiangiogenic activity of TSP-1 is potentially conferred by the amino acid sequence, CSVTCG, located in the third type I repeats. The CSVTCG motif can interact with CD36 on the surface of endothelial cells and induce endothelial cell apoptosis via the activation of tyrosine kinase p59^{fyn}, caspase-3-like proteases, and p38 MAPK *(145)*. Elevated expression of bax, decreased expression Bcl-2, and enzymatic processing of caspase-3 are also found in endothelial cells exposed to TSP-1 *(146)*.

The role of TSP-1 and 2 in angiogenesis is further solidified by genetic studies. TSP-1 and 2 affect angiogenesis during wound healing, although neither protein is essential for embryonic angiogenesis. Mice homozygous null for *TSP-1* or *2* exhibit increased vascularity in skin wounds that may result in accelerated wound healing *(147,148)*. Targeted expression of *TSP-1* in transgenic basal keratinocytes results in delayed wound healing that may be attributed, at least in part, to impaired endothelial cell proliferation and reduction in vascular density *(149)*.

Compelling evidence of an antiangiogenic function for TSP-1 in tumors is provided by studies showing that activation of the angiogenic switch in Li-Fraumeni fibroblasts lacking wild-type (wt) *p53* is associated with diminished *TSP-1* expression *(150,151)*. Conversely, introduction of wt *p53* into BT549 human breast carcinoma cells generates an antiangiogenic activity involving upregulated *TSP-1* expression *(141)*. In addition, overexpression of *TSP-1* is antiangiogenic in tumor models *(152,153)*. Reports that endogenous *TSP-1* expression can be induced by wt *p53*, repressed by oncogenic signals such as c-*jun* overexpression, and silenced by DNA methylation imply that down-

Angiogenic Switch

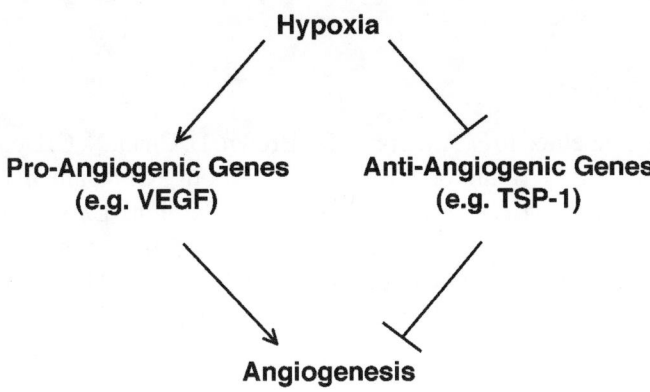

Fig. 3. Hypoxia causes a net increase in proangiogenic factors. Possible mechanisms include the upregulation of proangiogenic genes such as VEGF and the inhibition of antiangiogenic genes such as TSP-1.

regulation of TSP-1 contributes to oncogenesis *(150,154–157)*. As both *p53* and c-*jun* are inducible by hypoxia *(33,158)*, the regulation of TSP-1 expression by these proteins suggests that it is a hypoxia-responsive angiogenic inhibitor.

It has been shown that *TSP-1* gene expression is significantly reduced in rodent and human tumor cells under hypoxic conditions *(159,160)*. On the other hand, ectopic expression of either *TSP-1* or *2* in several xenograft tumor models results in retarded tumor growth associated with apparent reduction in tumor vasculature density (*see* review by de Fraipont et al. *[144]*). However, neither treatment of human squamous cell carcinoma cells with antisense *TSP-1* mRNA *(161)* or blocking the TSP-1 receptor with antibodies in human breast cancer MDA-MB-231 cells *(162)* reduces tumor progression. Clinically, there is lack of an unequivocal correlation between TSP and cancer malignancy. These observations may reflect the wide range of potential interactions of the complex domain structure of this molecule with its microenvironment, as well as the variable distribution of TSP-1 receptors in tissues *(163,164)*. Nevertheless, repression of *TSP-1* or *2* expression in certain hypoxic solid tumors may play an important role in facilitating tumor progression and dissemination by switching the tumor microenvironment to proangiogenesis (**Fig. 3**).

6. Summary

Tumor hypoxia reduces the efficiencies of therapy and is inversely correlated with clinical outcomes. Hypoxia provides a favorable microenvironment for the clonal expansion of tumor cells that have lost tumor suppressor genes or gained activated oncogenes, which may, in turn, cause a reduction in the production of antiangiogenic factors and an increase in the output of proangiogenic factors. The challenge remains to understand the transcriptional and posttranscriptional response of tumor cells to hypoxia. This will require a systematic and multifaceted approach. The unanswered question is how important are the contributions of tumor suppressor genes to hypoxia-induced changes in gene expression for malignant progression. Elucidation of the molecular basis of cellular responses to hypoxia will benefit the development of efficacious therapies for the treatment of hypoxic tumors.

7. Acknowledgments

The authors are indebted to R. Alarcon, J. M. Brown, D. Chan, N. C. Denko, L. Fontana, R. Freiberg, S. Green, E. Hammond, F. Kaper, A. Koong, C. Koumenis, K. Laderoute, Q.-T. Le, H. Maecker, M. O'Neill, M. Powell, P. Sutphin, L. Swiersz, S. Welford, B. Wouters, Wayne Zundel, and D. Zinyk for their contribution to this review. This work was supported by National Institutes of Health (NIH) Grants CA88480 and CA67166, a grant from Aventis (A.J.G.), and NIH Cancer Biology Training Grant CA09302 (Z.Y.).

References

1. Wouters, B. G. and Brown, J. M. (1997) Cells at intermediate oxygen levels can be more important than the "hypoxic fraction" in determining tumor response to fractionated radiotherapy. *Radiat. Res.* **147**, 541–550.
2. Brizel, D. M., Sibley, G. S., Prosnitz, L. R., Scher, R. L., and Dewhirst, M. W. (1997) Tumor hypoxia adversely affects the prognosis of carcinoma of the head and neck. *Int. J. Radiat. Oncol. Biol. Phys.* **38**, 285–289.
3. Nordsmark, M., Overgaard, M., and Overgaard, J. (1996) Pretreatment of oxygenation predicts radiation response in advanced squamous cell carcinoma of the head and neck. *Radiother. Oncol.* **41**, 31–40.
4. Brizel, D. M., Scully, S. P., Harrelson, J. M., et al. (1996) Tumor oxygenation predicts for the likelihood of distant metastases in human soft tissue sarcoma. *Cancer Res.* **56**, 941–943.
5. Hockel, M., Schlenger, K., Aral, B., Mitze, M., Schaffer, U., and Vaupel, P. (1996) Association between tumor hypoxia and malignant progression in advanced cancer of the uterine cervix. *Cancer Res.* **56**, 4509–4515.
6. Jenkins, W. T., Evans, S. M., and Koch, C. J. (2000) Hypoxia and necrosis in rat 9L glioma and Morris 7777 hepatoma tumors: comparative measurements using EF5 binding and the Eppendorf needle electrode. *Int. J. Radiat. Oncol. Biol. Phys.* **46**, 1005–1017.
7. Adam, M. F., Dorie, M. J., and Brown, J. M. (1999) Oxygen tension measurements of tumors growing in mice. *Int. J. Radiat. Oncol. Biol. Phys.* **45**, 171–180.
8. Lord, E. M., Harwell, L., and Koch, C. J. (1993) Detection of hypoxic cells by monoclonal antibody recognizing 2-nitroimidazole adducts. *Cancer Res.* **53**, 5721–5726.
9. Arteel, G. E., Thurman, R. G., and Raleigh, J. A. (1998) Reductive metabolism of the hypoxia marker pimonidazole is regulated by oxygen tension independent of the pyridine nucleotide redox state. *Eur. J. Biochem.* **253**, 743–750.
10. Raleigh, J. A., Miller, G. G., Franko, A. J., Koch, C. J., Fuciarelli, A. F., and Kelly, D. A. (1987) Fluorescence immunohistochemical detection of hypoxic cells in spheroids and tumours. *Br. J. Cancer* **56**, 395–400.
11. Aebersold, D. M., Burri, P., Beer, K. T., et al. (2001) Expression of hypoxia-inducible factor-1alpha: a novel predictive and prognostic parameter in the radiotherapy of oropharyngeal cancer. *Cancer Res.* **61**, 2911–2916.
12. Beasley, N. J., Wykoff, C. C., Watson, P. H., et al. (2001) Carbonic anhydrase IX, an endogenous hypoxia marker, expression in head and neck squamous cell carcinoma and its relationship to hypoxia, necrosis, and microvessel density. *Cancer Res.* **61**, 5262–5267.
13. Olive, P. L., Aquino-Parsons, C., MacPhail, S. H., et al. (2001) Carbonic anhydrase 9 as an endogenous marker for hypoxic cells in cervical cancer. *Cancer Res.* **61**, 8924–8929.

14. Zhong, H., De Marzo, A. M., Laughner, E., et al. (1999) Overexpression of hypoxia-inducible factor 1alpha in common human cancers and their metastases. *Cancer Res.* **59**, 5830–5835.

15. Salnikow, K., Costa, M., Figg, W. D., and Blagosklonny, M. V. (2000) Hyperinducibility of hypoxia-responsive genes without p53/p21-dependent checkpoint in aggressive prostate cancer. *Cancer Res.* **60**, 5630–5634.

16. Airley, R., Loncaster, J., Davidson, S., et al. (2001) Glucose transporter glut-1 expression correlates with tumor hypoxia and predicts metastasis-free survival in advanced carcinoma of the cervix. *Clin. Cancer Res.* **7**, 928–934.

17. Wykoff, C. C., Beasley, N. J., Watson, P. H., et al. (2000) Hypoxia-inducible expression of tumor-associated carbonic anhydrases. *Cancer Res.* **60**, 7075–7083.

18. Giatromanolaki, A., Koukourakis, M. I., Sivridis, E., et al. (2001) Expression of hypoxia-inducible carbonic anhydrase-9 relates to angiogenic pathways and independently to poor outcome in non-small cell lung cancer. *Cancer Res.* **61**, 7992–7998.

19. Koukourakis, M. I., Giatromanolaki, A., Sivridis, E., et al. (2001) Hypoxia-regulated carbonic anhydrase-9 (CA9) relates to poor vascularization and resistance of squamous cell head and neck cancer to chemoradiotherapy. *Clin. Cancer Res.* **7**, 3399–3403.

20. Younes, M., Lechago, J., Chakraborty, S., et al. (2000) Relationship between dysplasia, p53 protein accumulation, DNA ploidy, and Glut1 overexpression in Barrett metaplasia. *Scand. J. Gastroenterol.* **35**, 131–137.

21. Clifford, S. C. and Maher, E. R. (2001) Von Hippel-Lindau disease: clinical and molecular perspectives. *Adv. Cancer Res.* **82**, 85–105.

22. Ivan, M. and Kaelin, W. G., Jr. (2001) The von Hippel-Lindau tumor suppressor protein. *Curr. Opin. Genet. Dev.* **11**, 27–34.

23. Cockman, M. E., Masson, N., Mole, D. R., et al. (2000) Hypoxia inducible factor-alpha binding and ubiquitylation by the von Hippel-Lindau tumor suppressor protein. *J. Biol. Chem.* **275**, 25733–25741.

24. Maxwell, P. H., Wiesener, M. S., Chang, G. W., et al. (1999) The tumour suppressor protein VHL targets hypoxia-inducible factors for oxygen-dependent proteolysis [see comments]. *Nature* **399**, 271–275.

25. Ohh, M., Park, C. W., Ivan, M., et al. (2000) Ubiquitination of hypoxia-inducible factor requires direct binding to the beta-domain of the von Hippel-Lindau protein [see comments]. *Nat. Cell. Biol.* **2**, 423–427.

26. Wykoff, C. C., Pugh, C. W., Maxwell, P. H., Harris, A. L., and Ratcliffe, P. J. (2000) Identification of novel hypoxia dependent and independent target genes of the von Hippel-Lindau (VHL) tumour suppressor by mRNA differential expression profiling. *Oncogene* **19**, 6297–6305.

27. Ripley, B. D. (1996) *Pattern Recognition and Neural Networks.* Cambridge University Press, Cambridge, UK.

28. Wu, Y., Denhardt, D. T., and Rittling, S. R. (2000) Osteopontin is required for full expression of the transformed phenotype by the ras oncogene. *Br. J. Cancer* **83**, 156–163.

29. Thalmann, G. N., Sikes, R. A., Devoll, R. E., et al. (1999) Osteopontin: possible role in prostate cancer progression. *Clin. Cancer Res.* **5**, 2271–2277.

30. Tuck, A. B., Arsenault, D. M., O'Malley, F. P., et al. (1999) Osteopontin induces increased invasiveness and plasminogen activator expression of human mammary epithelial cells. *Oncogene* **18**, 4237–4246.

31. Greenblatt, M. S., Bennett, W. P., Hollstein, M., and Harris, C. C. (1994) Mutations in the p53 tumor suppressor gene: clues to cancer etiology and molecular pathogenesis. *Cancer Res.* **54**, 4855–4878.

32. Donehower, L. A., Harvey, M., Slagle, B. L., et al. (1992) Mice deficient for p53 are developmentally normal but susceptible to spontaneous tumours. *Nature* **356**, 215–221.

33. Graeber, T. G., Peterson, J. F., Tsai, M., Fornace, A. J., Jr., and Giaccia, A. J. (1994) Hypoxia induces the accumulation of p53 protein, but the activation of a G1-phase checkpoint by low oxygen conditions is independent of p53 status. *Mol. Cell. Biol.* **14**, 6264–6277.

34. Graeber, T. G., Osmanian, C., Jacks, T., et al. (1996) Hypoxia mediated selection of cells with diminished apoptotic potential in solid tumors. *Nature* **379**, 88–91.

35. Giaccia, A. J. and Kastan, M. B. (1998) The complexity of p53 modulation: emerging patterns from divergent signals. *Genes Dev.* **12**, 2973–2983.

36. Prives, C. (1998) Signaling to p53: breaking the MDM2-p53 circuit. *Cell* **95**, 5–8.

37. Levine, A. J. (1997) p53, the cellular gatekeeper for growth and division. *Cell* **88**, 323–331.

38. Maki, C. G. and Howley, P. M. (1997) Ubiquitination of p53 and p21 is differentially affected by ionizing and UV radiation. *Mol. Cell. Biol.* **17**, 355–363.

39. Mosner, J., Mummenbrauer, T., Bauer, C., Sczakiel, G., Grosse, F., and Deppert, W. (1995) Negative feedback regulation of wild-type p53 biosynthesis. *EMBO J.* **14**, 4442–4449.

40. Avantaggiati, M. L., Ogryzko, V., Gardner, K., Giodano, A., Levine, A. S., and Kelly, K. (1997) Recruitment of p300/CBP in p53-dependent signal pathways. *Cell* **89**, 1175–1184.

41. Lill, N. L., Grossman, S. R., Ginsberg, D., DeCaprio, J., and Livingston, D. M. (1997) Binding and modulation of p53 by p300/CBP coactivators. *Nature* **387**, 823–827.

42. Chen, X., Farmer, G., Zhu, H., Prywes, R., and Prives, C. (1993) Cooperative DNA binding of p53 with TFIID (TBP): a possible mechanism for transcriptional activation [published erratum appears in *Genes Dev.* 1993 Dec; 7(12B):2652]. *Genes Dev.* **7**, 1837–1849.

43. Farmer, G., Colgan, J., Nakatani, Y., Manley, J. L., and Prives, C. (1996) Functional interaction between p53, the TATA-binding protein (TBP), andTBP-associated factors in vivo. *Mol. Cell. Biol.* **16**, 4295–4304.

44. Farmer, G., Friedlander, P., Colgan, J., Manley, J. L., and Prives, C. (1996) Transcriptional repression by p53 involves molecular interactions distinct from those with the TATA box binding protein. *Nucleic Acids Res.* **24**, 4281–4288.

45. Haupt, Y., Maya, R., Kazaz, A., and Oren, M. (1997) Mdm2 promotes the rapid degradation of p53. *Nature* **387**, 296–299.

46. Kubbutat, M. H., Jones, S. N., and Vousden, K. H. (1997) Regulation of p53 stability by Mdm2. *Nature* **387**, 299–303.

47. Alarcon, R., Koumenis, C., Geyer, R. K., Maki, C. G., and Giaccia, A. J. (1999) Hypoxia induces p53 accumulation through MDM2 down-regulation and inhibition of E6-mediated degradation. *Cancer Res.* **59**, 6046–6051.

48. Pomerantz, J., Schreiber-Agus, N., Liegeois, N. J., et al. (1998) The Ink4a tumor suppressor gene product, p19Arf, interacts with MDM2 and neutralizes MDM2's inhibition of p53. *Cell* **92**, 713–723.

49. Zhang, Y., Xiong, Y., and Yarbrough, W. G. (1998) ARF promotes MDM2 degradation and stabilizes p53: ARF-INK4a locus deletion impairs both the Rb and p53 tumor suppression pathways. *Cell* **92**, 725–734.

50. An, W. G., Kanekal, M., Simon, M. C., Maltepe, E., Blagosklonny, M. V., and Neckers, L. M. (1998) Stabilization of wild-type p53 by hypoxia-inducible factor 1alpha. *Nature* **392**, 405–408.

51. Agani, F., Kirsch, D. G., Friedman, S. L., Kastan, M. B., and Semenza, G. L. (1997) p53 does not repress hypoxia-induced transcription of the vascular endothelial growth factor gene. *Cancer Res.* **57**, 4474–4477.

52. Wenger, R. H., Camenisch, G., Desbaillets, I., Chilov, D., and Gassmann, M. (1998) Up-regulation of hypoxia-inducible factor-1alpha is not sufficient for hypoxic/anoxic p53 induction. *Cancer Res.* **58**, 5678–5680.

53. Banin, S., Moyal, L., Shieh, S., et al. (1998) Enhanced phosphorylation of p53 by ATM in response to DNA damage. *Science* **281**, 1674–1677.

54. Canman, C. E. and Kastan, M. B. (1998) Small contribution of G1 checkpoint control manipulation to modulation of p53-mediated apoptosis. *Oncogene* **16**, 957–966.

55. Shieh, S. Y., Ikeda, M., Taya, Y., and Prives, C. (1997) DNA damage-induced phosphorylation of p53 alleviates inhibition by MDM2. *Cell* **91**, 325–334.

56. Siliciano, J. D., Canman, C. E., Taya, Y., Sakaguchi, K., Appella, E., and Kastan, M. B. (1997) DNA damage induces phosphorylation of the amino terminus of p53. *Genes Dev.* **11**, 3471–3481.

57. Hammond, E. M., Denko, N. C., Dorie, M. J., Abraham, R. T., and Giaccia, A. J. (2002) Hypoxia links ATR and p53 through replication arrest. *Mol. Cell. Biol.* **22**, 1834–1843.

58. Meek, D. W. Multisite phosphorylation and the integration of stress signals at p53. *Cell Signal.* **10**, 159–166.

59. Chen, X., Ko, L., Jayaraman, L., and Prives, C. (1996) p53 levels, functional domains, and DNA damage determine the extent of the apoptotic response of tumor cells. *Genes Dev.* **10**, 2438–2451.

60. Zhu, J., Zhou, W., Jiang, J., and Chen, X. (1998) Identification of a novel p53 functional domain that is necessary for mediating apoptosis. *J. Biol. Chem.* **273**, 13030–13036.

61. Caelles, C., Helmberg, A., and Karin, M. (1994) p53-dependent apoptosis in the absence of transcriptional activation of p53 target genes. *Nature* **370**, 220–223.

62. Wagner, A. J., Kokontis, J. M., and Hay, N. (1994) Myc-mediated apoptosis requires wild-type p53 in a manner independent of cell-cycle arrest and the ability of p53 to induce p21waf-1/cip1. *Genes Dev.* **8**, 2817–2830.

63. Ryan, K. M. and Vousden, K. H. (1998) Characterization of structural p53 mutants which show selective defects in apoptosis but not cell cycle arrest. *Mol. Cell. Biol.* **18**, 3692–3698.

64. Sakamuro, D., Sabbatini, P., White, E., and Prendergast, G. C. (1997) The polyproline region of p53 is required to activate apoptosis but not growth arrest. *Oncogene* **15**, 887–898.

65. Murphy, M., Hinman, A., and Levine, A. J. (1996) Wild-type p53 negatively regulates the expression of a microtubule- associated protein. *Genes Dev.* **10**, 2971–2980.

66. Roperch, J. P., Alvaro, V., Prieur, S., et al. (1998) Inhibition of presenilin 1 expression is promoted by p53 and p21WAF-1 and results in apoptosis and tumor suppression. *Nat. Med.* **4**, 835–838.

67. Venot, C., Maratrat, M., Dureuil, C., Conseiller, E., Bracco, L., and Debussche, L. (1998) The requirement for the p53 proline-rich functional domain for mediation of apoptosis is correlated with specific PIG3 gene transactivation and with transcriptional repression. *EMBO J.* **17**, 4668–4679.

68. Walker, K. K. and Levine, A. J. (1996) Identification of a novel p53 functional domain that is necessary for efficient growth suppression. *Proc. Natl. Acad. Sci. USA* **93**:15335–15340.

69. Murphy, M., Ahn, J., Walker, K. K., et al. (1999) Transcriptional repression by wild-type p53 utilizes histone deacetylases, mediated by interaction with mSin3a. *Genes Dev.* **13**, 2490–2501.

70. Koumenis, C., Alarcon, R., Hammond, E., et al. (2001) Regulation of p53 by hypoxia: dissociation of transcriptional repression and apoptosis from p53-dependent transactivation. *Mol. Cell. Biol.* **21**, 1297–1310.

71. Probst, H., Schiffer, H., Gekeler, V., et al. (1988) Oxygen dependent regulation of DNA synthesis and growth of Ehrlich ascites tumor cells in vitro and in vivo. *Cancer Res.* **48**, 2053–2060.

72. Wilson, R. E., Keng, P. C., and Sutherland, R. M. (1990) Changes in growth characteristics and macromolecular synthesis on recovery from severe hypoxia. *Br. J. Cancer* **61**, 14–21.

73. Gekeler, V., Epple, J., Kleymann, G., and Probst, H. (1993) Selective and synchronous activation of early S-phase replicons of Ehrlich ascites cells. *Mol. Cell. Biol.* **13**, 5020–5033.

74. Riedinger, H. J., Gekeler, V., and Probst, H. (1992) Reversible shutdown of replicon initiation by transient hypoxia in Ehrlich ascites cells. Dependence of initiation on short-lived protein. *Eur. J. Biochem.* **210**, 389–398.

75. Amellem, O., Stokke, T., Sandvik, J. A., and Pettersen, E. O. (1996) The retinoblastoma gene product is reversibly dephosphorylated and bound in the nucleus in S and G2 phases during hypoxic stress. *Exp. Cell Res.* **227**, 106–115.

76. MacManus, J. P., Koch, C. J., Jian, M., Walker, T., and Zurakowski, B. (1999) Decreased brain infarct following focal ischemia in mice lacking the transcription factor E2F1. *Neuroreport* **10**, 2711–2714.

77. Krtolica, A., Krucher, N. A., and Ludlow, J. W. (1998) Hypoxia-induced pRB hypophosphorylation results from downregulation of CDK and upregulation of PP1 activities. *Oncogene* **17**, 2295–2304.

78. Danielsen, T., Hvidsten, M., Stokke, T., Solberg, K., and Rofstad, E. K. (1998) Hypoxia induces p53 accumulation in the S-phase and accumulation of hypophosphorylated retinoblastoma protein in all cell cycle phases of human melanoma cells. *Br. J. Cancer* **78**, 1547–1558.

79. Krtolica, A., Krucher, N. A., and Ludlow, J. W. (1999) Molecular analysis of selected cell cycle regulatory proteins during aerobic and hypoxic maintenance of human ovarian carcinoma cells. *Br. J. Cancer* **80**, 1875–1883.

80. Green, S. L., Freiberg, R. A., and Giaccia, A. J. (2001) p21(Cip1) and p27(Kip1) regulate cell cycle reentry after hypoxic stress but are not necessary for hypoxia-induced arrest. *Mol. Cell. Biol.* **21**, 1196–1206.

81. Amellem, O., Sandvik, J. A., Stokke, T., and Pettersen, E. O. (1998) The retinoblastoma protein-associated cell cycle arrest in S-phase under moderate hypoxia is disrupted in cells expressing HPV18 E7 oncoprotein. *Br. J. Cancer* **77**, 862–872.

82. Ludlow, J. W., Howell, R. L., and Smith, H. C. (1993) Hypoxic stress induces reversible hypophosphorylation of pRB and reduction in cyclin A abundance independent of cell cycle progression. *Oncogene* **8**, 331–339.

83. Semenza, G. L., Roth, P. H., Fang, H.-M., and Wang, G. L. (1994) Transcriptional regulation of genes encoding glycolytic enzymes by hypoxia-inducible factor 1. *J. Biol. Chem.* **269**, 23757–23767.

84. Goldberg, M. A., Dunning, S. P., and Bunn, H. F. (1988) Regulation of the erythropoietin gene: evidence that the oxygen sensor is a heme protein. *Science* **242**, 1412–1415.

85. Goldberg, M. A., Glass, G. A., Cunningham, J. M. and Bunn, H. F. (1987) The regulated expression of erythropoietin by two human hepatoma cell lines. *Proc. Natl. Acad. Sci. USA* **84**, 7972–7976.

86. Shweiki, D., Itin, A., Soffer, D., and Keshet, E. (1992) Vascular endothelial growth factor induced by hypoxia may mediate hypoxia-initiated angiogenesis. *Nature* **359**, 843–845.

87. Graham, C. H., Forsdike, J., Fitzgerald, C. J., and Macdonald-Goodfellow, S. (1999) Hypoxia-mediated stimulation of carcinoma cell invasiveness via upregulation of urokinase receptor expression. *Int. J. Cancer* **80**, 617–623.

88. Imagawa, S., Goldberg, M. A., Doweiko, J., and Bunn, H. F. (1991) Regulatory elements of the erythropoietin gene. *Blood* **77**, 278–285.

89. Semenza, G. L., Nejfelt, M. K., Chi, S. M., and Antonarakis, S. E. (1991) Hypoxia-inducible nuclear factors bind to an enhancer element located 3' to the erythropoietin gene. *Proc. Natl. Acad. Sci. USA* **88**, 5680–5684.

90. Wang, G. L. and Semenza, G. L. (1995) Purification and characterization of hypoxia-inducible factor 1. *J. Biol. Chem.* **270**, 1230–1237.

91. Goldberg, M. A. and Schneider, T. J. (1994) Similarities between the oxygen-sensing mechanisms regulating the expression of vascular endothelial growth factor and erythropoietin. *J. Biol. Chem.* **269**, 4355–4359.

92. Huang, L. E., Arany, Z., Livingston, D. M., and Bunn, H. F. (1996) Activation of hypoxia-inducible transcription factor depends primarily upon redox-sensitive stabilization of its alpha subunit. *J. Biol. Chem.* **271**, 32253–32259.

93. Wenger, R. H., Kvietikova, I., Rolfs, A., Gassmann, M., and Marti, H. H. (1997) Hypoxia-inducible factor-1 alpha is regulated at the post-mRNA level. *Kidney Int.* **51**, 560–563.

94. Richard, D. E., Berra, E., Gothie, E., Roux, D., and Pouyssegur, J. (1999) p42/p44 mitogen-activated protein kinases phosphorylate hypoxia-inducible factor 1alpha (HIF-1alpha) and enhance the transcriptional activity of HIF-1. *J. Biol. Chem.* **274**, 32631–32637.

95. Baek, S. H., Lee, U. Y., Park, E. M., Han, M. Y., Lee, Y. S., and Park, Y. M. (2001) Role of protein kinase Cdelta in transmitting hypoxia signal to HSF and HIF-1. *J. Cell Physiol.* **188**, 223–235.

96. Mazure, N. M., Chen, E. Y., Laderoute, K. R., and Giaccia, A. J. (1997) Induction of vascular endothelial growth factor by hypoxia is modulated by a phosphatidylinositol 3-kinase/Akt signaling pathway in Ha-ras-transformed cells through a hypoxia inducible factor-1 transcriptional element. *Blood* **90**, 3322–3331.

97. Zundel, W., Schindler, C., Haas-Kogan, D., et al. (2000) Loss of PTEN facilitates HIF-1-mediated gene expression. *Genes Dev.* **14**, 391–396.

98. Pugh, C. W., O'Rourke, J. F., Nagao, M., Gleadle, J. M., and Ratcliffe, P. J. (1997) Activation of hypoxia-inducible factor-1; definition of regulatory domains within the alpha subunit. *J. Biol. Chem.* **272**, 11205–11214.

99. Srinivas, V., Zhang, L. P., Zhu, X. H., and Caro, J. (1999) Characterization of an oxygen/redox-dependent degradation domain of hypoxia-inducible factor alpha (HIF-alpha) proteins. *Biochem. Biophys. Res. Commun.* **260**, 557–561.

100. Jiang, B. H., Zheng, J. Z., Leung, S. W., Roe, R., and Semenza, G. L. (1997) Transactivation and inhibitory domains of hypoxia-inducible factor 1alpha. Modulation of transcriptional activity by oxygen tension. *J. Biol. Chem.* **272**, 19253–19260.

101. Ema, M., Taya, S., Yokotani, N., Sogawa, K., Matsuda, Y., and Fujii-Kuriyama, Y. (1997) A novel bHLH-PAS factor with close sequence similarity to hypoxia-inducible factor 1alpha regulates the VEGF expression and is potentially involved in lung and vascular development. *Proc. Natl. Acad. Sci. USA* **94**, 4273–4278.

102. Flamme, I., Frohlich, T., von Reutern, M., Kappel, A., Damert, A., and Risau, W. (1997) HRF, a putative basic helix-loop-helix-PAS-domain transcription factor is closely related to hypoxia-inducible factor-1 alpha and developmentally expressed in blood vessels. *Mech. Dev.* **63**, 51–60.

103. Hogenesch, J. B., Chan, W. K., Jackiw, V. H., et al. (1997) Characterization of a subset of the basic-helix-loop-helix-PAS superfamily that interacts with components of the dioxin signaling pathway. *J. Biol. Chem.* **272**, 8581–8593.

104. Tian, H., McKnight, S. L., and Russell, D. W. (1997) Endothelial PAS domain protein 1 (EPAS1), a transcription factor selectively expressed in endothelial cells. *Genes Dev.* **11**, 72–82.

105. Gu, Y. Z., Moran, S. M., Hogenesch, J. B., Wartman, L., and Bradfield, C. A. (1998) Molecular characterization and chromosomal localization of a third alpha-class hypoxia inducible factor subunit, HIF3alpha. *Gene Expr.* **7**, 205–213.

106. Huang, L. E., Gu, J., Schau, M., and Bunn, H. F. (1998) Regulation of hypoxia-inducible factor 1alpha is mediated by an O_2-dependent degradation domain via the ubiquitin-proteasome pathway. *Proc. Natl. Acad. Sci. USA* **95**, 7987–7992.

107. Salceda, S. and Caro, J. (1997) Hypoxia-inducible factor 1alpha (HIF-1alpha) protein is rapidly degraded by the ubiquitin-proteasome system under normoxic conditions. Its stabilization by hypoxia depends on redox-induced changes. *J. Biol. Chem.* **272**, 22642–22647.

108. Gassmann, M., Chilov, D., and Wenger, R. H. (2000) Regulation of the hypoxia-inducible factor-1 alpha. ARNT is not necessary for hypoxic induction of HIF-1 alpha in the nucleus [In Process Citation]. *Adv. Exp. Med. Biol.* **475**, 87–99.

109. Kamura, T., Koepp, D. M., Conrad, M. N., et al. (1999) Rbx1, a component of the VHL tumor suppressor complex and SCF ubiquitin ligase. *Science* **284**, 657–661.

110. Kibel, A., Iliopoulos, O., DeCaprio, J. A., and Kaelin, W. G., Jr. (1995) Binding of the von Hippel-Lindau tumor suppressor protein to Elongin B and C. *Science* **269**, 1444–1446.

111. Lonergan, K. M., Iliopoulos, O., Ohh, M., et al. (1998) Regulation of hypoxia-inducible mRNAs by the von Hippel-Lindau tumor suppressor protein requires binding to complexes containing elongins B/C and Cul2. *Mol. Cell. Biol.* **18**, 732–741.

112. Pause, A., Lee, S., Worrell, R. A., et al. (1997) The von Hippel-Lindau tumor-suppressor gene product forms a stable complex with human CUL-2, a member of the Cdc53 family of proteins. *Proc. Natl. Acad. Sci. USA* **94**, 2156–2161.

113. Pause, A., Peterson, B., Schaffar, G., Stearman, R., and Klausner, R. D. (1999) Studying interactions of four proteins in the yeast two-hybrid system: structural resemblance of the pVHL/elongin BC/hCUL-2 complex with the ubiquitin ligase complex SKP1/cullin/F-box protein. *Proc. Natl. Acad. Sci. USA* **96**, 9533–9538.

114. Deshaies, R. J. (1999) SCF and Cullin/Ring H2-based ubiquitin ligases. (1999) *Annu. Rev. Cell. Dev. Biol.* **15**, 435–467.

115. Koepp, D. M., Harper, J. W. and Elledge, S. J. (1999) How the cyclin became a cyclin: regulated proteolysis in the cell cycle. *Cell* **97**, 431–434.

116. Tyers, M. and Jorgensen, P. (2000) Proteolysis and the cell cycle: with this RING I do thee destroy. *Curr. Opin. Genet. Dev.* **10**, 54–64.

117. Duan, D. R., Pause, A., Burgess, W. H., et al. (1995) Inhibition of transcription elongation by the VHL tumor suppressor protein. *Science* **269**, 1402–1406.

118. Ohh, M., Takagi, Y., Aso, T., et al. (1999) Synthetic peptides define critical contacts between elongin C, elongin B, and the von Hippel-Lindau protein. *J. Clin. Invest.* **104**, 1583–1591.

119. Kamura, T., Sato, S., Iwai, K., Czyzyk-Krzeska, M., Conaway, R. C., and Conaway, J. W. (2000) Activation of HIF1alpha ubiquitination by a reconstituted von Hippel- Lindau (VHL) tumor suppressor complex. *Proc. Natl. Acad. Sci. USA* **97**, 10430–10435.

120. Tanimoto, K., Makino, Y., Pereira, T., and Poellinger, L. (2000) Mechanism of regulation of the hypoxia-inducible factor-1 alpha by the von Hippel-Lindau tumor suppressor protein. *EMBO J.* **19**, 4298–4309.

121. Bruick, R. K. and McKnight, S. L. (2001) A conserved family of prolyl-4-hydroxylases that modify HIF. *Science* **294**, 1337–1340.

122. Ivan, M., Kondo, K., Yang, H., et al. (2001) HIFalpha targeted for VHL-mediated destruction by proline hydroxylation: implications for O_2 sensing. *Science* **292**, 464–468.

123. Jaakkola, P., Mole, D. R., Tian, Y. M., et al. (2001) Targeting of HIF-alpha to the von Hippel-Lindau ubiquitylation complex by O_2-regulated prolyl hydroxylation. *Science* **292**, 468–472.

124. Ema, M., Hirota, K., Mimura, J., et al. (1999) Molecular mechanisms of transcription activation by HLF and HIF1alpha in response to hypoxia: their stabilization and redox signal-induced interaction with CBP/p300. *EMBO J.* **18**, 1905–1914.

125. Gu, J., Milligan, J., and Huang, L. E. (2001) Molecular mechanism of hypoxia-inducible factor 1alpha -p300 interaction. A leucine-rich interface regulated by a single cysteine. *J. Biol. Chem.* **276**, 3550–3554.

126. Kung, A. L., Wang, S., Klco, J. M., Kaelin, W. G., and Livingston, D. M. (2000) Suppression of tumor growth through disruption of hypoxia-inducible transcription. *Nat. Med.* **6**, 1335–1340.

127. Lando, D., Peet, D. J., Whelan, D. A., Gorman, J. J., and Whitelaw, M. L. (2002) Asparagine hydroxylation of the HIF transactivation domain a hypoxic switch. *Science* **295**, 858–861.
128. Arbiser, J. L., Moses, M. A., Fernadez, C. A., et al. (1997) Oncogenic H-ras stimulates tumor angiogenesis by two distinct pathways. *Proc. Natl. Acad. Sci. USA* **94**, 861–866.
129. Jiang, B. H., Jiang, G., Zheng, J. Z., Lu, Z., Hunter, T., and Vogt, P. K. (2001) Phosphatidylinositol 3-kinase signaling controls levels of hypoxia-inducible factor 1. *Cell Growth Differ.* **12**, 363–369.
130. Cantley, L. C. and Neel, B. G. (1999) New insights into tumor suppression: PTEN suppresses tumor formation by restraining the phosphoinositide 3-kinase/AKT pathway. *Proc. Natl. Acad. Sci. USA* **96**, 4240–4245.
131. Blancher, C., Moore, J. W., Robertson, N., and Harris, A. L. (2001) Effects of ras and von Hippel-Lindau (VHL) gene mutations on hypoxia-inducible factor (HIF)-1alpha, HIF-2alpha, and vascular endothelial growth factor expression and their regulation by the phosphatidylinositol 3′-kinase/Akt signaling pathway. *Cancer Res.* **61**, 7349–7355.
132. Laughner, E., Taghavi, P., Chiles, K., Mahon, P. C., and Semenza, G. L. (2001) HER2 (neu) signaling increases the rate of hypoxia-inducible factor 1alpha (HIF-1alpha) synthesis: novel mechanism for HIF-1-mediated vascular endothelial growth factor expression. *Mol. Cell. Biol.* **21**, 3995–4004.
133. Zhong, H., Chiles, K., Feldser, D., et al. (2000) Modulation of hypoxia-inducible factor 1alpha expression by the epidermal growth factor/phosphatidylinositol 3-kinase/PTEN/AKT/FRAP pathway in human prostate cancer cells: implications for tumor angiogenesis and therapeutics. *Cancer Res.* **60**, 1541–1545.
134. Alvarez-Tejado, M., Alfranca, A., Aragones, J., Vara, A., Landazuri, M. O., and del Peso, L. (2002) Lack of evidence for the involvement of the phosphoinositide 3-kinase/Akt pathway in the activation of hypoxia-inducible factors by low oxygen tension. *J. Biol. Chem.* **277**, 13,508–13,517.
135. Brown, J. M. and Giaccia, A. J. (1998) The unique physiology of solid tumors: opportunities (and problems) for cancer therapy. *Cancer Res.* **58**, 1408–1416.
136. Hanahan, D. and Folkman, J. (1996) Patterns and emerging mechanisms of the angiogenic switch during tumorigenesis. *Cell* **86**, 353–364.
137. Naik, P., Karrim, J., and Hanahan, D. (1996) The rise and fall of apoptosis during multistage tumorigenesis: down-modulation contributes to tumor progression from angiogenic progenitors. *Genes Dev.* **10**, 2105–2116.
138. Boehm, T., Folkman, J., Browder, T., and O'Reilly, M. S. (1997) Antiangiogenic therapy of experimental cancer does not induce acquired drug resistance. *Nature* **390**, 404–407.
139. Bornstein, P. and Sage, E. H. (1994) Thrombospondins. *Meth. Enzymol.* **245**, 62–85.
140. Fidler, I. J. and Ellis, L. M. (1994) The implications of angiogenesis for the biology end therapy of cancer metastasis. *Cell* **79**, 185–188.
141. Volpert, O. V., Stellmach, V., and Bouck, N. (1995) The modulation of thrombospondin and other naturally occurring inhibitors of angiogenesis during tumor progression. *Breast Cancer Res. Treat.* **36**, 119–126.
142. Volpert, O. V., Tolsma, S. S., Pellerin, S., et al. (1995) Inhibition of angiogenesis by thrombospondin-2. *Biochem. Biophys. Res. Commun.* **217**, 326–332.
143. Adams, J. C. (2001) Thrombospondins: multifunctional regulators of cell interactions. *Annu. Rev. Cell Dev. Biol.* **17**, 25–51.
144. de Fraipont, F., Nicholson, A. C., Feige, J. J., and Van Meir, E. G. (2001) Thrombospondins and tumor angiogenesis. *Trends Mol. Med.* **7**, 401–407.
145. Jimenez, B., Volpert, O. V., Crawford, S. E., Febbraio, M., Silverstein, R. L., and Bouck, N. (2000) Signals leading to apoptosis-dependent inhibition of neovascularization by thrombospondin-1. *Nat. Med.* **6**, 41–48.

146. Nor, J. E., Mitra, R. S., Sutorik, M. M., Mooney, D. J., Castle, V. P., and Polverini, P. J. (2000) Thrombospondin-1 induces endothelial cell apoptosis and inhibits angiogenesis by activating the caspase death pathway. *J. Vasc. Res.* **37**, 209–218.

147. Kyriakides, T. R., Tam, J. W. and Bornstein, P. (1999) Accelerated wound healing in mice with a disruption of the thrombospondin 2 gene. *J. Invest. Dermatol.* **113**, 782–787.

148. Lawler, J., Sunday, M., Thibert, V., et al. (1998) Thrombospondin-1 is required for normal murine pulmonary homeostasis and its absence causes pneumonia. *J. Clin. Invest.* **101**, 982–992.

149. Streit, M., Velasco, P., Riccardi, L., et al. (2000) Thrombospondin-1 suppresses wound healing and granulation tissue formation in the skin of transgenic mice. *EMBO J.* **19**, 3272–3282.

150. Dameron, K. M., Volpert, O. V., Tainsky, M. A. and Bouck, N. (1994) Control of angiogenesis in fibroblasts by p53 regulation of Thrombospondin -1. *Science* **265**, 1582–1584.

151. Volpert, O. V., Dameron, K. M. and Bouck, N. (1997) Sequential development of an angiogenic phenotype by human fibroblasts progressing to tumorigenicity. *Oncogene* **14**, 1495–1502.

152. Bleuel, K., Popp, S., Fusenig, N. E., Stanbridge, E. J., and Boukamp, P. (1999) Tumor suppression in human skin carcinoma cells by chromosome 15 transfer or thrombospondin-1 overexpression through halted tumor vascularization. *Proc. Natl. Acad. Sci. USA* **96**, 2065–2070.

153. Streit, M., Velasco, P., Brown, L. F., et al. (1999) Overexpression of thrombospondin-1 decreases angiogenesis and inhibits the growth of human cutaneous squamous cell carcinomas. *Am. J. Pathol.* **155**, 441–452.

154. Li, Q., Ahuja, N., Burger, P. C., and Issa, J. P. (1999) Methylation and silencing of the Thrombospondin-1 promoter in human cancer. *Oncogene* **18**, 3284–3289.

155. Mettouchi, A., Cabon, F., Montreau, N., et al. (1994) SPARC and thrombospondin genes are repressed by the c-jun oncogene in rat embryo fibroblasts. *EMBO J.* **13**, 5668–5678.

156. Slack, J. L. and Bornstein, P. (1994) Transformation by v-src causes transient induction followed by repression of mouse thrombospondin-1. *Cell Growth Differ.* **5**, 1373–1380.

157. Tikhonenko, A. T., Black, D. J., and Linial, M. L. (1996) Viral Myc oncoproteins in infected fibroblasts down-modulate thrombospondin-1, a possible tumor suppressor gene. *J. Biol. Chem.* **271**, 30741–30747.

158. Ausserer, W. A., Bourrat-Floeck, B., Green, C. J., Laderoute, K. R. and Sutherland, R. M. (1994) Regulation of c-jun expression during hypoxic and low glucose stress. *Mol. Cell Biol.* **14**, 5032–5042.

159. Laderoute, K. R., Alarcon, R. M., Brody, M. D., et al. (2000) Opposing effects of hypoxia on expression of the angiogenic inhibitor thrombospondin 1 and the angiogenic inducer vascular endothelial growth factor. *Clin. Cancer Res.* **6**, 2941–2950.

160. Tenan, M., Fulci, G., Albertoni, M., et al. (2000) Thrombospondin-1 is downregulated by anoxia and suppresses tumorigenicity of human glioblastoma cells. *J. Exp. Med.* **191**, 1789–1798.

161. Castle, V., Varani, J., Fligiel, S., Prochownik, E. V., and Dixit, V. (1991) Antisense-mediated reduction in thrombospondin reverses the malignant phenotype of a human squamous carcinoma. *J. Clin. Invest.* **87**, 1883–1888.

162. Wang, T. N., Qian, X. H., Granick, M. S., et al. (1996) Inhibition of breast cancer progression by an antibody to a thrombospondin-1 receptor. *Surgery* **120**, 449–454.

163. Bornstein, P. (1995) Diversity of function is inherent in matricellular proteins: an appraisal of thrombospondin 1. *J. Cell Biol.* **130**, 503–506.

164. Roberts, D. D. (1996) Regulation of tumor growth and metastasis by thrombospondin-1. *FASEB J.* **10**, 1183–1191.

32

Hormonal and Differentiation Agents in Cancer Growth Suppression

Mikhail V. Blagosklonny

1. What Do Hormones and Differentiation Agents Have in Common?

Although chemotherapeutic agents are a cornerstone of therapy in medical oncology, hormonal and differentiating therapies play an increasingly important role. There is no clear-cut margin between chemotherapy, hormonal, and differentiating modalities. Although chemotherapeutic agents are intended to kill cancer cells, they may induce differentiation. On the other hand, the induction of apoptosis by differentiating and hormonal agents is also an important mode of their action.

There are several common features in hormonal and differentiating therapies.

1. Physiologic relevance: hormones and differentiating agents (or their prototypes) are physiological regulators of cell growth and differentiation.
2. Cellular pathways and targets: many hormones (androgens, estrogens, glucocorticoids) and their antagonists (antihormones) as well as many differentiating agents (retinoids, vitamin D) act by binding to nuclear transcription factors that serve as their receptors. However, the differentiating agents are a broader group than the ligands of transcription factors/receptors, and include inhibitors of histone deacetylases, of DNA methylation, activators of kinases, and so on.
3. Tissue selectivity: there is selectivity in the ability of hormones and differentiating agents to affect certain tissues. Tissue selectivity is an especially important advantage compared with standard chemotherapy, since it allows drugs not to attack other tissue, thus diminishing side effects. For example, prostate epithelial and testicular cells depend selectively on androgens for survival and proliferation.
4. Nonapoptotic effects: although these agents can also induce apoptosis, there is an emphasis on their ability to control cell growth without apoptosis. In fact, many side effects of chemotherapy are due to apoptosis of normal cells. Lymphoid and hematopoietic organs, intestinal epithelia, hair follicules, and the testis are especially vulnerable to nonselective apoptosis-inducing drugs. Thus, utilizing tissue-selective differentiation-promoting agents offers an important advantage.

From: *Methods in Molecular Biology, Vol. 223: Tumor Suppressor Genes: Regulation, Function, and Medicinal Applications.* Edited by: Wafik S. El-Deiry © Humana Press Inc., Totowa, NJ

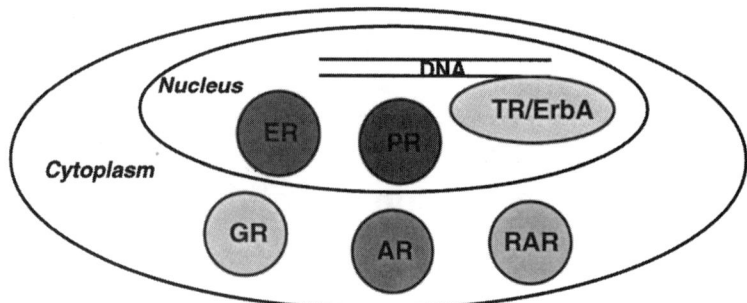

Fig. 1. Cellular location of unligated receptors of steroid hormone receptors superfamily. Thyroid hormone receptor (TR), estrogen receptor (ER), progesterone receptor (PR), glucocorticoid receptor (GR), retinoic acid receptor (RAR), androgen receptor (AR). Upon ligand binding, GR, RAR, and AR translocate to the nucleus.

2. Hormonal Therapy

2.1. Steroid Hormone Receptors Superfamily

Physiologic and therapeutic activities of steroid hormones are mediated by the super-family of nuclear receptors, which include those for steroid and thyroid hormones as well as retinoids. These receptors regulate nearly all aspects metazoan life *(1)*. Steroid/nuclear receptors are ligand-dependent transcription factors that can stimulate gene expression and regulate proliferation, differentiation, and specific functions of target tissues. Unligated receptors exist in inactive complexes with chaperone proteins such as heat-shock proteins (HSPs) *(2)*. In hormone-free cells, estrogen and progesterone receptors are localized predominantly in the nucleus, whereas androgen (AR), retinoic acid (RAR), and glucocorticoid (GR) receptor *(2,3)* are localized predominantly in the cytoplasm (**Fig. 1**). In the hormone-free GR, NLS is masked by the hormone-binding domain. Upon ligand binding, the GR translocates to the nucleus. In fact, the unliganded GR is a complex of a ligand-binding protein, HSP90, HSP70, HSP40, p23, and HOP *(4)*. It has been demonstrated, however, that only Hsp90 and HSP70 are required for the activation of glucocorticoid receptors in the presence of steroids *(5)*. Upon binding of glucocorticoid to the receptor, the complex moves along the microtubules to the nucleus. The HSP90-specific inhibitor geldanamycin impedes hormone-dependent GR translocation *(6)*. The geldanamycin-inhibited movement of the GR required intact microtubules *(7)*, and when cytoskeleton was disrupted the translocation became geldanamycin-insensitive. Once in the nucleus, GR binds to a specific site in DNA, and thus transmits signal(s). HSP70 also works as a molecular chaperone when the ligand-binding protein enters the nucleus *(8)*. Unlike most receptors of the superfamily (**Fig. 1**), the unliganded thyroid hormone receptor and highly related viral oncoprotein v-erbA associate with a co-repressor complex containing NCoR, SIN3, and histone deacetylase, bind DNA, and repress transcription. In the presence of ligand, the thyroid hormone receptor undergoes a conformational change that weakens interactions with the co-repressor complex while facilitating the recruitment of transcriptional co-activators such as p300 and PCAF that possess histone acetyltransferase activity *(9)*.

Steroid receptors interact with other transcription factors. For example, dexamethasone-activated GR inhibits p53-dependent transactivation of Bax and p21[WAF1/CIP1], cell cycle arrest, and apoptosis. GR forms a complex with p53, resulting in cytoplasmic sequestration of both p53 and GR. In neuroblastoma cells, cytoplasmic retention and inactivation of wild-type p53 involves GR *(10)*.

2.2. Tissue-Specific Hormonal Therapy

2.2.1. Antihormonal Therapy in Prostate Cancer

Normal prostate epithelial cells depend on androgens for their growth and survival. Androgen ablation induces apoptosis of androgen-dependent prostatic glandular cells, with 20% rate per day between d 2 and 5 after castration. p53 gene expression is not required, since the same degree of cell death occurred in prostates and seminal vesicles after castration of wild-type and p53-deficient mice *(11)*.

In 1941, it was established that the initial growth of prostate cancer is dependent on testosterone. Androgen suppression in the form of medical or surgical castration with inhibition of adrenal androgens became the standard primary treatment for advanced prostate cancer *(12)*. Following androgen suppression, recurrent prostate carcinomas are able to avoid apoptosis in the androgen-deprived environment. The clinical progression of prostate cancer during androgen withdrawal is associated with increased cell proliferation and decreased apoptosis *(13)*. Overexpression of Bcl-2 can enable prostate cancer cells to remain viable despite castrate levels of androgen, thus selecting for upregulation of Bcl-2 in prostate cancer cells *(14)*. During the course of prostate cancer progression, cells convert from an androgen-dependent to an androgen-independent growth status. Prostate cancer cells that have acquired the ability to survive and grow in a low-androgen environment might be activating the AR pathway using growth factors, cytokines, and steroids other than androgens *(15)*. Once a relapse occurs following primary endocrine treatment, metastatic prostate cancer is one of the most therapy-resistant human neoplasms *(16)*.

Like prostate cells, male testicular germ cells depend on testosterone for their survival *(17)*. Unlike prostate cancer, testicular cancer is highly curable by apoptosis-inducing chemotherapy *(18)*.

2.2.2. Antihormonal Therapy in Breast Cancer

The breast is a target organ for estrogens and progesterone. These hormones control several functions of the normal and abnormal mammary epithelium, including cell proliferation *(19)*. Estrogen is the major steroid mitogen for the luminal epithelial cell population. As demonstrated by postmenopausal hormone replacement therapy, estrogens plus progestines induce proliferation of breast epithelial cells localized to the terminal duct-lobular unit of the breast, which is the site of development of most breast cancers *(20)*.

Estrogens may exert its effects on cell proliferation indirectly via paracrine and/or autocrine secretion of GF (**Fig. 2**). According to some studies, there is dissociation between the population of luminal epithelial cells expressing the estrogen receptor and those that proliferate *(21)*. It has been speculated that the ER-negative proliferating cells represent a precursor cell population that differentiates to ER-containing, nonproliferative cells. In turn, these ER-positive cells act as "estrogen sensors" and transmit positive or negative paracrine growth signals to the precursor cells, depending on the prevailing hormonal environment *(21)*. It has been shown that autocrine PDGF mediates the estrogen mitogenic effect via stimulation of cyclin D1 in breast cancer cells *(22)*.

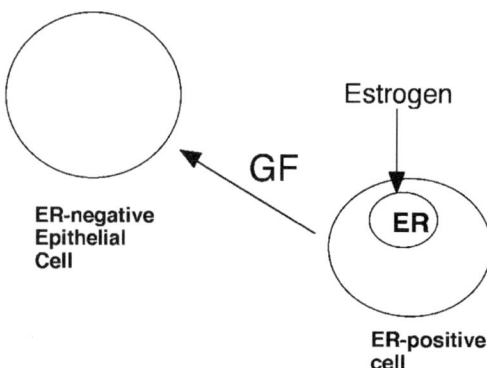

Fig. 2. Paracrine model of estrogen-stimulated proliferation of breast epithelia cells.

Since estrogens are mitogenic for breast epithelial cells, antiestrogens are used for therapy of breast cancer. Chemotherapy alone are insufficient for the younger age group, and these patients should strongly consider additional endocrine therapies (tamoxifen or ovarian ablation) if their tumors express estrogen receptors *(23)*.

Aromatase inhibitors are now used in postmenopausal women to block the *in situ* conversion of adrenal androgens to estrogens. This approach can only be successful if ER in a particular tumor is not directly stimulated by adrenal androgens. Thus, adrenal androgens must also be considered as risk factors in postmenopausal women *(24)*. Finally, loss of hormone dependency of breast cancer cells limits the antiestrogen therapy.

2.2.3. Endometrial Cancer

Endometrial glandular proliferation is inhibited by endogenous progesterone and is markedly reduced in premenopausal women receiving a synthetic progestin and in untreated postmenopausal women *(25)*. In endometrial carcinomas, treatment with progesterone suppresses cell proliferation and induce differentiation without altering apoptosis. This results in a shift in tissue kinetics toward a relative predominance cancer cell loss *(26)*. After progesterone therapy, in the response group of endometrial carcinoma, squamous differentiation of tumor cells was associated with downregulation of Bcl-2, ER, and PR *(27)*.

2.2.4. Steroid Hormones and Leukemia

Glucocorticoid-induced apoptosis is a well-recognized physiologic regulator of lymphocyte death and proliferation. DNA binding of the GR and transcriptional regulation of specific genes is required for glucocorticoid-induced apoptosis. However, it is not clear whether the activation or repression of responsive genes is essential *(28)*. Peripheral blood T-cells show sensitivity to glucocorticoid-induced apoptosis soon after the proliferative response to a mitogenic stimulation, and are also sensitive to spontaneous (i.e., growth factor deprivation-dependent) apoptosis. CD8+ T-cells were more sensitive than CD4+ T-cells. Interleukin-2, -4, and -10 protected T-cells from apoptosis *(29)*.

Glucocorticoids induced-apoptosis in cells of the lymphoid lineage at certain stages of differentiation has been exploited to a great extent in the therapy of malignant lymphoproliferative disorders. Glucocorticoids (dexamethasone and prednisone) are included in almost all treatment regimens for childhood acute lymphoblastic leukemia. Prolonged, i.e., 28-d, glucocorticoid therapy may be unnecessary, as exposure to glucocorticoid induces downregulation of glucocorticoid receptors *(30)*. Leukemic blast sensitivity to glucocorticoids correlates with sensitivity to chemotherapeutic agents and with outcome after multiagent therapy, suggesting overlapping mechanisms of action, and focusing attention on the determinants of the threshold for apoptosis *(30)*. In-vitro and in-vivo response of leukemic cells to glucocorticosteroids is highly predictive of outcome *(31)*.

Sensitivity of leukemic blasts to steroid therapy is one of the prognostic factors in acute lymphoblastic leukemia (ALL). Apoptosis was most pronounced in the first 6 h of treatment with prednisone and showed a good correlation with the clinical response. The number of steroid receptors on the blast cells was less informative than the in-vivo steroid response itself *(32)*. High levels of glucocorticoid receptors in leukemic cells may predict response to combination chemotherapy in acute lymphoblastic leukemia patients *(33)*.

Effect of glucocorticoid therapy is, however, hampered by the occurrence of resistant clones evolving under selective glucocorticoid pressure. At relapse, loss of in-vitro sensitivity to glucocorticosteroids is common and out of proportion to the loss of sensitivity to other agents. Glucocorticoid-induced apoptosis does not require p53. Glucocorticosteroid resistance is most commonly linked to altered receptor number or function leukemic cell lines *(31)*. Common mechanisms of glucocorticoid resistance may involve decreased expression of glucocorticoid receptor, mutant protein with low ligand-binding affinity, abnormal HSP expression, and increased Bcl-2 and Bcl-xL *(8,34)*.

The role of Bcl-2: Exogenous Bcl-2 protects pre-B- and immature B-cells from dexamethasone-induced cell death *(35)*. Bcl-2 is highly expressed in CD43+ B-cell precursors (pro-B-cells) and mature B-cells but downregulated at the pre-B- and immature B-cell stages of development. Susceptibility to apoptosis correlates with the levels of Bcl-2 protein *(35)*. High levels of Bcl-2 and Bcl-x renders intestinal intraepithelial lymphocytes resistant to apoptosis by glucocorticoids *(36)*. Up to about 48 h, exogenous Bcl-2 almost completely protected T-cell leukemia blasts from apoptosis. However, when the cells were cultured for another 24 h in the continuous presence of glucocorticoids, they underwent massive apoptosis that was associated with DEVDase activity and PARP cleavage. Bcl-2 did not markedly affect glucocorticoid-mediated growth arrest, thereby separating the antiproliferative from the apoptosis-inducing effect of glucocorticoid. Moreover, Bcl-2 did not prevent the dramatic reduction in the levels of several mRNAs observed during glucocorticoid treatment, including the transgenic Bcl-2 mRNA. Thus, Bcl-2 can be placed upstream of effector caspase activation, but downstream of other glucocorticoid-regulated events, such as growth arrest and the potentially critical repression of steady-state levels of multiple mRNA *(37)*.

2.2.5. High Doses of Sex Steroids and Antihormones

At micromolar concentrations, many sex steroids and their antagonists are cytotoxic in a receptor-independent manner. For example, tamoxifen at higher doses inhibits cell proliferation of estrogen receptor-negative cancer and leukemia cells. IC_{50}s for growth inhibition of several prostate cancer cell lines ranged from 5.5 to 10 μM, and were not affected by estrogen. Tamoxifen treatment was associated with inhibition of PKC, induction of p21[WAF1/CIP1], Rb dephosphorylation, and G1-phase cell cycle arrest *(38)*. In ER-negative lung cancer cells, tamoxifen induced G1 growth arrest, but the expression of G1 cyclins (including D1, 2, 3, and E) was not altered. The expression of p21 and p27, but not p57, was strongly activated by tamoxifen, leading to hypophosphorylated Rb *(39)*.

Although tamoxifen is one of the most potent inhibitors of cell growth, other sex steroids cause growth inhibition in receptor-negative manner. The sensitivity of 10 leukemia cell lines to sex steroids did not correlate with their sensitivity to glucocorticoids *(40)*. The order of potency was tamoxifen = diethylstilbestrol (DES) > progesterone > testasterone > estradiol *(40)*. Although cytotoxic doses of sex steroids are quite high compared with their physiologic concentrations, these levels are attainable, particular in the case of tamoxifen and DES (IC50 = 4 μM). Similarly, cytostatic effect was obtained with doses of 5 μM for estrogen and 10 μM for androgen. Growth inhibition was associated with a cell cycle arrest of U937 leukemia cells in G2/M phase *(41)*. Two structurally distinct antiestrogens, tamoxifen and ICI 182,780, inhibited the proliferation of all multiple myeloma cell lines and patient-derived cells with IC$_{50}$ 2–4 μmol/L. Furthermore, antiestrogens triggered cell death in both ER-α-positive as well as ER-α-negative multiple myeloma cell lines and patient MM cells *(42)*. Based on all these studies, high doses of tamoxifen was introduced for patients with hormone-refractory prostate cancer. All study patients had failed prior treatment with combined androgen blockade. The average steady-state plasma concentration of tamoxifen was 2.96 ± 1.32 μM *(43)*. Combined partial response/stable disease response rate was 23%, and grade 3 neurotoxicity was observed in 29% of patients *(43)*.

3. Differentiating Therapy

3.1. Retinoic Acid

3.1.1. Receptors

Retinoids are derivatives of vitamin A that include all-*trans*-retinoic acid (ATRA), 13-*cis*-retinoic acid (13-*cis*-RA), and fenretinide (4-HPR). Retinoids are essential for the maintenance of epithelial differentiation. Since high levels of either ATRA or 13-*cis*-RA can arrest growth and induce differentiation of numerous cell types, they are used in differentiation therapy *(44)*. Retinoids control gene expression through retinoic acid receptors (RARs)-α, -β, and -γ and 9-*cis*-retinoic acid receptors-α, -β and -γ, which bind with high affinity the natural ligands all-*trans*-retinoic acid and 9-*cis*-retinoic acid, respectively. RAR-RXR heterodimers bind retinoic acid response elements and act as transcription factors. In addition, RARs can repress transcription of genes that do not contain RAR-binding elements, due to competition for co-activators such as CBP *(45)*. In this regard, there is a stricking analogy between RARs and p53. p53 also activates transcription in a promoter-dependent manner but represses transcription by competiting for p300/CBP co-activators.

3.1.2. Acute Myelogenic Leukemia

RAR and AML1 transcription factors are found in leukemias as fusion proteins with PML and ETO, respectively. Association of PML-RAR and AML1-ETO with the nuclear co-repressor (N-CoR)/histone deacetylase (HDAC) complex is required to block hematopoietic differentiation *(46)*. PML-RAR exist in vivo within high-molecular-weight nuclear complexes, reflecting their oligomeric state. Oligomerization requires

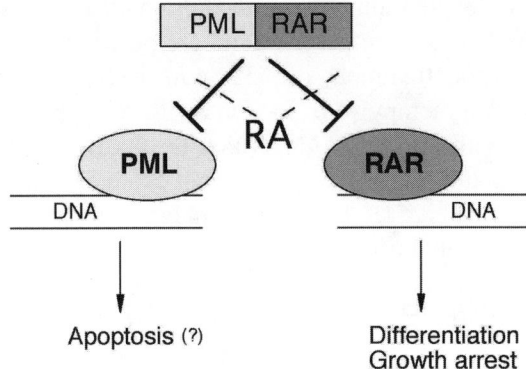

Fig. 3. Retinoic acid in acute promylocytic leukemia. The fusion protein PML-RAR blocks myeloid differentiation by direct transcriptional inhibition of RAR target genes and inhibits PML-dependent apoptotic pathways. In the presence of high concentrations of retinoic acid (RA), both PML- and RAR-dependent apoptotic and differentiation programs are reactivated.

PML coiled-coil regions and is responsible for abnormal recruitment of N-CoR, transcriptional repression, and impaired differentiation of primary hematopoietic precursors. Oligomerization of the transcription factor, imposing an altered interaction with transcriptional co-regulators, represents a mechanism of oncogenic activation *(47)*.

3.1.3. PML-RAR in Acute Promyelocytic Leukemia (APL)

Acute promyelocytic leukemia (APL) is caused by one of four genetic lesions that disrupt the α-receptor for retinoic acid, RAR-α *(48)*. The fusion protein responsible for more than 99% of APL cases, PML-RAR-α, inhibits PML-dependent apoptotic pathways in a dominant-negative fashion and blocks myeloid differentiation by direct transcriptional inhibition of retinoic acid target genes (**Fig. 3**). This transcriptional inhibition is mediated by recruitment of co-repressor proteins and resultant deacetylation of histones in the promoter regions of genes that control promyelocyte development. This protein functions as a dominant negative against the function of both PML and RARs, and its overexpression is able to re-create the phenotypes of the disease in transgenic mice. PML, a putative transcription factor, can suppress growth and induce apoptosis *(49)*. In the presence of high levels of all-*trans*-retinoic acid (ATRA), both PML-dependent apoptotic mechanisms and myeloid-specific gene expression programs are reactivated (**Fig. 3**).

This explains why induction of differentiation in promyelocytes and consequent remission of APL following retinoid therapy depends on expression of a chimeric PML-RAR-α fusion protein resulting from a t(15;17) chromosomal translocation.

Differentiation therapy of acute promyelocytic leukemia (APL) is based on the ability of retinoic acid to induce differentiation of leukemic promyelocytes. All-*trans*-retinoic acid ATRA therapy has been shown to reduce tumor burden by stimulating differentiation of the leukemic cells, induce long-term clinical remission when administered in conjunction with chemotherapy, and lower the incidence of consumptive coagulopathies in patients with APL. The treatment of APL has improved dramatically in the past decade, which has greatly enhanced the prognosis of patients with this disease. First remission rates have increased to greater than 85% worldwide, the incidence of disseminated intravascular coagulation has declined dramatically, and 60–70% of patients with APL have achieved long term survival and are potentially cured *(50)*. The "standard-of-care" for induction treatment of APL now includes ATRA plus concurrent anthracycline-based chemotherapy *(48)*. Patients with relapsed, retinoid-resistant APL are now being treated with arsenic oxide, which results in apoptosis of the leukemic cells. Combined arsenic and retinoic acid treatment enhances differentiation and apoptosis in arsenic-resistant cells *(51)*. A high percentage of relapsed patients can achieve second remissions with arsenic trioxide.

3.1.4. Retinoids and Non-APL Leukemia and Solid Tumors

Thus, differentiating therapy overcomes the site-specific transcriptional repression by dominant fusion leukemogenic proteins characteristic of APL and other forms of acute myelogenous leukemia (AML). These approaches are, in general, non-cross-resistant and are well tolerated *(52)*. Although M0/M1 leukemic cells are not sensitive to differentiating agents, M2 and M4 leukemic blast cells undergo monocytic differentiation with both RA and vitamin D. Retinoids inhibits the proliferation of Epstein-Barr virus (EBV)-immortalized lymphoblastoid B-cell lines. Vitamin D and retinoids cooperate to inhibit proliferation and induce differentiation of human non-APL myelomonocytic U937 cells, presumably through TGF-β autosecretion.

Also, retinoids inhibit the growth of human ovarian cancer cells both in vivo and in vitro and may play a role as an ovarian cancer chemotherapeutic agent *(53)*. In advanced non-small-cell lung cancer, the differentiation therapy with retinyl palmitate were administered between chemotherapy cycles. In high-risk neuroblastoma, therapy with 13-*cis*-RA given after completion of intensive chemoradiotherapy (with or without autologous bone marrow transplantation) significantly improves event-free survival *(54)*.

3.2. Inhibitors of Histone Deacetylase

3.2.1. Mechanisms

Histone acetylation regulates transcription by affecting the interaction between DNA and histones. Histone deacetylases (HDAC) is a class of enzymes consisting of at least two subfamilies with at least seven members *(55)*. HDAC inhibitors cause acetylated histones to accumulate in both tumor and normal tissues, and this accumulation can be used as a marker of the biologic activity of the HDAC inhibitors. These compounds act on gene expression, altering the expression of about 1–2% of the genes expressed in cultured tumor cells *(56)*. HDAC inhibitors can also acetylate proteins other than histones, and this may play a particular role in apoptosis.

HDAC inhibitors such as butyrate, trichostatin A (TSA), oxamflatin, and FR901228 vary widely in their potency *(56–60)*. Hydroxamic acid-based hybrid polar compounds induce differentiation at micromolar or lower concentrations *(56)*. FR901228 (NSC 630176, depsipeptide) is currently undergoing clinical trials (Sandor et al., submitted). The compound is cytotoxic at nanomolar concentrations in several in-vitro and in-vivo models *(60,61)*, showing 100,000-fold higher activity than butyrate.

Mechanism of G1-arrest: HDAC inhibitors including butyrate, trichostatin A (TSA), oxamflatin, suberoylanilide hydroxamic acid, and FR901228 upregulate p21 *(57–63)*. FR901228 downregulated cyclin D1 and upregulated CDK inhibitor p21$^{WAF1/CIP1}$, resulting in inhibition of CDK activity and Rb dephosphorylation and G1 arrest *(60)*.

Cells lacking p21 did not undergo G1 arrest, continued DNA synthesis, and were arrested in G2/M phase of the cell cycle *(60)*. Similarly, HDAC inhibitor oxamflatin increased expression of cyclin E and p21 and decreased expression of cyclin D1 *(64)*. It has been reported that growth arrest by butyrate was mediated by p21 in HCT116 cells *(65)*. However, p21 induction was dispensable for G1 arrest caused by butyrate in mouse fibroblasts *(58)*. In the later study, butyrate treatment decreased cyclin E. When cyclin D1 is downregulated and p21 is upregulated, CDKs are inhibited, resulting in Rb dephosphorylation. Dephosphorylated Rb complexed with E2F actively blocks cyclin E expression, thus preventing the next step of cell cycle progression *(66)*. The complex of dephosphorylated Rb and E2F associates with a histone deacetylase to effect transcriptional repression *(66)*. FR901228, by inhibiting histone deacetylase may prevent cyclin E downregulation. FR901228 treatment resulted in a unique combination: increased p21 and cyclin E accompany decreased cyclin D1 *(60)*. Following FR901228 treatment, increased p21 counterbalanced increased cyclin E in HCT116 cells, and G1 arrest, which is initially triggered by downregulation of cyclin D1, thus depends on p21 expression *(60)*. In contrast, p21 induction was dispensable for G1 arrest by butyrate in MEF cells, in which butyrate downregulated cyclin E *(58)*. Thus, the opposing effects on the levels of cyclin E may determine the requirement for p21 induction. However, neither G2/M arrest nor cytotoxicity of HDAC inhibitors depends on p21 *(60)*.

3.2.2. HDAC Inhibitors and Acute Leukemia

As we discussed, nuclear receptor co-repressor (N-CoR)-histone deacetylase (HDAC) complex is required for the biologic activities of the retinoic acid receptor fusion proteins (e.g., PMA-RAR) of acute promyelocytic leukemias. Therefore, HDAC inhibitors synergize with retinoic acid to stimulate leukemia cell differentiation *(46,59,67)*.

The t(8;21) translocation, found in adult acute myelogenous leukemia (AML), fuses ETO (eight-twenty-one or MTG8) to the acute myelogenous leukemia 1 (AML1) transcription factor, results in the formation of an AML1/ETO chimeric transcription factor. AML1 upregulates a number of target genes critical to normal hematopoiesis, whereas the AML1/ETO fusion interferes with this *trans*-activation. ETO interacts via its zinc-finger region with a conserved domain of the co-repressors N-CoR and SMRT and recruits HDAC *(68,69)*. The fusion protein AML1-ETO retains the ability of ETO to form stable complexes with N-CoR/SMRT and HDAC. Formation of a stable complex with CoR-HDAC is crucial to the activation of the leukemogenic potential of AML1 by ETO and suggests that aberrant recruitment of co-repressor complexes is a general mechanism of leukemogenesis *(68,69)*.

Since ETO recruits a transcription repression complex that includes the histone deacetylase (HDAC1) enzyme, inhibitors of HDAC such as trichostatin A (TSA) and phenylbutyrate partially reversed ETO-mediated transcriptional repression. Phenylbutyrate is also able to induce partial differentiation of the AML1-ETO cell line, Kasumi-1 *(70)*. TSA induced differentiation in erythroid cell lines by itself, whereas it synergistically enhanced the differentiation that was directed by all-*trans*-retinoic acid or vitamin D3 in U937, HL60, and NB4 cells. The combined treatment of HDAC inhibitors with ATRA induced p21, p16, and differentiation in ATRA-resistant HL60 and NB4 cells *(59)*. Furthermore, 34% of clinical samples from AML patients ranging from M0 to M7 also displayed both phenotypic and morphologic changes by the treatment with TSA *(59)*.

3.3. Inhibitors of Methylation

The methylation of DNA is an epigenetic modification that can play an important role in the control of gene expression in mammalian cells *(71)*. The enzyme DNA methyltransferase modifies a base that is found mostly at CpG sites in the genome. DNA methylation, a fundamental aspect of neoplasia, often causes silencing of tumor suppressors and CDK inhibitors such as p16, p27, and p57. The inhibitor of DNA methylation, 5-aza-2′-deoxycytidine (5-asa-CdR) reactivates the expression most of tumor suppressor genes in human tumor cell lines *(71)*.

Methylated DNA associates with transcriptionally repressive chromatin characterized by the presence of deacetylated histones. Hypermethylated genes p15 and p16 cannot be transcriptionally reactivated with TSA (a HDAC inhibitor) alone. Following minimal demethylation and slight gene reactivation in the presence of low-dose 5-aza-2′deoxycytidine (5Aza-dC), TSA treatment results in robust reexpression of each gene. Thus, although DNA methylation and histone deacetylation appear to act as synergistic layers for the silencing of genes in cancer, dense CpG island methylation is dominant for the stable maintenance of a silent state at these loci *(72)*. The AR promoter contains specific CpG methylation "hot spots" that are markers for gene silencing. AR methylation may represent a phenotype important in the development of hormone independence in a subset of advanced prostate cancer in which AR expression is lost *(73)*. Retinoic acid (RA) resistance in breast cancer cells has been associated with irreversible loss of RAR-β expression. Methylation in the promoter region of RAR in several primary breast tumors has been observed *(74)*.

Remarkably, breast cancer cells lacking final σ expression showed increased number of chromosomal breaks and gaps when exposed to γ-irradiation *(75)*. The mRNA of 14-3-3 final σ was undetectable in 45 of 48 primary breast carcinomas *(75)*. Hypermethylation of CpG islands in the final σ gene was detected in 91% of breast tumors and was associated with lack of gene expression. Hypermethylation of final σ is functionally important, because treatment of final σ-nonexpressing breast cancer cell lines with the drug 5-aza-2′-deoxycytidine resulted in demethylation of the gene and synthesis of final σ mRNA.

3.4. Activators of PKC

Protein kinase C (PKC) regulates fundamental cellular functions including proliferation, differentiation, tumorigenesis, and apoptosis. PKC-mediated activation of the Raf-1/MAPK pathway may simultaneously induce cyclin D1 and p21. Induction of p21 determines growth arrest caused by the PKC activator phorbol ester (PMA) (**Fig. 4**). Leukemia cells lacking p53 are especially sensitive to PMA-induced p21 *(76,77)*. Also, PMA induces growth arrest in SKBr3 breast cancer and LNCaP cancer cells *(78–80)*. Intriguingly, it has been shown that all-*trans*-retinoic acid interacts directly with PKC isozymes, suggesting the existence of a general mechanism for regulation of PKC activity during exposure to retinoids, as in retinoid-based cancer therapy *(81)*. PMA has been reluctantly considered for clinical use due to its potential side effects, including a possible tumor promoter activity. Recently, a remarkable potency of PMA in the treatment of refractory leukemia and its low toxicity for the patients has been demonstrated *(82)*.

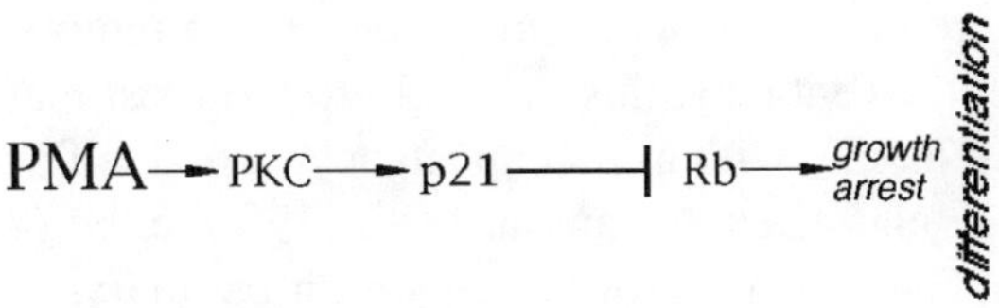

PMA→ PKC → p21 ——⊣ Rb → *growth arrest* | *differentiation*

Fig. 4. Phorbol ester (PMA)-induced growth arrest and differentiation.

4. Integration of Different Modalities

4.1. Chemotherapy and Differentiation

Although use of chemotherapeutic agents is intended to induce apoptosis, they also can induce differentiation and cell senescence. DNA-damaging drugs induce wt p53 and p21. On the other hand, p21 is also induced by many differentiating agents, in a p53-independent manner.

More important, DNA-damaging agents may induce therapy regardless of p53 status. For example, anthracycline antitumor drugs, particularly doxorubicin and aclacinomycin, are potent inducers of erythroid differentiation in human leukemic K562 cells (p53-null). Aclacinomycin could exert their erythroid-differentiating activity by modulating the expression of transcription factors that specifically regulate the transcription of erythroid genes *(83,84)*. Actinomycin D, the drug of choice in the treatment of rhabdomyosarcoma, induces differentiation in rhabdomyosarcoma cell line *(85)*.

4.2. Apoptotic Effects of Hormones and Differentiating Agents

As we discussed, glucocorticoids and other hormones can induce apoptosis. Intriguingly, dexamethasone can transcriptionally induced KILLER/DR5 mRNA *(86)*.Two synthetic retinoids, 4 HPR R(fenretinide) and AHPN/CD437, induced apoptosis that did not require the activation of classical retinoid receptors *(87,88)*. Therefore, CD437 induces in vitro the rapid apoptosis of both RA-sensitive and -resistant APL cells. 4-HPR achieves multi-log cell kills in neuroblastoma cell lines that are resistant to ATRA and 13-*cis*-RA *(51)*.

4.3. Thyroid Cancer Model of Tissue-Specific and Differentiating Therapy

Early thyroid tumor development is closely correlated with mutation of five alternative genes, Ras, Ret, Trk, Gsp, and the TSH receptor *(89)*. The hormone thyrotropin (TSH), the physiologic regulator of thyroid cells, stimulates proliferation, thyroid-specific gene expression (differentiation) of thyroid follicle cells. The mitogenic effects of TSH require Ras, which, via PI3K, confers hormone-independent proliferation *(90)*. In addition to its direct effects, TSH induces secretion of secretory Alzheimer amyloid precursor (sAPP), which may mediate the proliferative effect of TSH on neighboring thyroid epithelial cells *(91)*.

TSH suppression therapy is supported by most clinical studies *(92,93)*. Thus, for the treatment of differentiated thyroid cancer, radioiodine therapy and TSH-suppressive thyroxine application is a well-established and successful therapeutic approach. In one-third of cases, dedifferentiation is observed, giving rise to tumors that are refractory to conventional treatment. Eventually, this may lead to the most malignant human tumor, anaplastic thyroid carcinoma, with a life expectancy of only a few months.

Thyroid cancers are classified as papillary (82%), follicular (8%), medullary (9%), and anaplastic (1%). The prognosis of patients with papillary, the most differentiated thyroid cancer, is favorable and depends on tumor differentiation and the ability to take up radioactive iodine. Uptake of radioactive iodine, a normal function of thyroid cells, allow its accumulation to 2000 times serum levels, thus killing the thyroid cell selectively. The overall cure rate of ^{131}I (radioiodine) therapy after 12 yr was 50% (papillary 65% vs follicular 23%). Differentiation grade were significant prognostic factors *(94)*. ^{131}I can be an effective treatment in patients with advanced differentiated thyroid cancer at all sites and can cure, on average, 50% of all patients with advanced differentiated thyroid cancer *(94)*. Due to the loss of thyroid-specific functions associated with dedifferentiation, these tumors are inaccessible to standard therapeutic procedures such as radioiodine therapy and thyroxine-mediated thyrotrophin suppression. Medullary thyroid carcinomas are also highly aggressive *(95)*.

Among novel approaches for the treatment of dedifferentiated thyroid carcinomas, retinoic acid therapy was evaluated *(95)*. Cell culture experiments in thyroid carcinoma lines show that RA treatment caused partial redifferentiation and increased iodine uptake *(96)*. RA can induce radioiodine uptake in some patients with radioiodine-negative thyroid carcinoma tumor sites *(97)*. In clinical pilot studies, about 40% of the patients responded to RA application with an increased radioiodine uptake. Forty percent of patients exhibited stabilization in tumor size in addition to enhanced radioiodine transport *(95)*.

Anaplastic thyroid cancer cells are characterized by the absence of expression of thyroid-specific genes (thyroglobulin, thyroperoxidase, and TSH receptor). This results in their inability to incorporate radioactive iodine, and precludes this tissue-selective therapeutic modality.

One of the striking differences between differentiated and undifferentiated thyroid cancer is the frequencies of mutations found in p53. p53 abnormalities are associated with more advanced thyroid carcinomas and especially with anaplastic carcinomas. Inactivation of p53 did not result directly in acquiring of anaplastic phenotype, although lost of some tissue-specific markers was reported *(89)*. The presence of p53 mutations almost exclusively in poorly differentiated thyroid tumors and thyroid cancer cell lines suggests that inactivation of p53 may confer these neoplasms with loss of differentiated function and triggers the progression from differentiated to anaplastic carcinoma in the human thyroid gland *(89,98,99)*. Moreover, restoration of wt p53 may induce differentiation in thyroid cancer cell lines *(98,100)*. It has been shown that stably transfected wt p53 papillary carcinoma cells reexpressed thyroid peroxidase (TPO) *(98)*. In another study, ARO cells stably transfected with p53 reacquired the ability to respond to thyrotropin (TSH) stimulation, showing an increased expression of thyroid-specific genes (thyroglobulin, thyroperoxidase, and TSH receptor) *(100)*. It should be taking into account, however, that the ARO cells were not anaplastic and already expressed some thyroid-specific genes. We did not find reexpression of thyroid-specific TSH and TG in anaplastic Ad-p53-infected cells *(101)*. However, doxorubicin, a DNA-damaging drug, not only potentiated p53-mediated cytotoxicity but also induced morphologic alteration in an anaplastic cell line, consistent with more differentiated phenotype *(101)*. As discussed, chemotherapy

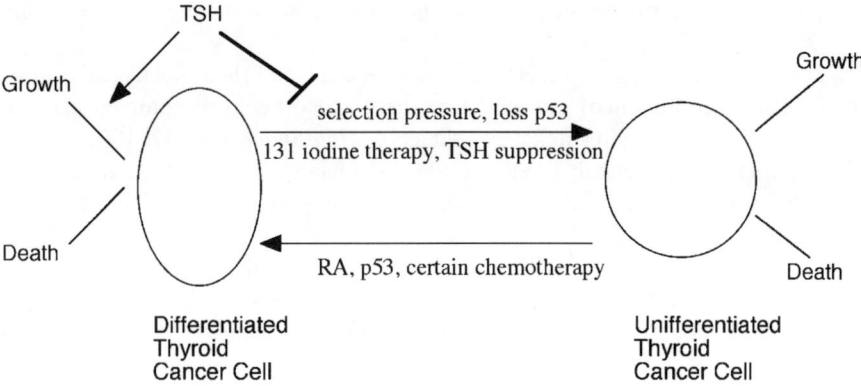

Fig. 5. Tissue-specific and differentiating therapy in thyroid cancer.

by itself can induce differentiation. Therefore combination of chemotherapy and radioiodine therapy may be warranted. In fact, a conversion of non-iodine-concentrating thyroid carcinoma metastasis into iodine-concentrating foci after anticancer chemotherapy has been reported. In this study, a patient with metastatic papillary carcinoma was treated with cisplatin and doxorubicin. Repeat ^{131}I imaging after three cycles of chemotherapy showed significant ^{131}I uptake in previously non-iodine-concentrating lesions *(102)*. This example illustrates a potential of both tissue-specific cytotoxic and differentiation therapies (**Fig. 5**). Inhibition of TSH and treatment with radioiodine are highly tissue-specific therapies that are effective in differentiated thyroid cancer (and normal) cells. Suppression of TSH and radioiodine inhibit growth and cause cell death, respectively. On the other hand, they target the most differentiated cells, and eventually may select for undifferentiated cancer. This cancer is insensitive to tissue-specific therapy. However, certain modalities such as retinoid acid, p53, and DNA-damaging drugs may induce these cells to differentiate. Although differentiation is an important goal by itself, even more important, by reactivating tissue-specific genes, it permits tissue-specific therapy (**Fig. 5**).

References

1. Owen, G. I. and Zelent, A. (2000) Origins and evolutionary diversification of the nuclear receptor superfamily. *Cell. Mol. Life Sci.* **57,** 809–827.
2. Tsai, M.-J. and O'Malley, B. W. (1994) Molecular mechanisms of action of steroid/thyroid hormone receptor superfamily members. *Ann. Rev. Biochem.* **63,** 451–486.
3. Poukka, H., Karvonen, U., Yoshikawa, N., Tanaka, H., Palvimo, J. J., and Janne, O.A. (2000) The RING finger protein SNURF modulates nuclear trafficking of the androgen receptor. *J. Cell. Sci.* **113,** 2991–3001.
4. Pratt, W. B., Silverstein, A. M., and Galigniana, M. D. (1999) A model for the cytoplasmic trafficking of signalling proteins involving the hsp90-binding immunophilins and p50cdc37. *Cell. Signal.* **11,** 839–851.
5. Rajapandi, T., Greene, L. E., and Eisenberg, E. (2000) The molecular chaperones Hsp90 and Hsc70 are both necessary and sufficient to activate hormone binding by glucocorticoid receptor. *J. Biol. Chem.* **275,** 22597–22604.
6. Czar, M. J., Galigniana, M. D., Silverstein, A. M., and Pratt, W. B. (1997) Geldanamycin, a heat shock protein 90-binding benzoquinone ansamycin, inhibits steroid-dependent translo-

cation of the glucocorticoid receptor from the cytoplasm to the nucleus. *Biochemistry* **36,** 7776–7785.

 7. Galigniana, M. D., Scruggs, J. L., Herrington, J., et al. (1998) Heat shock protein 90-dependent (geldanamycin-inhibited) movement of the glucocorticoid receptor through the cytoplasm to the nucleus requires intact cytoskeleton. *Mol. Endocrinol.* **12,** 1903–1913.

 8. Kojika, S., Sugita, K., Inukai, T., et al. (1996) Mechanisms of glucocorticoid resistance in human leukemic cells: Implication of abnormal 90 and 70 kDa heat shock proteins. *Leukemia* **10,** 994–999.

 9. Wolffe, A. P., Collingwood, T. N., Li, Q., Yee, J., Urnov, F., and Shi, Y. B. (2000) Thyroid hormone receptor, v-ErbA, and chromatin. *Vitam. Horm.* **58,** 449–492.

10. Sengupta, S., Vonesch, J. L., Waltzinger, C., Zheng, H., and Wasylyk, B. (2000) Negative cross-talk between p53 and the glucocorticoid receptor and its role in neuroblastoma cells. *EMBO J.* **19,** 6051–6064.

11. Berges, R. R., Furuya, Y., Remington, L., English, H. F., Jacks, T., and Isaacs, J. T. (1993) Cell proliferation, DNA repair, and p53 function are not required for programmed death of prostatic glandular cells induced by androgen ablation. *Proc. Natl. Acad. Sci. USA* **90,** 8910–8914.

12. Reese, D. M. (2000) Choice of hormonal therapy for prostate cancer. *Lancet* **355,** 1474–1475.

13. Koivisto, P., Visakorpi, T., Rantala, I., and Isola, J. (1997) Increased cell proliferation activity and decreased cell death are associated with the emergence of hormone-refractory recurrent prostate cancer. *J. Pathol.* **183,** 51–56.

14. Bruckheimer, E. M., Brisbay, S., Johnson, D. J., Gingrich, J. R., Greenberg, N., and McDonnell, T. J. (2000) Bcl-2 accelerates multistep prostate carcinogenesis in vivo. *Oncogene* **19,** 5251–5258.

15. Jenster, G. (2000) Ligand-independent activation of the androgen receptor in prostate cancer by growth factors and cytokines. *J. Pathol.* **191,** 227–228.

16. Middleman, M. N., Lush, R. M., Sartor, O., Reed, E., and Figg, W. D. (1996) Treatment approaches for metastatic cancer of the prostate based on recent molecular evidence. *Cancer Treat. Rev.* **22,** 105–118.

17. Erkkila, K., Henriksen, K., Hirvonen, V., et al. (1997) Testosterone regulates apoptosis in adult human seminiferous tubules in vitro. *J. Clin. Endocrinol. Metab.* **82,** 2314–2321.

18. Einhorn, E. H. (1997) Testicular cancer: an oncological success story. *Clin. Cancer Res.* **3,** 2630–2632.

19. Fanelli, M. A., Vargas-Roig, L. M., Gago, F. E., Tello, O., Lucero De Angelis, R., and Ciocca, D. R. (1996) Estrogen receptors, progesterone receptors, and cell proliferation in human breast cancer. *Breast Cancer Res. Treat.* **37,** 217–228.

20. Hofseth, L. J., Raafat, A. M., Osuch, J. R., Pathak, D., Slomski, C. A., and Haslam, S. Z. (1999) Hormone replacement therapy with estrogen or estrogen plus medroxyprogesterone acetate is associated with increased epithelial proliferation in the normal postmenopausal breast. *J. Clin. Endocrinol. Metab.* **84,** 4559–4565.

21. Anderson, E., Clarke, R. B., and Howell, A. (1998) Estrogen responsiveness and control of normal human breast proliferation. *J. Mammary Gland Biol. Neoplasia* **3,** 23–35.

22. Lu, R. and Serrero, G. (2001) Mediation of estrogen mitogenic effect in human breast cancer MCF-7 cells by PC-cell-derived growth factor (PCDGF/granulin precursor). *Proc. Natl. Acad. Sci. USA* **98,** 142–147.

23. Aebi, S., Gelber, S., Castiglione-Gertsch, M., et al. (2000) Is chemotherapy alone adequate for young women with oestrogen-receptor-positive breast cancer? *Lancet* **27,** 1869–1874.

24. Maggiolini, M., Donze, O., Jeannin, E., Ando, S., and Picard, D. (1999) Adrenal androgens stimulate the proliferation of breast cancer cells as direct activators of estrogen receptor alpha. *Cancer Res.* **59,** 4864–4869.

25. Moyer, D. L. and Felix, J. C. (1998) The effects of progesterone and progestins on endometrial proliferation. *Contraception* **57,** 399–403.

26. Saegusa, M. and Okayasu, I. (1998) Progesterone therapy for endometrial carcinoma reduces cell proliferation but does not alter apoptosis. *Cancer* **83,** 111–121.

27. Saegusa, M. and Okayasu, I. (1997) Down-regulation of bcl-2 expression is closely related to squamous differentiation and progesterone therapy in endometrial carcinomas. *J. Pathol.* **182,** 429–436.

28. Planey, S.L. and Litwack, G. (2000) Glucocorticoid-induced apoptosis in lymphocytes. *Biochem. Biophys. Res. Commun.* **279,** 307–312.

29. Brunetti, M., Martelli, N., Colasante, A., Piantelli, M., Musiani, P., and Aiello, F.B. (1995) Spontaneous and glucocorticoid-induced apoptosis in human mature T lymphocytes. *Blood* **86,** 4199–4205.

30. Gaynon, P. S. and Lustig, R. H. (1995) The use of glucocorticoids in acute lymphoblastic leukemia of childhood. Molecular, cellular, and clinical considerations. *J. Pediatr. Hematol. Oncol.* **17,** 1–12.

31. Gaynon, P. S. and Carrel, A. L. (1999) Glucocorticosteroid therapy in childhood acute lymphoblastic leukemia. *Adv. Exp. Med. Biol.* **457,** 593–605.

32. Csoka, M., Bocsi, J., Falus, A., et al. (1997) Glucocorticoid-induced apoptosis and treatment sensitivity in acute lymphoblastic leukemia of children. *Pediatr. Hematol. Oncol.* **14,** 433–442.

33. Iacobelli, S., Marchetti, P., De Rossi, G., Mandelli, F., and Gentiloni, N. (1987) Glucocorticoid receptors predict response to combination chemotherapy in patients with acute lymphoblastic leukemia. *Oncology* **44,** 13–16.

34. Strasser-Wozak, E. M., Hattmannstorfer, R., Hala, M., et al. (1995) Splice site mutation in the glucocorticoid receptor gene causes resistance to glucocorticoid-induced apoptosis in a human acute leukemic cell line. *Cancer Res.* **55,** 348–353.

35. Merino, R., Ding, L., Veis, D. J., Korsmeyer, S. J., and Nunez, G. (1994) Developmental regulation of the Bcl-2 protein and susceptibility to cell death in B lymphocytes. *EMBO J.* **13,** 683–691.

36. Van Houten, N., Blake, S. F., Li, E. J., et al. (1997) Elevated expression of Bcl-2 and Bcl-x by intestinal intraepithelial lymphocytes: resistance to apoptosis by glucocorticoids and irradiation. *Int. Immunol.* **9,** 945–953.

37. Hartmann, B. L., Geley, S., Loffler, M., et al. (1999) Bcl-2 interferes with the execution phase, but not upstream events, in glucocorticoid-induced leukemia apoptosis. *Oncogene* **18,** 713–719.

38. Rohlff, C., Blagosklonny, M. V., Kyle, E., et al. (1998) Prostate cancer cell growth inhibition of tamoxifen is associated with inhibition of protein kinase C and induction of p21waf1/cip1. *Prostate* **37,** 51–59.

39. Lee, T. H., Chuang, L. Y., and Hung, W. C. (1999) Tamoxifen induces p21(WAF1) and p27(KIP1) expression in estrogen receptor-negative lung cancer cells. *Oncogene* **18,** 4269–4274.

40. Blagosklonny, M. V. and Neckers, L. M. (1994) Cytostatic and cytotoxic activity of sex steroids against human leukemia cell lines. *Cancer Lett.* **76,** 81–86.

41. Mossuz, P., Cousin, F., Castinel, A., et al. (1998) Effects of two sex steroids (17 beta estradiol and testosterone) on proliferation and clonal growth of the human monoblastic leukemia cell line, U937. *Leukemia Res.* **22,** 1063–1072.

42. Treon, S. P., Teoh, G., Urashima, M., et al. (1998) Anti-estrogens induce apoptosis of multiple myeloma cells. *Blood* **92,** 1749–1757.

43. Bergan, R. C., Reed, E., Myers, C. E., et al. (1999) A Phase II study of high-dose tamoxifen in patients with hormone- refractory prostate cancer. *Clin. Cancer Res.* **5,** 2366–2373.

44. Hansen, L. A., Sigman, C. C., Andreola, F., Ross, S. A., Kelloff, G. J., and De Luca, L. M. (2000) Retinoids in chemoprevention and differentiation therapy. *Carcinogenesis* **21,** 1271–1279.

45. Kamei, Y., Xu, L., Heinzel, T., et al. (1996) A CBP integrator complex mediates transcriptional activation and AP-1 inhibition by nuclear receptors. *Cell* **85,** 403–414.

46. Lin, R. J., Nagy, L., Inoue, S., Shao, W., Miller, W. H., and Evans, R. M. (1998) Role of the histone deacetylase complex in acute promyelocytic leukemia. *Nature* **391,** 811–814.

47. Minucci, S., Maccarana, M., Cioce, M., et al. (2000) Oligomerization of RAR and AML1 transcription factors as a novel mechanism of oncogenic activation. *Mol. Cell* **5,** 811–820.

48. Slack, J. L. and Rusiniak, M. E. (2000) Current issues in the management of acute promyelocytic leukemia. *Ann. Hematol.* **79,** 227–238.

49. Wang, Z. G., Ruggero, D., Ronchetti, S., et al. (1998) PML is essential for multiple apoptotic pathways. *Nat. Genet.* **20,** 266–272.

50. Randolph, T. R. (2000) Acute promyelocytic leukemia (AML-M3)—Part 1: pathophysiology, clinical diagnosis, and differentiation therapy. *Clin. Lab. Sci.* **13,** 98–105.

51. Gianni, M., Koken, M. H. M., Chelbi-Alix, M. K., et al. (1998) Combined arsenic and retinoic acid treatment enhances differentiation and apoptosis in arsenic-resistant NB4 cells. *Blood* **91,** 4300–4310.

52. Waxman, S. (2000) Differentiation therapy in acute myelogenous leukemia (non-APL). *Leukemia* **14,** 491–496.

53. Zhang, D., Holmes, W. F., Wu, S., Soprano, D. R., and Soprano, K. J. (2000) Retinoids and ovarian cancer. *J. Cell. Physiol.* **185,** 1–20.

54. Matthay, K. K., Perez, C., Seeger, R. C., et al. (1999) Treatment of high-risk neuroblastoma with intensive chemotherapy, radiotherapy, autologous bone marrow transplantation, and 13-*cis*-retinoic acid. Children's Cancer Group. *N. Engl. J. Med.* **341,** 1165–1173.

55. Weidle, U. H. and Grossmann, A. (2000) Inhibition of histone deacetylases: a new strategy to target epigenetic modifications for anticancer treatment. *Anticancer Res.* **20,** 1471–1485.

56. Marks, P. A., Richon, V. M., and Rifkind, R. A. (2000) Histone deacetylase inhibitors: inducers of differentiation or apoptosis of transformed cells. *J. Natl. Cancer Inst.* **92,** 1210–1216.

57. Sowa, Y., Orita, T., Minamikawa, S., et al. (1997) Histone deacetylase inhibitor activates the WAF1/Cip1 gene promoter through the SP1 sites. *Biochem. Biophys. Res. Commun.* **241,** 142–150.

58. Vaziri, C., Stice, L., and Faller, D. V. (1998) Butyrate-induced G1 arrest results from p21-independent disruption of retinoblastoma protein-mediated signals. *Cell Growth Diff.* **9,** 465–474.

59. Kosugi, H., Towatari, M., Hatano, S., et al. (1999) Histone deacetylase inhibitors are the potent inducer/enhancer of differentiation in acute myeloid leukemia: a new approach to anti-leukemia therapy. *Leukemia* **13,** 1316–1324.

60. Sandor, V., Senderowicz, A., Mertins, S., et al. (2000) P21-dependent G1 arrest with down-regulation of cyclin D1 and upregulation of cyclin E by the histone deacetylase inhibitor FR901228. *Br. J. Cancer* **83,** 817–825.

61. Wang, R., Brunner, T., Zhang, L., and Shi, Y. (1998) Fungal metabolite FR901228 inhibits c-Myc and Fas ligand expression. *Oncogene* **17,** 1503–1508.

62. Richon, V. M., Sandhoff, T. W., Rifkind, R. A., and Marks, P. A. (2000) Histone deacetylase inhibitor selectively induces p21WAF1 expression and gene-associated histone acetylation. *Proc. Natl. Acad. Sci. USA* **97,** 10014–10019.

63. Huang, L., Sowa, Y., Sakai, T., and Pardee, A. B. (2000) Activation of the p21WAF1/CIP1 promoter independent of p53 by the histone deacetylase inhibitor suberoylanilide hydroxamic acid (SAHA) through the Sp1 sites. *Oncogene* **19,** 5712–5719.

64. Kim, Y. B., Lee, K. H., Sugita, K., Yoshida, M., and Horinouchi, S. (1999) Oxamflatin is a novel antitumor compound that inhibits mammalian histone deacetylase. *Oncogene* **18,** 2461–2470.

65. Archer, S. Y., Meng, S., Shei, A., and Hodin, R. A. (1998) p21WAF1 is required for butyrate-mediated growth inhibition of human colon cancer cells. *Proc. Natl. Acad. Sci USA* **95,** 6791–6796.

66. Zhang, H. S., Postigo, A. A., and Dean, D. C. (1999) Active transcriptional repression by the Rb-E2F complex mediates G1 arrest triggered by p16INK4a, TGFbeta, and contact inhibition. *Cell* **97,** 53–61.

67. Grignani, F., De Matteis, S., Nervi, C., et al. (1998) Fusion proteins of the retinoic acid receptor-alpha recruit histone deacetylase in promyelocytic leukemia. *Nature* **391,** 815–818.

68. Wang, J., Hoshino, T., Redner, R. L., Kajigaya, S., and Liu, J. M. (1998) ETO, fusion partner in t(8;21) acute myeloid leukemia, represses transcription by interaction with the human N-CoR/mSin3/HDAC1 complex. *Proc. Natl. Acad. Sci. USA* **95,** 10860–10865.

69. Gelmetti, V., Zhang, J., Fanelli, M., Minucci, S., Pelicci, P. G., and Lazar, M. A. (1998) Aberrant recruitment of the nuclear receptor co-repressor-histone deacetylase complex by the acute myeloid leukemia fusion partner ETO. *Mol. Cell. Biol.* **18,** 185–191.

70. Wang, J., Saunthararajah, Y., Redner, R. L., and Liu, J. M. (1999) Inhibitors of histone deacetylase relieve ETO-mediated repression and induce differentiation of AML1-ETO leukemia cells. *Cancer Res.* **59,** 2766–2769.

71. Momparler, R. L. and Bovenzi, V. (2000) DNA methylation and cancer. *J. Cell. Physiol.* **183,** 145–154.

72. Cameron, E. E., Bachman, K. E., Myohanen, S., Herman, J. G., and Baylin, S. B. (1999) Synergy of demethylation and histone deacetylase inhibition in the re- expression of genes silenced in cancer. *Nat. Genet.* **21,** 103–107.

73. Kinoshita, H., Shi, Y., Sandefur, C., et al. (2000) Methylation of the androgen receptor minimal promoter silences transcription in human prostate cancer. *Cancer Res.* **60,** 3623–3630.

74. Bovenzi, V., Le, N. L., Cote, S., Sinnett, D., Momparler, L. F., and Momparler, R. L. (1999) DNA methylation of retinoic acid receptor beta in breast cancer and possible therapeutic role of 5-aza-2'-deoxycytidine. *Anticancer Drugs* **10,** 471–476.

75. Ferguson, A. T., Evron, E., Umbricht, C. B., et al. (2000) High frequency of hypermethylation at the 14-3-3 sigma locus leads to gene silencing in breast cancer. *Proc. Natl. Acad. Sci. USA* **97,** 6049–6054.

76. Steinman, R. A., Hoffman, B., Iro, A., Guillouf, C., Liebermann, D. A., and El-Houseini, M. E. (1994) Induction of p21 (WAF-1/CIP1) during differentiation. *Oncogene* **9,** 3389–3396.

77. Zeng, Y. X. and El-Deiry, W. S. (1996) Regulation of p21WAF1/CIP1 expression by p53-independent pathways. *Oncogene* **12,** 1557–1565.

78. Blagosklonny, M. V., Prabhu, N. S., and El-Deiry, W. S. (1997) Defects in p21WAF1/CIP1, Rb, c-myc signaling in phorbol ester-resistant cancer cells. *Cancer Res.* **57,** 320–325.

79. Blagosklonny, M. V. (1998) The mitogen-activated protein kinase pathway mediates growth arrest or E1A-dependent apoptosis in SKBr3 human breast cancer cells. *Int. J. Cancer* **78,** 511–517.

80. Mitchell, K. O. and El-Deiry, W. S. (1999) Overexpression of c-myc inhibits p21WAF1/CIP1 expression and induces S-phase entry in 12-O-tetradecanoylphorbol-13-acetate (TPA)-sensitive human cancer cells. *Cell Growth Differ.* **10,** 223–230.

81. Radominska-Pandya, A., Chen, G., Czernik, P. J., et al. (2000) Direct interaction of all-trans-retinoic acid with protein kinase C (PKC). Implications for PKC signaling and cancer therapy. *J. Biol. Chem.* **275,** 22324–22330.

82. Han, Z. T., Zhu, X. X., Yang, R. Y., et al. (1998) Effect of intravenous infusions of 12O-tetradecanoylphorbol-13-acetate (TPA) in patients with myelocytic leukemia: Preliminary studies on therapeutic efficacy and toxicity. *Proc. Natl. Acad. Sci. USA* **95,** 5357–5361.

83. Trentesaux, C., Nyoung, M. N., Aries, A., et al. (1993) Increased expression of GATA-1 and NFE-2 erythroid-specific transcription factors during aclacinomycin-mediated differentiation of human erythroleukemic cells. *Leukemia* **7,** 452–457.

84. Aries, A., Trentesaux, C., Ottolenghi, S., Jardillier, J. C., Jeannesson, P., and Doubeikovski, A. (1996) Activation of erythroid-specific promoters during anthracycline-induced differentiation of K562 cells. *Blood* **87,** 2885–2890.

85. Melguizo, C., Prados, J., Aneiros, J., Fernandez, J. E., Velez, C., and Aranega, A. (1995) Differentiation of a human rhabdomyosarcoma cell line after antineoplastic drug treatment. *J. Pathol.* **175,** 23–29.

86. Meng, R. D. and El-Deiry, W. S. (2001) p53-Independent upregulation of KILLER/DR5 TRAIL receptor expression by glucocorticoids and interferon-gamma. *Exp. Cell Res.* **262,** 154–169.

87. Gianni, M. and de The, H. (1999) In acute promyelocytic leukemia NB4 cells, the synthetic retinoid CD437 induces contemporaneously apoptosis, a caspase-3-mediated degradation of PML/RAR alpha protein and the PML retargeting on PML-nuclear bodies. *Leukemia* **13,** 739–749.

88. Sun, S. Y., Yue, P., Wu, G. S., et al. (1999) Mechanisms of apoptosis induced by the synthetic retinoid CD437 in human non-small cell lung carcinoma cells. *Oncogene* **18,** 1357–2365.

89. Wynford-Thomas, D. (1997) Origin and progression of thyroid epithelial tumours: cellular and molecular mechanisms. *Hormone Res.* **47,** 145–157.

90. Cass, L. A. and Meinkoth, J. L. (2000) Ras signaling through PI3K confers hormone-independent proliferation that is compatible with differentiation. *Oncogene* **19,** 924–932.

91. Pietrzik, C. U., Hoffmann, J., Stober, K., et al. (1998) From differentiation to proliferation: the secretory amyloid precursor protein as a local mediator of growth in thyroid epithelial cells. *Proc. Natl. Acad. Sci. USA* **95,** 1770–1775.

92. Clark, O. H. (1996) Predictors of thyroid tumor aggressiveness. *West. J. Med.* **165,** 131–138.

93. Pujol, P., Daures, J. P., Nsakala, N., Baldet, L., Bringer, J., and Jaffiol, C. (1996) Degree of thyrotropin suppression as a prognostic determinant in differentiated thyroid cancer. *J. Clin. Endocrinol. Metab.* **81,** 4318–4323.

94. Pelikan, D. M., Lion, H. L., Hermans, J., and Goslings, B. M. (1997) The role of radioactive iodine in the treatment of advanced differentiated thyroid carcinoma. *Clin. Endocrinol. (Oxf).* **47,** 713–720.

95. Schmutzler, C. and Kohrle, J. (2000) Retinoic acid redifferentiation therapy for thyroid cancer. *Thyroid* **10,** 393–406.

96. Van Herle, A. J., Agatep, M. L., Padua, D. N., et al. (1990) Effects of 13 cis-retinoic acid on growth and differentiation of human follicular carcinoma cells (UCLA R0 82 W-1) in vitro. *J. Clin. Endocrinol. Metab.* **71,** 755–763.

97. Grunwald, F., Menzel, C., Bender, H., et al. (1998) Redifferentiation therapy-induced radioiodine uptake in thyroid cancer. *J. Nucl. Med.* **39,** 1903–1906.

98. Fagin, J. A., Tang, S. H., Zeki, K., Di Lauro, R., Fusco, A., and Gonsky, R. (1996) Reexpression of thyroid peroxidase in a derivative of an undifferentiated thyroid carcinoma cell line by introduction of wild-type p53. *Cancer Res.* **56,** 765–771.

99. Donghi, R., Longoni, A., Pilotti, S., Michieli, P., Della Porta, G., and Pierotti, M. A. (1993) Gene p53 mutations are restricted to poorly differentiated and undifferentiated carcinomas of the thyroid gland. *J. Clin. Invest.* **91,** 1753–1760.

100. Moretti, F., Farsetti, A., Soddu, S., et al. (1997) p53 re-expression inhibits proliferation and restores differentiation of human thyroid anaplastic carcinoma cells. *Oncogene* **14,** 729–740.

101. Blagosklonny, M. V., Giannakakou, P., Wojtowicz, M., et al. (1998) Effects of p53-expressing adenovirus on the chemosensitivity and differentiation of anaplastic thyroid cancer cells. *J. Clin. Endocrinol. Metab.* **83,** 2516–2522.

102. Morris, J. C., Kim, C. K., Padilla, M. L., and Mechanick, J. I. (1997) Conversion of non-iodine-concentrating differentiated thyrid carcinoma metastasis into iodine-concentrating foci after anticancer chemotherapy. *Thyroid* **7,** 63–66.

33

Regulation of NF-κB by Oncoproteins and Tumor Suppressor Proteins

Lee V. Madrid and Albert S. Baldwin, Jr.

1. Introduction

The transcription factor nuclear factor-κB (NF-κB) has been demonstrated to control cellular proliferation and oncogenesis. Consistent with this, NF-κB activity is stimulated by various mitogenic stimuli and by the action of numerous oncoproteins. This role of NF-κB in controlling oncogenesis is highlighted by observations that many oncoprotein signaling networks lead to the activation of NF-κB and that NF-κB is required for oncogene-induced cellular transformation *(1)*. Furthermore, it has also been demonstrated that various tumor suppressor gene products can function to negatively regulate NF-κB activity. However, evidence has been provided that NF-κB activity is required for the ability of p53 to induce cell death and that NF-κB can regulate p53 gene expression. This review highlights the complex roles that NF-κB subunits play in cellular transformation and analyzes mechanisms of oncoprotein-induced activation and tumor suppressor gene regulation of NF-κB. What is clear is that the role of NF-κB in regulating oncogenic mechanisms is complex, due to stimulus-specific, cell type-specific, or subunit-specific responses.

2. The Roles of NF-κB in Oncogenesis and Cell Survival

Like many transcription factors, NF-κB consists of a family of related proteins. Presently there are five known human Rel/NF-κB family members, p50, p52 (p50 and p52 are posttranslationaly derived from p100 and p105, respectively), p65/RelA, c-Rel, and RelB. NF-κB functions through hetero- or homodimerization of the majority of subunits to lead to distinct as well as redundant functions for the different forms *(2,3)*. The classic NF-κB transcription factor is a p50-p65 heterodimer which is considered to be widely responsible for controlling many of the transcription functions of NF-κB.

NF-κB is the prototypical, posttranslationally regulated transcription factor *(2,3)*. In unstimulated cells, the majority of NF-κB is found in the cytoplasm, associated with inhibitory molecules known as the IκB proteins. In response to a wide variety of stimuli, including cytokine or growth factor stimulation, NF-κB is induced to accumulate in the

From: *Methods in Molecular Biology, Vol. 223: Tumor Suppressor Genes: Regulation, Function, and Medicinal Applications*. Edited by: Wafik S. El-Deiry © Humana Press Inc., Totowa, NJ

nucleus. The canonical NF-κB activation mechanism involves the phosphorylation of IκB on two critical serine residues by the IκB kinase (IKK) signalosome complex *(4–7)*. Phosphorylated IκB is then targeted for ubiquitination and subsequent degradation by the 26S proteosome, which allows liberated NF-κB to accumulate in the nucleus where it activates transcription of NF-κB-responsive genes *(2,3)*. Genes regulated by NF-κB include those encoding cytokines, cytokine receptors, antiapoptotic proteins, and c-Myc. Although the induced nuclear translocation of NF-κB has been highly regarded as the principal method to activate NF-κB-dependent gene expression, an alternative mechanism of NF-κB activation is emerging that involves the phosphorylation of the NF-κB subunits. For example, it has been shown that the proinflammatory cytokines TNF and IL-1β lead to the phosphorylation of RelA/p65 and the subsequent stimulation of NF-κB transactivation through pathways apparently distinct from induced nuclear translocation *(8,9)*. These studies indicate that dual controls exists for NF-κB, with mechanisms controlling induction of nuclear translocation of NF-κB, as well as regulating the inherent transcriptional activity of NF-κB. Oncoprotein-controlled signaling pathways have been demonstrated to lead to the activation of NF-κB at the level of nuclear accumulation and transactivation stimulation, which is important for controlling the cellular transformation process (see below).

One of the surprising findings about NF-κB was that it is a powerful regulator of apoptosis. This was first revealed when it was shown that the p65/RelA knockout dies at d 16 of embryonic development from widespread liver apoptosis *(1)*. Subsequently, it was shown that the activation of NF-κB suppresses the apoptotic potential of TNF through the upregulation of certain TRAF and IAP proteins to block caspase-8 activation. Additionally, it has been shown that NF-κB regulates other antiapoptotic genes, such as Bcl-xL *(1)*.

Observations indicate that NF-κB controls aspects of growth control and proliferation in response to oncogenic stimuli *(1)*. Indeed, oncoproteins such as Ras, Akt, Bcr-Abl, the Epstein Bar viral oncoprotein LMP1, the HTLV-1 Tax protein, and Her-2/neu have been demonstrated to lead to the activation of NF-κB *(1)*. In the case of oncogenic Ras, our laboratory has shown that expression of oncogenic H-Ras V12 leads to transcriptional activation of a transfected, NF-κB-dependent luciferase reporter *(10,11)*. While oncogenic Ras has been demonstrated to lead to NF-κB nuclear translocation and DNA binding in transient Ras expression assays, fibroblasts transformed by Ras show hyperactivated NF-κB transcriptional activity, rather than increased NF-κB nuclear translocation and DNA binding. In this regard, Ras appears to utilize a signal transduction pathway or pathways to target the resident nuclear NF-κB for enhanced transactivation potential. Similarly, oncogenic Raf targets the transcriptional activation function of NF-κB in fibroblasts through a mechanism involving an autocrine factor induced by Raf *(11)*. Sonenshein and colleagues have shown that Ras- or Raf-transformed epithelial cells exhibit enhanced nuclear levels of NF-κB *(12)*. Consistent with these overall results, the Ras effector molecule PI3K has been shown to lead to the activation of NF-κB through the activation of Akt *(13)*. Interestingly, the strong antiapototic function of Akt can be explained partly through its ability to activate NF-κB. The mechanism of Akt-mediated NF-κB activation occurs by either enhancing NF-κB nuclear translocation *(14–16)* or by stimulating the transactivation domain of NF-κB independent of signaling pathways that lead to nuclear accumulation and DNA binding *(9,13)*. Recently, we have shown that the ability of Akt to enhance the transactivation potential of NF-κB requires IKK through a mechanism involving the stress-activated kinase p38 *(17)*.

Since NF-κB is activated by oncoproteins, it was important to determine if NF-κB is required for cellular transformation. It was shown that expression of IκBα inhibited the ability of Ras to induce focus formation *(1)* and others showed that inhibitors of IKK blocked the ability of Ras or Raf to transform epithelial cells *(12)*. NF-κB appears to be required for tumor formation, since expression of a modified form of IκB blocked tumor growth initiated by Bcr-Abl and blocked squamous cell carcinoma tumor formation *(1)*. In addition to the experimental evidence supporting the role of NF-κB in oncogenesis, NF-κB, or components of its regulation, have been demonstrated to be deregulated in human cancers. These abnormalities in NF-κB regulation are frequently seen in Hodgkin's and B-cell lymphomas, leukemias, and solid tumors *(1)*. Evidence for a role for NF-κB in these cancers comes from studies on Hodgkin's lymphoma, in which the inhibition of NF-κB blocked growth and cell survival *(18)*. The high NF-κB activities in cancer cells are due to perturbations in the signaling pathways leading to NF-κB activation, highlighting the importance of the regulation of this transcription factor *(1,19)*.

What is the role of NF-κB in oncogenesis? In-vitro experimentation revealed that NF-κB is required for cellular transformation mediated by H-Ras V12 to overcome Ras-induced apoptosis *(20)*. This suppression of apoptosis may represent an early event in oncogenesis, since inhibition of NF-κB in late-stage solid tumor models does not induce cell death. Although NF-κB plays a role in cell survival functions, evidence has also been presented that NF-κB controls factors involved in growth control and proliferation. In this way, mitogenic stimuli, such as growth factors, can lead to the activation of NF-κB and subsequently stimulate cell cycle progression. In this regard, NF-κB regulates expression of the cyclin D1 gene by binding to NF-κB sites located in the promoter *(21,22)*. This positive regulation of cyclin D1 by NF-κB also leads to increased G1-to-S-phase cell cycle transition. Likewise, inhibition of NF-κB reduces cyclin D1 activity and the subsequent hyperphosphorylation of the Rb tumor suppressor protein, resulting in delayed cell cycle progression *(21,22)*. Additional data indicate that NF-κB may control metastasis and angiogenesis *(23)*. Thus, these data taken as a whole suggest that aspects of NF-κB regulation or the induced transcriptional activity of NF-κB are crucial for the regulation of cellular growth control and oncogenesis.

3. Regulation of NF-κB by Tumor Suppressor Protein Function

Since NF-κB has been demonstrated to provide cell survival functions and promote cellular transformation, it is not surprising that there is evidence that tumor suppressor proteins can regulate NF-κB activity. In this way, suppression of NF-κB activity may relate to the induction of apoptosis and to the inhibition of transformation potential that are known to be controlled by tumor suppressor activities.

p53, the most widely studied tumor suppressor protein, promotes apoptosis and suppresses growth through the upregulation of proteins such as the cyclin-cdk inhibitor p21. Interestingly, there are reports that p53 and NF-κB are mutually antagonistic. Two reports indicate that p53 induction leads to the downregulation of NF-κB activity by competing for the limited pool of the co-activator CBP/p300 *(24,25)*. Likewise, in response to TNFα stimulation, increased NF-κB activity leads to the downregulation of p53 activity, again suggesting that these two transcription factors compete for co-activators and are mutually antagonistic *(24)*. Two other reports indicate that p53 induction inhibits NF-κB and NF-κB-regulated genes *(26,27)*. Consistent with this, papilloma virus E6 and E7 proteins, which inhibit p53 and Rb proteins, upregulate NF-κB-responsive genes in cervical keratinocytes *(28)*. Additionally, evidence using a nutrient-deficiency model indicates that NF-κB antagonizes the ability of p53 to induce apoptosis *(29)*. These studies provide evidence that p53 and NF-κB are functionally antagonistic. However, other evidence indicates that NF-κB may be required for the ability of p53 to induce apoptosis (*see* **Subheading 4.**).

An interesting protein that has been implicated in inhibiting NF-κB activity is the E2F transcription factor. The properties of E2F are complex, with evidence that this transcription factor is associated with cell growth, the induction of apoptosis, and tumor formation as well as tumor inhibition. Perturbed E2F regulation has been observed in a number of human tumors leading to the loss of Rb, upregulation of cyclin D1, and loss of the p16INK4a CDK inhibitor activity *(30)*. E2F-1 has been implicated in the induction of apoptosis, and mutant E2F-1 can efficiently rescue the induced apoptosis of Rb-null cells *(31,32)*. It has been reported that E2F-1-inducible cell lines can lead to the inhibition of NF-κB DNA-binding activity in response to TNFα as well as the inhibition of NF-κB-regulated genes. In addition, E2F-1 expression potentiates TNFα-mediated apoptosis, and this can be rescued by overexpressing the p50/p65 subunits of NF-κB *(33)*. Consistent with this, it was shown recently that E2F1 interacts with NF-κB to suppress its activity *(34)*. These results suggest that one mechanism of E2F-1-mediated apoptosis is by the inhibition of NF-κB activity and is consistent with the role of NF-κB in providing cell survival functions.

Disruption of the p16INK4a tumor suppressor protein is frequently detected in human cancer *(30)*. The p16INK4a protein contains ankryin repeats, similar to the IκB proteins, and certain ankyrin repeats containing proteins are known to bind to RelA family members. It was demonstrated that RelA/p65 and p16INK4a expression in 293 cells lead to an interaction between these two proteins *(35)*. Furthermore, stimulation of cells with TNFα, a potent inducer of IκB degradation, lead to an increase in the endogenous RelA/p65-p16INK4a interaction, consistent with the idea that p16INK4a binds RelA/p65 via its ankyrin repeat region. Finally, it was also demonstrated that the expression of p16INK4a decreased NF-κB transcriptional activity in response to TNFα. These results suggest that the p16INK4a protein can negatively modulate NF-κB activity by interacting with NF-κB and inhibiting its activity. Additionally, p16INK4a can lead to a G1 block and growth arrest in cells, and its ability to inhibit NF-κB activity may contribute to this function.

An interesting and recently identified protein that is frequently deleted in human cancer is the PTEN (MMAC) tumor suppressor. The loss of PTEN expression has been demonstrated to occur in a large percentage of advanced human prostate cancers *(36–38)*. Furthermore, mechanistic studies have revealed that the PTEN protein inactivates the PI3K/Akt pathway due to its inherent phosphatase activity *(39)*. PTEN dephosphorylates phosphatidlyinositol 3-OH lipids (PIP3), reducing these lipids to the PIP2 form. Since Akt requires PIP3 lipids for its activation, PTEN can effectively block Akt kinase activity. In addition, since Akt potently activates NF-κB (see above), the PTEN protein is hypothesized to lead to the inactivation of NF-κB. Recently it has been reported that in response to IL-1β, U251 glioma cells overexpressing PTEN are defective in NF-κB transcriptional activity in comparison to cells that are defective for PTEN expression *(40)*. This inactivation of NF-κB does not involve inhibition of degradation of IκB and is not due to decreased RelA/p65 nuclear translocation, but rather targets the phosphorylation

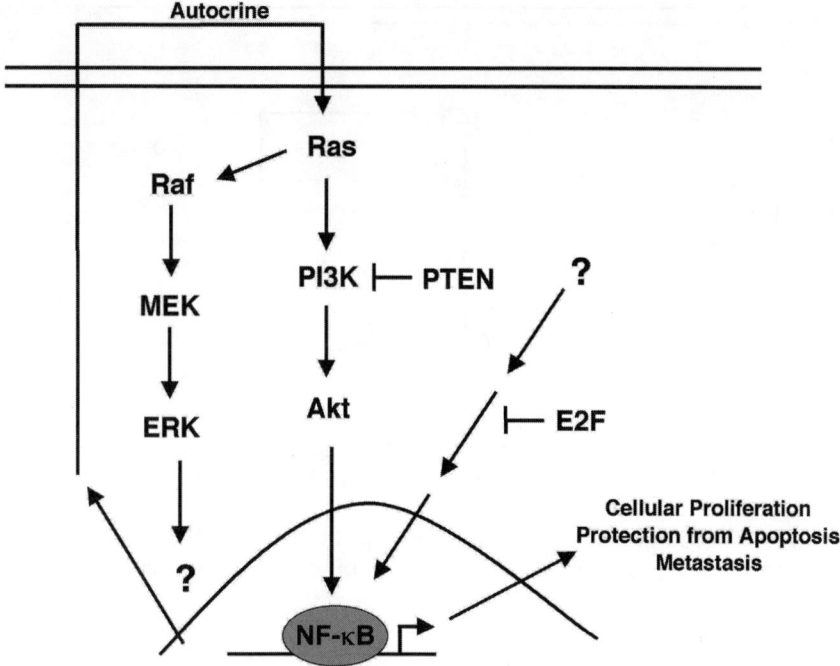

Fig. 1. The role of NF-κB in cellular proliferation and protection from apoptosis. Oncogenic Ras and PI3K-mediated activation of NF-κB and protection from apoptosis is inhibited by the tumor suppressor protein PTEN. Additionally, Raf-mediated NF-κB activation occurs through an undefined mechanism leading to autocrine activation of NF-κB. In addition, the tumor suppressor protein E2F inhibits NF-κB activity by an undefined mechanism.

of NF-κB subunits *(40,41)*. These results support the notion that the PI3K/Akt pathway does not modulate NF-κB activity by a nuclear translocation-mediated mechanism but rather targets the phosphorylation and inherent transcriptional activity of NF-κB (**Fig. 1**).

4. Potential Antioncogenic and Proapoptotic Roles for NF-κB

Although there is compelling evidence that NF-κB functions in prooncogenic and antiapoptotic roles, there is evidence that NF-κB is required to inhibit oncogenesis in certain settings and that NF-κB can function proapoptotically. Thus, expression of a modified form of IκBα in the skin epidermis led to a hyperplastic response *(42)* and caused the appearance of epithelial cancer *(43)*. Additionally, it has been shown that IκB blocked the induction of apoptosis induced by hydrogen peroxide *(44)*. Thus there is evidence that NF-κB can serve antioncogenic and proapoptotic functions. In these regards one must consider the stimulus, the cell type, the different NF-κB subunits that may be involved in the response, as well as the possibility that IκB expression may have functions distinct from blocking NF-κB.

The role of NF-κB subunits in inducing cellular proliferation is well established, but reports describing the action of the c-Rel NF-κB subunit suggest a divergent role for NF-κB subunits in cell survival and proliferation. It has been shown that c-Rel expression in HeLa cells leads to growth arrest at the G1–S phase transition *(45)*. Additionally, c-Rel expression lead to the inhibition of E2F DNA binding and an accumulation

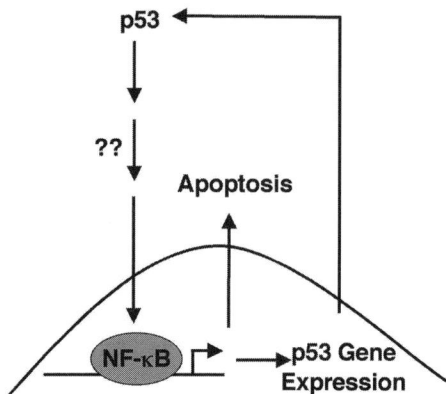

Fig. 2. The role of p53-mediated NF-κB activation. p53 overexpression can lead to the activation of NF-κB, causing an apoptotic response. Likewise, NF-κB transcription function controls p53 gene expression.

of the hypophosphorylated form of Rb. Interestingly, these authors also demonstrate that c-Rel expression leads to increased stability of the p53 protein *(45)*. These data, although inconsistent with the role of NF-κB in cell survival and pro-proliferation, demonstrate that the diverse NF-κB subunits may have opposing effects. Additionally, care must be taken with interpretations regarding expression of an isolated NF-κB subunit in the absence of a physiological signal.

The p53 tumor suppressor protein is frequently lost and/or mutated in a variety of tumors and has clear growth-inhibitory and proapoptotic function. Recently it was shown that expression of p53 leads to a subsequent increase in NF-κB DNA-binding activity *(46)*. Additionally, these authors showed inhibition of NF-κB blocked the ability of p53 to induce apoptosis. These results suggest that p53 expression leads to the posttranslational increase in NF-κB DNA-binding activity and that this activity is required for p53-induced cell death. These data are in apparent contrast with the reports described above regarding the potential mutual antagonism between NF-κB and p53. Adding to the complexity of the relationship between p53 and NF-κB are reports indicating that NF-κB can regulate p53 gene expression *(47)* (**Fig. 2**).

5. Conclusions

Traditionally, NF-κB has been implicated in the positive control of cellular growth and proliferation, yet the studies outlined here demonstrate that the function of NF-κB is not as straightforward. Whereas there are solid data indicating a causative role in NF-κB in promoting cancer and cell cycle progression, other data indicate the complex nature of NF-κB signaling and responses. The data concerning the c-Rel subunit of NF-κB suggests that at least one NF-κB subunit leads to the growth arrest of cells at the G1–S phase transition. This mechanism of growth arrest could be due to the ability of c-Rel to modulate components of the cell cyle that are known to function in the G1–S phase transition such as, E2F, Rb, and CDK2, and through the increased stability of p53. However, it should be pointed out that the phenotype of the c-Rel knockout mice indicates that c-Rel plays a positive role in B- and T-cell mitogenic responses *(3)*. It remains to be seen what the true physiologic function of c-Rel, or perhaps other Rel subunits, is in regulating growth arrest.

Perhaps the most confusing relationship is between p53 and NF-κB. The role p53 plays in the activation of NF-κB is not apparent, and although p53 has been shown to be mutually antagonistic, the inducible expression of p53 clearly leads to increased NF-κB DNA-binding functions *(46)*. It is interesting to note that the measurement of NF-κB transcription function has not been demonstrated in this inducible system, and although high levels of NF-κB were induced in the nucleus, it remains to be seen whether those complexes are transcriptionally active. Finally, it appears that cellular context is also a factor in the regulation between these two transcription factors and cell type specificities may play a role in determining the balance of activation vs repression of NF-κB and p53 *(48)*.

The regulation of NF-κB activity by the tumor suppressor PTEN may turn out to be the most interesting in terms of mechanisms for activation of NF-κB. There is considerable controversy concerning the mechanism of Akt-mediated activation of NF-κB *(49)*, and Akt has been shown to activate NF-κB DNA binding as well as stimulating transactivation function. The data suggested by PTEN inhibition of NF-κB activity suggests that the Akt pathway predominantly stimulates NF-κB activity by targeting transcriptional functions of NF-κB independent of a nuclear translocation-mediated mechanism. The role of IKK in this process has recently been demonstrated, and it appears as if IKK may be functioning as a RelA/p65 kinase leading to transcriptional activation of NF-κB *(17)*. Regardless of the mechanism, clearly one function of PTEN is to inhibit NF-κB activity, and this may contribute to its tumor suppressor function.

It remains to be seen what the true physiologic consequences of tumor suppressor-induced activation or inhibition NF-κB are. The diverse roles that NF-κB plays in cancer progression or inhibition may be more closely determined upon the analysis of endogenous tumor suppressor gene deletions found in human cancers or in model genetic systems. These analyses may lead to a clearer understanding of NF-κB function in the regulation of tumor suppressor mechanisms and oncogenesis in general.

References

1. Baldwin, A. S. (2001) Control of oncogenesis and cancer therapy resistance by the transcription factor NF-kappaB. *J. Clin. Inv.* **107**, 241–246.
2. Baldwin, A. S. (1996) The NF-kappa B and I-kappa B proteins: new discoveries and insights. *Annu. Rev. Immunol.* **14,** 649–681.
3. Ghosh, S., May, M. J., and Kopp, E. B. (1998) NF-kappa B and Rel proteins: evolutionarily conserved mediators of immune responses. *Annu. Rev. Immunol.* **16**, 225–260.
4. DiDonato, J., Mercurio, F., Rosette, C., et al. (1996) Mapping of the inducible IkappaB phosphorylation sites that signal its ubiquitination and degradation. *Mol. Cell. Biol.* **16**, 1295–1304.
5. Mercurio, F., Zhu, H., Murray, B. W., et al. (1997) IKK-1 and IKK-2: cytokine-activated IkappaB kinases essential for NF-kappaB activation. *Science* **278**, 860–866.
6. Zandi, E., Rothwarf, D. M., Delhase, M., Hayakawa, M., and Karin, M. (1997) The IkappaB kinase complex (IKK) contains two kinase subunits, IKKalpha and IKKbeta, necessary for IkappaB phosphorylation and NF-kappaB activation. *Cell* **91**, 243–252.
7. Zandi, E. and Karin, M. (1999) Bridging the gap: composition, regulation, and physiological function of the IkappaB kinase complex. *Mol. Cell. Biol.* **19**, 4547–4551.

8. Wang, D. and Baldwin, A. S., Jr. (1998) Activation of nuclear factor-kappaB-dependent transcription by tumor necrosis factor-alpha is mediated through phosphorylation of RelA/p65 on serine 529. *J. Biol. Chem.* **273**, 29411–29416.

9. Sizemore, N., Leung, S., and Stark, G. R. (1999) Activation of phosphatidylinositol 3-kinase in response to interleukin-1 leads to phosphorylation and activation of the NF-kappaB p65/RelA subunit. *Mol. Cell. Biol.* **19**, 4798–4805.

10. Finco, T. S., Westwick, J. K., Norris, J. L., Beg, A. A., Der, C. J., and Baldwin, A. S., Jr. (1997) Oncogenic Ha-Ras-induced signaling activates NF-kappaB transcriptional activity, which is required for cellular transformation. *J. Biol. Chem.* **272**, 24113–24116.

11. Norris, J. L. and Baldwin, A. S., Jr. (1999) Oncogenic ras enhances NF-kappaB transcriptional activity through raf- dependent and raf-independent mitogen-activated protein kinase signaling pathways. *J. Biol. Chem.* **274**, 13841–13846.

12. Arsura, M., Mercurio, F., Oliver, A., Thorgeirsson, S., and Sonenshein, G. (2000) Role of the Ikappa B complex in oncogenic Ras- and Raf-mediated transformation of rat liver epithelial cells. *Mol. Cell. Biol.* **20**, 5381–5391.

13. Madrid, L. V., Wang, C. Y., Guttridge, D. C., Schottelius, A. J., Baldwin, A. S., Jr., and Mayo, M. W. (2000) Akt suppresses apoptosis by stimulating the transactivation potential of the RelA/p65 subunit of NF-kappaB. *Mol. Cell. Biol.* **20**, 1626–1638.

14. Kane, L. P., Shapiro, V. S., Stokoe, D., and Weiss, A. (1999) Induction of NF-kappaB by the Akt/PKB kinase. *Curr. Biol.* **9**, 601–604.

15. Ozes, O. N., Mayo, L. D., Gustin, J. A., Pfeffer, S. R., Pfeffer, L. M., and Donner, D. B. (1999) NF-kappaB activation by tumour necrosis factor requires the Akt serine-threonine kinase. *Nature* **401**, 82–85.

16. Romashkova, J. A. and Makarov, S. S. (1999) NF-kappaB is a target of AKT in anti-apoptotic PDGF signalling. *Nature* **401**, 86–90.

17. Madrid, L., Mayo, M., Reuther, J., and Baldwin, A. (2001) Akt stimulates the transactivation potential of the RelA/p65 subunit of NF-kB through utilization of the IκB kinase and activation of the mitogen-activated protein kinase p38. *J. Biol. Chem.* **276**, 18934–18940.

18. Bargou, R., Emmerich, F., Krappmann, D., et al. (1997) Constitutive activation of NF-κB-RelA is required for proliferation and survival of Hodgkin's disease tumor cells. *J. Clin. Invest.* **100**, 2961–2969.

19. Yamamoto, Y. and Gaynor, R. B. (2001) Therapeutic potential of inhibition of the NF-kappaB pathway in the treatment of inflammation and cancer. *J. Clin. Invest.* **107**, 135–142.

20. Mayo, M. W., Wang, C. Y., Cogswell, P. C., et al. (1997) Requirement of NF-kappaB activation to suppress p53-independent apoptosis induced by oncogenic Ras. *Science* **278**, 1812–1815.

21. Guttridge, D. C., Albanese, C., Reuther, J. Y., Pestell, R. G., and Baldwin, A. S., Jr. (1999) NF-kappaB controls cell growth and differentiation through transcriptional regulation of cyclin D1. *Mol. Cell. Biol.* **19**, 5785–5799.

22. Hinz, M., Krappmann, D., Eichten, A., Heder, A., Scheidereit,C., and Strauss, M. (1999) NF-kappaB function in growth control: regulation of cyclin D1 expression and G0/G1-to-S-phase transition. *Mol. Cell. Biol.* **19**, 2690–2698.

23. Huang, S., DeGuzman, A., Bucana, C., and Fidler, I. (2000) NF-κB activity correlates with growth, angiogenesis, and metastasis of human melanoma cells in nude mice. *Clin. Cancer Res.* 6, 2573–2581.

24. Ravi, R., Mookerjee, B., van Hensbergen,Y., et al. (1998) p53-mediated repression of nuclear factor-kappaB RelA via the transcriptional integrator p300. *Cancer Res.* **58**, 4531–4536.

25. Webster, G. A. and Perkins, N. D. (1999) Transcriptional cross talk between NF-kappaB and p53. *Mol. Cell. Biol.* **19**, 3485–3495.

26. Shao, J., Fujiwara, T., Kadowaki, Y., et al. (2000) Overexpression of wild-type p53 inhibits NF-κB activity and synergizes with aspirin to induce apoptosis in human colon cancer cells. *Oncogene* **10**, 726–736.

27. Bartke, T., Siegmund, D., Peters, N., et al. (2001) p53 upregulates cFLIP, inhibits transcription of NF-κB-regulated genes and induces caspase-8-independent cell death in DLD-1 cells. *Oncogene* **20**, 571–580.

28. Nees, M., Geoghegan, J., Hyman, T., Frank, S., Miller, L., and Woodworth,C. (2001) Papillomavirus type 16 oncogenes downregulate expression of interferon-responsive genes and upregulate proliferation-associated and NF-κB-responsive genes in cervical keratinocytes. *J. Virol.* **75**, 4283–4296.

29. Holmes-McNary, A., Baldwin, A., and Zeisel, S. (2001) Opposing regulation of choline deficiency-induced apoptosis by p53 and NF-κB. *J. Biol. Chem.* **276**, 41197–41204.

30. Sherr, C. J. and Roberts, J. M. (1999) CDK inhibitors: positive and negative regulators of G1-phase progression. *Genes Dev.* **13**, 1501–1512.

31. Pan, H., Yin, C., Dyson, N. J., Harlow, E., Yamasaki, L., and Van Dyke, T. (1998) Key roles for E2F1 in signaling p53-dependent apoptosis and in cell division within developing tumors. *Mol. Cell* **2**, 283–292.

32. Tsai, K. Y., Hu, Y., Macleod, K. F., Crowley, D., Yamasaki, L., and Jacks, T. (1998) Mutation of E2F-1 suppresses apoptosis and inappropriate S phase entry and extends survival of Rb-deficient mouse embryos. *Mol. Cell* **2**, 293–304.

33. Phillips, A. C., Ernst, M. K., Bates, S., Rice, N. R., and Vousden, K. H. (1999) E2F-1 potentiates cell death by blocking antiapoptotic signaling pathways. *Mol. Cell* **4**, 771–781.

34. Tanaka, H., Matsumura, I., Ezoe, E., et al. (2002) E2F1 and c-Myc potentiate apoptosis through inhibition of NF-κB activity that facilitates MnSOD-mediated ROS elimination. *Mol. Cell* **9**, 1017–1029.

35. Wolff, B. and Naumann, M. (1999) INK4 cell cycle inhibitors direct transcriptional inactivation of NF-kappaB. *Oncogene* **18**, 2663–2666.

36. Li, J., C., Yen, C., Liaw, D., et al. (1997) PTEN, a putative protein tyrosine phosphatase gene mutated in human brain, breast, and prostate cancer. *Science* **275**, 1943–1947.

37. Di Cristofano, A., Pesce, B., Cordon-Cardo, C., and Pandolfi, P. P. (1998) PTEN is essential for embryonic development and tumour suppression. *Nat. Genet.* **19**, 348–355.

38. Whang, Y. E., Wu, X., Suzuki, H., et al. (1998) Inactivation of the tumor suppressor PTEN/MMAC1 in advanced human prostate cancer through loss of expression. *Proc. Natl. Acad. Sci. USA* **95**, 5246–5250.

39. Stambolic, V., Suzuki, A., de la Pompa, J. L., et al. (1998) Negative regulation of PKB/Akt-dependent cell survival by the tumor suppressor PTEN. *Cell* **95**, 29–39.

40. Koul, D., Yao, Y., Abbruzzese, J., Yung, W., and Reddy, S. (2001) Tumor suppressor MMAC/PTEN inhibits cytokine-induced NF-κB activation without interfering with the IkappaB degradation pathway. *J. Biol. Chem.* **276**, 11402–11408.

41. Mayo, M., Madrid, L., Westerheide, S., et al. (2002) PTEN blocks TNF-induced NF-kB-dependent transcription by inhibiting the transactivation potential of the p65 subunit. *J. Biol. Chem.* **277**, 11116–11125.

42. Seitz, C. S., Lin, Q., Deng, H., and Khavari, P. A. (1998) Alterations in NF-kappaB function in transgenic epithelial tissue demonstrate a growth inhibitory role for NF-kappaB. *Proc. Natl. Acad. Sci. USA* **95**(5), 2307–2312.

43. van Hogerlinden, M., Rozell, B., Ahrlund-Richter, L., and Toftgard, R. (1999) Squamous cell carcinomas and increased apoptosis in skin with inhibited Rel/NF-κB signaling. *Cancer Res.* **59**, 3299–3303.

44. Dumont, A., Hehner, S. P., Hofmann, T. G., Ueffing, M., Droge, W., and Schmitz, M. L. (1999) Hydrogen peroxide-induced apoptosis is CD95-independent, requires the release of mitochondria-derived reactive oxygen species and the activation of NF-kappaB. *Oncogene* **(18)**3, 747–757.

45. Bash, J., Zong, W. X., and Gelinas, C. (1997) c-Rel arrests the proliferation of HeLa cells and affects critical regulators of the G1/S-phase transition. *Mol. Cell. Biol.* **17**, 6526–6536.

46. Ryan, K. M., Ernst, M. K., Rice, N. R., and Vousden, K. H. (2000) Role of NF-kappaB in p53-mediated programmed cell death. *Nature* **404**, 892–897.

47. Benoit, V., Hellin, A. C., Huygen, S., Gielen, J., Bours, V., and Merville, M. P. (2000) Additive effect between NF-kappaB subunits and p53 protein for transcriptional activation of human p53 promoter. *Oncogene* **19**(41), 4787–4794.

48. Perkins, N. D. (2000) The Rel/NF-kappa B family, friend and foe. *Trends Biochem. Sci.* **25**, 434–440.

49. Delhase, M., Li, N., and Karin, M. (2000) Kinase regulation in inflammatory response. *Nature* **406**, 367–368.

34

Blocking Survivin to Kill Cancer Cells

Dario C. Altieri

1. Introduction

It has long been appreciated that dysregulation of apoptosis, or programmed cell death, plays a critical role in the onset and progression of cancer *(1)*. For the complexity of cell death pathways, this may result from deregulated overexpression of apoptosis inhibitors or loss/underrepresentation of apoptosis inducers. The net effect for both conditions is an aberrantly extended cell viability, which facilitates the acquisition of transforming mutations and the insurgence of chemoresistance *(2)*. Because of the tremendous progress recently made in elucidating apoptotic mechanisms and their relevance to human diseases, concrete therapeutic efforts are now underway to manipulate cell death pathways and eliminate survival signals that may contribute to keeping cancer cells alive *(3)*. To make an ideal drug target, an apoptosis regulator should be preferentially expressed in tumor cells but not in normal tissues, and interference with its expression/function should be sufficient to facilitate cell death, either alone, or in combination with chemotherapeutic drugs or ultraviolet/γ-irradiation. Data recently appeared in the literature suggests that the human *survivin* gene may fulfill both of these prerequisites. Furthermore, initial evidence in vitro and in vivo has suggested that targeting survivin may provide a viable approach to kill cancer cells selectively.

1.1. Survivin and the Inhibitor of Apoptosis (IAP) Gene Family

Survivin is an intracellular molecule of 16.5 kDa that belongs to the Inhibitor of Apoptosis (IAP) gene family *(4)*. IAP proteins are found in the genome of many organisms from yeast to mammalian cells and are structurally characterized by the presence of one to three copies of a ~70 amino acid zinc finger fold, designated Baculovirus IAP Repeat (BIR) *(5)*. Other structural modules found in certain IAP proteins include a —COOH-terminus RING finger, a caspase-recruitment domain (CARD), or a ubiquitin-conjugating domain *(5)*. Several members of the IAP family block apoptosis by physically associating with both initiator and effector caspases and prevent their proteolytic processing and enzyme catalytic activity *(5)*. The structural requirements of these interactions involving both initiator caspase-9 and down-

From: *Methods in Molecular Biology, Vol. 223: Tumor Suppressor Genes: Regulation, Function, and Medicinal Applications.* Edited by: Wafik S. El-Deiry © Humana Press Inc., Totowa, NJ

stream effector caspase-3 have been elucidated in considerable details, and discrete molecular regions involved in docking IAP proteins to terminal effector caspases have been determined at the atomic level *(6)*. The paradigm of an IAP–caspase complex that protects cells against apoptosis is highly evolutionary conserved, and a similar model has been demonstrated in model organisms, including *Drosophila (7)*. There are several features that distinguish survivin from other members of the IAP family. First, survivin is structurally unique for its highly stable and organized dimeric architecture, which has been demonstrated for both the human and mouse protein *(8–10)*. Second, survivin is the sole IAP family protein described so far to be expressed in mitosis in a strict cell cycle-dependent manner *(11)*, and to localize to discrete compartments of the mitotic apparatus, including spindle microtubules, centrosomes, and the central spindle midzone *(11–13)*.

1.2. Expression of Survivin in Cancer

One of the most significant features of survivin is its differential expression in cancer vs normal tissues *(4)*, with a pattern reminiscent of "onco-fetal" antigens. Although strongly expressed in embryonic and fetal tissues, including kidney, lung, brain, endocrine pancreas, endometrial glands, and epidermis *(14)*, survivin was conspicuously undetectable in terminally differentiated adult tissues, by Northern blotting, immunohistochemistry, and *in situ* hybridization *(4)*. As demonstrated by *in situ* hybridization, immunohistochemistry, and RT-PCR survivin became abundantly expressed in tumors of lung *(15)*, breast *(16)*, colon *(17–19)*, stomach *(20)*, esophagus *(21)*, pancreas *(22)*, bladder *(23)*, uterus *(24)*, large-cell non-Hodgkin's lymphoma *(25)*, acute myelogenous leukemia *(26)*, gliomas *(27)*, neuroblastoma *(28,29)*, melanoma *(30)*, and nonmelanoma *(31)* skin cancers and pediatric renal tumors *(32)* (**Table 1**). In genome-wide searches, survivin was identified as the fourth top "transcriptome" in cancers of colon, lung, brain, breast, and melanoma, but was undetectable in the corresponding normal tissues *(33)*. In addition to tumor cells, the endothelium during angiogenesis becomes intensely positive for expression of survivin in vivo, and after exposure to vascular endothelial cell growth factor, fibroblast growth factor, and angiopoietin-1, in vitro *(34,35)*.

1.3. Prognostic Significance and Impact of Survivin Expression in Cancer

Various retrospective trials looked at a potential relationship between survivin expression and cancer prognosis. Cancer patients expressing survivin consistently exhibited abbreviated survival rates *(15–18,21,25,27)*, association with negative predictive markers of disease progression *(26,28,32,36)* and accelerated rates of recurrences *(23)* (**Table 1**).

Because of its selective expression in cancer, the possibility of detecting survivin or survivin-related material in biologic fluids has also been investigated as a diagnostic tool. Consistent with this idea, patients with cancers of colon and lung, but not healthy volunteers, were found to contain circulating antibodies against survivin *(37)*. The fact that cancer patients recognize survivin as a "non-self" protein, and mount an immune response to it, may have practical implications for potential vaccination strategies in cancer treatment. This hypothesis has been experimentally validated in two independent studies in which a vigorous T-cell cytolytic response to survivin peptides was elicited in

Table 1
Survivin Expression in Human Cancer

Cancer series	n	Percentage survivin-positive	Detection method used[a]	Clinico-pathologic correlation	Reference
Stomach	174	34	IHC	Association with bcl-2, low apoptotic index	Lu et al. *(20)*
Colorectal	171	53	IHC	Association with bcl-2; abbreviated survival	Kawasaki et al. *(17)*
Colorectal	144	63	RT-PCR	Association with advanced stage	Sarela et al. *(18)*
Esophagus	51		RT-PCR	Abbreviated survival in high expressors	Kato et al. *(21)*
Lung	83	85.5	RT-PCR	Abbreviated survival	Monzo et al. *(15)*
Breast	167	70%	IHC	Association with bcl-2; abbreviated survival for cases with low apoptotic index	Tanaka et al. *(16)*
Large cell non-Hodgkin's lymphoma	222	60%	IHC	Abbreviated survival; independent unfavorable prognostic factor	Adida et al. *(25)*
Acute myeloblastic leukemia	167	60%	IF and IHC	Independent unfavorable prognostic factor.	Adida et al. *(26)*
Neuroblastoma	72	47%	IHC	Association with unfavorable negative prognostic factors	Adida et al. *(28)*
Nueroblastoma	34	Expression in advanced grade	NB	Association with unfavorable negative prognostic factors	Islam et al. *(29)*

[a]IF, immunofluorescence; IHC, immunohistochemistry; NB, Northern blot; RT-PCR, reverse-transcription polymerase chain reaction.

vitro and in vivo *(38,39)*. A second effort for diagnostic detection of survivin in biologic fluids involved analysis of urine from patients with bladder cancer, in which survivin is abundantly overexpressed and correlates with advanced grade and accelerated rates of recurrences *(23)*. This resulted in the implementation of a simple, one-step immunodetection method for measuring urine survivin, which exhibited encouraging specificity and sensitivity for bladder cancer detection *(40)*. In a retrospective trial, there was no detection of survivin in the urine of normal volunteers, or from patients with non-neoplastic genitourinary disease, or other genitourinary cancers except bladder. In contrast, all patients with new or recurrent bladder cancer tested positive for urine survivin, whereas patients with treated bladder cancer became negative for urine survivin *(40)*. If validated in larger patient series, a urine survivin test, potentially in combination with other urine markers, may provide a simple and inexpensive first-line diagnostic approach to monitor bladder cancer patients at risk of recurrent disease.

1.4. Induction of Apoptosis by Targeting Survivin Expression/Function

Because of its selective expression in cancer, survivin may also provide an ideal therapeutic target. Despite its structural organization as an IAP protein *(5)*, the actual role of survivin in the apoptosis cascade, and its relationship with caspases in particular, has been intensely debated. In earlier studies, a physical interaction between survivin and effector caspase-3 and -7 was demonstrated in vivo, and, similarly to other mammalian IAP proteins *(41)*, evidence for survivin-dependent inhibition of caspase activity was also provided *(42)*. These findings were disputed in more recent structure–function studies *(10)*, only to be more recently independently confirmed biochemically using a purified protein system *(43)*. Regardless, a substantial body of evidence independently generated in various laboratories has unambiguously demonstrated a role of survivin in cytoprotection and modulation of the apoptotic machinery *(44,45)*. In this context, the use of "molecular antagonists" of survivin, including antisense oligonucleotides or plasmids or point mutants of survivin, resulted in massive induction of cell death, which was characterized by all the biochemical and morphologic hallmarks of apoptosis *(12,46–48)*. One of the most significant aspects of these studies was that interference with the survivin pathway was sufficient to cause spontaneous apoptosis in cancer cells, without the need of other proapoptotic, death-inducing stimuli. This has prompted the model that survivin may counteract a "default" apoptotic checkpoint in cancer cells, increasing their antiapoptotic threshold during cell cycle progression *(11)*. Conversely, interference with expression/function of survivin may restore this checkpoint function and cause caspase-dependent apoptosis of cancer cells. Although it remains to be elucidated if survivin cytoprotection couples to upstream caspases, effector caspases, or both *(42,43)*, recent evidence has linked the survivin pathway to the modulation of the intrinsic, mitochondrial-dependent initiation of apoptosis. In this context, interference with survivin phosphorylation on Thr^{34} resulted in dissociation of a survivin-caspase-9 complex, mislocalization of caspase-9 from the mitotic apparatus, and caspase-dependent apoptosis *(49)*. This pathway has been shown to have profound consequences for cancer cell viability in both in-vitro and in-vivo studies *(49,50)*. In recent experiments, regulated expression of a phosphorylation-defective survivin Thr^{34}→Ala mutant interfered with phosphorylation of endogenous survivin, caused spontaneous apoptosis in vitro, and prevented tumor formation in a xenograft model of human melanoma in immunoincompetent mice *(50)*. When activated in the context of already established tumors, in vivo, expression of survivin Thr^{34}→Ala induced apoptosis and inhibited tumor growth by ~60–70% *(50)*. The next section reviews the use of molecular antagonists of survivin and their potential impact in novel therapeutic strategies in cancer. Determination of cancer cell apoptosis in these experiments included morphologic analysis of β-galactosidase-expressing cells, hypodiploid DNA content by propidium iodide staining and flow cytometry, internucleosomal DNA fragmentation by TUNEL staining and immunocytochemistry, nuclear morphology of apoptosis by DAPI staining and fluorescence microscopy, and increased caspase-3 activity by hydrolysis of the fluorogenic substrate Ac-DEVD-AMC (N-acetyl-Asp-Glu-Val-Asp-aldehyde).

2. Materials

1. Sense and antisense phosphorothioate-modified oligonucleotides.
2. Restriction enzymes.
3. Agarose and DNA analysis reagents.
4. Lipofectin reagent for mammalian cell transfection.
5. pcDNA3, pML1, and pTet-Splice/pTet-TTa constitutive and inducible expression vector.
6. Electroporator.
7. Fluorescence microscope.
8. Lipfectamine reagent for mammalian cell transfection.
9. G418 (neomycin).
10. pAlter DNA mutagenesis system.
11. $ZnSO_4$ and tetracycline.

3. Methods

3.1. Antisense Oligonucleotides

Forty 2'-O-methoxyethyl chimeric phosphorthothioate oligonucleotides have been screened for inhibition of survivin mRNA expression in T24 bladder carcinoma cells by quantitative real-time RT-PCR. The oligonucleotide 5'-<u>TGTG</u>CTATTCTGTG<u>AATT</u>-3' was identified as one of the most active oligonucleotides, and a scrambled control oligonucleotide with the sequence 5'-<u>TAAG</u>CTGTTCTATG<u>TGTT</u>-3' was synthesized for use in cell culture assays. Oligonucleotides contain uniform phosphorothioate linkages, and are fluorescein-conjugated to monitor transfection efficiencies by fluorescence microscopy. Underlined nucleosides correspond to 2'-O-methoxyethyl nucleosides. Cervical carcinoma HeLa cells (American Type Culture Collection) expressing endogenous survivin *(11)*, or angiogenically stimulated endothelial cells *(34)* are transfected with increasing concentrations (50–400 nM) of control or survivin antisense oligonucleotide by lipofectin. After a 24-h culture at 37°C, cells are harvested and analyzed for survivin or GAPDH RNA expression by Northern hybridization or by Western blotting with an antibody to survivin.

3.2. Antisense cDNA Plasmid Construct

A 708-nucleotide (nt) SmaI-EcoRI of the survivin cDNA is cloned in the reverse orientation in mammalian cell expression vectors pcDNA3 or PML-1 *(46)*. The PML-1 vector is derived from the episomal mammalian expression vector pCEP4 by replacing the cytomegalovirus promoter cassette with the mMT1 promoter, directing Zn^{2+}-dependent expression of recombinant proteins in mammalian cells. For transient transfection experiments, 10×10^6 HeLa cells are transfected with 10 µg of control empty pcDNA3 or the survivin antisense cDNA plus 50 mg of salmon sperm DNA by electroporation using a Gene-Pulser apparatus (Bio-Rad) with a single electric pulse at 350 V at 960 µF. Alternatively, HeLa cells in 6-well tissue culture plates are transfected with 1 µg of *LacZ* reporter plasmid and 4 µg of the various sense or antisense survivin constructs in pcDNA3. For selection of stable transfectants, 48 h after transfection, HeLa cells are diluted, plated onto 100-mm-diameter tissue culture dishes and selected in complete growth medium containing 0.4-mg/mL hygromycin for 4 wk.

Modulation of survivin expression in HeLa cells stably transfected with the inducible survivin antisense cDNA is assessed by Western blotting. For these experiments, HeLa cells are seeded in 24-well tissue culture plates in complete growth medium, induced with 200 μM $ZnSO_4$ and harvested after an additional 24–48 h culture at 37°C. For Western blotting, HeLa cells are solubilized in a lysis buffer containing 0.5% Triton X-100, 0.5% NP-40, 0.05 M Tris-HCl, 0.15 M NaCl, plus protease inhibitors. Alternative lysis procedures include solubilization of total cellular content in 2% sodium dodecyl sulfate (SDS), 50 mM Tris-HCl, and 10% glycerol. Protein-normalized aliquots of the various cell extracts are separated on a 5–15% polyacrylamide SDS gel and immunoblotted with an antibody to full-length recombinant survivin (commercially available from Novus Biologicals, Littleton, CO) or with an antibody to the survivin peptide A^3-I^{19}, followed by chemiluminescence and autoradiography.

3.3. Dominant Negative cDNA Constructs

Two point mutations in the human survivin cDNA have been recently shown to cause spontaneous apoptosis when overexpressed in transformed cell types, in vitro and in vivo. The first mutation is the substitution of Cys^{84} with Ala. Cys^{84} comprises the —COOH-terminus Cys contributing the Zn^{2+} coordination sphere and mutation of this residue to Ala is expected to induce profound conformational disruption of the survivin BIR. The second mutation is the substitution of Thr^{34} to Ala. Thr^{34} is located in a knuckle protruding from the acidic patch in the crystal structure of survivin and functions as a phosphorylation site for the main mitotic kinase complex $p34^{cdc2}$–cyclin B1. Overexpression of a phosphorylation-defective survivin mutant may also function as a dominant negative mutant for its ability to associate physically with $p34^{cdc2}$–cyclin B1, thus preventing phosphorylation of endogenous survivin on Thr^{34}.

The Thr^{34}→Ala mutation is introduced by site-directed mutagenesis into the 1.6-kb human survivin cDNA using the oligonucleotide 5 CATTGCGCCTGCgCCCCG-GAGCGGATG3 and the GeneEditor system (Promega, Madison, WI) according to the manufacturer's instructions, and cloned. The Cys^{84}→Ala mutation was inserted using the pAlter vector (Promega) annealed to the mutated oligonucleotide (underlined) 5 CATTCGTCCGGTgcCGCTTTCCTTTC3. Constructs are entirely sequenced to confirm the introduction of the desired mutation and inserted into the green fluorescent protein (GFP) marker plasmid pEGFPc1 (Clontech Laboratories, Palo Alto, CA). Productively transfected cells are identified by GFP expression by fluorescence microscopy or by flow cytometry and analyzed for apoptosis using the methods described above.

3.4. Inducible Expression of Survivin Thr^{34}→Ala Dominant Negative Mutant

To generate stable cell lines expressing survivin Thr^{34}→Ala under the control of an inducible, regulated promoter, the survivin Thr^{34}→Ala cDNA is cloned into the *EcoR*I of the pTet-splice mammalian expression vector (Stratagene, La Jolla, CA) downstream of the regulatory sequences of the tetracycline (tet)-resistance operon. A second plasmid, pTA-Neo, containing the tet-controlled transactivator sequence downstream of the tet operon, and a neomycin-resistance gene is also commercially available from Strategene or can be obtained by Dr. D. Schatz (Yale University School of Medicine). In this tandem plasmid system, tet prevents transactivator binding to the tet operon and transcription of the transgene; in the absence of tet, the transactivator upregulates its own transcription and the transgene is expressed. To generate stable tetracycline-regulated transfectants, the melanoma cell line YUSAC2 is transfected in 6-well plates by the addition of 0.8 µg pTet-splice containing the survivin Thr^{34}→Ala cDNA plus 0.8 µg pTA-Neo, 0.5 µg tet (Sigma, St. Louis, MO), and 5 µL Lipofectamine (Life Technologies, Gaithersburg, MD) per well. After 9 h, the transfection medium is aspirated and replaced with serum-containing medium in the presence of 0.5-µg/mL tet. Forty-eight hours after transfection, cells are trypsinized, washed, and replated at low density in 15 × 150-mm plates in fresh medium containing 1.5 mg/mL G418 (Life Technologies), 2 mM sodium hydroxide, and 0.5 µg/mL tet. This selection medium is changed every 6 d, and after 3 wk colonies are transferred to U-bottom microtiter wells for expansion and screening on the basis of tet-regulated differential growth. Clones stably transfected with survivin Thr^{34}→Ala are maintained in selection medium in the presence of G418 and tet, and analyzed for inducible expression of survivin Thr^{34}→Ala by Western blotting. For these experiments, YUSAC-2 transfectants are harvested 12, 24, 36, and 48 h after withdrawal of tetracycline, and 100 µg/lane of whole cell extract prepared as described above is separated by electrophoresis on a 5–15% SDS polyacrylamide gel. Filters are incubated with a rabbit antibody raised against full-length recombinant survivin, and reactive bands in the presence or absence of tetracycline are visualized by chemiluminescence and autoradiography.

Acknowledgment

This work was supported by National Institutes of Health Grants CA78810, CA90917 and HL54131.

References

1. Hanahan, D. and Weinberg, R. A. (2000) The hallmarks of cancer. *Cell* **100**, 57–70.
2. Evan, G. I. and Vousden, K. H. (2001) Proliferation, cell cycle and apoptosis in cancer. *Nature* **411**, 342–348.
3. Nicholson, D. W. (2000) From bench to clinic with apoptosis-based therapeutic agents. *Nature* **407**, 810–816.
4. Ambrosini, G., Adida, C., and Altieri, D. C. (1997) A novel anti-apoptosis gene, survivin, expressed in cancer and lymphoma. *Nat. Med.* **3**, 917–921.
5. Deveraux, Q. L. and Reed, J. C. (1999) IAP family proteins-suppressors of apoptosis. *Genes Dev.* **13**, 239–252.
6. Huang, Y., Park, Y. C., Rich, R. L., Segal, D., Myszka, D. G., and Wu, H. (2001) Structural basis of caspase inhibition by XIAP: differential roles of the linker versus the BIR domain. *Cell* **104**, 781–790.
7. Wang, S. L., Hawkins, C. J., Yoo, S. J., Muller, H.-A. J., and Hay, B. A. (1999) The *Drosophila* caspase inhibitor DIAP1 is essential for cell survival and is negatively regulated by HID. *Cell* **98**, 453–463.

8. Chantalat, L., Skoufias, D. A., Kleman, J. P., Jung, B., Dideberg, O., and Margolis, R. L. (2000) Crystal structure of human survivin reveals a bow tie-shaped dimer with two unusual alpha-helical extensions. *Mol. Cell* **6**, 183–189.

9. Muchmore, S. W., Chen, J., Jakob, C., et al. (2000) Crystal structure and mutagenic analysis of the inhibitor-of-apoptosis protein survivin. *Mol. Cell* **6**, 173–182.

10. Verdecia, M. A., Huang, H., Dutil, E., Kaiser, D. A., Hunter, T., and Noel, J. P. (2000) Structure of the human anti-apoptotic protein survivin reveals a dimeric arrangement. *Nat. Struct. Biol.* **7**, 602–608.

11. Li, F., Ambrosini, G., Chu, E. Y., et al. (1999) Control of apoptosis and mitotic spindle checkpoint by survivin. *Nature* **396**, 580–584.

12. Li, F., Ackermann, E. J., Bennett, C. F., et al. (1999) Pleiotropic cell-division defects and apoptosis induced by interference with survivin function. *Nat. Cell Biol.* **1**, 461–466.

13. Uren, A. G., Wong, L., Pakusch, M., et al. (2000) Survivin and the inner centromere protein INCENP show similar cell-cycle localization and gene knockout phenotype. *Curr. Biol.* **10**, 1319–1328.

14. Adida, C., Crotty, P. L., McGrath, J., Berrebi, D., Diebold, J., and Altieri, D. C. (1998) Developmentally regulated expression of the novel cancer anti-apoptosis gene survivin in human and mouse differentiation. *Am. J. Pathol.* **152**, 43–49.

15. Monzo, M., Rosell, R., Felip, E., et al. (1999) A novel anti-apoptosis gene: re-expression of survivin messenger RNA as a prognosis marker in non-small-cell lung cancers. *J. Clin. Oncol.* **17**, 2100–2104.

16. Tanaka, K., Iwamoto, S., Gon, G., Nohara, T., Iwamoto, M., and Tanigawa, N. (2000) Expression of survivin and its relationship to loss of apoptosis in breast carcinomas. *Clin. Cancer Res.* **6**, 127–134.

17. Kawasaki, H., Altieri, D. C., Lu, C. D., Toyoda, M., Tenjo, T., and Tanigawa, N. (1998) Inhibition of apoptosis by survivin predicts shorter survival rates in colorectal cancer. *Cancer Res.* **58**, 5071–5074.

18. Sarela, A. I., Macadam, R. C., Farmery, S. M., Markham, A. F., and Guillou, P. J. (2000) Expression of the antiapoptosis gene, survivin, predicts death from recurrent colorectal carcinoma. *Gut* **46**, 645–650.

19. Gianani, R., Jarboe, E., Orlicky, D., et al. (2001) Expression of survivin in normal, hyperplastic, and neoplastic colonic mucosa. *Hum. Pathol.* **32**, 119–125.

20. Lu, C. D., Altieri, D. C., and Tanigawa, N. (1998) Expression of a novel antiapoptosis gene, survivin, correlated with tumor cell apoptosis and p53 accumulation in gastric carcinomas. *Cancer Res.* **58**, 1808–1812.

21. Kato, J., Kuwabara, Y., Mitani, M., et al. (2001) Expression of survivin in esophageal cancer: Correlation with the prognosis and response to chemotherapy. *Int. J. Cancer* **95**, 92–95.

22. Asanuma, K., Moriai, R., Yajima, T., et al. (2000) Survivin as a radioresistance factor in pancreatic cancer. *Jpn. J. Cancer Res.* **91**, 1204–1209.

23. Swana, II. S., Grossman, D., Anthony, J. N., Weiss, R. M., and Altieri, D. C. (1999) Tumor content of the antiapoptosis molecule survivin and recurrence of bladder cancer. *N. Engl. J. Med.* **341**, 452–453.

24. Saitoh, Y., Yaginuma, Y., and Ishikawa, M. (1999) Analysis of Bcl-2, Bax and Survivin genes in uterine cancer. *Int. J. Oncol.* **15**, 137–141.

25. Adida, C., Haioun, C., Gaulard, P., et al. (2000) Prognostic significance of survivin expression in diffuse large B-cell lymphomas. *Blood* **96**, 1921–1925.

26. Adida, C., Recher, C., Raffoux, E., et al. (2000) Expression and prognostic significance of survivin in de novo acute myeloid leukaemia. *Br. J. Haematol.* **111**, 196–203.

27. Chakravarti, A., Noll, E., Black, P. M., et al. (2002) Quantitatively determined survivin expression levels are of prognostic value in human gliomas. *J. Clin. Oncol.* **20**, 1063–1068.

28. Adida, C., Berrebi, D., Peuchmaur, M., Reyes-Mugica, M., and Altieri, D. C. (1998) Anti-apoptosis gene, survivin, and prognosis of neuroblastoma. *Lancet* **351**, 882–883.

29. Islam, A., Kageyama, H., Hashizume, K., Kaneko, Y., and Nakagawara, A. (2000) Role of survivin, whose gene is mapped to 17q25, in human neuroblastoma and identification of a novel dominant-negative isoform, survivin- beta/2B. *Med. Pediatr. Oncol.* **35**, 550–553.

30. Grossman, D., McNiff, J. M., Li, F., and Altieri, D. C. (1999) Expression and targeting of the apoptosis inhibitor, survivin, in human melanoma. *J. Invest. Dermatol.* **113**, 1076–1081.

31. Grossman, D., McNiff, J. M., Li, F., and Altieri, D. C. (1999) Expression of the apoptosis inhibitor, survivin, in nonmelanoma skin cancer and gene targeting in a keratinocyte cell line. *Lab. Invest.* **79**, 1121–1126.

32. Takamizawa, S., Scott, D., Wen, J., et al. (2001) The survivin: fas ratio in pediatric renal tumors. *J. Pediatr. Surg.* **36**, 37–42.

33. Velculescu, V. E., Madden, S. L., Zhang, L., et al. (1999) Analysis of human transcriptomes. *Nat. Genet.* **23**, 387–388.

34. O'Connor, D. S., Schechner, J. S., Adida, C., et al. (2000) Control of apoptosis during angiogenesis by survivin expression in endothelial cells. *Am. J. Pathol.* **156**, 393–398.

35. Papapetropoulos, A., Fulton, D., Mahboubi, K., et al. (2000) Angiopoietin-1 inhibits endothelial cell apoptosis via the Akt/survivin pathway. *J. Biol. Chem.* **275**, 9102–9105.

36. Islam, A., Kageyama, H., Takada, N., et al. (2000) High expression of Survivin, mapped to 17q25, is significantly associated with poor prognostic factors and promotes cell survival in human neuroblastoma. *Oncogene* **19**, 617–623.

37. Rohayem, J., Diestelkoetter, P., Weigle, B., et al. (2000) Antibody response to the tumor-associated inhibitor of apoptosis protein survivin in cancer patients. *Cancer Res.* **60**, 1815–1817.

38. Schmitz, M., Diestelkoetter, P., Weigle, B., et al. (2000) Generation of survivin-specific CD8+ T effector cells by dendritic cells pulsed with protein or selected peptides. *Cancer Res.* **60**, 4845–4849.

39. Andersen, M. H., Pedersen, L. O., Becker, J. C., and Straten, P. T. (2001) Identification of a cytotoxic T lymphocyte response to the apoptosis inhibitor protein survivin in cancer patients. *Cancer Res.* **61**, 869–872.

40. Smith, S. D., Wheeler, M. A., Plescia, J., Colberg, J. W., Weiss, R. M., and Altieri, D. C. (2001) Urine detection of survivin and diagnosis of bladder cancer. *J. Am. Med. Assoc.* **285**, 324–328.

41. Deveraux, Q. L., Leo, E., Stennicke, H. R., Welsh, K., Salvesen, G. S., and Reed, J. C. (1999) Cleavage of human inhibitor of apoptosis protein XIAP results in fragments with distinct specificities for caspases. *EMBO J.* **18**, 5242–5251.

42. Tamm, I., Wang, Y., Sausville, E., et al. (1998) IAP-family protein survivin inhibits caspase activity and apoptosis induced by Fas (CD95), Bax, caspases, and anticancer drugs. *Cancer Res.* **58**, 5315–5320.

43. Shin, S., Sung, B. J., Cho, Y. S., et al. (2001) An anti-apoptotic protein human survivin is a direct inhibitor of caspase-3 and -7. *Biochemistry* **40**, 1117–1123.

44. Altieri, D. C. and Marchisio, P. C. (1999) Survivin apoptosis: an interloper between cell death and cell proliferation in cancer. *Lab. Invest.* **79**, 1327–1333.

45. Reed, J. C. and Bischoff, J. R. (2000) BIRinging chromosomes through cell division—and survivin' the experience. *Cell* **102**, 545–548.

46. Ambrosini, G., Adida, C., Sirugo, G., and Altieri, D. C. (1998) Induction of apoptosis and inhibition of cell proliferation by survivin gene targeting. *J. Biol. Chem.* **273**, 11177–11182.

47. Olie, R. A., Simoes-Wust, A. P., Baumann, B., et al. (2000) A novel antisense oligonucleotide targeting survivin expression induces apoptosis and sensitizes lung cancer cells to chemotherapy. *Cancer Res.* **60**, 2805–2809.

48. Chen, J., Wu, W., Tahir, S. K., et al. (2000) Down-regulation of survivin by antisense oligonucleotides increases apoptosis, inhibits cytokinesis and anchorage-independent growth. *Neoplasia* **2**, 235–241.
49. O'Connor, D. S., Grossman, D., Plescia, J., et al. (2000) Regulation of apoptosis at cell division by p34^{cdc2} phosphorylation of survivin. *Proc. Natl. Acad. Sci. USA* **97**, 13103–13107.
50. Grossman, D., Kim, P. J., Schechner, J. S., and Altieri, D. C. (2001) Inhibition of melanoma tumor growth in vivo by survivin targeting. *Proc. Natl. Acad. Sci. USA* **98**, 635–640.

35

Targeting the Mitochondria to Enhance Tumor Suppression

Eyal Gottlieb and Craig B. Thompson

1. Overview
1.1. Mitochondria in Cell Growth and Survival

A continuous supply of biochemical energy is required in order for cells to perform their physiologic functions as well as maintain their own homeostasis. Cellular energy is determined largely by the ratio of intracellular ATP to ADP. A high ratio of ATP/ADP is produced primarily by the oxidation of glucose. This process can be divided into two major steps. The first, which takes place in the cytosol, is glycolysis, which produces 2 mol of ATP per 1 mol of glucose utilized. The end product of glycolysis, pyruvate, is either converted to lactate and expelled from the cells or is utilized by the second step, which takes place in the mitochondria, and includes the citric acid cycle and oxidative phosphorylation (OXPHOS). Further oxidation of pyruvate in the mitochondria leads to the production of reducing equivalents (NADH and $FADH_2$). Under aerobic conditions these compounds fuel OXPHOS, producing between 29 and 36 mol of ATP per 1 mol of glucose in addition to the 2 mol produced by glycolysis alone. An average adult converts 50–75 kg of ADP into ATP per day, which is used mainly to regulate the homeostasis of cells (mostly by regulating Na^+/K^+ channels) and to enable motion (muscle contraction) and anabolic reactions (mostly macromolcule synthesis). The latter is important for promoting cell growth and proliferation, as cells must synthesize proteins, DNA, and other macromolcule in order to divide (1).

Mitochondria, in addition to being the source of energy that supports life under aerobic conditions, can also be the source of death signals that can initiate physiologic cell death (apoptosis) (2). Mitochondria can contribute to cell death in several ways, such as by decreasing energy production and by generating oxygen-derived free radicals, termed reactive oxygen species (ROS). Since the discovery that during apoptosis, cytochrome c is released from mitochondria to activate a catalytic cascade of proteolysis (3), it became apparent that mitochondria are essential components of the cell death machinery (4,5). Mitochondria contain key regulators of caspases, a family of proteases that are major factors in many apoptotic processes. Upon release of cytochrome c from mito-

From: *Methods in Molecular Biology, Vol. 223: Tumor Suppressor Genes: Regulation, Function, and Medicinal Applications*. Edited by: Wafik S. El-Deiry © Humana Press Inc., Totowa, NJ

chondria a cytosolic complex, an "apoptosome," is formed. The apoptosome, comprising cytochrome c, apoptosis activating factor-1 (Apaf-1), and caspase-9, is crucial for activating caspase-9, which in turn can activate other downstream caspases. Caspases can be negatively regulated by a set of inhibitors of apoptosis proteins (IAP) that can bind caspases and block their activity. Interestingly, during mitochondria-regulated apoptosis another proapoptotic factor, Smac/Diablo, is released from mitochondria. This protein binds to IAPs to release active caspases from their inhibitors *(6–8)*.

1.2. Cancer and Mitochondrial Functions

Apoptosis and cancer have been tightly linked opponents since the discovery that overexpression of the Bcl-2 oncogene contributes to neoplasia by blocking apoptosis *(9,10)* and that the tumor suppressor protein p53 can reduce tumor formation by activating apoptosis *(11)*. The major mechanisms that can lead to cancerous cells' death include immune system-mediated killing, signaling through death-inducing receptors, and nutrient or oxygen deprivation and all activate apoptosis. Furthermore, ionizing radiation and chemotherapy treatments also activate apoptotic cascades. Since suppression of tumorigenesis requires apoptosis and since mitochondria are essential for the apoptotic process, it is important that the connections between mitochondrial function and the formation or prevention of cancer be delineated.

1.2.1. Oxidative Phosphorylation and Glycolysis Rates in Cancer Cells

Otto Warburg was the first to discover a link between mitochondrial function and cancer formation, almost 80 years ago *(12)*. In his study, Warburg described that in many cancer cells, glycolysis rate is increased dramatically compared to normal cells, and even though oxygen is not limited, the rate of OXPHOS is decreased or remains unaltered. The increased rate of glycolysis can contribute to as much as 60% of ATP production in highly proliferating cancer cells *(13)*. This phenomenon, known as "aerobic glycolysis," is, to say the least, unexpected considering the fact that OXPHOS, rather than glycolysis, is the efficient way for producing ATP and that cancer cells have higher metabolic requirements due to a higher proliferation rate. Based on these observations, Warburg suggested that cancer cells evolve from cells defective in their mitochondrial ATP production *(14)*. Moreover, when some cells are grown under conditions that reduce OXPHOS *(15)*, they become more tumorigenic when transplanted into SCID mice *(16)*. This is probably due to a mitochondria to nucleus pathway that increases transcription of several nuclear genes as a result of depleted respiration in mitochondria *(16)*. Although it is still not clear what triggers aerobic glycolysis in cancer cells, it is believed to be regulated by the expression levels and activity of the glycolytic enzymes hexokinase, phosphofruktokinase-1, and pyruvate kinase *(17,18)*.

It is of great interest that recently it was discovered that many tumors contain mutations in their mitochondrial DNA (mtDNA) (reviewed in **ref. *19***). Human mtDNA is a 16.5-kb circular molcule that is present in between one and tens of copies in each mitochondrion. mtDNA alone does not support the mitochondria, since the nuclear DNA encodes most of the genes required for their functions. All the structural genes that are encoded by mitochondria are subunits of the different respiration complexes. In addition, the rRNA and tRNA genes, required for translation of these structural genes within mitochondria, are also encoded by the mtDNA *(20)*. mtDNA mutations in tumors have been found at high frequency in colorectal cancer *(21)*, bladder, head and neck and lung tumors *(22)*, as well as in breast *(23)* and hepatocellular carcinomas *(24)*. Most of these mitochondrial mutations are homoplasmic, meaning that all the mitochondria in a cell harbor the same mtDNA mutation(s). Moreover, all the cells within a given tumor contain the same mutated mtDNA. This surprising finding suggests that there is a selection for certain mitochondrial mutations within the cells. Although the authors did not report any major physiologic changes in these mitochondria, it is likely that the consequence of such mutations in mtDNA will be a decrease in the rate of OXPHOS. This is in line with Warburg's vision of the origin of cancer cells *(14)* and adds a new venue for investigating the basis of the aerobic glycolysis phenomenon described above.

The advantage of glycolysis-derived ATP is that it is oxygen-independent, thus cancer cells can grow under lower oxygen supply. Moreover, the higher rate of glycolysis results in an increase in macromolcule precursors, which are required for cell growth and division. In line with the fact that mitochondria are part of the apoptotic machinery, it is conceivable that reducing mitochondrial activity by preferring aerobic glycolysis may reduce the susceptibility of cancer cells to apoptosis. In addition, it is possible that, were the high-energy requirements of cancer cells to be fulfilled by extensive OXPHOS, the result would be extremely high ROS production that would lead to apoptosis.

1.2.2. Mitochondria as the Target for Members of the Bcl-2 Family and Other Oncoproteins

The different Bcl-2-like proteins are key regulators of apoptosis. Some members of the family are antiapoptotic (such as Bcl-2 and Bcl-x$_L$), while others are proapoptotic (such as Bax, Bak, Bad, and Bid). There are many variations and different degrees of similarities between these family members *(25)*. However, a common characteristic of all the Bcl-2 proteins is that they all appear to be able to localize to the outer mitochondrial membrane. In most cases, the antiapoptotic members of the Bcl-2 family reside permanently at the outer mitochondrial membrane, while the proapoptotic members translocate from the cytosol to the mitochondria following an apoptotic signal. The chief known role of the Bcl-2 proteins is to regulate mitochondria-mediated apoptosis. They are clearly important regulators of cytochrome C release from mitochondria—the proapoptotic Bcl-2 proteins promote cytochrome C release, while the antiapoptotic members block it. What is still under debate is the actual biochemical mechanism of cytochrome c release and how exactly does each member of the Bcl-2 family enable or encumber this process (for further reading, *see* **refs. 26–29**).

Recently it was shown that Bcl-2 proteins have a direct connection to mitochondrial physiology, particularly the ability of mitochondria to perform OXPHOS. Bax depends on an intact mitochondrial respiratory chain in order to induce toxicity in yeast *(30–32)*. Bcl-x$_L$ can alleviate the early reduction in mitochondrial respiration rate that is observed following treatment with tumor necrosis factor α (TNFα) *(33)*. This reduction in OXPHOS is attributed to alterations in metabolite exchange (particularly ADP and ATP) through the mitochondrial outer membrane *(34)*. Bcl-x$_L$ sustains the activity of the voltage-dependent anion channel (VDAC), the most abundant protein in the outer mitochondrial membrane, under conditions that favor VDAC closure *(35)*, and therefore maintains exchange of metabolites between the mitochondria and the cytosol.

There is an increasing amount of evidence describing the translocation of proteins other than Bcl-2 family members, into mitochondria under apoptotic conditions. Interestingly, these proteins are known to play a role in cancer progression, being well-described oncogenes or tumor suppressors. This list includes protein kinase Cδ *(36,37)*, stress-activated protein kinase (SPAK/JNK) *(38)*, von Hippel-Lindau protein *(39)*, and p53 *(40,41)*. All the translocations into mitochondria occur during apoptosis, and the effect is to regulate cytochrome C release either directly or indirectly. However, information about the exact mitochondrial role of each of these proteins in apoptosis is still lacking. It is apparent that mitochondria still hold in store a few apoptosis-regulation surprises.

1.2.3. Effects of Cancer Treatments on Mitochondria

There are two major pathways that can send a cell to death by apoptosis. One is to activate death-inducing receptors of the TNF receptor (TNFR) family and their downstream cascade (cell-extrinsic pathway), while the other is to activate the mitochondria-dependent cascade (cell-intrinsic pathway). The distinction between these pathways is not absolute, since mitochondria can also participate in the extrinsic pathway, most likely by assisting its amplification *(33)*. Although some treatments such as chemotherapeutic agents have the potential to induce both pathways of apoptosis, it seems that the intrinsic pathway is the one by which most cellular-damaging agents induce cell death *(42)*. Mitochondria are evidently important for the cell death induced by many drugs, including paclitaxel, etoposide, and arsenic trioxide *(43–45)*, even though the exact mechanism by which these agents affect mitochondria is still unknown. Once the mitochondrial pathway is initiated, the most common cascade observed is the release of cytochrome C and the activation of the apoptosome and caspase-3. Recently, it was shown that the mitochondrial pathway is also important for the activation of caspase-8 following chemotherapy *(46)*. Thus, even caspases-8 activation can occur in response to either the extrinsic pathway through TNFR family members or the intrinsic pathway involving the mitochondria.

Although what triggers cytochrome C release from mitochondria following chemotherapy is not fully understood, we nevertheless have some important clues. It is well established that p53 activation is a major outcome of many chemotherapeutic treatments. p53 affects mitochondrial functions indirectly by activating the transcription of genes that increase both the production of ROS by mitochondria and the susceptibility of cells to this redox stress, leading to peroxidation of mitochondrial membrane lipids *(47)*. As mentioned above, p53 may also have a direct effect by translocating into mitochondria. Another protein to mediate mitochondrial response to chemotherapeutics is the proapoptotic member of the Bcl-2 family, Bax. Bax translocates to mitochondria following an apoptotic signal. Being a target for p53 transcriptional activation on the one hand, and further modified and relocated on the posttranslational level on the other, suggests that Bax is an important mitochondrial executor of chemotherapy *(48–51)*. Since the p53 pathway is mutated in more than 50% of human cancers and the Bcl-2 proteins also go awry in many malignant processes (e.g., Bcl-2 overexpression in follicular lymphoma and Bax mutations in colon carcinoma), much attention is given in this chapter to drugs that can directly affect mitochondrial functions to induce the intrinsic apoptotic cascade (*see* **Subheading 2.**).

1.2.4. The Redox State of Cancer Cells and its Effects on Treatment

Oxidative stress can damage DNA, RNA, proteins, and lipids and can lead to cell death *(52–54)*. Mitochondria are central regulators of the redox state of cells via several mechanisms. They are the major source of cellular ROS that are generated when molecular oxygen (O_2) is partially reduced during respiration (electron leak). Thermodynamically, this can happen at any component of the electron transport chain since they all have reducing potential sufficient to reduce O_2, but it seems that the major site of partial oxygen reduction in the mitochondria is in the ubiquinone cycle *(52)*. Forms of ROS include superoxide anion, hydrogen peroxide, and the hydroxyl radical.

Mitochondria can also regulate the redox state of cells indirectly; the efficiency of utilization of glycolytic products for energy production by mitochondria can dictate the amount of glucose available for the pentose phosphate shunt in the cytosol. This pathway produces reducing agents such as NADPH and therefore reduced glutathione (GSH). Another important effector of the redox state is thioredoxin, a protein that is reduced (on cysteine residues) by NADPH and can protect other proteins from oxidative damage. In effect, the redox state of the cell is a consequence of the balance between the levels of oxidizing agents (such as ROS) and reducing agents (such as GSH and reduced thioredoxin).

The basal redox state of cells and their ability to respond to stimuli that alter the redox state are important factors in determining the potency of several apoptosis inducers. The situation in cancer cells is complex. On the one hand, ROS produced by cancer cells can induce damage to neighboring cells and tissues, enhancing the penetration of malignant cells into normal tissues and providing a counterattack on the immune system *(55)*. Furthermore, moderate ROS levels can be a positive proliferative signal through the activation of many cellular mitogenic factors *(52)*. In cancer cells that are homoplasmic for mtDNA mutations (see above), low levels of ROS may be constitutively produced by the partially defective mitochondria, producing a mitogenic signal that conveys a proliferative advantage over neighboring cells. Also, in many cancer cells, lowered levels of enzymatic antioxidants are observed, resulting in increased ROS levels compared to normal cells *(53)*. On the other hand, cellular GSH levels and thioredoxin activity may determine whether cells are resistant to chemotherapy *(54,55)*. Overexpression of Bcl-2 results in higher levels of GSH, which correlates with resistance to treatment. Reducing the levels of GSH restores chemosensitivity to Bcl-2-expressing cells *(55,56)*. This can be explained by the fact that most anticancer treatments, including ionizing radiation and many chemotherapeutic agents, induce oxidative stress. Manipulating the redox state of cancer cells is being investigated in the hope of achieving better treatment efficacy. Moreover, tipping the redox balance toward the oxidized part of the equation can reduce cell survival directly by facilitating the apoptotic machinery, as high levels of ROS accelerate apoptosis *(33)*.

2. Manipulating Mitochondrial Functions: Approaches to Cancer Therapy

Many mitochondria-related genes are deregulated in cancer cells. Overexpression of the antiapoptotic gene Bcl-2 occurs in many types of tumors *(57)*, and some evidence suggests that a proapoptotic member of the Bcl-2 family, Bax, is a tumor suppressor *(58–63)*. Moreover, high levels of mitochondria-associated hexokinase II are observed in many cancer cells *(64)*, and the levels of some other mitochondrial proteins can also increase in cancer cells. These include adenine nucleotide transporter (ANT), mitochondrial creatine kinase, and mitochondrial (or peripheral) benzodiazepine receptor (mBzR/PBR) *(65)*. These changes, alongside the physiologic changes of mitochondria during apoptosis, and the role of mitochondria in regulating apoptosis make mitochondria an attractive target for cancer treatment.

2.1. Cytosolic Acidification due to Aerobic Glycolysis

Many cancer cells have a high rate of glycolysis, which is not accompanied by increased OXPHOS. This could be due to upregulation of glycolytic enzymes or to homoplasmic mtDNA mutations. In either case, the consequences would be a higher dependency of cancer cells on glucose and a higher production of lactate in these cells. Assuming that cancer cells depend on extensive glycolysis, it is likely that treatments that reduce the rate of glycolysis or inhibit lactate clearance from the cells would deliver higher toxicity to cancer cells than to normal cells. Forcing lactate to accumulate in the cells would have a dual effect. First, it would acidify the cytosol, which can by itself induce or accelerate apoptosis *(66)*. Second, it would block (or even reverse) the activity of lactate dehydrogenase and consequently cause the depletion of cytosolic NAD^+, an essential co-factor for glycolysis. Lonidamine (LND), a derivative of indazole-3-carboxylic acid, disrupts energy production and bring about the accumulation of lactate in the cells *(67,68)*. LND can block aerobic glycolysis and is thought to prevent the removal of lactate from cells *(67,69)*. It is not clear, though, if blocking lactate transport is the direct mechanism by which LND-induced metabolic disruption occurs, since it has been shown to affect mitochondria directly, causing a drop in membrane potential and inducing cytochrome C release and ROS production *(70)*. This drug is used in combination with other treatments to enhance their effectiveness *(71)*, and is now being evaluated in Phase III human trials.

2.2. Mitochondrial-Specific Drugs

Many proapoptotic, mitochondria-specific drugs have been developed during the last few years (for recent comprehensive reviews, *see* **refs. *65*** and ***72***). Some of these drugs are already being tested in clinical trials. One such drug is MKT-077, a cationic, lipophilic rhodacyanine that incorporates into mitochondria in a membrane potential-dependent manner. Since some cancer cells show higher membrane potential than normal cells (a still poorly understood phenomenon), this drug accumulates selectively in these cancer cells to inhibit respiration and growth *(73,74)*. MKT-077 is now being evaluated in early clinical trials.

Other possible drugs that may affect mitochondrial functions are under early investigation. Any chemical compound that works directly on mitochondria has the potential to change the redox state of the cells by increasing ROS levels and therefore may be used for combined treatment together with other chemotherapeutic drugs. A recent experiment showed that an estrogen derivative, 2-methoxyoestradiol, inhibits the activity of superoxide dismutase (SOD) and induces apoptosis selectively in hematopoietic malignant cells that have a relatively low basal activity of SOD *(75)*. SOD is part of the enzymatic antioxidant machinery and upon its inhibition, cellular ROS levels increase. Drugs such as 2-methoxyoestradiol have the potential to modify the redox state of cancer cells and thus may affect their response to chemotherapy.

Mitochondria-directed drugs could also have the potential to induce apoptosis directly, by inducing cytochrome C release and by bypassing the upstream regulatory factors that might be altered in cancer cells. One such potential drug is $_D$(KLAKLAK)$_2$, a peptide that contains a mitochondrial membrane-disrupting domain *(76)*. This peptide can incorporate into some tumor cells and induce mitochondrial swelling and death. Impressively, it was shown to prolong survival of nude mice xenografted with cells derived from human breast carcinoma *(76)*.

2.3. Bcl-2-Related Treatments

Considerable attention is also being directed at drugs that can target Bcl-2. Specific compounds that inhibit Bcl-2 can overcome its antiapoptotic effects, particularly in Bcl-2-overexpressing cancer cells such as follicular lymphoma. One straightforward approach is to use antisense oligonucleotides. In this way, Bcl-2 levels can be lowered, doing the same to the apoptotic threshold. Although many problems with oligonucleotide treatments have yet to be solved (especially efficient uptake of these drugs by the cells), some promising preclinical experiments in animal models have been reported and a clinical trial with such drugs has been initiated *(72,77)*.

A second approach to targeting Bcl-2 functions is a structural one. All Bcl-2-family proteins contain up to four Bcl-2-homology (BH) domains, and most contain at least a BH3 domain. Some proapoptotic Bcl-2 proteins, such as Bad and Bid, have BH3 as their sole BH domain. Through the NMR and X-ray crystalography studies, the three-dimensional structure of Bcl-x$_L$ has been determined *(78,79)*. It was shown that a hydrophobic pocket mediates the protein–protein interactions between different Bcl-2 family members. The pocket is comprised of the BH1, BH2, and the buried BH3 domains.

It seems that an exposed BH3 domain confers proapoptotic activity. The antiapoptotic proteins Bcl-2 and Bcl-x$_L$ can bind via the pocket, to the BH3 domain of the proapoptotic family members. Although different family members exert their effects in various ways, it is believed that the BH3-only proteins perform their proapoptotic function solely by binding and disarming the antiapoptotic Bcl-2 proteins. This raises the possibility of designing peptides (or other molcules) that will mimic the proapoptotic activity of BH3-only-containing proteins by interfering with the antiapoptotic activity of Bcl-2. Two such molcules have already been described. HA14-1 was selected from a chemical compound database for its structure and theoretical ability to interact with Bcl-2 *(80)*. HA14-1 induces apoptosis in the etoposide-resistant cell line, HL-60, but whether this death is indeed the result of interactions with Bcl-2 still needs to be determined. The second group of molcules is antimycin A. This is a well-described drug that blocks mitochondrial electron transport at complex III. A recent report shows that antimycin A can also interact specifically with Bcl-2 (competes with a BH3 peptide), induce swelling of isolated mitochondria in a Bcl-2-dependent fashion, and accelerate death of Bcl-2-expressing cells *(81)*. An antimycin A subtype, such as antimycin A$_3$, that can bind Bcl-2 but not complex III, has potential therapeutic ability.

The third potential way to block Bcl-2 antiapoptotic activity is in a functional manner. The biochemical function of Bcl-2 is gradually being unveiled and is likely to be related to the activity of VDAC, which is important in regulating metabolite exchange through the outer mitochondrial membrane. Recently, it was suggested that upon the binding of proapoptotic Bcl-2 proteins to VDAC a pore, large enough to release cytochrome C from the intermembrane space, is formed *(82)*. Tsujimoto and co-workers also described an inhibitory effect of Bcl-2 on VDAC activity (closing the channel) and suggested that this is the antiapoptotic mechanism by which Bcl-2 prevents cytochrome C release. In our laboratory, Bcl-x$_L$ exhibited an opposite effect of on VDAC *(35)*, showing that under conditions that induce VDAC closure, VDAC activity is maintained by recombinant Bcl-x$_L$ both in artificial lipid membranes and in isolated mitochondria. This biochemical activity of Bcl-x$_L$ is in line with its ability to maintain metabolite exchange through the outer mitochondrial membrane during apoptosis *(34)*. Whatever the precise biochemical mechanism of action, the intimate relationship between Bcl-2 proteins and VDAC has initiated another approach toward overcoming Bcl-2 activity in cancer cells. By affecting VDAC activity directly, the antiapoptotic function of Bcl-2 may be circumvented. Candidates for affecting VDAC may be mBzR/PBR ligands. As mentioned earlier, high levels of mBzR/PBR are observed in many tumors. There is some evidence that this receptor associates with VDAC on the mitochondrial outer membrane and alters the transport of many metabolites into the mitochondria *(83)*. High-affinity ligands, such as PK11195, which mimic the endogenous ligand protoporphyrin IX, decrease mitochondrial membrane potential and release cytochrome C in isolated mitochondria *(72)*. Moreover, given in concert with some chemotherapeutic drugs, PK11195 has the ability to enhance their antitumor action. Whether PK11195 and mBzR/PBR work by regulating VDAC activity is not yet clear, but they can exert their activity even in the presence of high levels of Bcl-2 *(84)*.

2.4. Summary

It is still too early to know whether mitochondria will end up being the "Trojan horse" of cancer cells. Nevertheless, taking into account all the functions that are regulated by these organelles and are related to tumor formation and survival, it is an important subject of research. This research combines knowledge from diverse biologic fields, starting from basic bioenergetics and genetics through molcular, cellular, and structural biology, all the way to pharmacology. We described in this chapter several possible approaches to treat cancer based on physiologic changes of mitochondria that occur in tumor cells. Cytosolic changes, such as increased glycolysis and possible acidification by lactate, are characteristics of neoplastic cells that can be used for potential treatment. The lower mitochondrial respiration rate and higher levels of ROS in cancer cells can be pushed further to induce metabolic catastrophe and apoptosis. These treatments may be used in combination with current therapy to facilitate the action of other drugs in chemoresistant cells. The last approach we described is to inhibit Bcl-2 activity physically. Based on the growing understanding of the structural and functional relationship between different Bcl-2 proteins, the potential to design specific drugs to target cells overexpressing Bcl-2 is becoming a reality. The study of mitochondria is advancing cancer therapy research, since the mechanisms of available therapeutics often involve mitochondrial effects. The potential of designing drugs that will specifically induce mitochondria to initiate apoptosis of tumor cells holds great promise for the future.

Acknowledgments

We thank Ayala King, David Brooks, and Marian Harris for excellent editorial and scientific advice, and Giuseppe Attardi and Narayan Avadhani for helpful discussion.

EG is supported by a special fellowship from the Leukemia and Lymphoma society.

References

1. Rosenwald, I. B. (1996) Deregulation of protein synthesis as a mechanism of neoplastic transformation. *Bioessays* **18**, 243–250.
2. Gottlieb, R. A. (2000) Mitochondria: execution central. *FEBS Lett.* **482**, 6–12.
3. Liu, X., Kim, C. N., Yang, J., Jemmerson, R., and Wang, X. (1996) Induction of apoptotic program in cell-free extracts: requirement for dATP and cytochrome c. *Cell* **86**, 147–157.
4. Green, D. R. and Reed, J. C. (1998) Mitochondria and apoptosis. *Science* **281**, 1309–1312.
5. Hengartner, M. O. (2000) The biochemistry of apoptosis. *Nature* **407**, 770–776.
6. Du, C., Fang, M., Li, Y., Li, L., and Wang, X. (2000) Smac, a mitochondrial protein that promotes cytochrome c-dependent caspase activation by eliminating IAP inhibition. *Cell* **102**, 33–42.
7. Verhagen, A. M., Ekert, P. G., Pakusch, M., et al. (2000) Identification of DIABLO, a mammalian protein that promotes apoptosis by binding to and antagonizing IAP proteins. *Cell* **102**, 43–53.
8. Green, D. R. (2000) Apoptotic pathways: paper wraps stone blunts scissors. *Cell* **102**, 1–4.
9. Vaux, D. L., Cory, S., and Adams, J. M. (1988) Bcl-2 gene promotes haemopoietic cell survival and cooperates with c-myc to immortalize pre-B cells. *Nature* **335**, 440–442.
10. McDonnell, T. J., Deane, N., Platt, F. M., et al. (1989) bcl-2-immunoglobulin transgenic mice demonstrate extended B cell survival and follicular lymphoproliferation. *Cell* **57**, 79–88.
11. Yonish-Rouach, E., Resnitzky, D., Lotem, J., Sachs, L., Kimchi, A., and Oren, M. (1991) Wild-type p53 induces apoptosis of myeloid leukaemic cells that is inhibited by interleukin-6. *Nature* **352**, 345–347.
12. Warburg, O., Wind, F., and Neglers, E. (1930) On the metabolism of tumors in the body, in *Metabolism of Tumours* (Warburg, O., ed.). Constable & Co., London, UK, pp. 254–270.
13. Nakashima, R. A., Paggi, M. G., and Pedersen, P. L. (1984) Contributions of glycolysis and oxidative phosphorylation to adenosine $5'$-triphosphate production in AS-30D hepatoma cells. *Cancer Res.* **44**, 5702–5706.
14. Warburg, O. (1956) On the origin of cancer cells. *Science* **123**, 309–314.
15. King, M. P. and Attardi, G. (1989) Human cells lacking mtDNA: repopulation with exogenous mitochondria by complementation. *Science* **246**, 500–503.
16. Amuthan, G., Biswas, G., Zhang, S. Y., Klein-Szanto, A., Vijayasarathy, C., and Avadhani, N. G. (2001) Mitochondria-to-nucleus stress signaling induces phenotypic changes, tumor progression and cell invasion. *EMBO J.* **20**, 1910–1920.
17. Mazurek, S., Boschek, C. B., and Eigenbrodt, E. (1997) The role of phosphometabolites in cell proliferation, energy metabolism, and tumor therapy. *J. Bioenerg. Biomembr.* **29**, 315–330.
18. Dang, C. V. and Semenza, G. L. (1999) Oncogenic alterations of metabolism. *Trends Biochem. Sci.* **24**, 68–72.
19. Bianchi, N. O., Bianchi, M. S., and Richard, S. M. (2001) Mitochondrial genome instability in human cancers. *Mutation Res.* **488**, 9–23.
20. Anderson, S., Bankier, A. T., Barrell, B. G., et al. (1981) Sequence and organization of the human mitochondrial genome. *Nature* **290**, 457–465.
21. Polyak, K., Li, Y., Zhu, H., et al. (1998) Somatic mutations of the mitochondrial genome in human colorectal tumours. *Nat. Genet.* **20**, 291–293.
22. Fliss, M. S., Usadel, H., Caballero, O. L., et al. (2000) Facile detection of mitochondrial DNA mutations in tumors and bodily fluids. *Science* **287**, 2017–2019.
23. Richard, S. M., Bailliet, G., Paez, G. L., Bianchi, M. S., Peltomaki, P., and Bianchi, N. O. (2000) Nuclear and mitochondrial genome instability in human breast cancer. *Cancer Res.* **60**, 4231–4237.

24. Nishikawa, M., Nishiguchi, S., Shiomi, S., et al. (2001) Somatic mutation of mitochondrial DNA in cancerous and noncancerous liver tissue in individuals with hepatocellular carcinoma. *Cancer Res.* **61**, 1843–1845.

25. Gross, A., McDonnell, J. M., and Korsmeyer, S. J. (1999) BCL-2 family members and the mitochondria in apoptosis. *Genes Dev.* **13**, 1899–8911.

26. Vander Heiden, M. G. and Thompson, C. B. (1999) Bcl-2 proteins: regulators of apoptosis or of mitochondrial homeostasis? *Nat. Cell. Biol.* **1**, E209–E216.

27. Crompton, M. (2000) Bax, Bid and the permeabilization of the mitochondrial outer membrane in apoptosis. *Curr. Opin. Cell Biol.* **12**, 414–419.

28. Desagher, S. and Martinou, J. C. (2000) Mitochondria as the central control point of apoptosis. *Trends Cell Biol.* **10**, 369–377.

29. Harris, M. H. and Thompson, C. B. (2000) The role of the Bcl-2 family in the rgulation of outer mitochondrial membrane permeability. *Cell Death Differ.* **7**, 1182–1191.

30. Matsuyama, S., Xu, Q., Velours, J. and Reed, J. C. (1998) The mitochondrial F0F1-ATPase proton pump is required for function of the proapoptotic protein Bax in yeast and mammalian cells. *Mol. Cell* **1**, 327–336.

31. Harris, M. H., Vander Heiden, M. G., Kron, S. J., and Thompson, C. B. (2000) Role of oxidative phosphorylation in Bax toxicity. *Mol. Cell. Biol.* **20**, 3590–3596.

32. Gross, A., Pilcher, K., Blachly-Dyson, E., et al. (2000) Biochemical and genetic analysis of the mitochondrial response of yeast to BAX and BCL-X(L). *Mol. Cell. Biol.* **20**, 3125–3136.

33. Gottlieb, E., Vander Heiden, M. G., and Thompson, C. B. (2000) Bcl-x(L) prevents the initial decrease in mitochondrial membrane potential and subsequent reactive oxygen species production during tumor necrosis factor alpha-induced apoptosis. *Mol. Cell. Biol.* **20**, 5680–5689.

34. Vander Heiden, M. G., Chandel, N. S., Li, X. X., Schumacker, P. T., Colombini, M., and Thompson, C. B. (2000) Outer mitochondrial membrane permeability can regulate coupled respiration and cell survival. *Proc. Natl. Acad. Sci. USA* **97**, 4666–4671.

35. Vander Heiden, M. G., Li, X. X., Gottlieb, E., Hill, R. B., Thompson, C. B., and Colombini, M. (2001) Bcl-xL promotes the open configuration of VDAC and metabolite passage through the outer mitochondrial membrane. *J. Biol. Chem.* **276**, 19,414–19,419.

36. Li, L., Lorenzo, P. S., Bogi, K., Blumberg, P. M., and Yuspa, S. H. (1999) Protein kinase Cdelta targets mitochondria, alters mitochondrial membrane potential, and induces apoptosis in normal and neoplastic keratinocytes when overexpressed by an adenoviral vector. *Mol. Cell. Biol.* **19**, 8547–8558.

37. Majumder, P. K., Pandey, P., Sun, X., et al. (2000) Mitochondrial translocation of protein kinase C delta in phorbol ester-induced cytochrome c release and apoptosis. *J. Biol. Chem.* **275**, 21793–21796.

38. Kharbanda, S., Saxena, S., Yoshida, K., et al. (2000) Translocation of SAPK/JNK to mitochondria and interaction with Bcl-x(L) in response to DNA damage. *J. Biol. Chem.* **275**, 322–327.

39. Shiao, Y. H., Resau, J. H., Nagashima, K., Anderson, L. M., and Ramakrishna, G. (2000) The von Hippel-Lindau tumor suppressor targets to mitochondria. *Cancer Res.* **60**, 2816–2819.

40. Marchenko, N. D., Zaika, A., and Moll, U. M. (2000) Death signal-induced localization of p53 protein to mitochondria. A potential role in apoptotic signaling. *J. Biol. Chem.* **275**, 16202–16212.

41. Sansome, C., Zaika, A., Marchenko, N. D., and Moll, U. M. (2001) Hypoxia death stimulus induces translocation of p53 protein to mitochondria. Detection by immunofluorescence on whole cells. *FEBS Lett.* **488**, 110–115.

42. Kaufmann, S. H. and Earnshaw, W. C. (2000) Induction of apoptosis by cancer chemotherapy. *Exp. Cell Res.* **256**, 42–49.

43. Perkins, C. L., Fang, G., Kim, C. N., and Bhalla, K. N. (2000) The role of Apaf-1, caspase-9, and bid proteins in etoposide- or paclitaxel-induced mitochondrial events during apoptosis. *Cancer Res.* **60**, 1645–1653.

44. Shen, Z. Y., Shen, J., Cai, W. J., Hong, C., and Zheng, M. H. (2000) The alteration of mito-chondria is an early event of arsenic trioxide induced apoptosis in esophageal carcinoma cells. *Int. J. Mol. Med.* **5**, 155–158.

45. Seol, J. G., Park, W. H., Kim, E. S., et al. (2001) Potential role of caspase-3 and -9 in arsenic trioxide-mediated apoptosis in PCI-1 head and neck cancer cells. *Int. J. Oncol.* **18**, 249–255.

46. Ferreira, C. G., Span, S. W., Peters, G. J., Kruyt, F. A., and Giaccone, G. (2000) Chemother-apy triggers apoptosis in a caspase-8-dependent and mitochondria-controlled manner in the non-small cell lung cancer cell line NCI-H460. *Cancer Res.* **60**, 7133–7141.

47. Polyak, K., Xia, Y., Zweier, J. L., Kinzler, K. W., and Vogelstein, B. (1997) A model for p53-induced apoptosis. *Nature* **389**, 300–305.

48. Guo, B., Yin, M. B., Toth, K., Cao, S., Azrak, R. G., and Rustum, Y. M. (1999) Dimeriza-tion of mitochondrial Bax is associated with increased drug response in Bax-transfected A253 cells. *Oncol. Res.* **11**, 91–99.

49. Sawa, H., Kobayashi, T., Mukai, K., Zhang, W., and Shiku, H. (2000) Bax overexpression enhances cytochrome c release from mitochondria and sensitizes KATOIII gastric cancer cells to chemotherapeutic agent-induced apoptosis. *Int. J. Oncol.* **16**, 745–749.

50. Rahman, K. M., Aranha, O., Glazyrin, A., Chinni, S. R., and Sarkar, F. H. (2000) Translo-cation of Bax to mitochondria induces apoptotic cell death in indole-3-carbinol (I3C) treated breast cancer cells. *Oncogene* **19**, 5764–5771.

51. Zhang, L., Yu, J., Park, B. H., Kinzler, K. W., and Vogelstein, B. (2000) Role of BAX in the apoptotic response to anticancer agents. *Science* **290**, 989–992.

52. Kamata, H. and Hirata, H. (1999) Redox regulation of cellular signalling. *Cell Signal.* **11**, 1–14.

53. Mates, J. M. and Sanchez-Jimenez, F. M. (2000) Role of reactive oxygen species in apopto-sis: implications for cancer therapy. *Int. J. Biochem. Cell Biol.* **32**, 157–170.

54. Davis, W., Jr., Ronai, Z., and Tew, K. D. (2001) Cellular thiols and reactive oxygen species in drug-induced apoptosis. *J. Pharmacol. Exp. Ther.* **296**, 1–6.

55. Voehringer, D. W. (1999) BCL-2 and glutathione: alterations in cellular redox state that reg-ulate apoptosis sensitivity. *Free Radic. Biol. Med.* **27**, 945–950.

56. Mirkovic, N., Voehringer, D. W., Story, M. D., McConkey, D. J., McDonnell, T. J., and Meyn, R. E. (1997) Resistance to radiation-induced apoptosis in Bcl-2-expressing cells is reversed by depleting cellular thiols. *Oncogene* **15**, 1461–1470.

57. Kroemer, G. (1997) The proto-oncogene Bcl-2 and its role in regulating apoptosis. *Nat. Med.* **3**, 614–620.

58. Rampino, N., Yamamoto, H., Ionov, Y., et al. (1997) Somatic frameshift mutations in the BAX gene in colon cancers of the microsatellite mutator phenotype. *Science* **275**, 967–969.

59. Yagi, O. K., Akiyama, Y., Nomizu, T., Iwama, T., Endo, M., and Yuasa, Y. (1998) Proapop-totic gene BAX is frequently mutated in hereditary nonpolyposis colorectal cancers but not in adenomas. *Gastroenterology* **114**, 268–274.

60. Ouyang, H., Furukawa, T., Abe, T., Kato, Y., and Horii, A. (1998) The BAX gene, the pro-moter of apoptosis, is mutated in genetically unstable cancers of the colorectum, stomach, and endometrium. *Clin. Cancer Res.* **4**, 1071–1074.

61. Meijerink, J. P., Mensink, E. J., Wang, K., et al. (1998) Hematopoietic malignancies demon-strate loss-of-function mutations of BAX. *Blood* **91**, 2991–2997.

62. Shibata, M. A., Liu, M. L., Knudson, M. C., et al. (1999) Haploid loss of bax leads to accel-erated mammary tumor development in C3(1)/SV40-TAg transgenic mice: reduction in pro-tective apoptotic response at the preneoplastic stage. *EMBO J.* **18**, 2692–2701.

63. Knudson, C. M., Johnson, G. M., Lin, Y., and Korsmeyer, S. J. (2001) Bax accelerates tumorigenesis in p53-deficient mice. *Cancer Res.* **61**, 659–665.

64. Mathupala, S. P., Rempel, A., and Pedersen, P. L. (1997) Aberrant glycolytic metabolism of cancer cells: a remarkable coordination of genetic, transcriptional, post-translational, and

mutational events that lead to a critical role for type II hexokinase. *J. Bioenerg. Biomembr.* **29**, 339–343.

65. Costantini, P., Jacotot, E., Decaudin, D., and Kroemer, G. (2000) Mitochondrion as a novel target of anticancer chemotherapy. *J. Natl. Cancer Inst.* **92**, 1042–1053.

66. Matsuyama, S., Llopis, J., Deveraux, Q. L., Tsien, R. Y., and Reed, J. C. (2000) Changes in intramitochondrial and cytosolic pH: early events that modulate caspase activation during apoptosis. *Nat. Cell. Biol.* **2**, 318–325.

67. Ben-Horin, H., Tassini, M., Vivi, A., Navon, G., and Kaplan, O. (1995) Mechanism of action of the antineoplastic drug lonidamine: 31P and 13C nuclear magnetic resonance studies. *Cancer Res.* **55**, 2814–2821.

68. Mardor, Y., Kaplan, O., Sterin, M., et al. (2000) Noninvasive real-time monitoring of intracellular cancer cell metabolism and response to lonidamine treatment using diffusion weighted proton magnetic resonance spectroscopy. *Cancer Res.* **60**, 5179–5186.

69. Fanciulli, M., Valentini, A., Bruno, T., Citro, G., Zupi, G., and Floridi, A. (1996) Effect of the antitumor drug lonidamine on glucose metabolism of adriamycin-sensitive and -resistant human breast cancer cells. *Oncol. Res.* **8**, 111–120.

70. Ravagnan, L., Marzo, I., Costantini, P., et al. (1999) Lonidamine triggers apoptosis via a direct, Bcl-2-inhibited effect on the mitochondrial permeability transition pore. *Oncogene* **18**, 2537–2546.

71. Floridi, A., Bruno, T., Miccadei, S., Fanciulli, M., Federico, A., and Paggi, M. G. (1998) Enhancement of doxorubicin content by the antitumor drug lonidamine in resistant Ehrlich ascites tumor cells through modulation of energy metabolism. *Biochem. Pharmacol.* **56**, 841–849.

72. Fennell, D. A. and Cotter, F. E. (2000) Controlling the mitochondrial gatekeeper for effective chemotherapy. *Br. J. Haematol.* **111**, 52–60.

73. Petit, T., Izbicka, E., Lawrence, R. A., Nalin, C., Weitman, S. D., and Von Hoff, D. D. (1999) Activity of MKT 077, a rhodacyanine dye, against human tumor colony-forming units. *Anticancer Drugs* **10**, 309–315.

74. Tatsuta, N., Suzuki, N., Mochizuki, T., et al. (1999) Pharmacokinetic analysis and antitumor efficacy of MKT-077, a novel antitumor agent. *Cancer Chemother. Pharmacol.* **43**, 295–301.

75. Huang, P., Feng, L., Oldham, E. A., Keating, M. J., and Plunkett, W. (2000) Superoxide dismutase as a target for the selective killing of cancer cells. *Nature* **407**, 390–395.

76. Ellerby, H. M., Arap, W., Ellerby, L. M., et al. (1999) Anti-cancer activity of targeted proapoptotic peptides. *Nat. Med.* **5**, 1032–1038.

77. Nicholson, D. W. (2000) From bench to clinic with apoptosis-based therapeutic agents. *Nature* **407**, 810–816.

78. Muchmore, S. W., Sattler, M., Liang, H., et al. (1996) X-ray and NMR structure of human Bcl-xL, an inhibitor of programmed cell death. *Nature* **381**, 335–341.

79. Petros, A. M., Medek, A., Nettesheim, D. G., et al. (2001) Solution structure of the anti-apoptotic protein bcl-2. *Proc. Natl. Acad. Sci. USA* **98**, 3012–3017.

80. Wang, J. L., Liu, D., Zhang, Z. J., et al. (2000) Structure-based discovery of an organic compound that binds Bcl-2 protein and induces apoptosis of tumor cells. *Proc. Natl. Acad. Sci. USA* **97**, 7124–7129.

81. Tzung, S. P., Kim, K. M., Basanez, G., et al. (2001) Antimycin A mimics a cell-death-inducing Bcl-2 homology domain 3. *Nat. Cell. Biol.* **3**, 183–191.

82. Tsujimoto, Y. and Shimizu, S. (2000) VDAC regulation by the Bcl-2 family of proteins. *Cell Death Differ.* **7**, 1174–1181.

83. Krueger, K. E. (1995) Molcular and functional properties of mitochondrial benzodiazepine receptors. *Biochim. Biophys. Acta* **1241**, 453–470.

84. Hirsch, T., Decaudin, D., Susin, S. A., et al. (1998) PK11195, a ligand of the mitochondrial benzodiazepine receptor, facilitates the induction of apoptosis and reverses Bcl-2-mediated cytoprotection. *Exp. Cell Res.* **241**, 426–434.

36

Novel Approaches to Screen for Anticancer Drugs Using *Saccharomyces cerevisiae*

Julian A. Simon and Timothy J. Yen

1. Introduction

The development of anticancer drugs has relied primarily on two traditional approaches. Synthetic or natural compounds are routinely screened for anticancer activities using a cell-based assay. The National Cancer Institute (NCI) has relied on a panel of human tumor cell lines to search for compounds that inhibit cell growth *(1,2)*. Inhibitors identified in this screen are then further characterized for toxicity and antitumor activity in animal models. A strategy that is highly popular in industry is high-throughput screens for compounds that inhibit the in-vitro activities of specific enzymes or proteins (kinases, phosphatases, etc.) *(3)*. This approach relies on the establishment of a robust in-vitro assay for the protein of interest. Although kinases are the preferred substrates because many of the existing chemical libraries were designed to identify kinase inhibitors, screens for other cellular targets are limited only by the development of an appropriate assay.

Each of these approaches has unique strengths and disadvantages. For cell-based assays, compounds that inhibit cell growth or induce apoptosis can be readily identified. Nevertheless, finding the cellular target of the inhibitor and thus the mechanism of inhibition is difficult. In some cases, the chemical structure of the compound can provide clues as to its mechanism of action. For screens that rely on in-vitro assays, careful consideration must be given to justify the choice of the particular enzyme or protein. In this case, the choice might be made based on prior knowledge of the biologic function of the protein of interest. The in-vitro system is also limited by practical issues such as the retention of biochemical activity of the protein.

The discovery of the novel antimitotic compound, monastrol, illustrates a case where both cell-based and in-vitro assays were used in combination to identify the drug and its target *(4)*. By modifying and incorporating an immunocytologic screen into the cell-based assay, compounds that selectively blocked cells in mitosis were identified. Detailed characterization of the organization of chromosomes and the spindle in drug-arrested cells revealed similarities to those found when cells were defective for the spindle-associated kinesin-like microtubule motor, Eg5 *(5,6)*. The existence of an in-vitro assay for microtubule motors led to the confirmation that Eg5 was the target of monas-

From: *Methods in Molecular Biology, Vol. 223: Tumor Suppressor Genes: Regulation, Function, and Medicinal Applications.* Edited by: Wafik S. El-Deiry © Humana Press Inc., Totowa, NJ

trol. In this example, the ability to ascribe a particular growth-arrest phenotype to specific cellular pathways provides important clues as to the target of a drug. In many cases, this may be difficult when the outcome is cell death or growth arrest in interphase.

The goal of this chapter is to describe the benefits of using the budding yeast *Saccharomyces cerevisiae* to screen for novel anticancer drugs. In principle, this system incorporates the desirable features of the two traditional screening approaches described above. This system also offers added benefits such as screens for drug targets and context-based screens to identify drugs that act preferentially on cells with defined molecular defects, as is the case with human cancers. It is important to state at the outset that the yeast-based system is not intended to be a substitute but rather to complement and enhance existing methods for screening drugs.

2. Practical Considerations for Using Yeast in High-Throughput Screens

The use of microbes in drug discovery has existed for over 50 years *(7)* but has not been exploited for systematic and large-scale efforts to screen for anticancer drugs. One reason was the absence of any obvious connections between microbes and human cancers. This early misperception changed radically once we realized that nearly all cancers have in common defects in maintaining genome stability. At the most basic level, all cancers share the characteristic of deregulated growth. The mechanisms that lead to uncontrolled proliferation are diverse but are generally thought to arise from accumulation of multiple mutations in genes that are critical for normal cell growth and differentiation *(8)*. As many of these genes and pathways are evolutionarily conserved, it becomes clear that organisms such as yeast can serve as a valuable model for understanding the molecular basis of human cancers *(9)*. For example, genes that are essential for DNA repair, replication, cell cycle kinases and their inhibitors, checkpoint proteins, and components of signal transduction pathways that are mutated in many types of cancers are conserved in yeast. On the other hand, yeast does not have obvious homologs of the tumor suppressors such as p53 and Rb, two genes that are frequently mutated in cancers. This, however, does not reduce the importance of yeast as a model organism for cancer, because both p53 and Rb are known to control pathways in human cells that are conserved in yeast. Although it is true that yeast is not a model for the disease state of human cancers, it is fully adequate as a model for studying the molecular basis for genome instability that leads to cancer. From this perspective, yeast is a highly suited organism for screening anticancer drugs, because the objective of such screens is to identify compounds that inhibit pathways that are essential for cell growth *(8)*. We will describe target-based and context-based screens for compounds that exhibit anticancer properties. Additionally, we will discuss how to take advantage of existing yeast technology and apply these methods to drug screens.

From the practical point of view, yeast is an ideal organism for high-throughput cell-based screens *(10,11)*. Its rapid doubling time and its simple growth requirements make yeast highly cost-effective in both time and money. Typically, yeast grow in a defined liquid medium and growth is monitored by turbidity measurements. Its rapid growth rate means that drug screens can usually be completed after an overnight incubation, as opposed to week-long incubations for mammalian tissue culture cells. Compounds that inhibit growth can easily be detected by comparing the optical density of cultures grown in the presence and absence of drugs. And, as we will demonstrate, it is possible to design assay systems in yeast in which growth inhibition is not just an indicator of cytotoxicity.

Beyond these practical issues, yeast is a very robust organism that is amenable to a wide variety of experimental manipulations *(10)*. Yeast was the first eukaryote to have its entire genome sequenced *(12)*. This, coupled with sophisticated genetics, provides a very powerful system to perform target-based and context-based drug screens. Target-based screening is designed to identify compounds that will inhibit a specific protein or enzyme in an in-vivo setting rather than the standard in-vitro assays for inhibitors. Context-based screens are designed to identify drugs that act in a specific genetic context. As many types of cancer cells have loss-of-function mutations in genes that are important for DNA repair or checkpoint control, drug screens can be conducted using yeast that are deficient in these conserved genes. By comparing the sensitivities of a wild-type and an isogenic mutant to a panel of drugs, it is possible to identify drugs that exhibit selectivity for mutants. The use of "matched pairs," in which two cell lines differ only in defined genes that are the cause of the disease, is a powerful screening tool. This is especially important in anticancer drug development because many drugs in clinical use are cytotoxic to both normal and tumor tissues. Using matched pairs in cell-based assays, it is possible to identify drugs that exhibit selectivity toward cancer cells with defined mutations.

One potential disadvantage of using yeast in drug screens is that its cell wall and membrane acts an effective barrier to small molecules. Nevertheless, of 50 U.S. Food and Drug Administration (FDA)-approved nonbiologic, nonhormonal anticancer drugs that were tested for cytotoxity in yeast, about half of the drugs inhibited wild-type yeast with an IC_{50} less than 10 m*M*, and for most of these compounds the IC_{50} was in the mid-micromolar range *(11)*. The permeability barrier can be overcome by using mutants with weakened cell membrane and defective efflux. Although yeast express over 50 genes that are important for transporting molecules across the cell membrane, there are additional genes that affect small-molecule transport when mutated. ERG6, PDR1, and PDR3 are probably the best-characterized genes in terms of mutations that increase permeability to small molecules. ERG6 (*ergosterol biosynthesis*) is a nonessential gene that encodes a methytransferase that catalyzes the formation of ergosterol *(13)*, which is equivalent to cholesterol in mammalian cells. *erg6* mutants that lack ergosterol in their cell membranes have increased uptake of small molecules that is thought to be due to a defect in efflux *(14)*. PDR1 *(15)* and PDR3 *(16)* (*pleiotropic drug resistance*) are transcription factors that regulate the expression of a large number of transporters that are similar to the mammalian mdr1 transporters that are responsible for pumping small molecules out of the cell. Simultaneous deletion of ERG6, PDR1, and PDR3 did not affect viability or growth rate significantly but significantly increased sensitivity to a panel of drugs obtained from the NCI repository *(11)*. This strain was used in a pilot study to assess the utility of yeast in screening anticancer drugs.

3. Proof of Principle

A recent study was performed to validate the use of yeast in high-throughput drug screens *(11)*. In this pilot study, a panel of yeast mutants was tested for sensitivity toward a panel of nonhormonal, nonbiologic, FDA-approved anticancer agents. The goal of this study was to see whether it is possible to identify drugs that are toxic within a particular genetic context. Twenty-one yeast strains with specific mutations that affected DNA repair and checkpoint control were constructed by targeted gene deletion. Side-by-side comparison with isogenic wild-type controls revealed some drugs were more effective against certain mutations, whereas other drugs exhibited little specificity. This study included DNA damage-repair mutants (for review, *see* **ref.** *17*) that were defective in nucleotide excision repair (*rad1*, *rad14*), mismatch repair (*mlh1*, *pms1*), base excision repair (*mag1*, *apn1*), O^6-alkylguanine repair (*mgt1*), error-free (*rad18*), error-prone (*rev1*, *rev3*), and error-free and -prone (*rad6*) postreplication repair, and recombinational repair (*rad50*, *rad51*, *rad52*). The study also included mutants in DNA damage checkpoint (*rad9*, *rad17*) *(18)*, S-phase checkpoint (*mec1*, *mec2/rad53*) *(19)*, mitotic checkpoint (*mad1*, *mad3*) *(20)*, and homolog (*sgs1*) of human genes mutated in Bloom's and Werner's syndromes *(21)*. Twenty-three of the 44 FDA-approved drugs that were tested produced a measurable cytotoxic effect.

The results from this study revealed that DNA repair or damage pathways are an important determinant for sensitivity to many of the drugs that were tested. The authors were able to classify the drugs according to the degree of sensitivity that was exhibited by different mutants. Drugs such as cisplatin, mitoxantrone, cytarabine phosphate, camptothecin, and idarubicin exhibited a narrow spectrum of selectivity. Cisplatin, which forms DNA adducts and cross-links, was highly toxic to *rad6* and *rad18* mutants (**Fig. 1a**) that are required for a poorly understood postreplication damage-repair pathway *(22,23)*. Cytarabine phosphate, a nucleotide analog that is incorporated into DNA, was highly toxic to the *sgs1* mutants, as they were an order of magnitude more sensitive to this drug than the next most sensitive mutant. The basis for the selectivity of cytarabine for *sgs1* mutants is unclear but is likely due to the fact that both the drug and *sgs1* mutants increase the frequency of recombination in cells *(24,25)*. While increased recombination induced by either cytarabine or loss of Sgslp1 function is tolerated by the cell, the combination might increase recombination events to levels that are lethal. The third group of selective drugs that were identified in this study were inhibitors of topoisomerase I and II. Mutants (*rad50*, *rad51*, *rad52*) defective in double-stranded break repair (DSB) were highly sensitive to the topo I inhibitor, camptothecin, and the topo II inhibitors, mitoxantrone (**Fig. 1d**) and idarubicin. It is interesting that these two classes of topoisomerase poisons exhibited different sensitivities toward DNA damage (*mec1* and *mec2*) and S-phase checkpoint mutants (*rad9*, *rad17*). Mutants in these two checkpoint pathways were found to be highly sensitive to camptothecin because this drug is believed to induce DSBs as a consequence of DNA synthesis, when single-stranded breaks are converted into double-strand breaks. Loss of DNA damage checkpoint pathways in *mec1* and *mec2* mutants would cause camptothecin-treated cells to progress through S phase (rather than arrest and repair the damage) and thus induce lethal DSBs. In contrast, topoII inhibitors generate DSBs primarily during mitosis, when these checkpoints are no longer active. This would be consistent with the finding that these checkpoint mutants are not hypersensitive to topoII inhibitors. The different mechanisms by which the two types of topoisomerase poisons induce DSBs therefore lead to their differential sensitivities to checkpoint mutants.

The second category of drugs that was identified in this study were drugs that were effective against a broad number and types of DNA repair mutants. Unlike the highly selective drugs, for which there was usually an order-of-magnitude difference between the most sensitive and the next most sensitive mutants, the difference in sensitivity exhibited by different classes of DNA-repair mutants to the second class of drugs were much smaller and usually within severalfold of each other. Many of the drugs in this group, such as mitomycin C, thiotepa, carmustine, lomustine, mechlorethamine, and

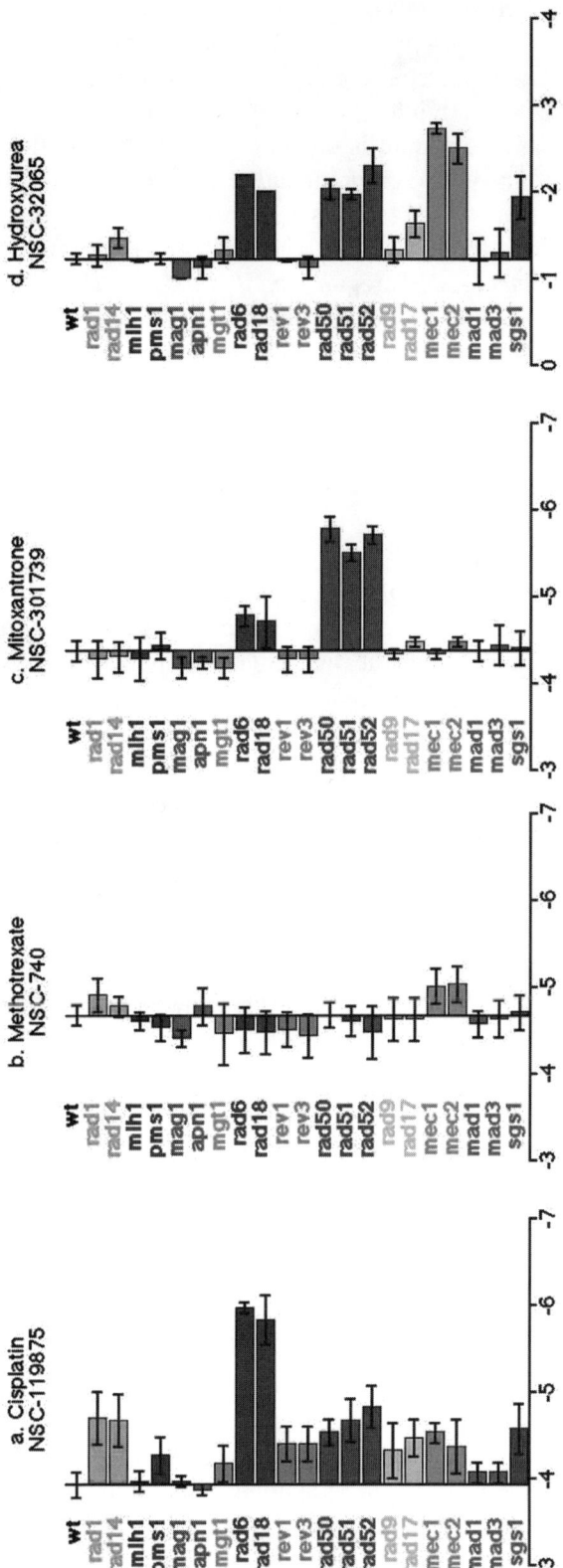

Fig. 1. Toxicity profile of anticancer drugs against a panel of yeast mutants. Comparison of the toxicity profiles of four FDA-approved anticancer drugs against wild-type yeast and a panel of yeast mutants that are defective for various DNA damage and repair genes. Genes that act along the same pathway are grouped by color. The histograms show the IC_{50} (log M) for each drug. The IC_{50} for the wild-type strain is the vertical bar. Log concentration of drug is plotted on the horizontal axis. All the strains were in an *erg6, pdr1, pdr3* background to increase uptake of drugs.

streptozocin, are alkylating agents that were found to exhibit broad selectivity for various repair mutants. The different levels of cytotoxicity exhibited by different groups of repair mutants by this class of drugs probably reflect the relative contribution of various repair pathways to removing and repairing DNA cross-links. For example, while the methylating agent streptozocin exhibited broad specificity, it was particularly effective against the *mgt1* mutant, as this mutant is unable to repair O^6-methylguanine *(26,27)*. Other agents belonging to this second group included hydroxyurea (**Fig. 1c**) and X-rays. Not surprisingly, hydroxyurea, which inhibits DNA replication and produces DSBs at stalled replication forks, was found to selectively kill cells that were defective for S-phase checkpoints (mec1, *mec2*), DSB repair (*rad50, rad51, rad52*), and repair after S phase (*rad6, rad18*). X-rays, which induce a spectrum of DNA damage including DSBs, were found to exhibit a similar profile of selectivity to what was found for hydroxyurea with the exception that *rad9* and *rad17* checkpoint mutants were hypersensitive to X-rays but not hydroxyurea. This difference can be accounted for by the fact that hydroxyurea induces strand breaks primarily during S phase, when the MEC1 checkpoint is operational. Thus, RAD9 and RAD17 functions are dispensable for DNA damage that is incurred during S phase as long as MEC1 remains functional. Because X-rays induce DSBs at all stages of the cell cycle, RAD9 and RAD17 checkpoint functions are probably required for arresting cells with DNA damage at other phases of the cell cycle.

The last class of drugs was classified as nonselective because they did not exhibit great differences in sensitivity between the different repair mutants and wild-type. These included methotrexate, trimetrexate, fluorouracil, and fluorodeoxyuridine, which are drugs that block biosynthesis of nucleotides and are thus expected to inhibit DNA synthesis. However, the nonselectivity of methotrexate against the panel of repair mutants (**Fig. 1b**) suggests that its mechanism of killing may not be based on interfering with DNA replication but possibly on RNA synthesis. On the other hand, trimetrexate, which is an analog of methotrexate, is more sensitive to the *mec1* and *mec2* checkpoint mutants, as might be expected for an inhibitor of DNA replication. Pentostatin, an inhibitor of adenine deaminase whose function is to modify RNA and thus is not expected to cause DNA damage, does not rely on DNA repair-deficient mutants for enhanced killing. Fluorouracil and fluorodeoxyuridine block the synthesis of dTMP and are expected to block DNA synthesis. Thus, it was unexpected that these drugs were not more cytotoxic to *mec1* and *mec2* mutants. It is unclear from these studies why some drugs that are known to inhibit DNA synthesis did not exhibit the selectivity as was seen for other drugs that also block DNA synthesis. One explanation may be that these drugs are killing cells through mechanisms other than DNA damage. This possibility is exemplified by the DNA damage-inducing drugs dacarbazine, doxorubicin daunorubicin, and actinomycin D. None of these drugs showed greater selectivity toward a particular class of repair-deficient mutants, as might be expected if these drugs were killing the yeast cells by induced lesions to DNA. However, these drugs are known to be capable of generating free radicals that can target membranes as well as DNA. The genetics indicate that drugs that do not exhibit strong preference for one class of mutants over another are likely to act through pathways that are not represented in the panel of mutants. In later sections we will discuss ways to identify drug targets using yeast. An important implication of this work is that the mechanism of action of a particular drug is best defined in the target cell, since compounds that appear to have a well-defined mechanism in vitro may be capable of causing many different forms of damage in vivo.

The results of this study showed that the drugs that were tested could be classified as being highly selective for a narrow group of DNA damage-repair mutants, broadly selective against many DNA damage-repair genes, or nonselective toward particular DNA damage-repair pathways or genes. The drugs that fall into the broad-selectivity category are likely to be those that can be used to treat a broad spectrum of cancers, because the toxicity of these drugs is unaffected by the genetic context of a particular cancer. The ability of this class of drugs to produce a response in a wide variety of cancers is offset by the strong likelihood that these drugs will also be cytotoxic to normal tissues. In contrast, drugs that are selective toward specific genes or pathways in yeast are likely to be highly selective for particular types of cancers that have similar defects. As different types of cancers are likely to arise from different genetic mutations, these genetic differences can be exploited as a major factor in selecting drugs with the highest probability of success.

This study also raised an important issue as to whether we should focus on making more drugs or on better diagnostics so that existing drugs can be used more effectively. For example, a demonstration of DSB repair defects in biopsy sample could be used to identify patients who are more likely to respond to topo I and II poisons. On the other hand, it is equally valuable to identify drugs that exhibit high selectivity based on the genetic profile of the cancer. This is especially true for profiles that do not sensitize cancer cells to any of the existing agents (e.g., mismatch-repair defects).

4. Context-Based Drug Screens

To identify novel anticancer drugs that exhibit selectivity toward cancer cells, a panel of yeast mutants that had similar genetic defects to those found in human cancers was used. A total of 100,000 compounds obtained from the Developmental Therapeutics Program at the NCI were tested against this panel of mutants. This panel of yeast mutants included *mlh1* (mismatch-repair defects), *mec2-1* (ATM and ATR kinases, checkpoint defects), *bub3* (spindle assembly checkpoint), and *CLN2* overexpression (cyclin D1 and E overexpression). All of these genes have been linked to genome instability and cancer in humans. In total, the screen yielded ~150 compounds that target these cancer-related defects and are currently being characterized.

This screen also used a different set of yeast mutants (*rad50*, *rad18*, *rad14*, *sgs1*, and *mgt1*), that were found in the earlier study to be hypersensitive to some of the FDA-approved anticancer drugs. This yielded another 140 compounds that were cytotoxic to the various mutants in this panel. The majority of compounds that targeted this set of strains act through mechanisms of action similar to those of the FDA-approved agents and are not being pursued. Fifty-four structurally novel compounds were found to selectively kill *rad50* mutants, which are defective in recombinational repair of DSBs. As *rad50* mutants are known to be hypersensitive to topo I and topo II inhibitors, the newly identified compounds were candidate novel topoisomerase poisons even though their chemical structures did not resemble bona-fide topo poisons. One of these compounds, NSC638432, was originally synthesized as an analog of a non-narcotic analgesic, yet it, like etoposide, selectively killed *rad50* or *rad52* mutants (**Fig. 2**). Given that *rad50* and *rad52* mutants are defective in DSB repair, NSC638432 was tested against cells that were defective for DSB repair. Cells derived from Ku70-knockout mice were chosen to test whether NSC638432 would act within the context of DSB repair-deficient cells despite the fact that the drug was found to sensitize mutations in other DSB-repair

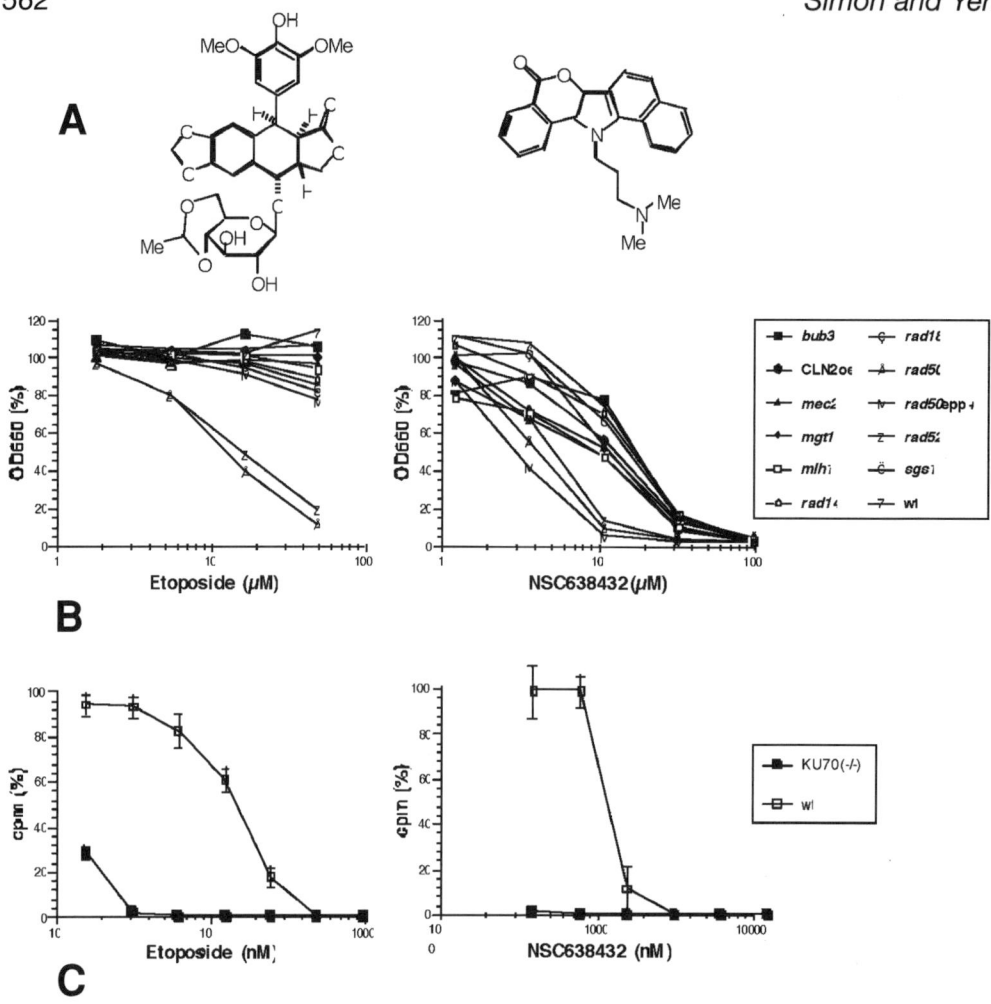

Fig. 2. Identification of NSC638432 as a new topoisomerase II poison. (**a**) Chemical structures of etoposide (*left*) and NSC638432 (*right*); (**b**) differential toxicity of etoposide and NSC638432 in *rad50* and *rad52* mutants; (**c**) differential toxicity of etoposide and NSC638432 in Ku70–/– mouse cells.

genes. Consistent with the yeast data, cells lacking Ku70 and thus crippled in the non-homologous-end rejoining DSB repair pathway were about an order of magnitude more sensitive to NSC638432 than wild-type cells. This difference in sensitivity was also observed with etoposide, suggesting that the target for NSC838432 might also be topo II. This prediction was confirmed when the drug was found to inhibit topo II activity in vitro. The discovery of NSC638432 validates the concept of context-based drug screens. Drugs such as NSC638432 may be useful against a variety of cancers that have defects in DSB-repair genes or pathways.

5. Target-Based Screens

Target-based screens are used to identify compounds that inhibit a specific protein or enzyme. Currently, target-based screens are carried out in vitro, and this approach is commonly used by pharmaceutical companies. Aside from the difficulties encountered with setting up an in-vitro assay for a particular protein, a greater concern is the inability to test for selectivity or specificity of a given compound in a meaningful way. In contrast, a cell-based assay would impose strict demands on selectivity of a drug for its intended target. Two examples will be given to illustrate the use of yeast in target-based screens for drugs that would specifically inhibit human proteins involved in cancer. These examples highlight the flexibility of yeast as a system to screen for drugs against nonyeast genes.

Many human cancers exhibit growth defects that result from deregulation of cell cycle kinases (cdks) *(28,29)*. This can be achieved by overexpression of cyclins, which are required for kinase activity of the cdks, or because of the loss of cdk inhibitors. Human cdk4, when overexpressed in yeast, will inhibit its growth *(30)*. Given that the cdk4-induced arrest can be rescued by co-expression of p16INK4A *(30)*, a specific inhibitor of cdk4/cyclinD1, and that the arrest is blocked by mutations in cdk4 that abolish its catalytic activity, the opportunity exists to use this model to screen for drugs that selectively inhibit human cdk4. As the inhibition of cdk4 restores cell growth, this positive selection (i.e., cells grow in the presence of an active compound) requires that cdk4 inhibitors do not inhibit the endogenous pool of kinases that are essential for growth. In practice, expression of cdk4 is driven from an inducible promotor. A popular choice is the GAL1 promotor that is repressed when cells are grown in the presence of glucose and is activated when the glucose is removed and the cells are inoculated into media containing galactose *(31,32)*. The screen is conducted with GAL:cdk4 yeast that are grown in the presence of glucose or galactose. Inhibitors that kill cells regardless of whether they are grown in glucose or galactose are likely nonspecific. On the other hand, compounds that allow yeast to grow in galactose media (in the presence of cdk4) and glucose media are likely to be specific inhibitors of cdk4 or components of the pathway throughout which cdk4 acts. The likelihood of the second outcome is low, given that the pathways in yeast that are usurped by human cdk4 are likely to be essential for normal growth. Compounds that inhibit transcription from the GAL1 promoter may also be isolated. Thus, secondary screens relying on expression from a different galactose-inducible promotor must be performed to eliminate false positives. An important aspect of a screen that selects for cell growth is that the compounds must inhibit only cdk4 kinase and not indigenous cdks *(33–37)* kinases *(12)*, or others that are essential for viability.

Yeast do not undergo apoptosis, yet they are sensitive to overexpression of the human proapoptotic BAX protein *(38,39)*. Importantly, this lethality can be rescued by expression of antiapoptotic BCL2. The ability to identify inhibitors of antiapoptotic proteins such as BCL2 is highly desirable, as BCL2 overexpression is thought to contribute to many tumors' ability to develop drug resistance *(40)*. Using BAX/BCL2 expression in yeast, it should be possible to design screen for drugs that inhibit BCL2. Unlike the situation with cdk4, in which inhibition of cdk4 allows yeast to grow, inhibitors of BCL2 should prevent it from neutralizing the toxicity of BAX overexpression. Therefore, compounds that inhibit BCL2 are identified based on their ability to prevent cell growth. To set up such a screen, BAX expression is put under GAL1 control while BCL2 is expressed from a constitutive promoter such as pADH (alcohol dehydrogenase). As before, parallel screens are performed in galactose and glucose media. Compounds that inhibit cell growth in both glucose and galactose will be nonspecifically toxic to yeast. Inhibitors of BCL2 will only kill cells grown in the presence of galactose when BAX expression is induced. An alternative way of targeting BCL2-expressing tumors was demonstrated with the identification of antimycin A methyl ether as an agent that is selectively toxic to BCL2-overexpressing mammalian cells. This agent appears to convert the normally antiapoptotic BCL2 protein into a proapoptotic protein. This observation could be used to set up a yeast-based screen in which the toxicity of candidate agents to BCL2-expressing cells is compared to that in wild-type cells.

6. Synthetic Lethal Screens

Synthetic lethality is a genetic term that defines the interactions between two genes. Typically, loss of either one of the two genes does not affect viability, but once one gene is inactivated, the second gene becomes essential. There are two ways that genes can be synthetically lethal. In the simplest example, two genes provide redundant functions so that one can compensate for the loss of the other. N-methionine aminopeptidase catalyzes the removal of the N-terminal methionine from newly translated proteins and is encoded by MAP1 and MAP2 in yeast *(41)*. Loss of either gene is not lethal because the remaining gene can supply sufficient enzyme for cell growth. However, cells cannot grow if both genes are lost.

Synthetic lethality can also occur between two unrelated genes because their gene products might act along redundant or overlapping pathways. To illustrate this point, we will describe the synthetic lethal interactions between a kinesin microtubule motor and a component of the mitotic checkpoint pathway. CIN8 (*C*hromosome *in*stability 8) is a member of the kinesin superfamily of microtubule motors that contributes in a nonessential way to spindle formation in yeast *(42)*. *cin8* mutants are viable but exhibit a delay in anaphase onset. This delay is likely imposed by the spindle checkpoint, whose function is to ensure that chromosomes are properly attached to the spindle before allowing cells to enter anaphase. The fact that *cin8* mutants exhibit a delayed anaphase implies that spindle formation must be retarded and the checkpoint delay provides the extra time required for these mutants to build a functional spindle. In yeast, the spindle checkpoint is specified by seven genes *(43)*, most of which are not essential unless there are problems with chromosomes attaching properly to the spindle. Thus, a *bub1* (*B*udding *U*ninhibited by *B*enzimidozole) mutant *(44)* that lacks an essential component of the spindle checkpoint can grow relatively normally until it encounters a defective spindle that might be brought about by the loss of CIN8. CIN8 and BUB1 exhibit synthetic lethal interactions because loss of either one is tolerated but cells cannot survive when both genes are lost. A *cin8 bub1* double mutant is nonviable because the loss of the checkpoint will cause the cells with defective spindles to exit mitosis with improperly aligned chromosomes. This leads to high rates of chromosome loss that cannot be tolerated by the cell. It is clear that CIN8 and BUB1 differ significantly in their biochemical properties (CIN8 is a microtubule motor, BUB1 is a protein kinase), and the synthetic lethal interactions cannot reflect redundancy activities. Knowing that CIN8 and BUB1 are involved in spindle formation and spindle checkpoint functions, their synthetic lethal interactions reveal connections between the two pathways along which these proteins act. Indeed, CIN8 exhibits synthetic lethal interactions with virtually all of the genes that are essential for spindle checkpoint functions *(45)*.

Both types of synthetic lethality can be exploited in context-based drug screens that are designed to identify compounds that inactivate a specific pathway. In this section we will discuss how the synthetic lethality between CIN8 and BUB1 can be used to identify drugs that target aneuploid tumor cells. A drug screen based on the redundancy of MAP1 and

MAP2 will be presented in the concluding section. It is perhaps noteworthy that certain human cancers that exhibit high rates of aneuploidy are thought to be defective in the spindle checkpoint pathway *(46)*. As many components of the spindle checkpoint are conserved between yeast and humans, it would be desirable to screen for drugs that selectively kill cells that are defective for checkpoint control. Using CIN8 and BUB1 as an example, a screen can be designed to search for drugs that inhibit components of the spindle checkpoint in yeast. The screen will search for compounds that exert chemical synthetic lethal interactions with CIN8 mutants. CIN8 mutants that fail to grow in the presence of a compound will likely be drugs that inhibit components of the checkpoint. Once such compounds are identified, they can be tested on human cancer cells that are defective for the spindle checkpoint. It is formally possible that compounds will be identified that inhibit other kinesin motor proteins that are known to provide overlapping functions with CIN8 *(47)*. Given that yeast can tolerate the loss of multiple kinesin family members, the probability of finding a drug that can inhibit multiple kinesins is lower than that of finding one that specifically inhibits any one of the seven proteins that are essential for the checkpoint.

A more useful application of the chemical synthetic lethal screens is to search for drugs that selectively kill mismatch-repair mutants. A large proportion of colorectal cancers are defective in DNA mismatch-repair as a result of mutations in hMLH1, hMSH2, hPMS1, or hMSH6 genes *(48,49)*. These mutations can be easily identified from tumor samples based on microsatellite instability, yet drugs that specifically target mismatch defective cells are unavailable. A search was therefore conducted to identify compounds that would be selectively cytotoxic to yeast *mlh1* mutants. A screen to identify compounds that exhibited synthetic lethal interactions with *mlh1* mutants was carried out as part of the NCI/DTP screen (see above). This approach yielded a very small number of hits, which are currently being characterized. Additional efforts will have to be applied to identify drug targets (see below). An alternative strategy was first to identify yeast genes synthetically lethal with an *mlh1* mutant. Once such genes are identified, a target-based approach can be applied to find inhibitors that specifically inhibited their gene products. An *mlh1* mutant strain that expressed a plasmid-borne wild-type MLH1 was first chemically mutagenized to isolate mutants that could no longer live in the absence of the wild-type MLH1-encoding plasmid. One gene identified from this screen was RNR1, which encodes the large subunit of the yeast ribonucleotide reductase that converts ribonucleotides into deoxyribonucleotides *(50,51)*. This RNR1 mutant allele was not only synthetically lethal with *mlh1* but was incompatible with other yeast mismatch-repair genes. The RNR mutant was found to contain a point mutation near the regulatory site of the enzyme. Because RNR1 is responsible for synthesizing all four dNTPs, the enzymatic activity is tightly regulated to ensure a balanced supply of the four nucleotides. Regulation is achieved by allosteric regulators ATP, dATP, dTTP, and dGTP that bind to the regulatory site and control which substrates the enzyme should process. It is conceivable that the RNR mutant is no longer sensitive to regulation and thus produces an imbalance in the ratio of dNTPs that would increase the rate of mismatches during replication or repair. These mismatches if not repaired, as in the case of an *mlh1* mutant, could result in lethality. Based on this analysis, it should be possible to use the *mlh1* mutant yeast to isolate a drug that specifically targets RNR1. One way to identify inhibitors of RNR1 that mimic the effect of the RNR1 mutation described above would be screen for compounds that show a selective toxicity to *mlh1* strains. A potential mechanism for achieving this selectivity is by altering the allosteric regulation of RNR1. However, other approaches such as structure-based drug design can be used to identify agents that bind to the allosteric regulatory site of RNR1 and disrupt the regulation of the enzyme.

Another genetic context that could be exploited to identify novel anticancer drugs is cyclin overexpression. In some human cancers, cyclin D1 and cyclin E overexpression is thought to disrupt the transition from G1 to S and thus increase genomic instability *(52,53)*. In yeast, CLN2 is a cyclin that associates with the cdc28 kinase and is responsible for regulating the G1-to-S transition *(54)*. As in mammalian cells, yeast cells overexpressing CLN2 accelerate the G1-to-S transition. From this perspective, CLN2 can be considered a functional homolog of human cyclinD1 and E *(55)*. Although overexpression of CLN2 is tolerated by yeast, it becomes lethal when combined with mutations that inactivate MEC1, CDC14, or NUP170. MEC1 is a member of a conserved family of PI3-related kinases that include human ATM. These protein kinases are essential components of the DNA damage and replication checkpoint response. Although MEC1 is an essential gene, specific mutant alleles (i.e., *mec1-1*) lack the nonessential checkpoint function and are viable. The synthetic lethality between CLN2 overexpression and *mec1-1* reinforces the notion that, in yeast as in mammals, accelerated G1-to-S progression causes DNA damage that requires cell cycle arrest for repair. CDC14 is a dual-specificity phosphatase that is required for cells to exit mitosis *(56)*. NUP170 is a component of the nuclear pore complex that is involved in mRNA export from the nucleus *(57)*. These genes are not essential for viability in yeast, and they do not appear to share overlapping functions. The molecular basis for why these genes are synthetically lethal with CLN2 overexpression is unknown. Nevertheless, this phenomenon may be used to identify drugs that inactivate MEC1, CDC14, or NUP170 and thus may be selectively toxic to CLN2 overexpressing yeast. Whether inhibition of the corresponding human orthologs in cancer cells overexpress cyclinD1 or cyclinE remains to be demonstrated.

It is important to emphasize that a context-based screen, as in the above example, is not restricted to a specific gene target. An inhibitor that is identified based on CLN2 overexpression may have the desirable quality that it acts not on CLN2 directly but rather within the context of inappropriate cell cycle timing. A drug that kills cells that accelerate through G1 inappropriately may have a broad spectrum of targets rather than cells that just overexpress cyclinD1 or cyclinE. Cancers with mutations in ras, p16INK, pRb, HER/neu, and other genes that exhibit an accelerated G1-to-S transition may also be targeted by a drug that was isolated in a context-based screen. As mentioned earlier, a screen for drugs that exhibit chemical synthetic lethality (also called conditional lethality) with a mutant gene (i.e., CLN2 overexpression) will not yield the target of the inhibitor. The next section describes genetic strategies that include genome-wide target identification to find drug targets in yeast.

7. Target Identification

Target identification is not an issue for target-based screens that are performed in vivo or in vitro. Drugs identified from context-based screens, however, require further work to reveal the drug target. Traditional approaches rely on using the drug as an affinity reagent to isolate the target protein from cell lysates. This approach can be costly and time-consuming if the target protein is not very abundant or has to be obtained from animal sources. Genetic approaches that rely on isolating resistant mutants require extensive efforts to identify the resistance gene. These efforts do not guarantee that the resistance is specified by mutations in target genes because resistance is frequently achieved by mutations in unrelated genes that indirectly modulate the toxicity of the drug. Yeast offers many of the same practical benefits for target identification as for drug screening. More important, yeast is the only eucaryote for which all of the protein-encoding genes have been identified and advanced genetic techniques exist for the manipulation of the expression level of specific proteins. This provides a genome-wide approach to target identification.

8. Suppressor Screens

One way to identify yeast genes that are targets for specific drugs is to overexpress the target gene as a way to decrease sensitivity of the cell to the drug. This strategy is a classic multicopy suppressor screen in which loss-of-function mutations are rescued by expression of the wild-type gene *(58,59)*. If a drug inhibits the function of a specific protein, this chemically induced loss of function should be complemented by overexpression of the target gene. Expression of the target gene would reduce the sensitivity of the cells to a drug. It is possible that a drug might enhance the activity of an enzyme or protein, as would a gain-of-function mutation. In this case, overexpressing the target gene would lead to increased sensitivity by the drug.

Drugs that target topoisomerase II are good examples of how drugs can act in a chemical genetic sense as loss of function or gain of function mutations. There are two classes of topoisomerase II-active compounds, which have different mechanisms of action. Topoisomerase II is an enzyme that induces transient DSBs in DNA to allow catenated DNA to become separated *(60)*. After strand passage through the break, the two ends of the broken DNA are ligated and repaired. ICRF159 and ICRF193 are inhibitors that block the enzymatic activity of topoisomerase II such that the enzyme fails to cut the DNA and thus prevent resolution of catenated DNAs. Cells treated with these drugs fail to decatenate DNAs and fail to separate sister chromatids in mitosis. These drugs mimic loss-of-function mutations, as overexpression of topoisomerase II will increase drug resistance. Etoposide and amsacrine, on the other hand, belong to a large family of topoisomerase II poisons, compounds that interact with topoisomerase II to stabilize what normally is a transient covalent interaction between topoisomerase II and DNA ends. In this case, topoisomerase II is trapped in the middle of its catalytic cycle and prevented from religating cleaved DNA, which leads to accumulation of DSBs. Overexpression of topoisomerase II does not relieve cells from the toxicity of etoposide, but rather increases their sensitivity. This is consistent with a gain-of-function mutation, as cells with increased levels of topoisomerase II accumulate higher levels of DSBs in the presence of etoposide.

Regardless of whether drug-treated cells behave as loss- or gain-of-function mutants, the effect of the drug can be detected by examining growth rates as a function of expression level of the target enzyme. To screen for a target, cells are transformed with a yeast cDNA or genomic library and then the transformants are replica-plated onto normal or drug-containing media. Transformation will introduce additional copies of genes into cells, resulting in a proportional increase in the amount of the corresponding proteins. The vast majority of transformants will express higher levels of an irrelevant gene and will retain the drug sensitivity of the untransformed parent. On the other hand, some transformants might change their growth rate as a result of overexpression of a potential target gene. Typically, transformants are scored based on increased drug resistance and one needs to rule out those plasmids that encode genes that affect drug uptake or metabolism.

9. Haplo-Insufficiency

A reduction in the level of a drug target might alter the growth of a strain in the presence of drug. Haplo-insufficiency refers to a heterozygous diploid organism in which loss of one copy of a specific gene leads to a reduction in the level of the encoded protein and thus a phenotype. By applying this idea on a genome-wide scale, it is possible to test a drug against a panel of yeast that are heterozygous diploid for each of the ~6200 protein-coding genes. An international consortium has deleted all the open reading frames (ORFs) in the yeast genome *(61)*. The panel of heterozygous diploids is publicly available and can be used to help identify drug targets.

Briefly, DNA sequences that flank the 5' and 3' ends of a target gene (usually 50 bp of homology is sufficient for targeting) are amplified by PCR along with an internal antibiotic-resistance gene (**Fig. 3**). This deletion cassette is transformed into yeast, and transformants that are resistant to the antibiotic are selected. This procedure works because yeast use homologous recombination as the major pathway for DNA repair. The deletion cassette is treated as a piece of DNA with two DBSs. Because of flanking homologous sequences, the deletion cassette is "repaired" by exchange with intervening sequences (ORFs) in the genome. To give each of the 6200 gene disruptions a unique identity, each disruption cassette contains two unique stretches of 20 nucleotides that flank the antibiotic-resistance marker. All of these unique "bar codes" are in turn flanked by another set of sequences that is common among all the cassettes. By using one set of PCR primers, it is possible to amplify any of the deletion cassettes from a mixed population of cells *(61)*. Further sequencing of the bar code or hybridization to oligonucleotide arrays identifies the specific gene that was disrupted in a strain.

The protocol will be simply to grow pools of the heterozygous diploid deletion mutants or as individuals in the presence and absence of drug. The dose of drug should be low enough to reduce the growth of the wild-type strain by only a small amount. At this dose of drug, a cell that expresses reduced levels of the target protein might exhibit enhanced growth inhibition. If pools of mutant yeast are tested, the growth of any mutant within this mixed population can be quantitated by comparing the amount of its DNA to that obtained from others. After an overnight incubation, DNA is isolated from the drug-treated and control cultures. The deletion cassettes from these two cultures are amplified using fluorescently labeled primers (**Fig. 4**). The DNAs from the drug-treated and control cells are distinguished by using different fluorescent primers for each of the cultures. The fluorescently labeled PCR products from the two populations are mixed together and hybridized to an oligonucleotide chip that contains all of the 6200 unique bar codes. Simultaneous detection of the fluorescent signals at each unique bar code will reflect the relative growth of a particular mutant in the presence and absence of drug. A drug that inhibits a specific enzyme might be more sensitive to cells have a reduced level of this enzyme. Thus, the reduction in growth rate of a sensitive strain in the presence of drug would produce fewer cells and thus lower amounts of cellular DNA should be recovered from these strains. The chip analysis would show that the bar code that represents a particular heterozygous mutant would be lower than the same bar code that was isolated from cultures grown in the absence of drug. This approach has been tested recently using a subset of the heterozygous diploid strains that were available at the time *(62)*.

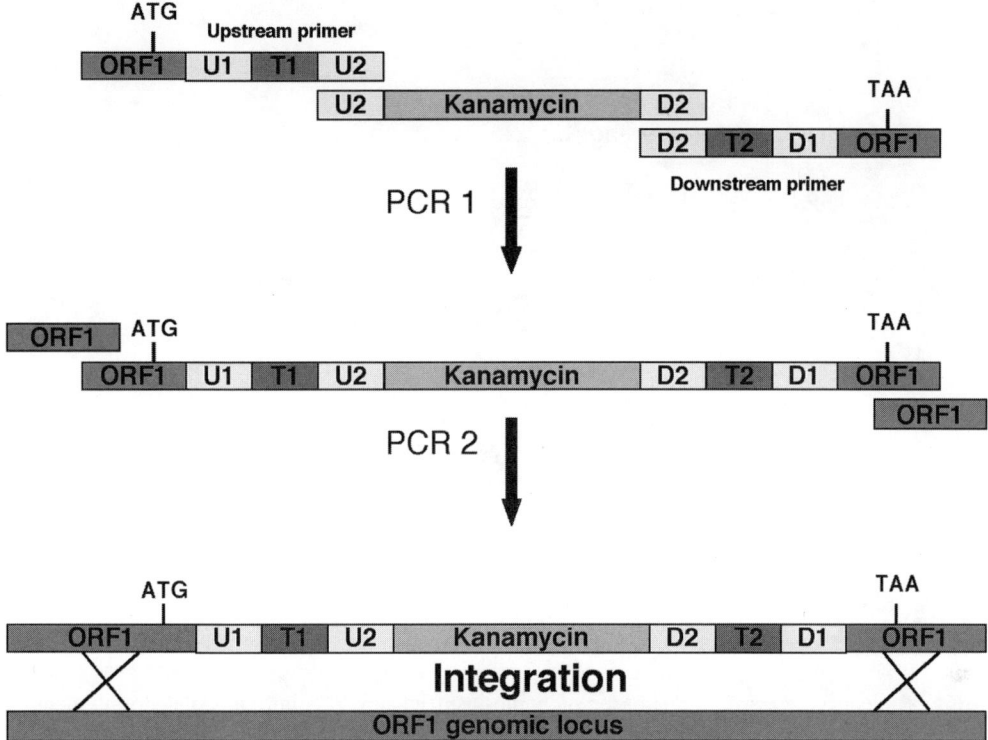

Fig. 3. Schematic for targeted deletion of all ORFs in the *S. cerevisae* genome. The breakdown of the various elements of the deletion cassette is shown at the top. U2 and D2 are 18 bp elements that flank the kanamycin resistance gene (Kan^R) and are used to facilitate PCR amplification of the deletion cassette using upstream and downstream primers. Upstream and downstream primers contain two sets of elements that uniquely describe an ORF in the genome. Each primer contains 18 bp of sequences that are homologous to sequences that lie immediately 5′ of the start codon (ATG) and 3′of the stop codon (TAA) of the target ORF, respectively. These elements are followed by U1 or D1, which are 18-bp elements that are common among all the upstream and downstream primers. T1 and T2 are 20-bp sequences that uniquely tag each of the ~6200 ORFs and act as bar codes. The U2 and D2 elements at the 3′ ends of each primer are used to generate a deletion cassette by PCR amplifying the kan^R gene. A second round of PCR is performed using gene-specific primers that extend the homology with the gene of interest from 18 to 45 bp. Each of the ~6200 deletion cassettes is transformed into yeast and Kan^R transformants are selected and screened for site-specific integration homologous recombination.

10. Drug Screens Using Yeast Three-Hybrid Technology

The yeast two-hybrid screen is a popular method to search for protein–protein interactions. This method relies on the modular design of the transcription regulators that recruit RNA polymerase to promoters. The transcriptional regulators contain a module that binds to a specific DNA sequence and an activation module that interacts with RNA polymerase to position it so that it can accurately initiate transcription. These two modules work together to recruit RNA polymerase to a specific promoter. The yeast two-hybrid approach takes advantage of the fact that the DNA-binding and activation modules do not have to be part of the same molecule as interactions that bring these two

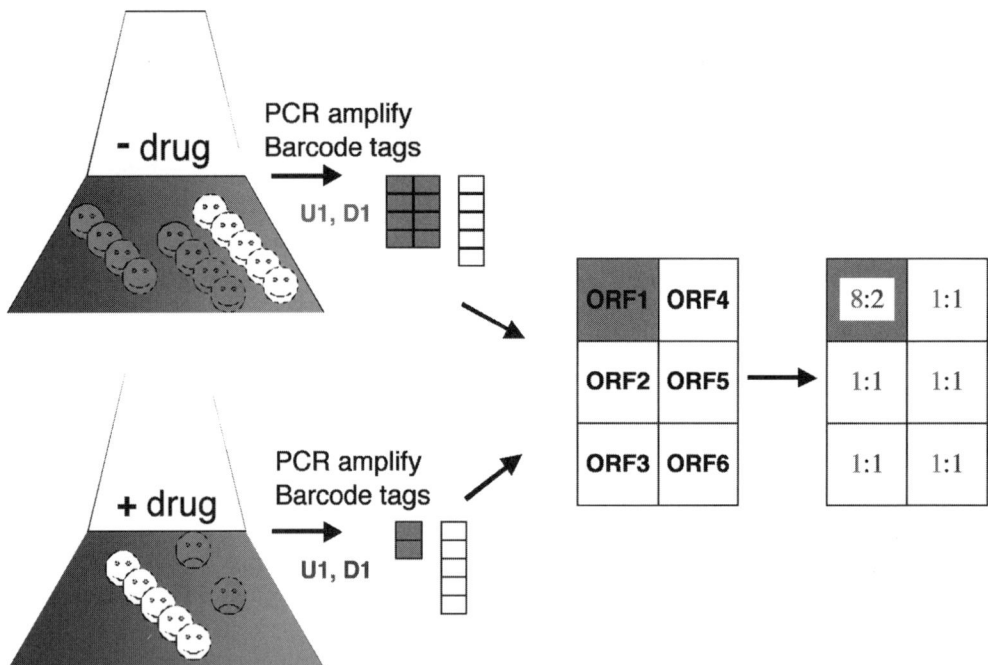

Fig. 4. Identifying a drug target based on haplo-insufficiency. A culture inoculated with identical number of cells from six different heterozygous deletion mutants (ORFs 1–6) . ORF1 mutant is sensitive to the drug, while the remaining five mutants are not. After ON incubation of the mixed cultures in the presence and absence of drug, genomic DNA is prepared from each culture and the deletion cassettes for all six mutants are simultaneously amplified using the U1 and D1 primers (*see* **Fig. 3** for primer information). The amount of each deletion cassette that is amplified should be proportional to the cell number of the corresponding mutant. To distinguish between the cassettes that were derived from control and drug-treated cultures, each culture is amplified with primers that have a different fluorescent tag. The amplified products are mixed together and hybridized to an oligonucleotide chip that contains the unique bar codes that were assigned to each gene. Simultaneous detection of the two fluorescent signals on each spot allows for direct comparison of the relative growth of a particular haplo-insufficient strain in the presence and absence of drug.

domains within close proximity to each other within a promoter will be sufficient to activate transcription of a reporter gene (**Fig. 5**). Typically, a protein of interest (called bait) is fused to an activation domain and expressed in yeast. A library of random cDNAs fused to a DNA-binding domain that binds to a specific sequence within the promoter of the reporter gene is transformed into the bait-containing yeast. If a cDNA encodes a protein that can bind to the bait, the reporter gene is activated.

Modification of the two-hybrid approach has led to a new strategy whereby the interaction between the DNA-binding and activation modules are mediated by a third component, which can be a drug or a third protein. To demonstrate the feasibility of the yeast three-hybrid screen, an experiment was designed to identify the protein target of the immunosuppressive drug FK506 *(63)*. Because the interaction of the DNA-binding and activation domains could only be brought together if both domains are bridged together by the drug, the FK506 was first chemically tethered to dexamethasone (**Fig. 5**). The DNA-binding domain was fused to the glucocorticoid receptor in order to bind to the dexamethasone moiety of the FK506 hybrid. Importantly, expression of the reporter

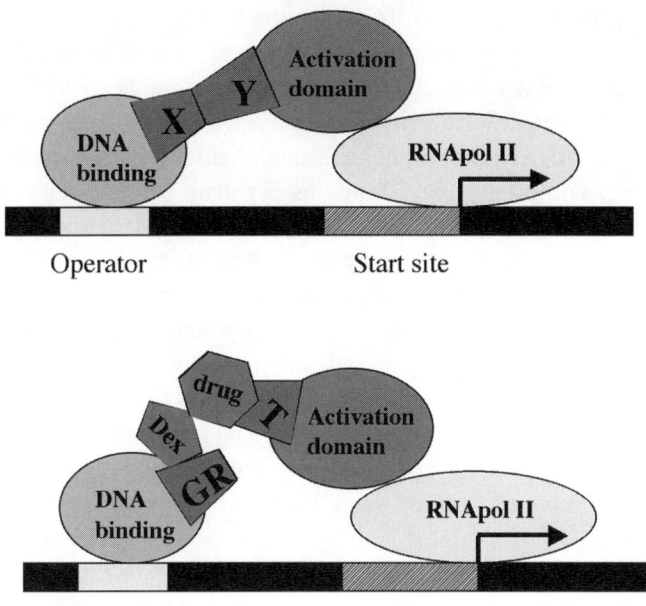

Fig. 5. Identification of drug targets using yeast three-hybrid screen. The top panel shows the principle behind the conventional yeast two hybrid system. This method is used to screen for proteins that interact with a protein of interest. The protein of interest "bait" (X) is fused to the cDNA-binding domain, which recognizes a specific sequence that lies within the promotor of a reporter gene. A cDNA library that is fused to the activation domain of a transcription factor is transformed into the strain expressing the "bait." The reporter gene is expressed only when the bait (X) is associated with its interactor (Y). The bottom panel depicts a three-hybrid system to screen for drug targets. The DNA-binding and activation domains do not interact physically with each other, so the reporter gene is inactive. However, a small molecule that tethers these domains and brings them within close proximity to each other should activate the reporter gene. The DNA-binding domain is fused to glutacorticoid receptor (GR) and the activation domain is fused to a cdna library that is being screened for drug targets. The drug is modified so that it carries a dexamethasone (Dex) moiety, which binds to the GR of the DNA-binding domain. Upon binding of the drug to its target (T), the transcription modules induce expression of the reporter gene.

construct was strictly dependent on the expression of an activation domain that was fused to FKBP12, the protein that binds FK506. More reassuring was the ability of the drug to identify FKBP12 from a cDNA library screen. Despite the promise of the three-hybrid approach to screen for drug targets, there are limitations. The main obstacle is the need to synthesize derivatives of drugs (i.e., linked to dexamethasone) and all of the caveats associated with modifying a drug. For drugs that can accommodate these modifications, the advantage is that the cDNA encoding the drug targets can be isolated very rapidly. Finally, the three-hybrid screen is not restricted to yeast genes, as cDNAs from all sources can be screened for drug binding.

11. Summary

Genome instability is one of the hallmarks of cancer and is primarily responsible for deregulation of the cell cycle that leads to uncontrolled cell growth. This fundamental difference between cancer and normal cells provides real opportunities to identify drugs that are selectively toxic to cancer cells. Given the highly conserved nature of the components and pathways that regulate the cell cycle, yeast serves as an ideal model system for such drug screens. Yeast is a cheap and rapid alternative to traditional cell-based screens that rely on tissue culture cells. The real advantage of yeast, however, is that it recapitulates many genetic changes that are associated with human cancers. This has led to two different but complementary strategies for drug screens. The principle behind context-based screens is to rely on yeast mutants that are defective for genes or pathways that are disrupted in human cancer cells to identify compounds that are selective toxic to cancer but not to normal cells. Drugs that selectively kill a specific yeast mutant are likely to act on cancer cells that share a similar genetic context as the yeast mutant. Once a drug is identified in a context-based screen, the target of the drug must be identified. Target genes can be identified by isolating multicopy suppressors that alter the sensitivity of the yeast to the drug, by testing drugs against a genome-wide panel of heterozygous diploid mutants or by yeast three-hybrid. These genetic approaches should facilitate characterization of protein–drug interactions using traditional biochemical approaches.

The goal of target-based screens is to identify inhibitors to a specific protein. Traditional target-based screens rely on in-vitro assays that require extensive effort to set up and are limited by the availability of large quantities of biochemically active protein. Target-based screens in yeast are cell-based screens that provide a high level of selectivity that is not possible for in-vitro assays. This screen is not limited to yeast genes but can accommodate target genes from other species. As long as a gene produces an easily recognizable phenotype in yeast, it is suitable for target-based drug screens. The added benefit of using a cell-based system for target screens is that the compounds must be highly specific for the target and must not inhibit endogenous proteins that may be structurally related to the target. In the example in which human cdk4 was used as target, inhibitors must be highly specific for cdk4, as compounds that lack specificity will inhibit a wide range of endogenous kinases that are essential for viability of yeast.

Although the focus of this chapter has been to use yeast to identify novel anticancer drugs, this is not its only application. In general, any yeast gene that exhibits structural or functional homologies with a human gene can be considered as a drug target. Angiogenesis, a process that is clearly absent in yeast, relies in part on enzymes that are conserved in yeast. Fumagillin and ovacilin are potent inhibitors of angiogenesis that target methionine aminopeptidase-2, an enzyme that removes the amino-terminal initiator methionine from newly made proteins. The substrates of this enzyme that are important for angiogenesis is not known. Yeast express two functionally redundant enzymes that are encoded by MAP1 and MAP2 genes. Neither gene is essential for yeast viability, but cells cannot grow when both genes are lost. Interestingly, MAP2 is highly homologous to the human enzyme and is inhibited by fumagillin and ovacilin *(64,65)*. In principle, drugs with similar specificity to fumagillin and ovacilin can be identified in a chemical synthetic lethal screen using a *map1* mutant as the host strain. The MAP2 gene is made essential by the deletion of MAP1, and any compound that is selectively toxic to the *map1* mutant strain but not the parental MAP1 strain should target MAP2. With the completion of the human genome sequence, we should expect increasing numbers of human genes that share similarities with yeast genes. These similarities provide new and expanded opportunities for using yeast as a model organism for drug discoveries that have broad impact on human diseases.

Acknowledgments

The authors would like to thank Roseanne Diehl for manuscript preparation. This work was supported by Grants PO1 CA75138, CA6927, and an Appropriation from the Commonwealth of Pennsylvania to T.J.Y. J.A.S. would like to thank Dr. John Lamb and Dr. Shlomo Handeli for helpful comments. J.A.S. was supported by grants from the National Cancer Institute and Merck.

References

1. Alley, M. C., Scudiero, D. A., Monks, A., et al. (1988) Feasibility of drug screening with panels of human tumor cell lines using a microculture tetraolium assay. *Cancer Res.* **48**, 589–601.
2. Paull, K. D., Shoemaker, R. H., Hodes, L., et al. (1989) Display and analysis of patterns of differential activity of drugs against human tumor cell lines: development of mean graph and COMPARE algorithm. *J. Natl. Cancer Inst.* **81**, 1088–1092.
3. Panek, R. L., Lu, G. H., Klutchko, S. R., et al. (1997) In vitro pharmacological characterization of PD 166285, a new nanomolar potent and broadly active protein tyrosine kinase inhibitor. *J. Pharmacol. Exp. Ther.* **283**, 1433–1444.
4. Mayer, T. U., Kapoor, T. M., Haggarty, S. J., et al. (1999) Small molecule inhibitor of mitotic spindle bipolarity identified in a phenotype-based screen. *Science* (1999) **286**, 971–974.
5. Blangy, A., Lane, H. A., D'Herin, P. D., Harper, M., Kress, M., and Nigg, E. A. (1995) Phosphorylation by p34cdc2 regulates spindle association of human eg5, a kinesin-related motor essential for bipolar spindle formation in vivo. *Cell* **83**, 1159–1169.
6. Sawin, K. E., Leguellec, K., Philippe, M., and Mitchison, T. J. (1992) Mitotic spindle organization by a plus-end-directed microtubule motor. *Nature* **359**, 480–481.
7. White, R. J. (1982) Microbiological models as screening tools for anticancer agents: potentials and limitations. *Ann. Rev. Microbiol.* **36**, 415–433.
8. Renan, M. J. (1993) How many mutations are required for tumorigenesis? Implications from human cancer data. *Mol. Carcinog.* **7**, 139–146.
9. Hartwell, L. H., Szankasi, P., Roberts, C. J., Murray, A. W., and Friend, S. H. (1997) Integrating genetic approaches into the discovery of anticancer drugs. *Science* **278**, 1064–1068.
10. Guthrie, C. and Fink, G. R. (1991) Guide to yeast genetics and molecular biology, in *Methods in Enzymology* (Abelson, J. N. and Simon, M. I., eds.). Academic Press, New York, Vol. 194.
11. Simon, J. A., Szankasi, P., Nguyen, D. K., et al. (2000) Differential toxicities of anticancer agents among DNA repair and checkpoint mutants of *Saccharomyces cerevisiae*. *Cancer Res.* **60**, 328–333.
12. Goffeau, A., Barrell, B. G., Bussey, H., et al. (1996) Life with 6000 genes. *Science* **274**, 5463–5467.
13. McCammon, M. T., Hartmann, M. A., Bottema, C. D., and Parks, L.W. (1984) Sterol methylation in *Saccharomyces cerevisiae*. *J. Bacteriol.* **157**, 475–483.
14. Gaber, R. F., Copple, D. M., Kennedy, B. K., Vidal, M., and Bard, M. (1989) The yeast gene ERG6 is required for normal membrane function but is not essential for biosynthesis of the cell-cycle-sparking sterol. *Mol. Cell. Biol.* **9**, 3447–3456.
15. Balzi, E. and Goffeau, A. (1995) Yeast multidrug resistance: the PDR network. *J. Bioenerg. Biomembr.* **27**, 71–76.
16. Kolaczkowski, M., Van der Rest, M., Cybularz-Kolaczkowska, A., Soumillion, J. P., Konings, W. N., and Goffeau, A. (1996) Anticancer drugs, ionophoric peptides, and steroids as substrates of the yeast multidrug transporter Pdr5p. *J. Biol. Chem.* **271**, 31543–31548.
17. Friedberg, E. C., Walker, G. C., and Siede, W. (1995) *DNA Repair and Mutagenesis*. American Society for Microbiology, Washington, DC.

18. Weinert, T. A. and Hartwell, L. H. (1988) The RAD9 gene controls the cell cycle response to DNA damage in *Saccharomyces cerevisiae*. *Science* **241**, 317–322.

19. Weinert, T. A., Kiser, G. L., and Hartwell, L. H. (1994) Mitotic checkpoint genes in budding yeast and the dependence of mitosis on DNA replication and repair. *Genes Dev.* **8**, 652–665.

20. Straight, A. F. and Murray, A. W. (1997) The spindle assembly checkpoint in budding yeast. *Meth. Enzymol.* **283**, 425–440.

21. Watt, P. M., Hickson, I. D., Borts, R. H., and Louis, E. J. (1996) SGS1, a homologue of the Bloom's and Werner's syndrome genes, is required for maintenance of genome stability in *Saccharomyces cerevisiae*. *Genetics* **144**, 935–945.

22. Prakash, L. (1981) Characterization of postreplication repair in *Saccharomyces cerevisiae* and effects of rad6, rad18, rev3, and rad52 mutations *Mol. Gen. Genet.* **184**, 471–478.

23. Prakash, S., Sung, P., and Prakash, L. (1993) DNA repair genes and proteins of *Saccharomyces cerevisiae*. *Annu. Rev. Genet.* **27**, 33–70.

24. McIntosh, E. M., Kunz, B. A., and Haynes, R. H. (1986) Inhibition of DNA replication in *Saccharomyces cerevisiae* by araCMP. *Curr. Genet.* **10**, 579–585.

25. Yamagata, K., Kato, J., Shimamoto, A., Goto, M., Furuichi, Y., and Ikeda, H. (1998) Bloom's and Werner's syndrome genes suppress hyperrecombination in yeast sgs1 mutant: implication for genomic instability in human diseases. *Proc. Natl. Acad. Sci. USA* **95**, 8733–8738.

26. Xiao, W., Derfler, B., Chen, J., and Samson, L. (1991) Primary sequence and biological functions of a *Saccharomyces cerevisiae* O6 methylguanine/O4 methylthymine DNA repair methyltransferase gene. *EMBO J.* **10**, 2179–2186.

27. Kokkinakis, D. M., Ahmed, M. M., Delgado, R., Fruitwala, M. M., Mohiuddin, M., and Albores-Saavedra, J. (1997) Role of O6-methylguanine-DNA methyltransferase in the resistance of pancreatic tumors to DNA alkylating agents. *Cancer Res.* **57**, 5360–5368.

28. Sherr, C. J. (1996) Cancer cell cycles. *Science* **274**, 1672–1677.

29. Sherr, C. J. (2000) The Pezcoller Lecture: cancer cell cycles revisited. *Cancer Res.* **60**, 3689–3695.

30. Moorthamer, M., Panchal, M., Greenhalf, W., and Chaudhuri, B. (1998) The p16(INK4A) protein and flavopiridol restore yeast cell growth inhibited by Cdk4. *Biochem. Biophys. Res. Commun.* **250**, 791–797.

31. Lorch, Y. and Kornberg, R. D. (1985) A region flanking the GAL7 gene and a binding site for GAL4 protein as upstream activating sequences in yeast. *J. Mol. Biol.* **186**, 821–824.

32. Johnston, M. (1987) A model fungal gene regulatory mechanism: the GAL genes of *Saccharomyces cerevisiae*. *Microbiol. Rev.* **51**, 458–476.

33. Nasmyth, K. A. and Reed, S. I. (1980) Isolation of genes by complementation in yeast: molecular cloning of a cell-cycle gene. *Proc. Natl. Acad. Sci. USA* **77**, 2119–2123.

34. Timblin, B. K., Tatchell, K., and Bergman, L. W. (1996) Deletion of the gene encoding the cyclin-dependent protein kinase Pho85 alters glycogen metabolism in *Saccharomyces cerevisiae*. *Genetics* **143**, 57–66.

35. Balciunas, D. and Ronne, H. (1995) Three subunits of the RNA polymerase II mediator complex are involved in glucose repression. *Nucleic Acids Res.* **23**, 4421–4425.

36. Lee, J. M. and Greenleaf, A. L. (1991) CTD kinase large subunit is encoded by CTK1, a gene required for normal growth of *Saccharomyces cerevisiae*. *Gene Exp.* **1**, 149–167.

37. Valay, J. G., Simon, M., and Faye, G. (1993) The kin28 protein kinase is associated with a cyclin in *Saccharomyces cerevisiae*. *J. Mol. Biol.* **234**, 307–310.

38. Tao, W., Kurschner, C., and Morgan, J. I. (1998) Bcl-xs and Bad potentiate the death suppressing activities of Bcl-xl, Bcl-2, and A1 in yeast. *J. Biol. Chem.* **273**, 23704–23708.

39. Xu, Q., Jurgensmeier, J. M., and Reed, J. C. (1999) Methods of assaying Bcl-2 and Bax family proteins in yeast. *Methods* **17**, 292–304.

40. Decaudin, D., Geley, S., Hirsch, T., et al. (1997) Bcl-2 and Bcl-XL antagonize the mitochondrial dysfunction preceding nuclear apoptosis induced by chemotherapeutic agents. *Cancer Res.* **57**, 62–67.

41. Li, X. and Chang, Y. G. (1995) Amino-terminal protein processing in *Saccharomyces cerevisiae* is an essential function that requires two distinct methionine aminopeptidases. *Proc. Natl. Acad. Sci. USA* **92**, 12357–12361.

42. Hoyt, M. A., He, L., Totis, L., and Saunders, W. S. (1993) Loss of function of *Saccharomyces cerevisiae* kinesin-related CIN8 and KIP1 is suppressed by KAR3 motor domain mutations. *Genetics* **135**, 35–44.

43. Straight, A. F. and Murray, A. W. (1997) The spindle assembly checkpoint in budding yeast. *Meth. Enzymol.* **283**, 425–440.

44. Hoyt, M. A., Totis, L., and Roberts, B. T. (1991) *S. cerevisiae* genes required for cell cycle arrest in response to loss of microtubule function. *Cell* **66**, 507–517.

45. Hardwick, K. G., Li, R., Mistrot, C., et al. (1999) Lesions in many different spindle components activate the spindle checkpoint in the budding yeast *Saccharomyces cerevisiae*. *Genetics* **152**, 509–518.

46. Cahill, D. P., Lengauer, C., Yu, J., et al. (1998) Mutations of mitotic checkpoint genes in human cancers. *Nature* **392**, 300–303.

47. Hoyt, M. A. and Geiser, J. R. (1996) Genetic analysis of the mitotic spindle. *Annu. Rev. Genet.* **30**, 7–33.

48. Parsons, R., Li, G. M., Longley, M. J., et al. (1993) Hypermutability and mismatch repair deficiency in RER+ tumor cells. *Cell* **75**, 1227–1236.

49. Fujiwara, T., Stolker, J. M., Watanabe, T., et al. (1998) Accumulated clonal genetic alterations in familial and sporadic colorectal carcinomas with widespread instability in microsatellite sequences. *Am. J. Pathol.* **153**, 1063–1078.

50. Elledge, S. J. and Davis, R. W. (1990) Two genes differentially regulated in the cell cycle and by DNA-damaging agents encode alternative regulatory subunits of ribonucleotide reductase. *Genes Dev.* **4**, 740–751.

51. Price, C., Nasmyth, K., and Schuster, T. (1991) A general approach to the isolation of cell cycle-regulated genes in the budding yeast, *Saccharomyces cerevisiae*. *J. Mol. Biol.* **218**, 543–556.

52. Delsal, G., Loda, M., and Pagano, M. (1996) Cell cycle and cancer: critical events at the G1 restriction point. *Crit. Rev. Oncogen.* **7**, 127–142.

53. Funk, J. O. (1999) Cancer cell cycle control. *Anticancer Res.* **19**, 4772–4780.

54. Richardson, H. E., Wittenberg, C., Cross, F., and Reed, S. I. (1989) An essential G1 function for cyclin-like proteins in yeast. *Cell* **59**, 1127–1133.

55. Reed, S. I., Dulic, V., Lew, D. J., Richardson, H. E., and Wittenberg, C. (1992) G1 control in yeast and animal cells. *Ciba Found. Symp.* **170**, 7–15; discussion 15–19.

56. Visintin, R., Craig, K., Hwang, E. S., Prinz, S., Tyers, M., and Amona, A. (1998) The phosphatase Cdc14 triggers mitotic exit by reversal of Cdk-dependent phosphorylation. *Mol. Cell* **2**, 709–718.

57. Marelli, M., Aitchison, J. D., and Wozniak, R. W. (1998) Specific binding of the karyopherin Kap121p to a subunit of the nuclear pore complex containing Nup53p, Nup59p, and Nup170p. *J. Cell Biol.* **143**, 1813–1830.

58. Rine, J., Hansen, W., Hardeman, E., and Davis, R. W. (1983) Targeted selection of recombinant clones through gene dosage effects. *Proc. Natl. Acad. Sci. USA* **80**, 6750–6754.

59. Barnes, G., Hansen, W. J., Holcomb, C. L., and Rine, J. (1984) Asparagine-linked glycosylation in *Saccharomyces cerevisiae*: genetic analysis of an early step. *Mol. Cell. Biol.* **4**, 2381–2388.

60. Ishida, R., Hamatake, M., Wasserman, R. A., Nitiss, J. L., Wang, J. C., and Andoh, T. (1995) DNA topoisomerase II is the molecular target of bisdioxopiperazine derivatives ICRF-159 and ICRF-193 in *Saccharomyces cerevisiae*. *Cancer Res.* **55**, 2299–2303.

61. Winzeler, E. A., Shoemaker, D. D., Astromoff, A., et al. (1999) Functional characterization of the *S. cerevisiae* genome by gene deletion and parallel analysis. *Science* **285**, 901–906.
62. Giaever, G., Shoemaker, D. D., Jones, T. W., et al. (1999) Genomic Profiling of drug sensitivities via induced haploinsufficiency. *Nat. Genet.* **21**, 278–283.
63. Licitra, E. J. and Liu, J. O. (1996) A three-hybrid system for detecting small ligand-protein receptor interactions. *Proc. Natl. Acad. Sci. USA* **93**, 12817–12821.
64. Griffith, E. C., Su, Z., Turk, B. E., et al. (1997) Methionine aminopeptidase (type 2) is the common target for angiogenesis inhibitors AGM-1470 and ovalicin. *Chem. Biol.* **4**, 461–471.
65. Sin, N., Meng, L., Wang, M. Q., Wen, J. J., Bornmann, W. G., and Crews, C. M. (1997) The anti-angiogenic agent fumagillin covalently binds and inhibits the methionine aminopeptidase, metap-2. *Proc. Natl. Acad. Sci. USA* **94**, 6099–6103.

37

Tumor Suppressor Gene Therapy

Jack A. Roth and Susan F. Grammer

1. Introduction

The preceding chapters in these two volumes on tumor suppressor genes (TSG) are a comprehensive compilation of what is currently known about the structure, function, activation, and regulation of TSGs, along with the roles they play in the myriad biochemical pathways that network to result in living cells. With recent developments in molecular genetics, new information is revealed daily, and it is likely that we have only scratched the surface of the base of knowledge about TSGs that will soon be available. Even so, the studies reported thus far have already led to the identification of genetic alterations in TSGs that are responsible, in whole or in part, for cancer, and subsequently to the development and implementation of formidable therapeutic strategies.

As has been described in previous chapters, the blueprint contained in the DNA in each cell of an organism normally directs the cell to grow when growth is required, to cease division when appropriate, to self-repair when a minor defect in DNA is detected, or to self-destruct when the defect cannot be repaired. Many known TSGs play prominent roles in the web of biochemical pathways governing these cellular functions, and mutations or genetic alterations in one or more of them can result in a single cell proliferating out of control, leading eventually to the development of cancer. This knowledge, and the possibility that correcting a genetic alteration in one or more TSGs can either halt the unregulated growth or make a cell more sensitive to other methods of destruction, has led to rapid development of gene therapy strategies for cancer.

1.1. History of Gene Therapy

Until the actual cause of a disease is known, treatment of the disease is necessarily palliative. For example, humans once suffered and often died from bacterial infections treated only with very toxic therapies such as arsenic and other heavy metals. It was not until Henle showed that a specific type of bacteria causes a specific disease and Koch subsequently demonstrated that a small amount of infected blood could transmit a specific infection from animal to animal that any type of treatment specific to the etiology of the disease could be considered. The development of antibiotics, along with vaccination strategies to enhance the body's ability to prevent infectious diseases on its own,

From: *Methods in Molecular Biology, Vol. 223: Tumor Suppressor Genes: Regulation, Function, and Medicinal Applications.* Edited by: Wafik S. El-Deiry © Humana Press Inc., Totowa, NJ

revolutionized the treatment of infectious diseases with highly targeted minimally toxic treatments, but they could not have been developed without first identifying the cause of disease.

So it is with the host of diseases that are now known to be, or suspected to be, caused by defects in one or more genes, rather than by infectious agents. The etiologies of these diseases were locked away in black boxes until the field of molecular genetics evolved sufficiently to develop methods for isolation and characterization of genetic material from patients. New tools of molecular biology allowed cell biologists to examine the workings of cells on the DNA level in order to confirm what was long suspected—that many diseases are caused, in whole or in part, by genetic mutations. Once the genetic material could be manipulated on a routine basis, the notion of treating disease at its core—the gene—became feasible. In spite of subsequent rapid development of tools for gene therapy during the last decade, this revolutionary field is truly in its infancy.

Early gene therapy research targeted inherited monogenic diseases caused by alteration of one gene. Although it initially appeared that the relatively simple strategy of inserting and expressing a normal copy of a mutated or missing gene might cure these diseases, gene therapy has not been so simple. Viral vectors first used to transfer genes were inefficient and unstable, and also stimulated the immune system of the recipient, leading to unacceptable toxicity. Consequently, vector design has been a critical factor in the development of the field of gene therapy from its inception, and will continue to be for some time.

When the first human gene therapy clinical trials began in 1990, cellular DNA was manipulated outside the patient, or ex vivo, and cells were subsequently transplanted back into patients. Some of the first trials involved treating an inherited immunodeficiency, and treating children and adults with high serum cholesterol levels. By the mid-1990s, in-vivo trials of gene therapy to treat cystic fibrosis had been reported, along with Phase I and Phase II clinical trials for adenosine deaminase (ADA), Gaucher's disease, hemophilia, Duchenne muscular dystrophy, and sickle cell anemia *(1–4)*. Gene therapy research soon expanded into acquired diseases with a genetic basis, including cardiovascular disease and cancer, with cancer rapidly becoming the number-one target. According to the minutes of the March 8, 2001, meeting of the Recombinant DNA Advisory Committee (RAC), which advises the NIH Director on protocols involving recombinant DNA techniques, of the 409 clinical protocols for gene transfer therapy approved between 1989 and 2001, 280 were for cancer.

1.2. History of Cancer Gene Therapy

Thanks to the rapid emergence of the field of molecular biology, gene mutations—either inherited, acquired, or both—are now known to be the root cause of cancer. Identification of the actual genes responsible for the many types of cancer has proved challenging though, because cancers are nearly always caused by multiple genetic lesions occurring sequentially over time, and the affected genes vary from one tumor to another. Because of this, many researchers at first doubted the feasibility of treating cancer by targeting a single damaged gene. In addition, due to the unique profile of genetic alterations in each tumor, the possibility of finding one gene that would affect a large subset of tumors seemed unlikely. Quite to the contrary, cancer has become the number-one application of gene therapy in recent years, largely because a common thread links all cancers—damage to the genes that control the cellular machinery common to all mammalian cells. In nearly all cases of cancer, at least one genetic alteration turns out to be in a gene involved with regulating cell proliferation, cell cycle, signal transduction, DNA transcription, apoptosis, cell migration, or angiogenesis.

Two gene families are commonly implicated as causal factors in cancers—protooncogenes *(5)* and tumor suppressor genes *(6)*. Alterations in members of each of these gene families leads to disruption of normal biochemical pathways in cells. Oncogenes, first identified when their normal counterparts, protooncogenes, were transformed by retroviruses, were relatively easy to identify for further study. The term "protooncogene" actually encompasses all normal genes that, when manipulated, result in an allele capable of transforming a cell *(5)*. Generally, protooncogenes play prominent roles in signal transduction and transcription. Because the existence of tumor suppressor genes is most evident when they are missing from the genome, research in this area lagged behind that of oncogenes until the development of more advanced gene manipulation technology *(6)*. TSGs are generally critical in governing proliferation by regulating transcription and the cell cycle and have also been implicated in induction of apoptosis.

Despite initial misgivings about the potential for successful gene replacement therapy for cancer, it is now apparent that correction of just one lesion—either deletion of an oncogene or insertion of a normal copy of a TSG—can be sufficient to eradicate the malignant potential of a tumor cell.

1.3. Tumor Suppressor Genes

One of the first TSGs described, *RB* (the retinoblastoma gene), has been implicated in retinoblastoma, osteosarcoma, and carcinomas of the bladder, lung, and breast. The TSG *p53* has been associated with carcinoma of the bladder, lung, and breast, along with astrocytoma and osteosarcoma. Other TSGs, such as *WT1* (Wilms tumor), *DCC* (associated with colon carcinoma), and *NF-1* (neurofibromatosis) were also described early in the history of TSG research and reviewed in 1991 by Robert Weinberg *(6)* and by Michael Bishop *(5)*. More recently, genes such as *PTEN (7)*, *BRCA-1*, and *p16* are being recognized as TSGs (see earlier chapters in this series) and their potential therapeutic uses are being explored.

Recent developments in molecular genetics have provided a rapidly expanding literature base on TSGs, far surpassing that which investigators of only a few years ago expected. In reviews of the literature published only 10 years ago, investigators spoke with cautious optimism about how the detection of tumor suppressor genes might someday be applied to cancer. J. Michael Bishop, in his 1991 review *(5)*, inquired, "how soon and in what ways will the revelations of molecular genetics become useful in the care of individuals with cancer?" Answering his own question, Bishop ventured forth, "There has been talk of restoring functional copies of tumor suppressor genes to tumors in which they are defective, but realization of this objective (if it is ever to come at all) seems many years distant." Detailed understanding of the roles that TSGs and their protein products play in normal cellular function, acquired in the ensuing decade and described in detail in these volumes, has allowed TSGs to make the transition from the laboratory bench to the patient's bedside in a few short years.

2. Clinical Application of TSG Replacement Therapy

2.1. Role of TSGs in Normal Cellular Function

The myriad roles TSGs play in normal cell function were described previously in this volume. Briefly, though, expression of growth factors, oncogenes, cyclins, and cyclin-dependent kinases (CDKs) drive the cell cycle toward proliferation. Expression of TSGs and other inhibitors of CDKs induce cell cycle arrest when appropriate. Two major G1 cell cycle-arrest pathways, the RB pathway and the P53 pathway, are tightly interwoven and regulated at the protein level by other TSGs and by oncogenes. In general, mammalian cell proliferation is under the control of these two pathways, with the Rb protein regulating maintenance of, and release from, the G1 phase, and the p53 protein monitoring cellular stress and DNA damage and effecting either growth arrest and repair or progression to apoptosis (8). When either of these interwoven pathways is disrupted by genetic damage to a single component, the cell, to avoid proliferating out of control, must find alternative means of regulating proliferation.

2.2. Rationale for Replacing TSGs in Cancer

Targeting specific genetic lesions responsible for tumorigenesis and cancer progression is an attractive strategy for developing more effective anticancer therapeutics and reducing treatment-related toxicity. To be successful enough to undergo trials for clinical benefit, though, a gene therapy strategy must first achieve three goals. Initially, the gene must be successfully delivered to the appropriate tissue; second, the gene must be relatively stably expressed in the target tissue; and finally, expression must be regulated. In the case of cancer, long-term expression is not generally necessary, as the goal of therapy is destruction of the tumor cells.

Successful transfer and expression of several TSGs has been documented in preclinical and clinical studies in several types of cancer. Although most cancer TSG gene therapy clinical trials to date have been Phase I trials, designed to document gene expression and toxicity data, several have already provided evidence of clinical benefits to study participants and given rise to larger Phase II trials, including several multicenter studies involving hundreds of patients.

As listed in the March 8, 2001, report of the RAC, (9) 25 protocols involving transfer of TSG DNA alone or in combination with conventional therapy were registered between 1989 and March 2001. Of these 25, five study protocols for *BCR-1* gene transfer in ovarian cancer were approved, along with 3 for *p16* (all prostate cancer) and one for the *Rb* gene. *p53* gene transfer was the most common TSG protocol approved (16 studies), including protocols for bladder (1), brain (1), breast (3), (liver (1), colon (1), prostate (1), general advanced cancers (1), and lung cancer (7).

2.2.1. TSGs in Combination with Conventional Therapy

In the case of many cancers, conventional treatments are not adequate for a majority of patients, because their tumors are resistant to chemotherapy and radiation, both of which target rapidly dividing cells by damaging DNA. The mechanism of action of

DNA-damaging agents has been linked to the induction of apoptosis—the cell must be able to detect the DNA damage and direct its own suicide through apoptotic pathways. Interestingly, the "guardian of the genome," the molecule that is responsible for detecting damage to DNA and either directing repair or destruction through apoptosis, is a tumor suppressor gene, *p53*. Early studies of *p53* gene transfer indicated that cells transduced in vitro with the gene became capable of inducing apoptosis, thereby destroying themselves and surrounding cells. This has also led to studies to determine whether this link to apoptosis might make TSG gene therapy beneficial in combination therapies with conventional DNA-damaging agents.

After several years of clinical trials, it appears that the rationale for using TSG gene replacement therapy in cancer is valid. Because most of the TSG gene therapy clinical data to date arises from *p53* gene replacement strategies, the following discussion of the medicinal applications of TSGs will focus on *p53*. Other TSGs, however, either alone or in combination with *p53*, have also shown promise as potentially beneficial therapeutic strategies.

2.3. Clinical Trials of p53 Gene Replacement

2.3.1. Rationale

Over 50% of cancers exhibit alteration of the TSG *p53*. Inactivation of this gene through deletion or mutation correlates with tumorigenesis, so it follows that replacement of a copy of a "wild-type" (wt; nonmutated) gene might restore normal cell growth and regulation of proliferation. In addition, *p53* in the normal cell is tasked with detection of damaged DNA and, when repair is not possible, induction of apoptosis (or "cell suicide") to ensure that the genetic defect is not carried on into daughter cells. For these reasons, replacing this particular gene has the potential to restore the normal apoptotic pathway to tumor cells, enhancing their sensitivity to conventional therapeutic strategies that require a cell to undergo apoptosis.

Normally, when DNA is damaged, p53 will cause G1 arrest until the damage is repaired, or will trigger apoptosis *(8)*. Loss of cell-cycle regulatory function, loss of the ability to self-repair when DNA is damaged, or unchecked expression of molecules normally regulated by the p53 protein, might make a cell more susceptible to transformation. It follows, then, that replacement of the *p53* gene and subsequent expression of p53 might allow induction of apoptosis, induction of tumor dormancy, or even prevention of progression of premalignant cells to the malignant phenotype. Restoration of p53 function to a cell also has the potential to suppress the tumorigenic influence of mutations in other genes critical to the functional program of the cell. For example, restoration of p53 activity might also correct a breakdown in the pathway caused by mutation of a different gene upstream of *p53*, obviating the need to develop therapy targeting that gene. Also, as discussed in the previous section, evidence is strong for a link between p53 expression and apoptosis *(10–12)*. Clinical trials reported to date in non-small-cell lung cancer (NSCLC) and head and neck cancer treated with wild-type p53 TSG replacement have consistently shown evidence of gene transduction and expression, mediation of apoptosis, and clinical responses including pathologic complete responses. More recent trials have suggested that gene transfer of p53 in combination with conventional strategies using DNA-damaging agents confers sensitivity to radiation and chemotherapy *(13)*.

2.3.2. Early Clinical Trials of p53 Gene Replacement Using Retroviral Vectors

Early gene therapy trials were carried out with a retroviral vector carrying wt-p53 cDNA driven by the actin promoter *(14)*. The vector was delivered to lung tumors by direct intratumoral injection. Tumors were biopsied before and after wt-p53 therapy and specimens were examined for evidence of apoptosis. In 6 of 9 cases, TUNEL staining (evidence of apoptosis) was greater in posttreatment biopsies than in pretreatment specimens, indicating that apoptosis had been induced by the treatment protocol. In addition, the percentage of cells staining positive in the TUNEL assay was greater than the percentage of cells expressing vector DNA in most specimens, indicating a bystander effect. Three of the 9 patients in this trial exhibited regression of tumors, and in 2 cases, no evidence of viable tumor was obtained 4 wk after administration of the retroviral vector. This study demonstrated, for the first time in a clinical setting, that replacement of a tumor suppressor gene can mediate tumor regression, and thus provided critical proof-of-principle. Importantly, vector-related toxicity was minimal and a putative mechanism of action of TSG therapy, apoptosis, was established.

2.3.3. p53 Delivery by Adenoviral Vectors

Retroviral vectors have limitations in gene therapy protocols in that they can transduce only dividing cells and are difficult to produce in the high titers required for gene therapy strategies. Adenoviral vectors, on the other hand, can transduce both dividing and nondividing cells, achieve higher expression levels, and can be produced with high titers and in large scale.

High-level gene transfer of p53 by an adenoviral vector into lung cancer cells was demonstrated by Zhang *(15)* and, based on these results, a Phase I clinical trial with 28 NSCLC patients whose cancers had not responded to conventional treatments was initiated *(16)*. Successful gene transfer was demonstrated during this trial by the presence of vector DNA in 80% of the evaluable patients. Vector-specific p53 DNA was detected in 46%, and apoptosis was demonstrated in all but one of the patients expressing the gene. Importantly, despite up to 6 injections per patient, there were no significant toxic effects related to the vector. Although this Phase I trial was designed to assess toxicity, gene transfer, and gene expression, antitumor activity was also observed; 2 patients achieved greater than a 50% tumor size reduction, and 1 patient remained free of tumor more than a year after the conclusion of therapy. One patient experienced nearly complete regression of an upper-lobe endobronchial tumor that had resisted chemotherapy, radiotherapy, and laser treatment.

In a Phase I study of 33 patients with bulky head and neck squamous cell carcinoma (HNSCC) *(17)*, the Adp53 construct was also found to cause little toxicity, and significant clinical response was again observed in 9 of 18 clinically evaluable patients. Interestingly, systemic Adp53 DNA was present transiently for under 48 h as demonstrated by presence in blood, urine, and sputum.

In Phase II studies of HNSCC with Adp53, over 200 patients were enrolled. Approximately 10% of patients with both recurrent or refractory head and neck cancer achieved complete or partial responses *(18,19)*. If patients with prolonged inhibition of tumor growth are included, then 60% of patients showed evidence of antitumor activity. The low toxicity seen with Adp53 administration, with less than a 5% incidence of serious adverse events, suggests that it can be readily combined with other anticancer treatments without significant increases in treatment-related toxicity *(20)*.

2.4. p53 Gene Therapy in Combination with Conventional Treatment

2.4.1. p53 and DNA-Damaging Chemotherapeutic Agents

As noted previously, many cancer patients fail conventional therapy because their tumors are resistant to DNA-damaging agents such as chemotherapy and radiation therapy. When apoptosis was implicated as the normal mechanism of cell destruction in response to these DNA-damaging agents, it followed that a defect in the normal apoptotic pathway might confer resistance to some tumor cells.

In-vitro studies *(10–12)* demonstrated that overexpression of p53 in cell lines transfected with wild-type p53-expressing plasmids could drive cells into apoptosis. Subsequent studies of apoptosis in tumor cells treated with chemotherapy agents or radiation in vitro or in animal models *(21–26)* supported a link between functional p53 and apoptosis induction, leading to the initiation of clinical trials combining Ad-p53 with the DNA-damaging agent, cisplatin.

Nemunaitis and colleagues *(27)* initiated a Phase I trial of *p53* gene transfer in sequence with cisplatin in 24 patients with non-small-cell lung carcinoma previously unresponsive to conventional treatments. Intravenous cisplatin was administered, and 3 d later *p53* was delivered by intratumoral injection; up to a total of 6 monthly courses were carried out. Seventeen patients remained stable for at least 2 mo, 2 achieved partial responses, and 4 continued with progressive disease. One patient was unevaluable due to progressive disease. When tumor biopsies were analyzed for apoptosis, 14% demonstrated no change, 7% showed a decrease in apoptosis, and 79% demonstrated an increased number of apoptotic cells. Of note is that 75% of the patients entered in the trial had previously demonstrated tumor progression on cisplatin or carboplatin-containing regimens.

2.4.2. p53 Gene Replacement in Combination with Radiation Treatment

Preclinical studies also suggested that p53 gene replacement might enhance the effect of radiation on some tumors *(28–32)*, leading to the initiation of a Phase II clinical trial of adenoviral mediated *p53* gene transfer in conjunction with radiation therapy. Preliminary data from this study *(33)* revealed that, of 19 patients with localized NSCLC, a complete response (CR) was achieved in one patient (1/19; 5%), partial response (PR) in 11 patients (11/19; 58%), stable disease (SD) in 3 patients (3/19; 16%), and progressive disease (PD) in 2 patients (2/19; 11%). Two patients (2/19; 11%) were nonevaluable because of tumor progression (patient #18) or early death (patient #6). Pathologic examination of biopsies 3 mo following completion of therapy revealed no viable tumor in 12 patients (12/19; 63%) and viable tumor in 3/19 (16%). Tumors of four patients (4/19; 21%) were not biopsied because of tumor progression (patients #10 and #18), early death (patient #6), or weakness (patient #19).This rate compared favorably to the 17% reported in studies of chemotherapy combined with radiation therapy *(34)*.

These results are encouraging and are the basis for a randomized clinical trial in patients with unresectable NSCLC that will compare a conventional therapeutic strategy of concurrent radiation and chemotherapy, with concurrent chemotherapy and radiation in conjunction with intratumoral injections of adenoviral *p53*.

2.5. Future Directions in TSG Gene Therapy

It is clear that, in spite of the promise apparent in these early trials, the current approach to gene therapy of cancer with TSG can be improved further. Promising avenues for investigation include improvement of gene delivery systems, enhanced induction of bystander effects, design of immunogene and antiangiogenesis gene therapies, and further adjuvant use of gene therapy with conventional chemotherapy, radiation therapy, and surgery.

2.5.1. Vector Development

Promising approaches to gene therapy used to date in clinical studies have utilized retroviral, adenoviral, and herpes vectors for gene delivery. However, due to various complications caused by injecting viral material with the genetic material, new modes of delivery using nonviral vectors and naked DNA are being explored.

2.5.2. Tissue Targeting

One readily apparent limitation to gene therapy thus far is the nonspecificity inherent in current gene delivery methods. So far, only tumors accessible by needle or by endoscopy are viable targets for gene therapy—tumors that are unreachable or undetectable are excluded by current gene therapy delivery methods. Thus far, recombinant adenovirus has proven the most efficient at systemic delivery of genes to less accessible tissues, such as heart and skeletal muscle. Several strategies have been tried to target recombinant adenovirus vectors to specific cell types by manipulating cell surface-binding properties of the viral particle.

One approach involved the use of a fusion protein consisting of an antibody fragment specific to the fiber protein of the virus and a ligand, in this case epidermal growth factor *(35)*. This approach uses the bifunctional fusion protein to block normal fiber binding to the coxsackie-adenovirus receptor, with concomitant targeting of the virus to the EGF receptor. Coated adenovirus particles could then target via the EGF receptor in vitro. Addition of the fusion proteins to the adenovirus enhanced the transduction efficiency of the epidermoid carcinoma cell line A431 16-fold at optimal conditions compared with infection with native adenovirus vector *(36)*. The results of this study indicated that this modification of the adenovirus could both block the native viral tropism and target gene delivery specifically to the EGF receptor.

2.5.3. Combining TSG Gene Therapy with Other Gene Therapy Strategies

Recent reports indicated that complementary tumor suppressor genes, delivered together, can cooperate to induce apoptosis. Combinatorial introduction of the tumor suppressor gene p16INK4 and wt-p53 demonstrated a synergistic effect on the induction of apoptosis in HuH7 hepatocellular carcinoma cells (mutated p53) and LOVO colon carcinoma cells (very low expression of wt-p53) in vitro. P16INK4, an inhibitor of cyclin-dependent kinase 4, indirectly regulates the activity of the tumor suppressor gene Rb. Overexpression of p16INK4 can induce cell cycle arrest, but only if the cell contains functional Rb protein *(37)*. The mechanisms by which p16INK4 and p53 cooperate to induce apoptosis are not yet fully understood.

3. Ethical Issues, Safety, Education and Debate

3.1. Ethical Issues

During the year before this chapter was written, two events of paramount importance to the future of health care and to gene therapy strategies in particular occurred. The first, the successful sequencing of the human genome, offers numerous possibilities for identification of targets for gene therapy and for the efficiency with which proven gene therapy strategies can be applied to new diseases. The second event was the restoration of normal immune systems to two children with severe combined immune deficiency disease (SCID) following infusions of stem cells containing the corrected gene *(38)*. This report, for the first time, provided proof of principle for the complete reversal of a disease state by gene therapy, prompting a vision for the future in which genetic analysis and gene therapy can become a standard practice in health care.

As these technologies are becoming a reality, it is a critical time for researchers, clinicians, patients, and the public to address ethical issues that loom on the horizon. Before genetic analysis and gene therapy can become standard health-care tools, issues surrounding privacy must be addressed and resolved. If employers and insurance companies have the option to refuse employment or insurance coverage to patients carrying a specific gene, gene therapy will never become economically feasible. The early consideration of these aspects by medical ethicists have led to laws already passed, and others in development, prohibiting discrimination by insurance providers and employers at this point.

3.2. Safety Issues

The safety of research study participants, and eventually of gene therapy patients, is of paramount importance, not only to individual patients, but also to the long-term future of gene therapy. Early studies demonstrated an excellent safety profile, and many gene therapy protocols currently in clinical trials have already proved safe enough to be administered on an outpatient basis. In spite of this excellent record, though, the first death clearly attributable to a gene therapy trial protocol occurred in the fall of 1999, when 18-year-old Jesse Gelsinger received systemic adenovirus at a dose that had already been tolerated by two other patients, and died. As *Nature Medicine*'s January 2000 editorial, "Gene Therapy—A Loss of Innocence," reads: "By openly discussing the limitations and problems of the technology, the gene therapy community has been able to carry a general acceptance and at times enthusiasm for a technique that has for a decade failed to deliver. But now that same community must also publicly acknowledge the first death clearly attributable to the experimental therapy, and a willingness to take on board the lessons that this first gene therapy tragedy offers" *(39)*. The salient point of the editorial is that it is "paramount that a single central agency with the resources and expertise to interpret the data, receive reports of all adverse events" to avoid missing the presence of rare but dangerous adverse events that will not be experienced and therefore noted in every clinical trial. Gelsinger's death prompted increased concerns over the documentation and reporting of all adverse events.

Common toxicity criteria guidelines are now accessible on-line *(40)*, and guidelines for reporting adverse events must be followed by all investigators receiving funding from the National Institutes of Health. Reports of side effects are submitted to the Office of Biotechnical Affairs (OBA) and to the RAC and the results of the investigations are made public. For example, of the 206 "serious or unexpected" reports to the OBA discussed in a March 2001 *(41)* RAC report, 160 were initial reports and 46 were follow-ups. The report showed that a clinical study participant had received *p53* gene transfer for ovarian cancer and died a week later. Although this was initially reported as "serious" when the preliminary autopsy noted severe peritonitis "possibly related" to the treatment, final autopsy attributed death to the extensive metastatic carcinoma, changing the event from "possibly related" to "unrelated." The careful monitoring of these events in order to establish the safety of gene therapy technologies can only help the gene therapy community to achieve a smooth transfer of technology from the laboratory bench to the bedside.

3.3. Public Perception of Gene Therapy

The future of gene therapy depends not only on good science and the ability to develop and test the necessary technology for gene transfer, but also on the acceptance of this new and sometimes intimidating technology by the public. Many consumers are currently wary of any type of genetic manipulation, as is evidenced by the public outcry against genetically modified crops and the controversy over the ethics of screening of in-vitro fertilized embryos for genetic diseases before embryo transfer, and the practice of conceiving children with the intention of transferring some of their healthy bone marrow to a terminally ill older sibling. This aura of discomfort is likely to linger until medical science has proven, not only to other scientists but also to the public, that responsible application of the potential benefits of gene technology is a far more likely scenario than one in which the technology is applied in an unethical way. In spite of the obvious promise evident in the results of the clinical trials of TSG gene transfer described in the previous section, it is critical to recognize that there are still gaps in knowledge and technology that must be filled before the most finely tuned gene therapy strategies can emerge. The most efficient method of successfully delivering the gift of gene therapy to health-care consumers will be to gain public acceptance of the technology through education in basic molecular genetics and its applications, through release of accurate information (both positive and negative) from clinical trials, and through honest discussion of the limitations—as well as the promises—of gene therapy. An official step in this direction was taken in 1974 when the Recombinant DNA Advisory Committee (RAC) was formed in answer to public concerns about the safety of gene manipulation. One of the RAC's early tasks was to develop and publish mandatory guidelines for investigators at institutions receiving NIH funds for research involving recombinant DNA. Thirty percent of the members of this committee do not have scientific expertise, but represent public interests or attitudes, and the periodic meetings of the RAC are open to the public. In the beginning one of the RAC's responsibilities was to review all gene therapy protocols. Later the committee functioned in more of an advisory than a regulatory capacity. An NIH Working Group on NIH Oversight of Clinical Gene Transfer Research recommended in July 2000 *(42)* that the RAC should once again take on a more regulatory function. They also noted:

> Because risks are not fully known, and unanticipated harms might occur, trial monitoring and oversight must be exemplary and investigators and institutions must be held accountable for this oversight. In addition, those agencies and institutions entrusted with this responsibility must work together to ensure proper oversight.

Public support for gene therapy was high as clinical trials began in the early 1990s. But because results were slow in coming, confidence that the field would eventually deliver what had been promised fell. In addition, negative publicity about genetically modified crops and genetic testing of in-vitro-fertilized embryos was more conspicuous than publicity from groups with the goal of responsibly educating the public in basic genetics and in ethics.

Even those who enthusiastically supported the theoretical potential of gene manipulation in curing disease realized by the mid-1990s that there were technological barriers still impeding progress. In 1995, a panel appointed by the director of the NIH to "assess the current status and promise of gene therapy and provide recommendations regarding future NIH-sponsored research" reported:

> ... while the expectations and the promise of gene therapy are great, clinical efficacy has not been definitively demonstrated at this time in any gene therapy protocol, despite anecdotal claims of successful therapy and the initiation of more than 100 RAC-approved protocols ... significant problems remain in all basic aspects of gene therapy ... basic studies of disease pathophysiology, which are likely to be critical to the eventual success of gene therapy, have not been given adequate attention.

The report also cautioned investigators and research sponsors

> ... not to oversell their results, lest the public continue the mistaken and widespread perception that gene therapy is further developed and more successful than it actually is. Such inaccurate portrayals threaten confidence in the integrity of the field and may ultimately hinder progress toward successful application of gene therapy to human disease.

Among the panel's many recommendations were (a) increase the emphasis on research dealing with the mechanisms of disease, (b) develop animal models of disease for preclinical gene therapy studies, and (c) adhere strictly to high standards of excellence in clinical protocols for all investigators. The NIH also urged scientists, clinicians, science writers, research advocates, institutions, industry, and the press to "inform the public of both the extraordinary promise of gene therapy, and its current limitations."

A 1996 article in *Scientific American (43)* described the unforeseen complications that produced roadblocks in early trials of gene therapy, but also suggested that the tide was turning and that "gene therapy may emerge a winner in this round, though the match will likely draw on for years to come." Soon, several successful Phase I trials were published and gene therapy moved forward rapidly for a few years. Too rapidly, in the eyes of some.

The unfortunate October 1999 death of 18-year-old gene therapy trial participant Jesse Gelsinger, however, quickly (and fittingly) heralded a new level of concern among both professionals and the public. Media reports suggested that, although safety of the adenoviral vectors had been well established in many trials in which they were delivered directly to tumors, investigators might have moved too quickly into trials involving systemic delivery. At the same time, concerns were raised in the media that some researchers and companies had been less than forthcoming with information about adverse events observed in clinical trials, as well as deaths of trial participants judged by investigators to be unrelated to the therapy protocol. Because patients undergoing experimental gene therapy are often in the late stages of their diseases, some researchers apparently did not report deaths that they judged were caused by the patient's underlying disease. News articles that soon followed, such as "Gene Therapy, Touted as a Breakthrough, Bogs down in Details" *(44)*, once again brought the more frightening aspects of gene therapy to the forefront for the public, and the resulting climate was one of criticism for most gene manipulation experiments.

Even top researchers in the field became cautious, realizing the implications of gene manipulation and the need for examination of emerging technologies. In "A Cure That May Cost Us Ourselves," published in *Newsweek* in January 2000 *(45)*, W. French Anderson, a pioneer of human genetic engineering, predicted, "within 30 years, there will be a gene-based therapy for most diseases." He also stated that he "feared the profound dangers" of his own work. An editorial in *Nature Medicine*, "A Loss of Innocence" *(39)*, maintained that, "by accepting the lessons that this death teaches, the gene therapy community could emerge stronger and better-prepared to advance this emerging field of medicine."

In response to these events, The Advisory Committee to the Director, Working Group on NIH Oversight of Clinical Gene Transfer Research (The Working Group) was established by NIH Director Harold Varmus. In July 2000 The Working Group issued recommendations *(42)* to change the process of recombinant DNA clinical trial protocol review and to increase professional and public education. They also reached a consensus on the need for a system of reporting adverse events, resulting in guidelines for reporting events to the NIH and FDA *(46)*.

3.4. Education

The Internet has made access to information on cancer clinical trials widely available to private physicians and to potential patients. In addition, the NIH Guidelines for Research Involving Recombinant DNA Molecules: RAC Guidelines (revised January 2001), prepared by the RAC, can be accessed on-line by physicans, scientists, patients, and the public *(47)*, and the RAC database listing all approved clinical trials, as well as minutes of their review meetings, are available on the NIH Office of Biomedical Affairs OBA/RAC website *(48)*. In addition, reports of adverse effects reported to the OBA are now accessible to other investigators and to the public in the Reports section of the RAC website *(49)*.

Information on availability of and participation in cancer clinical trials is increasingly more accessible to private physicians and to potential patients through various Internet sites, including the National Cancer Institute's "CancerTrials" website *(50)*, where links for physicians, researchers, and patients can be found. In addition, most cancer centers with ongoing clinical trials provide patient and physician education on their websites, several of which are listed under **Subheading 4.6.**

This is a critical time to educate the public about general genetics, and about the potential of the field of gene therapy to improve length and quality of life for all healthcare consumers. Education projects such as the NIH education kit, distributed free of charge on the Internet in 2001, will have enormous impact on the public perception of the potential usefulness of gene therapy. Only education of individuals will facilitate the exchange of ideas between the public, the medical community, and governing bodies as they work together to decide how best to use this new technology in an ethical, responsible, and fair manner. Public debate can only be effective when all the cards are on the table. The most flawless gene therapy technologies will be of little use to a public who does not understand, and is therefore not willing to accept, its gifts.

4. Clinical Trials: Design, Development, and Implementation

4.1. Gene Therapy Clinical Trials: Unique and Novel Issues

The first human gene transfer protocol was approved in 1988, and the number of protocols approved in subsequent years has increased dramatically—from 2 in 1990, to 43 in 1995, and then to 99 in 1999, with 70 more added in 2000. By February 2000, more than 4000 patients had participated in gene therapy studies, and of the 372 gene transfer clinical trials registered with the NIH at that time, 89% were Phase I studies, 10% were Phase II, and only 1% (3 protocols) had progressed to Phase III trials *(51)*. Although a majority of the trials are still at the stage of assessing toxicology and pharmacology, the number of potential applications of recombinant DNA technology to clinical medicine can only be expected to expand rapidly as successful outcomes are reported.

The transfer of recombinant DNA to humans in the last decade has raised new and complex scientific, medical, ethical, and social issues that justify enhanced oversight by regulatory agencies. For example, transfer of "foreign" DNA carries several theoretical risks, including risk of inadvertent transfer to reproductive cells, resulting in inheritable genetic changes, and risk that the technology could be used unethically. In addition, the most effective currently available technology for transferring DNA involves a backbone of viral DNA and, although all such viral vectors are disabled, there is still a risk that these molecules could cause limited infectious disease, exposing trial participants and their families to unexpected dangers. Also, many gene therapeutics are likely to require repeat administration over a long period of time in order to assure continuous expression of the therapeutic gene product, leading to risk of toxicity related to an immune response.

Because of enhanced real and theoretical risks, gene transfer clinical trials are subject to more regulatory and advisory oversight than many experienced clinical investigators may be used to encountering. This should not be perceived as a potential impediment to the progress of gene therapy research, but as a method of ensuring rapid progress in the development of gene therapy strategies that reflect safe, ethical, scientifically sound, and socially responsible medicine, fully primed to take its place in the standard of care for gene-based diseases, as soon as clinical trials are completed.

4.2. Clinical Trials: General Process and Structure

The following discussion of the general components of clinical trials is not meant to stand on its own as a handbook for developing a clinical trial. Instead, it is intended as a guide to electronic resources useful to investigators during the development of new clinical trial protocols, and to physicians seeking information on open clinical trials for their patients.

During recent decades, faced with rapid developments in all fields of medicine and the attending explosion of new potential treatment strategies, the medical community became convinced of the need for a formal, structured process for assessing all new therapies *(52)*. Structured and regulated clinical trials were born from the need to draw accurate conclusions as to whether a new treatment could be expected to help patients. Initial therapeutic attempts must now be designed, performed, analyzed, and reported according to guidelines for officially designated clinical trials. Generally, the first goal of assessing a potential new therapy is definition of the toxicology and pharmacology of the agent. Phase I clinical trials, during which the agent to be tested is administered in increasing doses to a small number of volunteers, usually at a single medical center, accomplishes this goal. Phase I trials for most drugs have, historically, been first carried out in normal, healthy volunteers, as the goal was to determine whether the drug was toxic and how it was processed normally. Clinical trials for cancer, however, have always involved drugs and procedures with a high level of risk and toxicity, so Phase I cancer clinical trials generally involve patients with a confirmed malignant disease for which there is no standard treatment or that has not satisfactorily responded to conventional therapy. In addition, the patient selection criteria outlined in a trial protocol normally dictates that the study participant must have normal organ function, an acceptable "performance status," and be expected to survive a sufficient period of time to allow observation of any potential toxic effects. Because the patient population in Phase I studies for cancer treatments is diseased, these trials can often provide some preliminary data regarding not only safety, but efficacy, with demonstration of measurable antitumor effects quite common, however rarely statistically significant due to the small sample size. In spite of their qualitative rather than quantitative nature, these observations can help to direct development of Phase II studies.

Phase II trials to assess whether an experimental protocol leads to a therapeutic effect often involve hundreds of patients at multiple medical facilities. Phase II trials should be painstakingly designed by a collaborative team including a member with expertise in statistical analysis.

Therapeutic benefits demonstrated in Phase II trials must then be compared, to benefits achieved with the current standard of care for the disease. In order to make any statistically significant difference between two beneficial therapies detectable, the sample size of Phase III clinical trials is large, often including hundreds or thousands of patients, at hundreds of different sites. Phase III trials are often sponsored by government agencies or drug companies.

In all levels of clinical trials there are critical features to be cognizant of when designing the protocol, including patient selection criteria, sample size, vector selection, requirements for biologic containment, public heath considerations, and informed consent. Clinical studies involving recombinant DNA must address these criteria during the FDA application process by answering the questions raised in Appendix M, "Points to Consider," of the NIH Guidelines *(53)*.

4.3. Gene Therapy Clinical Trials in the United States

Several government agencies have jurisdiction over all clinical trials in the United States involving recombinant DNA technology. The FDA and the NIH work together to oversee human gene transfer research, but have separate responsibilities. The FDA has statutory authority over approval of all Investigational New Drugs (INDs) that are intended for eventual distribution in the United States, while the NIH, through its Recombinant DNA Advisory Committee, conducts public scientific and ethical reviews on novel applications of gene transfer.

4.3.1. Negotiating the "Red Tape"

Clinical investigators intending to initiate gene therapy clinical trials with new or existing recombinant DNA constructs can access, on-line, the required regulatory agency paperwork, from protocol submission forms and instructions to documents required for follow-up of reports of adverse events occurring during the trial. In addition, resources helpful in navigating the maze of regulatory requirements are available, with updates posted periodically (see **Subheading 4.6.**).

A 1999 letter to all gene therapy Investigational New Drug (IND) Sponsors and Principal Investigators (P.I.) from the FDA's Center for Biologics Evaluation and Research (CBER) *(54)* outlined the appropriate process for submission of a gene therapy IND Application and any subsequent Adverse Event Report forms.

a. *Investigator submits NIH Guidelines, Appendix M-I, "Points to Consider in the Design and Submission of Protocols for the Transfer of Recombinant DNA Molecules into One or More Human Research Participants"* **(53)**, to the NIH/ORDA. Submission of Appendix M prior to step 2 will help to ensure that any novel issues requiring full public discussion will be added to the RAC agenda and be discussed in a timely manner.

b. *Investigator submits IND application to the FDA.* The IND application process is outlined on the FDA/CDER (Center for Drug Evaluation and Research) webpage *(55)*. A flowchart outlining the FDA review process for the IND, linked to a handbook for investigators, is also available on-line *(56)*, along with "Guidance for Institutional Review Boards and Clinical Investigators" *(57)*.

c. On receipt of the gene therapy IND, the FDA will notify NIH/ORDA, in order to assist NIH/ORDA in verifying compliance with the NIH Guidelines, and will complete review within 30 days of receipt.

d. Within 15 days after receipt by the NIH/ORDA of a complete submission, the RAC will make a determination regarding whether the protocol is novel and therefore requires public RAC review, or is exempt from review.

e. Once the RAC makes a decision regarding the need for full public review of a gene therapy protocol, the NIH will notify FDA within 1 working day.

f. If the RAC judges that full public review of the protocol is warranted, the FDA will request that the sponsors delay initiation of the protocol until RAC review process is complete.

g. Investigators and trial sponsors are required, under the U.S. *Code of Federal Regulations*, Title 21, part 312.32, "IND Safety Reports" and under the NIH Guidelines, Appendix M-VII-C "Adverse Event Reporting," *(53)* to report all serious adverse events to both the NIH and the FDA.

4.4. Regulatory Issues: The FDA

The FDA is tasked with protecting the public health by assuring that safe and effective medical products reach the public in a timely way. The agency is also tasked with monitoring those products for continued safety once they are in use, making the job of the agency a "blending of law and science" *(58)*. All products used by, or prescribed by, health professionals and marketed in the United States must be approved for sale and distribution by the FDA. The IND required for FDA approval of clinical trial protocols is essentially an application for exemption from FDA rules prohibiting distribution of non-FDA-approved drugs. Once clinical trials demonstrate the appropriateness of a new therapy, the FDA remains the oversight agency, helping to bring the drug to market and then monitoring its safety.

Regulation of human gene therapy products, including clinical trials, falls under the responsibility of the FDA's Center for Biologics Evaluation and Research (CBER). The CBER uses both the Public Health Service Act and the Federal Food Drug and Cosmetic Act as statutes for oversight.

4.5. Regulatory Issues: The NIH

The NIH established the Recombinant DNA Advisory Committee (RAC) in 1974 to consider the current state of knowledge and technology regarding DNA recombinants, their survival in nature, and their transferability to other organisms, and their societal impact to recommend guidelines for the safe and ethical conduct of recombinant DNA research. The RAC authored the *NIH Guidelines for Research Involving Recombinant DNA (47)*, which specify practices for constructing and handling recombinant DNA molecules, and organisms and viruses containing recombinant DNA molecules.

4.6 Who Must Comply with NIH Guidelines?

Compliance with the NIH Guidelines is mandatory for investigations conducted at or sponsored by any institution that receives any NIH support for recombinant DNA research, regardless of whether the protocol in question is funded by the NIH. Researchers subject to the NIH Guidelines are required to submit reports of serious adverse events to the NIH and the FDA immediately *(46)*.

Noncompliance may result in termination of the protocol if it is funded by the NIH, and, regardless of NIH funding status, in termination of funds for other recombinant DNA research ongoing at the institution, and a future requirement for prior NIH approval of any or all recombinant DNA projects at the institution.

Information concerning noncompliance with the NIH Guidelines may be reported by anyone and should be delivered to both NIH/OBA and the relevant institution. The institution, generally through the Institutional Biosafety Committee, will then take appropriate action and forward a complete report of the incident, recommending any further action, to the Office of Biotechnology Activities, National Institutes of Health.

Researchers not required to adhere to NIH Guidelines are encouraged to do so voluntarily by registering their protocols with the RAC and filing adverse event reports with the FDA and NIH.

4.7. Additional Electronic Resources

- *Code of Federal Regulations*, Title 45: Public Welfare, Department Of Health And Human Services, National Institutes Of Health, Office for Protection from Research Risks, Part 46: Protection of Human Subjects, http://ohrp.osophs.dhhs.gov/humansubjects/guidance/45cfr46.htm
- FDA Center for Biologics Evaluation and Research (CBER), http://www.fda.gov/cber/index.html
- Guidance for Institutional Review Boards and Clinical Investigators, http://www.fda.gov/oc/ohrt/irbs/default.htm
- Investigational New Drug Application Review Process, http://www.fda.gov/cder/handbook/ind.htm
- NIH/OBA Recombinant DNA and Gene Transfer, http://www4.od.nih.gov/oba/RDNA.htm (includes links to Latest News, RAC Roster, NIH Guidelines, Meetings and Conferences, Documents and the Human Gene Transfer Clinical Trials Database)
- NIH Guidelines, Appendix M. Points to Consider in the Design and Submission of Protocols for the Transfer of Recombinant DNA Molecules into One or More Human Research Participants, http://www4.od.nih.gov/oba/RAC/guidelines/appendix_m.htm

- NIH/OBA Human Gene Transfer Clinical Trials Database—Phase I, http://www4.od.nih.gov/oba/rac/clinicaltrial.htm
- RAC Documents Page, http://www4.od.nih.gov/oba/rac/documents1.htm (includes links to NIH Guidelines and Appendices, RAC Protocol List, Federal Registers, Principal Investigator List, RAC Roster, Serious Adverse Event Reporting Format, Reports from the RAC, Quarterly Data Management Reports, Amendments and Updates to Guidelines and appendixes)
- RAC Charter, http://www4.od.nih.gov/oba/rac/RACCharter.htm
- The Cancer Therapy Evaluation Program (CTEP), http://ctep.info.nih.gov/aboutctep/default.htm (CTEP is within the Division of Cancer Treatment and Diagnosis [DCTD] of the NCI; it plans, assesses, and coordinates all aspects of clinical trials)
- CTEP Investigator's Handbook: Part B—The Development of a Clinical Trial, http://ctep.info.nih.gov/handbook/handbook/Part_b.HTM (description of CTEP policies for each phase of clinical investigation with experimental drugs, outline of CTEP's scientific objectives as a drug sponsor, which physicians are eligible to study and administer investigational drugs)
- Common Toxicity Criteria, http://ctep.info.nih.gov/CTC3/default.htm (guide to recording Relevant Clinical Observations; links to definitions and descriptions of adverse events)
- NCI CancerNet homepage and Cancer Centers Program, http://cancernet.nci.nih.gov/
- NCI CancerTrials Resources for Researchers, http://cancertrials.nci.nih.gov/researchers/index.html
- NCI PDQ® searchable database for identifying clinical trials, http://cancernet.nci.nih.gov/trialsrch.shtml
- Cancer Center Websites: MDAnderson Cancer Center, http://www.mdanderson.org/; UCSF Cancer Center, http://cc.ucsf.edu/trials/index.html; USF Moffit Cancer Center, http://www.moffitt.usf.edu/medprof.htm; Memorial Sloan Kettering Cancer Center, http://www.mskcc.org/patients_n_public/clinical_trials/index.html; Fred Hutchinson Cancer Center, http://www.fhcrc.org/; Fox Chase Cancer Center, http://www.fccc.edu; University of Wisconsin, http://www.cancer.wisc.edu/index.html; UCLA, http://www.cancer.mednet.ucla.edu/; University of Michigan, http://www.cancer.mednet.ucla.edu/; UCSD, http://cancer.ucsd.edu/; University of Miami Sylvester Cancer Center, http://www.sylvester.org/; University of Pennsylvania, www.med.upenn.edu/ihgt/info/whatisgt.html)

4.8. Accessing Open Clinical Trials

Access to information for physicians and researchers regarding participation in most currently open clinical trials is available through Internet websites maintained by the NCI, other government agencies, and by individual cancer centers, some of them listed above.

4.9. Restructuring Clinical Trials at NCI

In 1998, in proposing a new, streamlined, more efficient, and simplified procedure for moving potential therapies for cancer into effective National Cancer Institute (NCI)-sponsored clinical trials, the NCI addressed three fundamental elements: assuring the best science, opening up access, and efficient, effective functioning *(52)*. The restructuring plan for NCI's clinical trials program was approved in 1998 and was implemented, in close collaboration with the clinical research community, beginning in 2000 *(59)*. The electronic resources available through the NCI (see above) will streamline the process for researchers and patients alike.

5. Summary

Although limitations still exist to the widespread application of gene therapy, the strategy has already been shown to be applicable in several clinical situations. Contrary to initial predictions, virus-assisted gene transfer has been shown to be more efficient in cancer cells than in normal tissue cells. Viral vectors appear to spread readily through a tumor and to encourage cell death via apoptosis. Initial concerns that the existence of multiple genetic lesions in cancer cells would prevent the application of gene therapy to cancer appear unfounded; on the contrary, correction of a single genetic lesion has, repeatedly, yielded significant tumor regression, and successful early clinical trials of *p53* gene replacement have provided information that will be useful in the design of future gene therapy strategies. Direct intratumoral injection has demonstrated low toxicity and thus can be readily combined with existing treatments. Postinjection gene expression can be documented and occurs in the presence of an antiadenovirus immune response. Very few complications have arisen from gene transfer to local tumors, and a number of patients have undergone treatments on an outpatient basis. Encouraging results in the context of combined modality protocols suggest that this mode of treatment should undergo further study.

Future research directions will include development of more efficient vectors, use of novel genes, and combined modality approaches. A large number of studies are now showing great potential for combining gene therapy and pharmaceutic, immunologic, and radiotherapeutic approaches to kill cells more effectively and in greater numbers. New vectors and targeting strategies will be critical to widening the applicability of gene therapy to systemic administration, and research is moving forward in those areas. Although there is a long road ahead before gene therapy strategies will approach their full potential, they are already taking their place as the next step in cancer treatment. And because gene therapy targets the etiology of the disease, these strategies may someday even be applied to cancer prevention.

Acknowledgments

This work was partially supported by grants from the National Cancer Institute and the National Institutes of Health (P01 CA78778-01A1) (JAR); SPORE (2P50-CA70970-04); by gifts to the Division of Surgery, from Tenneco and Exxon for the Core Laboratory Facility; by the UT M. D. Anderson Cancer Center Support Core Grant (CA 16672); by a grant from the Tobacco Settlement Funds as appropriated by the Texas State Legislature (Project 8), the W. M. Keck Foundation, and a sponsored research agreement with Introgen Therapeutics, Inc. (SR93-004-1).

References

1. Blaese, R. M., Culver, K. W., Miller, A. D., et al. (1995) T lymphocyte-directed gene therapy for ADA(-) SCID: initial trial results after 4 years. *Science* **270,** 475–480.
2. Crystal, R. G. (1995) Transfer of genes to humans: early lessons and obstacles to success. *Science* **270,** 404–410.
3. Kiem, H. P., von Kalle, D., Schuening, F., and Storb, R. (1995) Gene therapy and bone marrow transplantation. *Curr. Opin. Oncol.* **7,** 107–114.

4. Miller, N. and Vile, R. (1995) Targeted vectors for gene therapy. *FASEB J.* **9,** 190–199.
5. Bishop, J. M. (1991) Molecular themes in oncogenesis. *Cell* **64,** 235–248.
6. Weinberg, R. A. (1991) Tumor suppressor genes. *Science* **254,** 1138–1146.
7. Cantley, L. and Neel, B. (1999) New insights into tumor suppression: PTEN suppresses tumor formation by restraining the phosphoinositide 3-kinas/AKT pathway. *Proc. Natl. Acad. Sci. USA* **96,** 4240–4245.
8. Burns, T. and El-Deiry, W. (1999) The p53 pathway and apoptosis. *J. Cell. Physiol.* **181,** 231–239.
9. NIH/RAC (2001). Report of the March 8, 2001, Meeting of the Recombinant DNA Advisory Committee. NIH/OBA/RAC website, 3-8-2001. 6-27-2001, www4.od.nih.gov/oba/rac/minutes/march32001.pdf.
10. Ramqvist, T., Magnusson, K. P., Wang, Y., Szekeley, L., and Klein, G. (1993) Wild-type p53 induces apoptosis in a Burkitt lymphoma (BL) line that carries mutant p53. *Oncogene* **8,** 1495–1500.
11. Shaw, P., Bovey, R., Tardy, S., Sahli, R., Sordat, B., and Costa, J. (1992) Induction of apoptosis by wild-type p53 in a human colon tumor-derived cell line. *Proc. Natl. Acad. Sci. USA* **89,** 4495–4499.
12. Yonish-Rouach, E., Resnitzky, D., Lotem, J., Sachs, L., Kimchi, A., and Oren, M. (1991) Wild-type p53 induces apoptosis of myeloid leukemic cells that is inhibited by interleukin-6. *Nature* **352,** 345–347.
13. Swisher, S. G., Roth, J. A., Komaki, R., et al. (2001) Induction of pro-apoptotic mediators and tumor regression following intratumoral delivery of adenoviral p53 (RPR/INGN 201) and radiation therapy in patients with non-small cell lung cancer (NSCLC). *Proc. Am. Soc. Clin. Oncol.* **20,** 257a (abstr.).
14. Roth, J. A., Nguyen, D., Lawrence, D. D., et al. (1996) Retrovirus-mediated wild-type p53 gene transfer to tumors of patients with lung cancer. *Nat. Med.* **2,** 985–991.
15. Zhang, W. W., Fang, X., Mazur, W., French, B. A., Georges, R. N., and Roth, J. A. (1994) High-efficiency gene transfer and high-level expression of wild-type p53 in human lung cancer cells mediated by recombinant adenovirus. *Cancer Gene Ther.* **1,** 5–13.
16. Swisher, S. G., Roth, J. A., Nemunaitis, J., et al. (1999) Adenovirus-mediated p53 gene transfer in advanced non-small cell lung cancer. *J. Natl. Cancer Inst.* **91,** 763–771.
17. Clayman, G. L., El-Naggar, A. K., Lippman, S. M., et al. (1998) Adenovirus-mediated *p53* gene transfer in patients with advanced recurrent head and neck squamous cell carcinoma. *New Frontiers Res. Treat. Aerodigestive Tract Cancers* **41,** 109–110.
18. Goodwin, W. J., Esser, D., Clayman, G. L., Nemunaitis, J., Yver, A., and Dreiling, L. K. (1999) Randomized phase II study of intratumoral injection of two dosing schedules using a replication-deficient adenovirus carrying the p53 gene (AD5CMV-P53) in patients with recurrent/refractory head and neck cancer. *Proc. Am. Soc. Clin. Oncol.* **19,** 445a (abstr.).
19. Bier-Laning, C. M., VanEcho, D., Yver, A., and Dreiling, L. K. (1999) A phase II multi-center study of AD5CMV-P53 administered intratumorally to patients with recurrent head and neck cancer. *Proc. Am. Soc. Clin. Oncol.* **18,** 444a (abstr.).
20. Yver, A., Dreiling, L. K., Mohanty, S., et al. (1999) Tolerance and safety of RPR/INGN 201, an adeno-viral vector containing a p53 gene, administered intratumorally in 309 patients with advanced cancer enrolled in phase I and II studies world-wide. *Proc. Am. Soc. Clin. Oncol.* **19,** 460a (abstr.).
21. Dewey, W. C., Ling, C. C., and Meyn, R. E. (1995) Radiation induced apoptosis: relevance to radiotherapy. *Int. J. Radiat. Oncol. Biol. Phys.* **33,** 781–796.
22. Meyn, R. (1997) Apoptosis and response to radiation: implications for radiation therapy. *Oncology* **11,** 349–366.
23. Meyn, R. E., Stephens, LC., Hunter, N. R., and Milas, L. (1997) Apoptosis in murine tumors treated with chemotherapy agents. *Anticancer Drugs* **6,** 443–450.

24. Fujiwara, T., Cai, D. W., Georges, R. N., Mukhopadhyay, T., Grimm, E. A., and Roth, J. A. (1994) Therapeutic effect of a retroviral wild-type p53 expression vector in an orthotopic lung cancer model (Commentary). *J. Natl. Cancer Inst.* **86,** 1437–1438.

25. Nguyen, D., Wiehle, S., Koch, P., Roth, J. A., and Cristiano, R. (1996) Gene therapy for lung cancer: enhancement of tumor suppression by a combination of systemic cisplatin and adenovirus-mediated p53 gene transfer. *Proc. Am. Assoc. Cancer Res.* **37,** 347.

26. Hamada, M., Fujiwara, T., Hizuta, A., et al. (1996) The p53 gene is a potent determinant of chemosensitivity and radiosensitivity in gastric and colorectal cancers. *J. Cancer Res. Clin. Oncol.* **122,** 360–365.

27. Nemunaitis, J., Swisher, S. G., Timmons, T., et al. (2000) Adenovirus-mediated p53 gene transfer in sequence with cisplatin to tumors of patients with non-small cell lung cancer. *J. Clin. Oncol.* **18,** 609–622.

28. Spitz, F. R., Nguyen, D., Skibber, J., Meyn, R., Cristiano, R. J., and Roth, J. A. (1996) Adenoviral mediated p53 gene therapy enhances radiation sensitivity of colorectal cancer cell lines. *Proc. Am. Assoc. Cancer Res.* **37,** 347.

29. Jasty, R., Lu, J., Irwin, T., Suchard, S., Clarke, M. F., and Castle, V. P. (1998) Role of p53 in the regulation of irradiation-induced apoptosis in neuroblastoma cells. *Mol. Genet. Metab.* **65,** 155–164.

30. Feinmesser, M., Halpern, M., Fenig, E., et al. (1994) Expression of the apoptosis-related oncogenes bcl-2, bax, and p53 in Merkel cell carcinoma: can they predict treatment response and clinical outcome? *Hum. Pathol.* **30,** 1367–1372.

31. Broaddus, W. C., Liu, Y., Steele, L. L., et al. (1999) Enhanced radiosensitivity of malignant glioma cells after adenoviral p53 transduction. *J. Neurosurg.* **91,** 997–1004.

32. Sakakura, C., Sweeney, E. A., Shirahama, T., et al. (1996) Overexpression of bax sensitizes human breast cancer MCF-7 cells to radiation-induced apoptosis. *Int. J. Cancer* **67,** 101–105.

33. Swisher, S., Roth, J. A., Komaki, R., et al. (2000) A phase II trial of adenoviral mediated p53 gene transfer (RPR/INGN 201) in conjunction with radiation therapy in patients with localized non-small cell lung cancer (NSCLC). *Am. Soc. Clin. Oncol.* **19,** 461a (abstr.).

34. Le Chevalier, T., Arriagada, R., Quoix, E., et al. (1991) Radiotherapy alone versus combined chemotherapy and radiotherapy in nonresectable non-small-cell lung cancer: first analysis of a randomized trial in 353 patients. *J. Natl. Cancer Inst.* **83,** 417–423.

35. Shimizu, N., Chen, J., Gamou, S., and Takayanagi, A. (1996) Immunogene approach toward cancer therapy using epidermal growth factor receptor-mediated gene delivery. *Cancer Gene Ther.* **3,** 113–120.

36. Kawamoto, T., Sato, J. D., Le, A., Polikoff, J., Sato, G. H., and Mendelsohn, J. (1983) Growth stimulation of A431 cells by epidermal growth factor: identification of high-affinity receptors for epidermal growth factor by an anti-receptor monoclonal antibody. *Proc. Natl. Acad. Sci. USA* **80,** 1337–1341.

37. Lukas, J., Parry, D., Aagaard, L., et al. (1995) Retinoblastoma-protein-dependent cell-cycle inhibition by the tumour suppressor p16. *Nature* **375,** 503–506.

38. Cavazzana-Calvo, M., Hacein-Bey, S., de Saint Basile, G., et al. (2000) Gene therapy of human severe combined immunodeficiency (SCID)-X1 disease. *Science* **288,** 669–672.

39. *Nature* Editor (2000) Gene therapy—a loss of innocence. *Nature* **6,** 1.

40. NIH (2001) Common toxicity criteria. NIH Website, ctep.info.nih.gov/ctc3/default.htm.

41. NIH/RAC (2001) Report of the Recombinant DNA Advisory Committee. NIH/OBA/RAC website, 3-8-2001, 6-27-2001, www4.od.nih.gov/oba/rac/minutes/march32001.pdf.

42. NIH (2000) Report of The Working Group on Oversight of Clinical Gene Transfer Research. NIH, 7-12-2000.

43. Gibbs, W. W. (1996) Gene therapy. *Scientific American* website, www.sciam.com/explorations/101496explorations.html.

44. *Wall Street Journal* (1999) Gene therapy, touted as a breakthrough, bogs down in details. *Wall Street Journal*, October 27.

45. Anderson, W. F. (2001) A cure that may cost us ourselves. *Newsweek*, January 1.

46. NIH/RAC (2001) Reporting of adverse events in trials format. NIH, 6-1-2001, 7-27-2001, www4.od.nih.gov/oba/rac/SAEForm.rtf.

47. NIH/RAC (2001) NIH guidelines for research involving recombinant DNA molecules: RAC guidelines (revised Jan. 2001). NIH/RAC, http://www4.od.nih.gov/oba/rac/guidelines/guidejan01.htm.

48. NIH/OBA/RAC (2001) About Recombinant DNA Advisory Committee. NIH/OBA/RAC website, 7-27-2001, http://www4.od.nih.gov/oba/rac/aboutrdagt.htm.

49. NIH/OBA/RAC. (2001) Report of serious adverse effects. NIH/OBA/RAC website, 6–1-2001, 7-27-2001, http://www4.od.nih.gov/oba/rac/SAE_rpts/Mod0601s/Jun01_MODs.htm.

50. NCI (2001) NCI CancerTrials website, 6-1-2001, 6-27–2001, http://cancertrials.nci.nih.gov/researchers/index.html.

51. Patterson, A. (2000) Statement of Amy Patterson, MD, Director, NIH Office for Biotecholgy Activities, Before The Subcommittee on Public Health, Committee on Heath, Education, Labor and Pensions, U.S. Senate, Feb. 2, 2000. NIH/OBA/RAC website, 7-25-2001, www4.od.nih.gov/aba/rac/patterson2-00.pdf.

52. NCI (2000) Restructuring clinical trials. NIH/NCI, 7-28-2001, cancertrials.nci.nih.gov/researchers/restructuring/blueprint/intro.html.

53. NIH/OBA/RAC (2001) NIH Guidelines Appendix M: points to consider in the design and submission of protocols for the transfer of recombinant DNA molecules into one or more human research participants. NIH/OBA/RAC website, 1-1-2001, 7–27–2001, http://www4.od.nih.gov/oba/rac/guidelines/appendix_m.htm.

54. Zoon, KC. (1999) Letter to IND Sponsors and Prinicipal Investigators. RAC, www4.od.nih.gov/oba/rac/genetherapyltr.pdf.

55. FDA (2001) Investigational New Drug (IND) Application. FDA/CDER website, 1-1-2001. http://www.fda.gov/cder/regulatory/applications/ind_page_1.htm.

56. FDA (2001) *Handbook for Investigators—IND*. FDA/CDER website, 1-1-2001, www.fda.gov/cder/handbook/ind.htm.

57. FDA (2001) Guidance for institutional review boards and clinical investigators. FDA, 1-1-2001, www.fda.gov/oc/ohrt/irbs/default.htm.

58. FDA (2001) FDA's Health Professionals Page. FDA, 8-2-2001, 8–4-2001, http://www.fda.gov/oc/oha/default.htm.

59. NCI and NIH (2001) NCI clinical trials: a blueprint for the future. NCI-CancerTrials Website, 8-5-2001, cancertrials.nci.nih.gov/researchers/restructuring/blueprint/index.html.

38

Therapeutic Strategies Using Inhibitors of Angiogenesis

Michael S. O'Reilly

1. Introduction

Angiogenesis, the growth of new capillary blood vessels from preexisting vessels, is a fundamental process that is required for a wide variety of physiologic and pathophysiologic processes *(1,2)*. Examples of physiologic processes that require angiogenesis include wound healing, tissue repair, reproduction, and growth and development *(3)*. Disease states that are associated with malignant angiogenesis include cancer, ophthalmologic disorders such as diabetic retinopathy or macular degeneration, arthritis, psoriasis, and arteriosclerosis *(4)*. Leukemia has also been shown to be dependent on angiogenesis *(5)*.

Historically, the hyperemia and increased vascularity of malignant tissue as compared to normal tissue had been assumed to result from the dilation of preexisting host vessels in response to tumor necrosis and metabolic by-products. Neovascularization in association with a tumor was assumed to merely be a byproduct of inflammation and was not considered important. In 1971, Folkman first proposed that tumor growth was dependent on angiogenesis, based on his studies of tumor growth in isolated perfused organs *(6)*. Folkman's proposal that a tumor was dependent on the endothelial cell was met with skepticism and was widely criticized. However, numerous lines of direct evidence have now clearly established that the growth and expansion of tumors and their metastases is dependent on angiogenesis *(1)*.

It is now clear that both the growth of malignant tumors and the process of metastasis are dependent on the induction of angiogenesis to provide an adequate vasculature *(1,2)*. Any increase in tumor mass beyond a size that can be supported by diffusion of oxygen and nutrients (i.e., 200 mm^2) requires the induction of neovascularization *(6–10)*. Under physiologic conditions that require neovascularization, angiogenesis is a tightly regulated process *(11)*. However, in malignant angiogenesis the process is sustained and requires the continued production of stimulators by tumor and stromal cells in excess of inhibitors. The process of angiogenesis consists of multiple, sequential, and interdependent steps. In response to angiogenic stimuli, local degradation of the basement membrane surrounding capillaries begins allowing for endothelial cell invasion of the surrounding stroma. The endothelial cell migration and invasion is accompanied by

From: *Methods in Molecular Biology, Vol. 223: Tumor Suppressor Genes: Regulation, Function, and Medicinal Applications.* Edited by: Wafik S. El-Deiry © Humana Press Inc., Totowa, NJ

the proliferation of endothelial cells and their organization into three-dimensional structures that join with other similar structures to form a network of new blood vessels *(12)*. In order to control angiogenesis optimally as part of the treatment of pathologic conditions, strategies that target each of these steps will need to be developed. Thus, an improved understanding of the process of angiogenesis is required.

In addition to providing nutrients and oxygen and removing catabolites, angiogenesis can also be an important source of factors that have a paracrine effect for a malignant tumor. The proliferating endothelial cells found within the tumor itself and in the surrounding tissues can produce multiple growth factors that can promote tumor cell growth, invasion, and survival *(13–15)*. Angiogenesis, therefore, provides both a perfusion effect and a paracrine effect to a growing tumor. As a result, the tumor cells and endothelial cells can drive each other and perpetuate and amplify the malignant phenotype. These observations led Folkman to propose a two-compartment model of malignancy, consisting of endothelial cell and tumor cell populations *(16)*. Both of these compartments offer potential targets for anticancer strategies, and it will be prudent to target both the cancer cells and the angiogenic endothelium.

A number of antiangiogenic agents have now been described, and many of these are currently in clinical trials in cancer patients *(2,17,18)*. However, no angiogenesis inhibitor has yet been approved for clinical use. Before antiangiogenic agents can be successfully incorporated into clinical strategies to treat cancer and other angiogenesis-dependent diseases, a greater understanding of the process of angiogenesis and of the interaction between the endothelial cell and its microenvironment is required. The goal of this chapter is to provide an overview of the potential benefits of antiangiogenic therapy and discuss the challenges associated with the use of antiangiogenic agents.

2. Angiogenesis Is Regulated by a Balance of Stimulators and Inhibitors

The process of angiogenesis is regulated by a number of stimulators and inhibitors *(8)*, and it is the net balance of these factors that determine whether angiogenesis can occur. Endothelial cells are normally quiescent, and years may pass between cell divisions *(19)*. The onset of angiogenesis involves an alteration in the balance between proangiogenic and antiangiogenic molecules *(2,8,20)*. These molecules may mediate multiple steps in the process of angiogenesis and also may affect the function of diverse cell types not involved in angiogenesis. Several proangiogenic factors have now been described. These include the fibroblast growth factors (FGFs) such as acidic and basic FGF (aFGF, bFGF) *(21)*, vascular endothelial cell growth factor/vascular permeability factor (VEGF/VPF) *(22,23)*, interleukin-8 (IL-8) *(24,25)*, hepatocyte growth factor/scatter factor (HGF/SF) *(26)*, transforming growth factor beta (TGF-β) *(27,28)*, proliferin *(29)*, and erythropoietin *(30)*. A more refined definition of an angiogenic factor is a factor that selectively alters the characteristics of endothelial cells and associated perivascular structures (i.e., pericytes, vascular smooth muscle cells) but does not affect the function of other cell types *(8,20)*. For angiogenesis to occur, an increased production of stimulatory factors and/or a decreased production or generation of inhibitors is necessary *(8,31)*.

The studies of Bouck and her colleagues *(32,33)* support a model in which the net balance of the stimulators and inhibitors of angiogenesis controls tumor growth. Bouck and her colleagues demonstrated that when hamster cells were transformed to an angiogenic phenotype, they upregulated the production of angiogenesis stimulators but also downregulated the production of thrombospondin, an angiogenesis inhibitor, in order to form angiogenic lesions. Thrombospondin production is under the regulation of the wild-type p53 tumor suppressor gene *(34)*. A similar pattern was observed in malignant ocular angiogenesis in which VEGF/VPF expression increased and the expression of the angiogenesis inhibitor pigment epithelial-derived factor (PEDF) decreased *(35)*.

Under physiologic conditions that require neovascularization, angiogenesis is a tightly regulated process, and the balance in favor of stimulation is generally short-lived. As with physiologic angiogenesis, the neovascularization of solid tumors and their metastases is also regulated by the balance of stimulators and inhibitors. However, in malignant angiogenesis, the process is sustained and requires the continued production of stimulators by tumor and stromal cells in excess of inhibitors *(11,36)*. Further, a number of differences between malignant and physiologic angiogenesis are becoming apparent *(37)*. In malignant angiogenesis, angiogenic vessels are markedly distorted relative to their physiologic counterparts *(38,39)* and are characterized by a lack of pericytes and other support cells *(40)*, impaired and/or intermittent flow *(41–43)*, the presence of plasma proteins that function as a neostroma *(44–47)*, and an increase in vascular permeability *(48)*. Thus, strategies that target angiogenesis have clear clinical implications for a variety of conditions, both neoplastic and non-neoplastic, and the study of angiogenesis will help in the understanding of a number of disease states.

3. Proteinases and Their Inhibitors as Modulators of Angiogenesis

3.1. Proangiogenic Effects of Proteases

Increased protease activity is a critical step in both angiogenesis and tumor progression, and the extracellular matrix and its components play a critical role in angiogenesis (reviewed by Liotta *[49]* and Stetler-Stevenson *[50,51]*). A number of proteinases that are involved in the regulation of coagulation and fibrinolysis also play a role in the process of angiogenesis. Thrombin *(52–54)* is an important mediator of endothelial cell invasion and migration and can mobilize VEGF/VPF and other proangiogenic molecules. Plasmin is also important in the outgrowth of capillaries in the angiogenic response *(55,56)*. However, many other molecules that are involved with coagulation and fibrinolysis can inhibit angiogenesis *(57)*. The serpin plasminogen activator inhibitor-1 has been shown to inhibit angiogenesis possibly by blocking the interaction of $\alpha v \beta 3$ with vitronectin *(58,59)*. Taken together with the discovery that antithrombin *(60)*, another serpin, can also inhibit angiogenesis, these findings suggest that other serpins involved in the coagulation system may have effects on the endothelial cell. Further, the degradation of plasminogen by plasminogen activators in the presence of sulfhydryl donors can mobilize angiostatin *(61,62)*. Thus, many of the factors that are involved in the regulation of blood clotting also play a central role in angiogenesis and can function as stimulators or inhibitors of the process.

3.2. Inhibitors of Metalloproteinases

Several members of the tissue inhibitors of metalloproteinase (TIMP) family have been shown to inhibit angiogenesis, and the presence of metalloproteinase (MMP) inhibitors in cartilage may in part explain its resistance to angiogenesis *(63,64)*. However, cartilage also contains other factors, such as troponin I, that can inhibit angiogenesis *(65)*.

Small molecules that can inhibit MMP activity have been described, including minocycline and other tetracyclines *(66,67)* and COL-3 *(68)*. More recently, inhibitors that target the incorporation of zinc into the catalytic site of MMPs, such as marimistat and batimistat or the selective inhibitors BAY 12–9566 *(69,70)* and AG3340 *(71)*, have been developed to selectively target those MMPs that are important for tumor growth and angiogenesis. However, advanced clinical trials with many of these agents used alone or in combination with chemotherapy in cancer patients have been disappointing *(70)*, suggesting that a better understanding of their role in angiogenesis is needed.

3.3. Endogenous Angiogenesis Inhibitors Are Mobilized by Proteinases

Although the inhibition of protease activity may offer both a direct antitumor effect and an antiangiogenic effect in the treatment of cancer, there may be a potential disadvantage to administering a metalloproteinase inhibitor in some patients. Many of the endogenous inhibitors of angiogenesis are generated from the cleavage of other proteins by metalloproteinases and other enzymes. In our original studies of the suppression of tumor growth by tumor mass *(72)*, a variant of Lewis lung carcinoma (LLC-LM) was developed in which a primary tumor could completely suppress the growth of its metastases by generating the angiogenesis inhibitor angiostatin from plasminogen. Northern analysis of the LLC-LM cells revealed no evidence that the tumor cells expressed angiostatin or other fragments of plasminogen, suggesting that angiostatin was derived from proteolysis of plasminogen *(72,73)*.

In a series of studies of a variant of LLC-LM lung carcinomas in mice, tumors that suppressed the growth of their metastases were found to express high levels of GM-CSF *(74)*. In response to this GM-CSF *(75)*, tumor-infiltrating macrophages were induced to increase their expression of metalloelastase (MMP-12). Once activated, this MMP-12 then mobilized angiostatin from plasma-derived plasminogen and plasmin that was part of the tumor's neostroma. Thus, the production of angiostatin was dependent on the interaction of the tumor cells with their microenvironment and infiltrating stromal cells. These and other studies emphasize the close association of the immune system with angiogenesis. Stromal cells have also been shown to be a critical source for stimulators of angiogenesis in colon and other malignancies *(25)*.

In a separate series of studies using the original LLC-LM cells, active gelatinase A (MMP-2) was identified in the conditioned medium of the tumor cells. MMP-2 from these cells was able to process plasminogen into angiostatin. By specifically neutralizing the enzymatic activity, the mobilization of angiostatin from plasminogen by the tumor cells was blocked *(73)*. The processing of precursor or parent proteins into their bioactive form is not unprecedented. For example, MMPs are involved in the processing of the TNF-α precursor protein *(76,77)*, TGF-α *(78)*, the β-amyloid precursor protein *(79)*, the lymphocyte L-selectin adhesion molecule *(80–82)*, the IL-6 receptor ectodomain *(83)*, the human thyrotropin receptor ectodomain *(84)*, the FGFR1 ectodomain *(85)*, and EGF-like growth factor *(86)*. Taken together, these studies show that metalloproteinases can process a wide range of substrates, many of which play a role in angiogenesis.

Subsequent work has demonstrated that a variety of metalloproteinases and other enzymes that are important in tumor progression can cleave angiostatin from plasminogen *(61,87–89)*. An increasing number of studies now demonstrate that most angiogenesis inhibitors are fragments of larger parental molecules *(2)*. For example, an internal

16-kDa fragment of prolactin inhibits angiogenesis, whereas the parent molecule, intact prolactin, does not *(90–93)*. Other inhibitors of EC proliferation that are fragments of larger molecules include fragments of platelet factor 4 *(94)* with enhanced antiangiogenic activity as compared to the intact molecule *(95)*, thrombospondin *(96)*, EGF *(97)*, laminin *(98)*, fibronectin *(99)*, and endostatin *(100,101)*. Thus, metalloproteinases and other enzymes can play a central role in the regulation of tumor angiogenesis and potentially act in an inhibitory *(63,64)* or a stimulatory capacity. Further, the finding that angiogenesis inhibitors can be mobilized and/or activated by metalloproteinases suggests that therapeutic strategies for using inhibitors of metalloproteinase and other enzymatic activities must be carefully designed to consider both the clinical setting and the net effect on angiogenesis. Diagnostic and clinical strategies *(102,103)* for these inhibitors will need to be developed before they can be successfully incorporated into clinical practice.

The production of angiostatin by gelatinase A, metalloelastase, and other proteinases from the tumor or stromal cells helps to resolve the question as to why a primary tumor might be producing angiostatin. By increased production of gelatinase A, the tumor may become more locally invasive but might also mobilize more angiostatin *(73)*. It may therefore be prudent to consider gelatinase A and metalloproteinase inhibitors and metalloproteinases themselves as angiogenesis modulators instead of merely inhibitors or stimulators. Angiogenesis may therefore depend not only on the balance of endothelial stimulators and inhibitors but also on the balance of matrix-degrading proteases and their endogenous inhibitors. In addition, the increased permeability of the tumor vessels may lead to the sequestration of circulating plasminogen into the tumor's neostroma *(45,46)*, which could then be cleaved into active angiostatin.

It has been widely suggested that the limited clinical success of broad-spectrum MMP inhibitors is due, at least in part, to their lack of specificity and associated side effects. However, an alternate explanation is that the regulation of MMP activity may, in some cases, decrease the release of angiogenesis inhibitors. This possibility suggests that it will be important to define, for each different tumor type, the precise roles of MMPs and their inhibitors in the modulation of angiogenesis and malignancy. In these patients, the level of endogenous angiogenesis inhibitors could potentially decrease and would therefore need to be replaced.

The mobilization of angiogenesis inhibitors *(72)*, such as angiostatin, by enzymes produced by tumor cells or tumor-associated host cells *(73,74)* can inhibit the growth of distant metastases. Curative radiotherapy of primary tumors in rodents can be followed by the rapid growth of previously dormant metastases that kill the animal within 18 d after the completion of radiation therapy *(103)*. Systemic administration of recombinant angiostatin, however, prevented the growth of the metastases, suggesting that there may be a population of patients that would benefit from the addition of antiangiogenic therapy during and/or after radiation therapy to prevent the expansion of distant metastases.

4. Strategies to Control Tumor Growth by Blocking Angiogenesis

Many inhibitors of angiogenesis have now been described, and the discovery and characterization of these inhibitors has become an area of intense research. Many of the first angiogenesis inhibitors were substances with other functions whose antiangiogenic were discovered incidentally. More recently, strategies have been developed for the discovery of highly specific inhibitors of angiogenesis or for the specific targeting of factors that promote angiogenesis. In this section, some of the strategies to inhibit angiogenesis will be described in order to provide an outline of issues that are relevant to the translation of antiangiogenic therapy to the clinic.

4.1. Antiangiogenic Therapy Can Induce Tumor Dormancy

Prior to the discovery of angiostatin *(72)*, it had been generally assumed that antiangiogenic therapy would only produce cytostasis in tumors. Studies with angiostatin, however, demonstrated that antiangiogenic therapy can result in tumor regression *(72,104)*. The tumor dormancy that results from the suppression of angiogenesis is characterized by a balance of tumor cell proliferation and apoptosis *(72,104,105)*. However, in virtually all cases, antiangiogenic monotherapy is unable to eradicate microscopic disease completely. Since the discovery of angiostatin, several antiangiogenic agents, including endostatin *(100)*, antiangiogenic antithrombin *(60)*, and SU6668 *(106,107)*, an inhibitor of VEGF, bFGF, and PDGF receptor tyrosine kinase signaling, have been shown to induce regression of tumors. These agents and other angiogenesis inhibitors offer great potential, but before they can be integrated into clinical research a number of challenges will need to be overcome. In particular, a better understanding of their mechanism of action and their interaction with conventional modalities is required, and surrogates to monitor response to therapy must be developed.

4.2. Antivascular Agents and Vascular Targeting

Although the focus of this chapter is on antiangiogenesis, it is important to distinguish between antivascular and antiangiogenic therapy. Antiangiogenic therapy is directed against the neovasculature and thus does not have a significant effect on the resting vasculature. In contrast, antivascular therapy targets the existing vasculature and could therefore disrupt normal vessels. Given that in normal tissues endothelial cells generally remain quiescent for periods of months or years *(19)*, antiangiogenic strategies are not associated with significant toxicities. However, many agents will have both antivascular and antiangiogenic effects.

One strategy to target tumor vasculature selectively, is the use of antibodies directed against antigens expressed on some but not all endothelial cells conjugated to toxins *(108,109)*. Targeting of the tumor vasculature can also be achieved by using the tubulin-destabilizing agent combretastatin A4 *(10)*. Although the precise mechanism is not yet known, combretastatin A4 can target the established tumor vasculature, leading to the development of extensive tumor cell necrosis *(110)* without significant toxicity. Combretastatin A4 is now being evaluated in clinical trials.

The heterogeneity of the endothelium has recently been used for targeted antivascular therapies. Using phage display, specific peptide sequences that bind to receptors found on the endothelial cell surface in the capillary beds of different tissue beds have been identified *(37,111,112)*. Angiogenic endothelial cells can now be targeted by specific peptide sequences that bind selectively to proliferating endothelial cells but not to quiescent cells. Using these peptides conjugated to the cytotoxic agent doxorubicin, the antitumor activity of the drug could be enhanced several-fold in murine models of breast cancer *(37)*. Whether organ-specific peptides can home to sites of angiogenesis in selected tissues remains to be determined, but targeting the tumor vasculature is an attractive approach.

4.3. Agents That Block the Expression of Proangiogenic Molecules

Angiogenesis is controlled by the balance between stimulators and inhibitors *(8,113)*. For angiogenesis to proceed, there must be a shift in the balance of angiogenic factors in favor of stimulation. Tumors can achieve this by increasing the expression of proangiogenic molecules such as bFGF *(21,113)*, VEGF/VPF *(22,23)*, IL-8 *(24)*, or HGF/SF *(26)*, or by decreasing the expression of antiangiogenic molecules such as thrombospondin *(34,114)*. It is important to note that the source of angiogenic factors is often provided by both tumor cells and host cells *(25)*. Enzymes produced by tumor cells or stromal cells can mobilize proangiogenic molecules that can induce tumor angiogenesis. These enzymes can also mobilize antiangiogenic molecules *(73,74)*. Alternatively, tumor-derived growth factors can induce stromal cells to produce proangiogenic molecules. For example, the colonic mucosa adjacent to the colonic adenocarcinoma is responsible for the production of proangiogenic molecules that support tumor angiogenesis *(25)*. Thus, strategies to block the expression of angiogenic molecules offer great potential.

4.3.1. The Antiangiogenic Activity of the Interferons

One example of a molecule that can block the expression of proangiogenic factors is interferon. The interferon (IFN) family consists of IFN-α, IFN-β, and IFN-γ. IFNs regulate multiple biologic functions, have antiviral activities, and can also inhibit a number of steps in the angiogenic process and can block endothelial cell proliferation in vitro *(115–118)*. The mechanism of the antiangiogenic activity of IFN-γ is thought to be indirect and is mediated by the increased expression of the angiogenesis inhibitors IP-10 and MIG-1 *(119–121)*. IFN-α and IFN-β, but not IFN-γ, downregulate the expression of bFGF and metalloproteinases in human carcinoma cells *(9)*. Systemic administration of human IFN-α decreased the in-vivo expression of bFGF, decreased blood vessel density, and inhibited tumor growth of a human bladder carcinoma implanted orthotopically in nude mice *(122)*.

Systemic therapy using recombinant IFNs produces antiangiogenic effects in vascular tumors including life-threatening infantile hemangioma *(123–125)*, Kaposi's sarcoma *(126)*, giant cell tumor of the mandible *(127)*, and bladder carcinoma *(128)*. These tumors have also been documented as producing the high levels of bFGF often detectable in the urine or serum of these patients *(129,130)*.

In the original studies of the antiangiogenic effects of IFN-α, the relative activity was thought to be low *(131)*. However, by administering IFN at lower doses given more frequently, its antiangiogenic activity can be significantly enhanced and its toxicity diminished *(132)*. Studies of the effects of antiangiogenic IFN doses, both alone and in combination with other modalities, are now underway or will commence shortly at several centers throughout the world.

4.3.2. Receptor Tyrosine Kinase Blockade Can Inhibit Angiogenesis Indirectly and Directly

Many receptor tyrosine kinases are important for both tumor cells and endothelial cells. For example, the blockade of the EGF receptor targets both the tumor cell and tumor-associated endothelial cells *(133)*, leading to apoptosis and regression of the pancreatic cancer in mice. The blockade of these receptors works primarily by causing the downregulation of expression of proangiogenic factors by tumor and stromal cells. However, antagonists of EGF and its receptor may also have direct effects on the endothelial cells *(134)*. Recently, it has been shown that the expression of EGF or transforming growth factor alpha (TGF-⍺) by tumor cells results in the increased expression of the EGF-receptor by endothelial cells in vivo *(134)*. Platelet-derived growth factor (PDGF) may also have both indirect effects on the tumor vasculature, by increasing expression of proangiogenic molecules such as VEGF/VPF *(135,136)* and direct effects on endothelial cells that increase their expression of PDGF receptor in response to PDGF production by tumor cells *(137)*. Thus, these agents offer the potential for both indirect and direct effects on both the tumor and endothelial cell compartments.

Although strategies to target angiogenesis by blocking the proangiogenic molecules offer great promise, they are also associated with several potential limitations. The expression of receptor tyrosine kinases varies markedly among different tumor types, and not all tumor cells will express each receptor. It will therefore be critical first to screen tumor specimens for both the ligand and its receptor before initiating therapy with specific protein tyrosine kinase inhibitors. However, even tumors that express the growth factor receptor that is being targeted could eventually become resistant to therapy with the agent(s) by expressing different growth factors and/or their receptors. For example, insulin-like growth factor receptor 1 (IGFR-I) signaling through phosphoinositide 3-kinase was recently described as a mechanism of resistance to anti-EGFR therapy for some brain tumors *(138)*. Further, prolonged therapy of A431 human lung cancer with anti-EGFR antibodies in mice resulted in acquired resistance to further therapy targeting EGF-R in vivo and the emergence of tumors that produced increased amounts of VEGF/VPF and with increased angiogenic potential *(139)*. In addition, most advanced malignant tumors produce multiple angiogenic factors, and targeting only one may not be adequate for complete tumor control. For example, PTK787, an inhibitor of VEGF/VPF receptor tyrosine kinase phosphorylation, was able to reduce significantly the malignant pleural effusion associated with lung adenocarcinoma in mice but did not reduce the lung tumor burden *(140)*. One strategy to overcome the limitations of targeting individual growth factors is to design small molecules that can target multiple factors. SU6668, for example, is a small-molecule synthetic tyrosine kinase inhibitor of the receptors for VEGF/VPF (flk-1), FGF, and PDGF *(141)*. In vivo, SU6668 is a potent inhibitor of angiogenesis and is able to inhibit tumor growth and in some cases induce tumor regression when used alone in mice. As an alternative to using a single agent such as SU6668, it may also be possible to combine individual drugs that target different growth factors, such as EGF and VEGF/VPF *(142)*. Antiangiogenic therapy may thus have to be optimized on an individual basis.

Many receptor tyrosine kinases can affect the survival of both proliferating and quiescent endothelial cells. As a result, some of these agents could be associated with toxicity due to effects on and/or disruption of the normal vasculature. VEGF/VPF, for example, is critical not only for angiogenesis but also for endothelial cell integrity and function *(143–145)*. For example, VEGF/VPF produced by tumor cells is an proangiogenic factor but is also important for the survival of endothelial cells and prevents apoptosis and induces Bcl-2 expression *(144,146)*. In clinical trials with agents that target VEGF, hemorrhage and thrombosis have been observed as complications that may have been due to disruption of the normal vascular integrity. Thus, therapy with these agents will have to be monitored closely and optimized for each patient.

4.4. Endogenous Inhibitors of Angiogenesis

To promote angiogenesis, tumor cells produce a number of stimulatory factors and also downregulate the production of angiogenesis inhibitors such as troponin I *(65)* and thrombospondin, which is under the regulation of the wild-type p53 tumor suppressor gene *(34)*. Since thrombospondin also contains a domain that can stimulate angiogenesis and is upregulated in many tumors *(147)*, the therapeutic use of the intact molecule may have a mixed effect on angiogenesis. It may therefore be prudent to use fragments of thrombospondin from the antiangiogenic domains of the molecule rather than the intact molecule for cancer treatment. Indeed, several peptide fragments of thrombospondin have been generated that can effectively inhibit angiogenesis and are now being developed for clinical use *(96)*.

Several experimental rodent models suggest that the growth of a primary tumor can inhibit the production of distant metastases (reviewed by Gorelik *[148]* and by Prehn *[149]*). In one such model of concomitant resistance, in which a variant of Lewis lung carcinoma completely suppresses the growth of its metastases, angiostatin, a fragment of plasminogen, was purified from the urine of tumor-bearing mice *(72,105)*. Angiostatin potently inhibits endothelial cell proliferation *(150)* and migration in vitro and induces endothelial cell apoptosis *(151)* on a variety of endothelial cell types. The mechanism of the antiangiogenic activity of angiostatin remains unknown, but several lines of evidence suggest that it may have multiple targets. Angiostatin induces the activation of focal adhesion kinase *(152)* and has also been shown to interfere with several proteinases that are important in endothelial cell migration *(153)*. ATP synthase *(154)* and angiomotin *(155)* on the surface of endothelial cells have recently been described as receptors for angiostatin. Fragments of plasminogen that include at least the first three kringles and up to all five of the kringles have angiostatin activity.

Endostatin *(100,101)* is a 20-kDa carboxyl-terminal fragment of collagen XVIII and is a specific inhibitor of endothelial cell proliferation and migration *(156)*. In vitro, endostatin has no obvious effect on resting endothelial cells *(157)* or on a variety of nonendothelial cells *(100)*. The crystal structure of human and mouse endostatin have now been determined *(158,159)*. As with angiostatin, endostatin may have multiple mechanisms of action. Endostatin was shown to induce endothelial cell apoptosis *(160)* and a marked reduction in Bcl-2 and Bcl-Xl antiapoptotic protein and to inhibit migration and proliferation and cause G1 arrest of endothelial cells stimulated with bFGF or VEGF. It has also been shown to bind tropomyosin *(161)* on the surface of endothelial cells and to interact with endothelial integrins *(162)* and proteases that are important in endothelial cell shape *(163)* and motility *(164)*.

Endostatin is derived from the NC1 domain on collagen XVIII, and fragments from the NC1 domains of other collagens also inhibit angiogenesis. Restin, a C-terminal fragment of collagen XV that is homologous to collagen XVIII, has also been shown to be antiangiogenic *(165)*. As with endostatin, restin was able to potently inhibit the growth of human renal cell carcinoma xenografts. Further, tumstatin and other fragments from the NC1 domain of collagen IV, which can regulate endothelial cell migration *(166)*, have also been shown to be antiangiogenic *(167,168)*. Currently, endostatin and endo-

statin-like fragments derived from other collagens are being studied in order to elucidate their role in angiogenesis and are being developed for clinical use.

Several other endogenous angiogenesis inhibitors have now been described that are
fragments of proteins with distinct functions. One of the first such fragments to be
described is a 16-kDa fragment of the hormone prolactin produced from intact prolactin
in the pituitary *(92)*. Native and recombinant 16-kDa prolactin maintain biologic activity as prolactin agonists but, unlike intact prolactin, also inhibit endothelial proliferation
and capillary tube formation in vitro and angiogenesis in vivo and can have a potent antitumor effect *(169)*. Recently, N-terminal fragments of the human prolactin/growth hormone family and the intact molecules were found to have opposing effects on
angiogenesis *(93)*. Recombinant N-terminal fragments of prolactin, growth hormone,
placental lactogen, and growth hormone variant all inhibit angiogenesis, possibly by
blocking the activation of MAP kinase downstream of bFGF and VEGF or by increased
expression of plasminogen activator inhibitor-1 (PAI-1) in capillary endothelial cells
(91). A dual role in angiogenesis has also been described for proliferin, which is angiogenic, and proliferin-related peptide, which is antiangiogenic, in the placenta *(29)*. The
association between stimulatory and inhibitory effects on angiogenesis within the same
molecules may provide an efficient means of regulating physiologic angiogenesis. Further, a plasmin-derived internal fragment of the angiogenesis inhibitor platelet factor 4
(94) is up to 50-fold more potent *(95)* than the intact molecule. Synthetic fragments of
murine epidermal growth factor *(97)* and fragments of laminin *(170)* all inhibit endothelial cell proliferation and angiogenesis. Vasostatin *(171–173)*, a potent and specific
angiogenesis inhibitor, is an N-terminal fragment of calreticulin that was purified from
the supernatant of EBV-immortalized cell lines. The generation of vasostatin, along with
IP-10 and Mig, is in part responsible for angiogenesis suppression and tumor regression
induced by EBV-immortalized cells *(119)*. The cleaved conformation of antithrombin
has recently been shown to have potent antiangiogenic and antitumor activity *(60)*. The
intact native molecule did not have this effect. However, in the case of antithrombin it
is the change in conformation of the molecule, and not the cleavage *per se*, that accounts
for the antiangiogenic activity *(60,174,175)*.

When taken together with the discovery of angiostatin and endostatin, a theme has
emerged in which endogenous inhibitors of angiogenesis arise from larger proteins with
distinct and varied functions. If common proteases and/or similar cleavage patterns are
found, a new general mechanism of proteolytic events regulating the vascular system
may be revealed.

4.5. Targeted Therapy

Angiogenesis is regulated by a variety of stimulators of the process that interact with
growth factor receptors *(176,177)*. Many of these, such as VEGF/VPF and the FGFs,
are receptor tyrosine kinases. Recently, a new class of ligands for the receptor tyrosine
kinase Tie2, which is expressed almost exclusively by endothelial cells, has been
described *(178,179)*. The ligands for Tie2 include angiopoietin-1 (ang1) and angiopoi-
etin-2 (ang2), and together they function in the regulation of angiogenesis and vascular
development. Ang1 *(178,180)* recruits pericytes to newly formed capillary sprouts and
may help stabilize newly formed capillaries. Angiopoietin-2 *(179)* blocks the phosphorylation of Tie2 and can antagonize the effects of ang-1. Strategies to block receptor

tyrosine kinase function are currently being developed for use in the treatment of cancer, and several are now in clinical trial.

By using agents that target receptor tyrosine kinases that are involved in angiogenesis, the tumor vasculature can be targeted directly and/or indirectly *(2,181)*. One strategy involves directly blocking the binding of the growth factor to its target receptor using antibodies or small-molecule inhibitors. Monoclonal antibodies to angiogenic factors such as bFGF, VEGF/VPF *(182)*, epidermal growth factor (EGF) *(183,184)*, and Tie2 can potently inhibit angiogenesis and tumor growth for a wide variety of malignancies. Alternatively, the signaling of the growth factor receptor can be blocked using antibodies or by small-molecule receptor tyrosine kinase antagonists *(133,185–187)*. A variety of agents have also been produced that target autophosphorylation or the downstream signals of receptor tyrosine kinase activity. By selectively targeting the endothelial cell, these strategies have shown significant antitumor efficacy.

Soluble receptors for angiogenic factors can also be used to inhibit the activity of proangiogenic molecules. Circulating soluble receptors for flt-1, a receptor for VEGF/ VPF, and binding proteins for fibroblast growth factors have been described *(188,189)*. Their physiologic function is not known, but they may serve to limit the levels of angiogenic factors in the circulation and counteract the elevated levels of angiogenic factors seen in cancer patients. When delivered therapeutically in mice using adenoviral-mediated gene therapy, the regional delivery of a secreted form of the soluble flt-1 VEGF/VPF receptor inhibited tumor growth in experimental metastasis models of colon cancer *(190)*. Similar strategies using locoregional or systemic therapy with soluble receptors for VEGF/VPF and other growth factors may provide a means to control angiogenesis and tumor growth. Antisense targeting of angiogenic factors or their receptors, such as the VEGF/VPF receptor, can be used to inhibit angiogenesis and tumor growth *(191,192)*.

4.6. Gene Therapy and Angiogenesis

Gene therapy can be used to deliver both proangiogenic and antiangiogenic proteins in the treatment of angiogenesis-related diseases. The use of gene therapy to deliver proteins that regulate angiogenesis offers a number of potential benefits over systemic therapy (reviewed by Folkman *[193]*, Feldman *[194]* and Kong *[195]*). Several different angiogenesis stimulators have been used to promote the revascularization of coronary or peripheral arteries *(196)* as part of the treatment of arteriosclerosis *(197)*. Systemic therapy and locoregional therapy have both been employed *(198,199)*. However, a number of problems may arise with the systemic delivery of proangiogenic molecules, and proangiogenic factors will need to be targeted to the diseased vessels *(200)*. It is of interest that the atherosclerotic plaque is also dependent on angiogenesis. Moulton and colleagues *(4)* found that antiangiogenic therapy with endostatin or TNP-470 could actually prevent the development and progression of athersclerotic disease in mice. Thus, angiogenesis stimulators may one day be used to promote revascularization in patients with advanced peripheral vascular and/or coronary artery disease, whereas angiogenesis inhibitors could be used to prevent atherosclerosis or its recurrence.

Angiogenesis inhibitors require extended therapy for maximal efficacy and durability. As an alternative to systemic therapy, tumor cells can be transfected with the gene corresponding to an angiogenesis inhibitor. Cao et al. transfected an aggressive murine fibrosarcoma with a cDNA of murine angiostatin and observed an 80% inhibition of primary tumor growth (201). Further, metastases in 70% of mice implanted with the angiostatin-transfected tumors remained in a state of dormancy after primary tumor resection. The dormant metastases appeared as avascular cuffs around the normal lung capillaries and exhibited a high rate of proliferation balanced by apoptosis. Steeg and her colleagues transfected human breast carcinoma cells with thrombospondin-1 (202) and saw a significant inhibition of tumor growth in immunocompromized mice and a decreased microvessel density. The stable transfection of thrombospondin-1 into transformed endothelial cells restored a normal phenotype to these cells in vitro and suppressed tumorigenesis in vivo (203). Further, thrombospondin-1 transfection inhibited angiogenesis and tumor growth of human skin carcinomas injected subcutaneously in mice (204). These studies provide a basis for the use of gene therapy to deliver angiogenesis inhibitors.

Several angiogenesis inhibitors, including angiostatin (205–208), endostatin (209,210), a truncated VEGF receptor (211,212), and tie2 (213) have been delivered using gene therapy. In these studies, a potent inhibition of angiogenesis and tumor growth without evidence was observed. These studies suggest that gene therapy may provide a platform for the therapeutic delivery of angiogenesis factors. However, many of the antiangiogenic factors are internal fragments of proteins, and it may therefore be difficult to consistently deliver them in an active form using gene therapy.

4.7. Antiangiogenic Scheduling of Chemotherapy

Endothelial cell proliferation is an important component of angiogenesis, and dividing endothelial cells are highly sensitive to most chemotherapeutics. However, cytotoxic chemotherapy is typically administered at high doses given relatively infrequently, allowing for repopulation of endothelial cells. Thus, current strategies used to administer chemotherapy have only a limited effect on angiogenesis. By increasing the frequency of administration of these agents and by giving them over prolonged periods, the antiangiogenic activity of cytotoxic agents can be enhanced. For example, cyclophosphamide at low dose given every 6 d for prolonged periods can produce a greater antiangiogenic effect than higher doses given every 21 d (214). Antiangiogenic scheduling of chemotherapeutic agents was effective at controlling the growth of tumors that were almost completely resistant to the conventional scheduling. However, tumor growth resumed in many cases despite the continued administration of the antiangiogenic chemotherapy (214) when it was used alone.

The mechanism of this apparent resistance to the antiangiogenic chemotherapy is an increase in the production of proangiogenic molecules by the tumor cells in the treated mice, leading to an enhanced survival of endothelial cells (215). The treatment of neuroblastoma tumors in mice treated with an antiangiogenic schedule of vinblastine was associated with an increased expression of VEGF/VPF and hence increased survival of endothelial cells (215). By combining the vinblastine therapy with VEGF/VPF blockade, an almost complete and durable tumor control was achieved. An enhancement of the antitumor effects of the drug topotecan when given via antiangiogenic scheduling by the concurrent administration of antibodies directed against VEGF/VPF has also been observed in an experimental model of Wilms tumor (216). Similar enhancements of the antitumor effects of antiangiogenic scheduling have been reported for cyclophosphamide combined with TNP-470 (214) and carboplatinum and etoposide combined with the angiogenesis inhibitor PEX, a fragment of matrix metalloproteinase (MMP)-2 *(217)*. The various combinations of cytotoxic agents with antiangiogenic agents provided a means of overcoming the resistance to therapy with cytotoxic agents administered alone.

Given that cytotoxic drugs at conventional doses can produce rapid antitumor effects, an effective and rapid anticancer strategy might be to give cytotoxic agents at conventional doses but continue therapy using antiangiogenic doses of chemotherapeutics given chronically. Since the use of antiangiogenic schedules of chemotherapy can still be associated with toxicity to normal tissues, this strategy may only be useful until more specific antiangiogenic agents become available. It may therefore be prudent to combine conventional doses of chemotherapy (i.e., maximum tolerated dose) with antiangiogenic doses in order to achieve rapid and durable control of primary and metastatic disease. It is important to note that it is not the low dose of the chemotherapeutic agent that is responsible for the observed antiangiogenic effects. Instead, it is the more frequent and prolonged administration of the drug that results in the antiangiogenic effect. The lower doses are used to minimize the toxicity and mortality that would normally be associated with the frequent administration and prolonged course of chemotherapy.

5. Limitations of Antiangiogenic Monotherapy

Although antiangiogenic agents show great promise in preclinical cancer models, their use, particularly when administered as monotherapy, may be associated with a number of potential limits and problems in the clinic. In this section, some of these limitations are outlined.

5.1. Heterogeneity of the Endothelium

Recent work shows that endothelial cells in different organs express different surface receptors *(37)* and have different patterns of gene expression *(218)*. Angiogenic endothelial cells also express discrete surface receptors that were used effectively to target the tumor vasculature with adriamycin and other drugs. More recently, organ microenvironment has been shown to regulate the expression for the growth factors EGF and PDGF in mice *(134,137)*. These data suggest that endothelial cells in different tissue beds may respond differentially to antiangiogenic therapy. Thus, antiangiogenic agents may have to be used in combination with each other, and therapy may have to be optimized on an organ- and tissue-, and perhaps patient-, specific basis.

5.2. Potential Toxicities of Antiangiogenic Agents

In adults, endothelial cells typically maintain a state of prolonged quiescence *(19)*. The female menstrual cycle and wound healing are exceptions to this pattern. It had been assumed that antiangiogenic therapy would therefore impair wound healing. TNP-470 treatment of mice led to a decrease in the tensile strength of cutaneous wounds. However, this occurred only if the drug was administered within 24 h of the wounding *(219)*. TNP-470 has also been shown to impair fracture healing in the bones of rats *(220)*. More recently, endostatin has been tested for its effect on wound healing in mice. In one study, no effect was seen in mice receiving doses of endostatin that could completely suppress tumor and ocular angiogenesis *(221)*. In another study, only subtle histologic changes were seen in healing wounds of mice treated with endostatin *(222)*. Indeed, the decreased fibrosis observed in this model could be beneficial. In clinical trials of endostatin, no effect on wound healing has been observed *(223)*.

Although VEGF is a stimulator of angiogenesis, an equally important for the molecule is to maintain endothelial cell survival *(144)*. VEGF is also critical for normal vascular development, and VEGF knockout was associated with embryonic lethality in mice *(224)*. Thus factors that target the growth factor VEGF may cause a disruption of endothelial integrity and these factors may lead to a disruption of the normal vasculature architecture. In clinical trials, VEGF inhibitors that target the protein directly or target its receptor have been associated with thrombotic or hemorrhagic complications. Although observed in only a small percentage of patients, these data suggest that anti-VEGF strategies will have to be monitored closely, particularly if the agents are given at high dose for prolonged periods.

5.3. Delayed Tumor Response to Antiangiogenic Therapy

Angiostatin *(72)*, endostatin *(100)*, and antiangiogenic antithrombin (aaAT) *(60)* are endogenous angiogenesis inhibitors that are mobilized by proteolytic cleavage *(73,74,225,226)* of their parent proteins plasminogen, collagen XVIII, and antithrombin, respectively. All three are highly specific for microvascular endothelial cells, can induce a blockade of tumor angiogenesis, and have not caused any significant toxicity in preclinical and clinical use even after prolonged therapy. However, preclinical studies with all of these agents have demonstrated a number of potential therapeutic limitations that could arise with their use as monotherapy in the treatment of cancer. In several in-vivo models, all of these agents inhibited tumor growth but were associated with a delayed onset of activity *(101,104)*. In many of the animal models, tumors progressed by as much as 400% in the first several days after initiation of therapy. All of these models used rapidly growing tumors. Given that the doubling time of murine tumors is several-fold higher than is observed in the presentation of human cancer, this delay in the onset of activity could translate to several months in patients. For example, antiangiogenic therapy with IFN-α for the treatment of life-threatening pediatric hemangiomas *(123)* or giant cell tumors *(127)* required several months of continued therapy before a significant clinical response was observed. For patients with advanced metastatic and/or locally advanced disease, this delay in onset of activity may make the use of these agents unpractical.

Another potential limit for the use of antiangiogenic agents is the high dose over the prolonged treatment course that is required to obtain maximal efficacy. As a result, it may not be feasible and/or practical to produce the amount of material that would be required for widespread use. For angiostatin protein, a dose of up to 100 mg/kg/d *(104)* was required to regress tumors in mice, and treatment had to be continued for several months to prevent tumor recurrence. Although recent work suggests that the delivery of antiangiogenic agents via continuous infusion *(227,228)* or sustained release *(100,101)* may allow for a reduction of the bolus dose, the amount of protein needed for widespread use will still be challenging. By combining antiangiogenic agents with each other and conventional agents, it may be possible to reduce the dose required and the frequency of administration, and make their use more practical.

5.4. Microscopic Residual Disease and Antiangiogenic Monotherapy

Perhaps the most significant limitation of antiangiogenic monotherapy is the inability of these agents to eradicate disease completely, even after prolonged administration of high doses. For most angiogenesis inhibitors, tumor growth generally resumes within a short time *(104)* after cessation of therapy. Although prolonged therapy with endostatin, either by cycled administration of the drug *(101)* or by prolonged administration at high dose (O'Reilly et al., unpublished), did induce a self-sustained dormancy for a wide variety of tumors that persisted off therapy, residual microscopic disease was still present in all of the treated mice. Immunohistochemistry of the residual microscopic tumor revealed little or no evidence of angiogenesis but also revealed a high proliferative index of the tumor cells that was offset by a high apoptotic index.

If similar results are observed in patients treated with antiangiogenic agents, they might then require prolonged antiangiogenic therapy. Although little or no toxicity has been observed to date for angiostatin or endostatin, their prolonged use could result in late effects. Endothelial cells normally exist in a quiescent state, with a turnover rate of months to years *(19)*. The prevention of this turnover by prolonged therapy could lead to potential normal organ toxicities. Further, it would obviously be preferable to develop strategies using angiogenesis inhibitors in combination with more traditional modalities in order to eradicate the disease at the time of initial treatment in as many patients as possible.

6. Antiangiogenic Agents as Part of Combined-Modality Therapy

Angiogenesis inhibitors are currently being evaluated in the clinic in trials in patients with advanced cancer and other angiogenesis-dependent diseases. However, a number of obstacles must be overcome before they can be approved for more widespread clinical use. The limitations of antiangiogenic monotherapy will have to be overcome. In order to accomplish this, it will be necessary to combine antiangiogenic agents with each other and with existing and emerging modalities. In this section, the combination of antiangiogenic agents with other modalities used in the treatment of cancer is discussed.

6.1. Angiogenesis Inhibition and Immunotherapy

The use of angiogenesis inhibitors may also provide a platform for immunotherapy as part of the treatment of cancer. Recently, the combination of an integrin antagonist to inhibit angiogenesis and an antibody–cytokine fusion protein induced the complete regression of spontaneous liver metastases in a murine model of neuroblastoma *(229)*. Regressions were not seen with the two modalities when they were administered separately. Strategies that target VEGF/VPF may also be a valuable adjuvant in the immunotherapy of cancer. VEGF/VPF is a critical regulator of angiogenesis but can also inhibit the maturation and function of dendritic cells *(230,231)*. Gabrilovich and colleagues *(232)* observed that an anti-VEGF antibody significantly improved the number and function of lymph node and spleen dendritic cells in tumor-bearing animals and enhanced the effects of immunotherapy with dendritic cells pulsed with mutation-specific p53 peptides.

Further, some molecules may induce both an antiangiogenic and an immunomodulatory response. This is true of interleukin-12, which inhibits angiogenesis indirectly by upregulating interferon-γ *(121)*, that then induces interferon-inducible protein 10 and MIG-1 *(119)* via IFN-γ. Both are potent inhibitors of angiogenesis and tumor growth *(119,120,233)*. The potent efficacy of interleukin-12 against a wide variety of tumors may be explained by the combination of its effects on the immune system to induce a tumoricidal effect and its antiangiogenic effect.

Recently, the angiogenesis inhibitor calreticulin was linked to a tumor antigen, human papilloma virus type-16 (HPV-16) E7, for the development of a DNA vaccine. An enhanced antitumor activity was observed in the group of mice that received the combined antiangiogenic therapy with the DNA vaccine *(234)*. Gyorffy and colleagues *(235)* studied the intratumoral delivery of adenoviral vectors for angiostatin or IL-12 in a murine model of breast cancer and found that antiangiogenic therapy combined with immunotherapy could regress the original tumor and vaccinate the animal against rechallenge. Davidoff and colleagues *(236)* reported a gene therapy approach in which endostatin and the marker protein and potent immunogen, green fluorescent protein, were delivered to murine neuroblastoma cells prior to inoculation of the tumor cells into mice. Angiogenesis inhibition or immunomodulation alone resulted in only a modest delay in tumor growth, whereas the combination prevented the formation of appreciable tumors in the majority of the mice. A protection against subsequent tumor challenge with unmodified tumor cells was also observed. When taken together, these studies strongly suggest that antiangiogenic and immunotherapy strategies can act synergistically when used as part of a multimodality anticancer approach.

6.2. Combination of Antiangiogenic Agents and Chemotherapy

Teicher and her colleagues were among the first to show a synergistic effect between cytotoxic agents, such as tetrahydrocortisol, α-cyclodextrin tetradecasulfate, minocycline, and antiangiogenic agents for treatment of transplantable murine tumors *(237)*. In a similar set of experiments, a combination of the angiogenesis inhibitor TNP-470 (AGM-1470) and minocycline given alone or in combination with cytotoxic agents were administered to mice bearing Lewis lung carcinoma *(238)*. The angiogenesis inhibitor cortisone acetate has also been shown to enhance the activity of chemotherapy in a model of transitional cell carcinoma *(239)*. In a model of non-Hodgkin's lymphoma, the sequential administration of cytotoxic therapy followed by endostatin therapy induced stabilization of disease in mice *(240)*.

As described above, blockade of the EGF receptors has both a direct antitumor effect and an antiangiogenic effect. Thus, these agents are likely to be enhanced by the addition of conventional cytotoxic agents. ZD-1839 (Iressa), a quinazoline derivative that selectively inhibits the EGFR tyrosine kinase treatment of nude mice bearing established human GEO colon cancer xenografts, revealed a reversible dose-dependent inhibition of tumor growth *(241)*. However, tumors resumed the growth rate of controls at the end of the treatment. In contrast, the combination of cytotoxic agent treatment with ZD-1839 produced tumor growth arrest in all mice, and 50% of mice, treated with ZD-1839 plus topotecan, raltitrexed, or paclitaxel were still alive 10, 12, and 15 wk after cancer cell injection, respectively. Bruns and colleagues *(242)* used C225, an anti-EGF-R antibody *(243,244)*, in combination with gemcitabine in an orthotopic nude mouse model of pancreatic cancer. The combination of C225 and gemcitabine resulted in additive antitumor and antimetastatic effects and a dose-dependent decrease in expression of VEGF/VPF and IL-8. Further, agents that target EGF can be used to overcome tumor cell resistance to cytotoxic agents.

A similar enhanced antitumor activity has been observed when cytotoxic agents have been combined with a blockade of either the PDGF-R by SU101 *(245)* or with antibodies directed against VEGF/VPF *(215,216)*. Hormone ablation therapy and antiangiogenic therapy with anti-VEGF/VPF antibodies can also lead to an enhanced antitumor response *(246)*, which can be further improved by the addition of cytotoxic therapy with doxorubicin and cyclophosphamide. Androgen-dependent Shinogi male mouse mammary carcinomas were treated with anti-VEGFR-2 mAb, hormone ablation, and/or chemotherapy. In the groups that received combined-modality therapy, an increased intratumoral pO_2 and tumor growth arrest was observed. Taken together, these results provide a strong rationale for the combination of cytotoxic drugs with agents that target growth factors that are important in angiogenesis. Clinical trials are now being designed using antiangiogenic schedules of cytotoxic agents in patients with advanced malignancy that is resistant to conventional administration of cytotoxic agents *(247)*.

6.3. Combination of Antiangiogenic Agents and Radiation Therapy

Radiation therapy may have a systemic and/or local effect on angiogenesis. Studies of Canney and Dean *(248)* have demonstrated an increase in the level of TGF-β1, which is a potent stimulator of angiogenesis in vivo, after radiochemotherapy to the liver. More recently, Gorski et al. *(249)* demonstrated that the expression of VEGF/VPF, detected within tumors by both immunohistochemistry and Northern analysis, was increased after local irradiation. Radiation therapy may also inhibit systemic angiogenesis. In a recent study by Hartford et al. *(250)*, radiation therapy of a primary tumor in rodents was associated with an inhibition of angiogenesis seen in a cranial window at a distant site. An interesting finding was that the plasma levels of endostatin, an endogenous angiogenesis inhibitor, in the irradiated mice were twice those found in the nonirradiated mice. When taken together with prior studies *(103)*, these findings strongly suggest that radiation therapy may have a proangiogenic effect at a remote site by increasing the levels of angiogenesis stimulators such as VEGF/VPF or TGB-β or by decreasing the mobilization of angiogenesis inhibitors such as angiostatin. Therefore, the addition of antiangiogenic therapy to radiation therapy may restore the delicate balance of angiogenic factors and thereby control locoregional and metastatic tumor growth.

It had long been assumed that an angiogenesis inhibitor would impair the effect of ionizing radiation by inducing hypoxia in the tumor bed. Teicher et al. *(251)*, however, observed that antiangiogenic therapy given in combination with the angiogenesis inhibitors TNP-470 and minocycline, a weak inhibitor of metalloproteinase activity, improved tumor oxygenation and the antitumor effect of radiation therapy. To better understand the benefits of combining antiangiogenic agents with radiation therapy, Jain and colleagues examined intratumoral pO_2 during antiangiogenic therapy with anti-VEGF-R-2 antibodies. They observed that pO_2 initially decreased with therapy but then subsequently increased *(246)*. The antitumor effect of hypothermia is also enhanced by the combination with antiangiogenic therapy using TNP-470 *(252–254)*. Murata et al. *(255)*, however, observed that the concurrent treatment of mouse breast carcinoma xenografts with TNP-470 and fractionated radiation therapy resulted in a decrease in tumor oxygenation and a decrease in therapeutic response. The enhancement of the effect of radiation therapy by antiangiogenic therapy may also be dependent on the tumor microenvironment. Lund et al. *(256)* treated mice with xenografts of glioblastoma multiforme growing in the muscle of the thigh or in the brain with TNP-470 alone, radiation therapy alone, and in combination. The tumors reacted differently for undetermined reasons. It is tempting to speculate that differences in the capillary beds and the microenvironment of the brain and the musculature of the thigh may have contributed to the differences in response.

In the first study of the combination of a specific inhibitor of angiogenesis combined with radiation therapy, Weichselbaum and colleagues *(257)* demonstrated a synergistic effect when angiostatin and ionizing radiation were combined to treat a variety of transplantable tumors in mice. More recently, a number of agents with antiangiogenic activity have been used in combination with concurrent radiation therapy for the treatment of cancer in preclinical models. In studies conducted by Milas and colleagues, an enhancement of the response to radiation therapy was observed when it was combined with a selective inhibitor of the cyclooxygenase-2 enzyme *(258)*. A similar enhancement of the in-vivo antitumor activity of radiation therapy has been observed when it is combined with antibodies directed against the EGF receptor *(259,260)*. An increase in tumor growth delay and an augmentation of tumor curability was observed in the combined modality groups. A similar enhancement of the response to radiation therapy has also been demonstrated when it is combined with anti-VEGF, even for tumors that were markedly hypoxic *(261)*. Gorski and colleagues *(249)* showed that treatment of tumor-bearing mice with radiotherapy and antibodies against VEGF had a synergistic effect against the primary tumor. More recently, it has been shown that the combination of agents that prevent VEGF signaling with radiotherapy, tumor resistance to radiation could be reduced and response to therapy improved *(262)*. Kozin et al. *(263)* treated mice implanted with human small-cell lung cancer or glioblastoma multiforme with fractionated radiation therapy combined with antibodies directed against the VEGF receptor 2 and followed the surviving mice for at least 6 mo. The dose of radiation required to control 50% of tumors was diminished by at least 1.3-fold by the concurrent addition of the antiangiogenic agent.

In most of the studies to date, radiation therapy has been given concurrent with the antiangiogenic agents, and few studies have compared different sequences of radiation therapy and antiangiogenic agents. In the only published study to date comparing different sequencing of radiation therapy and antiangiogenic therapy *(264)*, the concurrent administration of radiation with antiangiogenic therapies produced the best therapy. However, the optimal strategies for combining antiangiogenic therapy with radiation therapy or other modalities used in the treatment of cancer will need to be determined with extensive preclinical and clinical testing.

7. Monitoring Antiangiogenic Therapy

Antiangiogenic agents, as with many other biologic therapies, may not result in the same pattern of regressions and responses that are seen with effective chemotherapeutic agents. As described above, tumor progression over a period of several weeks or months may occur when antiangiogenic agents are used to treat advanced malignancy. Thus, conventional strategies for monitoring anticancer therapies may not apply for antiangiogenic agents, and novel surrogates must be developed and validated.

Circulating levels of angiogenic factors have been used to determine prognosis for a wide variety of cancers in patients. In patients with early stages of disease, elevated levels of circulating proangiogenic factors are associated with an increased risk of locoregional failure and metastatic disease for a broad range of malignancies for VEGF/VPF *(265,266)*, bFGF *(267)*, and other proangiogenic molecules *(268)*. These strategies were based on the prognostic value to tumor microvessel density initially demonstrated for breast cancer *(269,270)*, which has since been demonstrated for many malignancies *(271)*. More recently, decreases in the circulating levels of proangiogenic molecules have been associated with a response to cytotoxic agents administered at conventional doses *(272,273)*. However, the use of circulating levels of these factors as surrogates for response to an antiangiogenic agent is less clear. In preclinical studies of an orthotopic model of lung cancer, the expression of VEGF/VPF and bFGF increased after effective therapy with endostatin *(274)*. The increased production of proangiogenic factors by tumors treated with an angiogenesis inhibitor such as endostatin is consistent with the theory that angiogenesis is regulated by a balance of pro- and antiangiogenic factors. Indeed, the increased expression of proangiogenic factors by tumor cells may provide a relative resistance of the tumor to the effects of antiangiogenic agents. However, since antiangiogenic agents that directly target the endothelial cell have not been associated with any toxicity, any resistance could be offset by an increase in dose. Thus, circulating levels of individual factors may not be adequate and angiogenic profiles that include both stimulators and inhibitors of angiogenesis may need to be developed.

Tumor biopsies may also provide information that can be used as surrogate markers for angiogenesis *(275–279)* and tissue expression of both pro- and antiangiogenic factors can be determined both before and after treatment with antiangiogenic agents. Tissues can also be studied for evidence of tumor and endothelial cell proliferation, apoptosis, and for microvessel density. However, these strategies have yet to be validated as a surrogate for response to antiangiogenic therapy, are impractical for many patients, and are invasive.

As an alternative to invasive strategies, imaging of tumor blood flow and tumor blood volume has been proposed to monitor angiogenesis. Ultrasound, MRI, CT, and PET are all currently being studied in a number of clinical trials of angiogenesis inhibitors; however, given that tumor angiogenesis is an inefficient process, there may be a period where tumor blood flow actually increases during effective antiangiogenic therapy. Jain *(280,281)* has proposed that the inhibition of angiogenesis may lead to tumor growth inhibition that would then result in a decrease in interstitial pressure and an apparent increase in blood flow. As with the use of tumor biopsies and bodily fluids, however, imaging modalities have not yet been validated as surrogates for response to antiangiogenic therapy. Thus, clinical trials of antiangiogenic agents need to be designed not only to determine if the agents are safe and have evidence of efficacy, but also to validate both invasive and noninvasive surrogates of response.

8. Concluding Remarks

Since its beginning in the early 1970s, the field of angiogenesis research has grown rapidly and has advanced our understanding of a number of biologic processes. Angiogenesis is a complex process that involves the interaction of endothelial cells with stromal cells and the organ microenvironment. In cancer and other angiogenesis-dependent diseases, the endothelial cell functions as a critical regulator of the malignant phenotype. Many endothelial targets have now been identified and many antiangiogenic agents are now being evaluated for clinical efficacy. These agents should have far-reaching applications in a variety of clinical settings and have the potential to improve efficacy and diminish toxicity in a variety of diseases. However, a better understanding of the process of angiogenesis is still needed before these agents can be successfully incorporated into clinical practice. To target the tumor vasculature effectively, therapies will have to be individualized and customized. Antiangiogenic therapy is likely to evolve on an organ- and tissue-specific basis and integrated with existing and emerging therapies for treating cancer.

Acknowledgment

This work was supported in part by Cancer Center Support Core Grant CA16672 from the National Cancer Institute, National Institutes of Health.

References

1. Folkman, J. (1990) What is the evidence that tumors are angiogenesis dependent? *J. Natl. Cancer Inst.* **82**, 4–6.
2. O'Reilly, M. S. (2000) Antiangiogenesis: basic principles, in *Principles and Practice of the Biologic Therapy of Cancer*, 3rd ed. (Rosenberg, S. A., ed.). Lippincott Williams & Wilkins, Philadelphia, pp. 827–843.
3. Folkman, J. and Shing, Y. (1992) Angiogenesis. *J. Biol. Chem.* **267**, 10931–10934.
4. Moulton, K. S., Heller, E., Konerding, M. A., Flynn, E., Palinski, W., and Folkman, J. (1999) Angiogenesis inhibitors endostatin or TNP-470 reduce intimal neovascularization and plaque growth in apolipoprotein E-deficient mice [see comments]. *Circulation* **99**, 1726–1732.
5. Perez-Atayde, A. R., Sallan, S. E., Tedrow, U., Connors, S., Allred, E., and Folkman, J. (1997) Spectrum of tumor angiogenesis in the bone marrow of children with acute lymphoblastic leukemia. *Am. J Pathol.* **150**, 815–821.
6. Folkman, J. (1971) Tumor angiogenesis: therapeutic implications. *N. Engl. J. Med.* **285**, 1182–1186.
7. Folkman, J., Watson, K., Ingber, D., and Hanahan, D. (1989) Induction of angiogenesis during the transition from hyperplasia to neoplasia. *Nature* **339**, 58–61.
8. Hanahan, D. and Folkman, J. (1996) Patterns and emerging mechanisms of the angiogenic switch during tumorigenesis. *Cell* **86**, 353–364.
9. Singh, R. K., Gutman, M., Bucana, C. D., Sanchez, R., Llansa, N., and Fidler, I. J. (1995) Interferons alpha and beta down-regulate the expression of basic fibroblast growth factor in human carcinomas. *Proc. Natl. Acad. Sci. USA* **92**, 4562–4566.
10. Dark, G. G., Hill, S. A., Prise, V. E., Tozer, G. M., Pettit, G. R., and Chaplin, D. J. (1997) Combretastatin A-4, an agent that displays potent and selective toxicity toward tumor vasculature *Cancer Res.* **57**, 1829–1834.
11. Folkman, J. (1995) Angiogenesis in cancer, vascular, rheumatoid and other disease, *Nature Med.* **1**, 27–31.
12. Fidler, I. J., Kerbel, R. S., and Ellis, L. M. (2001) Biology of cancer: angiogenesis, in *Cancer Principles and Practice of Oncology*, 6th ed. (DeVita, V. T., Hellman, S., and Rosenberg S. A., eds.). Lippincott Williams & Wilkins, Philadelphia, pp. 137–147.
13. Hamada, J., Cavanaugh, P. G., Lotan, O., and Nicolson, G. L. (1992) Separable growth and migration factors for large-cell lymphoma cells secreted by microvascular endothelial cells derived from target organs for metastasis. *Br. J. Cancer* **66**, 349–354.
14. Nicosia, R. F., Tchao, R., and Leighton, J. (1986) Interactions between newly formed endothelial channels and carcinoma cells in plasma clot culture. *Clin. Exp. Metastasis* **4**, 91–104.
15. Rak, J., Filmus, J., and Kerbel, R. S. (1996) Reciprocal paracrine interactions between tumour cells and endothelial cells: the "angiogenesis progression" hypothesis [Review]. *Eur. J. Cancer* **32A**, 2438–2450.

16. Folkman, J. (1996) Tumor angiogenesis and tissue factor. *Nat. Med.* **2**, 167–168.
17. Libutti, S. K. and Pluda, J. M. (2000) Antiangiogenesis: clinical applications, in *Principles and Practice of the Biologic Therapy of Cancer*, 3d ed. (Rosenberg, S. A., ed.) Lippincott Williams & Wilkins, Philadelphia, pp. 844–864.
18. Folkman, J. (1995) Clinical applications of angiogenesis research. *N. Engl. J. Med.* **333**, 1757–1763.
19. Hobson, B. and Denekamp, J. (1984) Endothelial proliferation in tumors and normal tissues: continuous labelling studies. *Br. J. Cancer* **49**, 405–413.
20. Fidler, I. J. and Ellis, L. M. (1994) Letter; comment. *Cell* **79**, 185–188.
21. Shing, Y., Folkman, J., Sullivan, R., Butterfield, C., Murray, J., and Klagsbrun, M. (1984) Heparin-affinity: purification of a tumor-derived capillary endothelial cell growth factor. *Science* **223**, 1296–1299.
22. Senger, D. R., Galli, S. J., Dvorak, A. M., Perruzzi, C. A., and Harvey, V. S. (1983) Tumor cells secrete a vascular permeability factor that promotes accumulation of ascites fluid. *Science* **219**, 983–985.
23. Ferrara, N. and Henzel, W. J. (1989) Pituitary follicular cells secrete a novel heparin-binding growth factor specific for vascular endothelial cells. *Biochem. Biophys. Res. Commun.* **161**, 851–858.
24. Koch, A. E., Polverini, P. J., Kunkel, S. L., et al. (1992) Interleukin-8 as a macrophage-derived mediator of angiogenesis [see comments]. Comment in: *Science* 1995 Apr 21; 268 (5209):447–448. *Science* **258**, 1798–1801.
25. Kuniyasu, H., Yasui, W., Shinohara, H., et al. (2000) Induction of angiogenesis by hyperplastic colonic mucosa adjacent to colon cancer. *Am. J. Pathol.* **157**, 1523–1535.
26. Grant, D. S., Kleinman, H. K., Goldberg, I. D., et al. (1993) Scatter factor induces blood vessel formation in vivo. *Proc. Natl. Acad. Sci. USA* **90**, 1937–1941.
27. Roberts, A. B., Sporn, M. B., Assoian, R. K., et al. (1986) Transforming growth factor type-beta: rapid induction of fibrosis and angiogenesis in vivo and stimulation of collagen formation in vitro. *Proc. Natl. Acad. Sci. USA* **83**, 4167–4171.
28. O'Mahony, C. A., Albo, D., Tuszynski, G. P., and Berger, D. H. (1998) Transforming growth factor-beta 1 inhibits generation of angiostatin by human pancreatic cancer cells. *Surgery* **124**, 388–393.
29. Jackson, D., Volpert, O., Bouck, N., and Linzer, D. (1994) Stimulation and inhibition of angiogenesis by placental proliferin and proliferin-related protein. *Science* **266**, 1581–1585.
30. Ribatti, D., Presta, M., Vacca, A., et al. (1999) Human erythropoietin induces a pro-angiogenic phenotype in cultured endothelial cells and stimulates neovascularization in vivo. *Blood* **93**, 2627–2636.
31. Iruela-Arispe, M. L. and Dvorak, H. F. (1997) Angiogenesis: a dynamic balance of stimulators and inhibitors. *Thromb. Haemost.* **78**, 672–677.
32. Good, D. J., Polverini, P. J., Rastinejad, F., et al. (1990) A tumor suppressor-dependent inhibitor of angiogenesis is immunologically and functionally indistinguishable from a fragment of thrombospondin. *Proc. Natl. Acad. Sci. USA* **87**, 6624–6628.
33. Rastinejad, F., Polverini, P. J., and Bouck, N. P. (1989) Regulation of the activity of a new inhibitor of angiogenesis by a cancer suppressor gene. *Cell* **56**, 345–355.
34. Dameron, K. M., Volpert, O. V., Tainsky, M. A., and Bouck, N. (1994) Control of angiogenesis in fibroblasts by p53 regulation of thrombospondin-1. *Science* **265**, 1582–1584.
35. Dawson, D. W., Volpert, O. V., Gillis, P., et al. (1999) Pigment epithelium-derived factor: a potent inhibitor of angiogenesis. *Science* **285**, 245–248.
36. Folkman, J. (1995) Tumour angiogenesis, in *The Molecular Basis of Cancer*. (Mendelsohn, P. M. H., Israel, M. A., and Liotta, L. A. ed.). Saunders, Philadelphia, 1995.
37. Arap, W., Pasqualini, R., and Ruoslahti, E. (1998) Cancer treatment by targeted drug delivery to tumor vasculature in a mouse model. *Science* **279**, 377–380.

38. Konerding, M. A., Malkusch, W., Klapthor, B., et al. (1999) Evidence for characteristic vascular patterns in solid tumours: quantitative studies using corrosion casts. *Br. J. Cancer* **80**, 724–732.

39. Jain, R. K. (2001) Normalizing tumor vasculature with anti-angiogenic therapy: a new paradigm for combination therapy [Review] [28 refs]. *Nat. Med.* **7**, 987–989.

40. Orlidge, A. and D'Amore, P. (1987) Inhibition of capillary endothelial cell growth by pericytes and smooth muscle cells. *J. Cell Biol.* **105**, 1455–1462.

41. Jain, R. K. and Baxter, L. T. (1988) Mechanisms of heterogeneous distribution of monocloncal antibodies and other macromolecules in tumors: significance of elevated interstitial pressure. *Cancer Res.* **48**, 7022–7032.

42. Jain, R. K. (1988) Determinants of tumor blood flow: a review. *Cancer Res.* **48**, 2641–2658.

43. Jain, R. K. (1989) Delivery of novel therapeutic agents in tumors: physiological barriers and strategies. *J. Natl. Cancer Inst.* **81**, 570–576.

44. Brown, L. F., Guidi, A. J., Schnitt, S. J., et al. (1999) Vascular stroma formation in carcinoma in situ, invasive carcinoma, and metastatic carcinoma of the breast *Clin. Cancer Res.* **5**, 1041–1056.

45. Dvorak, H. F. (1986) Tumors: wounds that do not heal. *N. Engl. J. Med.* **315**, 1650–1659.

46. Dvorak, H. F., Nagy, J. A., Dvorak, J. T., and Dvorak, A. M. (1988) Identification and characterization of the blood vessels of solid tumors that are leaky to circulating macromolecules. *Am. J. Pathol.* **133**, 95–109.

47. Nagy, J. A., Brown, L. F., Senger, D. R., et al. (1989) Pathogenesis of tumor stroma generation: a critical role for leaky blood vessels and fibrin deposition. *Biochim. Biophys. Acta* **948**, 305–326.

48. Gerlowski, L. E. and Jain, R. K. (1986) Microvascular permeability of normal and neoplastic tissues. *Microvasc. Res.* **31**, 288–305.

49. Liotta, L. A., Stetler-Stevenson, W. G., and Steeg, P. S. (1991) Cancer invasion and metastasis: positive and negative regulatory elements. *Cancer Invest.* **9**, 543–551.

50. Stetler-Stevenson, W. G. (1999) Matrix metalloproteinases in angiogenesis: a moving target for therapeutic intervention. *J. Clin. Invest.* **103**, 1237–1241.

51. Kleiner, D. E. and Stetler-Stevenson, W. G. (1999) Matrix metalloproteinases and metastasis. [Review]. *Cancer Chemother. Pharmacol.* **43** (Suppl.) S42–S51.

52. Mohle, R., Green, D., Moore, M. A., Nachman, R. L., and Rafii, S. (1997) Constitutive production and thrombin-induced release of vascular endothelial growth factor by human megakaryocytes and platelets. *Proc. Natl. Acad. Sci. USA* **94**, 663–668.

53. Haralabopoulos, G. C., Grant, D. S., Kleinman, H. K., and Maragoudakis, M. E. (1997) Thrombin promotes endothelial cell alignment in Matrigel in vitro and angiogenesis in vivo. *Am. J. Physiol.* **273**, C239–C245.

54. Tsopanoglou, N. E. and Maragoudakis, M. E. (1999) On the mechanism of thrombin-induced angiogenesis. Potentiation of vascular endothelial growth factor activity on endothelial cells by up-regulation of its receptors. *J. Biol. Chem.* **274**, 23969–23976.

55. Baker, E. A., Bergin, F. G., and Leaper, D. J. (2000) Plasminogen activator system, vascular endothelial growth factor, and colorectal cancer progression. *Mol. Pathol.* **53**, 307–312.

56. Brodsky, S., Chen, J., Lee, A., Akassoglou, K., Norman, J., and Goligorsky, M. S. (2001) Plasmin-dependent and -independent effects of plasminogen activators and inhibitor-1 on ex vivo angiogenesis. *Am. J. Physiol.—Heart Circ. Physiol.* **281**, H1784–H1792.

57. Browder, T., Folkman, J., and Pirie-Shepherd, S. (2000) The hemostatic system as a regulator of angiogenesis [Review] [67 refs]. *J. Biol. Chem.* **275**, 1521–1524.

58. Stefansson, S., Petitclerc, E., Wong, M. K., McMahon, G. A., Brooks, P. C., and Lawrence, D. A. (2001) Inhibition of angiogenesis in vivo by plasminogen activator inhibitor-1. *J. Biol. Chem.* **276**, 8135–8141.

59. Stefansson, S. and Lawrence, D. A. (1996) The serpin PAI-1 inhibits cell migration by blocking integrin avb3 binding to vitronectin. *Nature* **383**, 441–443.

60. O'Reilly, M. S., Pirie-Shepherd, S., Lane, W. S., and Folkman, J. (1999) Antiangiogenic activity of the cleaved conformation of the serpin antithrombin [see comments]. Comment in: *Science* 1999 Sep 17; 285(5435):1861–3, *Science* **285**, 1926–1928.

61. Stathakis, P., Fitzgerald, M., Matthias, L. J., Chesterman, C. N., and Hogg, P. J. (1997) Generation of angiostatin by reduction and proteolysis of plasmin. *J. Biol. Chem.* **272**, 20641–20645.

62. Gately, S., Twardowski, P., Stack, M. S., et al. (1997) The mechanism of cancer-mediated conversion of plasminogen to the angiogenesis inhibitor angiostatin. *Proc. Natl. Acad. Sci. USA* **94**, 10868–10872.

63. Moses, M. A., Sudhalter, J., and Langer, R. (1990) Identification of an inhibitor of neovascularization from cartilage. *Science* **248**, 1408–1410.

64. Moses, M. A. (1997) The regulation of neovascularization by matrix metalloproteinases and their inhibitors. *Stem Cells* **15**, 180–189.

65. Moses, M. A., Wiederschain, D., Wu, I., et al. (1999) Troponin I is present in human cartilage and inhibits angiogenesis. *Proc. Natl. Acad. Sci. USA* **96**, 2645–2650.

66. Tamargo, R. J., Bok, R. A., and Brem, H. (1991) Angiogenesis inhibition by minocycline. *Cancer Res.* **51**, 672–675.

67. Guerin, C., Laterra, J., Masnyk, T., Golub, L. M., and Brem, H. (1992) Selective endothelial growth inhibition by tetracyclines that inhibit collagenase. *Biochem. Biophys. Res. Commun.* **188**, 740–745.

68. Rudek, M. A., Figg, W. D., Dyer, V., et al. (2001) Phase I clinical trial of oral COL-3, a matrix metalloproteinase inhibitor, in patients with refractory metastatic cancer. *J. Clin. Oncol.* **19**, 584–592.

69. Gatto, C., Rieppi, M., Borsotti, P., et al. (1999) BAY 12–9566, a novel inhibitor of matrix metalloproteinases with antiangiogenic activity. *Clin. Cancer Res.* **5**, 3603–3607.

70. Zucker, S., Cao, J., and Chen, W. T. (2000) Critical appraisal of the use of matrix metalloproteinase inhibitors in cancer treatment [Review] [75 refs]. *Oncogene* **19**, 6642–6650.

71. Shalinsky, D. R., Brekken, J., Zou, H., et al. (1999) Marked antiangiogenic and antitumor efficacy of AG3340 in chemoresistant human non-small cell lung cancer tumors: single agent and combination chemotherapy studies. *Clin. Cancer Res.* **5**, 1905–1917.

72. O'Reilly, M. S., Holmgren, L., Shing, Y., et al. (1994) Angiostatin: a novel angiogenesis inhibitor that mediates the suppression of metastases by a Lewis lung carcinoma [see comments]. Comment in: *Cell* 1994 Oct 21; 79(2):185–8, *Cell* **79**, 315–328.

73. O'Reilly, M. S., Wiederschain, D., Stetler-Stevenson, W. G., Folkman, J., and Moses, M. A. (1999) Regulation of angiostatin production by matrix metalloproteinase-2 in a model of concomitant resistance. *J. Biol. Chem.* **274**, 29568–29571.

74. Dong, Z., Kumar, R., Yang, X., and Fidler, I. J. (1997) Macrophage-derived metalloelastase is responsible for the generation of angiostatin in Lewis lung carcinoma. *Cell* **88**, 801–810.

75. Dong, Z., Yoneda, J., Kumar, R., and Fidler, I. J. (1998) Angiostatin-mediated suppression of cancer metastases by primary neoplasms engineered to produce granulocyte/macrophage colony-stimulating factor. *J. Exp. Med.* **188**, 755–763.

76. McGeehan, G. M., Becherer, J. D., Bast, R. C. Jr., et al. (1994) Regulation of tumour necrosis factor-alpha processing by a metalloproteinase inhibitor. *Nature* **370**, 558–561.

77. Gearing, A. J., Beckett, P., Christodoulou, M., et al. (1994) Processing of tumour necrosis factor-alpha precursor by metalloproteinases. *Nature* **370**, 555–557.

78. Arribas, J., Coodly, L., Vollmer, P., Kishimoto, T. K., Rose-John, S., and Massague, J. (1996) Diverse cell surface protein ectodomains are shed by a system sensitive to metalloprotease inhibitors. *J. Biol. Chem.* **271**, 11376–11382.

79. Arribas, J. and Massague, J. (1995) Transforming growth factor-alpha and beta-amyloid precursor protein share a secretory mechanism. *J. Cell Biol.* **128**, 433–441.

80. Walcheck, B., Kahn, J., Fisher, J. M., et al. (1996) Neutrophil rolling altered by inhibition of L-selectin shedding in vitro. *Nature* **380**, 720–723.

81. Feehan, C., Darlak, K., Kahn, J., Walcheck, B., Spatola, A. F., and Kishimoto, T. K. (1996) Shedding of the lymphocyte L-selectin adhesion molecule is inhibited by a hydroxamic acid-based protease inhibitor. Identification with an L-selectin-alkaline phosphatase reporter. *J. Biol. Chem.* **271**, 7019–7024.

82. Bennett, T. A., Lynam, E. B., Sklar, L. A., and Rogelj, S. (1996) Hydroxamate-based metalloprotease inhibitor blocks shedding of L-selectin adhesion molecule from leukocytes: functional consequences for neutrophil aggregation. *J. Immunol.* **156**, 3093–3097.

83. Mullberg, J., Durie, F. H., Otten-Evans, C., et al. (1995) A metalloprotease inhibitor blocks shedding of the IL-6 receptor and the p60 TNF receptor. *J. Immunol.* **155**, 5198–5205.

84. Couet, J., Sar, S., Jolivet, A., Hai, M. T., Milgrom, E., and Misrahi, M. (1996) Shedding of human thyrotropin receptor ectodomain. Involvement of a matrix metalloprotease. *J. Biol. Chem.* **271**, 4545–4552.

85. Levi, E., Fridman, R., Miao, H. Q., Ma, Y. S., Yayon, A., and Vlodavsky, I. (1996) Matrix metalloproteinase 2 releases active soluble ectodomain of fibroblast growth factor receptor 1. *Proc. Natl. Acad. Sci. USA* **93**, 7069–7074.

86. Suzuki, M., Raab, G., Moses, M. A., Fernandez, C. A., and Klagsbrun, M. (1997) Matrix metalloproteinase-3 releases active heparin-binding EGF-like growth factor by cleavage at a specific juxtamembrane site. *J. Biol. Chem.* **272**, 31730–31737.

87. Gately, S., Twardowski, P., Stack, M. S., et al. (1996) Human prostate carcinoma cells express enzymatic activity that converts human plasminogen to the angiogenesis inhibitor, angiostatin. *Cancer Res.* **56**, 4887–4890.

88. Patterson, B. C. and Sang, Q. X. A. (1997) Angiostatin-converting enzyme activities of MMP-7 and MMP-9. *J. Biol. Chem.* **272**, 28823–28,825.

89. Falcone, D., Khan, K. M. F., Layne, T., and Fernandes, L. (1998) Macrophage formation of angiostatin during inflammation. *J. Biol. Chem.* **273**, 31480–31485.

90. Clapp, C., Martial, J. A., Guzman, R. C., Rentier-Delrue, F., and Weiner, R. I. (1993) The 16-kilodalton N-terminal fragment of human prolactin is a potent inhibitor of angiogenesis. *Endocrinology* **133**, 1292–1299.

91. D'Angelo, G., Struman, I., Martial, J., and Weiner, R. I. (1995) Activation of mitogen-activated protein kinases by vascular endothelial growth factor and basic fibroblast growth factor in capillary endothelial cells is inhibited by the antiangiogenic factor 16-kDa N-terminal fragment of prolactin. *Proc. Natl. Acad. Sci. USA* **92**, 6374–6378.

92. Ferrara, N., Clapp, C., and Weiner, R. I. (1991) The 16K fragment of prolactin specifically inhibits basal or FGF stimulated growth of capillary endothelial cells. *Endocrinology* **129**, 896–900.

93. Struman, I., Bentzien, F., Lee, H., et al. (1999) Opposing actions of intact and N-terminal fragments of human prolactin/growth hormone family members on angiogenesis: an efficient mechanism for the regulation of angiogenesis. *Proc. Natl. Acad. Sci. USA* **96**, 1246–1251.

94. Maione, T. E., Gray, G. S., Petro, J., et al. (1990) Inhibition of angiogenesis by recombinant human platelet factor-4 and related peptides. *Science* **247**, 77–79.

95. Gupta, S. K., Hassel, T., and Singh, J. P. (1995) A potent inhibitor of endothelial cell proliferation is generated by proteolytic cleavage of the chemokine platelet factor 4. *Proc. Natl. Acad. Sci. USA* **92**, 7799–7803.

96. Tolsma, S. S., Volpert, O. V., Good, D. J., Frazier, W. A., Polverini, P. J., and Bouck, N. (1993) Peptides derived from two separate domains of the matrix protein thrombospondin-1 have anti-angiogenic activity. *J. Cell Biol.* **122**, 497–511.

97. Nelson, J., Allen, W. E., Scott, W. N., Bailie, J. R., Walker, B., and McFerran, N. V. (1995) Murine epidermal growth factor (EGF) fragment (33–42) inhibits both EGF- and laminin-dependent endothelial cell motility and angiogenesis. *Cancer Res.* **55**, 3772–3776.

98. Grant, D. S., Tashiro, K.-I., Sequi-Real, B., Yamada, Y., Martin, G. R., and Kleinman, H. K. (1989) Two different laminin domains mediate the differentiation of human endothelial cells into capillary-like structures in vitro. *Cell* **58**, 933–943.

99. Homandberg, G. A., Williams, J. E., Grant, D., B., S., and Eisenstein, R. (1985) Heparin-binding fragments of fibronectin are potent inhibitors of endothelial cell growth. *Am. J. Pathol.* **120**, 327–332.

100. O'Reilly, M. S., Boehm, T., Shing, Y., et al. (1997) Endostatin: an endogenous inhibitor of angiogenesis and tumor growth. *Cell* **88**, 277–285.

101. Boehm, T., Folkman, J., Browder, T., and O'Reilly, M. S. (1997) Antiangiogenic therapy of experimental cancer does not induce acquired drug resistance [see comments]. Comment in: *Nature* 1997 Nov 27; 390(6658), 335–6. Comment in: *Nature* 1998 Jan 29; 391(6666), 450. Comment in: *Nature* 1998 May 14; 393(6681), 97. *Nature* **390**, 404–407.

102. Moses, M. A., Wiederschain, D., Loughlin, K. R., Zurakowski, D., Lamb, C. L., and Freeman, M. R. (1998) Increased incidence of matrix metalloproteinases in urine of cancer patients. *Cancer Res.* **58**, 1395–1399.

103. Camphausen, K., Moses, M. A., Beecken, W., Khan, M. K., Folkman, J., and O'Reilly, M. S. (2001) Radiation therapy to a primary tumor accelerates metastatic growth in mice. *Cancer Res.* **61**, 2207–2211.

104. O'Reilly, M. S., Holmgren, L., Chen, C., and Folkman, J. (1996) Angiostatin induces and sustains dormancy of human primary tumors in mice. *Nat. Med.* **2**, 689–692.

105. Holmgren, L., O'Reilly, M. S., and Folkman, J. (1995) Dormancy of micrometastases: balanced proliferation and apoptosis in the presence of angiogenesis suppression. *Nat. Med.* **1**, 149–153.

106. Laird, A. D., Vajkoczy, P., Shawver, L. K., et al. (2000). SU6668 is a potent antiangiogenic and antitumor agent that induces regression of established tumors. *Cancer Res.* **60**, 4152–4160.

107. Shaheen, R. M., Davis, D. W., Liu, W., et al. (1999) Antiangiogenic therapy targeting the tyrosine kinase receptor for vascular endothelial growth factor receptor inhibits the growth of colon cancer liver metastasis and induces tumor and endothelial cell apoptosis. *Cancer Res.* **59**, 5412–5416.

108. Burrows, F. J. and Thorpe, P. E. (1993) Eradication of large solid tumors in mice with an immunotoxin directed against tumor vasculature. *Proc. Natl. Acad. Sci. USA* **90**, 8996–9000.

109. Huang, X., Molema, G., King, S., Watkins, L., Edgington, T. S., and Thorpe, P. E. (1997) Tumor infarction in mice by antibody-directed targeting of tissue factor to tumor vasculature [see comments]. Comment in: *Science* 1997 Jan 24; 275(5299): 482–4. *Science* **275**, 547–550.

110. Tozer, G. M., Prise, V. E., Wilson, J., et al. (1999) Combretastatin A-4 phosphate as a tumor vascular-targeting agent: early effects in tumors and normal tissues. *Cancer Res.* **59**, 1626–1634.

111. Ellerby, H. M., Arap, W., Ellerby, L. M., et al. Anti-cancer activity of targeted pro-apoptotic peptides. *Nat. Med.* **5**, 1032–1038.

112. Pasqualini, R., Koivunen, E., Kain, R., et al. (2000) Aminopeptidase N is a receptor for tumor-homing peptides and a target for inhibiting angiogenesis. *Cancer Res.* **60**, 722–727.

113. Kandel, J., Bossy-Wetzel, E., Radvany, F., Klagsburn, M., Folkman, J., and Hanahan, D. (1991) Neovascularization is associated with a switch to the export of bFGF in the multistep development of fibrosarcoma. *Cell* **66**, 1095–1104.

114. Nickoloff, B. J., Mitra, R. S., Varani, J., Dixit, V. M., and Polverini, P. J. (1994) Aberrant production of interleukin-8 and thrombospondin-1 by psoriatic keratinocytes mediates angiogenesis. *Am. J. Path.* **144**, 820–828.

115. Heyns, A. D., Eldor, A., Vlodavsky, I., Kaiser, N., Fridman, R., and Panet, A. (1985) The antiproliferative effect of interferon and the mitogenic activity of growth factors are independent cell cycle events. Studies with vascular smooth muscle cells and endothelial cells. *Exp. Cell Res.* **161**, 297–306.

116. Friesel, R., Komoriya, A., and Maciag, T. (1987) Inhibition of endothelial cell proliferation by gamma-interferon. *J. Cell Biol.* **104**, 689–696.

117. Ruszczak, Z., Detmar, M., Imcke, E., and Orfanos, C. E. (1990) Effects of rIFN alpha, beta, and gamma on the morphology, proliferation, and cell surface antigen expression of human dermal microvascular endothelial cells in vitro. *J. Invest. Dermatol.* **95**, 693–699.

118. Hicks, C., Breit, S. N., and Penny, R. (1989) Response of microvascular endothelial cells to biological response modifiers. *Immunol. Cell Biol.* **67**, 271–277.

119. Angiolillo, A. L., Sgadari, C., Taub, D. D., et al. (1995) Human interferon-inducible protein 10 is a potent inhibitor of angiogenesis in vivo. *J. Exp. Med.* **182**, 155–162.

120. Sgadari, C., Angiolillo, A. L., Cherney, B. W., et al. (1996) Interferon-inducible protein-10 identified as a mediator of tumor necrosis in vivo. *Proc. Natl. Acad. Sci. USA* **93**, 13791–13796.

121. Voest, E. E., Kenyon, B. M., O'Reilly, M. S., Truitt, G., D'Amato, R. J., and Folkman, J. (1995) Inhibition of angiogenesis in vivo by interleukin 12. *J. Natl. Cancer Inst.* **87**, 581–586.

122. Dinney, C. P., Bielenberg, D. R., Perrotte, P., et al. (1998) Inhibition of basic fibroblast growth factor expression, angiogenesis, and growth of human bladder carcinoma in mice by systemic interferon-alpha administration. *Cancer Res.* **58**, 808–814.

123. Ezekowitz, R. A. B., Mulliken, J. B., and Folkman, J. (1992) Interferon alfa-2a therapy For life-threatening hemangiomas of infancy. *N. Engl. J. Med.* **326**, 1456–1463.

124. Orchard, P., Smith, C., Woods, W., Dehner, L. P., Day, D. L., and Shapiro, R. S. (1989) Treatment of hemangioendothiomas with alpha interferon. *Lancet* **2**, 565–567.

125. White, C. M., Sondheimer, H. M., Crouch, E. C., Wilson, H., and Fan, L. F. (1989) Treatment of pulmonary hemangiomatosis with recombinant interferon alfa-2a. *N. Engl. J. Med.* **320**, 1197–1200.

126. Mitsuyasu, R. T. (1991) Interferon alpha in the treatment of AIDS-related Kaposi's sarcoma. *Br. J. Haematol.* **79**(Suppl. 1), 69–73.

127. Kaban, L. B., Mulliken, J. B., Ezekowitz, R. A., Ebb, D., Smith, P. S., and Folkman, J. (1999) Antiangiogenic therapy of a recurrent giant cell tumor of the mandible with interferon alfa-2a [see comments]. *Pediatrics* **103**, 1145–1149.

128. Stadler, W. M., Kuzel, T. M., Raghavan, D., et al. (1997) Metastatic bladder cancer: advances in treatment. *Eur. J. Cancer* **33**(Suppl. 1), S23–S26.

129. Nanus, D. M., Schmitz-Drager, B. J., Motzer, R. J., et al. (1993) Expression of basic fibroblast growth factor in primary human renal tumors: correlation with poor survival. *J. Natl. Cancer Inst.* **85**, 1597–1599.

130. Nguyen, M., Watanabe, H., Budson, A. E., Richie, J. P., and Folkman, J. (1993) Elevated levels of the angiogenic peptide basic fibroblast growth factor in urine of bladder cancer patients. *J. Natl. Cancer Inst.* **85**, 241–242.

131. Brouty-Boye, D. and Zetter, B. R. (1980) Inhibition of cell motility by interferon. *Science* **206**, 516–518.

132. Slaton, J. W., Perrotte, P., Inoue, K., Dinney, C. P., and Fidler, I. J. (1999) Interferon-alpha-mediated down-regulation of angiogenesis-related genes and therapy of bladder cancer are dependent on optimization of biological dose and schedule. *Clin. Cancer Res.* **5**, 2726–2734.

133. Solorzano, C. C., Baker, C. H., Tsan, R., et al. (2001) Optimization for the blockade of epidermal growth factor receptor signaling for therapy of human pancreatic carcinoma. *Clin. Cancer Res.* **7**, 2563–2572.

134. Baker, C. H., McCarty, M. F., Tsan, R., and Fidler, I. J. (2001) Phenotypic diversity of organ-specific endothelial cells. *Proc. Amer. Assoc. Cancer Res.* **42**, 2186 (abstr.).

135. Tsai, J. C., Goldman, C. K., and Gillespie, G. Y. (1995) Vascular endothelial growth factor in human glioma cell lines: induced secretion by EGF, PDGF-BB, and bFGF. *J. Neurosurg.* **82**, 864–873.

136. Brogi, E., Wu, T., Namiki, A., and Isner, J. M. (1994) Indirect angiogenic cytokines upregulate VEGF and bFGF gene expression in vascular smooth muscle cells, whereas hypoxia upregulates VEGF expression only. *Circulation* **90**, 649–652.

137. Uehara, H., Kim, S. J., Karashima, T., Zheng, L., and Fidler, I. J. (2001) Blockade of the PDGF-R signaling by STI571 inhibits angiogenesis and growth of human prostate cancer cells in the bone of nude mice. *Proc. Amer. Assoc. Cancer Res.* **42**, 2192 (abstr.).

138. Chakravarti, A., Loeffler, J. S., and Dyson, N. J. (2002) Insulin-like growth factor receptor I mediates resistance to anti-epidermal growth factor receptor therapy in primary human glioblastoma cells through continued activation of phosphoinositide 3-kinase signaling. *Cancer Res.* **62**, 200–207.

139. Viloria-Petit, A., Crombet, T., Jothy, S., et al. (2001) Acquired resistance to the antitumor effect of epidermal growth factor receptor-blocking antibodies in vivo: a role for altered tumor angiogenesis. *Cancer Res.* **61**, 5090–5101.

140. Yano, S., Herbst, R. S., Shinohara, H., et al. (2000) Treatment for malignant pleural effusion of human lung adenocarcinoma by inhibition of vascular endothelial growth factor receptor tyrosine kinase phosphorylation. *Clin. Cancer Res.* **6**, 957–965.

141. Laird, A. D., Vajkoczy, P., Shawver, L. K., et al. (2000) SU6668 is a potent antiangiogenic and antitumor agent that induces regression of established tumors. *Cancer Res.* **60**, 4152–4160.

142. Ciardiello, F., Bianco, R., Damiano, V., et al. (2000) Antiangiogenic and antitumor activity of anti-epidermal growth factor receptor C225 monoclonal antibody in combination with vascular endothelial growth factor antisense oligonucleotide in human GEO colon cancer cells. *Clin. Cancer Res.* **6**, 3739–3747.

143. Jakkula, M., Le Cras, T. D., Gebb, S., et al. (2000) Inhibition of angiogenesis decreases alveolarization in the developing rat lung. *Am. J. Physiol.—Lung Cell. Mol. Physiol.* **279**, L600–L607.

144. Alon, T., Hemo, I., Itin, A., Pelee, J., Stone, J., and Keshet, E. (1995) VEGF acts as a survival factor for newly formed retinal vessels and has implications for retinopathy of prematurity. *Nat. Med.* **1**, 1024–1028.

145. Verheul, H. M., Hoekman, K., Lupu, F., et al. (2000) Platelet and coagulation activation with vascular endothelial growth factor generation in soft tissue sarcomas. *Clin. Cancer Res.* **6**, 166–171.

146. Nor, J. E., Christensen, J., Mooney, D. J., and Polverini, P. J. (1999) VEGF-mediated angiogenesis is associated with enhanced endothelial cell survival and induction of Bcl-2 expression. *Am. J. Pathol.* **154**, 375–384.

147. Taraboletti, G., Morbidelli, L., Donnini, S., et al. (2000) The heparin binding 25 kDa fragment of thrombospondin-1 promotes angiogenesis and modulates gelatinase and TIMP-2 production in endothelial cells. *FASEB J.* **14**, 1674–1676.

148. Gorelik, E. (1983) Concominant tumor immunity and the resistance to a second tumor challenge. *Adv. Cancer Res.* **39**, 71–120.

149. Prehn, R. T. (1991) The inhibition of tumor growth by tumor mass. *Cancer Res.* **51**, 2–4.

150. Hari, D., Beckett, M. A., Sukhatme, V. P., et al. (2000) Angiostatin induces mitotic cell death of proliferating endothelial cells. *Mol. Cell Biol. Res. Commun.* **3**, 277–282.

151. Lucas, R., Holmgren, L., Garcia, I., et al. (1998) Multiple forms of angiostatin induce apoptosis in endothelial cells. *Blood* **92**, 4730–4741.

152. Claesson-Welsh, L., Welsh, M., Ito, N., et al. (1998) Angiostatin induces endothelial cell apoptosis and activation of focal adhesion kinase independently of the integrin-binding motif RGD. *Proc. Natl. Acad. Sci. USA* **95**, 5579–5583.

153. Stack, M. S., Gately, S., Bafetti, L. M., Enghild, J. J., and Soff, G. A. (1999) Angiostatin inhibits endothelial and melanoma cellular invasion by blocking matrix-enhanced plasminogen activation. *Biochem. J.* **340**, 77–84.

154. Moser, T. L., Stack, M. S., Asplin, I., et al. (1999) Angiostatin binds ATP synthase on the surface of human endothelial cells. *Proc. Natl. Acad. Sci. USA* **96**, 2811–2816.

155. Troyanovsky, B., Levchenko, T., Mansson, G., Matvijenko, O., and Holmgren, L. (2001) Angiomotin: an angiostatin binding protein that regulates endothelial cell migration and tube formation [see comments]. Comment in: *J. Cell Biol.* 2001 Mar 19; 152(6):F35–6. *J. Cell Biol.* **152**, 1247–1254.

156. Yamaguchi, N., Anand-Apte, B., Lee, M., et al. (1999) Endostatin inhibits VEGF-induced endothelial cell migration and tumor growth independently of zinc binding. *EMBO J.* **18**, 4414–4423.

157. Read, T. A., Farhadi, M., Bjerkvig, R., et al. (2001) Intravital microscopy reveals novel antivascular and antitumor effects of endostatin delivered locally by alginate-encapsulated cells. *Cancer Res.* **61**, 6830–6837.

158. Hohenester, E., Sasaki, T., Olsen, B. R., and Timpl, R. (1998) Crystal structure of the angiogenesis inhibitor endostatin at 1.5 A resolution. *EMBO J.* **17**, 1656–1664.

159. Ding, Y. H., Javaherian, K., Lo, K. M., et al. (1998) Zinc-dependent dimers observed in crystals of human endostatin. *Proc. Natl. Acad. Sci. USA* **95**, 10443–10448.

160. Dhanabal, M., Ramchandran, R., Waterman, M. J. F., et al. (1999) Endostatin induces endothelial cell apoptosis. *J. Biol. Chem.* **274**, 11721–11726.

161. MacDonald, N. J., Shivers, W. Y., Narum, D. L., et al. (2001) Endostatin binds tropomyosin. A potential modulator of the antitumor activity of endostatin. *J. Biol. Chem.* **276**, 25190–25196.

162. Rehn, M., Veikkola, T., Kukk-Valdre, E., et al. (2001) Interaction of endostatin with integrins implicated in angiogenesis. *Proc. Natl. Acad. Sci. USA* **98**, 1024–1029.

163. Wickstrom, S. A., Veikkola, T., Rehn, M., Pihlajaniemi, T., Alitalo, K., and Keski-Oja, J. (2001) Endostatin-induced modulation of plasminogen activation with concomitant loss of focal adhesions and actin stress fibers in cultured human endothelial cells. *Cancer Res.* **61**, 6511–6516.

164. Huang, X., Wong, M. K., Zhao, Q., et al. (2001) Soluble recombinant endostatin purified from Escherichia coli: antiangiogenic activity and antitumor effect. *Cancer Res.* **61**, 478–481.

165. Ramchandran, R., Dhanabal, M., Volk, R., et al. (1999) Antiangiogenic activity of restin, NC10 domain of human collagen XV: comparison to endostatin. *Biochem. Biophys. Res. Commun.* **255**, 735–739.

166. Xu, J., Rodriguez, D., Petitclerc, E., et al. (2001) Proteolytic exposure of a cryptic site within collagen type IV is required for angiogenesis and tumor growth in vivo. *J. Cell Biol.* **154**, 1069–1079.

167. Petitclerc, E., Boutaud, A., Prestayko, A., et al. (2000) New functions for non-collagenous domains of human collagen type IV. Novel integrin ligands inhibiting angiogenesis and tumor growth in vivo. *J. Biol. Chem.* **275**, 8051–8061.

168. Colorado, P. C., Torre, A., Kamphaus, G., et al. (2000) Anti-angiogenic cues from vascular basement membrane collagen. *Cancer Res.* **60**, 2520–2526.

169. Bentzien, F., Struman, I., Martini, J. F., Martial, J., and Weiner, R. (2001) Expression of the antiangiogenic factor 16K hPRL in human HCT116 colon cancer cells inhibits tumor growth in Rag1(–/–) mice. *Cancer Res.* **61**, 7356–7362.

170. Sakamato, N., Iwahana, M., Tanaka, N. G., and Osaka, Y. (1991) Inhibition of angiogenesis and tumor growth by a synthetic laminin peptide, CDPGYIGSR-NH$_2$. *Cancer Res.* **51**, 903–906.

171. Pike, S. E., Yao, L., Jones, K. D., et al. (1998) Vasostatin, a calreticulin fragment, inhibits angiogenesis and suppresses tumor growth. *J. Exp. Med.* **188**, 2349–2356.

172. Pike, S. E., Yao, L., Setsuda, J., et al. (1999) Calreticulin and calreticulin fragments are endothelial cell inhibitors that suppress tumor growth. *Blood* **94**, 2461–2468.

173. Yao, L., Pike, S. E., Setsuda, J., et al. (2000) Effective targeting of tumor vasculature by the angiogenesis inhibitors vasostatin and interleukin-12. *Blood* **96**, 1900–1905.

174. Larsson, H., Sjoblom, T., Dixelius, J., et al. (2000) Antiangiogenic effects of latent antithrombin through perturbed cell-matrix interactions and apoptosis of endothelial cells. *Cancer Res.* **60**, 6723–6729.

175. Larsson, H., Akerud, P., Nordling, K., Raub-Segall, E., Claesson-Welsh, L., and Bjork, I. (2001) A novel anti-angiogenic form of antithrombin with retained proteinase binding ability and heparin affinity. *J. Biol. Chem.* **276**, 11996–12002.

176. Gale, N. W. and Yancopoulos, G. D. (1999) Growth factors acting via endothelial cell-specific receptor tyrosine kinases: VEGFs, angiopoietins, and ephrins in vascular development. *Genes Dev.* **13**, 1055–1066.

177. Klint, P. and Claesson-Welsh, L. (1999) Signal transduction by fibroblast growth factor receptors. *Frontiers Biosci.* **4**, D165–D177.

178. Davis, S., Aldrich, T. H., Jones, P. F., et al. (1996) Isolation of angiopoietin-1, a ligand for the TIE2 receptor, by secretion-trap expression cloning. *Cell* **87**, 1153–1155.

179. Maisonpierre, P. C., Suri, C., Jones, P. F., et al. (1997) Angiopoietin-2, a natural antagonist for tie2 that disrupts in vivo angiogenesis. *Science* **277**, 55–60.

180. Suri, C., Jones, P. F., Patan, S., et al. (1996) Requisite role of angiopoietin-1, a ligand for the TIE2 receptor, during embryonic angiogenesis. *Cell* **87**, 1153–1155.

181. Baker, C. H., Solorzano, C. C., and Fidler, I. J. (2001) Angiogenesis and cancer metastasis: antiangiogenic therapy of human pancreatic adenocarcinoma. *Int. J. Clin. Oncol.* **6**, 59–65.

182. Yuan, F., Chen, Y., Dellian, M., Safabakhsh, N., Ferrara, N., and Jain, R. K. (1996) Time-dependent vascular regression and permeability changes in established human tumor xenografts induced by an anti-VEGF/VPF antibody. *Proc. Natl. Acad. Sci. USA* **93**, 14765–14770.

183. Perrotte, P., Matsumoto, T., Inoue, K., et al. (1999) Anti-epidermal growth factor receptor antibody C225 inhibits angiogenesis in human transitional cell carcinoma growing orthotopically in nude mice. *Clin. Cancer Res.* **5**, 257–265.

184. Petit, A. M., Rak, J., Hung, M. C., et al. (1997) Neutralizing antibodies against epidermal growth factor and ErbB-2/neu receptor tyrosine kinases down-regulate vascular endothelial growth factor production by tumor cells in vitro and in vivo: angiogenic implications for signal transduction therapy of solid tumors. *Am. J. Pathol.* **151**, 1523–1530.

185. Sirotnak, F. M., Zakowski, M. F., Miller, V. A., Scher, H. I., and Kris, M. G. (2000) Efficacy of cytotoxic agents against human tumor xenografts is markedly enhanced by coadministration of ZD1839 (Iressa), an inhibitor of EGFR tyrosine kinase. *Clini. Cancer Res.* **6**, 4885–4892.

186. Drevs, J., Hofmann, I., Hugenschmidt, H., et al. (2000) Effects of PTK787/ZK 222584, a specific inhibitor of vascular endothelial growth factor receptor tyrosine kinases, on primary tumor, metastasis, vessel density, and blood flow in a murine renal cell carcinoma model. *Cancer Res.* **60**, 4819–4824.

187. Wood, J. M., Bold, G., Buchdunger, E., et al. (2000) PTK787/ZK 222584, a novel and potent inhibitor of vascular endothelial growth factor receptor tyrosine kinases, impairs

vascular endothelial growth factor-induced responses and tumor growth after oral administration. *Cancer Res.* **60**, 2178–2189.

188. Kendall, R. L. and Thomas, K. A. (1993) Inhibition of vascular endothelial cell growth factor activity by an endogenously encoded soluble receptor. *Proc. Natl. Acad. Sci. USA* **90**, 10705–10709.

189. Kendall, R. L., Wang, G., and Thomas, K. A. (1996) Identification of a natural soluble form of the vascular endothelial growth factor receptor, FLT-1, and its heterodimerization with KDR. *Biochem. Biophys. Res. Commun.* **226**, 324–328.

190. Kong, H. L., Hecht, D., Song, W., et al. (1998) Regional suppression of tumor growth by in vivo transfer of a cDNA encoding a secreted form of the extracellular domain of the flt-1 vascular endothelial growth factor receptor. *Hum. Gene Ther.* **9**, 823–833.

191. Pavco, P. A., Bouhana, K. S., Gallegos, A. M., et al. (2000) Antitumor and antimetastatic activity of ribozymes targeting the messenger RNA of vascular endothelial growth factor receptors. *Clin. Cancer Res.* **6**, 2094–2103.

192. Parry, T. J., Cushman, C., Gallegos, A. M., et al. (1999) Bioactivity of anti-angiogenic ribozymes targeting Flt-1 and KDR mRNA. *Nucleic Acids Res.* **27**, 2569–2577.

193. Folkman, J. (1998) Antiangiogenic gene therapy. *Proc. Natl. Acad. Sci. USA* **95**, 9064–9066.

194. Feldman, A. L. and Libutti, S. K. (2000) Progress in antiangiogenic gene therapy of cancer [Review] [147 refs]. *Cancer* **89**, 1181–1194.

195. Kong, H. and Crystal, R. G. (1998) Gene therapy strategies for tumor angiogenesis. *J. Natl. Cancer Inst.* **90**, 273–286.

196. Isner, J. M., Pieczek, A., Schainfeld, R., et al. (1996) Clinical evidence of angiogenesis after arterial gene transfer of phVEGF165 in patient with ischaemic limb [see comments]. Comment in: *Lancet* 1996 Nov 16; 348(9038), 1380–81, discussion 1381–1382. Comment in: *Lancet* 1996 Nov 16; 348(9038), 1381, discussion 1381–2. *Lancet* **348**, 370–374.

197. Isner, J. M., Vale, P. R., Symes, J. F., and Losordo, D. W. (2001) Assessment of risks associated with cardiovascular gene therapy in human subjects [Review] [94 refs]. *Circ. Res.* **89**, 389–400.

198. Isner, J. M. (2000) Angiogenesis: a "breakthrough" technology in cardiovascular medicine [Review] [26 refs]. *J. Invasive Cardiol.* **12** (Suppl. A), 14A–17A.

199. Sellke, F. W. and Simons, M. (1999) Angiogenesis in cardiovascular disease: current status and therapeutic potential [Review] [42 refs]. *Drugs* **58**, 391–396.

200. Rosengart, T. K., Lee, L. Y., Patel, S. R., et al. (1999) Angiogenesis gene therapy: phase I assessment of direct intramyocardial administration of an adenovirus vector expressing VEGF121 cDNA to individuals with clinically significant severe coronary artery disease. *Circulation* **100**, 468–474.

201. Cao, Y., O'Reilly, M. S., Marshall, B., Flynn, E., Ji, R. W., and Folkman, J. (1998) Expression of angiostatin cDNA in a murine fibrosarcoma suppresses primary tumor growth and produces long-term dormancy of metastases. *J. Clin. Invest.* **101**, 1055–1063.

202. Weinstat-Saslow, D. L., Zabrenetzky, V. S., VanHoutte, K., Frazier, W. A., Roberts, D. D., and Steeg, P. S. (1994) Transfection of thrombospondin 1 complementary DNA into a human breast carcinoma cell line reduces primary tumor growth, metastatic potential and angiogenesis. *Cancer Res.* **54**, 6504–6511.

203. Sheibani, N. and Frazier, W. A. (1995) Thrombospondin 1 expression in transformed endothelial cells restores a normal phenotype and suppresses their tumorigenesis. *Proc. Natl. Acad. Sci. USA* **92**, 6788–6792.

204. Bleuel, K., Popp, S., Fusenig, N. E., Stanbridge, E. J., and Boukamp, P. (1999) Tumor suppression in human skin carcinoma cells by chromosone 15 transfer or thrombospondin-1 overexpression through halted tumor vascularization. *Proc. Natl. Acad. Sci. USA* **96**, 2065–2070.

205. Tanaka, T., Cao, Y., Folkman, J., and Fine, H. A. (1998) Viral vector-targeted antiangiogenic gene therapy utilizing an angiostatin complementary DNA. *Cancer Res.* **58**, 3362–3369.
206. Griscelli, F., Li, H., Bennaceur-Griscelli, A., et al. (1998) Angiostatin gene transfer: inhibition of tumor growth in vivo by blockage of endothelial cell proliferation associated with a mitosis arrest. *Proc. Natl. Acad. Sci. USA* **95**, 6367–6372.
207. Liu, Y., Thor, A., Shtivelman, E., et al. (1999) Systemic gene delivery expands the repertoire of effective antiangiogenic agents. *J. Biol. Chem.* **274**, 13338–13344.
208. Gorrin-Rivas, M. J., Arii, S., Furutani, M., et al. (2000) Mouse macrophage metalloelastase gene transfer into a murine melanoma suppresses primary tumor growth by halting angiogenesis. *Clin. Cancer Res.* **6**, 1647–1654.
209. Blezinger, P., Wang, J., Gondo, M., et al. (1999) Systemic inhibition of tumor growth and tumor metastases by intramuscular administration of the endostatin gene. *Nat. Biotechnol.* **17**, 343–348.
210. Feldman, A. L., Restifo, N. P., Alexander, H. R., et al. (2000) Antiangiogenic gene therapy of cancer utilizing a recombinant adenovirus to elevate systemic endostatin levels in mice. *Cancer Res.* **60**, 1503–1506.
211. Kong, H. L., Hecht, D., Song, W., et al. (1998) Regional suppression of tumor growth by in vivo transfer of a cDNA encoding a secreted form of the extracellular domain of the flt-1 VEGF receptor. *Hum. Gene Ther.* **9**, 823–833.
212. Davidoff, A. M., Ng, C. Y., Brown, P., et al. (2001) Bone marrow-derived cells contribute to tumor neovasculature and, when modified to express an angiogenesis inhibitor, can restrict tumor growth in mice. *Clin. Cancer Res.* **7**, 2870–2879.
213. Lin, P., Buxton, J. A., Acheson, A., et al. (1998) Antiangiogenic gene therapy targeting the endothelium-specific receptor tyrosine kinase Tie2. *Proc. Natl. Acad. Sci. USA* **95**, 8829–8834.
214. Browder, T., Butterfield, C. E., Kraling, B. M., et al. (2000) Antiangiogenic scheduling of chemotherapy improves efficacy against experimental drug-resistant cancer. *Cancer Res.* **60**, 1878–1886.
215. Klement, G., Baruchel, S., Rak, J., et al. (2000) Continuous low-dose therapy with vinblastine and VEGF receptor-2 antibody induces sustained tumor regression without overt toxicity [see comments]. Comment in: *J. Clin. Invest.* 2000 Apr; 105(8), 1045–7. *J. Clin. Invest.* **105**, R15–R24.
216. Soffer, S. Z., Moore, J. T., Kim, E., et al. (2001) Combination antiangiogenic therapy: increased efficacy in a murine model of Wilms tumor. *J. Pediatr. Surg.* **36**, 1177–1181.
217. Bello, L., Carrabba, G., Giussani, C., et al. (2001) Low-dose chemotherapy combined with an antiangiogenic drug reduces human glioma growth in vivo. *Cancer Res.* **61**, 7501–7506.
218. St Croix, B., Rago, C., Velculescu, V., et al. (2000) Genes expressed in human tumor endothelium [see comments]. Comment in: *Science* 2000 Aug 18; 289(5482), 1121–2. *Science* **289**, 1197–1202.
219. Brem, H., Goto, F., Budson, A., Saunders, L., and Folkman, J. (1994) Minimal drug resistance after prolonged antiangiogenic therapy with AGM-1470. *Surg. Forum* **XLV**, 674–677.
220. Hausman, M. R., Schaffler, M. B., and Majeska, R. J. (2001) Prevention of fracture healing in rats by an inhibitor of angiogenesis. *Bone* **29**, 560–564.
221. Berger, A. C., Feldman, A. L., Gnant, M. F., et al. (2000) The angiogenesis inhibitor, endostatin, does not affect murine cutaneous wound healing. *J. Surg. Res.* **91**, 26–31.
222. Bloch, W., Huggel, K., Sasaki, T., et al. (2000) The angiogenesis inhibitor endostatin impairs blood vessel maturation during wound healing. *FASEB J.* **14**, 2373–2376.
223. Mundhenke, C., Thomas, J. P., Wilding, G., et al. (2001) Tissue examination to monitor antiangiogenic therapy a phase I clinical trial with endostatin. *Clin. Cancer Res.* **7**, 3366–3374.

224. Gerber, H. P., Hillan, K. J., Ryan, A. M., et al. (1999) VEGF is required for growth and sur-vival in neonatal mice. *Development—Supplement* **126**, 1149–1159.

225. Felbor, U., Dreier, L., Bryant, R. A., Ploegh, H. L., Olsen, B. R., and Mothes, W. (2000) Secreted cathepsin L generates endostatin from collagen XVIII. *EMBO J.* **19**, 1187–1194.

226. Wen, W., Moses, M. A., Wiederschain, D., Arbiser, J. L., and Folkman, J. (1999) The gen-eration of endostatin is mediated by elastase. *Cancer Res.* **59**, 6052–6056.

227. Drixler, T. A., Rinkes, I. H., Ritchie, E. D., van Vroonhoven, T. J., Gebbink, M. F., and Voest, E. E. (2000) Continuous administration of angiostatin inhibits accelerated growth of colorectal liver metastases after partial hepatectomy. *Cancer Res.* **60**, 1761–1765.

228. Kisker, O., Becker, C. M., Prox, D., et al. (2001) Continuous administration of endostatin by intraperitoneally implanted osmotic pump improves the efficacy and potency of ther-apy in a mouse xenograft tumor model. *Cancer Res.* **61**, 7669–7674.

229. Lode, H. N., Moehler, T., Xiang, R., et al. (1999) Synergy between an antiangiogenic inte-grin alphav antagonist and an antibody-cytokine fusion protein eradicates spontaneous tumor metastases. *Proc. Natl. Acad. Sci. USA* **96**, 1591–1596.

230. Oyama, T., Ran, S., Ishida, T., et al. (1998) Vascular endothelial growth factor affects den-dritic cell maturation through the inhibition of nuclear factor-kappa B activation in hemo-poietic progenitor cells. *J. Immunol.* **160**, 1224–1232.

231. Gabrilovich, D. I., Chen, H. L., Girgis, K. R., et al. (1996) Production of vascular endothe-lial growth factor by human tumors inhibits the functional maturation of dendritic cells [erratum appears in *Nat. Med.* 1996 Nov; 2(11), 1267]. *Nat. Med.* **2**, 1096–1103.

232. Gabrilovich, D. I., Ishida, T., Nadaf, S., Ohm, J. E., and Carbone, D. P. (1999) Antibodies to vascular endothelial growth factor enhance the efficacy of cancer immunotherapy by improving endogenous dendritic cell function. *Clin. Cancer Res.* **5**, 2963–2970.

233. Arenberg, D. A., Kunkel, S. L., Polverini, P. J., et al. (1996) Interferon-gamma-inducible protein 10 (IP-10) is an angiostatic factor that inhibits human non-small cell lung cancer (NSCLC) tumorigenesis and spontaneous metastases. *J. Exp. Med.* **184**, 981–992.

234. Cheng, W. F., Hung, C. F., Chai, C. Y., et al. (2001) Tumor-specific immunity and antian-giogenesis generated by a DNA vaccine encoding calreticulin linked to a tumor antigen. *J. Clin. Invest.* **108**, 669–678.

235. Gyorffy, S., Palmer, K., Podor, T. J., Hitt, M., and Gauldie, J. (2001) Combined treatment of a murine breast cancer model with type 5 adenovirus vectors expressing murine angio-statin and IL-12: a role for combined anti-angiogenesis and immunotherapy. *J. Immunol.* **166**, 6212–6217.

236. Davidoff, A. M., Leary, M. A., Ng, C. Y., et al. (2001) Autocrine expression of both endo-statin and green fluorescent protein provides a synergistic antitumor effect in a murine neu-roblastoma model. *Cancer Gene Ther.* **8**, 537–545.

237. Teicher, B. A., Sotomayor, E. A., and Huang, Z. D. (1992) Antiangiogenic agents potentiate cytotoxic cancer therapies against primary and metastatic disease. *Cancer Res.* **52**, 6702–6704.

238. Teicher, B. A., Holden, S. A., Dupuis, N. P., et al. (1995) Potentiation of cytotoxic thera-pies by TNP-470 and minocycline in mice bearing EMT-6 mammary carcinoma. *Breast Cancer Res. Treat.* **36**, 227–236.

239. Lee, K., Erturk, E., Mayer, R., and Cockett, A. T. (1987) Efficacy of antitumor chemother-apy in C3H mice enhanced by the antiangiogenesis steroid, cortisone acetate. *Cancer Res.* **47**, 5021–5024.

240. Bertolini, F., Fusetti, L., Mancuso, P., et al. (2000) Endostatin, an antiangiogenic drug, induces tumor stabilization after chemotherapy or anti-CD20 therapy in a NOD/SCID mouse model of human high-grade non-Hodgkin lymphoma. *Blood* **96**, 282–287.

241. Ciardiello, F., Caputo, R., Bianco, R., et al. (2000) Antitumor effect and potentiation of cytotoxic drugs activity in human cancer cells by ZD-1839 (Iressa), an epidermal growth factor receptor-selective tyrosine kinase inhibitor. *Clin. Cancer Res.* **6**, 2053–2063.

242. Bruns, C. J., Harbison, M. T., Davis, D. W., et al. (2000) Epidermal growth factor receptor blockade with C225 plus gemcitabine results in regression of human pancreatic carcinoma growing orthotopically in nude mice by antiangiogenic mechanisms. *Clin. Cancer Res.* **6**, 1936–1948.

243. Kawamoto, T., Sato, J. D., Le, A., Polikoff, J., Sato, G. H., and Mendelsohn, J. (1983) Growth stimulation of A431 cells by epidermal growth factor: identification of high-affinity receptors for epidermal growth factor by an anti-receptor monoclonal antibody. *Proc. Natl. Acad. Sci. USA* **80**, 1337–1341.

244. Mendelsohn, J. and Baselga, J. (2000) The EGF receptor family as targets for cancer therapy [Review] [155 refs]. *Oncogene* **19**, 6550–6565.

245. Strawn, L. M., Kabbinavar, F., Schwartz, D. P., et al. (2000) Effects of SU101 in combination with cytotoxic agents on the growth of subcutaneous tumor xenografts. *Clin. Cancer Res.* **6**, 2931–2940.

246. Hansen-Algenstaedt, N., Stoll, B. R., Padera, T. P., et al. (2000) Tumor oxygenation in hormone-dependent tumors during vascular endothelial growth factor receptor-2 blockade, hormone ablation, and chemotherapy. *Cancer Res.* **60**, 4556–4560.

247. Margolin, K., Gordon, M. S., Holmgren, E., et al. (2001) Phase Ib trial of intravenous recombinant humanized monoclonal antibody to vascular endothelial growth factor in combination with chemotherapy in patients with advanced cancer: pharmacologic and long-term safety data. *J. Clin. Oncol.* **19**, 851–856.

248. Canney, P. and Dean, S. (1990) Transforming growth factor beta: a promoter of late connective tissue injury following radiotherapy? *Br. J. Radiol.* **63**, 620–623.

249. Gorski, D. H., Beckett, M. A., Jaskowiak, N. T., et al. (1999) Blockage of the vascular endothelial growth factor stress response increases the antitumor effects of ionizing radiation. *Cancer Res.* **59**, 3374–3378.

250. Hartford, A. C., Gohongi, T., Fukumura, D., and Jain, R. K. (2000) Irradiation of a primary tumor, unlike surgical removal, enhances angiogenesis suppression at a distal site: potential role of host-tumor interaction. *Cancer Res.* **60**, 2128–2131.

251. Teicher, B. A., Dupuis, N., Kusomoto, T., et al. (1995) Antiangiogenic agents can increase tumor oxygenation and response to radiation therapy. *Rad. Oncol. Invest.* **2**, 269–276.

252. Ikeda, S., Akagi, K., Shiraishi, T., and Tanaka, Y. (1998) Enhancement of the effect of an angiogenesis inhibitor on murine tumors by hyperthermia. *Oncol. Rep.* **5**, 181–184.

253. Masunaga, S., Ono, K., Nishimura, Y., et al. (2000) Combined effects of tirapazamine and mild hyperthermia on anti-angiogenic agent (TNP-470) treated tumors-reference to the effect on intratumor quiescent cells [see comments]. Comment in: *Int. J. Radiat. Oncol. Biol. Phys.* 2000 Jun 1; 47(3), 549–0. *Int. J. Radiat. Oncol. Biol. Phys.* **47**, 799–807.

254. Nishimura, Y., Murata, R., and Hiraoka, M. (1996) Combined effects of angiogenesis inhibitor (TNP-470) and hyperthermia. *Br. J. Cancer* **73**, 270–274.

255. Murata, R., Nishimura, Y., and Hiraoka, M. (1997) An antiangiogenic agent (TNP-470) inhibited reoxygenation during fractionated radiotherapy of murine mammary carcinoma. *Int. J. Radiat. Oncol. Biol. Phys.* **37**, 1107–1113.

256. Lund, E. L., Bastholm, L., and Kristjansen, P. E. (2000) Therapeutic synergy of TNP-470 and ionizing radiation: effects on tumor growth, vessel morphology, and angiogenesis in human glioblastoma multiforme xenografts. *Clin. Cancer Res.* **6**, 971–978.

257. Mauceri, H. J., Hanna, N. N., Beckett, M. A., et al. (1998) Combined effects of angiostatin and ionizing radiation in antitumour therapy. *Nature* **394**, 287–291.

258. Milas, L., Kishi, K., Hunter, N., Mason, K., Masferrer, J. L., and Tofilon, P. J. (1999) Enhancement of tumor response to gamma-radiation by an inhibitor of cyclooxygenase-2 enzyme [see comments]. Comment in: *J. Natl. Cancer Inst.* 2000 Feb 16; 92(4), 346–7. *J. Nat. Cancer Inst.* **91**, 1501–1504.

259. Milas, L., Mason, K., Hunter, N., et al. (2000) In vivo enhancement of tumor radioresponse by C225 antiepidermal growth factor receptor antibody [see comments], *Clin. Cancer Res.* **6**, 701–708.

260. Huang, S. M. and Harari, P. M. (2000) Modulation of radiation response after epidermal growth factor receptor blockade in squamous cell carcinomas: inhibition of damage repair, cell cycle kinetics, and tumor angiogenesis. *Clin. Cancer Res.* **6**, 2166–2174.

261. Lee, C. G., Heijn, M., di Tomaso, E., et al. (2000) Anti-vascular endothelial growth factor treatment augments tumor radiation response under normoxic or hypoxic conditions. *Cancer Res.* **60**, 5565–5570.

262. Geng, L., Donnelly, E., McMahon, G., et al. (2001) Inhibition of vascular endothelial growth factor receptor signaling leads to reversal of tumor resistance to radiotherapy. *Cancer Res.* **61**, 2413–2419.

263. Kozin, S., Boucher, Y., Hicklin, D. J., Bohlen, P., Jain, R. K., and Suit, H. D. (2001) Vascular endothelial growth factor receptor-2-blocking antibody potentiates radiation-induced long-term control of human tumor xenografts. *Cancer Res.* **61**, 39–44.

264. Gorski, D. H., Mauceri, H. J., Salloum, R. M., et al. (1998) Potentiation of the antitumor effect of ionizing radiation by brief concomitant exposures to angiostatin. *Cancer Res.* **58**, 5686–5689.

265. Salven, P., Orpana, A., Teerenhovi, L., and Joensuu, H. (2000) Simultaneous elevation in the serum concentrations of the angiogenic growth factors VEGF and bFGF is an independent predictor of poor prognosis in non-Hodgkin lymphoma: a single-institution study of 200 patients. *Blood* **96**, 3712–3718.

266. Fontanini, G., Boldrini, L., Chine, S., et al. (1999) Expression of vascular endothelial growth factor mRNA in non-small-cell lung carcinomas. *Br. J. Cancer* **79**, 363–369.

267. Watanabe, H., Nguyen, M., Schizer, M., Li, V., Hayes, D. F., and Sallan, D. F. (1992) Basic fibroblast growth factor in human serum -a prognostic test for breast cancer. *Mol. Biol. Cell* **3**, 234a.

268. Poon, R. T., Fan, S. T., and Wong, J. (2001) Clinical implications of circulating angiogenic factors in cancer patients [Review] [196 refs]. *J. Clin. Oncol.* **19**, 1207–1225.

269. Weidner, N., Semple, J. P., Welch, W. R., and Folkman, J. (1991) Tumor angiogenesis and metastasis—correlation in invasive breast carcinoma. *N. Engl. J. Med.* **324**, 1–8.

270. Gasparini, G. (2001) Clinical significance of determination of surrogate markers of angiogenesis in breast cancer [Review] [184 refs]. *Crit. Rev. Oncol. Hematol.* **37**, 97–114.

271. Fox, S. B. and Harris, A. L. (1997) Markers of tumor angiogenesis: clinical applications in prognosis and anti-angiogenic therapy [Review] [174 refs]. *Invest. New Drugs* **15**, 15–28.

272. Lissoni, P., Fugamalli, E., Malugani, F., et al. (2000) Chemotherapy and angiogenesis in advanced cancer: vascular endothelial growth factor (VEGF) decline as predictor of disease control during taxol therapy in metastatic breast cancer. *Int. J. Biol. Markers* **15**, 308–311.

273. Kido, Y. (2001) Vascular endothelial growth factor (VEGF) serum concentration changes during chemotherapy in patients with lung cancer. *Kurume Med. J.* **48**, 43–47.

274. Boehle, A. S., Kurdow, R., Schulze, M., et al. (2001) Human endostatin inhibits growth of human non-small-cell lung cancer in a murine xenotransplant model. *Int. J. Cancer* **94**, 420–428.

275. Yano, S., Shinohara, H., Herbst, R. S., et al. (2000) Expression of vascular endothelial growth factor is necessary but not sufficient for production and growth of brain metastasis. *Cancer Res.* **60**, 4959–4967.

276. Bruns, C. J., Solorzano, C. C., Harbison, M. T., et al. (2000) Blockade of the epidermal growth factor receptor signaling by a novel tyrosine kinase inhibitor leads to apoptosis of endothelial cells and therapy of human pancreatic carcinoma. *Cancer Res.* **60**, 2926–2935.

277. Herbst, R. S., Yano, S., Kuniyasu, H., et al. (2000) Differential expression of E-cadherin and type IV collagenase genes predicts outcome in patients with stage I non-small cell lung carcinoma. *Clin. Cancer Res.* **6**, 790–797.

278. Kuniyasu, H., Troncosco, P., Johnston, D., et al. (2000) Relative expression of type IV collagenase, E-cadherin, and vascular endothelial growth factor/vascular permeability factor in prostatectomy specimens distinguishes organ-confined from pathologically advanced prostate cancers. *Clin. Cancer Res.* **6**, 2295–2308.

279. Kuniyasu, H., Ellis, L. M., Evans, D. B., et al. (1999) Relative expression of E-cadherin and type IV collagenase genes predicts disease outcome in patients with resectable pancreatic carcinoma. *Clin. Cancer Res.* **5**, 25–33.

280. Carmeliet, P. and Jain, R. K. (2000) Angiogenesis in cancer and other diseases [Review] [75 refs]. *Nature* **407**, 249–257.

281. Boucher, Y., Leunig, M., and Jain, R. K. (1996) Tumor angiogenesis and interstitial hypertension. *Cancer Res.* **56**, 4264–4266.

39

Isolation of p53 Inhibitors by Screening Chemical Libraries in Cell-Based Readout System

Andrei Gudkov and Elena Komarova

1. p53 as a Subject for Drug Targeting

p53 is a key mediator of cell response to a variety of stresses, inducing growth arrest or apoptosis, thereby eliminating damaged and potentially dangerous cells from the organism *(1,2)*. Once p53-dependent mechanisms are broken, conditions for rapid accumulation of genetic changes are established, leading to dramatic destabilization of the genome and acceleration of carcinogenesis. Indeed, in the majority of tumors, p53-mediated response is broken either by inactivation of the p53 gene itself *(3,4)*, by other members of the pathway (Arf) *(5)*, or by natural negative p53 regulators of cellular (Mdm2) *(6)* or viral (E6) *(7,8)* origin.

Tumor cells are usually very sensitive to wild-type p53 and respond to ectopic expression of p53 by apoptosis or growth arrest *(9)*. Frequent inactivation of p53 in cancer and high sensitivity of tumor cells to p53 determine the most straightforward p53-based therapeutic approach to cancer treatment: restoration or imitation of p53 function in p53-deficient tumors, resulting either in a direct (tumor growth inhibition) or indirect (sensitization to treatment) therapeutic benefit. This general strategy is being extensively explored through a variety of approaches (see below).

Lack of p53 function makes tumor cells genetically different from their cellular environment. This fact is used by another elegant therapeutic approach: "virotherapy" of p53-deficient cancers by replication-competent adenoviruses deficient in apoptosis suppression function and therefore capable of replication exclusively in p53-deficient cells *(10)*.

Although activation of p53 is generally viewed as a most direct and promising anti-cancer strategy, it is not a favorable event for normal tissues. p53 activation as a result of genotoxic stress associated with chemo- or radiation therapy was found to be responsible for massive apoptosis in several normal tissues known to be most sensitive to genotoxic stress (hematopoietic system, epithelia of digestive tract, etc.), possibly contributing to severe side effects of cancer treatment *(11,12)*. Based on these observations, we hypothesized that p53 is a mediator and a determinant of radiation and drug toxicity and, therefore, could be considered a target for therapeutic suppression to reduce cancer-treatment

From: *Methods in Molecular Biology, Vol. 223: Tumor Suppressor Genes: Regulation, Function, and Medicinal Applications.* Edited by: Wafik S. El-Deiry © Humana Press Inc., Totowa, NJ

side effects *(13)*. Obviously, such an approach should be applicable to the treatment of p53-deficient tumors, which form a major proportion of all cancers. To prove this principle, a chemical inhibitor of p53 named pifithrin (PFT) was isolated that, in fact, rescued p53 wild-type cells from apoptotic death induced by DNA damage *in vitro* and reduced lethality of mice from γ-radiation in vivo without a detectable increase in tumor incidence *(14)* and did not cause a protective effect on treatment sensitivity of p53-deficient tumors (unpublished observations). This result indicated that reversible repression of p53 is a valid approach to reduce cancer-treatment side effects and that p53 inhibitors could be useful drugs to be applied in combination with chemo- or radiation therapy. This work also showed that small molecules valuable both as protodrugs and as experimental tools can be effectively isolated in laboratory setup. In this chapter we describe principles and essential steps of chemical library screening using cell-based readout systems.

2. Approaches to p53 Modulators

So far, the majority of our efforts have been applied to restore p53 function in tumors. Development of p53-targeting therapeutic approaches takes advantage of the fact that this important signaling pathway is relatively well studied, making it possible to develop tools affecting individual components or interactors within the p53 pathway. p53 function is inactivated in tumors either by mutations/deletions in the p53 gene itself, by viral p53-inactivating proteins (i.e., E6 of papilloma virus), or through deregulation of other members of the pathway: inactivation of positive (Arf) and overexpression of negative (Mdm2) regulators. Consistently, modulation of the p53 pathway activity may target any of the above factors. The prospective pharmaceuticals being developed for such targeting include specific modifying peptides *(15–19)*, antisense oligonucleotides *(20)*, and small molecules *(14)*.

2.1. Mutant p53 as a Target for Pharmacological Rescue

The transcription regulatory and tumor suppressor activity of p53 is dependent on the ability of the protein to maintain the DNA-binding conformation *(21)*. The most frequently encountered mutations in p53 reduce the thermodynamic stability of the DNA-binding domain and weakens the contact between p53 and the DNA of p53-responsive genes. Activity of mutant p53 form might be restored, at least in part, by application of antibodies and peptides to a negative regulatory domain at the p53 COOH-terminus *(17,18)*. Moreover, restoration of p53 activity might be realized by stabilizing the active conformation of the DNA-binding domain by chemicals. Recently, this idea was confirmed by Foster et al. *(22)*, who used purified wild-type p53 DNA-binding domain that lost the active conformation and binding with wild-type-specific p53 antibodies immobilized in microtiter wells. A chemical library (100,000 compounds) was screened in this system, and active chemicals, which promoted conformational stability of wild-type p53 as judged by binding with mutant-specific antibodies, were found. These compounds were also effective in vivo, slowing tumor growth in mice.

2.2. p53 Regulatory Proteins as Targets

Besides p53, any other member of the p53 pathway (i.e., MDM2, E6, ARF) might be a target for screening molecules, modulating p53 activity, by a similar biochemical approach. Potential modulators may include peptides, antibodies, antisense oligonu-cleotides, ribozymes, and small molecules. Peptide-mimetic strategies have demon-strated a possibility of modulation of p53 function by affecting members of the p53 pathway. p53 protein degradation can be regulated by MDM2-dependent and JNK-dependent pathways. The development of small-peptide effectors that can inhibit MDM2 binding to p53 protein provides a potential drug target for reactivating the p53 pathway in cancers that overamplify MDM2 *(23)*. Similar inhibition of JNK binding to p53 by small peptides derived from the JNK/p53 interface can reduce JNK-dependent ubiquitination and degradation of p53 *(24)*.

p14Arf blocks the degradation of p53 by MDM2 through the inhibition of its ubiq-uitin ligase-associated function *(25)*. The use of small peptides derived from p14Arf, which map at the p14Arf /MDM2 interface, can activate p53, providing an additional target for modulating the MDM2-degradation pathway *(26)*.

Additionally, a new type of transcriptional transactivator of p53 termed a trabody (transcription-activating antibody) *(27)*, provides an antibody-based technology for p53 restoration. Unfortunately, practical applications of all these approaches, despite their high specificity, is restricted by the problem of in-vivo delivery of peptides.

Numerous attempts have been made to inactivate E6 papilloma virus protein in cer-vical carcinoma cells by antisense oligonucleotides and ribozymes *(28)*.

The use of already-known inhibitors of cellular factors that happen to be involved in p53 function is another approach to modulate p53 pathway. Thus, molecular chaperones, including HSP90, bind to mutant p53 in tumors and regulate p53 folding and stability. It was found that a reduction in mutant p53 levels after treatment of tumors with the HSP90-inhibiting agent geldanamycin (benzoquinone class) mediated the refolding of mutant p53 into the native, tetrameric conformation *(29)*.

2.3. Targeting the Entire p53 Pathway

Development of pharmacologic agents acting as p53 modulators is not limited to tar-geting known members of the pathway. There is another strategy that is less focused and is applicable to situations in which the exact molecular target cannot be defined. In such cases, selection of compounds involves a biologic assay that is indicative of the ultimate activity of the signaling cascade and is usually done in cell-based readout systems. The strong advantage of this approach is that it is aimed at a much broader target (the entire pathway), thus increasing the likelihood of successful screening. Specifically, in the case of p53, the activity of the potential modulator of p53 function could be monitored by its effect on the biologic outcome of p53 activation: growth arrest, apoptosis, or transacti-vation of p53-responsive genes. We used this approach to isolate small molecules inhibiting p53.

3. Cell-Based Readout System
3.1. General Requirements

The advantage of cell-based readout is that one can use a functional assay, thus bring-ing selection conditions closer to the final application of the compounds. It also allows combining in one assay testing of specific desirable activity of the compounds while filtering out cytotoxic ones. These considerations will be illustrated by our work on the isolation of p53 inhibitors.

Primary screening involves individual testing of numerous compounds from chemical libraries, the complexity of which is usually somewhere between 10^4 and 10^5 chemicals (see below). The purpose of primary screening is to greatly reduce (usually more than 100-fold) the complexity of the library by selecting a set of "hits" that would include molecules with the desired properties without losing any prospective candidate compounds. The sublibrary of primary "hits" may not necessarily be comprehensive; it is then subjected to a more accurate selection by passing it through additional assays or "filters" to allow further "noise" reduction and to increase the probability of isolation of the desired compounds. In order to achieve screening with reasonable labor and within a reasonable time frame (up to 10 wk), the assay should be robust (consisting of a small number of simple, easily automated steps) and reliable (highly reproducible, with high signal-to-noise ratio). The strictness of the assay is determined by a compromise between its selectivity and the necessity to retain the prospective compounds. We will now discuss how all these requirements were met in a cell-based readout system used for the isolation of p53 inhibitors.

3.2. Readout Assay

3.2.1. Real Biological Effects vs Surrogate Reporter

The ultimate goal of this work was to identify small molecules that would inhibit p53-mediated apoptosis occurring in normal tissues during genotoxic stress associated with anticancer therapy and cause cancer-treatment side effects. Therefore, the "ideal" readout would be to rescue normal mammalian cells that are normally highly sensitive to p53-dependent apoptosis from p53-dependent death. Epithelia of the digestive tract, splenocytes and thymocytes, are known to undergo rapid p53-dependent apoptosis after genotoxic stress in vivo and, in principle, could be viewed as candidate cells for a readout system. Among those, only thymocytes have been used as a cell model of p53-dependent apoptosis in vitro *(30)*. However, use of this system cannot be extended to high-throughput screening for a number of reasons; primary thymocytes can be maintained only as a short-term, fragile culture that is highly sensitive to variations in conditions of cultivation that unavoidably occur during large-scale screening. Moreover, it requires massive isolation of thymocytes from the animals, a laborious procedure that is hard to standardize. Facing the problems with the in-vivo-derived cells, we switched to stable cell lines, the majority of which have an inactivated p53 pathway and therefore could not be used for our purposes. Even those cells that maintain functional p53 pathway and could be maintained as a long-term culture do not usually have pronounced apoptotic response. One well-known exception is mouse embryo fibroblasts transformed with E1a+ras (line C8), which have been successfully used as a model of p53 wild-type tumor cells *(31)*. If properly maintained, these cells keep their wild-type p53 status associated with extremely high sensitivity to genotoxic stress, resulting in rapid (usually within 8–10 h) apoptosis. C8 cells, in combination with an apoptotic assay, were considered as a potential readout system for selection of p53 inhibitors. However, the extremely high sensitivity of these cells to even slight variations in growth conditions associated with a large-scale screening (temperature, pH, cell density, suboptimal concentrations of chemicals, etc.) made it difficult to use them for primary screening.

Thus, the use of the ultimate biologic effect as a readout assay may come into conflict with the necessity for a robust, high-throughput readout system. As a compromise, we substituted for growth-arrest or apoptotic assays in primary screening with another cell system that allowed determination of the p53-dependent transcriptional activation using a surrogate reporter. In this case, however, as an essential additional filtering step, the screening procedure should include identification and elimination of those compounds that are active specifically and exclusively in a surrogate-based readout but lack the desired effect on the pathway itself.

Being a DNA-binding transcription factor, p53 exerts many, if not all, of its functions through regulation of expression of p53-responsive genes *(1)*. Activation of p53-inducible marker would allow one to monitor the very last among direct functions of p53 that follows after stress signal, p53 activation, covalent modifications, stabilization, translocation to the nucleus and binding to regulatory sites within p53-responsive genes. Interference of library compounds with activation of reporter gene by activated p53 was chosen as a primary readout assay to screen for p53 inhibitors.

3.2.2. p53 Reporter Construct

The reporter construct should consist of a promoter that is specific for the transcription factor studied (in our case, p53) and that has a low basal level of expression. The reporter itself should produce an easily detectable product, ideally in a quantitative manner. Preferably, the combination of a specific promoter with a sensitive reporter should ensure high noise-to-signal ratio (preferably, at least 1:10).

To choose an adequate reporter promoter, the same rules can be applied to the creation of a reporter construct for p53-dependent transactivation. In principle it is possible to use natural regulatory regions of the genes normally showing p53-dependent transactivation, such as *p21/waf1*, *mdm2*, or *gadd45*, as a source of p53-responsive promoters. However, the high basal level of expression of such promoters in many cell types does not make then suitable for such a task. Therefore, different artificial constructs consisting of a minimal promoter with a series of p53-binding sites have been used in different studies. In our study, we used a reporter with a minimal *hsp70* promoter combined with two p53-binding sites representing a natural sequence from ribosomal gene (element A) *(21)* and an artificial sequence identified as a p53-binding consensus *(32)*. It has been successfully used before to monitor p53 activity in vitro and in vivo in transgenic animals *(12)*.

The reporter genes should be chosen and optimized to achieve better sensitivity of detection and lesser influence of the cell context and integration sites on the expression of a reporter. The list of popular reporters includes luciferase, several fluorescent proteins that vary in their colors, half-lives and cellular localization, alkaline phosphatase, and others (see, for example, www.clontech.com). The choice of a prospective reporter is dictated by the level of natural cellular background of the respective detectable activity (i.e., aoutofluorescence for luciferase or GFP) and the necessity to ensure rapid, high-throughput monitoring with minimal additional operations. Based on our experience, *LacZ* is a very suitable reporter gene for a high-throughput automated screening system because of its simplicity, sensitivity of detection, stability, specificity, and low background of expression. Activation of *LacZ* can be monitored by colorimetric assay followed by automated quantitation, which gives it advantages over other reporters (GFP, for example), quantitation of which is either difficult or impossible.

3.2.3. Cell Context

The choice of cells for the readout is a compromise between those systems that most adequately reflect the "disease" properties in vitro and the requirements of a high-throughput screening. Specifically for the purposes of isolation of p53 inhibitors, it would be advantageous to use normal (nontransformed) cells that represent tissues normally involved in cancer treatment side effects (i.e., epithelia of digestive tract, chematopoietic cells). All these requirements are difficult to combine with the practical need of screening (see above). Therefore, we used a mouse cell line, ConA, originating from fibroblasts, which did not undergo p53-dependent apoptosis, but was easy to handle in vitro and had well-preserved p53 transactivating function *(12)*. By choosing such a compromise, we took the risk of losing those particular small molecules that are species-specific (act only in human but not in mouse cells) or tissue-specific (suppress p53 in cells of epithelial but not of connective tissue origin).

3.2.4. p53-Activating Treatment

As a major stress-response factor, p53 is induced by a variety of stresses. Choosing one particular p53-inducing treatment for our readout (as well as a specific reporter), we risk isolating chemicals that may not have general applications, since they might be directed against one particular specific treatment. For example, chemicals that inactivate doxorubicin could be picked as "hits" in a readout that is based on p53 activation by this chemotherapeutic drug. In this regard, physical DNA damage by UV or γ-radiation should be considered as a "clean" way to induce p53 that has clear advantages prior to drug treatment. Nevertheless, technical difficulties in applying γ- and UV radiation in a high-throughput format prompted us to use drug-mediated genotoxic stress (doxorubicin treatment). Strength and duration of treatment were accurately calibrated to keep cells in the middle of the dose–response curve: p53 response lasts a limited period of time, and too strong overexpression of the reporter could make it difficult to detect inhibitory effects of the compounds and reduce sensitivity the readout system.

4. Chemical Libraries

Chemical libraries available from many vendors differ in their origin, composition, complexity, and purity. They consist of large sets of individual organic molecules with molecular weights ranging from 300 to 600, usually dissolved in dimethyl sulfoxide (DMSO) at concentrations around 10 m*M*. High purity of individual chemicals in the library is important to avoid selecting compounds because of the activity of their impurities. In the best available libraries, the quality of each library compound is confirmed by NMR analysis. Libraries are usually formatted to facilitate their transfer to a screening system (i.e., 96- or 384-well plates). Based on relatively limited accumulated experience, the minimal size of a diverged chemical library worth screening with a reasonable chance of isolating the desired "hits" is of the order of 10^4.

There are several types of chemical libraries used for screening biologically active compounds, including historical, combinatorial, and focused libraries.

Historical libraries are a combination of collections of handcrafted small molecules. For example, compounds in such collections (Prime-Collection 2000 TM library and DIVERSetTM library) distributed by ChemBridge Corporation (San Diego, CA) are specially preselected from around 1.5 million molecules potentially available from thousands of international sources. ChemBridge Corporation proved their success as a key lead-generation tool in a variety of screening programs *(33–35)*.

Combinatorial libraries represent a combination of small sublibraries produced by complex multistep synthetic schemes through a large number of proprietary templates, intermediates, and building blocks to optimize its diversity (i.e., PHARMACoreTM Library, ChemBridge).

Some of the companies producing chemical libraries offer their clients the ability to do "cherry picking," either from their collections or using their databases (i.e., EXPRESS-PickTM Database of ChemBridge's compounds) selected based on computationally prefiltered structures. Screening of such *"focused libraries"* is the first step in the development stage of small-molecule screening, since they allow us to optimize structure–function relationship of the isolated candidate chemicals.

5. Screening Procedure

5.1. Primary Screening

5.1.1. Pilot Screening

As with any technology, high-throughput screening of biologically active chemicals is a compromise between the scale and speed of screening on one hand and accuracy and reliability, on the other, between the ability to cover numerous individual compounds and almost unavoidable loss of some prospective "hits." Many challenges of a large-scale screening of chemicals could be resolved during "pilot prescreening," a relatively small-scale procedure involving 1000–5000 chemicals that are applied at varying conditions of testing to estimate and establish the exact parameters of the full-scale procedure.

Ideally, each compound should be subjected to several parallel tests in a range of concentrations. However, neither of these requirements is possible during real screenings involving tens to hundreds of thousands of compounds. Therefore, in the majority of cases chemicals are applied at one concentration based on specific features of the readout. In the case of cell-based readouts, the most important limiting parameters are concentration of the solvent (DMSO), toxicity rate and the frequency of primary "hits." Solvent should be used at a concentration that does not jeopardize the assay. The proportion of compounds that show toxicity within the time of screening should not exceed reasonable limits (usually within 1–2%). If these two requirements are satisfied and the system still generates too many "hits" (usually more than 1%), the acting concentrations of the compounds should be reduced to ensure picking most effective candidates for further characterization. In reality, compounds from the library are almost never added to test cells at concentrations exceeding 10 μM; this usually means that 1% DMSO is added to the cells.

5.1.2. Format and Scale of Screening, Library Delivery

A fraction of 1000–5000 chemicals randomly picked from the library for pilot screening are added to the cell medium in 96- or 384-well format plates to a final concentration of 1–10 μM (predetermined during pilot screening) and carefully mixed. The control rows in each plate are treated with the appropriate amount of pure DMSO, with and without a control chemical with known properties. In our screening for p53 inhibitors, we used sodium salicilate as an example of a compound with p53 suppressor activity *(36)*. A "positive control" drug is needed to determine the exact assay conditions (in the case of p53-responsive β-gal, the strength of p53-inducing treatment and the conditions of β-gal reaction). Pure solvent (usually DMSO) was used as a negative control. It is noteworthy that the peripheral rows of multiwell plates could be different from the rest of the plate due to slight variations in growth conditions and in some cases are omitted from the screening procedure.

In summary, the readout system and screening assays in the pilot stage are tuned in such a way as to make the frequency of "hits" not higher than 1–2%.

5.1.3. Toxicity and the Problem of Potential Loss of Effective "Hits"

Since the ultimate readout system is often based on a colorimetric or fluorescent reaction, cytotoxic compounds that abolish the color or fluorescent product development by virtue of killing the target cells may be among the identified "hits." Such false "hits" may form a significant proportion of the entire screening outcome. The rate of their appearance depends greatly on the specific properties of each given readout system and on the concentration of the compounds applied. This latter parameter should be estimated during preliminary pilot screening by microscopic analysis of cells in the wells with the inhibited color or fluorescent reaction. The cutoff rate for the selection of "hits" is established on the basis of the frequency of formal "hits" in several variants of pilot screening differing in compound concentrations. For example, twofold reduction in the library compound concentration (from 20 to 10 μM) in our readout used for the isolation of p53 inhibitors resulted in a dramatic drop of hit rate due to the reduction in the rate of acute toxicity.

5.2. Filtering Primary "Hits"

Even the most advanced cell-based readout systems cannot ensure isolation of a "pure" set of "hits" that is not contaminated with compounds possessing other than desired activities. The goal of primary screening is to enrich as much as possible the proportion of candidate molecules, without losing any valuable candidates. The number of compounds passing through primary screening should be low enough to be suitable for a more advanced, less robust screening to filter out the unwanted "false hits." The proportion of false "hits" may be especially high if the readout assay is based on suppression of a certain reaction (i.e., suppression of p53-mediated activation of β-gal), since those molecules that either kill the cell or inhibit the assayed product will be picked as formal "hits."

5.2.1. Toxicity

Microscopic examination of each "positive" well in multiwell plates to detect toxic compounds cannot be done within a high-throughput primary screening. Moreover, it gives only a vague, unreliable estimation of toxicity, since this approach does not rule out those compounds that kill the cells without altering their morphology. It is therefore recommended to subject all the primary "hits" to additional testing for toxicity, either on the same cells that served as a primary readout or on other cells more suitable for this type of testing. Thus, we performed p53 inhibitor toxicity assays, along with primary readout ConA cells, on E1a+ras-transformed mouse embryo fibroblasts, line C8, that are known to respond by a very quick apoptosis accompanied by cell detachment from substrate in a variety of toxic conditions. In such experiments, DMSO is used as a negative control and a range of concentration of some cytotoxic drugs (for example, adriamycin or 5-fluorouracil) is used as a positive control of killing. The choice of actual assay depends on the specific properties of the cells and may involve simple crystal violet or methylene blue staining (both measuring the numbers of attached cells; applicable to those cell types that easily detach from the substrate after toxic treatment) or more sophisticated detection of toxicity (i.e., MTT assay). It is important to avoid using an apoptosis detection system, since many drugs may induce necrotic rather than apoptotic death. To test toxicity, compounds are added to the cells usually (but not always) in the same conditions as were applied during primary screening.

5.2.2. Target Specificity Filter

Any screening procedure, especially if it involves a surrogate reporter as an indicator of certain pathway function, generates "hits" that may not necessarily target the studied pathway but instead the specific components of the readout itself. Since many readout systems are based on the activation or inactivation of certain reporters by target cells, compounds that inhibit the activity of such reporters (i.e., inhibitors of β-gal) could be among the identified "hits." To rule out such compounds, the primary "hits" should be tested for the ability to suppress the activity of the appropriate reporter expressed from a constitutive (i.e., CMV) promoter.

Another type of false hit coming out of the p53 inhibitor screening includes those chemicals that are directed against the p53-activating factor itself. Thus, in our first screening we used Adriamycin as a DNA-damaging agent inducing p53. Among the isolated hits that inhibited β-gal in the test cells, we found, besides p53 inhibitors, compounds with another type of activity: they inhibited p53 activation by stimulating P-glycoprotein-mediated efflux of Adriamycin, thus preventing the cells from drug-induced DNA damage. Although these compounds are useless as p53 inhibitors, they allowed us to describe a new, potentially important phenomenon in multidrug resistance field and opened an entirely new opportunity for targeting P-glycoprotein against certain specific toxins *(37)*. The p53-unrelated nature of biologic activity of these compounds became clear after we used alternative approaches to p53 activation (UV or γ-irradiation, use of DNA-damaging drugs that are not P-glycoprotein substrates) as filters for chemicals directed against p53-inducing agent rather than the p53 pathway.

The critical step in the screening procedure is to confirm that the "hits" isolated due to their effect on a surrogate reporter are in fact directed against the desired pathway. In the case of the p53-inhibitor screening, this was achieved in two steps. First, the isolated candidates were tested on endogenous p53-responsive genes (*p21, cyclin G, mdm2*). Then, the compounds inhibiting transactivation function of p53 were tested for their ability to suppress p53-dependent apoptosis. This involved the application of apoptotic assay on C8 cells and their derivative, C8-GSE56, with p53 activity suppressed by a strong dominant negative p53 mutant *(38)*. Both cell types were treated by apoptosis-inducing factors (UV, γ-irradiation, 5-fluorouracil, etc.) in combination with the selected "hits" and cell death/survival rates were quantitatively estimated by methylene blue assay for cell survival. Compounds that rescue C8 cells but do not affect the survival of C8-GSE56 cells, or enhance their killing, were picked for further analysis.

5.2.3. Cell-Type Specificity Filter

The outcome of a chemical library screening may be affected by specific biologic properties of the cells used as a reporter. Thus, p53 activity can be regulated differently in cells of different tissue origin *(13)*, and p53 inhibitors with selective tissue specific effect may be of limited value. In order to determine the general p53-suppressive effect of the isolated compounds, their suppressive effect on p53-mediated transactivation and cell growth were tested in isogenic pairs of cells differing in their p53 status. This testing included γ-radiation treatment of human diploid fibroblasts (p53 wild-type WI38 and p53-deficient MDAH041) *(39)* as a model of p53-dependent growth arrest and primary mouse thymocytes (from wild-type and p53-knockout mice) as a model of p53-mediated apoptosis.

6. From Candidate Compounds to Pharmaceuticals

6.1. Candidate Compounds as Research Tools

The compounds chosen as primary "hits" and passed through the toxicity and target specificity filters may serve as valuable research tools, providing insight into the mechanics of the process they modulate. For example, chemical inhibitors of p53 function can target any component and step in this long and divergent pathway, including many unknown factors. Determination of the mechanism of action of p53 inhibitors is an essential step toward their use for the analysis of p53 function and, hopefully, identification of new components of the pathway. Characterization of PFT showed that it possibly acts at the stage of nuclear accumulation of p53, thus providing a potential lead toward cellular control mechanisms of p53 nuclear export/import machinery. More detailed analysis of cellular effects of PFT indicated that this compound, in addition to p53, inhibits heat-shock response, presumably at the same level as nuclear accumulation of HSF-1 protein (Komarova, Neznanov, and Gudkov, in preparation), suggesting an intriguing hypothesis that these two important stress-signaling pathways share a common regulatory stage that is yet to be understood.

6.2. From Hits to Leads

Isolation of chemical compounds with desirable biologic properties targeting certain cellular targets is only the first, though essential, step in development of useful drugs. Converting of "hits" into "leads" suitable for preclinical and clinical trials is a long, complicated, and costly process, involving a number of essential stages that are outside the scope of this review. Very briefly, several most effective "hits" are subjected to additional filtering based on their in-vivo validation. In the case of p53 inhibitors, in-vivo toxicity and radioprotective effect are the key characteristics of the candidate compounds that have to be determined as a pass for further development process. This included determination of highest tolerable dose of the compound, followed by testing its activity in protecting mice from a single dose of γ-radiation. The advantages of γ-radiation before systemic genotoxic stress induced by chemotherapeutic drugs is that it is a short-lasting treatment that allows at this stage to avoid dealing with pharmacokinetics issues. Compounds showing the best dose/effect ratio are used for creation of focused libraries, which usually include from hundreds to thousands of chemicals picked by medicinal chemists on the basis of three-dimensional pharmacophor analysis. Focused libraries are passed through the same screening pipeline that supposedly results in generation of "strong hits" that work in the nanomolar range of concentrations. New "hits" are subjected to pharmaco-informatic analysis, classified according to their strength in projection on their structural similarities. This should hopefully result in the establishment of rules for structure–activity relationship that will rationalize further optimization of the "hits."

Pharmacologic properties of advanced candidates are then characterized by assays including determination of pharmacokinetics, pharmacodynamics, formulation, oral bioavailability, acute and chronic toxicity, and so on, in preparation for further preclinical testing and ultimate clinical trials.

6.3. Potential Value of p53 Inhibitors

p53 inhibitors that show protective effect against genotoxic stress in vivo could be prototype new drugs to reduce cancer-treatment side effects. There are other situations, besides anticancer therapy, when the human organism is exposed to stressful conditions known to involve p53 activation, which, in extreme cases, could result in the development of life-threatening diseases. p53-dependent tissue injuries might be developed associated with hypoxia (heart and brain ischemia) and hyperthermia (fever and burns) *(40)*. Thus, therapeutic application of the p53 inhibitors might be useful for reducing of pathologic outcome of the above stresses.

References

1. Gottlieb, T. M. and Oren, M. (1996) p53 in growth control and neoplasia. ***Biochim. Biophys. Acta*** **1287**, 77–102.
2. Prives, C. and Hall, P. A. (1999) The p53 pathway. ***J. Pathol.*** **187**, 112–126.
3. Hansen, R. and Oren, M. (1997) p53: from inductive signal to cellular effect. ***Curr. Opin. Genet. Dev.*** **7**, 46–51.
4. Levine, A. J. (1997) p53, the cellular gatekeeper for growth and division. ***Cell*** **88**, 323–331.
5. Sherr, C. J. (1998) Tumor surveillance via the ARF-p53 pathway. ***Genes Dev.*** **12**, 2984–2991.
6. Momand, J., Jung, D., Wilczynski, S., and Niland, J. (1998) The MDM2 gene amplification database. ***Nucleic Acids Res.*** **26**, 3453–3459.
7. Scheffner, M., Werness, B. A., Huibregtse, J. M., Levine, A. J., and Howley, P. M. (1990) The E6 oncoprotein encoded by human papillomavirus types 16 and 18 promotes the degradation of p53. ***Cell*** **63**, 1129–1136.
8. Thomas, M., Pim, D., and Banks, L. (1999) The role of the E6-p53 interaction in the molecular pathogenesis of HPV. ***Oncogene*** **18**, 7690–7700.
9. May, P. and May, E. (1999) Twenty years of p53 research: structural and functional aspects of the p53 protein. ***Oncogene*** **18**, 7621–7636.
10. Bischoff, J. R., Kirn, D. H., Williams, A., et al. (1996) An adenovirus mutant that replicates selectively in p53-deficient human tumor cells. ***Science*** **274**, 373–376.

11. Rogel, A., Popliker, M., Webb, C. G., and Oren, M. (1985) p53 cellular tumor antigen: analysis of mRNA levels in normal adult tissues, embryos, and tumors. *Mol. Cell Biol.* **5**, 2851–2855.

12. Komarova, E. A., Chernov, M. V., Franks, R., et al. (1997) Transgenic mice with p53-responsive lacZ: p53 activity varies dramatically during normal development and determines radiation and drug sensitivity in vivo. *EMBO J.* **16**, 1391–1400.

13. Komarova, E. A. and Gudkov, A. V. (1998) Could p53 be a target for therapeutic suppression? *Semin. Cancer Biol.* **8**, 389–400.

14. Komarov, P. G., Komarova, E. A., Kondratov, R. V., et al. (1999) A chemical inhibitor of p53 that protects mice from the side effects of cancer therapy. *Science* **285**, 1733–1737.

15. Ball, K. L., Lain, S., Fahraeus, R., Smythe, C., and Lane, D. P. (1997) Cell-cycle arrest and inhibition of Cdk4 activity by small peptides based on the carboxy-terminal domain of p21WAF1. *Curr. Biol.* **7**, 71–80.

16. Chene, P., Fuchs, J., Bohn, J., Garcia-Echeverria, C., Furet, P., and Fabbro, D. (2000) A small synthetic peptide, which inhibits the p53-hdm2 interaction, stimulates the p53 pathway in tumour cell lines. *J. Mol. Biol.* **299**, 245–253.

17. Hupp, T. R., Sparks, A., and Lane, D. P. (1995) Small peptides activate the latent sequence-specific DNA binding function of p53. *Cell* **83**, 237–245.

18. Selivanova, G., Iotsova, V., Okan, I., et al. (1997) Restoration of the growth suppression function of mutant p53 by a synthetic peptide derived from the p53 C-terminal domain. *Nat. Med.* **3**, 632–638.

19. Fuchs, S. Y., Adler, V., Buschmann, T., et al. (1998) JNK targets p53 ubiquitination and degradation in nonstressed cells. *Genes Dev.* **12**, 2658–2663.

20. Chen, L., Lu, W., Agarwal, S., Zhou, W., Zhang, R., and Chen, J. (1999) Ubiquitous induction of p53 in tumor cells by antisense inhibition of MDM2 expression. *Mol. Med.* **5**, 21–34.

21. Kern, S. E., Kinzler, K. W., Bruskin, A., et al. (1991) Identification of p53 as a sequence-specific DNA-binding protein. *Science* **252**, 1708–1711.

22. Foster, B. A., Coffey, H. A., Morin, M. J., and Rastinejad, F. (1999) Pharmacological rescue of mutant p53 conformation and function. *Science* **286**, 2507–2510.

23. Bottger, A., Bottger, V., Sparks, A., Liu, W. L., Howard, S. F., and Lane, D. P. (1997) Design of a synthetic Mdm2-binding mini protein that activates the p53 response in vivo. *Curr. Biol.* **7**, 860–869.

24. Fuchs, S. Y., Fried, V. A., and Ronai, Z. (1998) Stress-activated kinases regulate protein stability. *Oncogene* **17**, 1483–1490.

25. Lowe, S. W. (1999) Activation of p53 by oncogenes. *Endocr. Relat. Cancer* **6**, 45–48.

26. Midgley, C. A., Desterro, J. M., Saville, M. K., et al. (2000) An N-terminal p14ARF peptide blocks Mdm2-dependent ubiquitination in vitro and can activate p53 in vivo. *Oncogene* **19**, 2312–2323.

27. Venot, C., Maratrat, M., Sierra, V., Conseiller, E., and Debussche, L. (1999) Definition of a p53 transactivation function-deficient mutant and characterization of two independent p53 transactivation subdomains. *Oncogene* **18**, 2405–2410.

28. Alvarez-Salas, L. M., Arpawong, T. E., and DiPaolo, J. A. (1999) Growth inhibition of cervical tumor cells by antisense oligodeoxynucleotides directed to the human papillomavirus type 16 E6 gene. *Antisense Nucleic Acid Drug Dev.* **9**, 441–450.

29. Blagosklonny, M. V., Toretsky, J., and Neckers, L. (1995) Geldanamycin selectively destabilizes and conformationally alters mutated p53. *Oncogene* **11**, 933–939.

30. Lowe, S, Schmitt, E., Smith, S., Osborne, B., and Jacks, T. (1993) p53 is required for radiation-induced apoptosis in mouse thymocytes. *Nature* **362**, 847–849.

31. Lowe, S, Ruley, H., Jacks, T., and Housman, D. (1993) p53-dependent apoptosis modulates the cytotoxicity of anticancer agents. *Cell* **74**, 954–967.

32. Funk, W. D., Pak, D. J., Karas, R. H., Wright, W. E., and Shay, J. W. (1992) A transcriptionally active DNA binding site for human p53 protein complexes. *Mol. Cell. Biol.* **12**, 866–2871.
33. Cheung, A., Hollingworth, S, Baylor, S. M., Goldman, Y. E., Mitchison, T. J., and Straight, A. F. (2002) A small-molecule inhibitor of skeletal muscle myosin II. *Nat. Cell Biol.* **4**, 83–84.
34. Degterev, A., Lugovskoy, A., Cardone, M., et al. (2001) Identification of small-molecule inhibitors of interaction between the BH3 domain and Bcl-xL. *Nat. Cell Biol.* **3**, 173–182.
35. Su, G. H., Taylor, A. S., Byungwoo, R., and Kern, S. E. (2000) A novel histone deacetylase inhibitor identified by high-throughput transcriptional screening of a compound library. *Cancer Res.* **60**, 3137–3142.
36. Chernov, M. V. and Stark, G. R. (1997) The p53 activation and apoptosis induced by DNA damage are reversibly inhibited by salicylate. *Oncogene* **14**, 2503–2510.
37. Kondratov, R., Komarov P. G., Becker, Y, Ewenson, A., and Gudkov, A. (2001) Small molecules that dramatically alter multidrug resistance phenotype by modulating the substrate specificity of P-glycoprotein. *Proc. Natl. Acad. Sci. USA* **98**, 14078–14083.
38. Ossovskaya, V. S., Mazo, I. A., Chernov, M. V., et al. (1996) Use of genetic suppressor elements to dissect distinct biological effects of separate p53 domains. *Proc. Natl. Acad. Sci. USA* **93**, 10309–10314.
39. Agarwal, M. L., Agarwal, A., Taylor, W. R., and Stark, G. R. (1995) p53 controls both the G2/M and the G1 cell cycle checkpoints and mediates reversible growth arrest in human fibroblasts. *Proc. Natl. Acad. Sci. USA* **92**, 8493–8497.
40. Komarova, E. A. and Gudkov, A. V. (2001) Chemoprotection from p53-dependent apoptosis: potential clinical applications of the p53 inhibitors. *Biochem. Pharmacol.* **62**, 657–667.

Index